Bioorganic Chemistry:
Peptides and Proteins

Topics in Bioorganic and Biological Chemistry
A Series of Books in Support of Teaching and Research

Series Editors:
Sidney M. Hecht, University of Virginia
Richard L. Schowen, University of Kansas

Bioorganic Chemistry: Nucleic Acids
S. Hecht, editor

Bioorganic Chemistry: Peptides and Proteins
S. Hecht, editor

Bioorganic Chemistry: Carbohydrates
S. Hecht, editor

Enzyme Catalysis
R. Schowen

Steady State Enzyme Kinetics
P. Cook

Bioorganic Chemistry: Peptides and Proteins

Edited by
Sidney M. Hecht
University of Virginia

New York   Oxford
OXFORD UNIVERSITY PRESS
1998

Oxford University Press

Oxford   New York
Athens   Auckland   Bangkok   Bogota   Bombay   Buenos Aires
Calcutta   Cape Town   Dar es Salaam   Delhi   Florence   Hong Kong
Istanbul   Karachi   Kuala Lumpur   Madras   Madrid   Melbourne
Mexico City   Nairobi   Paris   Singapore   Taipei   Tokyo   Toronto   Warsaw

and associated companies in
Berlin   Ibadan

Copyright © 1998 by Oxford University Press, Inc.

Published by Oxford University Press, Inc.,
198 Madison Avenue, New York, New York 10016

Oxford is a registered trademark of Oxford University Press.

All rights reserved. No part of this publication may be reproduced,
stored in a retrieval system, or transmitted, in any form or by any means,
electronic, mechanical, photocopying, recording, or otherwise,
without the prior permission of Oxford University Press.

Library of Congress Cataloging-in-Publication Data
Bioorganic chemistry : peptides and proteins / edited by Sidney M.
   Hecht.
     p.   cm. — (Topics in bioorganic and biological chemistry)
   Includes bibliographical references and index.
   ISBN 0-19-508468-3 (cloth : alk. paper)
   1. Proteins. 2. Peptides. I. Hecht, Sidney M. II. Series.
QP551.B474   1998
572.6—dc21   97-27060

9 8 7 6 5 4 3 2 1

Printed in the United States of America
on acid-free paper

Contents

Preface, vii

Contributors, ix

1. Introduction to Peptides and Proteins, 1
   Milton J. Axley

2. Chemical Synthesis of Peptides, 27
   Victor J. Hruby and Jean-Philippe Meyer

3. Total Chemical Synthesis of Proteins, 65
   Michael C. Fitzgerald and Stephen B. H. Kent

4. Structural Analysis of Proteins, 100
   John E. Shively

5. Protein Structure, 153
   Charles W. Carter, Jr.

6. Protein Folding, 224
   Zhi-Ping Liu, Josep Rizo, and Lila M. Gierasch

7. Nucleic Acid Interactive Protein Domains That Require Zinc, 258
   Michael A. Massiah, Paul R. Blake, and Michael F. Summers

8. Understanding the Mechanisms and Rates of Enzyme-Catalyzed Proton Transfer Reactions to and from Carbon, 279
   John A. Gerlt

9. Site-Directed Mutagenesis, 312
   Paul J. Loida, Ronald A. Hernan, and Stephen G. Sligar

10. The Structural Basis of Antibody Catalysis, 335
    Donald Hilvert, Gavin MacBeath, and Jumi A. Shin

11. Peptide Hormones, 367
    Arno F. Spatola

12. Peptide Mimetics, 395
    Hiroshi Nakanishi and Michael Kahn

13   Use of Enzymes in Organic Synthesis, 420
     Zhen Yang and Alan J. Russell

14   Engineered Proteins in Materials Research, 446
     *David A. Tirrell, Jane G. Tirrell, Thomas L. Mason, and
     Maurille J. Fournier*

References, 473

Index, 523

Preface

This is the second volume of a series of books in bioorganic and biological chemistry. The first volume, dealing with nucleic acids, appeared in the spring of 1996. A third volume in the area of carbohydrates is in production and will appear shortly.

As noted in the preface to the first volume, the increasingly detailed understanding of the molecular basis by which biological systems operate has dramatically increased the range of studies now considered to be within the domain of organic chemistry. The "core" of expertise required to function effectively in this rapidly expanding field has increased correspondingly. This poses an ongoing educational challenge both for scientists presently working in the field and, especially, for students encountering the subject matter for the first time. This series of books is intended to support the teaching of graduate students in bioorganic chemistry.

In keeping with the format of the first volume, *Bioorganic Chemistry: Peptides and Proteins* consists of a set of 14 chapters, approximately equal to the number of weeks in a semester. The subject matter of each chapter is judged to be both representative of and central to an understanding of ongoing research activity in the field of peptides and proteins, as practiced by bioorganic chemists. Each chapter begins with an overview of basic principles and a summary of key findings that form the basis for current research activity in the specific area of focus in the chapter. The remainder of each chapter presents a limited number of examples of recent studies in greater depth. The chapters are thus organized in much the same fashion as typical lectures in special topics courses. A set of overheads corresponding to each of the figures in the book is available to aid classroom presentation.

In addition to my own favorable experience in teaching the subject matter of all three books in the set to graduate students at the University of Virginia, others who have used the nucleic acids book for a course have affirmed that it functions as intended. Gratifyingly, numerous colleges have commented that the book has been no less valuable as an educational tool within their own research laboratories.

I would like to thank the authors of this volume for their efforts in writing the chapters. Oxford University Press has continued to provide excellent support for this educational experiment; the ongoing help and advice of Bob Rogers, Senior Editor, has been invaluable. This volume was typed by Vickie Thomas, who provided uncommon technical support and much patience. Carolyn Esau assisted with verification of numerous literature references. I thank them both for their assistance. I also acknowledge with gratitude the contributions of many graduate and postdoctoral students who have participated enthusiastically in the special topic courses that I have given at the University of Virginia based on the material in this book.

Charlottesville   S. M. H.
October, 1997

Contributors

Milton J. Axley
Naval Medical Research Institute

Paul R. Blake
University of Maryland, Baltimore County

Charles W. Carter, Jr.
University of North Carolina, Chapel Hill

Michael C. Fitzgerald
The Scripps Research Institute

Maurille J. Fournier
University of Massachusetts, Amherst

John A. Gerlt
University of Maryland

Lila M. Gierasch
University of Massachusetts, Amherst

Ronald A. Hernan
The Scripps Research Institute

Donald Hilvert
The Scripps Research Institute

Victor J. Hruby
University of Arizona

Michael Kahn
Molecumetics

Stephen B. H. Kent
The Scripps Research Institute

Zhi-Ping Liu
University of Massachusetts, Amherst

Paul J. Loida
University of Illinois at Urbana-Champaign

Gavin MacBeath
The Scripps Research Institute

Thomas L. Mason
University of Massachusetts, Amherst

Michael A. Massiah
University of Maryland, Baltimore County

Jean-Philippe Meyer
University of Arizona

Hiroshi Nakanishi
Molecumetics

Josep Rizo
University of Texas, Southwestern

Alan J. Russell
University of Pittsburgh

Jumi A. Shin
The Scripps Research Institute

John E. Shively
Beckman Research Institute

Stephen G. Sligar
University of Illinois at Urbana-Champaign

Arno F. Spatola
University of Louisville

Michael F. Summers
University of Maryland, Baltimore County

David A. Tirrell
University of Massachusetts, Amherst

Jane G. Tirrell
University of Massachusetts, Amherst

Zhen Yang
University of Pittsburgh

Bioorganic Chemistry:
Peptides and Proteins

Introduction to Peptides and Proteins

Milton J. Axley

Proteins and peptides provide many of the chemical and physical processes that constitute life in biological organisms. These molecules are ubiquitous in all living creatures. Although certain types of proteins are shared by all organisms, each life form has a unique make-up of proteins that defines the characteristics of that organism. The synthesis of a protein within an organism is determined by the genetic make-up of the organism, and most of the genetic make-up of an organism is devoted to the expression of proteins.

The four main groupings of biological macromolecules each have their own functionalities. *Nucleic acids* store and manipulate the genetic information that details the make-up of an organism. *Carbohydrates* store energy and, in plants, serve structural purposes. *Lipid* assemblies make up the bulk of cellular membranes that compartmentalize cells and subcellular organelles. *Proteins* and *peptides* are the agents of action in a cell. Some of the biological roles in which proteins function include: chemical reaction catalysis, motility, physiological regulation, transport, structural composition, and defense. Proteins perform most of the activities and provide much of the structure that constitutes cells and organisms.

The Dutch chemist Gerardus Mulder coined the name "protein" in 1838, when he found proteins to be the major chemical constituent of cellular matter. The name is derived from the Greek word "proteis," which means "of primary importance." Proteins have more than lived up to their name throughout all the advances in understanding since Mulder's time, and today these molecules continue to be the center point for progress in diverse areas of scientific and industrial research.

Modern biotechnology is directed toward the exploitation of proteins and peptides. The ability to harness the vast potential of proteins provides a powerful tool for medicine, as proteins are involved in and can be used for the treatment of infectious, genetic, and traumatic diseases. Agriculture benefits from modern protein chemistry with improved crop yields, pest control, and animal husbandry. Chemistry has gained new and more efficient synthetic methods catalyzed by proteins.

Structure

Proteins and peptides are linear polymers consisting of amino acids. Short chains of amino acids (two to about forty) are known as peptides. Two amino acids linked by a peptide bond constitutes a dipeptide, while three amino acids linked together make up a tripeptide. Longer chains containing more than 30 to 50 amino acids are polypeptides. There is no single cutoff length that separates peptides from polypeptides; a 40-amino acid chain might be called either a peptide or a polypeptide. The individual amino acids within a peptide or polypeptide are called subunits or residues.

Proteins are polypeptides with a function. The terms "protein" and "polypeptide" are sometimes used interchangeably, although polypeptide strictly refers to an amino acid chain and protein refers to the overall entity, possibly including multiple polypeptide subunits and cofactors in a functional conformation. Enzymes are proteins that catalyze chemical reactions.

Amino Acids

In order to understand the nature of polypeptides, it is necessary to understand the characteristics of the amino acid subunits. Amino acids contain an amino group, a carboxy group, a hydrogen atom, and a side group all connected to a central (α) carbon atom (see Fig. 1-1). Since the central carbon is attached to four different groups, amino acids (except for glycine) are chiral molecules. All amino acids found in proteins and nearly all found in peptides are of the "L" (or S in the Cahn-Ingold-Prelog nomenclature) configuration. Amino acids in solution generally exist as zwitterions, as their amino and carboxy groups are both charged at pH values between 3 and 9.

As amino and carboxy groups are common to all amino acids, it is the side group that defines amino acids and determines their individual characteristics. The chemical and structural natures of the side groups determine the structure and function of the amino acid within the context of a protein or peptide. A variety of functional groups are available as side chains, including acidic, basic, hydrophobic, hydrophilic, redox-active, bulky, and compact chemical moieties. This range of functional groups provides the chemical basis for the vast array of biochemical properties displayed by proteins and peptides.

Twenty amino acids are commonly found as constituents of proteins and peptides in all living organisms. Figure 1-2 gives diagrams of the twenty "standard amino acids." Amino acids can be grouped into several types, based on the chemical properties of their side groups. Amino acids are often described as acidic, basic, polar, nonpolar, bulky, and so on. Other groupings are obviously also possible. Such descriptions based on side group properties are informative within the context of a polypeptide chain of amino acids, but may be inaccurate for the individual free amino acid. For example, alanine may be called a nonpolar or hydrophobic amino acid due to its methyl side group, but the free amino acid alanine by itself will have two charged groups at neutral pH due to the carboxy and amino groups. Groupings of amino acids allow comparison of the side group properties and prediction of the effect that amino acid sub-

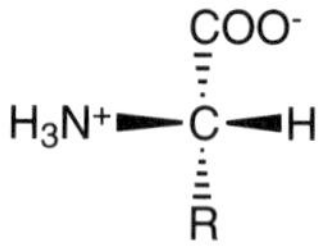

$$H_3N^+ - \overset{COO^-}{\underset{R}{C}} - H$$

Figure 1-1. Structure of an L-amino acid indicating absolute stereochemistry. The general structure of an L-amino acid consists of a central carbon atom surrounded by four different chemical groups: carboxylic acid and amino groups, a hydrogen atom, and a side chain. The side chain is here represented by R.

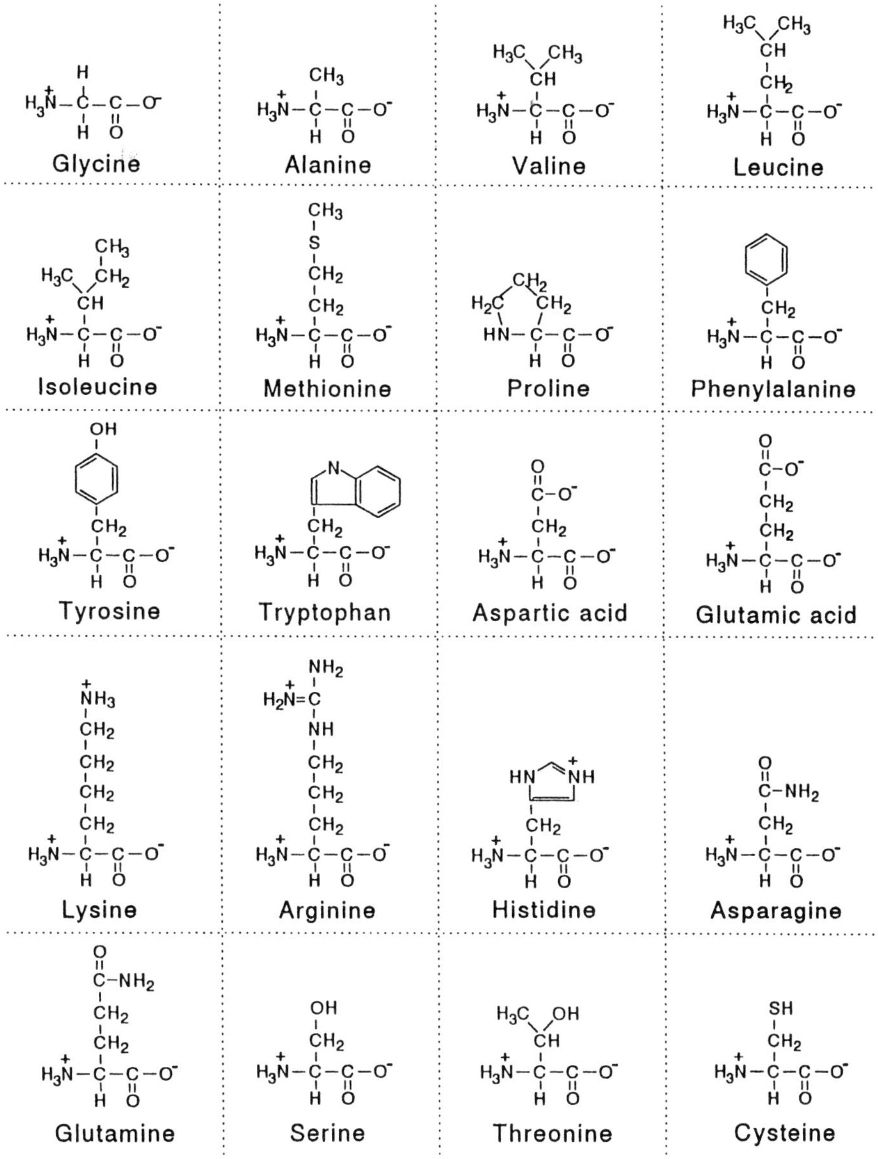

Figure 1-2. Chemical structures for the twenty common amino acids. The amino acids are presented in the order in which they are discussed in the text. Ionic charges reflect the situations at pH 7.

stitutions may have on the structure or biochemistry of polypeptides. For example, substitution of the amino acid aspartic acid for glutamic acid at the same position in a polypeptide might be expected to have little effect on the physical-chemical properties, as both amino acids contain carboxylic acid side groups. However, substitution of tryptophan, containing a hydrophobic side group, at the same position might have a more drastic effect on polypeptide structure and function.

Glycine is the smallest and simplest amino acid, as it has only a single hydrogen atom as a side group. Since the central carbon of glycine has two hydrogens attached to it, this is the only standard amino acid that is not chiral.

There are six amino acids with small, nonpolar side groups. Alanine has a methyl side chain, valine an isopropyl, leucine a 2-methylpropyl and isoleucine a 1-methylpropyl group. Methionine is one of two sulfur-containing amino acids, with a methylated thioether side group. Proline is unusual among the 20 standard amino acids, because its side chain is attached to the amino group to form a five-membered ring. The nitrogen group of proline is actually a secondary amine.

Three amino acids have bulky, uncharged, aromatic side groups. Phenylalanine has a benzyl functional group, while tyrosine has a phenol, and tryptophan has an indole moiety in its side group. Phenylalanine and tryptophan are considered nonpolar and hydrophobic, but the hydroxyl group of tyrosine gives its side chain some polar character.

Five amino acids have R groups that are charged at neutral pH. Aspartic acid and glutamic acid both have carboxyl-containing side groups, and they are, therefore, organic acids. Lysine, arginine, and histidine have basic side groups containing amino, guanidino, and imidazolium functional groups, respectively. The side group of histidine has a pK_a of about 6, and is the only amino acid of this group with a pK_a near physiological pH.

The side groups of five amino acids are small and polar. Asparagine and glutamine are the amide forms of aspartic and glutamic acids, respectively. Serine has a hydroxymethyl R group, and threonine has a 1-hydroxyethyl side group. Cysteine is a sulfur-containing amino acid, as its side group contains a thiol (-SH) functional group. The thiol of cysteine readily undergoes a reversible oxidation, and this reaction is important to many structural and functional properties of polypeptides. Specifically, the thiols of two cysteines can be oxidized to form a covalent disulfide bond between the two amino acids. The resulting derivatized amino acid, called cystine, stabilizes the three-dimensional structure of many polypeptides.

Amino acids are often referred to by three- or one-letter abbreviations. These are particularly useful for writing out the sequences of polypeptides. Table 1-1 presents the accepted three-letter and one-letter abbreviations for the 20 standard amino acids.

In addition to the 20 standard amino acids, several other amino acids are found naturally in certain proteins and peptides. In most such cases, amino acids are chemically modified after incorporation into a polypeptide. This posttranslational modification can be essential to the function of the polypeptide. For example, tropocollagen, the polypeptide component of collagen, contains a relatively large proportion (about 9%) of the amino acid hydroxyproline. The hydroxyproline is formed by hydroxylation of proline after synthesis of the polypeptide chain by an enzyme that requires ascorbic acid (vitamin C) for its activity. Without the activity of the hydroxylating enzyme, collagen is not properly formed. The human disease scurvy, characterized by improper collagen formation, is caused by vitamin C deficiency.

A number of proteins have been shown to contain phosphorylated derivatives of the hydroxyl-containing amino acids tyrosine, serine, and threonine. In these derivatives, a phosphate group is attached to the hydroxyl group as an ester to form phosphotyrosine, phosphoserine, and phosphothreonine, respectively. These phosphoester derivatives are formed posttranslationally, that is, after synthesis of the polypeptide by translation within a cell. Phosphorylation occurs at specific sites of certain proteins, and the reaction is reversible. Specific phosphorylation and dephosphorylation of certain proteins affect the activities of those proteins: increasing, decreasing, or otherwise alter-

Table 1-1 Amino Acid Abbreviations

Amino Acid	Three-letter Code	One-letter Code
Alanine	Ala	A
Arginine	Arg	R
Asparagine	Asn	N
Aspartic Acid	Asp	D
Cysteine	Cys	C
Glutamine	Gln	Q
Glutamic Acid	Glu	E
Glycine	Gly	G
Histidine	His	H
Isoleucine	Ile	I
Leucine	Leu	L
Lysine	Lys	K
Methionine	Met	M
Phenylalanine	Phe	F
Proline	Pro	P
Serine	Ser	S
Threonine	Thr	T
Tryptophan	Trp	W
Tyrosine	Tyr	Y
Valine	Val	V

ing the activities of the proteins. Thus, phosphorylation of proteins can act as a reversible control mechanism, switching proteins between active and inactive forms.

Most of the nonstandard amino acids found in natural proteins are produced by posttranslational modification of one of the twenty standard amino acids present in a peptide chain. However, this is not the case for selenocysteine, which is identical to cysteine except that a selenium atom replaces the sulfur atom of cysteine. Selenocysteine is found as an essential component of certain proteins in many classes of organisms. It is coded for in the DNA, and a specific tRNA directs its cotranslational insertion into proteins. For these reasons, selenocysteine has been called the "twenty-first amino acid" (Söll, 1988). In mammals, selenium is a required nutrient, and at least half a dozen proteins contain essential selenocysteine residues. Selenium deficiency causes a cardiomyopathy in humans and white-muscle disease in domestic livestock. Glutathione peroxidase is a mammalian selenoprotein that catalyzes the repair of cellular oxidative damage caused by free radicals and peroxides; therefore, this selenoprotein may play a role in prevention of cancer and aging.

Polypeptide Synthesis

Each polypeptide synthesized in a cell is encoded by a single gene. The DNA sequences of chromosomal genes directly encode the amino acid sequences of all proteins and most peptides. The information encoding each amino acid in a polypeptide is found in a "codon." The codons of a gene are arranged sequentially in the same order in which the amino acids appear in a polypeptide.

The linear arrangement of information found in the nucleotide sequence of DNA can be directly translated into the linear arrangement of amino acids in a polypeptide sequence through the use of the universal *genetic code*. The genetic code is thus a biological algorithm for the translation of information found in gene codons into amino acids in polypeptides.

A codon consists of three adjacent nucleotides within a gene, and the three nucleotides of a codon are sometimes called a triplet. Since there are four different nucleotide bases in DNA (and RNA), there are 64 (4^3) possible triplet codons. As there are only 20 standard amino acids for which to code, there is redundancy in the genetic code. That is, there is more than one codon for some of the amino acids. In addition, three of the codons do not code for amino acids. These codons are stop codons, also known as end or termination codons. Stop codons designate the end of the polypeptide coding sequence of a gene.

The genetic code is said to be universal. Virtually every tested species of living organism uses the universal genetic code. However, there are also exceptions where gene codons do not match the corresponding amino acids found in polypeptides. One type of exception is when the amino acid is chemically altered after polypeptide synthesis. Examples of this "posttranslational modification" include glycosylated and phosphorylated amino acids.

In other exceptions to the universal genetic code, the codons that normally code for termination instead code for amino acids. For example, in *Tetrahymena* the UAA and UAG codons designate glutamine insertion into polypeptide; in *Mycoplasma* and eukaryotic mitochondria the UGA codon specifies tryptophan, while in many eukaryotes and prokaryotes the UGA codon can also code for selenocysteine.

Synthesis of a polypeptide by a living cell occurs in two discrete steps, transcription and translation. In the first step, transcription, an RNA (ribonucleic acid) copy of the coding portion of the DNA (deoxyribonucleic acid) gene is synthesized. RNA consists of a linear array of four different ribonucleotide bases: adenine, uracil, guanine, and cytosine, abbreviated A, U, G, and C, respectively. DNA consists of a string of deoxyribonucleotides, which also contain the bases adenine, guanine, and cytosine; however, DNA contains thymine in place of uracil.

The RNA copy that is transcribed from the DNA gene is called the messenger RNA, or mRNA. Although DNA is the storehouse for coding information, it is actually the mRNA that is used during translation, the second major step of protein biosynthesis. During translation, the codons contained in the mRNA are translated sequentially via the genetic code, affording a specific polypeptide sequence. Two other macromolecular species are required for translation, namely transfer RNA and ribosomes. Transfer RNA (tRNA) molecules carry amino acids to the site of translation, which is within the ribosome. Each tRNA has a single anticodon, a three-base sequence complementary to an mRNA codon. There is a tRNA for every codon, and tRNAs are linked to (activated with) the amino acid corresponding with the tRNA anticodon.

The tRNA anticodon can anneal to (hybridize with) the corresponding codon through nucleotide base pairing interactions within the context of the ribosome. Ribosomes are very large macromolecular bodies, made up of numerous proteins and a few large RNA molecules (ribosomal RNA). During translation, ribosomes bind the mRNA, direct the codon-anticodon interactions between the mRNA and tRNAs and then link together the amino acids presented by the tRNAs.

Many proteins and peptides have value for pharmaceutical, agricultural, and industrial purposes, and understanding the process of polypeptide biosynthesis provides the ability to control this synthesis. Modern techniques of molecular biology have allowed the exploitation of the natural processes leading to protein biosynthesis in many beneficial ways.

DNA genes can be isolated and identified, and then expressed in foreign "hosts." Thus, the human gene for insulin has been expressed in bacterial hosts. Since bacteria can be produced in large quantities, the insulin expressed by such bacteria can be produced in relatively large quantities also. Purification of the insulin from the bacterial host proteins provides a ready supply of human insulin for people with diabetes.

Direct chemical synthesis of polypeptides is also possible. This is used primarily for the synthesis of shorter peptides, usually less than forty amino acids in length. The Merrifield method of protein synthesis provides a solid-phase approach in which amino acids are linked together using organic chemistry techniques. Peptides can be produced by chemical synthesis that are indistinguishable from biologically produced ones, or they can be synthesized with altered structures not found in nature. The chemical synthesis of peptides is described in more detail in Chapter 2.

Polypeptide Structure

The linkage between amino acids in proteins and peptides is known as the *peptide bond*. The peptide bond is created by a condensation reaction involving the carboxyl group of one amino acid and the amino group of another with concomitant formation of a H_2O molecule. The resulting amide C—N bond has some double bond character due to resonance, as shown in Figure 1-3. For this reason, there is not free rotation about the peptide bond, which is an important limitation to the degrees of freedom available to the structure of a polypeptide.

Proteins and peptides are synthesized by the stepwise linkage of amino acids in a specific sequence. The resulting chain of amino acids has two ends, one of which contains a free amino group, and the other a free carboxyl group. These are called, respectively, the amino and carboxy ends, or termini, of the polypeptide. By convention, a peptide or polypeptide structure is displayed with the amino terminus on the left and the carboxy terminus on the right.

Size is a distinctive characteristic of polypeptides, and polypeptides are often referred to on the basis of their molecular weight or number of amino acids. The number of amino acids in a polypeptide can be readily estimated from the molecular weight, since polypeptides are linear chains of about 20 different amino acids. The average weight of amino acid residues in polypeptides is about 110 Da. Using this figure, one can estimate that a polypeptide having a molecular weight of 33,000 Da would consist of about 300 amino acids.

The repetitive element of the linked amino acids within a polypeptide is called the peptide "backbone." In one view, the polypeptide can be seen as a series of side groups arrayed along a constant backbone. Two different polypeptides of equal length will share in common the properties of the backbone, while differing in the properties determined by the array of side groups displayed along the backbone.

A polypeptide is defined not only by its constituent amino acids, but also by the precise arrangement of these amino acids. The order in which amino acids are found in a polypeptide chain is known as the "amino acid sequence." The amino acid se-

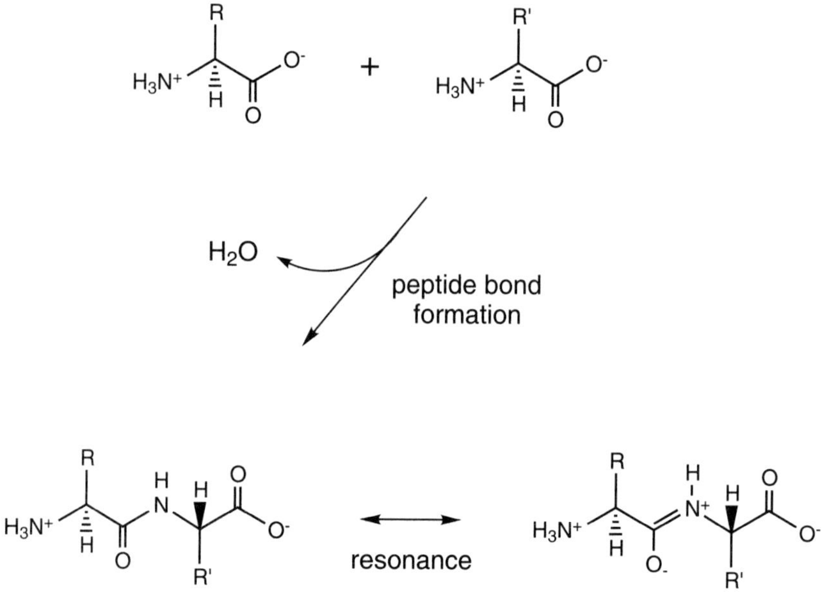

Figure 1-3. Peptide bond formation and resonance structures. A dehydration reaction links two amino acids with side groups R and R9. The resulting amide linkage is the peptide bond. Resonance structures reflect the partial double bond character of the C^N bond.

quence is also known as the *primary structure* of a polypeptide. The *secondary structure* of a protein refers to the folding of local regions into minimal-energy conformations which are stabilized by noncovalent bonds (such as hydrogen bonds) between nonadjacent residues. There are two major secondary structure motifs: the alpha helix and the beta sheet. A third motif is the turn, which describes a connection commonly found between different stretches of alpha helix or beta sheet.

The alpha helix structure describes a right-handed helix formed by the peptide chain, with about 3.6 amino acids per turn of the helix. Within an alpha helix, the amino hydrogen of an amino acid residue forms a hydrogen bond with the carbonyl oxygen of a residue in an adjacent turn of the helix. An amino acid residue of an alpha helix will form hydrogen bonds involving both its carbonyl and amino groups; therefore, it will be hydrogen-bonded to residues three amino acids away from it in both directions. The crosslinking involving multiple hydrogen bonds between residues results in high stability for the alpha helical structure.

The R groups of amino acids in an alpha helix are displayed along the outside of the helix. The alpha helices of some proteins contain hydrophobic amino acids along one side of the helix, while having charged groups on the other side. Such amphipathic helices are able to interact with hydrophobic surfaces (such as other hydrophobic polypeptides or cell membranes) on the one side and charged or polar surfaces (such as nucleic acids or the aqueous solvent) on the other.

The beta sheet describes a structure in which linear stretches of amino acids are aligned with other linear stretches of the polypeptide. The arrangement of hydrogen bonds is different from the alpha helix in that each amino acid residue is hydrogen bonded to an amino acid in a different stretch of the polypeptide chain. This structure has a sheetlike appearance, which forms the basis for its name. Within the beta sheet, the side groups of the amino acids will be arranged above and below the plane of the sheet. Beta sheets can be either parallel, where the two stretches have the same orientation of amino and carboxy ends, or antiparallel, where the orientations of the two stretches are opposite.

While peptides and localized regions of proteins adopt common structural motifs, the secondary structures of a protein are folded into a precise three-dimensional arrangement, or conformation. This *tertiary structure* is essential to the function of a protein. The tertiary structure is a minimum energy structure that is held together by multiple hydrogen bonds, as well as ionic and hydrophobic interactions. In some proteins this is stabilized by one or a few covalent bonds, usually disulfides formed by the oxidative crosslinking of two cysteine sulfhydryl groups. Side groups from various distant parts of a polypeptide can be juxtaposed in certain orientations to provide the chemical sites necessary for the functional activity of the protein. The tertiary structures of proteins, and the relationship to function, is described in more detail in Chapters 5 and 6.

The structure of many polypeptides can be subdivided into separate regions of distinct physical character. These regions are called *domains*. Domains often have distinct functional properties.

Many proteins consist of more than one polypeptide chain. The individual chains, or *subunits*, of such proteins are assembled into a *quaternary structure*. The polypeptide subunits are held together by multiple noncovalent interactions of the same type involved in tertiary structure. A protein with a single polypeptide chain is called a *monomer*, a two-polypeptide protein is called a *dimer*, a three-polypeptide protein is a *trimer*, and so on. The prefixes "homo" and "hetero" refer to multimeric proteins in which the subunits are all identical or different, respectively. For example, a protein that contains four nonidentical polypeptide subunits would be called a heterotetramer.

There is generally a single, low-energy conformation in which a polypeptide exists while functioning in its natural setting. This *native state* conformation is required for proper protein function. Disruption of the native structure, or *denaturation*, causes loss of activity (Fig. 1-4). Several physical factors can cause polypeptide denaturation, including heat, pH, ionic strength, and organic solvents.

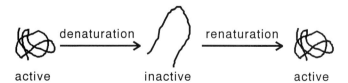

Figure 1-4. Protein denaturation. A protein must be folded into its proper three-dimensional conformation to be functional. Denaturation is the disruption or removal of the tertiary structure of the protein by chemical or physical processes, such as application of detergents or heat. In the absence of the denaturing agent, the denatured protein can, in some cases, be refolded into the functional state or renatured.

Cofactors

Important additions to the structure of many proteins are small chemical groups, known as protein *cofactors*. In general these cofactors are essential to the function of the proteins. A wide variety of cofactors are found within proteins, ranging from single metal atoms to complex organic biomolecules up to 1000 Da in molecular mass.

Although some cofactors are attached through covalent bonds, many are bound by noncovalent forces, especially hydrophobic and ionic interactions, and hydrogen bonds. Since the cofactor may be bound by amino acid side groups from disparate locations along the polypeptide chain, the tertiary structure of a polypeptide is essential for cofactor binding, and protein denaturation will release the cofactor from the polypeptide.

In some cases, cofactors can be released from the proteins under sufficiently mild conditions such that the tertiary structures of the polypeptides remain intact. The resulting cofactorless proteins are known as "apoproteins," or in the case of enzymes, "apoenzymes." Protein synthesis in the absence of the cofactor may also produce an apoprotein, although some proteins will not adopt their proper tertiary structure in the absence of essential cofactors. Many apoproteins can be reconstituted with their cofactors to form the "holoprotein" (or "holoenzyme"), the complete and fully functional form of the protein. Reconstitution of apoproteins with cofactor analogs can be used to probe the biochemistry of the structure and function of a cofactor within a holoenzyme.

The addition of cofactors to the arsenal of chemical groups available in polypeptides greatly extends the range of functional capabilities provided by proteins. Cofactors can have different functions depending on the proteins in which they are present. In hemoglobin and myoglobin, the cofactor heme is essential for the binding, transport, and release of the gases oxygen and carbon dioxide. However, in the enzyme catalase heme contributes to the catalytic conversion of hydrogen peroxide to water and oxygen.

A variety of metals are found in proteins as cofactors, and this is the reason many of them are essential nutrients. Iron is a component of heme, and it is also found within many oxidation-reduction enzymes within iron-sulfur clusters, which are involved in electron transfer. Zinc is a component of the digestive enzyme carboxypeptidase, and zinc is also required for the formation of unique secondary structures, know as "zinc fingers," within proteins involved in DNA binding and gene expression. Cobalt is present in the coenzyme cobalamine, which is derived from vitamin B_{12}.

Mixed Macromolecules

Some proteins function as complexes with other biological molecules, such as carbohydrates, lipids, and nucleic acids. Such combinations greatly extend the range of functions executed by proteins and peptides.

Carbohydrates are covalently attached to numerous polypeptides, particularly in eukaryotic cells. The carbohydrate moieties of proteins consist of long chains of simple sugars, and frequently these polysaccharide chains are branched at multiple points. The carbohydrates are covalently linked to proteins via either a hydroxyl group of serine or threonine, or an amido group of asparagine. Carbohydrate additions are therefore divided into O-linked and N-linked sugars, respectively. Carbohydrates can be substantial portions of the overall macromolecule; glycoproteins are known in which the carbohydrate comprises more than 20% of the overall molecular mass.

The carbohydrates are added soon after a newly synthesized polypeptide is exported across a cellular membrane, for example, within the Golgi apparatus just after transport from the cytoplasm. The carbohydrate additions function to localize proteins to specific cellular compartments. Most proteins which are secreted from eukaryotic cells contain carbohydrates.

A few proteins contain lipids linked to the polypeptide chain near the carboxy terminus. This linkage appears to be reversible within the cell, and the presence of the lipid alters the activity of the protein. The lipid portion is "dissolved" into a cell membrane, and this localizes the protein to the surface of the cell membrane. Without the lipid, the protein diffuses within the cytoplasm.

While most cellular proteins exist within the aqueous environment of the cytoplasm, many proteins are found embedded within the hydrophobic lipid membrane. Such membrane-bound proteins contain one or more hydrophobic stretches, regions consisting primarily of amino acids with nonpolar side groups, which are soluble in the lipid bilayer environment of the membrane. Membrane-bound proteins may also contain hydrophilic domains that are soluble in aqueous environments and, depending on the protein, such domains can project into the aqueous environment on one or both sides of the membrane. For example, many cellular receptors are membrane bound, with a hydrophilic domain for binding ligand. There are several types of membrane within a eukaryotic cell (nuclear, mitochondrial, Golgi, etc.) and each membrane type contains a specific set of membrane-bound proteins.

Nucleic acid-protein complexes are universal and highly conserved among species. Many nucleoprotein complexes are known, and they generally serve as components of essential cellular functions such as protein synthesis. The ribosome is a large nucleoprotein complex consisting of a few RNA molecules noncovalently associated with many different proteins. During protein synthesis, the ribosome also binds a messenger RNA, transfer RNAs, and the nascent polypeptide.

Analyzing Protein and Peptide Structure

In order to understand how proteins and peptides function, it is necessary to understand protein and peptide structure. The activity exhibited by a protein is manifested by its particular structure and composition. Proteins of similar function often share functional structures, and structures can be used for grouping proteins and determining relationships. There are several types of analysis available for discerning protein structures.

Gel Electrophoresis

Among the most informative and distinguishing characteristics of proteins are the number and sizes of their constituent polypeptides. For this reason techniques for size determination are used first and foremost in protein structure analysis. For most routine uses, gel electrophoresis is the method of choice for analyzing the basic subunit composition of proteins.

Although there are several different types of gel electrophoresis, they all work by the same basic principles. A protein sample is loaded onto a gel matrix, usually polyacrylamide for proteins, and an electrical potential is applied across the matrix. Since proteins are charged molecules, they will migrate due to the force of the electrical field. The gel matrix presents a frictional resistance to the migration.

In *denaturing gel electrophoresis*, the protein sample is first denatured with a detergent, usually sodium dodecyl sulfate (SDS), in the presence of a reducing agent, such as β-mercaptoethanol. The reducing agent breaks the disulfide bonds between cysteine residues that stabilize the polypeptide structure. In addition to denaturing the protein, the SDS also binds to and "coats" the denatured polypeptide. The amount of SDS that binds per unit length of polypeptide is nearly constant for all polypeptides, about one SDS molecule per two amino acid residues. Since SDS is a charged molecule, all proteins denatured and coated by SDS will have a nearly equal charge to mass ratio, and they will, therefore, all have the same attractive force applied to them when subjected to an electrical field. The electrophoretic migration velocity is due to the balance of attractive and resistive forces. The resistive force from the gel matrix is proportional to the size of the molecule. Since all SDS-denatured proteins in an electric field will be subjected to the same attractive force, a mixture of SDS-denatured polypeptides will differ in their electrophoretic migration velocities solely on the basis of size. In this way, the individual polypeptides in a mixture can be separated, as the smaller ones migrate faster than the larger ones. The SDS-denatured proteins are subjected to the electrical field for a set amount of time, then the gel is stained to make the protein bands visible. This technique is abbreviated "SDS-PAGE" for sodium dodecyl sulfate-polyacrylamide gel electrophoresis. Comparison of migration velocities with standards of known size allows a highly accurate determination of the sizes of polypeptides. Also, the number and sizes of the individual subunit polypeptides in a multimeric protein can be determined in this way.

Electrophoresis of proteins under conditions that retain the native conformation of the protein is called *native gel electrophoresis*. This method separates whole proteins on the basis of both size and shape.

Another commonly used electrophoretic technique is *isoelectric focusing*. In this technique, a pH gradient is placed across the gel matrix parallel to the direction of protein migration. The protein will migrate in the electrophoretic field until it reaches a point at which the pH of the solution matches the isoelectric point (pI) of the protein. At this point the protein has no net charge, and it will cease to migrate. Isoelectric focusing can be used to separate a mixture of proteins as well as determine the pI of an individual protein.

Two-dimensional electrophoresis allows even greater separation of protein mixtures. In this method, a mixture of proteins is first separated by isoelectric focusing. The resulting slab of protein bands is rotated 90° and then subjected to SDS-PAGE. This method allows resolution of more than a thousand proteins within a single gel, or about the number of proteins present in a bacterial cell. The method can be used to identify the slight differences in protein expression for cells grown under different conditions.

The *Western blot* is a powerful electrophoretic technique. In this method, proteins are first separated by SDS-PAGE. The gel is then placed in an electric field perpendicular to the plane of the gel, causing the proteins to migrate. A membrane, usually nitrocellulose, is placed next to the gel in the path of the migrating proteins. Upon leaving the gel, the proteins contact and bind tightly to the membrane. The membrane is removed from the electric field and incubated with a solution containing an antibody specific to a protein of interest. The antibody will bind the protein antigen stuck to the membrane. The antibody-antigen complex can be visualized in one of several ways.

For example, the antibody can be prelabeled with radioactive iodine-125 prior to incubation with the membrane, and subsequent exposure of the membrane to film reveals the places at which ^{125}I-antibody is bound to protein antigen. Western blots allow the specific detection and sizing of single proteins present in a mixture.

There are two other methods traditionally used for determining protein size. Analytical ultracentrifugation is employed for the study of the rate of migration of proteins in a centrifugal field. In the presence of a constant force provided by centrifugation, sedimentation rate is dependent on both the size and shape of a protein, due to the resistive force of the medium through which the protein passes. Many proteins have a globular shape, and therefore the differences in shape are minimized. The molecular mass of a protein can be determined by comparison of its sedimentation rate to the sedimentation rates of known proteins of similar shape. Gel filtration is a column chromatography method. As is true for analytical centrifugation, gel filtration separates on the basis of both size and shape.

Primary Structure Analysis

Amino acid analysis is the determination of the relative number of each type of amino acid that makes up a polypeptide chain. The polypeptide chain is hydrolyzed into its amino acid subunits with strong acids at high temperature. The resulting mixture contains amino acids, which can be separated and quantified. This gives the amino acid composition, or the ratio of each amino acid present in the polypeptide.

The amino acid sequence is a defining characteristic of a polypeptide, and amino acid sequence analysis is, therefore, an essential component of protein structure determination. Most methods for sequence determination involve the use of Edman degradation chemistry. In this reaction, a single amino acid residue is quantitatively removed from the amino terminus of a polypeptide. The identity of the amino acid released by this process can be determined subsequently. Removal of the amino terminal amino acid by Edman degradation leaves the remainder of the polypeptide intact, but with a new amino terminus (the second amino acid from the N-terminus in the original polypeptide). The process can be repeated to determine the identity of the new N-terminal amino acid. Repetitive processing of the polypeptide in this way allows for the amino acid sequence to be determined from the N-terminus. Machines have been developed that have automated this process. Such machines allow determination of amino acid sequence using as little as femtomole (10^{-15} mole) amounts of polypeptide.

With sufficient material, sequence can be determined more than 50 amino acids from an amino terminus. However, many polypeptides are hundreds of amino acids in length. In order to determine the entire sequence of longer polypeptides, the polypeptide chain must first be cleaved into smaller peptide fragments using specific chemical reagents or proteases (enzymes that cleave peptide bonds). The peptide fragments can then be isolated and sequenced individually.

Secondary and Tertiary Structure Analysis

Information about polypeptide secondary structure can be determined by circular dichroism. In this technique, the absorption of circularly polarized light by a protein or peptide is measured. The proportion of alpha helix and beta sheet structure within the polypeptide can be determined by this method. However, circular dichroism does not reveal the specific location of helical or sheet structures within the protein.

Tertiary and quaternary structures are determined through X-ray crystallography. A crystal of a pure protein placed in an X-ray beam will cause the beam to diffract in a unique pattern due to deflection of the X-rays from the atoms of the polypeptide. This diffraction pattern is used to calculate the three-dimensional distribution of atoms and groups of atoms within the protein. At highest resolution, X-ray diffraction can be used to assign the three-dimensional location of every atom in a protein crystal.

Physical Techniques for Structure Analysis

Spectrophotometry is used to determine the absorption of ultraviolet and visible light by proteins. The amino acids tryptophan, tyrosine, and phenylalanine all have useful absorption spectra. Measuring the absorption of light at 280 nanometers (nm) wavelength, which is due to tyrosine and tryptophan, is a simple way of determining protein concentration.

Many cofactors also have strong light absorption that can be monitored by spectrophotometry. The iron-sulfur centers of some redox enzymes absorb light at about 410 nm. Changes in the redox state of the centers can be monitored at this wavelength, as the different redox states have different light absorptivities. Heme cofactors have useful absorbance spectra that give much information about the cofactor environment and ligands.

A number of other physical techniques are available for protein and peptide analysis. Nuclear magnetic resonance (NMR) has recently been used for determining the tertiary structures of peptides and small proteins. Electron paramagnetic resonance (EPR) analysis provides information about the redox state of cofactors within some proteins. Mass spectrometry gives highly accurate determinations of protein molecular mass. Although mass spectrometry is used primarily for peptides and smaller polypeptides, recent advances have extended the range of protein sizes that can be determined by this technique.

Predicting Protein and Peptide Structure

When the rules governing adoption of certain structures are sufficiently understood, it is possible to predict structural characteristics of proteins based on rudimentary information. This is most clearly seen in the case of protein primary structure. As described earlier, the rules governing the amino acid sequence of a protein are simply those controlling the translation of DNA sequence by means of the genetic code. Determination of the DNA sequence for a gene encoding a protein reveals a series of DNA triplet codons, and the genetic code in turn reveals the amino acid corresponding to each codon. Thus, determination of the DNA sequence provides the amino acid sequence of the protein. This has become increasingly important in recent years, as methods for cloning, identifying, and determining the DNA sequence of genes have become more expedient than methods for the direct determination of complete amino acid sequences of proteins. However, some genes have protein coding sequences that are interrupted by *intervening sequences*, or *introns*, which complicates the prediction of protein sequence from DNA sequence.

Frequently, amino acid sequencing of proteins is used in tandem with cloning and DNA sequencing. A partial amino acid sequence of a protein provides some information about the DNA sequence that encodes it, as the genetic code describes which set of DNA codons are able to code for a particular amino acid sequence. By "reverse translation" of the genetic code, a partial amino acid sequence can be used to synthesize oligonucleotides that correspond to the DNA sequence of the gene. Such oligonucleotides can be used to clone and identify gene sequences, either by acting as labeled hybridization probes or through the use of the polymerase chain reaction (PCR). The DNA sequence of a cloned gene can, in turn, reveal the amino acid sequence of the polypeptide for which it codes.

It is quite possible to predict, with some certainty, secondary structures of proteins based on their amino acid sequence (Fasman, 1989). Modeling studies with synthetic peptides have shown that the 20 amino acids vary in their tendency to contribute to the formation of an alpha helical structure. This tendency has been quantified and is known as the "helicity" of an amino acid. Certain amino acids, such as proline, are known as "helix breakers," as helices do not form well around these amino acids in a polypeptide chain. Computer algorithms are available that aid in predicting the secondary structure based on primary sequence. Beta sheets and turns can also be predicted from primary structure. Note that knowledge of the DNA sequence of a gene for a protein allows prediction of the amino acid sequence, which in turn allows prediction of secondary structure.

The primary sequence of a protein contains the information necessary for the adoption of the final three-dimensional conformation of that protein. This was shown by studies involving the complete denaturation of enzyme proteins (Anfinsen, 1973). Denaturation removes the tertiary structure of a protein. Since the tertiary structure is necessary for enzymatic activity, the denatured enzyme is no longer enzymatically active. However, removal of the denaturing agent may allow an enzyme to refold, or renature, into its proper tertiary conformation, thus restoring enzymatic activity (Fig. 1-4).

These findings not only show that tertiary structure is essential for function, but also that the primary sequence of a protein can determine the secondary and tertiary conformations of that protein. This suggests that we might be able to predict protein structures based on their primary sequences, if we were able to "decode" the chemical information found within the primary sequence that directs adoption of higher order structure.

Although the information for tertiary structure is present within the primary sequence, decoding that information for the accurate prediction of tertiary structure is extremely difficult at present. The problem is that the number of potential folding conformations for a polypeptide is enormous, and the rules governing three-dimensional folding are not sufficiently understood. However, the rapidly increasing number of known protein structures should provide sufficient examples for the development of testable hypotheses concerning fundamental principles of higher order protein folding. This can then be coupled with advances in artificial intelligence computer progams to create future methods for the prediction of complete, three-dimensional protein structures based on primary sequences.

Certain structural features are found to be common among several different proteins, and these features are often correlated with particular functions. These recurring features are called *structural motifs*, and they can often be recognized from the primary sequence or secondary structure of a protein. For example, the helix-turn-helix motif has been found in several gene regulatory proteins from a variety of species, and it has been shown that in these proteins this motif binds to specific DNA sequences.

Structural motifs have been recognized for ATP binding, glycosylation sites, membrane interactions, and many other functions.

Protein Purification

The use of a protein, whether it is for scientific study or commercial purposes, usually requires isolation of the protein in a homogeneous state. A purified protein can be studied and characterized without interference from the presence of contaminants. Protein purification is the process whereby a protein is separated from unwanted molecules or contaminants. Ideally, a purified protein will be completely free of all other molecular components of the source from which it was isolated. The aim of protein purification is to obtain a homogeneous population of proteins, with all the individual molecules sharing the same polypeptide sequence, cofactors, physical characteristics, and activities.

Purification of a bacterial protein requires the physical separation of the protein from all the other bacterial molecules, including DNA, RNA, carbohydrates, lipids, cell wall material, as well as other bacterial proteins. Proteins are by far the most prevalent class of molecule within a cell, making up more than half the dry weight of a typical bacterial cell. Because of the differences in physical properties, it is usually a simple matter to separate the proteins from nonprotein cell constituents. However, isolation of a single protein from all the other proteins in a cell can be more difficult. The typical bacterial cell contains more than a thousand different protein species. Therefore, a single protein may constitute in the range of 0.1% of the total protein of a cell.

In order to purify a protein, one must first have an accurate and specific assay for the protein. An assay is simply a measurement of the amount of the protein of interest. Enzyme assays generally involve the reaction catalyzed by the enzyme; the rate at which a protein solution catalyzes the conversion of substrate to product can quite accurately reveal the concentration of a specific enzyme within it. Other proteins can be assayed based on physical characteristics. Several proteins have been purified solely on the basis of color, as they contained highly chromogenic cofactors which could be observed by the eye. If antibodies to a protein or peptide are available, these can be used to develop highly specific assays.

The quantitative result one obtains from an assay is known as the "activity." Although this term most often is used in reference to enzymes and enzyme assays, it can also be applied to the results of any assay. The extent to which a protein is purified can be measured quantitatively by determining the specific activity, which is the ratio of the amount of activity in a protein solution divided by the amount of total protein in the solution. Since purification removes contaminating proteins that have no assay activity, the specific activity increases as the protein becomes purer. The highest specific activity will be observed when a protein is completely pure, and the specific activity of the pure protein is a defining characteristic of that protein.

There are many ways in which proteins can be purified. Very few proteins can be isolated in a single step, and protein purifications generally involve multiple steps employing various techniques. Proteins all differ in their physical properties to some extent, and this is what allows proteins to be separated. However, the physical differences between proteins also require that the purification procedure for each protein be different; no single protocol works for the purification of all proteins.

The initial step in any purification is to prepare a cell-free solution of the protein. If the protein is present intracellularly, the cell must be broken open to release its contents. Membranes and other insoluble particles can be removed through centrifugation or other bulk methods. Soluble proteins will then be available in the resulting crude extract. Membrane-bound proteins must be released from the membranes, which can sometimes be accomplished through the use of detergents.

Proteins that are secreted into the extracellular medium can be much easier to purify than intracellular proteins. For such cases, the cells are not broken open; a simple filtration removes the cells, and all the intracellular proteins, from the secreted protein present in the medium. Thus a great deal of purification can be done quickly and efficiently. Many fungi are very efficient at secreting proteins, as most of their digestion of food sources occurs outside the cell. For this reason, fungi are being investigated as potential hosts for expression and secretion of foreign gene products.

Many purification protocols involve an early step in which the protein solution is subjected to a precipitation via the agency of organic alcohols, heat, or salt. In particular, ammonium sulfate precipitations are often used in protein purifications. This salt can be used in high enough concentrations to cause many proteins to precipitate without affecting their activity. Frequently, differential precipitations will be used, whereby a low level of precipitant will be used first to remove contaminants, followed by a higher level to precipitate the protein of interest.

The most powerful methods of purification usually involve some form of chromatography. A solution containing a set of proteins is pumped through a solid matrix contained within a cylindrical column. The matrix is a porous, insoluble material such as cellulose or agarose. The matrix material is chemically or physically modified such that it will have some affinity for the proteins, and this affinity will vary among the different proteins bound to the matrix. The proteins are eluted from the column using a solution which disrupts the binding to the matrix. A fraction collector automatically dispenses the flow-through solution into separate tubes with time. Thus, proteins that elute from the column at different times will be physically separated by collection into distinct fractions.

There are many types of matrices available for column chromatography of proteins and peptides, and the binding between the matrix and the protein defines the type of chromatography. In ion exchange chromatography, ionic interactions provide the affinity between protein and matrix. Hydrophobic interaction chromatography uses matrices modified to display hydrophobic moieties. A matrix can also be chemically modified to display chemical groups to which proteins bind through specific interactions. This is called affinity chromatography, and the types of chemical groups that can be used for this include enzyme substrates, protein ligands, and specific antibodies. In gel filtration or size exclusion chromatography, the matrix is physically altered to contain pores of a certain size that exclude molecules of larger size. Smaller proteins enter the pores and are, therefore, retained longer than larger proteins, which do not enter the pores. Thus, gel filtration chromatography separates proteins on the basis of size and shape.

An example of ion exchange chromatography is a matrix that is chemically modified so that it displays positively charged chemical groups. Negatively charged proteins will bind to such a matrix through ionic interactions, with affinities dependent on their charge. Proteins bound to such a matrix can be eluted with a salt solution, which disrupts the ionic interactions. The different proteins are bound to the column with different affinities, and therefore they will each be eluted by unique and specific salt concentrations.

A special type of column chromatography is called high performance liquid chromatography, or HPLC. This method involves pumping solutions through a column at higher pressures than with traditional, or open column, techniques. HPLC provides greatly improved resolution and speed, although it is limited in the quantity of material that can be handled.

Structural, Regulatory, and Transport Proteins

The functions performed by proteins and peptides are highly varied; some of these are described below. For the purposes of introduction, proteins are divided into four functional groupings: structural, regulatory, transport, and enzymatic. However, one should keep in mind that the diversity of protein functions is such that many proteins defy simple classification, while others could belong to more than a single group.

Structural Proteins

As the name implies, structural proteins provide much of the framework upon which a cell or multicellular organism is built. Within cells, the protein tubulin makes up much of the cytoskeleton. Although vertebrates have an inorganic skeleton, the bones are held together by the protein collagen and are articulated by the actions of the muscle proteins, actin and myosin. Skin, hair, nails, hooves, and feathers are other examples of biological features made up primarily of structural proteins. Blood clots are composed of the protein fibrin, which is derived from its precursor fibrinogen. Histones are proteins of eukaryotes that bind and give structure to chromosomal DNA.

Many structural proteins have a fibrous shape, whereas most nonstructural proteins are globular. A protein is said to be fibrous if the long axis of its native conformation is more than seven times its width. The fibrous shape underlies many of the functional characteristics of structural proteins.

The function of certain proteins is to store chemical components for future use by the organism. Ferritin is a storage protein that binds iron in the blood of mammals. Soybean storage proteins provide a reservoir of amino acids for use by the developing soybean seedling.

Regulatory Proteins

Regulatory proteins affect the activities expressed by other agents, including other proteins, DNA, organelles, cells, and tissues. Many proteins involved in the immune response fall in this category. Antibody proteins present in serum bind foreign antigens in order to inactivate them. The binding of foreign antigen to membrane receptors on certain immune system cells induces an activation of the cells, which causes the cells to become motile and cytolytic. Immune system cells communicate with each other using interleukins, a class of polypeptides that cause cellular activation, differentiation, and division. Interferon is a type of interleukin with antiviral properties, which is being investigated as a therapeutic drug.

Many hormones are peptides or polypeptides. Insulin is the classic example of a polypeptide hormone. This protein regulates cellular metabolism of glucose; inappropriate insulin production can result in diabetes. However, diabetes can also be manifested through faulty expression of the insulin receptor found at the cell surface. Although not all hormones are polypeptide in nature, all hormone receptors are proteins.

Expression of the genetic information contained within chromosomal DNA is controlled by the action of proteins. Transcriptional regulators bind specific DNA sequences in the vicinity of genes and cause the transcriptional expression of those genes to be up- or down-regulated. The activities of transcriptional regulators themselves respond to cellular conditions. For example, membrane-bound hormone receptors, upon binding a hormone ligand, cause an intracellular signal to be produced that is relayed to a transcriptional regulator, which in turn alters gene expression. Thus, membrane receptors may work in concert with DNA-binding transcriptional regulators to translate information from the extracellular environment into the appropriate genetically programmed response.

Certain antibiotics are peptide in nature. Fungi produce unusual peptide antibiotics, including cyclic peptides and peptides containing D-amino acids. Recently antibiotic peptides were discovered and isolated from frog skin. They have been named "magainins," and they show powerful activity toward certain bacterial pathogens with little toxicity to the mammalian host.

Transport Proteins

Transport proteins are involved in the movement of other chemical species within the cell, across cell membranes, or around multicellular organisms. The types of particles requiring transport (and examples) include energy sources (glucose), gases (oxygen), ions (Ca^{++}), and lipids (cholesterol).

Hemoglobin in blood carries oxygen from the lungs to the rest of the body tissues, and it carries carbon dioxide and protons from the body tissues to the lungs. Oxygen binds to the iron moiety of the heme cofactor of the protein. Binding of oxygen to hemoglobin is cooperative, meaning that binding of an initial molecule of oxygen to hemoglobin facilitates binding subsequent oxygen molecules. A marked hysteresis in hemoglobin binding to oxygen is observed upon slight changes in pH. This is thought to promote the transport activity of hemoglobin: hemoglobin binds oxygen more tightly in the higher pH environment of the lungs, whereas the lower pH of the external tissues promotes oxygen release from hemoglobin.

Enzymes

By far the most diverse set of proteins are the enzymes, proteins that catalyze specific biochemical reactions. Enzymes facilitate nearly every action performed by a biological organism, whether it is energy production and consumption, synthesis and decomposition, growth and reproduction, or thinking and breathing. As such, enzymes play a major role in determining and defining biological life as we know it.

Enzyme Biological Chemistry

As catalysts, enzymes cause a reduction in the activation energy of chemical reactions. Enzymes do not alter the thermodynamic equilibrium of a reaction, they only enhance the rate at which a reaction approaches equilibrium. Like other catalysts, enzymes are not modified or consumed in the overall process of reaction catalysis.

One of the most powerful aspects of enzyme catalysis is the high degree of specificity which enzymes have for reaction substrates. Most enzymes catalyze a single reaction. Enzymes are "stereoselective," that is, they will catalyze the reaction of only a single stereoisomer. Although other substrate stereoisomers may have the same overall thermodynamic equilibria, typically the reaction of only a single stereoisomer will be catalyzed by an enzyme.

The enhancement of reaction rates by enzymes is also quite impressive. Enzymes are able to enhance reaction rates by factors of 10^9 to 10^{12}, and many enzymes are known that catalyze reactions at rates of greater than one thousand substrate molecules converted to products per second per enzyme molecule. Under the same conditions, the reaction rates in the absence of the enzyme catalyst would be on the order of years. Further, this diverse array of reactions proceeds under the mild and limited environment of physiological conditions, generally limited to aqueous solutions with narrow ranges of pH and temperature.

Enzyme Nomenclature and Classification

In the early years of enzyme research, enzymes were named by their discoverers. The proliferation of enzymes found in the 1940s and 1950s led to confusion, as some enzymes were given several names by different workers, and some names were given to multiple enzymes.

In order to develop a logical and regulated system for naming enzymes, an International Commission on Enzymes was formed by the International Union of Biochemistry in 1955. The Committee on Enzymes formulated guidelines that allowed enzymes to be named in an orderly manner (Webb, 1992). Periodic revisions have kept this system updated.

The Enzyme Commission system of enzyme nomenclature is based upon catalytic function; that is, enzymes are named by the reactions they catalyze. Six major categories of catalytic reaction were recognized, and these are shown in Table 1-2. Each enzyme is assigned to one of the six categories, and further subdivisions of the categories specify each enzyme until a unique name is assigned. The specific reaction substrates are used as the final qualifiers for an enzyme name. Names determined in this manner are the systematic names.

Table 1-2 Enzyme Nomenclature

E.C. Number	Systematic Name	Reaction Type
1	Oxido-reductases	Oxidation-reduction
2	Transferases	Functional group transfer
3	Hydrolases	Hydrolysis
4	Lyases	Addition to double bonds
5	Isomerases	Isomerization
6	Ligases	ATP hydrolysis in bond formation

In addition to a name, each enzyme is also assigned a number specifying the category and subcategories to which the enzyme belongs. This number provides a convenient shorthand notation for identifying a unique enzyme type. The number is preceeded by the initials E.C., standing for "Enzyme Commission." The numerical designation consists of a series of numbers referring to categories and subcategories separated by periods. The first digit is a number between one and six referring to a major category. Subsequent numbers refer to branching subcategories.

While the IUB system of enzyme nomenclature provides a precise and unambiguous framework, these names can be cumbersome for regular use. For this reason, the Enzyme Commission gives "recommended" names, also known as "common" or "trivial" names. These are often abbreviated versions of the systematic name, providing limited information about the enzymatic reaction. Many enzymes have common names that predate the Enzyme Commission and have been retained, such as catalase, which catalyzes the conversion of hydrogen peroxide into oxygen and water.

Since enzymes do not affect the thermodynamic equilibrium of a reaction, the direction in which an enzyme-catalyzed reaction proceeds depends on the concentration ratios of substrates and products. Under normal conditions within a cell, most enzymatic reactions proceed in only a single direction, and enzyme names are assigned based on the physiologically relevant reaction direction. However, in some cases the reaction direction in vivo is not clear, or the identification of some enzymatic activities involves placing the enzyme under nonphysiological conditions for observation of the reaction proceeding in a direction opposite the physiological one. For this reason many enzymes are named by reactions that are the reverse reactions of the physiologically relevant ones.

Kinetics

The rate at which an enzyme-catalyzed reaction occurs is determined by a number of factors both inherent and external to the enzyme. Quantifying the factors that affect reaction rates is of obvious importance, and therefore methods have been developed to describe mathematically the various factors which determine catalytic rates. These methods are grouped under the term *kinetics*, as they quantify the rates of changes that occur in the chemical system during reaction. Understanding the kinetic properties of an enzyme allows one to predict activity in a given situation and also to compare and relate the kinetic properties of an enzyme to those of other enzymes.

The rate at which an enzyme catalyzes a reaction is also known as the "activity." Enzyme activity is generally measured by observing the rate at which the product of a reaction is produced or the substrate of a reaction is consumed in the presence of a given amount of enzyme.

The *steady-state* theory of enzyme kinetics was developed by L. Michaelis and M. Menten in 1913 (and it is often called Michaelis-Menten kinetics), with modification by G. Briggs and J. B. S. Haldane in 1925 (Dixon & Webb, 1979). This theory analyzes a reaction according to the following equation:

$$E + S \rightleftharpoons ES \rightarrow P \tag{1}$$

In this reaction, a single substrate, S, binds to the enzyme, E, to form a complex, ES. The ES complex can dissociate back into E and S, or else the substrate can be converted to product, P.

The Michaelis-Menten, or steady-state, theory is summarized by the following equation:

$$v = \frac{V_{max}*[S]}{K_m + [S]} \qquad (2)$$

This equation describes a reaction in which a single substrate goes to a single product, where v is the reaction velocity, [S] is the substrate concentration, K_m is the Michaelis-Menten constant, and V_{max} is the maximal velocity of reaction possible.

There are a few assumptions that underlie the steady-state theory, and these are important for understanding. The first assumption is that the enzyme and substrate form a complex, ES. The second assumption is that the concentration of the ES complex is in a steady state. The third assumption is that the concentration of S does not change significantly during the reaction. The first assumption has been demonstrated physically for a number of enzymes, while the other two are generally true under most enzyme assay conditions where S is in great excess relative to E and P. Reactions that involve two or more substrates or products can also be modeled using the steady-state approach, but for the purposes of simplified illustration the single substrate-single product reaction will be discussed here.

According to the steady-state equation, the reaction velocity, v, will increase with increasing S as shown by the plot in Figure 1-5. This shows a hyperbolic curve that ap-

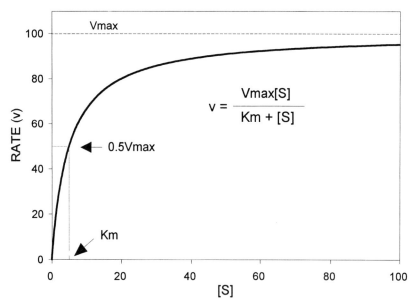

Figure 1-5. The effect of substrate concentration on enzyme reaction rate. The Michaelis-Menten equation predicts a plot of v versus [S] (rate versus substrate concentration) to be in the shape of a hyperbola. At low [S], small changes in [S] have relatively large effects on rate, while at very high [S], rates approach an asymptote of V_{max} at infinite [S]. The K_m is the substrate concentration which gives a rate of one-half the V_{max}. In this example, the rates and substrate concentrations are given in arbitrary units, the $V_{max} = 100$ and $K_m = 5$.

proaches a plateau value for v at infinite [S]. This plateau is the maximal velocity of reaction, or V_{max}. The V_{max} can also be defined as the reaction velocity when the enzyme is completely saturated with substrate. From the steady-state equation, it can be seen that the K_m is equal to the substrate concentration at half-maximal reaction velocity (when $v = 0.5*V_{max}$), and this is also indicated in the v versus [S] curve of Figure 1-5.

The steady-state equation describes the kinetic properties of an enzyme in terms of only two parameters, V_{max} and K_m. The V_{max} is a measure of the potential catalytic activity of an enzyme. This is sometimes expressed as the catalytic constant, k_{cat}, defined as the molecules of substrate converted to product per second per molecule of enzyme. The k_{cat} is given in units of sec^{-1}, and it is also known as the turnover number. The K_m is a measure of the affinity of an enzyme for substrate, and it is given in units of concentration (usually millimolar or micromolar). When comparing kinetic parameters of different enzymes, the K_m gives a quantitative denotation for substrate affinity: the lower the K_m, the lower the [S] required to produce a given submaximal reaction velocity.

In order to determine V_{max} and K_m, reaction velocities for an enzyme are determined at a variety of substrate concentrations, and the results are plotted as rate versus [S]. Currently, computer algorithms are readily available for determining the values of K_m and V_{max} that give the "best fit" of the data to a hyperbolic curve described by the steady-state equation. Simpler graphical methods are also available for determining these values. Although many different methods are available, the most common method traditionally used is the double-reciprocal, or Lineweaver-Burke, plot. Inversion of the steady-state equation results in the following:

$$\frac{1}{v} = \frac{K_m + [S]}{V_{max}*[S]} = \frac{K_m}{V_{max}}*\frac{1}{[S]} + \frac{1}{V_{max}} \quad (3)$$

By this equation, a plot of $1/v$ versus $1/[S]$ will give a straight line with a y-intercept of $1/V_{max}$, a slope of K_m/V_{max}, and an intercept at the x-axis (when $1/v = 0$) of $-1/K_m$ (see Fig. 1-6).

In some cases the binding of a substrate molecule changes the affinity of an enzyme toward the binding of further substrate molecules. This effect is known as cooperativity. If the affinity increases after the initial binding, the cooperativity is said to be positive, and if the affinity decreases after the initial binding, the cooperativity is negative.

Many factors external to the enzyme can play a role in determining the activity an enzyme will express. As described above, substrate concentrations play a major part. Environmental conditions, such as pH and temperature, are also important. Most enzymes display a pH optimum; that is, under the conditions of certain pH values an enzyme will show optimal activity. A graphical profile of pH versus activity is an important characterization of an enzyme, with some enzymes showing a sharp, distinct pH optimum and others having a broad pH optimum.

The activities of most enzymes increase with temperature. A general rule of thumb is that activity increases twofold for every ten degrees Celsius rise in temperature. However, this effect can only be seen to a certain point, as enzymes will begin to denature at some high temperature and then be subject to thermal inactivation.

Chemical substances that specifically reduce the activity of an enzyme are known as inhibitors. An inhibitor may bind to an enzyme in such a way as to diminish the

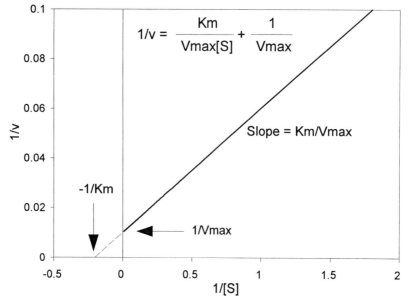

Figure 1-6. Double-reciprocal plot. Plotting the reciprocal of rate versus the reciprocal of substrate concentration gives a straight line, as predicted from inversion of the Michaelis-Menten equation. Values for V_{max} and K_m can be determined from y- and x-intercepts, respectively. As in Figure 1-5, the rates and substrate concentrations are given in arbitrary units, the $V_{max} = 100$ and $K_m = 5$.

binding of substrate or alter the enzyme-substrate interaction upon binding. If removal of the inhibitor from the enzyme restores full activity, the inhibition is reversible. An inhibitory agent that completely eliminates activity and is irreversible is called an inactivator.

Chemistry of Enzyme Mechanisms

Although enzymes exhibit diverse and powerful catalytic abilities that provide the tools for running complicated biological entities, it is important to bear in mind that enzymes are chemicals that obey physical laws. For this reason it is possible to study and learn the physical chemistry underlying enzyme mechanism. Such knowledge allows an understanding of enzyme catalysis at a molecular level, and this provides a predictive basis for the rational modification of the enzyme molecule.

Many enzymes employ acid-base catalysis as part of their catalytic mechanisms. Redox enzymes use internal electron transfer to pass electrons from a reducing substrate to an oxidizing substrate. Some enzymes involved in bond breakage induce a torsional strain to a substrate upon binding. This important area of research is expanded upon in Chapter 8.

Cofactors play an important role in the activities of many enzymes. Of special interest are the cofactors derived from B vitamins, the "coenzymes." These cofactors were originally isolated as small, heat-stable organic molecules which were required for the activities of certain enzymes. Figure 1-7 shows the structures of a few coenzymes. These organic compounds are involved in many enzymatic reactions, especially

Introduction to Peptides and Proteins / 25

oxidation-reduction and group-transfer reactions. In some cases coenzymes are covalently linked to enzymes, in which case the coenzyme is called a *prosthetic group*. Some enzymes bind coenzymes very tightly but noncovalently, while others bind loosely. Coenzymes can also be used as substrates by some enzymes.

Summary

The recent explosion in our knowledge and understanding of protein and peptide structure and function has come through a synergistic increase in techniques of molecular biology, cloning, and genetics. New avenues of analysis have been opened that allow

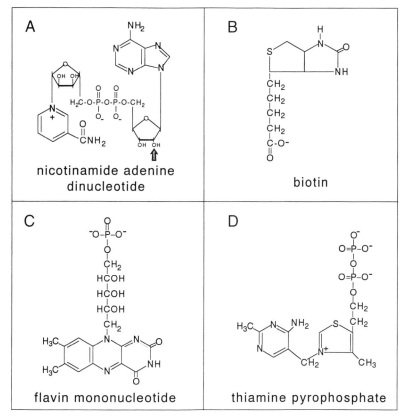

Figure 1-7. Chemical structures of four coenzymes. (A) Nicotinamide adenine dinucleotide (NAD) is derived from the vitamin niacin. Addition of a phosphate group at the 29-hydroxyl group marked by the arrow results in the coenzyme nicotinamide adenine dinucleotide phosphate (NADP). Both NAD and NADP are involved in oxidation-reduction reactions. (B) Biotin is crucial for certain reactions involving carboxy group transfer. (C) Flavin mononucleotide (FMN) is also called riboflavin 59-phosphate. In the coenzyme flavin-adenine dinucleotide (FAD), a phosphodiester bond links riboflavin phosphate to adenosine 59-diphosphate. FAD and FMN are cofactors and substrates for many electron-transfer reactions. (D) Thiamine pyrophosphate is a cofactor for decarboxylaton reactions. Thiamine deficiency can lead to the human disease beriberi.

increasing control over the understanding and manipulation of proteins and peptides. Certain techniques will continue to evolve and play a major role in protein chemistry research.

Site-directed mutagenesis (Chapter 9) is a method for altering specific codons in a cloned gene. Expression of such modified genes produces polypeptides altered at specific amino acids. This powerful technique allows the analysis of polypeptide chemistry at the amino acid level. Such mutagenesis also gives the ability to improve the activity of proteins in a directed manner.

Catalytic antibodies (Chapter 10) resemble enzymes in that they can act as catalysts for certain chemical reactions. They are made by using analogs of reaction transition states as immunogens to produce the antibodies. This technique can provide biological catalysts for reactions that have no known catalyst.

Enzyme immobilization is an active field of research, particularly for industrial uses. The ability to link an enzyme to a solid support while retaining its activity allows flow-through methods for the bioprocessing of substrates. The large-scale production of high-fructose corn syrup has been achieved through the use of immobilized glucose isomerase, which converts glucose to the sweeter sugar fructose.

Although not reviewed in this volume, an important future use for proteins and peptides will be as pharmaceuticals. The integral roles that they play in the activities and regulation of human health and disease will be harnessed increasingly for the improvement of human health. Every aspect of human health care will be affected, ranging from diagnostic tests to regulation of physiology to treatment of cancer and other diseases.

2

Chemical Synthesis of Peptides

Victor J. Hruby and Jean-Philippe Meyer

The synthesis of peptides, proteins, pseudopeptides, and peptidomimetics has become of central importance to essentially all fields of chemistry, biology, and biophysics. Proteins and peptides and their conjugates (glycopeptides, lipoproteins, etc.) are the major products of genes and posttranscriptional processing and thus at the center of the revolution in molecular biology. Despite their central importance, many chemists are unaware of the progress that has been made in this area. The synthesis of polypeptides, even complex polypeptides, is one of the most highly developed areas of synthetic chemistry, and its best practitioners can "routinely" synthesize peptides of molecular weight 3,000 to 10,000 or greater. The development of protecting groups and racemization-free condensation reactions in essentially quantitative yield (>99.5%) in optimized cases, has made it possible to address problems in the chemistry of living systems that would have been impossible even a decade ago. With the current growing interest in pseudopeptides, peptidomimetics, peptide and peptidomimetic libraries, peptide catalysts, and other related areas, there is a growing need for synthetic chemists to familiarize themselves thoroughly with the methods, tactics, strategies, and limitations of the current methods and technology.

The purpose of this chapter is to provide an overview of the synthetic methodology available both for solution phase peptide synthesis and for solid-phase peptide synthesis (SPPS). It is not possible in the limited space available to give a comprehensive treatment of the field. Rather, the chapter emphasizes those general considerations that are important in peptide synthesis, to introduce the reader to some current topics that seem of general interest, and to point the interested reader to more comprehensive treatments and other aspects of the subject in the literature.

General Considerations

The development of rapid, highly stereospecific and very high yield synthesis of polypeptides and proteins represents one of the major successes in organic chemistry in the past 30 years. Using the methods discussed in this chapter, it often is possible to synthesize and purify polypeptides of 30–100 amino acid residues (3,500–12,000 Da) in two to three weeks or less. This has allowed access of millions of peptides for evaluation of their conformational, catalytic, hormonal, neurotransmitter, antibiotic, and other chemical, physical, and biological properties. As a result, our understanding of such diverse chemical and biological problems as *de novo* peptide and protein secondary structure design; the protein folding problem; structure-biological activity relationships for hormones, neurotransmitters, cytokines, and so on; the development of peptide catalysts; enzyme inhibition; immune response regulation; and so on, have greatly expanded in recent years. The powerful synthetic methods that have been developed in the polypeptide area have had a dramatic effect not only on what can be synthesized, but also on what chemists, biologists, and biophysicists who need peptides for a wide variety of physical, chemical, and biological studies can do. Indeed, it might be argued that diverse peptides and their analogues are among the most readily available biologically active compounds. In any case, the remarkable achievements in synthetic peptide chemistry, including the powerful solid phase synthetic method of Merrifield (1963) and a variety of sophisticated methods for orthogonal synthesis (Atherton & Sheppard, 1989; Barany & Merrifield, 1979; Bodanszky et al., 1976; Bodanszky, 1993; Erickson & Merrifield, 1976; Fields & Noble, 1990; Jones, 1991; Stewart & Young, 1984) are impressive.

In general, there are two major impetuses for peptide synthesis. First, peptides, pseudopeptides, depsipeptides, and related compounds generally are biologically active and stimulate, control, and modulate most biological functions in living systems, and in addition can act as hormones, neurotransmitters, antibiotics, antifungal agents, anticancer agents, and so on. Hence, there is a need to synthesize the native ligands for use in biological studies, but also to design and synthesize analogues to improve on their properties in terms of potency, selectivity, stability, diminution of toxic side effects, and so forth. Second, the ability of scientists to predict on first principles the three-dimensional structure of a peptide or protein (the "second code") still needs much improvement. Hence there continues to be a need to design peptides with predicted conformations, and then to test these predictions by a variety of biophysical methods (NMR, CD, fluorescence, etc.). *De novo* design and synthesis of polypeptide ligands is an area of great current interest, and the development of excellent methods for peptide synthesis has had a critical impact on design development. The discussion begins with solution phase methods.

Solution Phase Synthesis

Numerous methods have been developed throughout the years for the solution phase synthesis of peptides (Bodanszky et al., 1976; Bodanszky, 1993; Jones, 1991). Before starting the synthesis of any peptide, one should be aware of the advantages, drawbacks and limitations of the various strategies that are available. The choice of strategy often will depend on the sequence of the peptide and its special properties. In general, however, all strategies will depend on several general considerations.

Choice of the N^α-Protecting Groups

Since peptides are usually synthesized from the C- to the N-terminus in order to minimize racemization, the choice of temporary α-amino protecting groups for the growing peptide is crucial, and will influence the coupling strategy, the side chain protection schemes, and the final deprotection step(s). Two urethane protecting groups, namely *t*-butoxycarbonyl (Boc) and fluorenylmethoxycarbonyl (Fmoc) are used most commonly (Fig. 2-1). They are cleaved, respectively, under relativity mild acidic and

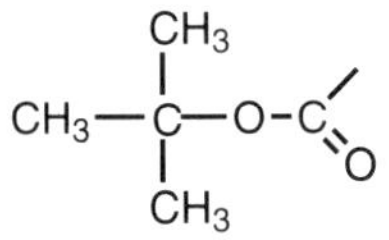

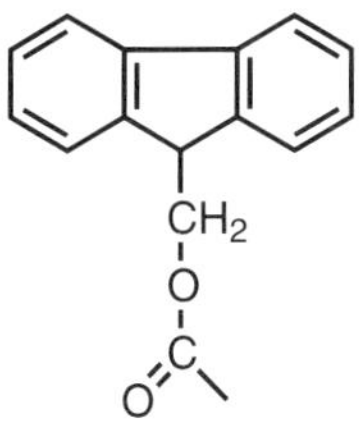

Figure 2-1. Structures of the *tert*-butoxycarbonyl (Boc, *left*) and 9-fluorenylmethoxycarbonyl (Fmoc, *right*) protecting groups.

basic conditions. Other protecting groups, removed, for instance, by thiols or by photolysis also are available. Comparing the Boc and Fmoc strategies for overall efficiency is rather difficult. Nevertheless, the latter is easier to handle in a sense because it does not require the use of special equipment for hydrofluoric acid or other strong acids that often are used for final side chain deprotection using the Boc strategy. The synthesis of difficult peptides containing highly hydrophobic sequences seems to demonstrate the superiority of the Fmoc strategy for this application, but in most cases the method of choice is not critical.

Side Chain Protection

These usually are chosen to be "orthogonal," that is, stable to the condition of deprotection of the temporary N^α-protecting group, but readily removable at the final deprotection step. Side chain protections orthogonal to the Boc and the Fmoc groups, respectively, often are referred to as "benzyl" and "*tert*-butyl" groups as they are usually related to them structurally. In some cases, only partial deprotection of the peptide is desired, and then special considerations arise. For example, the synthesis of a specific disulfide bridge between two cysteine residues in a peptide containing more than two cysteines can require partial deprotection that is regiospecific (three-dimensional orthogonal synthesis).

Coupling Methods

A large number of methods are known since amide bond formation is one of the most highly developed transformations in organic chemistry. The best methods tend to proceed rapidly to completion of coupling with no or few side reactions or racemization. Though the coupling of natural amino acids is usually straightforward and can be achieved by classical methods such as DCC-HOBt (Fig. 2-2), incorporation of sterically hindered residues or synthesis of hydrophobic sequences may require the use of more efficient reagents, such as those belonging to the BOP or HBTU families (Fig. 2-3).

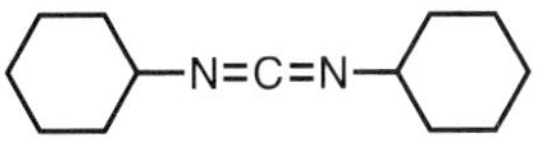

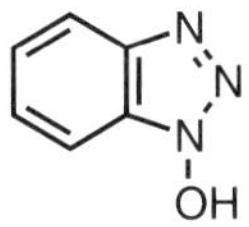

Figure 2-2. Structures of dicyclohexylcarbodiimide (DCC) and 1H-hydroxybenzotriazole (HOBt).

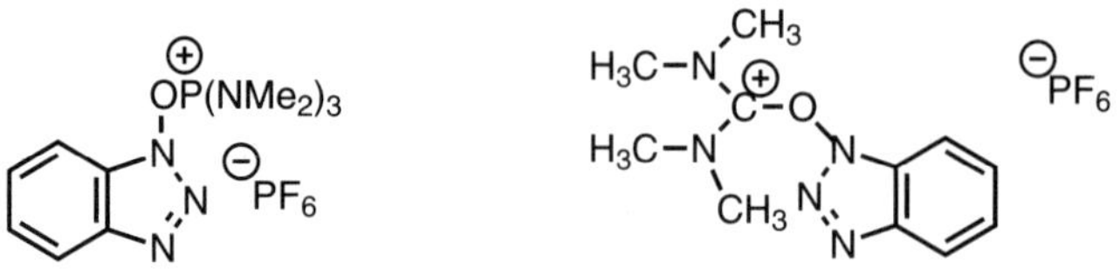

Figure 2-3. Structures of benzotriazole-1-yl-oxy-tris-(dimethylamino)phosphonium hexafluorophosphate (BOP, *left*) and 2-(1H-benzotriazol-1-yl)-1,1,3,3-tetramethyluronium hexafluorophosphate (HBTU, *right*).

Deprotection Strategies

The method of choice depends directly on the synthesis strategy used. If Boc-benzyl protecting groups are employed, the final deprotection is usually performed using HF or trifluoromethanesulfonic acid (TFMSA). On the other hand, the Fmoc-*tert*-butyl protecting groups usually are removed by trifluoroacetic acid (TFA). As already mentioned, partial deprotection can be accomplished for special purposes. Finally, the choice of scavengers, that is, additives necessary to avoid side reactions such as alkylation of the indole ring of tryptophan by reactive carbocations, is crucial.

Protection of the C-Terminal Carboxyl Group

Depending on what kind of C-terminal group is required (usually a carboxyl or an amide), the C-terminal amino acid involved generally will have this group protected as an ester (which can be removed by mild hydrolysis) or as an amide.

Stepwise and Fragment Condensation

Peptides can be synthesized either by stepwise addition and N^α-deprotection of each residue one after the other, or by fragments (containing two or more residues) of the peptide that have been synthesized separately before being ligated together.

All these decisions need to be made before attempting the synthesis of any peptide. Each of these topics will be discussed again in this chapter, in the context of coupling methods or protecting groups.

This section describes the main types of chemical reactions used to form peptide bonds. Though a fairly comprehensive discussion of numerous methods is attempted, the focus is mainly on those methods most widely used by peptide chemists. When two amino acids, one with a free carboxyl group and the other with a free amino group, are present in solution, they do not spontaneously form a peptide bond. The free carboxyl group must first be activated prior to aminolysis by the free amino group. Different methods can be used, and they mainly fall into two categories. The carboxyl component can be activated leading to the formation of a reactive acylating agent and immediately treated with the amine component, leading to the formation of the peptide bond. In this case, the protected amino acid acylating agent can be formed *in situ* by addition of the coupling reagent, allowing the condensation reaction to take place as soon as the reactive intermediate is formed. In the second case, if the acylating agent is stable, the activated amino acid derivatives can be purified and characterized before being reacted with the amine component. Though this classification is convenient, it will be shown later that in many cases the distinction between these classes is not clear.

The first coupling methods discussed in this section (carbodiimide, mixed anhydride, active ester, and azide methods) have been in use routinely for over three decades and have been reviewed extensively (Bodanszky, 1979; Meienhofer, 1979; Rich & Singh, 1979). Therefore, the discussion here will be brief, outlining the mechanistic aspects, drawbacks, and advantages of each. Coupling reagents and methods that have been introduced to peptide chemistry more recently, such as the BOP reagent (and related analogues), or the UNCA method also will be discussed.

Coupling Methods Involving the Use of Carbodiimides

Carbodiimides (Rich & Singh, 1979) were first used in peptide synthesis by Sheehan & Hess (1955). They used dicyclohexylcarbodiimide (DCC) to form a peptide bond between Z-Gly-OH and H-Gly-OMe leading to the protected Z-Gly-Gly-OMe as shown in Scheme 2-1. N,N'-dicyclohexylurea (DCU) is formed as a by-product. This coupling method gained immediate attention due to its efficiency and ease of use. DCC and other carbodiimides like diisopropylcarbodiimide or water soluble carbodiimides (Sheehan & Hlavka, 1956) are widely used for numerous solution syntheses. Moreover, most solid phase syntheses, whether automated or not, used DCC or a related compound to form the peptide bond in the growing peptide chain. It is important to note that DCC and other carbodiimides can be used directly either by adding them to a solution of N-protected amino acid and peptide or amino acid ester, or for the formation of activated N-protected amino acids that can be used subsequently for peptide bond formation. Compounds such as symmetrical anhydrides (Erickson & Merrifield, 1976) and activated esters like pentafluorophenyl or 1-hydroxybenzotriazole (HOBt) esters are easily obtained with DCC. These reactive amino acid derivatives are usually stable enough to allow purification and characterization prior to their use in peptide synthesis, thereby decreasing the possibility of side reactions.

Nevertheless, the carbodiimide method has some drawbacks. These include racemization (Anderson & Callahan, 1958; Hofmann et al., 1958), dehydration of some amino acids like asparagine and glutamine (Gish et al., 1956; Ressler, 1956), and formation of by-products such as N-acylureas. Many of these unwanted side reactions occur at the activation stage of the carbonyl component when the intermediate O-acylisourea

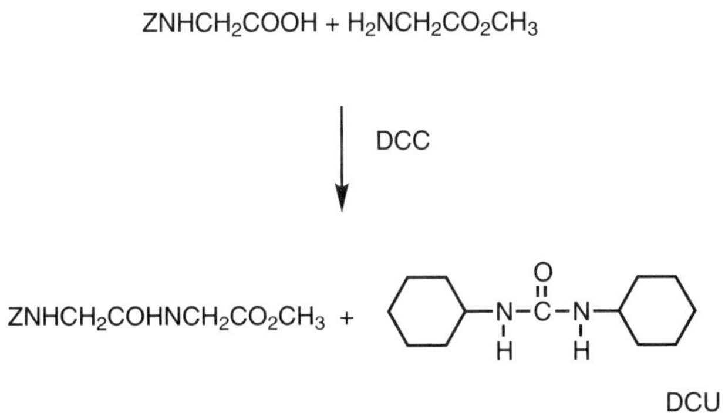

$$ZNHCH_2COOH + H_2NCH_2CO_2CH_3$$

$$\downarrow DCC$$

$$ZNHCH_2COHNCH_2CO_2CH_3 \quad + \quad \text{(Cy)}-\underset{H}{N}-\underset{\underset{O}{\|}}{C}-\underset{H}{N}-\text{(Cy)}$$

DCU

Scheme 2-1. Synthesis of Z-Gly-Gly-OMe.

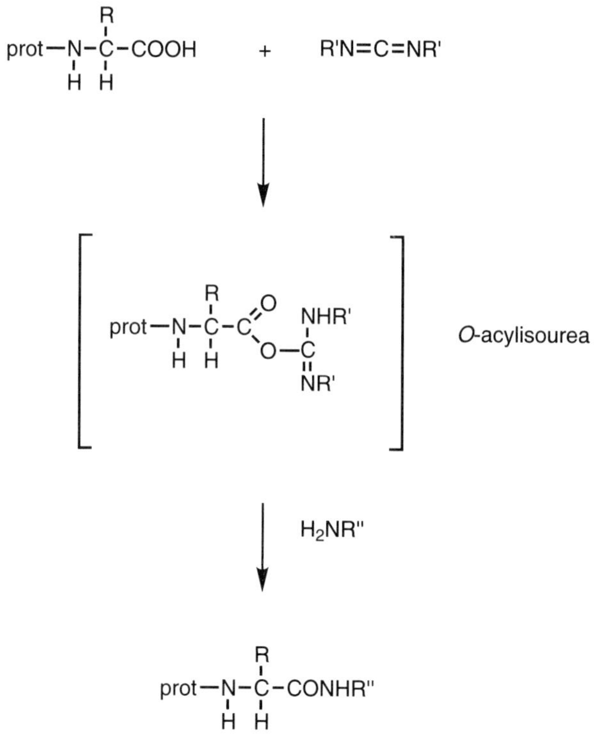

$$\text{prot}-\overset{H}{\underset{H}{N}}-\overset{R}{\underset{H}{C}}-\text{COOH} \quad + \quad R'N=C=NR'$$

$$\downarrow$$

$$\left[\text{prot}-\overset{H}{\underset{H}{N}}-\overset{R}{\underset{H}{C}}-C\underset{O-C(=NR')}{\overset{O}{\diagup}}\text{NHR'}\right] \quad O\text{-acylisourea}$$

$$\downarrow H_2NR''$$

$$\text{prot}-\overset{H}{\underset{H}{N}}-\overset{R}{\underset{H}{C}}-\text{CONHR''}$$

Scheme 2-2. Amide bond formation using carbodiimides.

(Scheme 2-2) is formed. These findings prompted the use of trapping agents like *N*-hydroxybenzotriazole (HOBt), *N*-hydroxysuccinimide (HOSu) and others.

In the usual case, the carboxyl component reacts first with DCC (or another carbodiimide) to yield the highly reactive acetylating agent, an *O*-acylisourea (Scheme 2-2). This intermediate has not been isolated and characterized due to its instability; it reacts readily with any available nucleophile. Ideally, in the presence of an amino acid or peptide ester, a peptide bond is formed. Depending on the reaction conditions and the amino acids and their mode of protection, several reactions can take place. For example, in the presence of an excess of the carbonyl component a symmetrical anhydride can be formed (Scheme 2-3). Symmetrical anhydrides are highly reactive species and in the presence of free N-terminal amino acids or peptides, the desired peptide bond is formed. Therefore, they often are used for automated solid phase peptide synthesis (SPPS).

As noted, side reactions can take place after formation of the *O*-acylisourea intermediate. They mainly occur when the reaction conditions favor intramolecular versus intermolecular nucleophilic attack. In order to suppress or strongly reduce the occurrence of these side reactions, trapping agents were introduced into peptide synthesis. These reagents react more quickly with the *O*-acylisourea than the intramolecular nucleophile and lead to the formation of activated amino acid esters suitable for the formation of peptide bonds. *p*-Nitrophenol (Bodanszky, 1956), pentachloro and pentafluorophenols (DeTar et al., 1966; Kisfaludy et al., 1967, 1970) fulfill these goals, but do

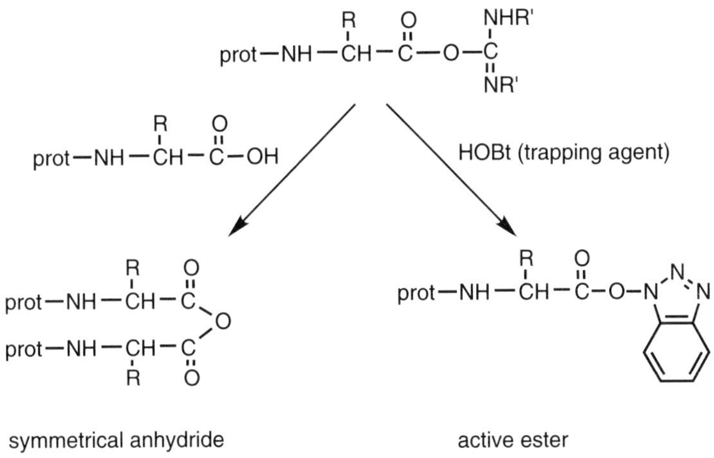

Scheme 2-3. Synthesis of reactive species from the O-acylisourea.

not react fast enough to suppress racemization completely through formation of the 5(4H) oxazalone. The use of N-hydroxysuccinimide (HOSu, Fig. 2-4) brought the first real improvement to the DCC method (Wünsch & Drees, 1966) by almost completely suppressing racemization and formation of N-acylurea. HOSu reacts readily with the O-acylisourea to form an activated ester before any side reaction can occur.

The most used trapping agent, HOBt (Scheme 2-3) and other related compounds (König & Geiger, 1970), were found to be very effective when used in combination with DCC for peptide bond formation. Furthermore, it suppressed racemization and N-acylurea formation efficiently. As for HOSu, HOBt reacts with DCC and the carboxyl component to form an activated ester that rapidly acylates amino acid peptide esters. The HOBt/O-acylisourea reaction is so fast that competing intramolecular reactions, even the one leading to dehydration of asparagine or glutamine, do not take place (Rich & Singh, 1979). Therefore, the method utilizing a combination of DCC and HOBt is one of the most widely effective and easy to use for peptide bond formation, as no major drawback has been found (Rich & Singh, 1979).

The Mixed Carbonic Anhydride Method

Mixed carbonic anhydrides (Meienhofer, 1979) are formed by reaction of N^α-protected amino acids with alkyl chloroformates according to Scheme 2-4. This method was introduced in 1951, and has been employed widely because of its simplicity and the high yields and purity of the peptides obtained (Boisonnas, 1951; Vaughan, 1951; Wieland & Bernhard, 1951). Nevertheless, a major drawback is racemization through the for-

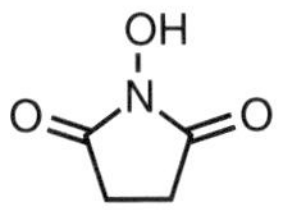

Figure 2-4. Structure of N-hydroxysuccinimide (HOSu).

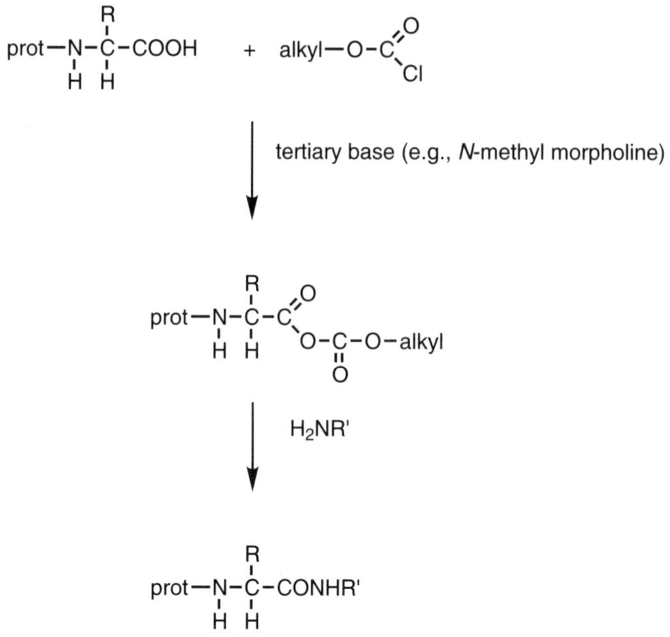

Scheme 2-4. Amide bond formation via a mixed anhydride.

mation of 5(4H) oxazolone (Kemp, 1979), due to the extremely high activation of the carboxyl carbonyl. However, appropriate amino acid protection and precisely defined reaction conditions strongly reduce this side reaction.

The activation step is performed by addition of an alkylchloroformate to a cold solution of the protected amino acid in an aqueous organic solvent in the presence of a tertiary base (Scheme 2-4). Typical activation conditions include the use of isobutylchloroformate as the activating agent and 1 equivalent of N-methylmorpholine (Anderson et al., 1967) (Fig. 2-5). This reaction should be carried out in solvents such as ethyl acetate, acetonitrile, tetrahydrofuran, or *tert*-butanol (Anderson et al., 1967; Steward, 1965; Wieland et al., 1962) at a temperature of −15 °C, for 1 or 2 minutes (Anderson, 1970). Carbamate N^α-protecting groups such as Boc or Z should be used to minimize the risk of racemization.

Generally, mixed anhydrides are not stable enough to be isolated and are usually reacted *in situ* with the amine component. The latter is dissolved in any organic solvent except primary alcohols and phenol and neutralized, if necessary, by the use of one equivalent of N-methylmorpholine prior to addition to the mixed anhydride (Anderson et al., 1967). The temperature should be maintained at −15 °C and, as the coupling is exothermic, the addition should be slow and careful.

Figure 2-5. Structures of isobutyl chloroformate and N-methylmorpholine.

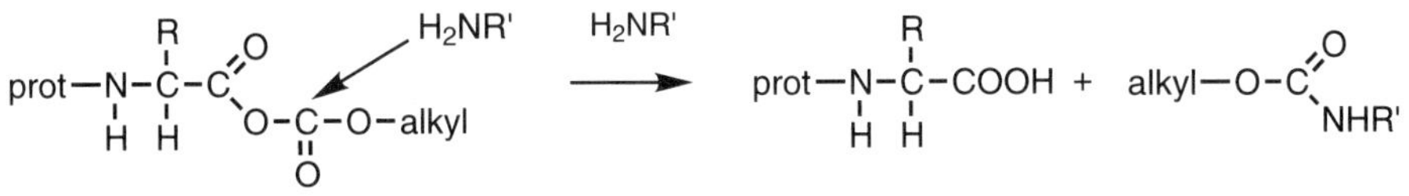

$$\text{prot}-\underset{H}{\underset{|}{N}}-\underset{H}{\underset{|}{C}}-\underset{R}{\overset{|}{C}}\overset{O}{\underset{O-\underset{\|}{C}-O-\text{alkyl}}{\diagdown}} \xleftarrow{H_2NR'} \xrightarrow{H_2NR'} \text{prot}-\underset{H}{\underset{|}{N}}-\underset{H}{\underset{|}{C}}-\underset{R}{\overset{|}{C}}-COOH + \text{alkyl}-O-\underset{NHR'}{\overset{O}{\diagup}}C$$

Scheme 2-5. Urethane formation.

If all the conditions above are respected, side reactions should be minimal. Urethane formation, which occurs when the amino group attacks the carbonyl of the alkyl chloroformate moiety of the mixed anhydride (Scheme 2-5), is usually minimal when isobutyl chloroformate is used (Meienhofer, 1979). The use of an excess of alkylchloroformate, which then competes with the mixed anhydride for reaction with the amine, leads to the formation of the same urethane. Racemization can take place during activation of the carboxyl group (Meienhofer, 1979). To avoid this problem, N^α-carbamate protection (Boc or Z) should be used, and the amount of base should be as close as possible to one equivalent. Solvents like those described previously should be used and the reaction time kept as short as possible. Though the mixed anhydride method is very efficient and gives excellent results in terms of yield and optical purity, it is less used than the DCC method.

The Active Ester Method

Active esters still are widely used in liquid phase peptide synthesis. Nevertheless, they are usually formed by addition of a trapping agent to a mixture of an N^α-protected amino acid and DCC (or a related carbodiimide); therefore they can be considered as part of the carbodiimide method. As already mentioned, the reagents most in use are HOBt (and its analogues) (König & Geiger, 1970), N-hydroxysuccinimide (Wünsch & Drees, 1966) and pentafluorophenol (Kisfaludy et al., 1967, 1970) because they lead to the formation of highly reactive amino acid esters capable of readily acetylating the amine component. Alternative routes of formation and the use of different esters have been reviewed (Bodanszky, 1979).

The Azide Method

Introduced by Curtius (1902), the azide method (Meienhofer, 1979) lost most of its interest when newer coupling methods like the DCC/HOBt combination and the BOP reagent started to be used routinely. Therefore, only the general mechanism of peptide bond formation through this method will be outlined here. This method is usually carried out in three steps (Scheme 2-6): hydrazide formation, azide formation, and reaction with the amine component.

An amino acid ester can be converted to the corresponding hydrazide upon reaction with hydrazine. This amino acid hydrazide can then be transformed into the azide by the use of sodium nitrite in an aqueous acid medium, and the amino acid azide then can undergo aminolysis by the amine component, leading to peptide bond formation. A detailed discussion of the method has been made by Meienhofer (1979).

BOP Reagent

The synthesis of benzotriazolyloxytris[dimethylamino]phosphonium hexafluorophosphate (better known as the BOP or Castro's reagent) was first described in 1975 (Cas-

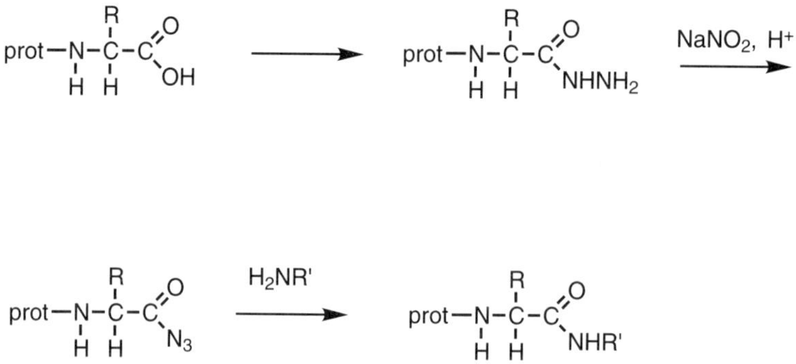

Scheme 2-6. Mechanism of amide bond formation using the azide method.

tro et al., 1975) (Scheme 2-7). It was preceded by the synthesis of coupling reagents of the general formula $^+XP(NMe_2)_3A^-$, where X is a halogen (Castro & Dormoy, 1972) or a pseudohalogen (Castro & Dormay, 1973a, b; Castro & Selve, 1971) and A^- is a nonnucleophilic anionlike perchlorate or hexafluorophosphate. In order to improve on the results obtained with these reagents, hydroxybenzotriazole (HOBt) was added to the coupling reagent bromotrisdimethylaminophosphonium. The favorable results (Castro & Dormay, 1972) that were obtained led Castro and his team to synthesize the oxyphosphonium salt of HOBt according to Scheme 2-7. This synthesis follows the usual technique for preparation of alkoxytrisdimethylaminophosphonium salts (Castro & Selve, 1971). The BOP reagent possesses the highly useful characteristics of being stable, nonhygroscopic, and easily soluble in organic solvents commonly used in peptide chemistry such as dimethylformamide and dichloromethane. It was used initially for the synthesis of model dipeptides that proved to be obtained quickly in high yields and with low racemization (Castro et al., 1975, 1977). The synthesis of the BOP reagent was latter improved in order to make its use more cost effective (Castro et al., 1976).

The BOP reagent has been used successfully for the syntheses of numerous peptides and other amino acid derivatives. One of the first applications was the synthesis of phenyl esters of N^α-Boc protected amino acids, according to Scheme 2-8 (Castro et al., 1977). This type of reaction proceeded in low yield when DCC was used. The BOP reagent has become used extensively in peptide chemistry, both in solid and solution

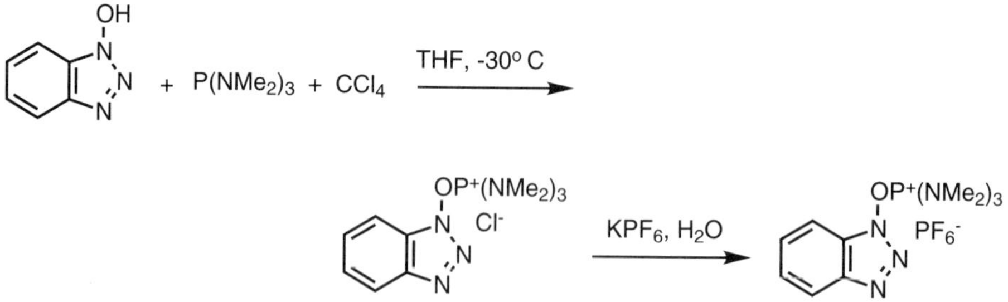

Scheme 2-7. First synthesis of benzotriazol-1-yloxytris-(dimethylamino)phosphonium hexafluorophosphate (BOP) reagent.

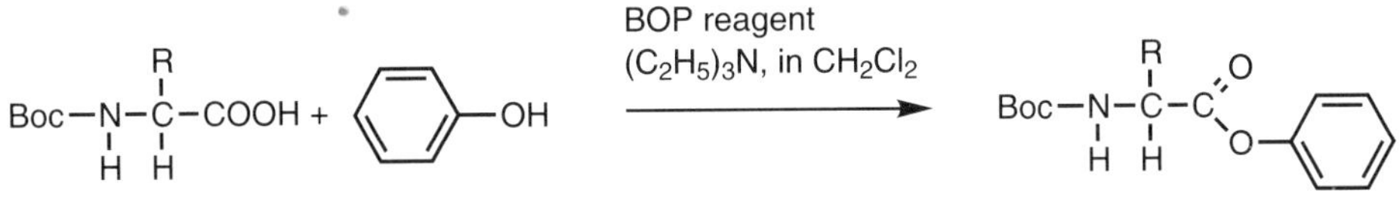

Chemical Synthesis of Peptides / 37

$$\text{Boc}-\overset{H}{\underset{H}{N}}-\overset{R}{\underset{H}{C}}-\text{COOH} + \text{C}_6\text{H}_5-\text{OH} \xrightarrow{\substack{\text{BOP reagent} \\ (\text{C}_2\text{H}_5)_3\text{N, in CH}_2\text{Cl}_2}} \text{Boc}-\overset{H}{\underset{H}{N}}-\overset{R}{\underset{H}{C}}-\overset{O}{\underset{}{C}}-\text{O}-\text{C}_6\text{H}_5$$

Scheme 2-8. Synthesis of phenyl ester derivatives of various amino acids using the BOP reagent.

phase syntheses. For example LH-RH was synthesized very efficiently by Rivaille et al. via a fragment condensation strategy with the BOP reagent (Rivaille et al., 1980; Seyer et al., 1990). The reagent also was used for the synthesis of renin substrates both in solution via a fragment condensation strategy (Le Nguyen et al., 1985b) and on solid phase using the classical stepwise elongation procedure (Le Nguyen et al., 1985a). In both cases, excellent yields and overall high purity of the peptides was obtained. A series of publications by Fournier and coworkers (Felix et al., 1988; Forest & Fournier, 1990; Fournier et al., 1988, 1989) exemplifies the advantages of this reagent for difficult coupling and on-resin side chain-to-side chain cyclization by lactam formation. The yields were superior and purities were comparable with single BOP coupling to those observed using multiple couplings via the DCC method. The BOP reagent also was shown to be much more efficient than the DCC/HOBt combination in the side chain-to-side chain cyclization of the same peptide (Felix et al., 1988). Similar results were obtained by Al-Obeidi et al. (1989) to form superpotent, prolonged acting cyclic lactam analogues of α-MSH.

However, the manufacture and utilization of this reagent involves the use or formation of hexamethylphosphoric triamide, whose toxicity (carcinogenicity) has been well established. This led to the preparation of a new coupling reagent, PyBOP (Fig. 2-6, where dimethylamino groups on the phosphorus are replaced with pyrrolidine groups), which exhibits equivalent or superior properties in peptide bond formation to BOP and is safer to use (Coste et al., 1990). Later, oxybenzotriazole-free peptide coupling reagents such as BrBOP and PyBroP, have been introduced (Coste et al., 1991a,b), and exhibit the same general characteristics and usefulness as the parent compound BOP. PyBOP and PyBroP are safer to use. PyBOP, PyBroP and BroP (Fig. 2-7) have been found to be even more effective than BOP for the difficult coupling of N^α-methylated amino acids and others like aminoisobutyric acid (Aib) (Coste et al., 1991a; Frérot et al., 1991).

HBTU Reagents

Though introduced in 1984 (Dourtoglou et al., 1984), 2-(1H-benzotriazol-1-yl)-1,1,3,3-tetramethyluronium hexafluorophosphate (HBTU) and its analogues did not gain recog-

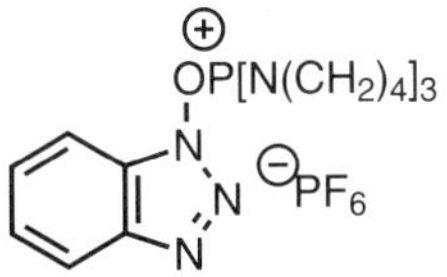

Figure 2-6. Structure of benzotriazole-1-yl-oxy-tris-(dimethylamino)phosphonium hexafluorophosphate (PyBOP).

[structures of PyBroP and BroP phosphonium hexafluorophosphate salts]

Figure 2-7. Structures of bromo-tris-pyrrolidinophosphonium hexafluorophosphate (PyBroP, left) and bromo-tris-dimethylaminophosphonium hexafluorophosphate (BroP, right) coupling reagents.

nition until more recently (Knorr et al., 1989). Uronium salts have been shown to quickly form the active species for coupling, even in polar solvents, which has applications in automated solid phase synthesis (Beck-Sickinger et al., 1991; Fields et al., 1991). Results obtained with those reagents are comparable to those obtained with BOP and related analogues. They have also proved to be efficient for the side chain-to-side chain cyclic lactam formation.

Amino Acid Halides

N-protected amino acid chlorides have been known since 1903 (Fisher & Otto, 1903), but have seldom been used in peptide synthesis due to their instability. Nevertheless, the synthesis of stable Fmoc-protected amino acid chlorides by treatment with thionyl chloride in dichloromethane (Scheme 2-9) has been reported more recently by Carpino et al. (1986). Reported yields are 80% and higher. These new activated amino acids were used by the same group for the rapid synthesis of short peptides. The coupling takes place in a mixture of immiscible solvents and is followed after removal of the excess starting material by quick deblocking of the amino group. This new method has proved to be highly efficient and usually faster than other solution phase coupling methods. Other advantages include ease of synthesis of the activated amino acids at a low cost. This new method could not be transposed easily to solid phase peptide synthesis. For example, due to the necessary addition of basic coreactants like DIEA, the Fmoc amino acid chlorides are converted to the corresponding oxazolones. Nevertheless, if HOBt is added in a 1:1 ratio relative to the base, acylation is again possible due to the intermediate formation of the HOBt activated ester (Carpino et al., 1991).

In the same way, protected amino acid fluorides have been synthesized (Scheme 2-10) by reacting the amino acids with cyanuryl fluoride in the presence of pyridine (Olah et al., 1973). Berthot et al. (1991) and Carpino et al. (1991) have reported the synthesis of Fmoc, Boc and Z protected fluoride derivatives of almost all naturally occurring amino acids. All of these derivatives appeared to be stable, though the ones bearing

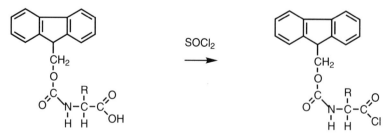

Scheme 2-9. Synthesis of Fmoc-protected amino acid chlorides.

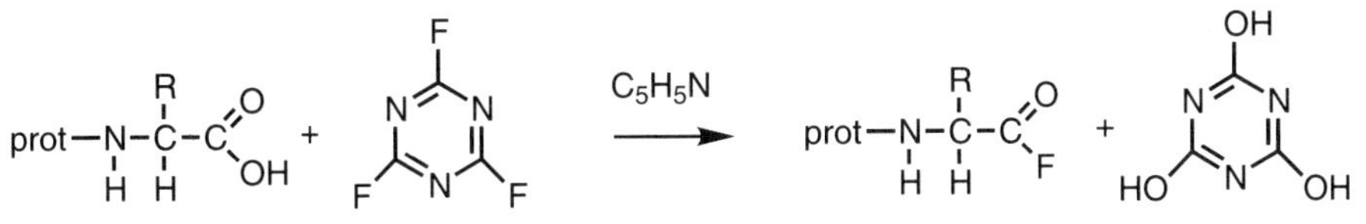

Scheme 2-10. Synthesis of protected amino acid fluorides.

N^α-Boc protection need to be kept cold. These derivatives were used for the facile synthesis of dipeptides with little racemization (Berthot et al., 1991).

Urethane Protected Amino Acid N-Carboxyanhydrides (UNCAs)

Amino acid N-carboxyanhydrides (NCAs) have been used since the late 1940s in stepwise elongation peptide synthesis (Katakai, 1975). This approach has been somewhat unsuccessful due to severe drawbacks and side reactions like the poor stability of NCAs and the possibility of multiple additions at each coupling step (Bailey, 1950; Kricheldorf, 1987; Hirschmann et al., 1967). Several groups reported attempts to protect these compounds with little success (Akiyama et al., 1979; Block & Cox, 1963; Kricheldorf, 1977). UNCAs (Fig. 2-8) were introduced in 1990 by Fuller et al. (1990) and show many promising characteristics. They are stable (in the absence of water), often crystalline, can be obtained as Boc, Fmoc, or Z derivatives, and are highly reactive toward nucleophiles like amines. Therefore, they form peptide bonds quickly and cleanly with carbon dioxide as the only by-product. Almost all derivatives are now commercially available with different urethane protection. UNCAs are prepared by the condensation of acyl halides, chloroformates, or anhydrides with NCAs (Fuller et al., 1990).

UNCAs have been used in both solution and solid phase synthesis with excellent results. Martinez and coworkers (Bourdel et al., 1993) reported the synthesis of C-terminal fragments of gastrin, bombesin, cholecystokinin, and neurotensin in solution employing Boc, Fmoc, and Z protected UNCAs in only a 10% excess. Overall yields ranged from 66 to 85% and a high degree of purity was obtained (75 to 95%). The authors also stressed the simplicity of the work-up in this method (Bourdel et al., 1993; Rodriguez et al., 1993). In their introductory paper, Fuller et al. (1990) reported the solid phase synthesis of the acyl carrier decapeptide-(65-74). After cleavage from the resin, the crude peptide was obtained in a very good yield (73%) and a high degree of purity without racemization. Swain et al. (1993) report the liquid and solid phase syn-

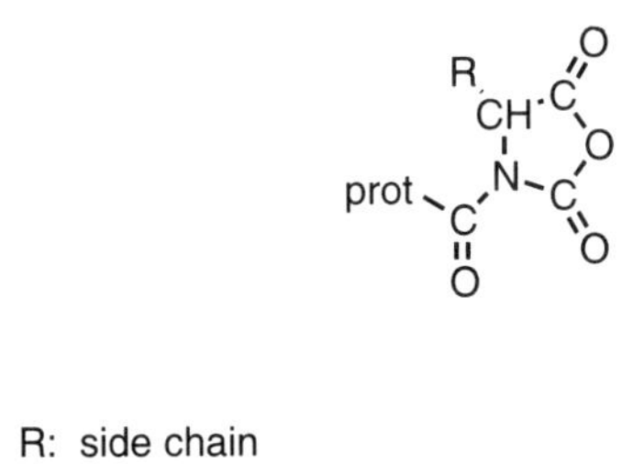

R: side chain
prot: BzlO, tert-BuO or FmO

Figure 2-8. Structure of urethane-protected amino acid N-carboxyanhydrides (UNCAs).

thesis of [D-Trp^6]LH-RH, without any major problem and in a high overall yield. In the solid phase synthesis of several peptides, Xue & Naider (1993) showed that, compared to the BOP and HBTU reagents, the results obtained with UNCAs were very similar in terms of yield, purity and degree of racemization. All authors reported that adding 0.5 to 1 equivalent of a tertiary base at the coupling stage increased the yields and final purity. UNCAs have proven to be very efficient in the coupling of highly sterically hindered amino acids and peptides or N-methylated amino acids, as shown by Spencer et al. (1992). Again, UNCAs in difficult coupling reactions gave about the same results as the coupling reagents HBTU and PyBroP, at room temperature. Finally the stability, reactivity and solubility of UNCAs, as well as their attachment to Wang, Rink and other resins have been reported (Fuller et al., 1993a, b). Due to their ease of use (no preactivation step is necessary) and commercial availability, and due to the excellent results obtained by several groups for difficult coupling steps, with no racemization and in good yields, UNCAs appear to be one of the best coupling methods.

Difficult Couplings

Several authors have compared the efficiencies of the BOP reagent and related compounds, the uronium salts and other amino acid derivatives like UNCAs and fluorides in the difficult coupling of hindered peptides or amino acids bearing N-terminal α,α-dialkylation or N-methylation (Frérot et al., 1991; Goodman et al., 1993; Spencer et al., 1992). Amino acids examined included Aib, N-MeAib, N-Me-α-Ac^5, and so forth. Results obtained for the synthesis of model peptides appear varied depending on the coupling considered. Nevertheless, UNCAs and amino acid fluorides, as well as the HBTU and PyBroP reagents, seemed to give the best results compared to the BOP reagent or methods involving the use of carbodiimides. Other reagents like TOPPipU and CIP also have also been reported to be useful in such difficult couplings (Akaji et al., 1993; Geiger & König, 1981; Pipkorn et al., 1993).

In peptide synthesis, amino acids have to be protected at three positions, the N-terminal amino group, the C-terminal carboxyl group and any reactive group present in the side chain (e.g., the ε-amino group of lysine and the hydroxyl groups of tyrosine, serine and threonine). A number of considerations should be kept in mind in choosing a protecting group strategy. These include the fact that peptides are usually synthesized from the C-terminus toward the N-terminus. To minimize racemization through formation of 5(4H) oxazalones (Scheme 2-11), urethane-type protection of the α-amino group is the most suitable. Finally, due to the increased use of solid phase peptide synthesis, C-terminal protection is determined by how the carboxyl group is linked to the

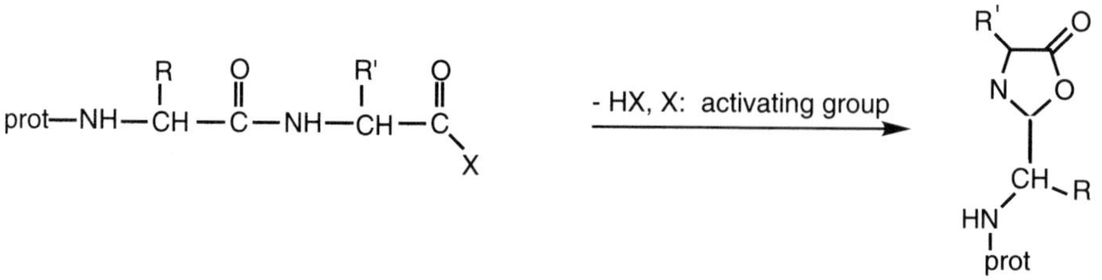

Scheme 2-11. 5(^{4}H)Oxazolone formation.

resin. Thus, this chapter focuses on urethane-type protection of the α-amino group and side chain protecting groups that are compatible with them. The most widely used N^α-Boc and N^α-Fmoc protecting groups are discussed first and then the newly introduced allyl and Alloc protection schemes are presented.

tert-Butyloxycarbonyl Protection

Introduced in 1957 (Anderson & McGregor, 1957; McKay & Albertson, 1957), the N^α-Boc protecting group (Geiger & König, 1981) is cleaved easily under acidic conditions but is stable to base, sodium in liquid ammonia and catalytic hydrogenation. In the 1960s and 1970s, simple methods for the synthesis of N^α-Boc-protected amino acids were published, allowing their commercialization and increased use in peptide synthesis (Itoh et al., 1974; Schnabel, 1967; Tarbell et al., 1972). The Boc group is usually removed with TFA-CH_2Cl_2 or with HCl, HBr and other acids. To avoid alkylation of some residues like tryptophan by the reactive *tert*-butyl carbocations, suitable scavengers such as anisole and ethanedithiol should be used (Chang, 1980; Dryland & Sheppard, 1986). In classical solid and liquid phase peptide synthesis (either stepwise elongation or fragment condensation), the side chain protecting groups that must be stable to the acid used during N^α-Boc deprotection are removed by liquid HF in the presence of appropriate scavengers (Barany et al., 1988; Stewart & Young, 1984). Cleavage from the resin is usually performed at this stage. Therefore, the classical side protection of N^α-Boc derivatives are the following (Gross & Meienhofer, 1981). The guanidyl group of Arg is usually protected with a tosyl group. Histidine can have the same side chain protecting group but this protecting group is removed by HOBt, which often is used. Dinitrophenyl (Dnp) is stable to the coupling conditions but also to HF cleavage. Its use, therefore, requires an additional deprotection step with thiophenol in DMF, prior to HF deprotection. The most suitable imidazole protecting group is the BOM group, which is cleanly removed by HF and TFMSA-TFA and minimizes racemization (Brown & Jones, 1981; Brown et al., 1982). The thiol group of cysteine and of related amino acids is usually protected by the *p*-methylbenzyl group. The Asp and Glu carboxyl groups usually are protected as cyclohexyl esters, whereas the ε-amino group of Lys usually is protected as a Z or substituted Z group. Methionine, if unprotected, can be oxidized to Met(O) during the cleavage step. This can be prevented by using appropriate scavengers during cleavage, or by introducing methionine in its oxidized form and reducing it by the low-high HF procedure (Tam et al., 1983) and other improved methods (Tam et al., 1986). The indole ring of tryptophan is protected by the formyl group, which is stable to HF. It can be removed prior to cleavage by the use of piperidine in DMF or during the low-high HF procedure (Tam et al., 1983). The hydroxyl groups of Ser, Thr, and Tyr are often protected as benzyl ethers. However in the case of tyrosine, strong acid removal of this protecting can lead to a considerable amount of benzylation of the aromatic ring by the Fries rearrangement. Hence a 2',6'-dichlorobenzyl or Br-Z group usually is employed for protection.

The 9-Fluorenylmethyloxycarbonyl (Fmoc) Group

The 9-fluorenylmethoxycarbonyl (Fmoc) amino protecting group (Fields & Noble, 1990) (Fig. 2-1) was introduced in 1970 (Carpino & Han, 1970, 1972). This protecting group can be removed cleanly under mild basic conditions by dilute liquid am-

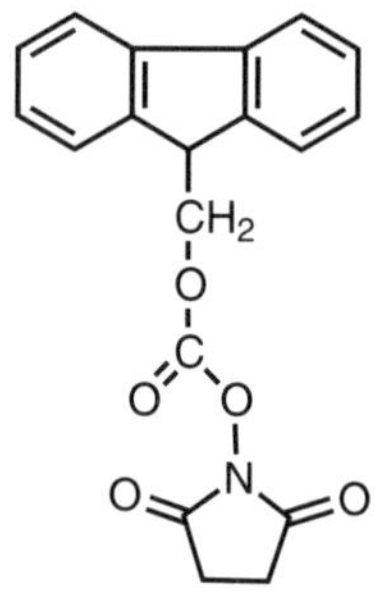

Figure 2-9. Structure of 9-fluorenylmethylsuccinimidylcarbonate (Fmoc-ONSu).

monia, ethanolamine, morpholine, or piperidine. The finding of the Fmoc protecting group was an important development since most amino protecting groups in use were removed by acids of various strengths. The Fmoc group can be introduced by reacting the amino acid with 9-fluorenylmethylchloroformate (Fmoc-Cl) in aqueous dioxane in the presence of sodium carbonate. The protected amino acid is then separated from by-products by classical extraction procedures and crystallized. The synthesis of Fmoc-protected amino acids with *tert*-butyl type side chain protection using Fmoc-Cl also was reported by Chang et al. (1980). Introduction of the Fmoc group by Fmoc-Cl leads to the undesirable formation of a small amount of oligopeptides (Sigler et al., 1983; Tessier et al., 1983). In order to overcome this problem, several groups proposed the use of either Fmoc-azidochloroformate (Tessier et al., 1983) as introduced by Carpino & Han (1972) or Fmoc-succinimide (Fmoc-ONSu, Fig. 2-9) (Milton et al., 1987; Ten Kortenaar et al., 1985). The latter leads to the formation of Fmoc derivatives of amino acids in high yields (73 to 100%) and with no racemization. Since Fmoc-ONSu is commercially available, this method should be used for the preparation of N^α-Fmoc amino acids.

The N^α-Fmoc group is removed during the course of peptide synthesis under mild basic conditions, such as 20% piperidine in dimethylformamide; the cleavage of the peptide from the resin (in solid phase synthesis) then generally is done with TFA. Hence side chain protecting groups should be stable to organic bases, and removed by TFA. Though several kinds of protecting groups fit these criteria and are commercially available, only the more easily and commonly used will be discussed here. For arginine, the PMC group introduced by Ramage and Green (1987) is more TFA labile than any other guanidine protecting group. Asparagine and glutamine generally can be used unprotected. However, if dehydration is possible, the trityl or benzhydryl groups are the most suitable as amide protecting groups. The corresponding acids (aspartic and glutamic) are protected as *tert*-butyl esters, and tyrosine as a *tert*-butyl ether. Cysteine is usually protected with the trityl group which, under acidic treatment by TFA, yields the free thiol. Other groups like Acm (acetamidomethyl) are stable under these conditions and therefore allow the differential deprotection of cysteine residues, thus permitting the selective formation of disulfide bridges. The imidazole ring of histidine is protected with the trityl group, which suppresses racemization at this residue. Lysine is used as its N^α-Boc derivative. Finally, methionine and tryptophan can be used without protection. However, the Boc-indole derivative of tryptophan seems to be more suitable and leads to fewer side reactions.

N^α-Allyloxycarbonyl Type Protection

As previously mentioned, many undesired side reactions can occur during the deprotections-cleavage step(s) due to the presence of carbocations and other active species generated by the use of strong acids like HF and TFA (Birr, 1990; King et al., 1990). In order to overcome these problems a new strategy based on allyl type protection has been developed. The N^α-allyoxycarbonyl group (Alloc) (Fig. 2-10) (Stevens & Watanabe, 1950) is suitable for the protection of amines and alcohols. In addition carboxylic acids can be protected as their allyl ester derivatives while aryl alcohols are protected as ethers (Hey & Arpe, 1973; Stevens & Watanabe, 1950). These protecting groups can be introduced by an esterification reaction or via reaction with the commercially available allyloxycarbonyl chloride. Essentially all functions groups, whether the α-amino or a side chain functionality usually present in amino acids, can be protected by these new protecting groups. Thus they can prove useful in peptide synthesis provided that an easy deprotection method is available; these groups are orthogonal to Fmoc, Boc or other commonly used protecting groups, and functionalized resin of the same type can be synthesized and used in solid phase peptide synthesis.

Following the studies of Tsuji and Trost on the complexation of allyl groups by Pd(0) (Trost, 1980; Tsuji, 1980), Guibé and coworkers have developed a new and extremely mild hydrogenation method to deprotect amines, alcohols and acids blocked as their Alloc and allyl derivatives (Scheme 2-12) using Pd(0) as a catalyst and tributyl tin as the hydrogen donor (Dangles et al., 1987; Four & Guibé, 1982; Guibé & Saint M'Leux, 1981; Guibé et al., 1986a, b; Zhang et al., 1988). This reaction proceeds in high yield and over a short time course. These findings led Loffet and Zhang (Loffet & Zhang, 1992, 1993), Kunz and coworkers (Kunz, 1987; Kunz & Dombon, 1988; von dem Bruch & Kunz, 1990) and Dangles et al. (1987) to synthesize all allyl and Alloc side chain-protected amino acids as well as the N^α-Alloc protected amino acids. They also investigated the stability of these derivatives under acidic and basic conditions (Dangles et al., 1987; Kunz, 1987; Kunz & Dombon, 1988; Loffet & Zhang, 1992, 1993; von dem Bruch & Kunz, 1990). Their results showed that all allyl-based derivatives, whether N^α or side chain protected, could be synthesized and that allyl and Alloc groups were orthogonal to Boc and t-butyl groups. They also found that Z and Bzl groups were stable to the hydrogenation conditions used to remove protecting allyl and Alloc groups and that Alloc side chain-protected cysteine and histidine residues were not stable toward basic conditions and, therefore, were not fully orthogonal to the Fmoc group.

The successful synthesis of glycopeptides (Kunz, 1987; Kunz & Dombon, 1988), various model peptides and bioactive peptides like substance P (Dangles et al., 1987; Kunz, 1987; Kunz & Dombon, 1988) with N^α-protected amino acids both in solution and solid phase peptide synthesis has demonstrated the usefulness of these types of protection. The development of allylic anchor groups (resin linkers) added the final touch to this new synthetic method so that the synthesis of protected fragments could

Figure 2-10. Structures of the Allyloxycarbonyl (Alloc) and allyl (All) protecting groups.

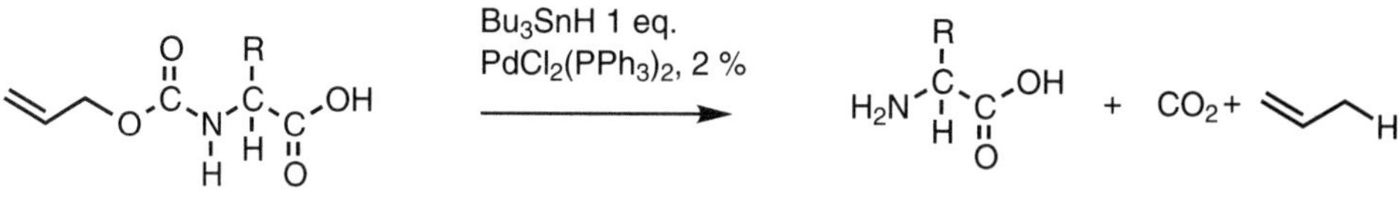

Scheme 2-12. Example of deprotection of an allyloxycarbonyl-protected amino acid derivative.

be achieved, as well as cleavage of a peptide from the resin under neutral and extremely mild conditions (Blankemeyer-Menge & Frank, 1988; Guibé et al., 1986a, b; Kunz & Dombon, 1988).

Solid Phase Peptide Synthesis

General Strategy

Since solid phase peptide synthesis (SPPS) was introduced by Merrifield (1963), the chemistry associated with this technique has developed tremendously. New methods of coupling and of deprotection, new protecting groups, new resins and new resin "handles" have been developed. Nevertheless, the basic synthetic tactics used now generally are still the same as those outlined by R. B. Merrifield more than three decades ago.

In SPPS, the resin is functionalized with a chloromethyl (Merrifield's original resin), an aminomethyl, or some other group. The protected amino acid (temporarily N^α and permanently side chain protected) C-terminal carboxyl group is attached covalently to the resin, either directly or through a bifunctional handle or spacer placed between the resin and the amino acid. The N^α amino group is then liberated by removal of the temporary protection and a second amino acid is added by creation of an amide or peptide bond. The N^α deprotection-coupling procedure is repeated until all amino acid residues have been introduced. The side chain protecting groups are then removed and the peptide cleaved from the resin (Scheme 2-13). Depending on the strategy followed, these two can be accomplished in sequence or in one step (Atherton & Sheppard, 1989; Barany & Merrifield, 1979; Bodanszky, 1993; Bodanszky et al., 1976; Erikson & Merrifield, 1976; Fields & Noble, 1990; Jones, 1991; Stewart & Young, 1984). As the growing peptide is covalently linked to an insoluble support, unreacted amino acids and coupling reagents can be removed at each step by filtration and washing without loss of peptide. In order to push coupling reactions to completion, and therefore obtain the peptide in high yield and purity, a two- to fourfold excess of reagents often is used. Finally, due to the ease of use, the speed, and the repetitive aspect of the deprotection-coupling cycles, SPPS can be automated easily.

In SPPS, general strategic considerations are the same as these outlined for the solution phase synthesis (choice of the protecting groups, the coupling methods, etc.). In addition, the choice of the solid support, usually know as the resin, is extremely important at different stages of the synthesis. For instance, linkage of the first amino acid, functionalization of the C-terminal group of the peptide after cleavage, and more generally the physicochemical properties such as swelling in different solvents, resistance to pressure and to chemicals, all are affected by the nature of the resin.

Though each paragraph following on SPPS addresses a specific topic, it is important to note that a rigid classification is very hard to achieve. For instance, oxime resins are important for the synthesis of fully protected peptides and therefore could appear under "SPPS by fragment condensation," but also under "orthogonal protection and cleavage methods," or in the "synthesis of cyclic peptides." Thus the reader should consider the section on SPPS as a whole to attain an overview.

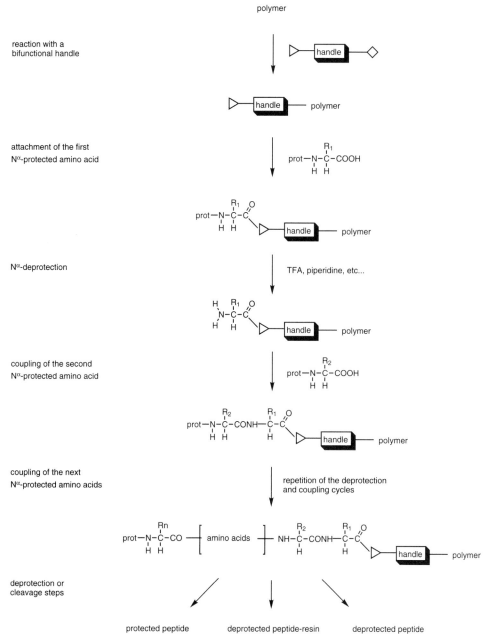

Scheme 2-13. General strategy in solid phase peptide synthesis.

Solid Support

The solid support is usually a polymer of styrene 1% crosslinked with *m*-divinylbenzene that can be modified in numerous ways (Birr, 1978; Merrifield, 1986). The extent of the crosslinking of a resin is important as it controls some of its physicochemical parameters such as rigidity and physical stability, but also its ability to swell in the solvents that are used during the synthesis and, therefore, the accessibility of the reacting groups. Merrifield performed the first SPPS of a pentapeptide on a chloromethylated 2% crosslinked polystyrene resin (Merrifield, 1963). Peptide chemists soon noticed that crosslinking should be reduced to about 1% for the synthesis of longer peptides, while a 0.5% crosslinking was found to render the resin too fragile (Birr, 1973). Other supports using different polymers have also been described, of which the polyamide supports are the most widely used (Arshady et al., 1979; Atherton et al., 1975, 1977; Birr, 1973; Butwell et al., 1988).

The 1% crosslinked polystyrene resin type is still the most widely used both in the Fmoc and Boc synthetic strategies. Often spacer molecules or so-called resin handles or linkers have been added as links between the amino acid and the resin (Barany & Merrifield, 1979; Barany et al., 1987; Bodanszky, 1993). The modifications can increase the swelling properties of the resin and decrease the interactions between peptide chains and the polymer, thus facilitating the synthesis of longer peptides. Handles also have the advantage of allowing different anchoring and cleavage techniques that can be used for various orthogonal synthesis schemes, thus increasing the versatility of the method.

Solid Supports and Resins for SPPS

Perhaps the most widely used resin is still the polystyrene polymer crosslinked with *m*-divinylbenzene and functionalized with a chloromethyl group, or functionalized with bifunctional spacers or handles or otherwise modified.

If a handle is used, it often is attached by one of its functional groups to the resin, and the first amino acid of the peptide is covalently linked via its carboxyl group to the second function group of the handle. This approach has prompted the development of a wide variety of different handles that have allowed the improved synthesis of peptides with higher yields, increased optical purity, and better control of cleavage conditions. Intensive efforts have been made to synthesize handles that were fully orthogonal to the protecting groups used in SPPS. Resins bearing these different handles and suitable for diverse purposes are described in the following.

There are several resins used in the Fmoc-*t*-butyl strategy. The "Wang" resin (Wang, 1973) contains a *p*-alkoxybenzyl ester linkage (Fig. 2-11). The handle, 4-hydroxybenzyl alcohol, is reacted with a Merrifield resin to yield the Wang resin. The first amino acid is usually attached to the alcohol group by a carbodiimide (DCC or DIC)-mediated coupling in the presence of *N,N*-dimethylaminopyridine (Wang et al., 1974). Since some problems were noted with this method (Wang & Kulesha, 1975; Wang et al.,

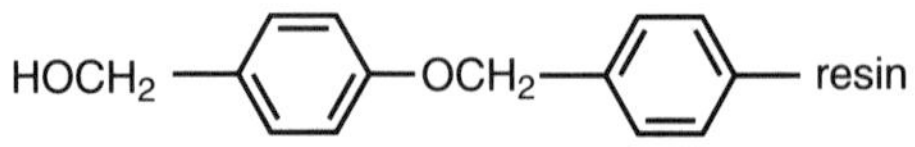

Figure 2-11. Structure of the "Wang" resin.

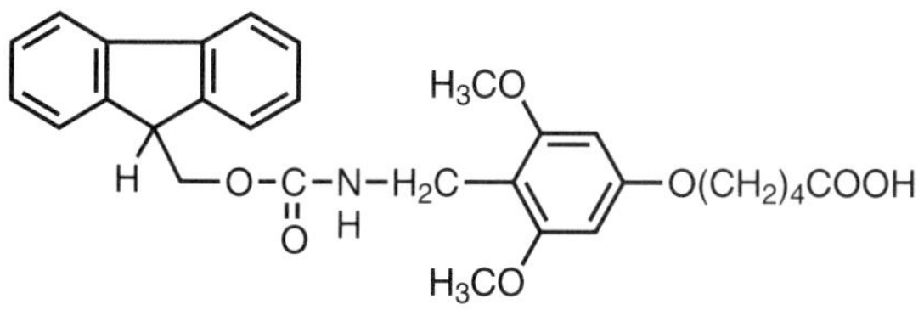

Figure 2-12. Structure of the peptide amide linker ("PAL") handle.

1981), Albericio and Barany (1984, 1985) have investigated new ways of anchoring the amino acids to the handle. Deprotection and cleavage of the peptide are performed in 50% TFA in dichloromethane and lead to a peptide acid. Since then, several different analogues of this handle have been synthesized including new "PAL" (for Peptide Amide Linker) handles (Fig. 2-12) (Albericio et al., 1990). They often are linked through the free carboxyl group to a *p*-methylbenzhydrylamine (MBHA) resin. The cleavage of the Fmoc group yields a free amino group on which the peptide is elaborated. Cleavage of the peptide from the resin is performed with 70 to 90% TFA in dichloromethane and leads to peptides with a C-terminal primary amide.

Other resins used in the Fmoc-*t*-butyl strategy include the "Rink" resin (Rink, 1987), a 4-(2′,4′-dimethoxyphenylhydroxymethyl (or amino acid)) phenoxy resin (Fig. 2-13) introduced by Rink and Sieber (Rink, 1987; Sieber, 1987). It allows the synthesis of fully protected peptides after facile cleavage from the resin under mild acidic conditions such as 10% acetic acid or 0.2% TFA in dichloromethane at room temperature. The "Sasrin" resin (Mergler et al., 1988a, b), in common with the Rink resin, is sensitive to dilute acids yielding protected peptide acids (Scheme 2-14). Cleavage takes place at room temperature with 0.5% TFA in dichloromethane. It is obtained by reaction of a disubstituted phenol with the Merrifield resin.

There are also resins used with a Boc-benzyl protection strategy. These include the *p*-methylbenzhydrylamine resin (Fig. 2-14) which is used for the synthesis of peptide amides with Boc-protected amino acids and HF or TFMSA in TFA cleavage conditions (Fujii et al., 1987; Tam et al., 1983). The "PAM" resin (Mitchell et al., 1978; Pietta et al., 1974; Sparrow, 1976), a phenylacetamidomethyl resin (Fig. 2-14), and the oxymethyl-PAM resin are used for obtaining peptide acids. These resins have replaced oxymethylpoly(styrene) based resin, since the PAM resins have a high resistance to acidolysis during the course of peptide synthesis (Gutte & Merrifield, 1971; Ragnarsson et al., 1970).

Oxime resins (Fig. 2-15) have been developed by DeGrado and Kaiser (1980, 1982) for the SPPS of fully protected peptides. After synthesis, the peptide is cleaved from the resin by aminolysis or by hydrazinolysis. More recently, the cyclization of protected peptides performed on oxime resins also have been described (Nishino et al., 1993; Taylor & Osapay, 1990).

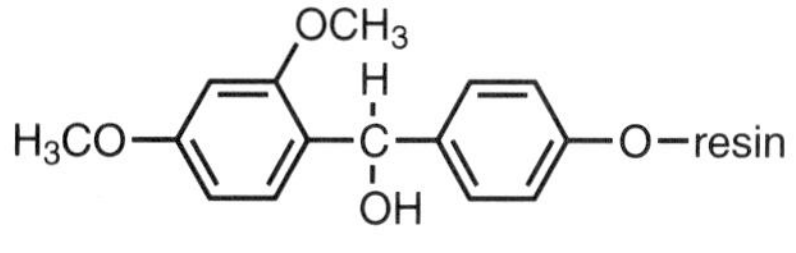

Figure 2-13. Structure of the "Rink" resin.

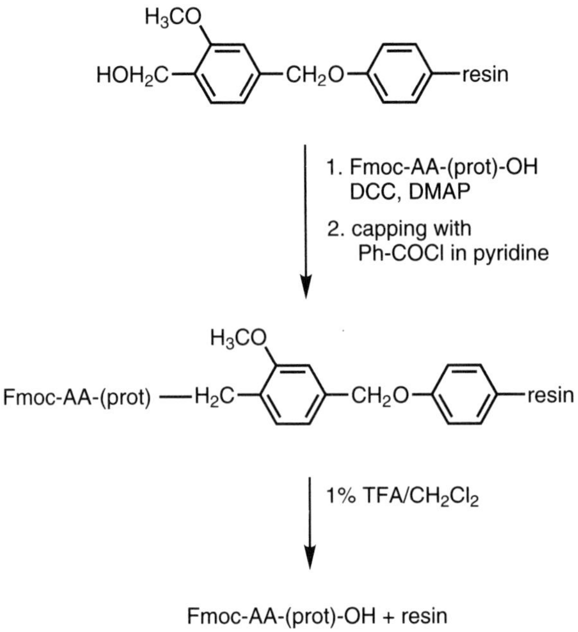

Scheme 2-14. Structure of the Sasrin resin. Attachment conditions for amino acids to the resin and conditions for removal of the formed protein.

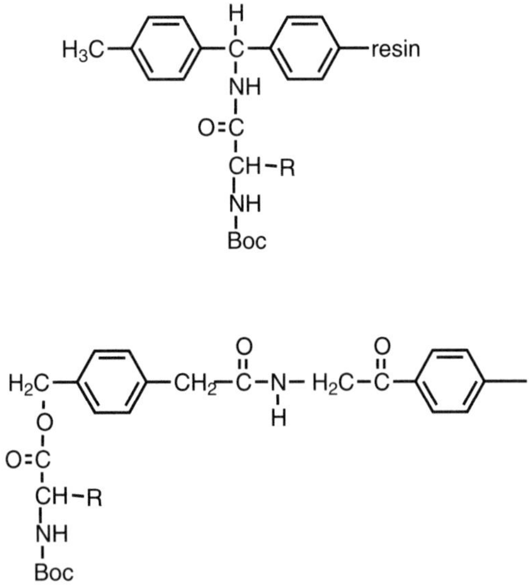

Figure 2-14. Structures of *tert*-butoxycarbonylaminoacyl-*p*-methylbenzhydrylamine (MBHA) and 4-(oxymethyl)phenylacetamidomethyl (oxymethyl PAM) resins.

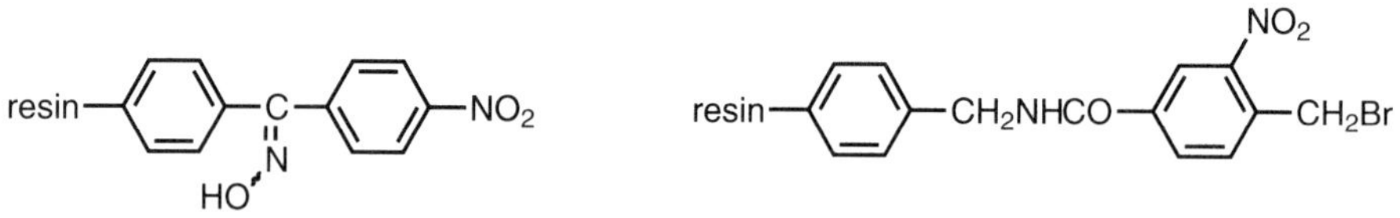

Figure 2-15. Structures of *p*-nitrobenzophenone oxime ("oxime") and 3-nitro-4-bromomethylbenzoylamide ("NBB") resins.

Peptides elaborated on nitrobenzyl and related (Fig. 2-15) are removed from the resin by photolysis at 350 nm. These resins were first developed for the synthesis of protected peptide fragments, as photolysis does not affect the usual protecting groups used in SPPS (Giralt et al., 1981, 1982; Rich & Gurwara, 1975; Tam et al., 1980). Therefore they add a third dimension to orthogonal synthetic strategies and have proven to be very useful (Barany & Albericio, 1985).

Resins with allylic handles have been developed as counterparts of the Alloc and allyl protective groups. Hycram resin (Fig. 2-16) and others have been used for the synthesis of protected peptide fragments, the Alloc group being used for the temporary protection of the α-amino group (Blankenmeyer-Menge & Frank, 1988; Guibé et al., 1986a, b). They provide the possibility of cleaving peptide resins under neutral and mild conditions. In common with nitrobenzyl resins, allylic resins add a third dimension to the orthogonal synthesis of peptides. They also can be used for the synthesis of fully deprotected peptides, the Boc group, or others then being used for the temporary protection of the α-amino group.

Deprotection Methods

Once the assembly of the protected peptide linked to the resin is completed, one must remove the side chain protecting groups and cleave the peptide from the resin. Different approaches can be taken depending on the final result one wishes to obtain. For the synthesis of peptides for fragment condensation, this step is performed so as to cleave the peptide from the resin while leaving the permanent side chain protecting groups unaltered. The fragments that are obtained are then condensed with other fragments. In the case of an orthogonal synthetic-deprotection strategy, it is possible to remove first some selected side chain protecting groups, perform a specific reaction on the peptide-resin (e.g., cyclization) and then remove the remaining protecting groups and cleave the peptide from the resin.

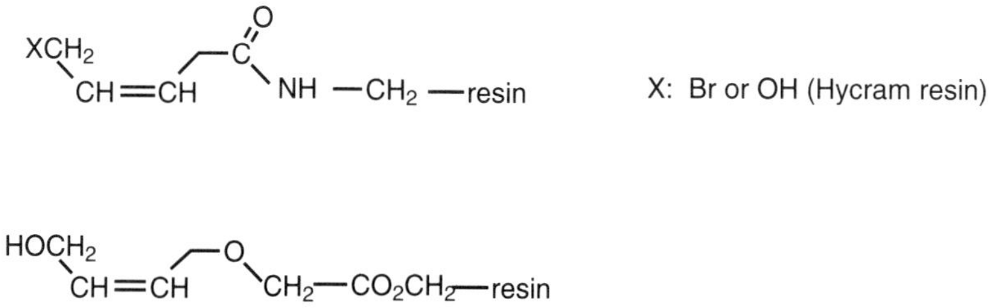

Figure 2-16. Structures of the allylic type resin.

In the most common case, removal of all protecting groups and cleavage from the resin are performed at the same time. Classical cleavage-deprotection methods have been widely used and reviewed. Strong acids such as liquid HF (Sakakibara & Shimonishi, 1965; Tam et al., 1983), TFMSA in TFA, or HBr in HOAc are used when Boc-benzyl protected amino acids are utilized. On the other hand, when a Fmoc-*tert*-butyl protection scheme is followed, TFA in dichloromethane is generally used for the last step. In order to prevent side reaction to the extent possible, due for instance to the generation of highly reactive carbocations generated in acidic media, mixtures of scavengers also are added to the cleavage milieu (Tam et al., 1986). The most widely used scavengers are anisole, ethanedithiol, dimethylsulfide, indoles, and others.

Though very efficient, the use of strong acids such as HF in the cleavage-deprotection step is accompanied by drawbacks, such as danger to the operator and the use of special equipment. Acids also can be deleterious to some peptides or amino acid side chains and often can induce side reactions. Therefore, milder methods like direct or phase-transfer hydrogenation and photolysis have been developed. These methods are often used for the sake of orthogonal synthetic strategies or fragment synthesis as shown in several examples (Anwer et al., 1983; Barany & Albericio, 1985; Colombo, 1981; Jones, 1977).

Some Current Topics

Peptide Libraries

The recognition of peptides and proteins that participate in all aspects of the chemistry of living systems has had a dramatic impact on biology and chemistry. In particular, their ability to recognize and to interact with virtually all chemical classes and as a result to control, modulate, and mediate most biochemical and physiological effects in living systems have made peptides and proteins and their (glyco and lipo) conjugates and analogues perhaps the central natural products of modern biology. As such, they are critical structures and targets for understanding and controlling biological processes.

Despite the widespread recognition of the central importance of peptides and proteins, in many cases it has been difficult to translate this insight into practical applications in diagnostic and therapeutic medicine. The major reasons for this are threefold. First, peptides and proteins are very complex structures with numerous functional groups, complex conformations and dynamics, and often multiple biological effects. Second, many chemists and biologists have little knowledge regarding peptide and protein synthesis, structure and conformation and, except for enzymes, little knowledge of the enormous diversity of functions they serve in living systems. Finally, the complexity of peptide and protein structure and conformation and the enormous possibilities for structural and conformational modification have made it difficult to develop a rational, systematic approach to structure-biological function analysis. Thus methods have been sought recently to overcome these problems by searching for more rapid ways to examine the diversity of structures, conformations, and other properties of polypeptides and their potential application in particular biological problems.

The complexity of the problem is illustrated by simply considering a pentapeptide. Even for just the common 20 amino acids found in most proteins, there are 20^5 (3,200,000) possible pentapeptides and most of these can possess numerous conformations to add to the complexity. Further, there are many other amino acids available as well. In addition, developing assays (binding, second messenger, functional whole tissue-cell, or whole animal) that can detect only those specific species that are interacting with the primary binding site for the macromolecular acceptor (enzyme, receptor, antibody, etc.) can be difficult because many interactions are possible that are not relevant to a particular receptor site. We will concentrate our discussion on the chemical synthesis of polypeptide and pseudopeptide-peptidomimetic libraries, though it should be pointed out that it is possible to use molecular biological and microbiological methods to create very diverse libraries of polypeptides (Huse et al., 1989; Scott & Smith, 1990; Smith, 1985). These methods are outside the scope of this section, but are highly complementary to the synthetic organic methods to be discussed here, and are being used by many laboratories. Of course one can also consider other classes of organic compounds that possess potential structural complexity and diversity in terms of physicochemical properties such as aromatics, heteroaromatics, heterocyclics, nucleic acids, sugars, alkaloids, steroids, and so forth that could lead to complex mixtures of organic structures that can be screened for possible biological activities.

The creation of complex mixtures of peptides was generally viewed during the development of synthetic peptide chemistry as a "problem" rather than an "opportunity." Problems of incomplete coupling, racemization, side reactions, double couplings, and the like often led to peptide products other than the desired product, and required extensive and often very tedious separations that were difficult to accomplish. Even as the methods to prevent racemization and other side reactions were developed, suggestions to make mixtures of peptides for purposes of exploring properties of peptides or structure-activity relationships generally were frowned upon by most synthetic chemists. Thus development of methods to examine large arrays of peptide structure for structure-activity studies were retarded because it was considered to be bad chemistry and bad science. It is interesting to note, that though the same criticisms can be made about many of the peptide and nonpeptide libraries being examined today, that is, critical comments regarding quality control in terms of synthetic and analytical chemistry, they seldom are discussed since combinatorial chemistry is a "hot" area of science. This is unfortunate since a critical view certainly will be important to the scientific development of this area. Thus, it is distressing that many synthetic chemists doing combinational chemistry do not provide a method for evaluating the quality of the synthetic methods being used. Since much of this chemistry is being done on solid supports, on surfaces, or on a variety of polymers, and so forth, an evaluation of the quality of the chemistry before constructing very diverse libraries will be important in the long term. Though ligands with important biological properties may nonetheless be discovered, good quality control is needed.

Construction of large diverse peptide libraries is certainly viable, since optimized solid phase peptide chemistry (Merrifield, 1986) often can be accomplished with 99.9% fidelity for each extension of the peptide chain. Though this kind of fidelity cannot be obtained for every peptide synthesized in a large peptide library, the use of optimized peptide synthetic methods can ensure some reasonable measure of quality for the library. In the succeeding paragraphs several approaches are outlined briefly for constructing peptide libraries that appear to work well.

The multiple synthetic method developed by Geysen and coworkers (Bray et al., 1990, 1991a; Geysen et al., 1984) was perhaps the first method that allowed rapid construction of relatively large numbers of peptides (hundreds to thousands) in a few days in a form suitable for exploring peptide receptor-acceptor interactions while the peptide was still attached to a polymer support. Subsequently, methods were developed to release the peptides into solution (Bray et al., 1990, 1991a, b). The methodology involves the synthesis of peptides on the head of polyacrylic and grafted polyethylene rods arranged in a microplate format or other device suitable for the simultaneous synthesis of ninety or more separate peptides. The peptides are attached to the resin in such a way that both the N^α and side chain protecting groups can be removed independently and without releasing the peptide from the support. In this way a single peptide on the surface of a pin can be prepared following synthetic assembly and utilized to explore its interactions with acceptor molecules, or alternatively released into solution for various other binding and biological assays. Both continuous and discontinuous epitopes can be evaluated with this method, though of course much large peptides with increased diversity of structure must be prepared in such cases; Geysen and coworkers have suggested an approach for this (Geysen et al., 1985).

Another early approach to preparing large numbers of peptides for screening was the "tea bag" method developed by Houghten (1985). In this method, porous polypropylene containers (tea bags) each containing a small amount of a solid phase resin support are identified and separated for coupling reactions where diversity is desired, but combined for common coupling or other reactions where the same structure is to be maintained or for other common reactions such as removal of N^α-protecting groups or specific side chain protecting groups. Eventually large numbers of specific peptides can be prepared with each tea bag possessing a unique peptide. Each peptide can then be purified and used in the usual manner for bioassays and so forth. More recently this approach was developed further (Houghten et al., 1991) using a combinatorial approach that allows the preparation of large numbers (hundreds of thousands to millions) of peptides in solution. These can be used to screen for binding or for some other biochemical or pharmacological end point. As generally practiced thus far, the fractions that contain activity are then resynthesized in a reduced combinatorial format such that by a process of iteration the "best" peptide sequence is obtained. This approach can be used to screen millions of peptides, and for some optimization of a specific lead.

In a somewhat different approach referred to as the one peptide, one bead method, Lam and coworkers (Lam et al., 1991, 1992) have developed a methodology based on the Merrifield solid phase synthesis procedures that can produce large (10^5 to 10^8) peptide libraries in small volumes in a form suitable for screening the peptides on each bead for a particular interaction with a particular acceptor molecule of interest. Those peptide beads that interact with the acceptor are then identified using a color reaction, fluorescent detection, or some other "reporter" reaction. Those individual beads then can be isolated, and the structure of the peptide on the bead determined by microsequencing methods. As currently practiced, with 50–100 micron beads, there are about 1,000,000 peptides per 1–2 mL of peptide resin. Each bead possesses sufficient peptide for all aspects of the methodology including preview analysis to assess the purity of the peptide on the bead ($\pm 0.5\%$). The rapid synthesis of large diverse mixtures requires the use of statistical combinatorial methods similar to those of Houghton, as first introduced by Furka et al. which they termed "proportioning-mixing" (Furka et al., 1988, 1991). Lam et al. termed it "split synthesis." In this approach, instead of using "tea bags" to separate peptides for individual coupling reactions, a mixture of beads is divided into equal portions at each coupling step, and each portion is individually coupled with a specific amino acid. Then all the peptides are mixed again for removal

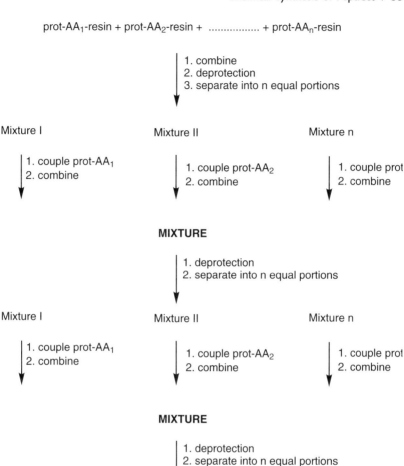

Figure 2-17. General procedure for the creation of a peptide library (Prot: protecting group, AA: amino acid).

of the N^α-protecting group, reproportioned for the next coupling step, and so on (Fig. 2-17). Note that at each synthesis step a single amino acid residue is being added to the growing peptide in each reaction. In this manner a complex mixture of peptides with a specific statistical distribution of peptides that can be calculated by well-established mathematical methods can be prepared. Since each bead contains a unique peptide structure, those peptides that interact with a specific receptor molecule can be determined directly, and information such as consensus sequences rapidly determined, often from a single library since each individual peptide has a unique spatial location on a particular bead in the library. Subsequently, it has been demonstrated that partial release of peptides from individual beads can be accomplished, which allows for solution assays. In general, this method is relatively simple and optimized methods of solid phase peptide synthesis can be applied readily.

A technologically sophisticated method for construction of peptide libraries on solid support in a manner that allows addressing a particular peptide in a unique spatial location on the surface was developed by Fodor and coworkers (Dower & Fodor, 1991; Fodor et al., 1991). This method also used solid phase peptide synthesis, in this case on a surface, in combination with photolithography. For this methodology N^α-protecting groups that can be removed photochemically are needed. Because of the high spatial resolution of photolithography, in principle as many as 40,000 peptides can be made and detected on a 10 mm surface, though to date many fewer peptides can be made on such a surface because the chemistry, especially the photochemistry, has not been optimized. In view of the potential of this methodology, synthetic chemists undoubtedly will expend considerable energy to optimize a variety of chemistries that are compatible with the method. However, to date very little has appeared in the literature regarding steps being taken to improve the chemistry needed for this approach to work well.

Peptide libraries constitute an interesting and exciting area of research that has attracted considerable attention. Several reviews already have appeared (Jung & Beck-Sickinger, 1992; Pavia et al., 1993; Scott, 1993) regarding the foregoing and other approaches that have been taken to prepare chemically diverse peptide mixtures. Much still needs to be done to evaluate the synthetic chemistry and the diverse way in which this approach might be utilized to examine chemical structures and reactivities that depend on molecular recognition processes. In addition, D-amino acids, unusual amino acids, cyclic peptides, pseudopeptides, peptoids, peptidomimetics, and other structures can be incorporated into such chemical libraries. The development of quantitative synthetic organic chemistry that will meet the challenge is needed and will provide for much work for the synthetic chemist for the foreseeable future. Similar work by chemists and biologists to create chemically clean binding assays and other bioassays for this technology is also needed. Overall these approaches can provide chemists the opportunity to participate in biological discovery using chemical methods similar to those discussed here (Hruby, 1996).

Stepwise" and "Fragment or Segment Condensation" Strategies

Whether in solution or by solid phase synthesis, peptides can be obtained by a stepwise strategy by applying repetitive and standardized synthesis cycles that include coupling of the residue onto the growing peptide and deprotection of the N^α-protecting group. This method is usually used for the synthesis of shorter peptides (up to 30–40 residues). Though easily applicable to automated solid phase peptide synthesis, this method presents some drawbacks for solution synthesis including insolubility of the growing peptide once it has reached a critical size or hydrophobicity, and difficult purifications of the intermediates. In solid phase synthesis, the major problems are aggregation of the peptide, leading to poor coupling and deprotection, and synthesis of peptide by-products missing one or more amino acids (deletion phenomenon), which are difficult to separate from the desired peptide after the final deprotection step. Numerous examples of solution and solid phase stepwise syntheses of peptides can be found in the literature.

In order to address some of these problems and to be able to synthesize peptides incorporating over a hundred residues, improved approaches for fragment (or segment) condensation peptide synthesis have been under development. In these approaches, segments of the peptide are synthesized in a protected form by classical stepwise methods in solution or on a solid support. The different fragments are then coupled together, yielding the desired peptide. This method suffers three primary limitations, including the fact that poor coupling may occur between the different fragments, racemization may take place at the activated C-terminal amino acid of the fragment, and the low solubility of some fragments may limit coupling. Nonetheless, examples using fragment condensation are reported to provide a sense of what can be done.

Semenov and Lomonosova (1993) have reported the gram-scale synthesis of salmon calcitonin-(1–16). Four fragments were synthesized in solution by the DCC/HOBt coupling method. They were then condensed using the same method without detectable racemization. For the synthesis of small peptides like Leu-enkephalin, Miyazawa et al. (1992) reported that the coupling of the fragments proceeded with higher yields and lower racemization levels when copper (II) chloride was added to the DCC and HOBt reagents. Other coupling reagents have been used as well. Gobbo et al. (1991) reported the solution synthesis of a glyco-hexapeptide by the fragment and stepwise procedure using a combination of BOP and HOBt for the couplings. Poulos et al. (1991) used carboxylic-phosphoric mixed anhydrides for the fragment peptide synthesis of protected analogues of substance P using similar methods. In a very important series of experiments to overcome the problems of solubility and slow coupling in fragment condensation reactions, Sakakibara and coworkers have developed a variety of mixed solvent systems such as hexafluoropropanol-chloroform that not only solubilize peptides previously insoluble even in DMSO or N-methylpyrrolidone, but also help to promote reactions of peptide fragments that are sparingly soluble in their solvent systems (Kimura et al., 1991; Kuroda et al., 1992). They have used this approach very successfully (Kuroda et al., 1992, 1993) for several syntheses of large (20–84 residues) polypeptides and small proteins.

The solid phase synthesis of a cyclic analogue of neuropeptide Y has been published by Bouvier & Taylor (1992). In this case, a cyclic fragment was condensed onto a linear fragment in 60% yield using DCC and ethyl 2-(hydroxoimino)-2-cyanoacetate. Excellent yields at the condensation stage were obtained using the BOP reagent in the presence of HOBt for the synthesis of human gastrin (Kneib-Cordonier et al., 1990).

Instead of utilizing classical coupling reagents at the fragment condensation stage, different groups report the use of enzymes like papain and α-chymotrypsin as condensation reagents for the peptide fragments. For example, such methods have been used in the synthesis of alamethicin (Slomczynaska et al., 1992) and an insect neuropeptide (Xaus et al., 1992) in yields of 62 and 82%, respectively. In the latter case DIC-HOBt and BOP-HOBt also were used and compared. Yields proved to be lower and racemization levels much higher using the chemical methods.

Cyclization of Peptides in Solution and on Solid Supports

Cyclization of peptides can lead to analogues with disulfide, lactam, and other functional groups. The fully deprotected or partially protected peptide can be transformed by intramolecular cyclization into its cyclic analogue. For the solution synthesis of cyclic peptides incorporating side chain-to-side chain, backbone-to-side chain, backbone-to-backbone linkages, or multiple disulfide bridges, there is a need for orthogonal protection of the different reactive amino, carboxyl, sulfhydryl, or other functions groups in order to ensure selective cyclization. The major drawback of solution cyclization is the risk of forming peptide polymers through intermolecular reactions. In

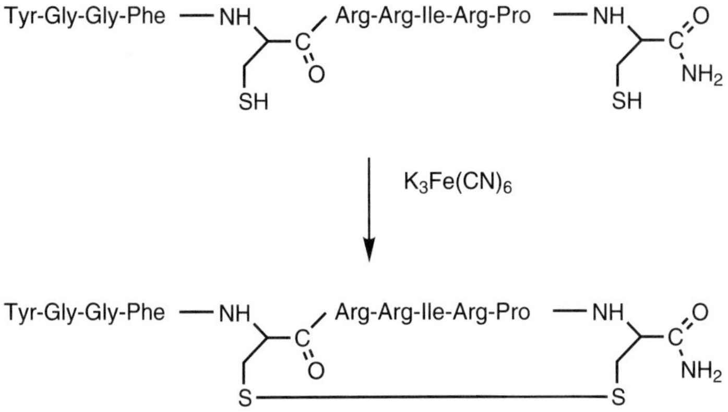

Scheme 2-15. Solution cyclization of [Cys^{5,11}]Dyn A-(1-11)-NH$_2$.

order to avoid that outcome, high dilutions (10^{-3} to 10^{-4} M) of the substrate are necessary so as to favor intramolecular bond formation. As an example, monocyclic disulfide analogues of dynorphin A (Scheme 2-15) have been synthesized by Kawasaki et al. (1993). The peptides were elaborated stepwise on a MBHA resin with Boc-benzyl protected amino acids. Once the synthesis was finished, the peptide was cleaved from the resin and fully deprotected with HF and appropriate scavengers. The peptide could then either be diluted with water to a concentration of 10^{-4} M at pH 8 and oxidized (K$_3$Fe(CN)$_6$) over a 30-minute period, or could be dissolved in a much smaller volume (about 50 mL) of water and oxidized by the very slow addition of K$_3$Fe(CN)$_6$, usually over a period of 24 hours. The excess oxidant was removed by addition of an ion exchange resin, and the solution was lyophilized. The latter method gave better results in terms of yield and purity of the peptide. Air oxidation (i.e., without any added oxidant) and other oxidants also have been used for the disulfide bridge formation (Hiskey, 1981). When large proteins containing multiple disulfide bonds are to be formed, the choice of solvent, reaction conditions, and oxidant often can be critical to success. In many cases mixed solvent systems are needed so that the properly folded structure can be obtained (Kuroda et al., 1993). Similar considerations need to be made when the formation of lactams or other macrocyclic rings are done in solution.

The basis of cyclization on a solid support (Hruby et al., 1997) is the availability of orthogonal schemes of synthesis with temporary and semipermanent protecting groups. Some of these schemes are described in the section on orthogonal synthesis. One of the most common example is the mixed Boc-Fmoc strategy for on-resin lactam formation (Scheme 2-16) (Hruby et al., 1990). The peptide is elaborated using N^α-Boc protected amino acids; Fmoc or OFm esters are used for the side chain protection of the amino and carboxyl groups to be incorporated in the lactam bridge. Though giving good results, this synthesis scheme and others relies on the use of HF for the final deprotection-cleavage step, with the risk of unwanted side reactions in some cases. Therefore, new protecting groups and resin handles that can be removed and cleaved under milder conditions are now currently in use (Barany et al., 1988). These include allyl (Loffet & Zhang, 1992) or trimethylsilylethyl esters (Marlowe, 1993) (removed by Pd(0) (Lloyd-Williams et al., 1991) and fluoride ion, respectively) and "Rink" (Rink, 1987) or "Sasrin" (Mergler et al., 1988a) resin handles cleaved with dilute TFA.

Chemical Synthesis of Peptides / 57

Though on-resin cyclization has been reviewed thoroughly (Hruby et al., 1997), two examples will be outlined here. Osapay and Taylor (1990) report the solid phase synthesis on an oxime resin of two 21-mer model peptides constrained by three side chain-to-side chain lactam bridges. Oxime or "Kaiser's" resins have been used previously for the synthesis of fully protected peptide fragments (DeGrado & Kaiser, 1980; Findeis & Kaiser, 1989; Osapay & Taylor, 1992). The two peptides have a general structure of tricyclo(3–7, 10–13, 17–21)-H-[Lys-Leu-Lys-Glu-Leu-Lys-Xxx]$_3$-OH, where Xxx is ether Asp or Glu. The dipeptide Boc-Lys(2-ClZ)-Xxx-OPac was linked to the

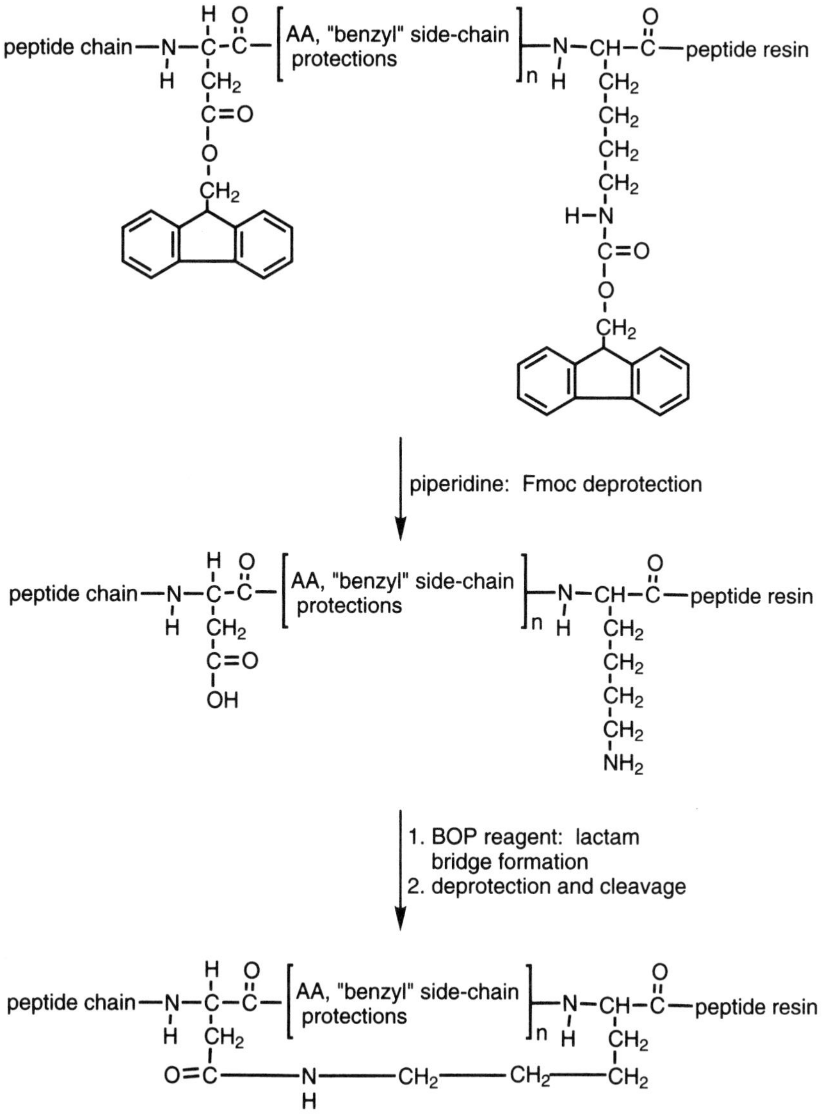

Scheme 2-16. On-resin lactam bridge formation.

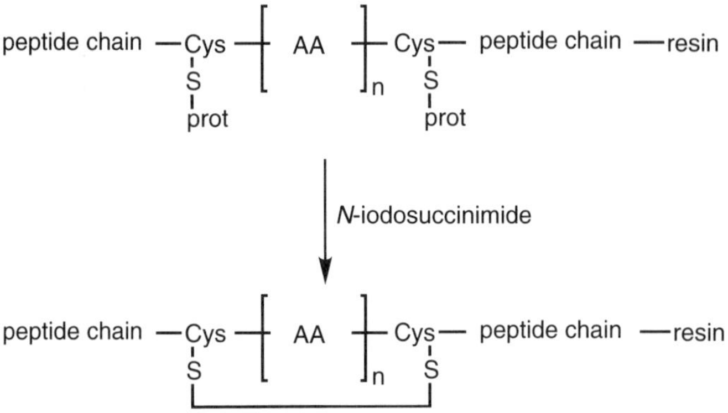

Scheme 2-17. On-resin disulfide bridge formation using N-iodosuccinimide.

oxime resin through the carboxyl group of the side chain of Xxx. The peptide was elaborated with all but one amino acid temporarily protected by groups resistant to TFA. Lysine residues 10 and 17 had their amino groups protected as trityl derivatives. Once the 7-mer was synthesized, the trityl group was removed by the action of TFA. Upon liberation of the amino group by DIEA, intrachain cyclization in the presence of AcOH released the protected cyclic heptapeptide from the resin. Three segments were then condensed, deprotected, and purified to give the 21-residue peptide incorporating the three lactam bridges in high overall yield. In this case, the authors took advantage of the active ester character of the peptidyl oxime resin linkage to perform the cyclization reaction and the cleavage of a protected peptide segment from the solid support at the same time.

The other example concerns on-resin disulfide bridge formation (Shih, 1993). N-Halosuccinimides have been found to convert S-protected cysteine (Acm, p-MeOBz or p-MeBz protected), or unprotected cysteine into cystine in good yields. Shih reported the on-resin deprotection of cysteines and their concomitant intramolecular reaction to form a disulfide bridge using N-iodosuccinimide (NIS) (Scheme 2-17). This method has been used for the synthesis of oxytocin and [Arg^8] vasopressin. These successful preparations of peptides prompted the author to synthesize the cyclic [Cys^5, Cys^{12}] analogue of the human growth hormone-releasing factor [hGHRF-(1-29)]. The [Cys^5, Cys^{12}] hGHRF-(5-29) sequence was assembled on a MBHA resin on a fully automated peptide synthesizer. Boc-benzyl protected amino acids were used and the thiol groups of the cysteine residues were protected with p-MeOBz groups. Deprotection and cyclization were performed in the synthesizer by addition of N-iodosuccinimide to the peptide. Amino acids 1 to 4 were then added stepwise and the peptide was deprotected and cleaved with HF. The overall yield for the synthesis of this 29-residue cyclic peptide was 15.6%. The same fully automated synthetic scheme was applied to the synthesis of apamin, a bicystine peptide.

These two examples outline the advantages of on-resin versus solution cyclization, which include higher overall yields and processes that often can be automated on a synthesizer. Other advantages include reduction of polymer formation through intermolecular reactions due to the "pseudo-dilution effect," the facile removal of cyclization reagents from the peptide resin by filtration and washing, and the ability to perform all major types of cyclizations on solid supports.

Orthogonal Protection and Synthesis

Two protecting groups or resin handles are said to be orthogonal when, under a specific reaction condition, one of them is removed while the other is stable (Barany & Merrifield, 1977). A classical example of orthogonal protecting groups is the combination of the N^α-Fmoc group (cleaved under basic conditions), and side chain protecting tert-butyl or trityl groups, which are cleaved by TFA. The handle linking the peptide to the resin can also be included in this strategy, adding a third dimension to the orthogonality (Fig. 2-18).

Hill et al. (1990) report the solid phase synthesis of a bicyclic (lactam and disulfide bridges) analogue of oxytocin: the analogue was elaborated on a MBHA resin (cleaved by HF), using N^α-Boc protected amino acids (cleaved by TFA). The side chain groups were protected by benzyl-type groups (also removed by HF), except for the amino acids destined to be part of the lactam bridge (glutamic or aspartic acid, and lysine or ornithine) which were protected by OFm or Fmoc groups, respectively. Once the synthesis was complete, the peptide resin was treated with dilute piperidine to yield a peptide containing one free carboxyl and one free amino side chain group that could be condensed to form the lactam bridge. The monocyclic peptide resin was then cleaved for the resin and the side chain protecting groups were removed by treatment with HF. The disulfide bridge was finally formed by classic oxidation methods ($Fe_3(CN)_6$). Munson et al. (1993) reported the solid phase synthesis of the parallel dimer of deamino-oxytocin. Deamino-oxytocin incorporates one cysteine residue and L-β-mercaptopropionic acid (β-Mpa, or deamino-cysteine). The goal was to avoid intramolecular cyclization between the thiol groups, as well as polymer formation. This was achieved by differential protection of the thiol groups. Cysteine was protected by the Acm group (removed by thallium(III) trifluoroacetate), while β-Mpa was protected by the novel

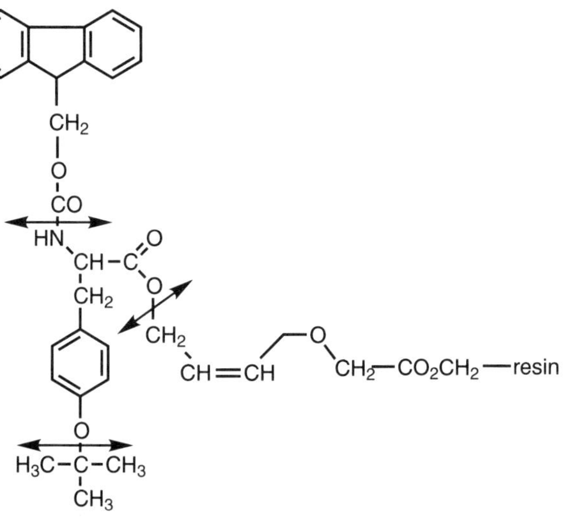

Figure 2-18. An example of three-dimensional orthogonal protection-cleavage scheme: the temporary a-amino Fmoc protecting group is removed with piperidine, whereas the permanent tert-butyl (and related) side chain protecting groups are removed by treatment with TFA. The peptide is cleaved from the resin by chemical hydrogenation.

Tmob (2,4,6-trimethoxybenzyl) group which is removed under very mild acidic conditions (in this case 7% TFA). The peptide was elaborated on a tris(alkoxy)benzylamine (PAL) resin (cleavage by 90 to 95% TFA) and amino acids were introduced as Fmoc derivatives. Once the peptide synthesis was finished, the Tmob groups were removed and the first intermolecular disulfide bridge was created. The Acm groups were then cleaved by Tl(TFA)$_3$ and the second bridge between the cysteines was formed. The peptide was finally cleaved from the resin by concentrated TFA.

Finally, an elegant, mild, three-dimensional orthogonal protection scheme has been proposed (Barany & Albericio, 1985; Barany & Merrifield, 1980). The dithiasuccinoyl (Dts) group (cleaved by thiols) was used for the temporary protection of the amino acids (Barany & Merrifield, 1979) while their side chains were protected with *tert*-butyl groups. The first amino acid was attached to the resin through an *o*-nitrobenzyl ester linkage that could be cleaved by photolysis at 350 nm. The authors applied this scheme to the synthesis of Leu-enkephalin (Barany & Albericio, 1985). The utilization of orthogonal protecting strategies also has proven to be very useful in fragment condensation as exemplified by the synthesis of human gastrin I under mild conditions (Kneib-Cordonier et al., 1990). N^α-Amino and side chain protection were provided in this study by Fmoc and *tert*-butyl groups, respectively, and a series of different resin handles cleavable under mild conditions were used. The overall yield was 30%, comparing favorably with those from stepwise assembly.

Pseudopeptides: Amide Bond Replacements

Backbone modifications usually involve the peptide bond. Nevertheless, they also can take place at the αCH group of one or more amino acids in the peptide chain. Backbone modifications by isosteric peptide bond replacement are being used increasingly in peptide chemistry. They are employed, for example, to increase the stability of biologically active peptides towards proteolytic enzymes to develop peptide enzyme inhibitors, to achieve receptor selectivity, and to obtain peptide antagonists (Spatola, 1983, 1993). Among the numerous replacements the ψ[NH—CO], ψ[CH$_2$—S], ψ[CH(OH)—CH$_2$] and ψ[CH$_2$—NH] functionalities have received considerable attention and are used routinely. The symbol ψ[ ] between two amino acids is now commonly accepted: ψ denotes the absence of the amide bond CO—NH, while the structure that replaces it is within the brackets. Several examples of peptide bond replacements are given below as well as an indication of their importance in terms of biological activities and stability of the analogues that were obtained.

The ψ[CH$_2$—NH] amide bond replacement, also known as the reduced peptide bond, has been used widely for the last decade. Most commonly the reduced bond is synthesized via the amino acid aldehyde prepared according to the method developed by Fehrentz and Castro (1983) (see also Guichard et al., 1993). The aldehyde is then condensed (Scheme 2-18) in the presence of sodium cyanoborohydride as described (Doulut et al., 1992; Coy et al., 1988; Hocart et al., 1990; Martinez et al., 1985; Qian et al., 1989; Rodriguez et al., 1987). The method used routinely was later modified (Ho et al., 1993). This replacement has been incorporated into sequences and analogues of many biologically active peptides including tetragastrin, cholecystokinin, bombesin, somatostatin, opioid peptides, substance P, neurotensin, secretin, and growth hormone-releasing factor. In many of these studies, peptide bonds were replaced one-by-one and the analogues tested for their biological activities with varied results. Some analogues

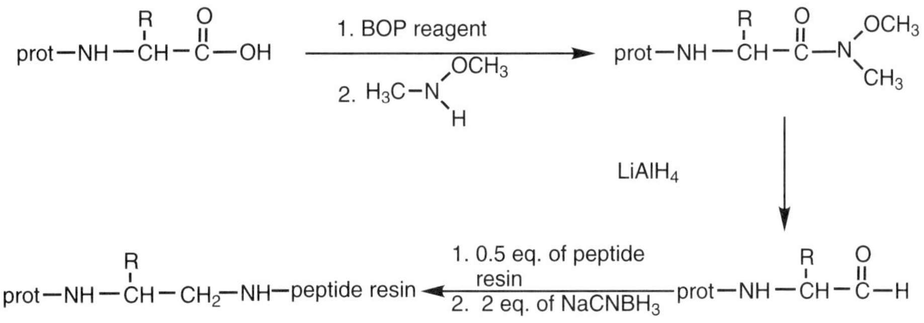

Scheme 2-18. Aminoaldehyde synthesis and reduced bond (CH$_2$NH) formation. The protecting group (Prot) can be a Boc or an Fmoc group.

totally lost activity, whereas other analogues of tetragastrin, bombesin, some opioids, substance P, secretin, and growth hormone-releasing factor turned out to be antagonists, or weak or partial agonists at their receptors. These analogues also showed improved resistance to enzymatic degradation. The results clearly show the importance of the backbone and the amide bond for interactions between the peptide and its receptor.

In many cases, instead of forming the amide bond replacement *in situ* by a stepwise elongation technique (as for ψ[CH$_2$—NH]), the dipeptide isostere bearing the peptide bond modification is synthesized prior to its incorporation into the peptide. This is the case for the carba (ψ[CH$_2$-CH$_2$]) (Rodriguez et al., 1990a, b), the ψ[CH(CN)NH] (Herranz et al., 1991), the ψ[CH$_2$-S] (Fehrentz & Castro, 1983), the *trans* olefinic (ψ[*trans* CH=CH]) (Cox et al., 1980a, b; Hann et al., 1982; Holladay & Rich, 1983; Tourwe et al., 1989, 1990), the hydroxyethylene and ketomethylene (Cox et al., 1980a, b) replacements, and for many other amide bond replacements as well.

In terms of size and spatial configuration, a *trans* carbon double bond should represent a good surrogate for the nearly planar and quite rigid amide bond. Several reports of the synthesis of Glyψ[*trans*CH=CH]Gly, and Xxxψ[*trans*CH=CH]Gly dipeptides (Hann et al., 1982; Tourwe et al., 1989, 1990) and the same dipeptides, as well as a general scheme allowing the synthesis of Xxxψ[*trans* CH=CH]Yyy surrogates have appeared. However, these nonstereospecific routes lead to a mixture of diastereoisomeric "dipeptides." These analogues or the peptides in which they are incorporated must, therefore, be separated from one another.

Enkephalin and substance P analogues have also been prepared (Fig. 2-19). According to Hann et al. (1982), [Tyr^{1}ψ[CH=CH]Gly^2 Leu^5] enkephalin (with the natural S isomer at Tyr^1) exhibits the same binding affinity as the parent compound, whereas for Cox et al. (1980a), the same analogue in racemic form had activities equal to 300% and 24% those of enkephalin in the GPI and MVD bioassays. Tourwe et al. (1989, 1990) reported that the cyclic analogue Tyr-D-Cys-Glyψ[CH=CH]Phe-Cys-OH was quite potent. On the other hand the [Gly^{2}ψ[CH=CH]Gly^3 Leu^5]enkephalin analogue had only about 0.1% activity in the same bioassay (Cox et al., 1980a). These results seem to indicate that the amide bonds between Tyr and Gly, and Gly and Phe are not important per se for recognition by the opioid receptor, whereas the Gly^Gly bond is important. The authors also reported an increase in stability to enzymatic degradation.

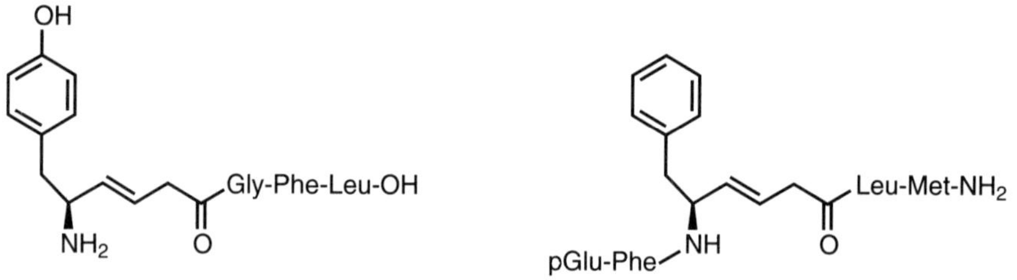

Figure 2-19. Structure of Leu-enkephalin (*top*) and substance P (*bottom*) analogues incorporating a *trans*-olefinic dipeptide isostere.

A modified C-terminal fragment of substance P, pGlu-Phe-Phe-Gly-Leu-Met-NH$_2$ was prepared incorporating Phe$^3\psi$[CH=CH]Gly^4 and Phe$^2\psi$[CH=CH]Phe^3 as racemic mixtures (Fig. 2-19) (Cox et al., 1980a, b). The latter analogue proved to be equipotent with the parent hexapeptide in the GPI bioassay. These results show that, except for the racemization and synthesis problems, *trans* olefinic replacements can be useful for the synthesis of stable and potently active peptides.

Modifications at the αCH Group

The replacement of the αCH group of one or more amino acid residues in the peptide chain by a nitrogen atom leads to azapeptides (Fig. 2-20). The introduction of such modifications anywhere in a peptide is quite easy synthetically, as shown and reviewed by Gante (1989). The synthesis of various analogues of angiotensin (Hess et al., 1963), lulibern (LH-RH) (Dutta et al., 1978; Ho et al., 1984), ACE inhibitors (Ulm et al., 1982) and enkephalin (Dutta et al., 1977; Gacel et al., 1988) exemplify such modifications. Some of the most active and stable analogues are currently being used as drugs (Furr et al., 1983; Ulm et al., 1982), indicating the utility of this approach.

The Allyl-UNCA Strategy

Recent and useful advances have involved the introduction of UNCAs as powerful acylating agents (Lofflet et al., 1993a, b) and the development of allyl-based α-amino and side chain protecting groups and resin handles; the latter allow deprotection under mild and neutral conditions, avoiding the use of strong acids at the final deprotection-cleavage stage.

It therefore seemed reasonable to combine both methods in order to develop a new strategy in peptide synthesis. The synthesis of several model peptides and LH-RH were carried out on different supports, including one bearing an allylic handle while amino

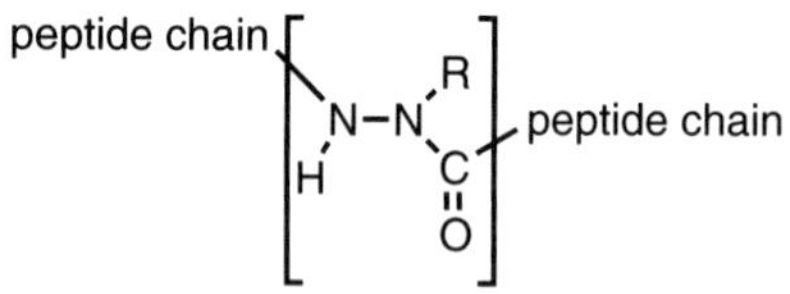

Figure 2-20. General structure of an azapeptide.

acids were introduced as Fmoc or Boc-protected NCAs. The results that have been obtained show that this new strategy is highly efficient for the purpose of peptide synthesis, very easy to set up and use, and can give excellent results in terms of yield and purity.

Continuous Flow Synthesis

In this mode, amino acids, reagents, and solvents are forced through a resin packed into columns. It has been postulated that as the peptide-resin is always solvated in this mode, SPPS should be carried out much more efficiently (Lukas et al., 1981; Scott et al., 1972). The continuous flow strategy has become much easier to use with the introduction of Fmoc-*t*-butyl protected amino acids and Fmoc resin handles, which allow deprotection and cleavage under mild conditions.

One of the first problems encountered is that resins used for batch SPPS are usually too fragile; they are not flow stable and, therefore, collapse after a few synthesis cycles. Continuous flow SPPS has been made possible by the use of resins composed of inorganic matrixes with high permeability to solvents. Examples include those obtained by polymerization of poly-dimethylacrylamide gel, polyethylene glycol or polystyrene onto supports of Kieselguhr or crosslinked polystyrene (Atherton et al., 1981; Dryland & Sheppard, 1986; Fields & Noble, 1990). Meldal has reported on the synthesis of the difficult acyl carrier protein-(65-74)-sequence on a flow-stable polyethyleneglycol dimethylacrylamide (PEGA) copolymer (Meldal, 1992). This resin is chemically stable, has good swelling properties in polar solvents and, furthermore, does not absorb in the aromatic region, thus allowing the spectrophotomertric monitoring of the coupling reaction within the resin. Once the copolymeric resin is synthesized, it is packed into a glass column and derivatized with the "Rink linker," 4-Fmoc-amino(2,4-dimethoxyphenyl)methylphenoxyacetic acid. The test peptide is then synthesized using classical coupling-deprotection methods and cleaved from the resin with concentrated TFA containing the appropriate scavengers. The reaction time was shorter than that previously observed for the synthesis of the same peptide on a different flow-stable polymeric support. The synthesis of several other peptides including insulin and FOS oncogene protein fragments have been reported by the continuous flow synthesis method (Bayer et al., 1985; Daniels et al., 1989).

Summary

Peptide synthesis has become a critical aspect of many areas of chemistry, biology, and medicine. In this chapter we have sought to provide both old-timers and newcomers to the field of peptide research with a broad overview of how to synthesize peptides successfully, avoid making common mistakes, and solve common and less common problems. Obviously, not all aspects of peptide synthesis could be covered, but we have made an effort to include the most common and generally useful methods.

The examples given here should allow the reader to choose from among the multitude of resins, protecting groups, synthetic strategies, and so forth, the ones best suited and adapted to his or her specific problems and needs, whether for solid phase or solution phase synthesis of linear and cyclic peptides.

No peptide is ever trivial to synthesize, and unexpected problems can and often will arise, but carefully planned syntheses, the use of common sense, and reading the extensive literature (such as this book and the references provided herein) will make your experience as a peptide chemist enjoyable and rewarding.

Acknowledgements

This work was supported by grants from the U.S. Public Health Service, the National Institute of Drug Abuse, and the National Science Foundation. We thank Guoxia Han, Charlene Morgan, and Margie Colie for their help in preparing this manuscript.

3

Total Chemical Synthesis of Proteins

Michael C. Fitzgerald and Stephen B. H. Kent

The total chemical synthesis of proteins has been a long-standing goal of organic chemistry, dating from the early years of the twentieth century (Fisher, 1906). Notable achievements in the history of total chemical synthesis of proteins include the synthesis of insulin (Sieber et al., 1974) and several syntheses of the enzyme ribonuclease (Gutte & Merrifield, 1969; Hirschmann et al., 1969; Yajima & Fujii, 1981). The passage of time has shown that much of this early work was premature in the sense that none of these pioneering synthetic efforts led to the productive use of total chemical synthesis for the study of protein function. Therefore, the goal of undertaking successfully the total chemical synthesis of proteins has continued to be one of the driving forces of the development of synthetic organic chemistry (Inui et al., 1996). Until recently these efforts have met with only limited success (Bayer, 1991; Hirschmann, 1991). That situation has now changed dramatically.

In this chapter, a set of techniques will be presented that give practical access to the world of proteins by total chemical synthesis. The application of total chemical synthesis to the study of the chemical basis of protein function will be described in several case studies. Finally, a new approach for systematically exploring the chemical basis of protein function will be described. This approach involves a unique combination of total protein synthesis and novel analytical chemical techniques.

Chemical Synthesis of Peptides

Proteins are molecules that owe their specific chemical, biochemical, and biological properties to the defined three-dimensional structure formed by the unique fold of a polypeptide chain (Anfinsen, 1973; Yue & Dill, 1995). They are not simply large polypeptides. However, the total chemical synthesis of a protein begins with construction of its polypeptide chain.

Principles of Solid Phase Peptide Synthesis

In recent years, the chemical synthesis of polypeptides has been carried out mostly by stepwise solid phase peptide synthesis (SPPS) (Kent, 1988; Merrifield, 1963) (Fig. 3-1). SPPS is carried out on a so-called insoluble or solid resin support, which is actually a swollen gel phase polymer (Kent et al., 1992). The C-terminal amino acid of the target polypeptide chain is attached to the resin support by a covalent bond that is sta-

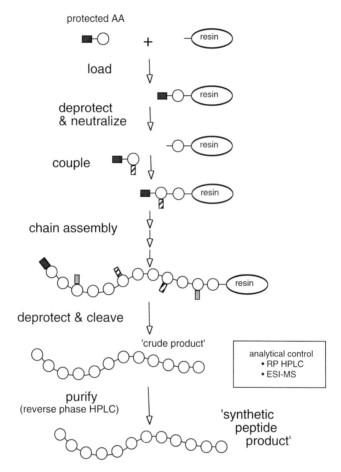

Figure 3-1. Overview of stepwise SPPS (Merrifield, 1963, 1986). The C-terminal amino acid residue of the target polypeptide chain is covalently attached to a resin support. The polypeptide chain is assembled in stepwise fashion from C- to N-terminus. Each cycle involves removal of the temporary α-amine protecting group, followed by coupling of the next N^α-protected, carboxyl-activated amino acid to the resin-bound peptide chain. Excess reagents and coproducts are removed by filtration and washing. Finally, side chain protecting groups are removed and the target polypeptide is cleaved from the resin support for work-up.

ble to the conditions used to assemble the peptide chain. All of the chemical reactions involved in building the protected peptide chain occur *within* the swollen beads. A cycle of deprotection and coupling, with any necessary washes in between, is repeated with the appropriate amino acids until the target polypeptide chain has been assembled in a linear, stepwise fashion on the resin support. The genius of the solid phase synthesis technique (Merrifield, 1986) is that excess reactants and coproducts can be simply washed away and removed by filtration, leaving the desired peptide product attached to the polymeric support.

After chain assembly is complete, the product peptide-resin is "deprotected," that is, protecting groups are removed from any side chain functionalities. The polypeptide chain is "cleaved" from the resin support, releasing the crude peptide into solution. If the chemistry of SPPS chain assembly has been done well enough, it is then possible to purify the target peptide in a straightforward fashion and obtain a homogeneous synthetic peptide product. The synthetic peptide is then characterized for purity, typically by analytical reverse phase HPLC, and for molecular structure by electrospray mass spectrometry.

Optimized SPPS

Recently, we described a set of optimized, generally applicable procedures for the efficient chemical synthesis of peptides (Schnolzer et al., 1992a). The method that we described incorporates several key improvements to earlier SPPS methods. The improvements were based on a detailed understanding of the physical and organic chemical principles of resin-supported chemical synthesis (Kent et al., 1992). Problems observed, and their effects on peptide synthesis, are summarized in Table 3-1. The features of the resulting technique are described in Figure 3-2. This highly optimized stepwise SPPS method is routinely applicable to the chemical synthesis of peptides 60 to 80 residues in length.

The highly optimized Boc chemistry (Schnolzer et al., 1992a) summarized in Figure 3-2 has eliminated both the sequence-dependent difficulties (Milton et al., 1990) and the common side reactions (Kent, 1988; Kent et al., 1992) in chain assembly. The remaining by-products in stepwise SPPS now arise from two principal sources. The first of these derives from impurities in the reagents used for the synthesis, primarily the protected amino acids (Kent et al., 1992). Most of these by-products are actually acetyl-peptides, arising from the presence of traces of acetic acid, which in turn are derived from residual traces of ethyl acetate used in bulk production of the protected amino acids. Such terminating impurities can be a major problem in the chemical synthesis of peptides. Rigorous control of the purity of resins, solvents, protected amino acids, and other reagents is essential for the unambiguous synthesis of peptides by SPPS (Kent et al., 1992). In addition, chemical side reactions occur in the final deprotection-cleavage steps, giving rise to complex mixtures of products. Considerable effort has been expended to understand and minimize these problems (Tam et al., 1983). Ultimately, the key is precise analytical evaluation of peptide products.

Table 3-1 Features of Optimized Stepwise Solid Phase Peptide Synthesis (Boc chemistry)

Problem	Solution
Chemical Side Reactions	
Chain loss	TFA-stable peptide-resin
N^α-trifluoroacetylation	$HOCH_2$-free resin; TFA-stable peptide-resin
Late initiation	$HOCH_2$-free resin; stable peptide-resin
Deletions	Aldehyde-free resin
Termination	Pure Boc-amino acids, solvents and reagents
Physical Chemical Problem	
"Difficult sequences" (peptide chain aggregation)	*in situ* neutralization coupling protocols

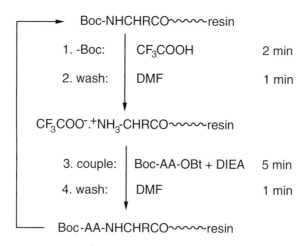

Figure 3-2. A highly optimized SPPS protocol, using *in situ* neutralization and *tert*-butoxycarbonyl (Boc) chemistry (Schnolzer et al., 1992). The key difference from standard Boc chemistry protocols is combining the neutralization and coupling steps into one operation.

Analytical Control of Peptide Synthesis

For the past twenty years, the principal method for the analysis of the products of stepwise SPPS has been reversed phase high pressure liquid chromatography (HPLC), using gradients of organic solvents containing ion-pairing agents such as trifluoroacetic acid (Hearn, 1991). Since its introduction as a practical, high resolution tool for the analysis of peptides, reversed phase HPLC has been invaluable in the characterization and purification of synthetic peptides, and is still used routinely today. Reversed phase HPLC is rapid, highly resolving for even complex mixtures of peptides, sensitive and quantitative, with a wide dynamic range. In about one hour, HPLC analysis can give a complete profile of the number and amounts of peptide components present in the total crude product from a synthesis. It has also been used routinely to guide the purification of the principal product from contaminating coproducts (Clark-Lewis et al., 1989).

However, until recently, information about the *covalent structure* of these isolated coproducts has been difficult to obtain. In the past few years, new mass spectrometry (MS) techniques have revolutionized the analysis and covalent characterization of synthetic polypeptides (Chait & Kent, 1992). The most generally useful of these techniques for the total chemical synthesis of polypeptides and proteins has been electrospray MS. The electrospray ionization process involves the evaporative ionization, at atmospheric pressure, of dilute solutions of the polypeptide analyte (Fenn, 1989). The evaporative ionization process gives rise to multiple ionized species for a polypeptide chain depending on the number of excess protons on each molecule. After ionization, the desolvated, charged polypeptide species are drawn into the high vacuum region of a quadrupole mass spectrometer and separated according to their mass-to-charge ratio. In the resulting mass spectrum, each molecular species is represented by multiple peaks corresponding to different charge states. This allows the molecular weight of each polypeptide to be measured with great precision (Mann et al., 1989). For a typical polypeptide, the MW can be measured to 1 part in 10,000: for example, 5019.5 ± 0.5 Da. Such high precision MW measurements place severe constraints on structural hypotheses and, in conjunction with a detailed understanding of the synthetic chemistry used, can serve as a means of directly assessing the covalent structure of synthetic peptides products and coproducts.

When used in a hands-on fashion by the synthetic protein chemist, electrospray MS can be an indispensable tool for controlling the chemical synthesis of peptides (Schnolzer et al., 1992b). Most importantly, the total crude peptide products from stepwise solid phase peptide synthesis are directly amenable to analysis by electrospray MS. Typically, an aliquot of the dissolved product mixture is analyzed by reverse phase HPLC and another much smaller aliquot from the same solution is diluted and analyzed by electropsray MS. As illustrated in Figure 3-3, these techniques can be used together to determine the composition of a crude peptide product. MS analysis of the crude product from a synthesis takes only a few minutes and is very sensitive, so peaks from the analytical HPLC separation of the same material can also be collected and subsequently analyzed by MS. The exact mass data, together with knowledge of the synthetic chemistry used, gives the researcher a very complete picture of the total synthetic products within a short time. Ultimately, this information is used to guide the purification of the desired peptide by preparative HPLC, where electrospray MS is also useful to assess the overall purity of individual and pooled fractions (Milton et. al., 1993).

This immediate feedback about the chemical composition of the total crude product generated in a single synthesis has transformed our ability to prepare peptides routinely by total chemical synthesis (Schnolzer et al., 1992a, b). It has substantially reduced the time it takes to assess the quality of a synthesis. Thus, if a synthesis fails for any reason, both the presence and the nature of the chemical errors are known immediately, allowing the problem to be fixed or reaction conditions to be adjusted (Muir et al., 1995). Armed with this knowledge, a repeat synthesis invariably gives useful yields of the purified target ~60-residue polypeptide within one or two days. What was once a prolonged and uncertain undertaking (Hodges & Merrifield, 1975) is now much more straightforward. This combination of advanced analytical and synthetic chemistries has made the total synthesis of large (~60 amino acid) unprotected polypeptides routine.

Purification of Synthetic Peptides

In addition to the almost ubiquitous use of preparative reversed phase HPLC (Clark-Lewis & Kent, 1989), other separation techniques can be very useful in isolating the target peptide from a synthesis. These include ion exchange chromatography (Alpert, 1991), which can be performed under nondenaturing conditions; gel permeation (size-exclusion) chromatography (Gooding & Freiser, 1991), which can be performed under native or denaturing conditions and which separates based on size (hydrodynamic volume); and, functional separation techniques such as affinity chromatography (Scopes, 1993) in which the folded polypeptides are separated based on their relative

A
human pth(1-84)   crude product
(*before* lyophilization)

analytical (reverse phase) hplc

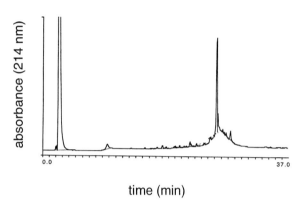

B
human pth(1-84)   crude product

electrospray
mass spectrometry

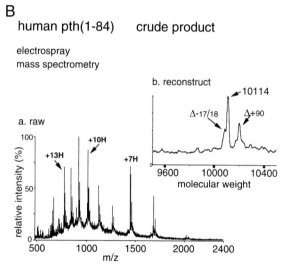

Figure 3-3. Crude synthetic human parathyroid hormone (1–84) from the stepwise SPPS. (A) Reversed-phase HPLC analysis of the crude synthetic product. (B) Electrospray mass spectrum of crude synthetic product showing multiple charge states. (The inset represents the reconstruction of the spectrum displayed as a single charge state.) (C) Electrospray mass spectrum of the main peak in the HPLC chromatogram shown in (A).

affinity for a specific ligand. In conjunction with the improved synthetic techniques and with analytical control for correct covalent structure of the purified product, the techniques that use the functional, folded form of the synthetic product can play an important role in high-yield isolation of the target polypeptide from a synthesis.

C

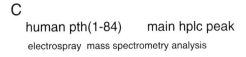

human pth(1-84)   main hplc peak

electrospray mass spectrometry analysis

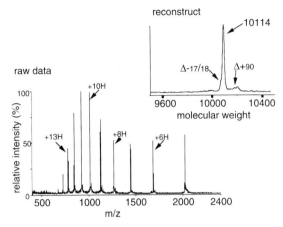

mass: observed   10114 ±2 Dalton
calculated 10117.2 (av. isotopes)

Figure 3-3. (Continued)

Chemical Synthesis of Human Parathyroid Hormone(1–84)

An example of current capabilities in the chemical synthesis of large polypeptides is given in Figure 3-3. This shows the total crude product from the stepwise solid phase synthesis of the 84-amino acid residue polypeptide human parathyroid hormone (Keutmann et al., 1978), a large polypeptide having MW ~10,000. Although PTH is larger than some of the proteins that will be discussed later in this chapter, it is only a polypeptide, because it does not owe its activity to tertiary structure (a defined, stable three-dimensional fold of the polypeptide chain).

The total crude product obtained from the stepwise SPPS of human PTH(1–84) was subjected to both analytical HPLC and electrospray MS analysis. The analytical HPLC (Fig. 3-3A) showed a major component, and a number of other species in smaller amounts. The electrospray MS analysis of the total crude products (Fig. 3-3B) displayed a multiplicity of charge states for each molecular species present. The data from these charge states was combined (Fig. 3-3B, inset) to show that the target 84-residue peptide chain at 10,114 Da was the predominant species present.

The high sensitivity of electrospray MS permitted direct MS analysis of the main peak in the analytical HPLC (Fig. 3-3A). From the results shown in (Fig. 3-3C) it can be seen that even with a very steep (low resolving power) gradient of 2% acetonitrile per minute in the HPLC separation, most of the coproducts have been separated from the target peptide giving the highly purified 84-residue polypeptide chain. Preparative reverse phase HPLC is invariably more highly resolving, and with a shallow (high resolving power) gradient, it was straightforward to purify the 84-residue peptide from the crude synthetic product.

General Utility of Peptide Synthesis

In the past, chemical synthesis of peptides has invariably been a slow and painstaking process, where no effort was spared in an attempt to assure that a single synthesis gave the correct product (Hodges & Merrifield, 1975). For long polypeptide chains, this has not worked as an approach to reliable synthesis of products of reproducible high purity (Stewart & Young, 1984). By contrast, the approach just described represents a quite different philosophy: rapid synthesis, with optimization *between* syntheses. Because of rapid, direct structural feedback from electrospray MS, this has been far more effective as a general approach to the chemical synthesis of large peptides. It has allowed the rapid and reliable synthesis of virtually every ~60 residue peptide attempted in our lab in the last several years.

Total Chemical Synthesis of Proteins

As detailed previously, reliable procedures exist for the total chemical synthesis of polypeptides. But where does this leave us in terms of making proteins? Unfortunately, our ability to synthesize long polypeptides chemically, even using highly optimized SPPS methods, has been limited to polypeptide chains about 70-amino acid residues in length as shown in Figure 3-4. It is important to define what is meant by this size limit. It is the practical limit (i.e., routinely useful, reproducible size) for the preparation of polypeptide chains by total chemical synthesis in the most highly skilled laboratories. In rare instances, larger polypeptide chains have been prepared, by our own laboratory and by others. For example, several laboratories in Japan have done beau-

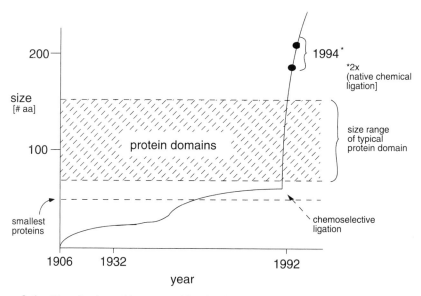

Figure 3-4. Size of polypeptides prepared by chemical synthesis in the twentieth century, starting with Emil Fischer in 1906 (first chemical synthesis of peptides), illustrating the impact of chemical ligation from 1992 on.

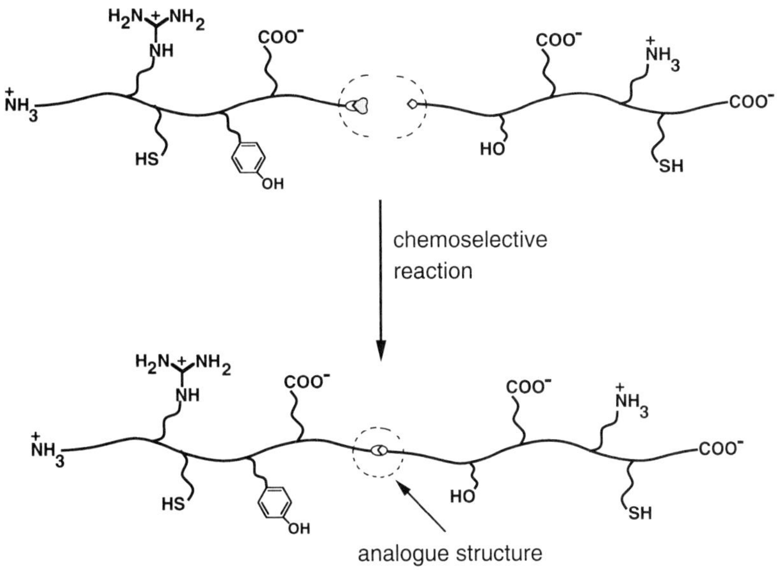

Figure 3-5. Principles of chemical ligation. Unprotected peptide segments are covalently linked together through chemoselective reaction of the C-terminal of one segment with the N-terminal of the other, resulting in the formation of an analogue (nonpeptide bond) structure joining the two fragments. The ligation reaction relies on the presence of unique functionalities, one in each of the two segments, that are mutually reactive with one another yet unreactive with other functionalities present in the reacting peptides.

tiful work with solution synthesis using classical synthetic organic chemistry applied to the preparation of polypeptide chains larger than 70 residues (Inui et al., 1996; Yajima & Fujii, 1981). But in terms of routine, reproducible, and practical syntheses that can be completed in weeks or months by an individual researcher, the limit in even the best chemical laboratories was for many years somewhere between 50 and 70 residues.

The size of the smallest proteins is ~50-amino acid residues, as indicated in Figure 3-4. So our ability to make long polypeptide chains by stepwise SPPS or classical methods has permitted synthetic access to only the smallest proteins. Moreover, the size of typical functional domains in proteins is ~130 ± 40 residues (Berman et al., 1994), so the study of structure-function relationships in these molecular building blocks has in most cases still been beyond the realm of total chemical synthesis.

In 1991, we presented a new approach for preparing large polypeptides (>70 amino acids in length) (Kent et al., 1992). This new approach was based on "chemical ligation," the stitching together of large unprotected synthetic peptide segments by chemoselective reaction. The impact of chemical ligation on our ability to prepare polypeptides and proteins by total chemical synthesis is indicated on the curve shown in Figure 3-4.

Chemical Ligation

Chemoselective reaction is one of the oldest principles of synthetic organic chemistry (see Vogel, 1989). The essential features of the chemical ligation approach to the chemical synthesis of polypeptides are illustrated in Figure 3-5. It involves the reaction of two unprotected peptide chains having unique functionalities that are mutually reactive with one another, yet nonreactive (under the reaction conditions used) with all the other functionalities present in both peptide segments (Schnolzer & Kent, 1992). The two unprotected peptide segments are thus joined efficiently by a covalent bond. The essence of the technique is that the chemistries used give a *nonpeptide* bond at the ligation site, at least as the initial product (see the following). For example, one peptide might contain a uniquely reactive nucleophile, which attacks an electrophilic center present in the second peptide. A reaction such as this will form an analogue (i.e., nonpeptide bond) structure linking the two peptides. The nonnative structure arises because in order to get selective reaction it is necessary to use functionalities that are not found in the rest of the molecule. In most cases, this rules out the direct formation of peptide bonds from activated carboxyl and amino groups (see Jones, 1991).

Chemical ligation based on the principle of chemoselective reaction has become a simple and practical approach to the total synthesis of very large polypeptide chains. The principles of the technique were first described in 1991 and 1992 (Kent et al., 1992; Schnolzer & Kent, 1992). Since developing this technique, we have used chemical ligation exclusively for the synthesis of peptides larger than about 60-amino acids in size. Peptide segments ~40 to 60 amino acids in length can be easily used in ligation reactions. Such unprotected peptides are straightforward to make by highly optimized stepwise SPPS; because the peptides are unprotected they can be purified by straightforward means such as preparative HPLC, and characterized by simple techniques such as analytical HPLC and electrospray mass spectrometry. The chemical ligation reactions can be carried out in the presence of chaotropic (solubility enhancing) agents such as the 6 M guanidine · HCl or 8 M urea. This allows the use of very high concentrations of peptide reactants (typically 2 to 5 mM), so that ligation reactions occur rapidly and in high yield; reactions take minutes or hours, and the ligation product is typically obtained directly in final, deprotected form.

Chemical ligation is a useful and practical way of connecting large peptide segments. Examples of the protein systems to which the technique has been successfully applied are given in Table 3-2. The protein targets prepared by total synthesis using chemical ligation in the few years since the inception of the method cover a wide range of molecular weights and a variety of protein types.

Thus, chemical ligation has solved what was an almost intractable problem: the inability to carry out routine total chemical syntheses of proteins. This problem had arisen from the inability to apply classical synthetic organic chemistry effectively to the preparation of maximally protected peptide segments, together with the inability to join maximally protected segments to form large polypeptides. The synthesis and reaction of large maximally protected peptide segments has not worked as a useful route to the preparation of proteins, largely for practical reasons. These include chronic, poor solubility of protected peptide segments, leading to very low reaction rates (Kiyama et al., 1984); racemization of amino acids in the carboxy-activated segments (Jones, 1991); and the inability to handle, purify, and characterize the intermediate protected peptide products (Carreno et al., 1995). In addition, it was generally observed that the larger the protected peptide segments involved, the lower the resulting reaction yields. This became codified as an inherent obstacle to the chemical synthesis of large polypeptide chains.

The chemical ligation approach has overcome these problems. In our hands, it has been used routinely for the synthesis of polypeptides over 100-amino acid residues in length. No handling problems have been observed, even for ligation products of over MW 20,000 (Baca et al., 1995; Canne et al., 1995). The chemical ligation of large un-

Table 3-2 Total Chemical Protein Synthesis Target Systems

Protein	Size (No. amino acid residues)	Reference
HIV-1 Protease	2 × 99	Schnölzer & Kent, 1992
10Fn3 module	97	Williams et al., 1994
Interleukin 8	72	Dawson et al., 1994
Integrin2b3a 'Neoprotein'	122	Muir et al., 1994
b/HLH/z transcription factors	182	Canne et al., 1995
HIV-1 PR tethered dimer	202	Baca et al., 1995
FIV PR	2 × 116	Muir & Kent, unpublished
Barnase analogues	110	Dawson et al., 1997a
OMTKY3	51	Lu et al., 1996
Eglin C	63	Lu & Kent, unpublished
β2-microglobin	99	Wilken & Kent, unpublished
Human secretory PLA2	124	Hackeng et al., 1997
Human SOD	153	Simon et al., unpublished

protected peptide segments is a general solution to the total chemical synthesis of the large polypeptide chains found in proteins (Muir, 1995).

Ligation Chemistries

A number of chemistries can be used for the chemical ligation of unprotected peptides. Originally, a thioester-forming ligation was used (Schnolzer & Kent, 1992). The oxime-forming ligation developed by Rose (1994) is also very useful. More recently, several ligation chemistries have been based on entropy-activated reactions, a principle most clearly enunciated by Max Brenner (Brenner, 1967). These chemistries include the "thiol capture" method (Fotouhi et al., 1989) and a multistep capture-rearrangement-deprotection approach (Liu & Tam, 1994a, b). We have also adapted our original thioester-forming ligation chemistry to the straightforward formation of native peptide bonds at the ligation site (Fig. 3-6). In "native chemical ligation," as this technique is called (Dawson et al., 1994), the initial thioester-forming ligation reaction was designed to give an intermediate which spontaneously rearranges through intramolecular five-membered ring attack (Mandolini, 1978) to give a native amide bond. The rearrangement step is irreversible under the conditions used, so the reaction strongly favors the native peptide bond-linked product. This results in selective reaction at the N-terminal cysteine even in the presence of other thiol side chains (Dawson, et al., 1994).

This is a very simple trick. It turns out that the exact chemistry described here was used for connecting two amino acids by Wieland and coworkers in 1953 (Wieland et al., 1953). Of course the chemistry had not been applied to joining peptide segments; indeed, Brenner's original suggestion (Brenner, 1967) had been made in the context of reacting amino acids, not segment ligation.

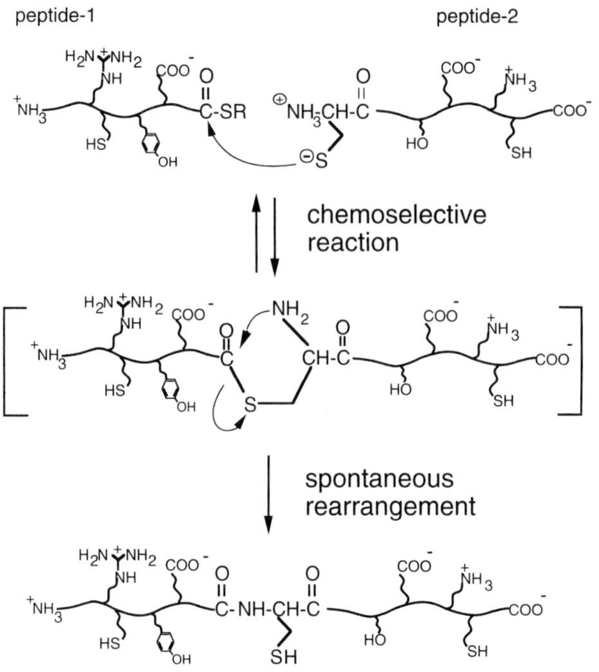

Figure 3-6. Mechanism of native chemical ligation. The -SH moiety in Cys residues undergoes a thiol exchange reaction with the thioester at the C-terminal of the peptide segment. This exchange is *reversible* under the conditions used. Uniquely for the N-terminal Cys, the initial reaction product spontaneously rearranges through a five-membered ring intermediate to give a native amide bond. This step is *irreversible* under the conditions used, so the native amide-linked product is formed.

Total Synthesis of Human IL8 by Native Chemical Ligation

Native chemical ligation has been applied to the total synthesis of a small protein, the chemokine human IL8, which contains 72 amino acids in its polypeptide chain (Dawson et al., 1994). The synthesis of [Ala^{33}]IL8 is shown in Figure 3-7A. The segment (1-33)$^\alpha$COSR was reacted with the completely native, unprotected segment 34–72 (con-

Figure 3-7. Total chemical synthesis of IL-8 by native chemical ligation. (A) The synthetic strategy for the construction of [Ala^{33}]IL8 is outlined (*left*). Segment 1–33 was prepared in unprotected form as the α-thiobenzyl ester, and treated with the completely native, unprotected segment 34–72 (containing an N-terminal Cys residue). Analytical HPLC analyses of the purified synthetic peptide fragments used in the ligation reaction are shown (*upper right*), along with the HPLC chromatograms obtained for the unfolded and folded ligation products. The two segments reacted overnight in ~65% yield; this reaction time has since been substantially reduced by the addition of thiol catalysts. Preparative HPLC gave the purified 72-amino acid polypeptide chain in reduced form (containing four Cys residues), with exact mass 8319.8 Da as determined by electrospray MS (ESMS). The reduced product was folded and oxidized by stirring in 1 M guanidine • HCl under atmospheric oxygen, to give the target IL8 protein; the exact mass of 8315.6 Da (4.2 Da loss) showed that the protein contained two disulfide bonds. (B) The biological activity of IL8 prepared by chemical ligation was identical to that of authentic samples.

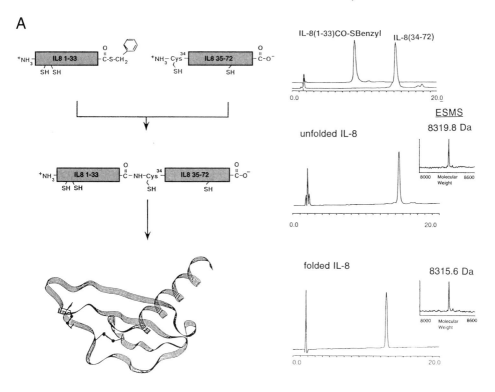

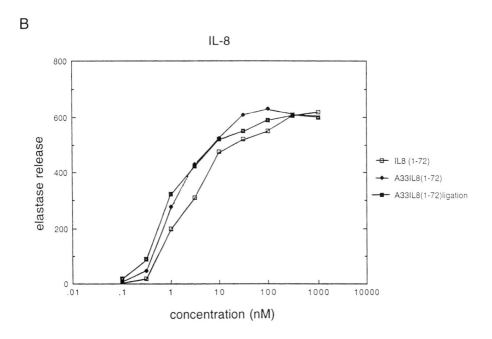

taining an N-terminal Cys residue). These two segments reacted to give the target 72-amino acid polypeptide chain in reduced form, containing four Cys residues; after purification, this was folded and oxidized to give the target IL8 protein containing two disulfide bonds. The folded product displayed full biological activity (Fig. 3-7B). Formation of the correct disulfide linkages in IL8 is essential for activity (Rajarathnam et al., 1994). Therefore, rearrangement of the initial thioester ligation product to the native amide must have occurred to release the Cys^{34} side chain for disulfide formation.

Extending the Applicability of Native Chemical Ligation

We have recently extended the native chemical ligation approach to link peptide segments at sites not adjacent to Cys residues (Canne et al., 1996). The approach is shown in Figure 3-8; in essence it amounts to "moving" the Cys side chain from the α-carbon to the α-amino group of the second peptide segment. The N^α-linked O-ethyl mercaptan auxiliary functionality reacts initially to form a thioester-linked product, which spontaneously rearranges to give an amide-linked peptide product with a cleavable N-linked substituent on the amide nitrogen at the ligation site. The pendant alkoxy group can be easily removed by zinc-acid reduction; repurification by HPLC gives the desired native product.

Because the thioester produced initially in this generalized native chemical ligation approach rearranges through a six-membered ring cyclic intermediate, a significantly reduced rate of the S-to-N rearrangement is observed for α-substituted amino acids on either side of the ligation site; for α-substituted amino acids on *both* sides of the ligation site, *no* rearrangement is observed (Canne et al., 1996). This does limit the applicability of the more generalized native ligation to X-Gly and Gly-X sites. However, it works well for joining peptides at a number of pairs of amino acids and represents a useful enhancement of the versatility of this ligation reaction to an estimated 50 dipeptide sequences (of the 400 possible). Since there is considerable leeway in the choice of ligation site in a given protein system, the two "native ligation" strategies outlined above are sufficient in most cases for applying the chemical ligation approach to the total chemical synthesis of proteins.

Other Ligation Chemistries

There are several other ligation chemistries that give nonnative (analogue) structures at the ligation site when unprotected peptides are joined by chemoselective reaction. These include oxime-forming ligation, originally introduced in protein semisynthesis, that is, the recombining of peptide segments generated by fragmentation of a naturally derived protein (Rose, 1994); stable amide bond formation by an O-to-N acyl shift (Liu & Tam, 1994a, b); and, thioether formation (Englebretsen et al., 1995). Under the appropriate conditions, these ligation chemistries can be compatible with thioester-forming ligations, and can be used consecutively in linear or convergent syntheses.

Convergent Chemical Ligation Strategies

The repertoire of ligation chemistries can be used as the basis of a classic convergent strategy for the synthesis of larger polypeptide chains (Fig. 3-9). For example, in order to make a 200-residue polypeptide chain, one might assemble four unprotected ~50-residue polypeptide segments. Initially, the segments are separately reacted in pairs to give ~100-residue intermediate products. Then another class of unique, mutually reactive functionalities (compatible with the first ligation chemistry used and already built into the reacting segments) is used to link the two ~100-residue peptides in a second chemoselective reaction, resulting in production of the target ~200-amino acid residue polypeptide chain.

The covalent bonds formed at the ligation sites in convergent chemical ligation strategies need to have the following general characteristics. In our experience, in-

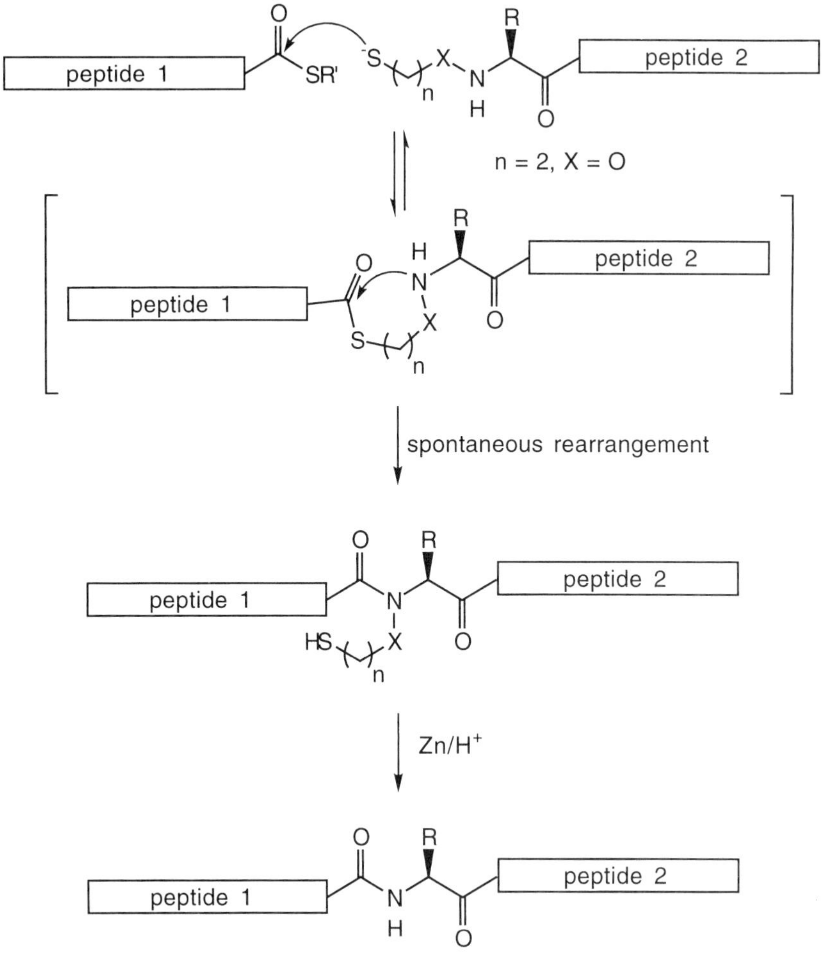

Figure 3-8. Extending the applicability of native ligation. This more generalized approach in essence amounts to moving the Cys side chain from the a-carbon to the nitrogen (a-amino group) of the second peptide segment. The resulting species has an $HSCH_2CH_2O-$ moiety on the a-amino group. This can react (by thioester exchange) to form a thioester-linked initial product, which rearranges through a six-membered ring. The product is an amide-linked polypeptide with a $HSCH_2CH_2O-$ substituent on the amide nitrogen (i.e., an O-alkyl hydroxamate). The pendant alkoxy group can be removed by zinc-acid reduction, which cleaves the (amide) N—O(alkyl) bond cleanly and quantitatively (Canne et al., 1996).

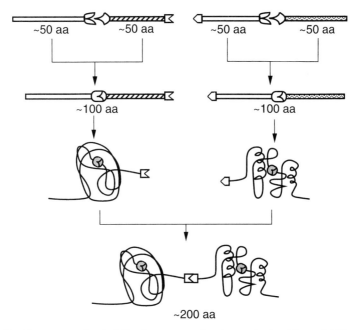

Figure 3-9. Convergent chemical ligation strategy for the synthesis of large (>120 amino acid) polypeptide chains. For example, in order to make a 200-residue protein consisting of two domains, one starts with four ~50 residue polypeptide segments, all unprotected. The segments are separately reacted in pairs to give ~100 residue intermediate products; these contain a second class of unique, mutually reactive functionalities compatible with the first ligation chemistry used. Thus, either before or after folding of the domains, the two large ~100 residue peptides can be linked together by a further chemoselective reaction to give the target (two domain) protein.

tradomain linkages should have native-like structure, (e.g., the native peptide bond itself or a thioester surrogate, which we have found very useful and minimally perturbing). *Interdomain* regions (loop structures) are usually more tolerant of unnatural structures. In such instances, the oxime-forming ligation chemistry (Rose, 1994) is a convenient and versatile method for joining two protein domains.

Total Chemical Synthesis of a cMyc-Max Covalent Construct

An example of a convergent chemical ligation strategy is provided by the total chemical synthesis of covalent heterodimers of the dimerization-DNA-recognition domain of the b/HLH/Z family of transcription factors. These transcription factors are involved in the regulation of gene expression in mammalian cells (Amati et al., 1993). It is believed that the activation or repression of a gene is controlled by the noncovalent association of cMyc with other members of the family (Mad, Max, or Mixi1) to form noncovalent heterodimeric transcription factors (Lahoz et al., 1994). The first three-dimensional structure of a member of this class of transcription factors, the homodimer of a recombinant polypeptide corresponding to the Max b/HLH/Z domains bound to its cognate DNA, was reported recently (Ferre D'Amare et al., 1993).

In order to study heterodimeric forms of the b/HLH/Z domains of these transcription factors, a convergent ligation total chemical synthesis strategy was designed (Canne et al., 1995) (Fig. 3-10). The individual b/HLH/Z polypeptides were synthesized by thioester-forming ligation. These two ~10 kDa segments were then reacted by oxime-forming ligation to give the target construct of MW ~20,000. The actual oxime-forming final ligation reaction is shown in Figure 3-11. The product was well resolved from the reactants by reversed phase HPLC; the full length, target protein was isolated easily by preparative HPLC. Electrospray MS analysis of the final product indicated that the material was of the correct molecular weight and of good purity.

The synthetic covalent heterodimer also recognizes cognate DNA in a specific fashion, as shown by the gel shift binding data (Canne et al., 1995). Interestingly, the covalently linked b/HLH/Z synthetic dimer is preorganized, as shown by CD measurements, whereas the noncovalent Max dimer is only partly organized in the absence of its cognate DNA. In the presence of the cognate DNA, the noncovalent Max dimer has a CD spectrum typical of a helical coiled-coil protein and consistent with the X-ray structure. Even in the absence of cognate DNA, the covalently linked Myc-Max, joined together at the C-termini of the two domains, shows the CD signature of a fully formed, coiled coil helical protein (Ferre D'Amare et al., unpublished).

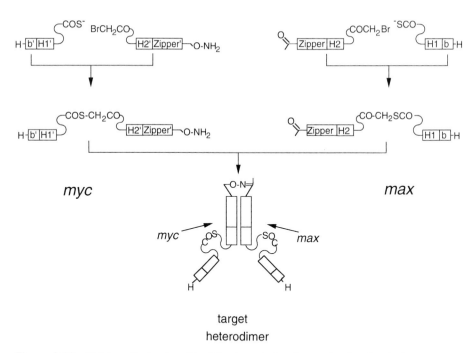

Figure 3-10. Total synthesis of a cMyc-Max transcription factor-related molecule, by convergent chemical ligation. A ~35 residue segment is ligated to a ~55 residue segment by thioester-forming ligation to give a product ~90 amino acids in length. This has a C-terminal aminooxy functionality. The two corresponding segments from the other sequence are similarly ligated to give a ~90 residue product containing a C-terminal keto function (on a levulinic acid). These two ~10 kDa segments are reacted by oxime-forming ligation to give the target construct having MW ~20 kDa (Canne et al., 1995).

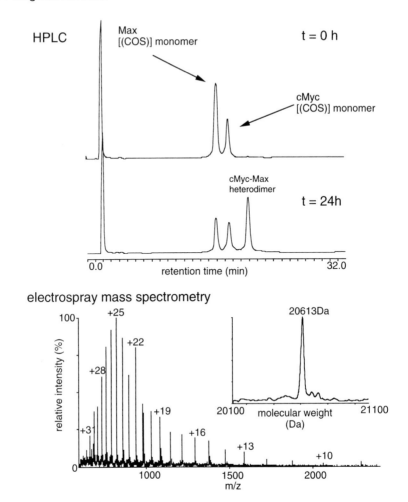

Figure 3-11. Oxime-forming ligation reaction from the scheme shown in Figure 3-10. The two ~90 residue monomers reacted overnight to give the desired ligation product. (*upper*) Analytical HPLC traces show the progress of the reaction, which proceeded to completion. (*lower*) Electrospray MS of the ligation product, showing raw and reconstructed data; observed mass 20613 6 3 Da (calculated mass 20,614 Da, average isotope composition).

The ability to prepare obligate (covalent) heterodimeric forms of the b/HLH/Z transcription factors in unambiguous fashion is enormously useful to the biological research community. These molecular constructs have previously been inaccessible. It was possible to express Myc and Max recombinantly, mix the heterologous species, and obtain the heterodimeric forms of the molecule. However, a mixture of molecular species was obtained and crystallization of the desired product was not possible. Using the convergent chemical ligation strategy previously described, we have made by total chemical synthesis, in "cassette" fashion, a number of heterodimeric b/HLH/Z constructs. This is permitting us to study, in collaboration with Stephen Burley (Rockefeller University), the chemical basis of gene regulation.

General Applicability of Chemical Ligation

The syntheses described in the preceding studies and in other published work (Baca et al., 1995) show that it is possible to handle in a routine, practical fashion not only ~5 kDa segments and their ~10 kDa ligation products, but also to manipulate and react these ~10 kDa segments and their ~20 kDa ligation products. Moreover, in combination with native ligation chemistries that can be easily used to generate ~10 kDa unprotected peptide segments, the strategies outlined should be easily extended to the total chemical synthesis of covalent constructs in the 40+ kDa range. The median size of proteins in a cell is ~30–35 kDa. Thus, the bulk of proteins are rendered accessible to total synthesis by use of the chemical ligation approach.

The chemical ligation approach to the total synthesis of proteins has been used for a number of projects in our laboratory (Table 3-2). A variety of different classes of proteins has been synthesized. Typically, a project involves the total synthesis of the parent form of a protein molecule, and the subsequent design and synthesis of a number of analogue protein molecules. The biological properties of the various proteins are measured and compared, to draw inferences about the molecular origin of protein function. These projects can be undertaken successfully by individual graduate students or postdoctoral fellows and completed within periods of time measured in months. Some examples are discussed below.

The Chemical Basis of Protein Function

An important goal is to complement the rapidly growing body of descriptive information on proteins with corresponding information on how proteins work. Thanks to recent innovations in the methodology of X-ray crystallography (computer-controlled area detectors with direct digital acquisition of diffraction data; molecular replacement for the solution of molecular structures; advances in computer-aided molecular graphics), there is an abundance of excellent new information on the three-dimensional structures of protein molecules (see Hendrickson & Wutrich, 1996). Many of these structures are of considerable beauty in themselves, and in some cases even suggest the molecular basis of biological phenomena (Branden & Tooze, 1991). However, as chemists, we would like to gain a more fundamental understanding of these phenomena at the molecular level. Our goal, then, is to develop a useful understanding of the chemical basis of protein function, including the most fundamental property of all: the ability of the linear polypeptide chain to fold into the defined three-dimensional structure of the protein molecule (Anfinsen, 1973).

Total chemical synthesis has a unique and important role to play in elucidating the chemical basis of protein function. Most importantly, total synthesis can be used to ask, and answer, very precise questions about the chemical basis of the function of proteins. An example follows.

Catalytic Contribution of Backbone-Substrate Hydrogen Bonds in HIV-1 Protease

The enzyme HIV-1 protease is a homodimer of two 99-residue subunits (Fig. 3-12). HIV-1 protease belongs to the aspartyl proteinase class of proteolytic enzymes, which use the side chain carboxyl functionalities of aspartic acid residues to catalyze the hy-

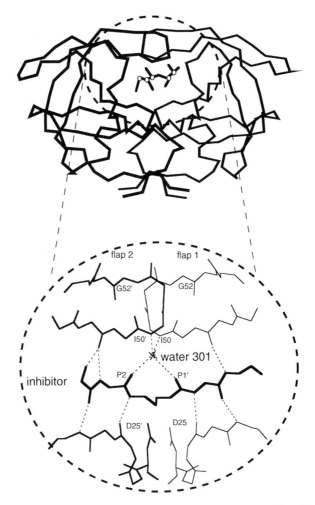

Figure 3-12. (*upper*) The enzyme HIV-1 protease is a homodimer of two 99-residue subunits. The substrate-derived inhibitor is shown as white spheres connected by rods, near the center of the molecule. (*lower*) H- bonds are transmitted from the enzyme molecule's "flaps" to the substrate via an internal, tetrahedrally coordinated water molecule, "water 301."

drolysis of peptide bonds (Davies, 1990). The catalytic apparatus lies at the bottom of the cleft created at the interface between the two domains (monomers). One essential aspartic acid is contributed from each domain (Pearl & Taylor, 1987). Using enzyme prepared by total chemical synthesis (Schneider & Kent, 1988), the three-dimensional structure of the HIV-1 protease molecule bound to a substrate-derived peptide inhibitor was elucidated (Miller et al., 1989).

A unique feature of the (surrogate substrate)-enzyme complex first observed in that work was an internal tetrahedrally-coordinated water molecule, transmitting H-bonds from backbone amide NH moieties at the tip of each HIV-1 protease flap, through the water molecule to the carbonyl oxygen on either side of the scissile bond (the substrate amide bond being broken by the enzyme) (Fig. 3-12). Subsequent studies have shown that this internal water molecule ("water301") is always found in these types of (substrate-derived) inhibitor-enzyme complexes (Wlodawer & Erickson, 1993). Many such complexes have now been studied by X-ray crystallography in the successful effort to develop HIV protease inhibitors as AIDS drugs. In the context of enzyme function, it is of great interest to establish whether these (protein backbone)-substrate hydrogen bonds contribute to the catalytic activity of the enzyme.

We have used total chemical synthesis to establish the role of the water301-flap hydrogen bonds in HIV-1 protease catalysis (Baca & Kent, 1993). The design of the experiment is shown in Figure 3-13A. The internal, tetrahedrally coordinated water molecule is hydrogen bonded to the two "flaps" of the HIV protease, one from each domain of the protein. The hydrogen bonds from the flaps to the substrate are donated by the amide bond NH of Ile^{50} of each domain. We asked what would happen if each of these Gly^{49}—Ile^{50} peptide bonds were modified so that it could *not* donate an H-bond?

Using total synthesis by chemical ligation, such an experiment was straightforward. The synthetic approach is shown in Figure 3-13B. Thioester-forming ligation was used to react two peptides to give the 99-residue HIV-1 PR monomer polypeptide chain, modified at a single backbone atom: the α-N of Ile^{50} was converted to an S atom. This polypeptide was then purified by HPLC and folded to give the two-domain protein molecule. The synthetic backbone-engineered protein was characterized by size exclusion chromatography and by native electrospray MS, both of which confirmed the homodimeric nature of the analogue protein.

The enzymatic properties of this backbone engineered [(COS)$^{49-50}$]HIV-1 protease were investigated (Baca & Kent, 1993). This modified enzyme analogue has been shown to have exactly the same substrate specificity as the native HIV-1 protease (Schneider & Kent, 1988); that is, it cleaves all of the standard sites of proteolytic processing observed in the maturation of the virus (Oroszlan, 1989). This is illustrated in Figure 3-14A for the matrix-capsid (MA-CA) cleavage site, and the capsid-nucleocapsid (CA-NC) cleavage sites 1 and 2, which are processed in the normal fashion. However, the *rate* of substrate cleavage was significantly reduced (Fig. 3-14B). For the backbone-engineered [(COS)$^{49-50}$]HIV-1 protease, the turnover number (k_{cat}) was *reduced* by a factor of about 3,000. This corresponds to an ~5 kcal/mole *increase* in the relative energy of the rate-limiting step of catalysis.

The rate-limiting step in the catalysis of peptide bond hydrolysis by HIV-1 protease is the breakdown of the tetrahedral intermediate (Hyland et al., 1991a, b). Thus, the 3,000-fold reduction observed in k_{cat} is consistent with our expectations based on the crystal structure of the enzyme bound to a substrate-derived inhibitor, that is, there is participation of two backbone-substrate H-bonds, one from the NH of Ile^{51} in each flap mediated by water301 to the carbonyls on either side of the scissile bond, in the breakdown of the tetrahedral intermediate (Baca & Kent, 1993).

This illustrates how one can use the unique approach of backbone engineering through total chemical synthesis to address very precise questions about enzyme mechanism. Chemistry can be used to explore such questions from the perspective of the protein molecule itself, not just the low molecular weight ligand that binds to the active site of an enzyme. Such studies apply the full power of organic chemistry to the study of the protein molecule and thereby can contribute importantly to understanding the chemical basis of protein function.

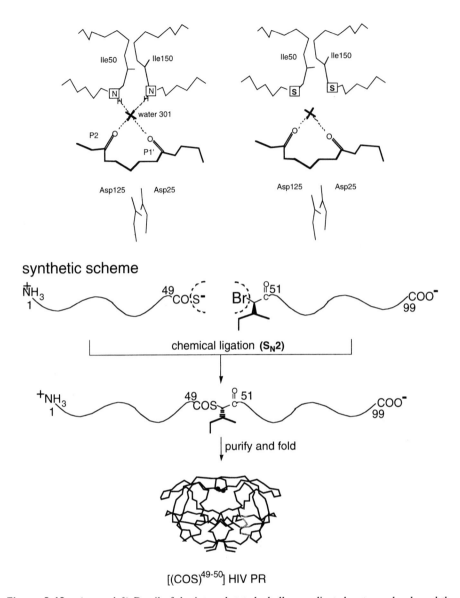

Figure 3-13. (*upper left*) Detail of the internal, tetrahedrally coordinated water molecule and the two "flaps" of the HIV-1 protease, one from each subunit of the protein. The hydrogen bonds from the flaps to the substrate are donated by the amide bond NH of Ile[50] of each subunit. We asked what would happen if these Gly[49]-Ile[50] amides were replaced with thioesters? That is, (*upper right*) if the NH which acted as an H-bond donor was replaced with a S atom (which cannot donate an H-bond). (*lower*) Synthetic scheme. Stepwise SPPS was used to make a 49-residue HIV-1 protease (1-49)-^{a}COSH unprotected peptide segment. The complementary segment [α-bromo′Ile′[50]]HIV-1 protease (51-99) was similarly prepared as an unprotected peptide. The α-bromo acid at the N-terminal of this second 49 residue segment corresponds to an Ile residue (converted to the bromo derivative), with the stereochemistry of a D-amino acid. In the S_N2 ligation reaction, epimerization (inversion of configuration) occurs at the a-carbon of ′Ile′[50]. The target 99-residue polypeptide chain is obtained, with a sulfur replacing the -NH- of ′Ile′[50], and with the correct stereochemistry. This polypeptide was then purified by HPLC and folded to give the homodimer protein molecule.

Scope and Limitations

The range of protein systems amenable to the total chemical synthesis approach is illustrated by the list of some current projects in our laboratory, as given in Table 3-2. In addition to work on the HIV-1 protease, we are studying the corresponding enzyme from feline immunodeficiency virus, the FIV protease, which is a dimeric 116-residue enzyme. Again, the defining initial work on this enzyme has been carried out using total chemical synthesis (Elder et al., 1993; Schnolzer et al., 1996). Other work includes the total chemical synthesis of covalent heterodimeric b/HLH/z transcription factors [(Canne et al., 1995) and outlined above] and of the chemokine IL-8 (Dawson et al., 1994). Artificial protein constructs have been created to investigate key aspects of signal transduction, such as the neoprotein models of the cytoplasmic domains of the cell-

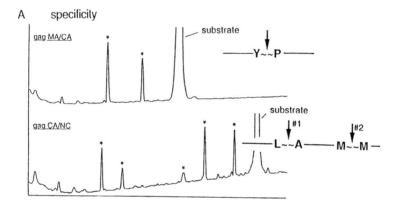

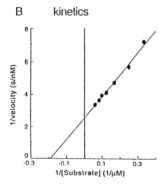

Figure 3-14. Enzymatic activity of the backbone-engineered HIV-1 protease. (A) Specificity. Synthetic peptide substrates corresponding to the MA-CA cleavage site, and the CA-NC cleavage sites 1 and 2, were processed in the normal fashion. The * peaks are cleavage products. (B) Rate of cleavage. An HPLC assay of the action of the enzyme on synthetic peptide substrates was used to give an extended dynamic range of $\sim 10^5$ for rate measurements. The kinetic data is shown as a double reciprocal plot (1/v versus 1/[S]). Analysis of the data revealed that the turnover number (k_{cat}) of the backbone-engineered [(COS)$^{49-50}$]HIV-1 protease, is reduced by a factor of about 3,000 compared to the wild-type enzyme, while the K_m was unchanged.

surface receptor integrin 2b3a (Williams et al., 1994). The oligomeric enzyme 4-oxalocrotonate tautomerase has also been prepared by total chemical synthesis. In this case, both the D- and L-protein forms of the enzyme have been prepared and shown to have equal activities on the achiral substrate, although acting on protons from different faces of the planar substrate, as expected (Fitzgerald et al., 1995). We have also developed an efficient one-step chemical synthesis of analogues of the enzyme "barnase," an important model system for protein folding (Dawson et al., 1997).

The application of total chemical synthesis to a wide range of proteins of biological interest has been undertaken successfully. However, there are molecules that are refractory to current chemical synthesis approaches. In fact, although we have been able to make the 116-residue FIV protease polypeptide chain both stepwise and by several segment ligation strategies, the FIV protease is not a protein amenable to synthesis in a useful, reproducible fashion. The reason for this difficulty is that the 116-residue polypeptide chain itself and several C-terminal segments are so insoluble that they cannot be manipulated in unfolded form. However, with only one or two exceptions, in the more than 20 systems that have been studied in the past few years, we have been able to apply total synthesis in a useful fashion to each protein that we set out to study.

The utility of the total chemical synthesis approach to the in-depth study of a protein has been exemplified most thoroughly by our work on HIV-1 protease, which is summarized in Table 3-3. Since 1987, we have successfully undertaken the chemical synthesis of this enzyme many times. More importantly, we have realized *six generations* of improved synthetic technology, while at the same time answering specific questions

Table 3-3 Six Generations of Chemical Synthesis of the HIV-1 Protease

Year	Synthetic Technology Achievement	Reference
1988–89	Stepwise, Boc-chemistry, double coupling [Aba[67,95,167,195]]HIV-1 protease crystal structures—empty; (PR inhibitor complexes)	Schneider & Kent, 1988 Wlodawer et al., 1989 Miller et al., 1989 Swain et al., 1990 Jaskowlski et al., 1991
1990	Stepwise, single coupling [+12 recouples] [BTD[17-18,117-118]]HIV-1 protease	Baca & Kent, 1993
1991	Stepwise, single coupling [*in situ* neutralization] D-enzyme native MS ternary complex	Milton et al., 1992 Baca & Kent, 1992
1992–93	Chemical ligation crystal structure D-enzyme backbone-engineered enzyme	Miller et al., 1996 Baca & Kent, 1993
1994	Tethered dimer (21,500 Da) [convergent ligation] chemical synthesis flap H-bond deleted [single-flap protease]	Baca et al., 1995 Baca & Kent, 1996
1995	Native chemical ligation NMR probe nuclei	Smith et al., 1996

about the activities of this interesting molecule. For example, in an extension of studies of the mechanistic role of enzyme flap-substrate H-bonding, we have recently developed a convergent ligation approach to the total chemical synthesis of a tethered dimer form of the HIV-1 protease (Baca et al., 1995), and used that to delete the H-bonding potential of just one flap in the enzyme molecule. The resulting enzyme analogue retained full catalytic activity, suggesting that the HIV-1 protease uses only one flap-substrate H-bond in its enzymatic action (Baca & Kent, unpublished). This mode of action would be similar to that of cell-encoded aspartyl proteinases, which are single polypeptide, two-domain molecules that contain only one flap structure (Davies, 1990).

The extensive studies of the HIV-1 protease carried out over the past few years have led to an increasingly precise description of the molecular basis of the action of this enzyme. However, we still lack fundamental insight into many aspects of HIV-1 protease action. For example, although the HIV-1 protease is an enzyme of exquisite specificity in the processing of polyprotein substrates, it is not well understood how the enzyme chooses its substrates. Furthermore, attempts to alter the substrate specificity of HIV-1 protease by rational alteration of the molecular structure of the enzyme have led to unexpected results: these altered enzyme constructs continued to cleave the original substrates with only slightly reduced activity.

As the example of this well-studied enzyme molecule illustrates, there is a need for better, more systematic approaches to gaining fundamental insights into the molecular basis of protein function. Such an approach is outlined in the next section.

Protein Signature Analysis: A Type of Combinatorial Protein Chemistry

In this section we explore a useful new way to elucidate the relationship between the folded, functional form of a protein domain and its covalent structure. Protein structural domains are the fundamental units of function in the biological world; they are typically $\sim 130 \pm 40$ amino acids in size (Berman et al., 1994). Domains are independent units of folding and function, and as such are the fundamental building blocks of the protein world. Therefore, it is sufficient for the chemist to seek to understand protein structure and function at the domain level. Conveniently, the synthetic tools described above provide easy and general access to such protein domains.

We have made use of combinatorial synthesis of defined arrays of analogue molecules corresponding to a protein domain to explore structure-function relationships. The essence of the approach is *to build into the protein molecule a systematic variation of the covalent structure, coupled to a latent readout chemistry*. This is illustrated schematically in Figure 3-15. An analogue unit is placed at each relevant position in the polypeptide chain, giving a family of related molecules each of which contains only one structural alteration. After a suitable functional selection, such as folding or ligand-binding, the molecules in each population (functional & nonfunctional) are then identified using the built-in readout chemistry. This gives a signature relating the covalent structure to the function of the protein domain.

Chemical Synthesis of Protein Analogue Arrays

The goal is to obtain directly a defined array of protein analogues as a single mixture containing unique, single modifications in each polypeptide chain. Phil Dawson in this laboratory has devised a simple, ingenious procedure by which this can be done in a

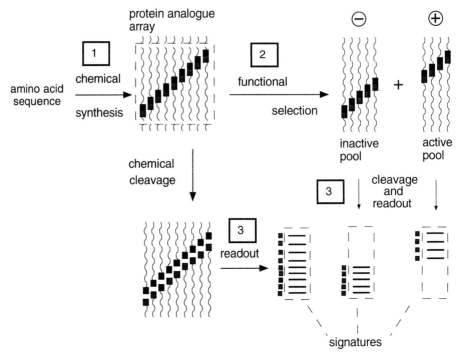

Figure 3-15. Protein signature analysis. This new technique involves three steps. (1) Synthesis. An array of protein analogues of defined molecular composition is prepared in a single chemical synthesis. (2) Selection. The array of analogues is separated into *active* and *inactive* pools, based on functional selection. (3) Readout. Each "self-encoded" array of analogues is chemically cleaved, and the resulting characteristic fragments give a "signature" defining the composition of each pool. The positions of the analogue structure in the analogue molecules in the active and inactive pools is defined by this signature. This information allows correlation of the chemical nature of the analogue structure with effects on protein function over a region of the polypeptide chain.

single synthesis (Dawson et al., 1997b), using the modified stepwise solid phase procedure shown in Figure 3-16.

Stepwise synthesis of the target polypeptide chain is carried out in normal fashion (Schnolzer et al., 1992a) until the region to be studied is reached. For the introduction of the analogue structures, two reaction vessels are used. Identical chain extension chemistry is carried out in both vessels. But, after addition of each amino acid in vessel A, a sample is taken; for a nine-component analogue mixture each sample should be one-tenth on a mole fraction basis, because there should be some of the unmodified target sequence left in vessel A at the end of the synthesis. The one-tenth mole fraction of peptide-resin is placed in a small auxiliary reaction vessel where the analogue unit is added by suitable chemical reaction. Meanwhile, the peptide-resin in vessel A is extended by another amino acid residue, and another (one residue longer) one-tenth mole fraction sample of peptide-resin is placed in a second auxiliary vessel.

For an analogue unit that replaces two amino acids in the target sequence, after *two cycles* of synthesis in vessel A, the modified peptide-resin sample in the first auxiliary reaction vessel is placed in vessel B, and the remainder of the target amino acid sequence is added by exactly the same stepwise SPPS synthetic operations as in vessel A. This procedure is followed throughout the region to be modified. The remaining amino acids in the target sequence are then added to the peptide-resin in vessel A and to the mixture of peptide-resins in vessel B in stepwise fashion using standard SPPS protocols.

At the end of the stepwise SPPS of the full target polypeptide sequence, vessel B will contain, in this case, the nine different full-length resin-bound polypeptide analogues. This product mixture in vessel B consists of nine distinct molecular species where each species has a precisely defined change at only one of the nine positions at which the analogue structure was introduced.

Analogue Units

Several of the analogue units that we have successfully incorporated into peptides and proteins are shown in Figure 3-17A. These dipeptide mimics contain side chain deletions and modified polypeptide backbones. One important feature that each analogue unit also contains is a chemically cleavable bond. The ester bonds in each of the analogues can be specifically cleaved in an aminolysis reaction as shown in Figure 3-17B. The cryptic cleavage site in each dipeptide mimic is a key feature of the analogue unit and serves to self-encode the members of a given protein analogue array as described later.

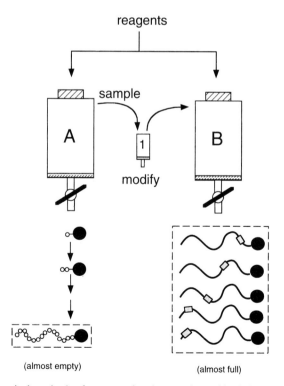

Figure 3-16. Chemical synthesis of an array of analogue polypeptide chains, using a modified split-resin procedure (Dawson et al., 1997b).

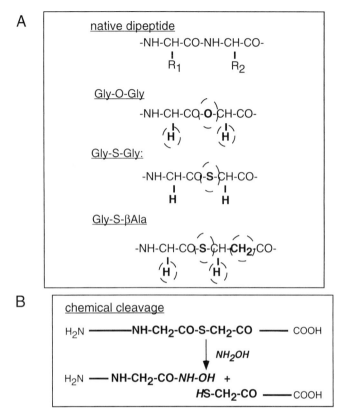

Figure 3-17. Analogue units. (A) A native dipeptide is compared with three analogue structures that have been used as replacements in protein signature analysis. (B) Chemical cleavage of the thioester functionality of the analogue dipeptide unit by hydroxylamine at neutral pH.

Readout of Protein Analogue Arrays

The composition of a mixture of protein analogues can be read out directly by a novel self-encoding method (Fig. 3-18). This method takes advantage of synthetic chemistry to build a latent cleavage site into each polypeptide chain. This reactive moiety allows us to cleave the polypeptide chain when desired (Fig. 3-18), to obtain a characteristic mixture of fragments. After mass spectrometric analysis (Chait & Kent, 1992), the signature of the parent array of protein analogues takes the form of a "sequence ladder" (Chait et al., 1993) (Fig. 3-18) that defines the amino acid sequence of the continuous region of the polypeptide chain that was subject to introduction of the analogue unit (Fig. 3-18).

The cCrk N-Terminal SH3 Domain

The utility of this signature analysis method, which combines combinatorial generation of protein analogues with a functional selection and subsequent readout of the self-encoded arrays, was illustrated by studies on a small protein domain, a src homology type 3 unit (SH3 domain) (Dawson et al., 1997b; Muir et al., 1996). SH3 domains are one of the more ubiquitous components of proteins in signal transduction pathways and are currently of great interest to biologists (Cohen et al., 1995); cCrk is one of the so-called adaptor proteins in signal transduction, and has two SH3 domains and one SH2 domain. The three-dimensional structure of the N-terminal SH3 domain of cCrk is given in Figure 3-19. The molecule is shown complexed with a synthetic Pro-rich peptide derived from the protein C3G, a natural ligand of cCrk (Wu et al., 1995).

The 58-residue amino acid sequence of the murine cCrk amino terminal SH3 domain, consisting of residues 134-191 of the cCrk open reading frame, is shown in Figure 3-20. Initially, a ten-amino acid region in the middle of the polypeptide chain was selected for investigation. An analogue unit (-Gly-[COS]-βAla-) was placed systematically at each possible position across the nine dipeptide units of the ten amino acid sequence, as indicated in Figure 3-20. This array of SH3 domain analogues was prepared using the previously described chemical synthesis procedure.

After deprotection and cleavage from the resin, the mixture of 58-residue long synthetic polypeptide chains, consisting of an array of nine analogue-containing subfam-

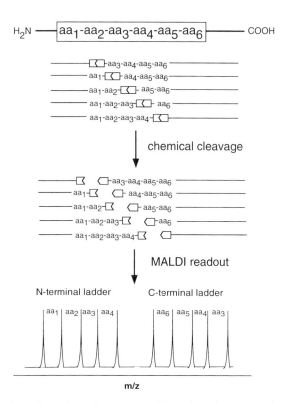

Figure 3-18. Readout of protein analogue arrays. The polypeptide chains of the members of the array are cleaved at the position of the analogue unit. The resulting characteristic fragments are identified by MALDI mass spectrometry, to give sequence ladders (Chait et al., 1993) that form a characteristic pattern defining the *positions* of the analogue units in the protein array.

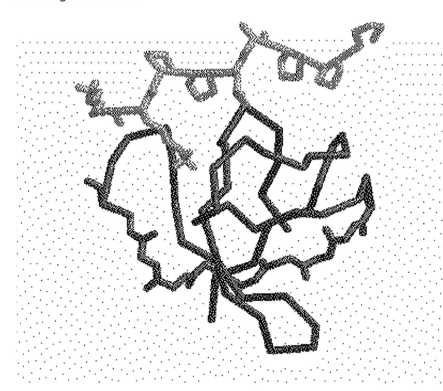

Figure 3-19. The folded backbone three-dimensional structure of the cCrk N-terminal SH3 domain (Wu et al., 1995). The C3G-derived Pro-rich peptide ligand is shown (light color) at the top of the SH3 molecule.

ilies, was folded in native buffer and then passed over the {C3G-derived Pro-rich peptide}-affinity column. Two populations of proteins were obtained: those that specifically recognized the Pro-rich peptide and bound to the affinity column; and those that were eluted by a salt wash that removed any material that was adsorbed nonspecifically. In this way, the array of synthetic SH3 protein analogues was separated into a pool of functional molecules and another pool containing the nonfunctional molecules. The members of each pool have the analogue unit modification at different positions of the polypeptide chain.

The composition of the self-encoded array of protein analogues was read out by chemical cleavage of the polypeptide chains at the analogue unit, followed by MALDI mass spectrometry of an aliquot of each pool (Fig. 3-20). Thus, when the nine-membered SH3 protein analogue array was synthesized, a thioester bond was built into the -Gly-[COS]-βAla- dipeptide analogue unit in the 58-residue polypeptide chain. This thioester backbone moiety was stable to normal handling. On treatment with hydroxylamine at neutral pH, the peptide chains were selectively cleaved at the thioester bond. The resulting fragments were read out by MALDI mass spectrometry to give a "signature" characteristic of the (parent) array of polypeptide analogues (see Fig. 3-20).

Profile of Structure-Activity in the SH3 Domain

The composition of the binding and nonbinding pools of analogues were read out in a similar fashion (Fig. 3-20). Five of the SH3 analogue proteins did *not* bind to the affinity column, three showed specific binding, and one exhibited partial binding, under the near-physiologic binding conditions used. The signatures obtained were quite unambiguous, and the whole process did not take long, especially the one-step readout which involves about 30 minutes.

The precise *nature* of the chemical changes introduced into each protein analogue is determined during the course of synthesis, while the *positions* of the analogue unit in the active and inactive pools have been determined as described. That information

```
1                                                                               58
AEYVRALFDFNGNDEEDLPFKKGDILRIRDKPEEQWWNAEDSEGKRGMIPVPYVEKYR
```

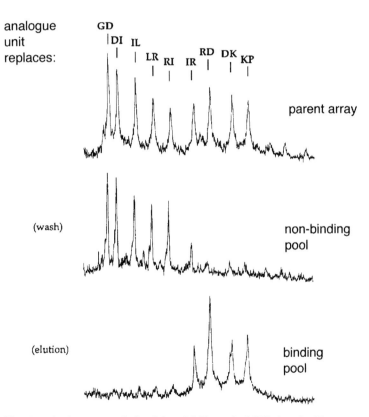

Figure 3-20. Protein signature analysis of the cCrk N-terminal SH3 domain. The sequence 23–32 (highlighted) of the 58-residue protein was explored by preparation of a nine-membered array in which each dipeptide sequence was replaced by an analogue unit. The MALDI mass spectrometric signature of the parent array, after chemical cleavage at the analogue units, is shown (*top*). Signatures of non-binding and binding pools, arising from affinity selection, are shown beneath.

can be simply correlated with effects on the function of the protein molecule, in this case binding activity. In this way, testable hypotheses can be investigated, for example, "lack of binding in the inactive analogues is due to the extra -CH_2- in the backbone." To test this hypothesis, the array of protein analogues was resynthesized with the corresponding dipeptide mimic *without* the extra -CH_2-; that is, using a - Gly{COS}Gly- analogue unit. After functional separation on the peptide-affinity column, the pool of specifically bound protein analogues was read out by chemical cleavage—MALDI mass spectrometry. In this case all nine of the analogue proteins showed specific binding activity (Muir et al., 1996). This is consistent with the nonbinding components of the original array of analogues arising from the effects of the backbone -CH_2- moiety at certain specific positions in the polypeptide chain.

These types of experiments can be run iteratively as necessary, based on tentative interpretations of the molecular basis of the observed effects on function. At each cycle of signature analysis, the signatures are correlated with chemical changes at *all* sites being studied in the molecule. Such a process will reveal the chemical basis of the effects, even if it is not the same at all sites within the protein being studied.

These chemical signatures of protein function can also be correlated with more conventional ways of looking at the molecular basis of protein function, such as the three-dimensional protein structures generated by structural biologists. An example of this is shown in Figure 3-21, where a characteristic signature was obtained for a nonlinear binding epitope on the SH3 domain. It should be emphasized that the protein signature analysis method does *not* rely on prior knowledge of the three-dimensional structure of a protein domain in order to develop an understanding of the chemical basis of function. The patterns of structure-function relationships revealed by the method tell directly how changes in the chemical make-up of a protein affect the properties of a polypeptide chain as it tries to behave as a protein. Even more can be learned by varying the stringency of the functional selection and by looking at effects of temperature on function.

Future Applications of Protein Signature Analysis to Understanding the Chemical Basis of Enzyme Action

One of the most important aspects of proteins is their ability to act as enzymes, that is, to catalyze reactions with great efficiency and precision. The protein signature analysis technique just described promises to be a useful way to examine the chemical basis (molecular origins) of the action of an enzyme. This technique is currently being used in collaborative studies of the enzyme 4-oxalocrotonate tautomerase (4-OT) with Chris Whitman (University of Texas). This enzyme is active as a hexamer of identical 62-amino acid subunits (Roper et al., 1994). Both the D and the L-amino acid forms of this enzyme have been prepared by total chemical synthesis (Fitzgerald et al., 1995). "Native" electrospray mass spectrometry studies have also been used to correlate the activity of several 4-OT analogues with their oligomer state (Fitzgerald et al., 1996).

Several features of the 4-OT molecule make it suitable for probing catalysis using this signature technique. The approach we have designed is illustrated in Figure 3-22. First, routine access to 4-OT by total chemical synthesis using stepwise SPPS (Fitzgerald et al., 1995) allows the introduction of any desired chemical perturbation into the molecule. An array of analogues is prepared in this fashion. Pro[1] is kept invariant in

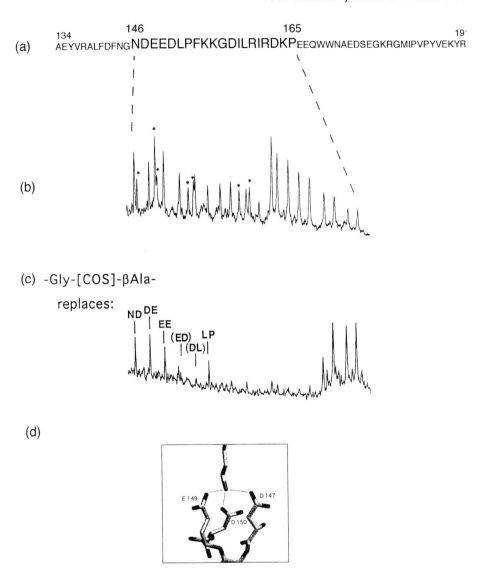

Figure 3-21. Nonlinear (conformational) epitopes. Protein signature analysis of the cCrk N-terminal SH3 domain, residues 146–165 (a). Affinity selection from the parent array of protein analogues (b) for binding to a C3G-derived Pro-rich peptide gave (c) the binding pool shown. (d) This signature showed that the effect of Asp[150] on binding is dominant under the conditions used (Wu et al., 1995).

the members of the 4-OT family of analogues, because the imino nitrogen of Pro[1] appears to be the catalytic general base in the mechanism of action of the enzyme on its substrates (Fitzgerald et al., 1996; Stivers et al., 1996a, b). When the 62-amino acid polypeptide folds correctly, it forms a hexamer. Therefore, folded and unfolded molecules can be distinguished by size and separated conveniently using gel filtration. The use of a suicide substrate (inhibitor) to probe for 4-OT enzyme activity (Stivers et al.,

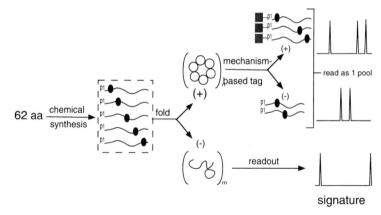

Figure 3-22. Probing the chemical basis of catalysis in 4-oxalocrotonate tautomerase (4-OT). A family of 4-OT analogues is synthesized; a specific analogue structure is systematically placed throughout a region of the polypeptide chain, such that each member of the family contains one analogue unit at a specific and defined position. The protein analogue array is formed by folding this family of polypeptides, and the folded and unfolded populations are separated into two pools by size exclusion chromatography. The catalytically active members of the folded population are then reacted with a suicide substrate. The components of each pool (unfolded; folded-inactive; folded-catalytically active) can be read out by MALDI mass spectrometry after a chemical cleavage step (as shown in Fig. 3-18).

1996c) in the pool of folded protein analogues gives rise to two subpools; in one of these the catalytically active proteins will add the suicide substrate as a covalent modification at the N-terminal of the 62-amino acid polypeptide chain, causing a distinctive mass shift in that signature. The other *inactive* pool of folded proteins will not contain such a modification.

This gives three pools of 4-OT analogues, namely (1) those members of the array that did not fold, (2) the folded-active members of the array, and (3) the folded-inactive members. Each of these self-encoded pools is read out by mass spectrometry following a chemical cleavage step. By correlating the folding and activity signatures, it should be possible to distinguish to some extent the structural from catalytic effects of analogue substitution in the polypeptide chain of this enzyme. It is anticipated that protein signature analysis will be very useful for looking at the chemical basis of both protein folding and enzyme catalysis in this system.

Summary

Probably one of the more significant recent developments in the application of chemistry to the study of proteins has been the development of the chemical ligation approach to total synthesis. For the first time this has provided for the routine, reproducible preparation of proteins by chemical synthesis. The amounts obtained, typically several tens of milligrams of protein from the *smallest* convenient laboratory scale of synthesis, are ideal for the application of modern physical techniques that are used for the study of proteins such as NMR and X-ray crystallography. Once synthetic access has been established for a particular system, the tools of chemistry allow complete control over every aspect of the covalent structure of the protein molecule (Muir & Kent, 1993). The chemical ligation approach provides access to novel proteins with unlimited chemical variations. Chemical ligation also permits the application of protein signature analysis to structural and functional domains in the context of larger protein molecules. This combination of techniques will allow the systematic elucidation of the chemical basis of protein function.

What will be the impact of the techniques described in this chapter? First and foremost, it will now be possible to study fundamental aspects of protein folding, that is, how the linear information of the biological world, contained in DNA and expressed on the ribosome as a nascent polypeptide chain, determines a three-dimensional protein structure. Initially, these studies will be carried out at the level of protein domains, the fundamental building blocks of the protein world. It will also be possible to look at how proteins function, particularly in binding to hormones and other potent biological effector molecules. Perhaps most importantly, chemical synthesis and the self-encoding principle can be used to systematically investigate protein-protein interactions, by building a latent footprinting chemistry into the protein molecule. Eventually, it is hoped that it will be possible to use these combinatorial chemistry approaches for protein design, and in this way to both build new functions into existing protein scaffolds and to design proteins *de novo*.

4

Structural Analysis of Proteins

John E. Shively

This chapter presents a systematic approach to the determination of the primary structure of peptides and proteins. While it is true that one can apply general principles toward the determination of protein structures, it should also be realized that every protein studied has unique properties that may make analysis difficult, challenging the researcher to try alternative approaches or to modify existing ones to suit the problem at hand. Proteins provide such a challenge because of their extremely diverse properties, a consequence of the almost limitless combinations of 20 amino acids; even for a protein with 75 residues there are 3×10^{97} possible combinations. In addition, their properties may change dramatically depending on the conditions (pH, ionic strength, solvents, denaturants) of purification and structural analysis. Thus, a water soluble protein may become insoluble after the reduction of disulfide bonds and alkylation of cysteine residues; a membrane protein may become water soluble after brief treatment with proteases; and a detergent soluble glycoprotein may be first rendered water soluble by treatment with a lipase (to release a GPI lipid anchor), and then become water insoluble again after treatment with glycosidases (to release glycosyl units). Furthermore, it is often true that one may determine 90% of the sequence of a protein in a relatively short time (say 1 to 3 weeks for a small protein, and 1 to 3 months for a medium size protein), but the last 10% may take a very long time. The reason for this truism is an inherent problem with either the alignment of peptides (every set must be overlapped to complete the structure) or the sequence analysis of one or more difficult sections. Such sections may be due to a run of hydrophobic amino acids, or the presence of highly modified residues (posttranslational modifications). Everyone knows that the simple way around this problem is to combine protein sequencing with gene cloning. The two techniques are complementary. A little bit of protein sequence information can dramatically speed up cloning of the corresponding gene by either providing oligonucleotide probe information, or confirming the sequence of existing cDNA clones.

With these facts in mind, it is clear that there is an important role for protein chemists in molecular biology. Interesting bands on sodium dodecyl sulfate (SDS) or 2D gels can provide the protein chemist with sufficient starting material to determine partial amino acid sequence information. This information can lead to the cloning and positive identification of the gene of interest. More complete protein sequence information can provide crucial information on protein processing, a rather common event, which is evident from an examination of the molecular sizes of the gene products, and which may not agree well with the molecular size predicted from the gene sequence. In this respect, all co- or posttranslational changes that occur are the special domain of the protein chemist, since none of this information can be learned from examining a gene sequence. Certainly, one can predict that a certain asparagine residue (in the context of the asn-xxx-ser/thr sequence) *may* be glycosylated, but in the end analysis, only a careful structural analysis can prove whether or not such an event did occur, and if it did, the structure of the glycosyl unit.

The types of posttranslational modifications which proteins undergo are amazing, and many are still being discovered. The N-terminus may be myristylated to anchor it to the membrane, or the C-terminus may be linked to a glycosylphosphatidylinositol (GPI) moiety which links it to the membrane via diacylglcerol. N-terminal glutamine may be converted to a pyroglutamic acid residue, while other N-terminal amino acids may be acetylated or removed by specific proteases. The protein may be cleaved internally by specific proteases before or after it leaves its point of synthesis. The protein may be phosphorylated, sulfated, *N*- or *O*-glycosylated, derivatized with γ-carboxylglutamic acid, esterified, amidated, crosslinked, or a combination of all of these possibilities. In the past, the determination of these posttranslational modifications was a slow and tedious process, but with the advent of ionization techniques for mass spectrometers compatible with high molecular weight biomolecules, these studies have become easier and possible with increasingly smaller amounts of material. As a result of a number of recent breakthroughs in this area, mass spectrometers have become one of the most potent tools of protein structural analysis available to the protein chemist. Although the day has not yet arrived where the entire structure of proteins is routinely determined by mass spectrometry, it may come soon, as more and more labs begin to depend on these powerful techniques.

This chapter is written for researchers who are not already expert protein chemists. It is written for researchers who have diverse interests in biological sciences, recognizing that at some stage in their research they are likely to identify a protein as a central character, and desire to participate to some degree in its structural analysis. By understanding the steps through which this analysis must proceed, they are more likely to help the process along, and in the end, may become, if not experts, at least knowledgeable protein chemists. For graduate students who are contemplating a career in this area, this chapter should help them gain an appreciation of the tools and approaches available. The current status of the field is one in which the tools are being rapidly perfected, leading to exciting discoveries almost daily. Throughout the text the most sensitive methods available are emphasized, since these are the methods which most researchers *require* for their projects and which present the greatest challenge to master.

Key Elements of Protein Structural Analysis: An Overview

Protein Purity

Before embarking on the structural analysis of a protein, it is essential that one establish its purity. If it was identified on the basis of its biological activity, its specific activity should increase during the purification process, but this parameter alone is not an absolute criterion of purity. The sample should be analyzed by a number of methods to ascertain its purity: a single band on an SDS gel, a single spot on a 2D gel, or a single symmetrical peak on several types of chromatography (ion exchange, hydrophobic, reverse phase, etc.) or capillary electrophoresis. Many researchers have been fooled by using SDS gel electrophoresis alone as a criteria for purity. Many samples may comigrate, or many bands may be unintentionally ignored: for example, low molecular contaminants may be run off the gel or be poorly stained; high molecular weight contaminants may not enter the gel or be dismissed as aggregated protein. For those who only feel comfortable with gel techniques as a method for determining sample purity, the following strategy should be used. The sample should be run at several loadings and on several different types of gels stained with both Coomassie Blue and silver stains. SDS gels can be run over a range of percent polyacrylamide (e.g., from 5 to 20%) and under both reducing and nonreducing conditions. Native gels may reveal subunit structure or contaminants missed by SDS gels. Acid-urea (with or without Triton X-100) or 2D gels may provide new information. The use of many gel techniques is also encouraged because often the sample *is* at a stage of sufficient purity to proceed with sequence analysis, but the investigator is doubtful due to the presence of multiple bands. Many proteins inherently give multiple bands on gels due to post-translational modifications, and these modifications may not interfere with the analysis of the amino acid sequence. For proteins that exist with multiple subunits or isoforms, multiple bands are the *expected* result. In the case where subunits are noncovalently linked, it is essential to run native gels that will not dissociate the protein into subunits, and in the case where they are linked by sulfide bonds, to run the sample on SDS gels with and without reducing agent.

What about other criteria of purity? The emphasis should be on methods that use only small amounts of the sample, since there should be something left for structural analysis. Small amounts (1 to 100 pmoles) can be spared for amino acid analysis or precise mass measurements by mass spectrometry. The pros and cons of these methods will be discussed in later sections. Most labs have access to HPLC (high performance liquid chromatography) equipment, which can give high resolution separations of samples applied to ion exchange or reversed phase columns. The current trend is toward packed capillary columns (i.d. <1 mm) which permit high sensitivity analysis (ca. 10 pmoles). If these columns are not available, the use of columns with an inner diameter of 1 to 2 mm and, in general, shorter columns, make it less likely that the sample will be nonspecifically absorbed. Samples eluted from these columns may often be sequenced directly. A single N-terminal sequence can also be used as a criterion of purity, especially if the yield is in the range of 30 to 60% of theoretical. Most proteins do not give quantitative yields of their N-terminal amino acid, where 50% yield is considered very good (as discussed later). As mentioned earlier, some proteins have a blocked N-terminus, and only under special conditions can they be sequenced from the N-terminus. It is estimated that 50% of the proteins in nature have a blocked N-terminus.

How pure should the sample be? As a general rule, it should be more than 95% pure. Samples with substantial amounts of impurity (even 5%) may complicate the analysis at the stage of N-terminal sequence analysis or peptide mapping, especially if there is a single contaminating protein and it has a lower molecular mass than the main species. If the main species is blocked at its N-terminus, you may observe only the sequence of the contaminant, and if you have not quantitated the sample, you may believe that the contaminant sequence *is* the main protein. In sequencing peptides from a peptide map, some of the peptides sequenced will be from the contaminant and will greatly confound peptide alignment for the major protein. An even worse scenario is that the researcher will clone a gene based on the sequence of the contaminating protein, and will incorrectly conclude that it is the gene of interest. In devising a strategy to sequence a protein, then, it is important to keep track of the contaminants. One can run an analytical SDS gel and analyze the major protein by carefully taking a region from the gel or blot, as described later, and performing either sequence analysis or peptide mapping experiments.

Protein Quantitation

Many researchers try to quantitate their proteins on SDS gels by judging the amount of staining of their protein by eye compared to several standards. To say the least, this is not a precise method, and even in the best of circumstances, may be off by more than 200%. Proteins do not take up stains in a uniform manner and may not be compared accurately to a different protein standard. To obtain accurate estimates of a protein amount by this method, you must first determine the concentration of your protein by another method, run a series of dilutions on an SDS gel, and establish a calibration curve by scanning each lane with a densitometer. The best method for quantitating the sample is by amino acid analysis. Good analyzers can give an accurate analysis of as little as 0.1 to 1.0 μg of sample, an amount that most researchers can spare for this important information. The sample cannot be analyzed directly from gels, because gels are usually contaminated with large amounts of glycine and Tris, both of which severely interfere with amino acid analysis. If the gel sample can be transferred to a membrane, a "blot," it can be hydrolyzed and analyzed, but since such transfers are never quantitative, the estimate will be on the low end (say 20 to 70%). If the sample can be analyzed directly from HPLC, a more accurate analysis is possible, as long as the buffer does not contain amino groups or large amounts of certain salts or detergents. In many cases, the sample must be desalted before either amino acid analysis or sequence analysis. A later section will deal with these strategies.

Amino acid analysis is a key method for any protein chemistry laboratory because it is one of the few quantitative methods available. However, it is also a demanding method, because the sample and all of the materials which contact it during the analysis must be free of amino acid contamination. Thus, the investigator must demonstrate through careful controls that the procedure itself gives acceptable levels of background amino acids. Even so, most samples will have elevated levels of glycine and serine which must be subtracted from the analysis in order to give an accurate weight analysis. If during purification the sample begins to give a reproducible amino acid composition, this also may be taken as evidence that the sample is very pure.

How much sample should be accumulated before starting structural analysis? Some protein chemists will claim that they can obtain good sequence information with only 1 or 2 pmoles of sample. However, it is unlikely that a single analysis at this level will be trustworthy. The cautious investigator will attempt to purify at least 10 to 20 pmoles of sample before beginning significant structural studies, and if multiple peptide maps and posttranslational information are desired, 100 to 200 pmoles are required. In spite of these realities, the ultimate goal would to be able to obtain significant structural information from a single spot on a 2D gel (about 10 ng of sample). Excellent progress towards reaching this goal is being made in many laboratories, but it should be stressed here that the techniques and protocols for achieving these sensitivities are still in the research stage and cannot be achieved routinely. Table 4-1 provides a guide to sample

Table 4-1 Sample Amounts Required for Routine Analyses

Sample Amount (weight)	Sample Amount (moles)	Procedure
Protein, 10 kDa		
10 ng	1 pmole	Minimum for silver staining
		Minimum for mass analysis by MS
100 ng	10 pmole	Minimum for CB[a] staining
		Minimum for amino acid analysis
		Minimum for peptide map, sequencer
Protein, 100 kDa		
10 ng	100 fmole	Minimum for silver staining
100 ng	1 pmole	Minimum for mass analysis by MS
		Minimum for CB[a] staining
		Minimum for amino acid analysis
1 µg	10 pmole	Minimum for peptide map, sequencer

[a] CB, Coomassie blue. Other stains are Ponceau S, Amido black, and calconcarboxylic acid.

amounts for routine analyses of two protein sizes, 10 kDa and 100 kDa. Proteins of intermediate or larger sizes can be extrapolated from this information.

Peptide Mapping

Once sufficient quantities of pure protein are available (i.e., preliminary mass and N-terminal sequence analyses have been performed), it is highly desirable to obtain a peptide map. While some proteins in their native state are readily digested with a variety of proteases, most will require thorough denaturation before digestion. Usually, treatment of proteins in their native state with proteases will result in cleavage between domains and at other protease sensitive sites. However, the goal of peptide mapping is to obtain a complete peptide map with cleavage occurring at every possible site. Maps of this sort yield peptides that are amenable to complete sequence analysis and that can cover the entire protein sequence. In spite of this goal, it is not unusual for a protein to have a protease resistant core (even though it contains potential cleavage sites), and it is necessary to use alternative proteases or conditions for which the core is accessible. If the goal is to obtain complete sequence information, many peptide maps are required in order to align individual peptides within the protein sequence.

Most proteases have very strict conditions for activity, including consideration of pH, ionic strength, detergents, and denaturants such as urea and guanidine HCl. A good strategy will include matching the last purification step with the best conditions for protease digestion. Since many investigators prefer sequencing samples from gels, strategies have been devised to either sequence the sample directly from a gel (*in-gel digests*) or after electrotransfer to a membrane (*blot digests*). One of the reasons that these strategies often work is that the protein has already been thoroughly denatured by SDS gel electrophoresis. However, this advantage is somewhat offset by the tendency of SDS to inhibit protease activity. If the protein contains cysteine residues, it is a good idea to reduce and S-alkylate the protein before running the gel. This strategy will open up the sample for a more complete digestion and allow positive identification of cysteine residues which may be otherwise difficult to identify. Large proteins (>50 kDa) will often give too many peptides, complicating the separation and identification procedure. In these cases it may be necessary to use multidimensional analysis (several HPLC columns or LC-MS-MS).

Chemical cleavage methods have the advantage that they do not require prior denaturation of the protein. The most popular method, CNBr cleavage at methionine residues, has the disadvantage that the peptides produced are often large and hydrophobic, and may be even more difficult to analyze than the starting sample. If the CNBr fragments are large enough to be resolved on an SDS gel, they may be further digested with proteases to yield complete sequence information. Occasionally, the CNBr fragments will be separable on reversed phase HPLC columns, the usual method for smaller peptides, but more often they will elute together in broad peaks with low yields. In the latter case, it may be necessary to digest the sample with a protease before the peptide mapping step is attempted. Other chemical cleavage methods will be discussed later.

Peptide Sequencing Methods

Once peptide maps are obtained, the next goal is sequence analysis. The most popular method is automated Edman chemistry (run on a *sequencer*), a method which removes one amino acid at a time from the N-terminus, resulting in the sequential liberation of phenythiohydantoin (PTH) amino acids which are identified by on-line HPLC analysis. Sample amounts required are in the 5 to 20 picomole range, and cycle times are about 40 to 50 minutes. Thus, it may be possible to analyze about 2 to 3 peptides per day. A peptide map containing 50 peptides may require up to one month of sequencer time. Most peptides of length 3–30 amino acids can be sequenced completely, but some may "wash out" early or have regions that are hard to identify. In any given sequencer run, especially those that have problematic sequences or amino acids, there may be a few questionable residues. Thus, it is not unusual for protein chemists to repeat their analyses (an automatic outcome of sequencing different peptide maps from the same protein), or to rely on complementary methods for confirmatory information.

The most popular confirmatory method is *mass analysis* by mass spectrometry. Mass analysis for peptides is commonly performed on time of flight (TOF), quadrupole or magnetic sector instruments using a variety of ionization techniques. SIMS (secondary ion mass spectrometry) or FAB (fast atom bombardment) are the oldest techniques and are usually available with the quadrupole and magnetic sector instruments. SIMS or FAB analysis involves putting 5 to 10 pmoles of the sample in a liquid matrix on a sample probe. While the results are excellent for single peptides above m/z (mass-to-charge ratio) 300, the technique often fails for very small (m/z <300), very large (m/z >4000), or very hydrophilic peptides. In addition, peptide mixtures may give only a selected distribution (i.e., some will inhibit the ionization of others). This technique is sensitive to the presence of salts and detergents. Electrospray inoization (ESI) is an ionization technique in which liquid sample can either be directly infused into the instrument or coupled to an HPLC (LC/MS (liquid chromatography/mass spectrometry)). Like SIMS (or FAB), ESI sources are usually found on quadrupole or magnetic sector instruments. Since direct infusion ESI is very sensitive to salts, coupling to an LC offers the most versatile approach, allowing mass analysis of entire peptide mixtures. Matrix assisted laser desorption ionization (MALDI) is one of the newest and least expensive options for mass analysis, since it is easily coupled to the simple-to-operate TOF instruments. Samples (1 to 2 pmoles) are mixed with a UV-absorbing matrix, allowed to dry on a sample stage, and ionized with a UV laser. The matrix is relatively salt insensitive and usually gives good peaks even for peptide mixtures. With the advent of less expensive MALDI-TOF mass spectrometers, the goal of mass analysis of all peptides from peptide maps is now within reach.

It is also possible to sequence peptides by mass spectrometry alone. In order for this approach to be successful, the peptide must be fragmented in the mass spectrometer and reanalyzed in a second stage of analysis. This technique requires several analyzers linked in tandem (hence the terms tandem MS or MS-MS). Previously, instruments yielding the most complete sequence information have been four-sector instruments, costing over one million dollars, and employing high-energy collisions for breaking peptide bonds. Triple quadrupole instruments, costing up to $500,000, have also been utilized extensively, but produce less complete sequence information since they rely on low-energy collisions to fragment peptide bonds. TOF instruments with post source decay that provide fragmentation data are now available from several vendors, as are tandem TOF instruments. Tandem instruments can give complete or partial peptide sequence information on samples in the range of 1 to 10 pmoles. With the advent of microspray (flow rates <1 μL/min) the sensitivity can be pushed below 1 pmole (Davis et al., 1995). It is not unusual for these methods to leave gaps in the sequence due to an uneven probability of breaking all peptide bonds. Accordingly, some fragments may be completely absent, or other fragments may predominate the series. This approach is also handicapped by the lack of good computer programs, which can take the tedium out of identifying all of the peaks in an MS-MS spectrum. However, there are several good programs available for predicting fragments for known sequences (Lee & Vemuri, 1990) and for searching data bases (Eng et al., 1990). Tandem instruments employing either SIMS or ESI sources can be used to sequence peptides. Tandem mass spectrometers are the instruments of choice for identifying posttranslational modifications in proteins.

Whether or not mass spectrometric methods can completely replace Edman sequencers for peptide sequence analysis is frequently debated; in fact, both methods are heavily used and, since the sensitivity of MS methods are similar to that of Edman chemistry, both methods will continue to be used. On the one hand, since MS methods cannot determine the N- or C-terminal sequences of *intact proteins*, there will be a continued need for the Edman sequencer. On the other hand, MALDI and ESI-MS methods can give accurate molecular masses for *intact proteins*, which when compared to their predicted sequences can verify a given structure, giving confidence to the N- and C-terminal sequence predicted from the peptide maps.

Peptide Alignment and Data Bases

Once multiple peptide maps have been sequenced for a given protein, it is necessary to align the individual sequences properly. This can be done by inspection or by protein sequence alignment algorithms found in most data base packages. Once a trial sequence is obtained, then programs such as MacProMass (Lee & Vemuri, 1990) will allow one to test if each peptide mass fits the aligned sequence. Programs which identify and align sequences by searching databases with MS/MS data are now available (Eng et al., 1990). It is also possible to search data bases for individual masses and check to see if they fit known proteins (or peptide sequences in open reading frames).

Examples of this type of program are given by Henzel et al. (1993), Mann et al. (1993), and Pappin et al. (1993). These powerful approaches save considerable time and effort, since it is not unusual to find that the "new" protein that you have just begun to sequence is already in the data base.

Posttranslational Modifications

Evidence for posttranslational modifications may be obtained during protein isolation and characterization, and during the sequence analysis. Holes in the peptide sequence may be seen at the level of Edman chemistry or mass analysis (as *mass discrepancies* compared to the predicted sequence). In rare cases the nature of the posttranslational change may be obtained directly from Edman chemistry, but more often, further analysis is required. As mentioned above, the method of choice is MS-MS analysis. Even MS-MS may not provide sufficient detail, and it may be necessary to perform enzymatic, chemical, or other spectral analyses which elucidate the exact details of the modification. Mass analysis is usually the first clue that a posttranslational modification is present. In the case of peptides, the MS method may allow one to measure the mass of both the modified and unmodified peptides. In the case of the intact protein, the mass discrepancy may give a clue as to the nature of the modification (e.g., plus 80 for sulfate or phosphate) or the mass heterogeneity profile may be characteristic of some glycoproteins. In many cases, the heterogeneity of the posttranslational modification is so great as to preclude accurate mass analysis of any given species. These problems may be solved by the sequential removal of the modifications by enzymatic or chemical treatment until a more tractable sample is obtained.

N-Terminal Sequence Analysis of Intact Proteins

Sample Preparation

The sample may be spotted directly onto a glass fiber filter or a polyvinylidene difluoride (PVDF) membrane that is placed in the sequencer. If it contains substantial amounts of salts and detergents, a high background may be observed on the first cycles, making identification of these cycles difficult. Buffers containing amino groups such as Tris, Tricine, or glycine will give high backgrounds, since the Edman reagent reacts with them, giving products that interfere in the subsequent HPLC analysis. This problem may be obviated by washing the sample with organic solvent in the sequencer (glass fiber or PVDF) or water outside the sequencer (PVDF only). Samples may be adsorbed onto PVDF membranes from buffers by immersing a PVDF strip into the sample buffer for a period of 1 to 2 hr and then rinsing with water to remove excess buffer. The hydrophobic PVDF membranes must be prewetted with methanol or acetonitrile prior to immersion. Most proteins bind very strongly to PVDF and can only be eluted with strong acids such as trifluroacetic acid (TFA)-water (e.g., 50:50). The presence of large amounts of detergent may prevent absorption of the sample to PVDF and thus should be removed prior to this step. SDS can be removed by extraction into an organic solvent such as ethyl acetate with a phase transfer salt such as triethylamine (e.g., extract 100 μL of sample with 200 μL of ethyl acetate and 1 μL of TEA, five times). If the sample is precipitated with trichloroacetic acid (TCA) (see the following), it may be extracted with acetone-acetic acid-TEA (90:5:5; see Henderson et al., 1979) Alternatively, SDS may be removed by running the sample over a reversed phase column with an inverse gradient, as described by Simpson and coworkers (1989a) or over an ion exchange column as described by Kawasaki & Suzuki (1990). This latter method was described primarily for tryptic digests, but also works well for intact proteins. Briefly, the sample is run over a diethylaminoethyl (DEAE)-type resin (DEAE-Toyopearl 650S, or a micro SDS trap column from Michrom Bioresources) which is connected in-line to a small reversed phase HPLC column. The protein is eluted from the dual column system with a linear gradient from 100% of 0.1% TFA to 100% of TFA-water-acetonitrile (0.1:9.9:90). The SDS elutes in the void volume and the protein elutes during the gradient (see later description of methods to collect the protein). The sample may also be concentrated onto a PVDF membrane using a modified Millipore Ultrafree concentrator as described by Sheer et al. (1990). After sample concentration, the membrane may be washed with water directly in the Ultrafree concentrator. It appears that the Centricon concentrators from Amicon could be used in a similar manner.

PVDF Membranes

In our own lab, we sequence all of our proteins routinely on PVDF membranes because of the obvious advantages in using an absorptive membrane that can be washed with water to remove salts prior to sequence analysis. Since we build our own sequencers and use continuous flow reactors for holding the PVDF membranes (Calaycay et al., 1991), our advice may not translate well to every commercial sequencer. Sequencers from Applied Biosystems and Porton were originally designed for glass fiber filters, but have been modified to accept PVDF membranes ("problott" cartridge). The Hewlett-Packard instrument has a biphasic cartridge containing ion exchange and reversed phase packing (Slattery & Harkins, 1993). This arrangement allows on-line sample cleanup, since the sample is unlikely to elute from both phases when either organic or water washes are applied. This cartridge has also been modified to accept PVDF strips (Reim & Speicher, 1993a, b).

Many samples are obtained from SDS gels. If they are eluted from the gel by a method such as that described by Kurth & Stoffel (1990) (0.1 M sodium acetate, pH 8.5, 0.1% SDS, 37°C, overnight), they may be spotted directly onto a PVDF membrane and washed with water. In general, protein samples are difficult to elute from gels after they have been fixed and stained. Direct elution methods may be further improved if the protein is located by a staining a guide strip, then eluting the unstained portion of the gel. Since the gel contains large amounts of Tris, glycine and SDS, a washing step is mandatory prior to sequence analysis. If the gel has been stained and dried, it may be better to perform an in-gel digestion with a protease rather than attempting to elute the intact protein (see later). If an SDS gel sample is electroblotted onto a PVDF membrane, it may be stained, washed, and sequenced according to standard protocols (see the following for an example protocol). Although extremely popular, these methods rarely give transfer yields above 50%, and since the usual yield for the N-terminal amino acid is also around 50%, overall sequencing yields in the range of 10 to 20% are common. There is good evidence to believe that electroblot yields are related to the SDS concentration in the gel, which if too high will cause the sample to pass through the PVDF membrane, and if too low, will leave the sample in the gel. In addition, it has been observed that proteins with high isoelectric points, or proteins at their isoelectric points, transfer poorly. Thus, it is necessary to check the gel for untransferred protein by staining the gel, and if protein is left behind, it may be necessary to raise the pH of the electrotransfer buffer.

Gel slices may *not* be sequenced directly because of their instability to the sequencer chemicals. The same is true for samples on membranes such as nylon and nitrocellulose. Recent experiments with Zitex (porous Teflon) or Teflon tape suggest that Teflon substrates may be suitable for sequence analysis without the addition of Polybrene (Burkhart et al., 1995). Zitex has very different properties from PVDF. Samples that are dried directly onto Zitex remain immobilized during sequence analysis, but are eluted easily with aqueous buffers. Electrotransfer onto Zitex or Teflon tape is possible, but only under conditions in which organic solvents such as methanol, acetonitrile, or 2-propanol are added to the electrotransfer buffer or used to prewet the membrane (Burkhart et al., 1995). Samples on Zitex, Teflon tape, or PVDF can be mass analyzed by MALDI-TOF if they are prewetted with 2-propanol. The addition of cationic polybrene to membranes such as PVDF can increase the yields of electrotransfer (Xu & Shively, 1988), but is unpopular because of its interference in the subsequent staining-destaining steps (most protein stains are anionic and are destained poorly from Polybrene).

Stains for PVDF Blots

Membranes stained with Coomassie blue will give artifact peaks on peptide mapping due to the stain, and there is evidence for protein loss during the destaining procedure (Reim & Speicher, 1993). Excess Coomassie blue may be removed by extracting the membrane with 1:4:1 water-methanol-chloroform (Wessel & Flugge, 1984). Alternative stains are amido black or Ponceau S (Fernandez et al., 1992). These stains give fewer background peaks during peptide mapping. A general procedure for staining-destaining nitrocellulose or PVDF membranes is the following.

1. Stain in 0.1% Ponceau S in 1% acetic acid for 2 to 3 min.
2. Destain in 5% acetic acid for 5 to 10 min.
3. Wash with water 2 to 3 times.
4. Further destain with 200 μM NaOH/20% MeCN.
5. Wash with water 2 to 3 times.

In spite of the previously mentioned problems, N-terminal sequence analysis from PVDF blots has become very popular, and if carefully performed can yield valuable sequence information on the protein of interest. For sample amounts in the range of 10 to 20 picomoles, it is not unusual to obtain the first 10 to 20 amino acids, and with some optimization, assignments through 30–50 residues are obtainable. If the protein fails to give a sequence, it may be blocked at its N-terminus. In such cases, the PVDF strip should not be discarded. It may be possible to deblock the sample by enzymatic or chemical treatment, or failing that, digest it with CNBr to obtain internal sequence information.

Treatment of N-Blocked Proteins on PVDF Membranes

Strategies for deblocking samples on PVDF strips have been described by Tsunasawa & Hirano (1993). Proteins blocked with a pyroglutamic acid group may be deblocked directly with the enzyme pyroglutamyl peptidase after first treating the membrane with PVP-40 (polyvinylpyrrolidone-40). The PVP-40 step, originally described by Aebersold et al. (1987), blocks sites on the PVDF which would otherwise bind the pyroglutamyl peptidase and prevent it from digesting the sample, which is also bound to the membrane. The procedure described by Tsunasawa and Hirano is summarized here:

1. Block with 200 μL of 0.5% PVP-40 in 0.1 M acetic acid for 30 min.
2. Remove excess liquid and wash 10 times with 1 mL of water.
3. Add 100 μL of 0.1 M phosphate, pH 8.0, 5 mM DTT, 10 mM EDTA.
4. Add 5 μg of pyroglutamyl peptidase for 24 hr at 30 °C.
5. Wash with water, dry, and apply membrane to sequencer.

Tsunasawa & Hirano (1993) also described methods for treating samples on PVDF with acid to remove N-formyl and N-acetyl groups. The yields for these methods may be low or result in internal peptide bond cleavage. Nonetheless, they are worth trying. The method for removal of N-acetyl groups involves treatment of the membrane with vapor phase TFA at 60 °C for 30 to 60 min. The method for removal of N-formyl groups involves treatment with 0.6 M HCl at 25 °C for 24 hr. Miyatake et al. (1993) have described a method for treatment of pyroglutamyl or N-formyl blocked proteins on PVDF membranes with hydrazine vapor. The recommended procedure is −5°C, 8 hr, for pyroglutamyl-blocked proteins and 20 °C, 4 hr for N-formyl blocked proteins. After treatment, the membrane is dried and applied to the sequencer.

Reduction-Alkylation of Cysteine Residues

Cysteine residues form unstable derivatives during Edman chemistry and may give blank cycles or be misinterpreted as serine residues, since both degrade to dehydroalanine derivatives. Cystines in disulfide bonds may not be released by Edman chemistry unless they are reduced, or until the second Cys residue is reached. These problems can be overcome by reduction and S-alkylation of the sample prior to sequence analysis. If the sample is going to be run on an SDS gel and then sequenced directly or from a PVDF blot, it may be reduced and alkylated at this stage. Briefly, the sample is mixed with an equal volume of 2% SDS, 20% glycerol, 20 mM dithiothreitol (DTT), 0.004% bromophenol blue in 120 mM Tris-HCl pH 6.8 buffer, heated at 95°C for 3 to 5 min, and mixed with 1 μL of 4-vinylpyridine for 30 min. The SDS gel is run in the usual way. Alternatively, the sample may be reduced and alkylated on a glass fiber filter or PVDF membrane by vapor phase reduction and alkylation, essentially as described by Amons (1987). Briefly, the PVDF (or glass fiber) strip is placed in a glass tube with a restriction, below which is a mixture of 10:10:2:2 water-pyridine-4-vinylpyridine-tributylphosphine; the tube is evacuated and sealed, and heated for 2 hr at 60°C. Andrews & Dixon (1987) described a similar method performed in the sequencer, but it appears to lead to higher backgrounds than the vapor phase method. Reduction with tributylphosphine (TBP) and S-carboxamidomethylation of samples on PVDF membranes is reported to give no increase in background even at the 10 pmole level (Atherton et al., 1993). Samples are first treated with tributylphosphine (equal amounts of 2% TBP in acetonitrile and 40% acetonitrile-0.4 M N-ethyl morpholineacetate, pH 8.0) for 1 hr at 55°C, dried, and then treated with 10 mM iodoacetamide in 40% acetonitrile-0.4 M N-ethylmorpholine acetate (pH 8.0) for 15 min at room temperature in the dark. Another alkylating agent, which has certain advantages, is 3-bromo-aminopropane (Jue & Hale, 1993). The advantages include conversion of cysteine residues to derivatives which can be cleaved with trypsin and identified easily by sequence or amino acid analysis.

Samples from HPLC Columns

Samples that are collected from HPLC columns may be spotted directly onto PVDF membranes for sequence analysis. If the sample is in a volatile buffer such as 0.1% TFA in acetonitrile, it will dry without residual salt. If the sample was eluted from an ion exchange column, it may contain small amounts of salt, which will not interfere with sequence analysis or which can be washed off with water as described earlier. If the sample is in a large volume (>20 μL), it may be easier to immerse the PVDF strip (previously wetted with methanol or acetonitrile) directly into the sample tube and allow it to absorb the sample over a period of 30 to 60 min. As mentioned earlier, detergents may prevent the absorption, necessitating a cleanup step. Integral membrane proteins are often isolated in the presence of detergents and 5 to 10% glycerol, a combination which is sure to prevent sample absorption to the PVDF membrane. These samples may be quickly desalted on a packed capillary column containing a large pore size reversed phase packing such as Polymer Labs PLRP-S (Amherst, MA). The sample is loaded onto the column and rinsed with 0.1% TFA in water until the baseline is reestablished, and then eluted with a rapid (<10 min) gradient to 100% solvent B (0.1:9.9:90 TFA-water-acetonitrile). The sample should be collected into a polypropylene tube containing 2 μL of DMSO to prevent adsorption to the walls of the tube. For a typical capillary column (i.d. 0.5 mm or less) the volume collected should be less than 100 μL. The sample may be further concentrated on a vacuum centrifuge to a volume of 10 μL before spotting onto the PVDF membrane. It is never a good idea to dry protein samples completely, a procedure that may cause them to adsorb strongly to the walls of the sample tube. If the sample inadvertently dries, it may be possible to desorb it by treatment with 20 μL of hexafluoroacetone (HFA) for 1 to 2 hr at 50°C.

Several strategies have been developed to prevent loss of sample by adsorption to tube walls during sample collection. One is to collect the sample into 5 to 10 μL of a material known to prevent sample adsorption. An example for a membrane protein was described earlier. Other proteins may prefer a small amount of phosphate buffer (10 μL of 1 M), guanidine HCl (10 μL of 1 to 6 M), or urea (10 μL of 1 to 8 M). The important point to remember is that even though a large UV peak corresponding to the sample is observed (a sure fact that it eluted from the column), it may immediately adsorb to the tube walls during the collection step. This can be checked by reinjecting an aliquot of the sample. This procedure can also be used to determine what is the best additive to prevent the sample from adsorbing irreversibly to the walls of the sample tube. This is perhaps the single most frustrating problem in sequencing a protein, namely, losing it at the last step. This problem may be further reduced by either collecting samples into Teflon tubes (tubes of the right size are not commercially available), or collecting the samples directly onto PVDF membranes. Samples may be hand-collected onto PVDF as they elute or automatically streaked on a long strip of PVDF (Murata et al., 1993).

Samples from 2D Gels

Depending on the size of the 2D gel, samples may be Coomassie blue or silver stained. Large gels, 1.5 mm thick and 15 cm long, can accommodate as much as 100 μg of loaded protein, and resolve individual spots containing about 1 μg of protein. Sample amounts in this range correspond to 10 to 100 pmoles of protein, depending on the molecular weight (100 to 10 kDa). These are sufficient amounts to plan a sequencing strategy such as described (Aebersold et al., 1987; Bauw et al., 1989; Henzel et al.,

1993; Simpson et al., 1989b). Small 2D gels, for example, Pharmacia Phast gels, which should be silver stained, can accommodate sample loads in the 1 to 10 μg range. These gels have as little as 10 ng sample per spot (0.1 to 1.0 pmole, for proteins of size 100 and 10 kDa, respectively), an amount that severely challenges the current capability of sequence analysis, especially in light of the low recoveries expected from the gel during electroblotting or in gel digestion.

Precipitation with Cold TCA

The most common obstacle preventing a successful sequence run is that the sample is dilute and in the presence of interfering buffers and detergents. Several strategies for overcoming these problems have already been discussed along with the sample isolation method. Any of the methods may fail because of excessive amounts of glycerol or detergents, nonspecific adsorption to tube walls during the desalting step, or the inherent physical problem of trying to concentrate the sample from an unusually large volume (i.e., how a 1 to 2 mL sample can be applied to a small strip of PVDF). Therefore, many researchers ask if there is a single method that can accomplish both steps with a high degree of success. One of the oldest methods of protein precipitation is with cold TCA (trichloroacetic acid), a step that usually (but not always) will precipitate the protein without also precipitating salts and detergents. If this procedure is performed carefully, it almost always works. The following protocol should be tried first on a small aliquot of the sample, with sufficient material to run on an SDS gel (preferably silver stained) to judge whether or not the method is appropriate for a sample.

1. Chill sample and TCA (e.g., 60:40 TCA-water on ice for 30 min).
2. Add a volume of TCA to sample to achieve a final percentage of 10 to 20% TCA.
3. Mix and let stand on ice for 1 to 2 hr.
4. Centrifuge at 4°C at 10,000 to 30,000 rpm for 10 to 20 min in a microfuge.
5. Remove the supernatant with a micropipetter, taking care not to remove the last few microliters.
6. Wash the precipitate three times with 100 μL of cold acetone.

It is important not to touch the bottom of the sample tube with the hands or the precipitate may warm up and redissolve. The washed precipitate may not be visible to the eye, but one should proceed with the subsequent steps. The precipitate must be redissolved immediately for further analysis or it may "set up" and become insoluble. In the case of running an SDS gel, simply add SDS sample buffer and analyze in the usual way. If the bromphenol blue turns yellow, then not all of the TCA was removed, and it will be necessary to add a few microliters of 1 M Tris to neutralize it (otherwise the gel will not run properly). In the case of sequence analysis, the precipitate may be dissolved in a few microliters of HFA, 70% aqueous TFA, hot 0.1% SDS, or DMSO. It may be necessary to try several solvents to determine which is best for your sample.

Edman Chemistry

The general scheme for Edman chemistry is shown in Figure 4-1. A protein or peptide with a free amino group is reacted with PITC (phenylisothiocyanate) under basic conditions to form a phenylthiocarbamyl peptide (PTC-peptide) in a step termed *cou-*

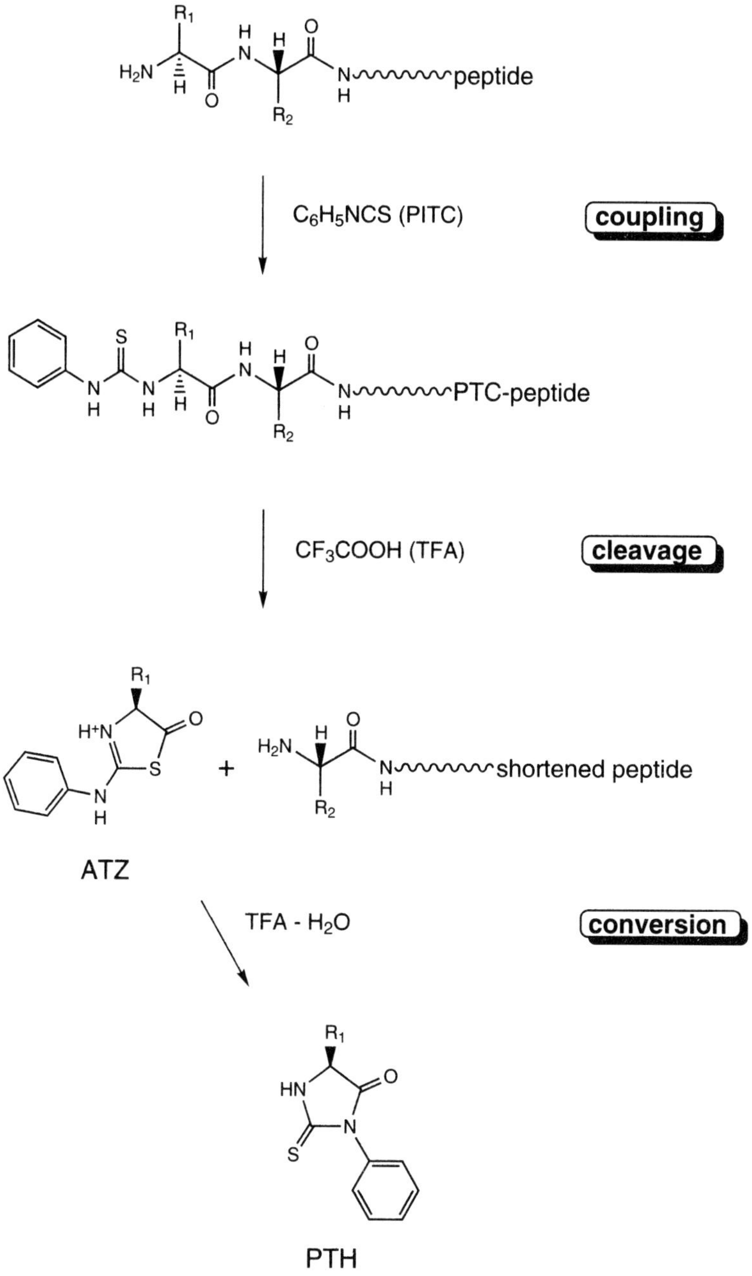

Figure 4-1. Edman Chemistry. The free α-amino group of a peptide or protein reacts with PITC to form a PTC-peptide in the coupling step. The first amino acid is cleaved from the peptide chain by anhydrous TFA in the cleavage step forming an ATZ amino acid derivative. The ATZ amino acid derivative is converted to a PTH amino acid derivative in the conversion step.

pling. The sample is then treated with anhydrous acid, usually TFA, in a step called the *cleavage* reaction, which releases the first amino acid as an anilinothiazolinone derivative (ATZ) and the shortened peptide. Since ATZs are chemically unstable, they are further treated with aqueous TFA and converted into phenythiohydantoin (PTH) derivatives in a step termed the *conversion* reaction. The PTH amino acids are separated by reversed phase HPLC and identified on the basis of their retention times. The steps are performed in an instrument called a protein sequencer, which incorporates all of the chemical steps and an on-line HPLC analyzer for identification of the PTH amino acid derivatives. The cycle time for one round of Edman chemistry is 40 to 60 min (accommodating both the chemistry and HPLC analysis). The sensitivity is limited by UV detection of the PTH amino acids, which for "narrow bore" (2.1 mm i.d.) columns is about 0.5 to 5 pmole at 269 nm. The success of Edman chemistry is based on high repetitive yields, often in the range of 93 to 96%. It can be easily calculated that one will have only about 50% of a sample left after 10 rounds of Edman chemistry if the average repetitive yield per step is 94%, that is, 0.94^{10}. The sensitivity is also limited by the initial yield of the first amino acid, usually in the range of 20 to 60%. Thus, if one starts with 10 pmoles of sample and sequences with an initial yield of 50% and a

Table 4-2 Amino Acid Single Letter Codes and Anhydro-masses[a]

Amino Acid	Single Letter Code	Monoisotopic Mass	Average Mass
Alanine	A	71.03712	71.079
Arginine	R	156.10112	156.188
Asparagine	N	114.04293	114.104
Aspartic acid	D	115.02695	115.089
Cysteine	C	103.00919	103.144
Glutamic acid	E	129.04260	129.116
Glutamine	Q	128.05858	128.131
Glycine	G	57.02147	57.052
Histidine	H	137.05891	137.142
Isoleucine	I	113.08407	113.160
Leucine	L	113.08407	113.160
Lysine	K	128.09497	128.174
Methionine	M	131.04049	131.198
Phenylalanine	F	147.06842	147.177
Proline	P	97.05277	97.117
Serine	S	87.03203	87.078
Threonine	T	101.04768	101.105
Tyrosine	Y	163.06333	163.170
Tryptophan	W	186.07932	186.213
Valine	V	99.06842	99.133

[a]From Biemann (1990)

repetitive yield of 94%, one may expect to identify as many as 10 to 20 residues. Often one can do much better than this, and it is possible to obtain longer sequences on similar amounts of material, or a similar number of cycles on less material. A further complication is that not all of the amino acids give high yields of their PTH derivatives, perhaps due to chemical breakdown (Ser or Thr), or poor extraction from the sequencer or adsorption to the HPLC column packing (His and Arg). The molecular masses and single letter codes for the 20 common amino acids are shown in Table 4-2.

The Sequencer

The first automated protein sequencer was designed by Edman & Begg (1967). The current generation of sequencers is based on the gas phase delivery of the coupling base and the cleavage acid, and the use of a sample cartridge with a glass fiber filter (Hewick et al., 1981). As previously mentioned, the cartridge has been modified in each of the commercial instruments to accept PVDF membranes. The continuous flow reactor described by us (Calaycay et al., 1991) can also be accommodated on the commercial instruments. A general schematic of a gas phase sequencer is shown in Figure 4-2. The amount of reagents and solvents delivered to the cartridge or conversion flask is controlled by their timed flow through zero-dead volume, chemically inert valves. The sample is spotted on a porous, inert support such as glass fiber or PVDF, and dried in the sequencer at 50 to 55°C in a stream of nitrogen or argon. The sample is then exposed to gas phase base (aqueous trimethylamine, triethylamine, or diisopropylethylamine) followed by delivery of PITC. Since PITC is delivered in a solvent such as heptane, excess solvent is removed by drying. The coupling reaction may require 5 to 10 min for complete reaction. The sample is then washed with solvents such as ethyl

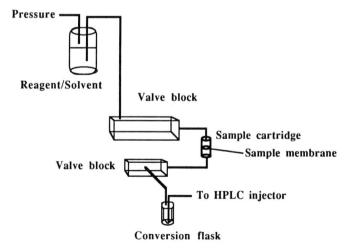

Figure 4-2. A schematic of an Edman sequencer. Reagents and solvents that are maintained under gas pressure are connected to a valve block that controls their delivery to a sample cartridge. Products from the sample cartridge are either delivered to waste (not shown) or to a conversion flask via a second valve block. The ATZ amino acid derivatives are converted to their respective PTH derivatives in the conversion flask and then delivered to an on-line HPLC.

Table 4-3 Reagents and Solvents in a Gas Phase Edman Sequencer[a]

Reagent or Solvent	Composition	Usage per Cycle
R1	5% PITC in heptane	2 µL
R2	2% TEA in water[b]	gas
R3	TFA	gas
R4	1:30:70 TFA-water-MeOH	60 µL
S1	not used	
S2	15:85 MeCN-EA	200 µL
S3	15:85 MeCN-BuCl	200 µL
S4	10:90 Water-MeCN	100 µL

[a]Abbreviations: PITC (phenylisothiocyanate), TEA (triethylamine), MeOH (methanol), MeCN (acetonitrile), TFA (trifluroroacetic acid), EA (ethyl acetate), BuCl (n-butyl chloride). The amounts used are for the minisequencer described by Calaycay et al. (1991), but are similar for most instruments.
[b]Other bases used are 12.5% trimethylamine, or 3.3% diisopropylethylamine (DIEA), both in water. The HP biphasic reactor sequencer uses 10:60:30 DIEA-1-propanol-water.

acetate to remove excess PITC and base. The cleavage reaction is accomplished with the delivery of gas phase TFA for 5 to 10 min, after which the ATZ derivative is extracted into butyl chloride and delivered to the conversion flask. While the ATZ is being converted to a PTH derivative by treatment with 5 to 25% aqueous TFA, the shortened peptide is either treated further with TFA (to remove the last traces of the first amino acid) or the next cycle begun. The conversion step requires 5 to 10 min at 55 to 60°C for complete conversion of an ATZ to a PTH derivative. Excess acid is removed by drying, and the PTH derivative is dissolved in water-acetonitrile and injected onto the on-line HPLC. The time for a complete cycle must be adjusted for the time for a complete HPLC separation, including column wash, and re-equilibration to initial conditions (40 to 50 min, depending on the program). A summary of the reagents and solvents used in Edman chemistry is given in Table 4-3.

The original gas phase instrument used a 1.2 cm diameter glass fiber disk in a glass cartridge. The glass fiber disk was coated with Polybrene, a poly(quaternary amine), originally shown by Tarr et al. (1978) and Klapper et al. (1978) to retain small peptides in spinning cup sequencers. It was first thought that the Polybrene together with a small peptide needed to be precycled with Edman chemistry in order to reduce background; however, this is rarely done today. In theory, proteins do not require Polybrene as a carrier, but in practice, the Polybrene is always added whether the sample is a protein or a small peptide. The size of the original glass fiber filter was appropriate for sample volumes up to 200 µL. With the advent of narrow bore and capillary LC, smaller sample volumes are usually applied (1 to 20 µL), and there has been a corresponding effort to reduce the size of the glass fiber disk and cartridge (Totty et al., 1992). The use of PVDF membranes has had an effect on the design of the original cartridge, which was much too large and often gave poor results for small strips of PVDF (1 × 10 mm). Three types of sample cartridges in common use are shown in Figure 4-3.

The sequence analysis of a standard protein is shown in Figure 4-4. The first 16 cycles of a 20 pmole sample of horse apomyoglobin are shown (Fig. 4-4A). The identi-

Structural Analysis of Proteins / 117

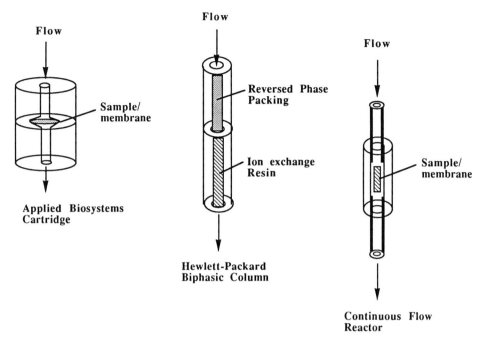

Figure 4-3. Sample chambers-reactors for Edman chemistry. (*left*) The glass cartridge from Applied Biosystems has a 1.2 cm glass fiber filter and a Zitex seal. (*center*) The Hewlett Packard biphasic column is made from polyethylene and has two types of packing to prevent the sample from eluting from the column during Edman chemistry. The sample may be desalted on the column prior to sequence analysis. (*right*) The continuous flow reactor is made from Teflon tubing, is self-sealing, and is ideal for PVDF membrane strips.

fied residues are shaded. Background peaks include DPTU, DPU, DEPTU (see later) and two peaks eluting at 20 and 21 min. The initial yield was 60% with a repetitive yield of 92% (Fig. 4-4C). The separation of the 20 common PTH amino acid derivatives is also shown (Fig. 4-4B).

Common Problems

The most common problems encountered in sequencing include low initial yield, low repetitive yield, and high background peaks. If any one of these problems is observed with a standard protein, it may be necessary to change reagents and solvents or reoptimize the sequencer program. Most sequencers have a large background peak, diphenythiourea (DPTU), caused by the hydrolysis of PITC to aniline and its subsequent reaction with PITC. This peak or its oxygen analog, diphenylurea (DPU), may interfere with the analysis of PTH-Trp, and thus should be minimized. Strategies for reducing DPTU include lowering the concentration of PITC in R1, or lowering the amount delivered to the sample membrane. Another common background peak is the PTC derivative of dimethylamine (DMPTU) or diethylamine (DEPTU), due to the breakdown of the coupling bases in water. This problem may be minimized by using freshly made base or the hindered base diisopropylethylamine. Sample contaminants

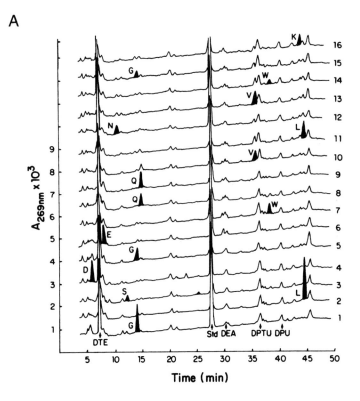

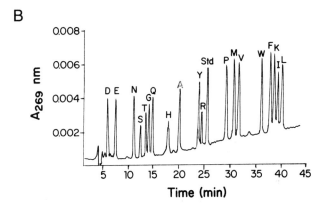

Figure 4-4. Sequence analysis of horse apomyoglobin. (A) HPLC traces of the first 16 cycles of 20 pmoles of horse apomyoglobin sequenced on a PVDF strip in a continuous flow reactor. The internal standard is PTH-aminoisobutyric acid. The background peaks are indicated at the bottom of cycle one. The expected peaks for each cycle are labeled and shaded. (B) HPLC separation of 19 of the PTH amino acid derivatives (C is omitted since it must be identified as an appropriate S-alkyl derivative) on a C-18 reversed phase column (2.1 × 25 cm). (C) Yields of PTH amino acids at each cycle. The yields for each amino acid shown in (A) are plotted opposite cycle number (as identified by the expected amino acid). The initial yield was 60% and the repetitive yield was 92%.

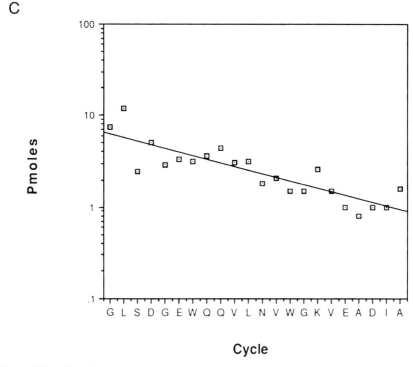

Figure 4-4. (Continued)

such as Tris or glycine will cause large background peaks, which may obscure large areas of the chromatograms for the first few cycles. They may be removed by strategies already discussed, such as washing the glass fiber filter or PVDF membrane with ethyl acetate or water, respectively. Another source of background peaks (in this case actual PTH-amino acids) is the protein itself. In general, the larger the protein, the higher the background, with increasing cycle number. Some proteins are worse than others, and it is likely that the background is due to peptide bond cleavage during Edman chemistry. This problem may be lessened by lowering the temperature for coupling and cleavage, a strategy that works well until a proline residue is reached. Proline presents a special problem in proteins, since its peptide bonds are cleaved only slowly by Edman chemistry (the problem is lessened for small peptides where there is less steric hindrance). Several strategies have been devised for sequencing through Pro bonds. The most general approach is to raise the temperature during every cleavage step just in case a Pro residue is encountered. A more specific approach is to sequence the protein at least twice, once to identify the Pro residues, and on the second run to perform special Pro cycles only at those residues. Special Pro cycles may involve raising the temperature during cleavage only on those cycles, or treating the protein with o-phthaldialdehyde (OPA) just before the coupling step at the Pro cycle. Since OPA does not react with proline, this approach will block all of the α-amino groups in the sample, leaving the N-terminal Pro sequence unblocked. This strategy will clean up a sequence run that is ragged because of the occurrence of multiple proline residues (Brauer et al., 1984).

Low repetitive yields may also be due to *sample washout*. This problem is especially prevalent for small hydrophobic peptides, or peptides that end in hydrophobic sequences. Washout occurs because the sample has sufficient solubility in the solvent wash (especially ethyl acetate) to be extracted from the Polybrene-support matrix. The problem occurs for both glass fiber filters and PVDF membranes as a support matrix. If the Polybrene is omitted, the peptides will wash out rapidly. Sequencer programs have been designed to minimize washout, but in the final analysis, one may wonder if the end of the peptide has been reached, since yields tend to drop off dramatically at the end of a peptide run. If you are sequencing a tryptic fragment, you may conclude that the end has been reached if the "last" cycle shows a Lys or an Arg. However, not all tryptic peptides end in Lys or Arg (e.g., the C-terminus, or cleavage at anomalous residues), or may contain internal Lys or Arg residues followed by Pro and other hydrophobic residues. Thus, it is important to verify the sequence of the peptide by another technique such as mass analysis. If it is certain that the peptide has washed out before completing the sequence analysis, it may be necessary to perform solid phase Edman chemistry or verify the C-terminal sequence by another technique (MS-MS, carboxypeptidase, or by a chemical C-terminal sequence analysis).

Solid Phase Edman Chemistry

Solid phase chemistry involves the covalent attachment of the peptide (or protein) to a solid support. The first solid phase sequencer was described by Laursen (1971). Peptides were attached by their carboxyl groups to amino-glass supports using water-soluble carbodiimide. Major problems with this approach are low yields of covalent attachment, and in the case of Asp- and Glu-containing peptides, attachment at these residues decreasing the overall sequence yield when the residues are encountered. Ideally, the peptide should only be attached through the C-terminal carboxyl group, but in reality, this goal may never be achieved. Coupling reactions with water soluble carbodiimides (e.g., N-ethyl-N'-(3-dimethylaminopropyl)carbodiimide; EDC) must be performed at or near the pH optimum, pH 4.5, and acetate buffers must be avoided for obvious reasons. A good buffer is 0.1 M pyridine HCl, pH 4.5. In addition, water-soluble carbodiimides will slowly rearrange to their N-acyl derivatives after reaction with carboxyl groups. Unless this reaction is suppressed, a major portion of the peptide may never react with the solid phase amino groups. A good strategy is to perform the coupling reaction at 4°C with the addition of equivalent amounts of carbodiimide and sulfo-N-hydroxysuccinimide; this converts the carboxyl group to an active ester that is more stable than the O-acyl urea derivative. The best solid supports have arylamino groups such as described by Aebersold et al. (1990) or Liang & Laursen (1990).

Cyanogen bromide peptides ending in a homoserine lactone may be coupled directly to amino-glass. The peptides are first treated with TFA to ensure conversion to the lactone, dried, dissolved in TEA-DMF-formamide (1:3:4) and added to amino-glass (45°C, 2 hr; Liang & Laursen, 1990). Proteins may also be sequenced on a solid phase. Unlike small peptides, proteins may be attached to the solid phase through their amino groups, since proteins contain many lysine residues. This is the preferred method of solid phase attachment for proteins, since it avoids the pitfalls of carbodiimide chemistry, and amino groups couple efficiently to solid phases such as DITC (1,4-

phenylenediisothiocyanate) glass (Liang & Laursen, 1990) or PVDF (Pappin et al., 1990). Proteins may be either attached to DITC-PVDF (Sequelon membranes from Millipore) or trapped indirectly by a combination of DITC and polyamines on underivatized PVDF (Sequenet approach). The sample is spotted onto PVDF, briefly treated with PITC to convert some of the amino groups of the protein to PTC-derivatives, and reacted with DITC and a polyamine such as polyallylamine. This procedure entraps the protein in a crosslinked net on the PVDF membrane. Solid phase sequence analysis for proteins usually gives equivalent results to those obtained for samples immobilized in Polybrene and, thus, has not become popular. However, there are some advantages to the covalent immobilization methods. The main advantage is that the sample may be washed with more polar solvents without threat of sample washout, and PTH derivatives may be eluted with more polar solvents such as TFA. Thus, there is the potential for longer sequencer runs, due to lower backgrounds, and polar PTH derivatives such as phospho-Ser and Thr may be identified; these derivatives bind strongly to Polybrene and are not eluted in conventional sequencers.

Unusual Amino Acids

Occasionally during sequence analysis, one will observe an unusual peak on PTH amino acid analysis, and will begin to suspect the presence of an unusual amino acid (i.e., not one of the standard 20 PTH-amino acid derivatives). A careful inspection of the HPLC separation of the 20 standard PTH amino acids (Fig. 4-4B) reveals little room for additional derivatives. Thus, the occurrence of an unusual amino acid (e.g., hydroxyproline, N-ε-methyl lysine, etc.) may or may not be noted, based on its unusual retention time alone. Furthermore, many amino acid derivatives are not detected by amino acid analysis, since they may coelute with or be hydrolyzed to a common amino acid. The solution to this problem is to perform mass analysis of each peptide. Any mass discrepancy between the mass predicted by the amino acid sequence (obtained by either amino acid or cDNA sequence analysis) will be a clue to the nature of the unusual amino acid. Table 4-2 gives the expected anhydro-masses for each of the common 20 amino acids. A review by Wold gives a thorough survey of the unusual amino acids encountered (Wold & Krishna, 1993).

Serine and Threonine

These amino acids are sometimes problematic since they tend to convert to their respective dehydro amino acid derivatives which then react with reducing agent (dithiothreitol, DTT) in the conversion flask. The elution position of their DTT adducts should be noted by sequencing a standard protein or peptide containing Ser and Thr. The dehydration product is more prevelant for Ser than for Thr; in some cases the DTT-Ser-adduct peak may be the only peak observed. Ser and Thr dehydration becomes more severe with decreasing sample amounts and with increasing amounts of TFA in R4 (Table 4-3). The best solution to the problem is to omit TFA from R4 entirely. Conversion is still efficient due to the residual TFA brought over with the ATZ during extraction from the cartridge (also, the conversion reaction depends more on temperature than on acid content). Nonetheless, yields of Ser during sequence analysis are often low, and for low picomole sequencing, may be missed entirely. Thr also dehydrates and forms DTT adducts, but due to the multiplicity of the peaks formed (isomers), the adducts are hard to identify and substantially lower the calculated yield of PTH-Thr.

Cysteine Derivatives

It has already been mentioned that unmodified cysteines cannot be positively identified during sequence analysis. The first clue may be a blank cycle (cystine) or the occurrence of a PTH-dehydroalanine. The sample should be resequenced following reduction and S-alkylation as described earlier. One should choose a preferred alkylating agent (iodoacetamide, iodoacetic acid, vinylpyridine, 3-bromopropane) and ascertain that the cysteine derivative chosen is separated from other PTH amino acid derivatives on HPLC.

Phosphorylated and Sulfated Amino Acids

Phospho-Ser and Thr also form dehydro amino acid derivatives in the sequencer. It has been noted that if sequencing conditions are chosen to minimize dehydration of regular Ser and Thr residues, then the occurrence of DTT-adduct peaks for Ser and Thr may be indirect evidence for their phospho-derivatives. Meyer et al. (1986) described a method in which phospho-Ser-containing peptides are treated under alkaline conditions with ethanethiol to produce thiol adducts. Phospho-Ser is identified as S-ethyl-Cys and phospho-Thr as β-methyl-S-ethyl-Cys (Meyer et al., 1991). Although the reaction will also occur for Cys residues, this problem can be prevented by previously S-alkylating all Cys residues. Phospho-Tyr is stable to Edman chemistry, and while the PTH derivative is not extracted in a conventional sequencer, it can be extracted on a solid phase sequencer. Even on a solid phase sequencer, however, positive identification requires altered chromatography conditions. Alternatively, if the phosphorylated sample is labeled with ^{32}P, the phosphorylated site may be identified by radioactivity determination of pieces of the glass fiber disks (Wang et al., 1988) on a conventional sequencer, or counting aliquots from the conversion flask on a solid phase sequencer. Phospho-amino acids can be detected by TLC analysis of protein hydrolysates from SDS gels (Neufeld et al., 1989) or PVDF membranes (Hildebrandt & Fried, 1989). Tyrosine may also be sulfated. Sulfo-Tyr is acid labile and is converted to Tyr during Edman chemistry. An indirect method of identification was devised by Henshen (1993). The sample is treated with tetranitromethane prior to sequence analysis, converting regular tyrosine residues to their nitro derivatives. Sulfo- and phospho-Tyr residues are protected, with sulfo-Tyr being identified as PTH-Tyr (due to its acid liability) and phospho-Tyr as a blank cycle (due to its acid stability). In addition, these peptides can be treated with arylsulfatase or alkaline phosphatase and resequenced before and after treatment with tetranitromethane.

Alternative Edman Reagents

Considerable effort has been expended to develop more sensitive Edman reagents. In general, most such reagents have drawbacks and have not come into popular use. PITC has optimal properties for both the coupling and cleavage steps. Fluorescent Edman reagents such as fluorescein isothiocyanate (FITC) couple slowly and lead to sequence overlaps (Muramoto et al., 1993). An interesting alternative is to preserve PITC chemistry through the generation of ATZ derivatives that can be reacted with amino-fluorescein in the conversion flask to produce fluorescent PTC-amide derivatives (Tsugita et al., 1989). This approach gives low yields for many amino acids that rapidly proceed from ATZ to PTH derivatives before they can react with amino-fluorescein. The yields for Asp are especially poor, and due to water quenching of the fluorescein derivatives during chromatography, the gain in sensitivity for most of the other amino acids is minimal. Recently, Farnsworth & Steinberg (1993b) have modified the conversion conditions to convert both ATZ and PTH amino acids to their PTC derivatives prior to their conversion to ATZ derivatives and reaction with amino-fluorescein. The procedure has been automated and claimed to give femtomole sensitivity for protein sequence analysis (Farnsworth & Steinberg, 1993a). The approach suffers from the lack of authentic standards, no yield for Asp, and fluorescent quenching as mentioned above. In spite of these problems, this is still an active area of investigation, and hybrid methods such as modified Edman reagents with mass spectrometric detection may ultimately gain femtomole sensitivity for Edman chemistry (Aebersold et al., 1992; Chait et al., 1993).

C-Terminal Sequence Analysis

It is highly desirable to obtain the C-terminal sequence of intact proteins. Together with N-terminal sequence information, this information verifies the start and stop points for a protein or gene, and may allow PCR cloning of a complete gene. While there are no C-terminal sequence methods in common use, several chemical methods are under development. The most successful strategy to date employs sequential removal of amino acids as their thiohydantoin derivatives (Inglis, 1991). Several of these methods are under commercial development and can give sequence information on 1 nmole of sample for 1 to 3 cycles. Enzymatic methods, such as the release of amino acids with carboxypeptidase, are generally unsuccessful because the C-terminus may not be readily available to the enzyme, and the kinetic analysis of the released amino acids is problematic (i.e., each amino acid is released at a different rate). Methods to identify the C-terminus include hydrazinolysis (Strydom, 1988) and tritium (tritiated water) incorporation after treatment with acetic anhydride to form the oxazolone (Matsuo et al., 1966). Both methods are insensitive and fraught with problems. Most researchers are forced to identify the C-terminus of their proteins by peptide mapping strategies (see the next section). This method is not quantitative and may either miss the correct C-terminal peptide or minor, but important C-terminal peptides.

Peptide Mapping

Sample Preparation

Three methods of sample preparation-digestion are discussed: in-gel digestion, blot digestion, and solution phase digestion. Since the primary goal of peptide mapping is to obtain a complete sequence, the first step should be to reduce and S-alkylate the protein, unless no cysteine residues are present. Reduction and S-alkylation must be carried out in the presence of strong denaturants such as urea, guanidine HCl, or SDS. In this way, all of the cysteine residues will be identified, and the protein will be completely denatured. While denaturation of a protein allows the digestion to proceed to completion, it may also cause the protein to precipitate, necessitating alternative strategies. If the reduced and S-alkylated protein is digested from a gel, then the latter problem is obviated. If the protein has no cysteine residue, the reduction-alkylation procedure may be skipped, but the protein may still be resistant to protease digestion due to its tightly folded native structure. The last goal of sequence determination may be to identify the positions of the disulfide bonds. In this case, the sample must be digested in its native state using more aggressive procedures, producing small peptides containing only a few residues around the disulfide bonds. In all cases, the peptide mapping should be performed by reversed phase HPLC, preferably using capillary columns for the highest possible sensitivity.

In-Gel Digests

A typical procedure has been described by Rosenfeld et al. (1992) with a slight modification that includes drying the gel slice in a Speed-Vac (Hellman et al., 1995). SDS gels are run in the normal way, stained with 0.2% Coomassie blue in 40:7.5:52.5 methanol-acetic acid-water for 20 min, destained in 30% methanol, washed twice with 150 µL of 0.2 M ammonium bicarbonate, pH 8.9-acetonitrile (1:1) for 20 min at 30°C, dried in a Speed-Vac for 20 min, and partially rehydrated with 3 to 5 µL 0.2 M ammonium bicarbonate, pH 8.9, containing 0.02% Tween-20. After addition of 0.5 µg of trypsin (3 to 5 µL), additional aliquots of buffer are added until the gel slices are completely rehydrated. The gel slices are then covered with buffer and digested for 4 hr at 30°C. Peptides are extracted twice with 100 µL of 0.1% TFA-60% acetonitrile at 30°C for 20 min. For hydrophobic peptides, 0.02% Tween-20 is added. The samples are concentrated on a Speed Vac and analyzed by reversed phase HPLC. This method appears to work well on low molecular weight samples in the 1 to 2 µg range (100 to 200 pmoles). The efficiency of the method for high molecular weight proteins (>50 kDa) is not known. It may be advisable to try the method on standards before proceeding with a larger protein; also it may be necessary to increase the sample size to 10 to 20 µg to maintain the yields. In order to perform mass analysis on peptides (see later discussion), the detergent must either be omitted from the protocol or removed prior to mass analysis.

On-Blot Digests

The original protocol for on-blot digestions was developed by Aebersold et al. (1987) for samples electrotransfered to nitrocellulose membranes. The nitrocellulose membranes are blocked with PVP-40 prior to protease treatment to avoid adsorption of the protease to the membrane. Recent modifications include electrotransfer to PVDF membranes and blocking with either PVP-40 (Aebersold et al., 1987) or reduced Triton X100 (Fernandez et al., 1992). The first use of PVDF as an electrotransfer membrane was by Matsudaira (1987). A general protocol taken from Henzel et al. (1993) is shown here.

1. Run SDS PAGE according to Laemmli (1970).
2. Electroblot to Immobilon PSQ membranes in a BioRad Trans-Blot transfer cell using 10 mM CAPS (3-(cyclohexylamino)-1-propanesulfonic acid), pH 11.0 to 20% methanol for 45 min at 250 mA, constant current.
3. Stain PVDF membrane with 0.1% Coomassie blue in 40% methanol, 0.1% acetic acid for 1 min; destain with acetic 10% acid-50% methanol for 2 to 3 min.
4. Wash with water, extract dye (wet strips with methanol, add 100 µL of water, 400 µL of methanol, and 400 µL of chloroform, vortex, discard liquid phase).
5. Reduce with 7 mM DTT in 100 µL of 0.5 M Tris HCl, pH 8.5, containing 10% acetonitrile and 5 mM EDTA for 1 hr at 45°C.
6. Alkylate with 10 µL of 0.2 M iodoacetic acid in 0.5 M NaOH for 20 min at 25°C in the dark.

7. Rinse with water, then incubate with 200 μL of 0.25% PVP-40 in 0.5 M acetic acid for 20 min; rinse with water and 20% acetonitrile.
8. Digest with 0.2 ug of Promega-modified trypsin or Lys-C (Wako) in 50 μL of 0.1 M ammonium bicarbonate, pH 8.0, containing 10% acetonitrile at 37°C for 24 hr.
9. Concentrate supernatant to 10 to 20 μL and analyze; strips may also be extracted with 20 μL of DMSO to remove hydrophobic peptides.

This protocol illustrates several key points: the sample must be reduced and alkylated to give good digestion, and the PVDF membrane must be blocked with an agent such as PVP-40 (or more recently, PVP-360). The investigator may use different SDS gel systems and apparatuses, different membranes, other stains such as amido black and Ponceau S (Fernandez et al., 1992), or calconcarboxylic acid (Hong et al., 1993), and reduce and alkylate with other alkylating agents such as iodoacetamide or vinyl pyridine, before or after applying the sample to the gel. Thus, there is a lot of room for variation, and there will be as many opinions on the correct combinations as there are investigators. The above procedure was chosen as an example because it has been used for both 1D and 2D gels.

There has been much experimentation with the choice of membranes. Nitrocellulose (NC) membranes are still a good choice, but have been largely displaced by PVDF since PVDF membranes can also be placed in a sequencer for N-terminal sequence analysis, whereas NC membranes are unstable in organic solvents. PVDF membranes can be purchased from several manufacturers and with different characteristics. Immobilon P and PSQ from Millipore are preferred by some investigators, while Aebersold and coworkers prefer Immobilon CD, which has a positive charge (Patterson et al., 1992). Staining is problematic with Immobilon CD, necessitating visualization by negative stains (3,3'-diethyloxacarbocyanine iodide), or by prestaining with dyes such as calconcarboxylic acid (Hong et al., 1993). The advantage of digesting from Immobilon CD is that the PVP-40 blocking step can be omitted. Zitex or Teflon tape may also be used (Burkhart et al., 1995).

Samples blotted to PVDF membranes may be treated with CNBr prior to protease digestion to improve digestion and subsequent elution from the membranes (Stone et al., 1992). This may be a good tactic if one wishes to avoid the reduction-alkylation step (e.g., for determination of disulfide linkages). The CNBr protocol of Stone et al. (1992) is given here.

1. Rinse stained PVDF strips twice with 500 μL of dry, ice-cold acetone.
2. Add 50 μL of 70% formic acid and 10 μL of 70 mg/mL CNBr in 70% formic acid; digest for 24 hr at room temperature.
3. Vacuum dry the supernatant and combine with 100 μL-rinses of 40% acetonitrile and 0.05% TFA in acetonitrile at 50°C; vacuum dry, and redry from 60 μL of water.

Stone et al. (1992) also reduce and alkylate their samples, and digest with protease. The advantage of the protocol is higher yields of peptides from the membrane. Possible alternatives to the procedure are to substitute 70% TFA for the formic acid (as the latter sometimes causes trouble with N-formylation) and to extract the digested membranes with DMSO.

Solution Phase Digests

Solution phase samples should be reduced and alkylated before digestion with proteases, but may be treated directly with CNBr if the reduction-alkylation step is undesirable or unsuccessful. If the sample is reduced and alkylated, it must be desalted before proceeding to protease digestion. A general protocol is given below, assuming a volume of 100 µL with the sample in an aqueous buffer in a 1.0 mL Pierce Reacti-Vial.

1. Adjust the pH of the solution to 8.5 with 1 M Tris, containing 0.1 M EDTA.
2. Add sufficient solid guanidine HCl (GuHCl) to the sample to bring the final concentration to 6 M.
3. Flush the vial with argon, add 1 µL of 100 mg/mL DTT and incubate for 1 hr at 30 to 50°C.
4. Add 1 µL of 4-vinylpyridine, and alkylate for 1 hr at 30 to 50°C.
5. Desalt either on an Aquapore butyl-300 column (Brownlee Labs; 2.1 × 30 mm) or on a PLRP-S column (Polymer Labs; 2.1 × 50 mm); inject sample and wash with solvent A (0.1% TFA) until baseline is reestablished. Elute with a linear gradient to 100% solvent B (0.1:9.9:90 TFA-water-acetonitrile) over 20 min; collect the sample in polypropylene tubes containing 10 µL of DMSO.
6. Concentrate samples to <50 µL in a vacuum centrifuge; add 50 µL of 0.1 M ammonium bicarbonate, pH 8.5, and 0.2 µg of Promega-modified trypsin or WAKO Endo Lys C, and digest for 24 hr at 37°C.

This protocol may need to be refined for a given protein. Refinements may include drying the sample prior to reduction-alkylation and using 100 µL of 6 M GuHCl-0.1 M Tris HCl, pH 8.5, using other reducing or alkylating agents, capillary columns for increased sensitivity, and other additives (besides DMSO) to prevent sample adsorption to the collection tube. Two controls should be run to test the procedure. One is a standard protein such as serum albumin (it has a number of disulfide bonds and elutes well from the desalting column). The other is a reagent control, with protease but no sample.

Proteases

The choice of protease depends on the sample and method. Some samples will not digest well with some proteases, and some proteases will not digest any protein well in the presence of organic solvents, denaturants, reducing agents, and detergents. Trypsin is perhaps the most commonly chosen protease because so much is known about its behavior. On the one hand, trypsin can tolerate small amounts of SDS (<0.1%), urea (<2 M), EDTA, and organic solvents (<20% methanol or acetonitrile). On the other hand, even low levels of GuHCl (1 to 2 M) or denaturants plus reducing agents can completely inactivate trypsin. Thus, before modifying a protocol and trying it on a sample, it is a good idea to try it first on a standard protein such as lysozyme or serum albumin. The main reasons for adding denaturants to the sample are to speed digestion and maintain sample solubility (although it should be noted that even insoluble samples may be digested with proteases). If poor digestion is noted with trypsin, it is worth adding 20% acetonitrile, 0.1% SDS, or 1 to 2 M urea, and trying again. If this is performed on the same sample, then additional trypsin should be added. In general, it is a good idea to maintain a trypsin-to-sample ratio of 1:50 to 1:100 (wt/wt), but it should not be overlooked that on a molar basis, it is easy to reach a point where there is more trypsin present than your sample. Thus, trypsin itself and its tryptic peptides (since it will autodigest) could contaminate the peptide map. *Running a protease control is essential.* It is also useful to have the sequence of trypsin (and other proteases) available to prevent one from accidentally assigning a protease sequence to a sample.

Table 4-4 summarizes the properties of a number of proteases useful for peptide mapping. The effect of organic solvents on most of these proteases has been studied in detail by Welinder (1988). In addition, most of these proteases can tolerate small amounts of SDS or reduced Triton X-100 (reduced Triton is preferred over its UV absorbing parent). For small proteins (<50 kDa), a good strategy is to start with trypsin or Endo Lys C. For larger proteins, trypsin may produce too many fragments, although this can be modified as shown below, and a more restricted protease such as Endo Lys C may be preferred. Producing a second map with a protease of different specificity such as Endo Glu C or Asp N protease is a good idea. Endo Glu C may be used at either pH 8.0 or 4.0 (at pH 4.0 in ammonium bicarbonate, it is reported to be specific for Glu bonds, but in phosphate at either pH, cleavage at Asp bonds is also observed). Endo Glu C activity may vary from batch to batch and can be used with 0.1% SDS if checked beforehand. Asp N protease can be used routinely with 0.1% SDS and is fairly specific for Asp bonds, with occasional cleavage at Glu bonds. Endo Arg C is a poorly active protease, but can occasionally produce good results. Another very specific protease that may be useful for peptide mapping is Asn-endopeptidase (Abe et al., 1993). Proteases such as chymotrypsin, elastase, thermolysin, and pepsin should be used with caution, since they have broad specificities and may produce peptide maps of high complexity. Nonetheless, these are important proteases and may be the key to obtaining a good map or a necessary overlap. A good strategy is to reduce the sample-protease ratio (<100:1) and time of digestion (to perhaps 1 to 4 hr, instead of 18 to 24 hr). Pepsin is of special interest since it is active at pH 2.0, and can be used with samples in 0.1% TFA. The best method to control pepsin digestion is to limit the time to 2 to 4 hr; otherwise the protein may be overdigested. Overdigestion results in low yields of overlapping small peptides, a situation that will frustrate the assembly of a complete sequence for a protein.

Table 4-4 Common Proteases Used in Protein Structural Analysis

Protease	Specific Activity	Source
Trypsin	110 U/mg	Bovine, pancreas
Chymotrypsin	90 U/mg	Bovine, pancreas
Endo Arg-C	220 U/mg	Murine, submaxillary
Endo Glu-C	20 U/mg	*S. aureus* V8
Endo Lys-C	30 U/mg	*Lysobacter enzymogenes*
Endo Asp-N		*Pseudomonas fragi*
Thermolysin	40 U/mg	*B. thermoproteolyticus*
Pepsin	2500 U/mg	Porcine gastric mucosa
Papain	30 U/mg	*Cariaca papaya*

Modifications of Trypsin

Since trypsin is so popular, investigators have worked out methods to increase its stability (Rice et al., 1977), and to limit its reactivity to either Lys only (Patthy & Smith, 1975) or Arg only (Tarr et al., 1983). If trypsin is reductively formylated, it becomes more resistant to autodigestion (Rice et al., 1977). Stabilized trypsin is commercially available from Promega. If highly purified trypsin is desired, it can be purchased from Boehringer, or purified by reversed phase HPLC (Titani et al., 1982). This is a good idea, since most commercial preparations of trypsin are contaminated with chymotrypsin and may lead to a number of unexpected (and perhaps unwanted) cleavages. Modification of proteins at Lys residues to limit trypsin cleavage to Arg bonds, can be performed by either treatment with acylating agents (Tarr et al., 1983) or isothiocyanates. For the latter approach, we prefer Braunitzer's reagent (4-sulfophenylisothiocyanate) since it is water soluble and produces peptides which have increased water solubility. We have used this reagent successfully to eliminate tryptic cores from membrane proteins (Botelho et al., 1982). The only problem with this approach is that modified Lys residues will not extract from polybrene in the sequencer, and thus will give blank cycles. A good acylating agent is succinic anhydride. The sample (1 to 10 μg) is treated at pH 8-9 (e.g., pH 8.0 sodium phosphate; avoid buffers with amino groups!) with 1 to 10 mg of succinic anhydride for 1 to 2 hr at 25°C, and acidified to pH 2 with TFA or acetic acid for 1 to 2 hr; the pH is then readjusted to 8.0 (e.g., with trisodium phosphate), and trypsin is added. Arginine residues may be reversibly modified with 1,2-cyclohexanedione (Patthy & Smith, 1975) or glyoxal (Glass & Pelzig, 1978). The 1,2-cyclohexanedione procedure is performed at pH 8-9 in borate buffer for 1 to 2 hr at 25 to 40°C. The sample can be either treated directly with trypsin or desalted first. The protecting group can be removed by treatment with hydroxylamine at pH 7 for 7 to 8 hr (Patthy & Smith, 1975).

Urea

The addition of urea to enhance the digestion of a protein is a good idea, but not without problems. Even ultrapure urea will contain small amounts of cyanate which can carbamylate the amino groups of peptides. This will result in both N-blocked and ε-amino-Lys substituted peptides. To prevent this possibility, the urea should be freshly deionized over a mixed bed ion exchanger. It is also essential to use an amino-containing digestion buffer such as Tris or ammonium bicarbonate and maintain the pH above 7.5. GuHCl cannot be easily substituted for urea, since it is more likely to inactivate proteases.

Reduced Triton X-100

Triton X-100 is a heterogeneous UV-absorbing detergent that will obscure large portions of a peptide map. This problem can be obviated by using reduced Triton X-100 (Fernandez et al., 1992), but certain cautions should be noted. If the peptides are going to be mass analyzed, the reduced Triton X-100 will contribute many peaks in the mass range of 500 to 1200. These peaks may obscure or suppress the peptide peaks of interest, especially if electrospray ionization is used.

Chemical Methods

Cleavage at methionine by CNBr was first described by Gross & Witkop (1962). Under strongly acidic conditions, cyanogen bromide cleaves on the C-terminal side of methio-

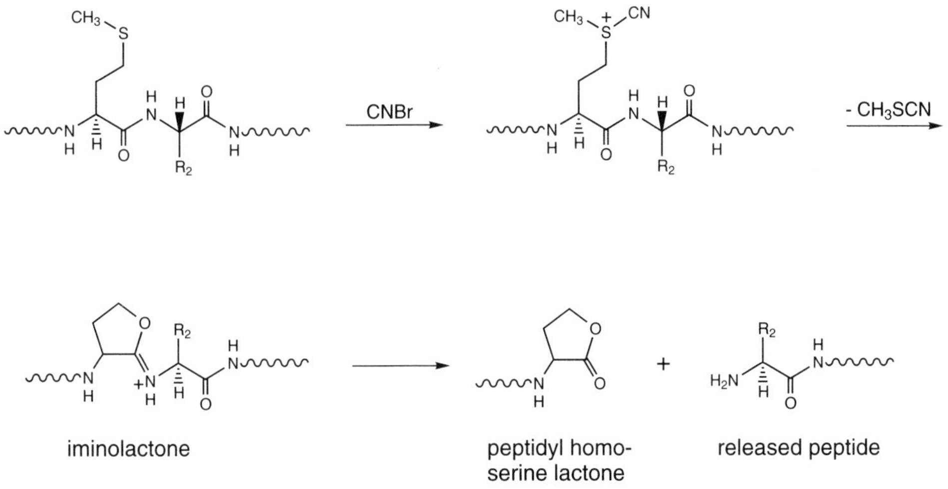

Figure 4-5. Mechanism of cyanogen bromide cleavage at methionine residues.

nine residues (Fig. 4-5). In the presence of acid, the C-terminal methionine is simultaneously converted to homoserine lactone; this can be hydrolyzed to homoserine under basic conditions. Since methionine sulfones (oxidized methionine) will not be cleaved by CNBr, it may be necessary to reduce the sample beforehand if oxidation of methionines is suspected. Although the original method calls for 70% formic acid, this acid should be avoided since it may formylate Ser and Thr residues (Goodlett et al., 1990). CNBr cleavage of PVDF-blotted samples has been previously described. A more general method for solution phase samples is to dry the sample in a Speed-Vac, add 100 μL of 70% TFA containing 2 mg of CNBr, digest for 2 to 12 hr at 25°C, dry, redissolve the sample in 10 to 20 μL of 70% TFA, and inject onto a reversed phase HPLC column. If the CNBr peptides are poorly resolved, the sample may be further digested with trypsin or other proteases after adding sufficient 1 M ammonium bicarbonate to adjust the pH to 8.0. Alternatively, the CNBr digested sample can be redissolved in DMSO or hexafluoroacetone and then mixed with digestion buffer. The dissolution step should be performed at 40 to 50°C for several hours to ensure that all of the peptides are in solution.

Cleavage at Tryptophan

Oxidative cleavage at the C-terminal side of tryptophan can be achieved with reagents such as N-chlorosuccinimide (Schechter et al., 1976), BNPS-skatole (Omenn et al., 1970), or iodosobenzoic acid (Mahoney & Hermodson, 1979). Like CNBr cleavage, the peptides formed are very large and may be difficult to separate by HPLC. In addition, the oxidizing agents may oxidize methionine and cysteine residues and destroy some tyrosine residues. Nonetheless, this is a potentially useful method. One example is given: the sample is dissolved in 10 to 50 μL of 50% acetic acid, and treated with 5 μL of 0.01 M N-chlorosuccinimide in DMF for 2 hr at 25°C; the reaction is terminated with 4 μL of 0.08 M N-acetylmethionine. The digest may be chromatographed directly, or dried and treated with protease if desired. Crimmins et al. (1990) give an example of the use of BNPS-skatole on proteins blotted onto PVDF membranes.

Cleavage at Cysteine

Cysteine residues may be either be methylated and treated with CNBr under acidic conditions (Yamauchi et al., 1977) or S-cyanylated with 2-nitro-5-thiocyanobenzoic acid (NTCB) and cleaved under alkaline conditions (Degani et al., 1970) to generate cleavages on the N-terminal side of cysteine (Fig. 4-6). The CNBr method requires acid hydrolysis to liberate the peptidyl dehydroalanine residue, and the S-cyanylation method produces N-terminally blocked peptides. Because of these problems, these methods have not been popular. However, if the blocked peptides are further digested with protease or analyzed by MS-MS techniques, the presence of an N-blocking group is not an obstacle. Conditions for specific methylation of cysteine residues include treatment with trimethylphosphate (Yamauchi et al., 1977) or methyltosylate (Heinrikson, 1970). Following methylation, CNBr cleavage is carried out in the usual way. Cysteine residues may be selectively methylated and cleaved in the presence of cystine residues. If cleavage is desired at cystine residues, they must be reduced prior to methylation, a procedure which may require strong denaturants such as 6 M GuHCl.

The S-cyanylation-cleavage procedure is described here. The sample (e.g., 1 to 10 μg in 100 μL of 0.05 M sodium phosphate, pH 8.0) is placed in a sealed 1.0 mL Reacti-Vial, purged with argon, reduced with 1 μL of 10 mg/mL DTT, treated with 5 μL of 50 mg/mL NTCB, and incubated for 2 to 12 hr at 37 to 50°C. The cleavage reaction can be monitored by SDS gel electrophoresis or analytical runs on HPLC. The major competing reaction, β-elimination of thiocyanate, can be reduced by maintaining the pH at 8.0.

Another method for cleavage at cysteine residues is to alkylate them with 1-bromo-3-aminopropane followed by cleavage with Lys C endoprotease (Jue & Hale, 1993). This is an attractive alternative to the older procedure, which utilized ethyleneimine, a dangerous carcinogen. The strategy is to convert cysteine residues into lysine lookalikes. The specificity depends on reagents which will only alkylate cysteine residues and the elimination of the ϵ-amino groups of lysine. A major advantage of this method is that the formation of N-blocked peptides is avoided. In order to be successful, the sample must be freshly reduced (see earlier discussion), and if cystine residues are to be modified, it may be necessary to perform the reduction in 6 M GuHCl. Since Endo Lys C will also cleave at Lys residues, the sample must be first treated with an acylating agent such as succinic anhydride (see earlier section, "Modifications of Trypsin.").

Cleavage at Aspartic Acid

Asp-Pro bonds are very susceptible to mild acid hydrolysis (Jauregui-Adell & Marti, 1975). Conditions may range from 0.1 M TFA for 1 to 2 hr at 50°C to 70% TFA at 25°C for 24 to 72 hr. Since Asp—Pro bonds are rare (most proteins have 1 or 2 at most), the procedure may produce only a few, potentially very large peptides. A more thorough digest can be obtained by acid cleavage at Asp bonds (Inglis, 1991). The major site of cleavage for mild acid hydrolysis is on the N-terminal side of Asp residues. Starting conditions are 0.1 M TFA at 100°C for 1 to 2 hr. This method must be monitored by analytical HPLC. If overdone, Asn and Gln residues will be deamidated, and some cleavage may occur at Glu, Gln, and Asn bonds. We have found that this cleavage method is protein dependent, with some proteins yielding excellent maps, and others giving poor results (Shively, 1986). It should also be mentioned that Asn—Gly

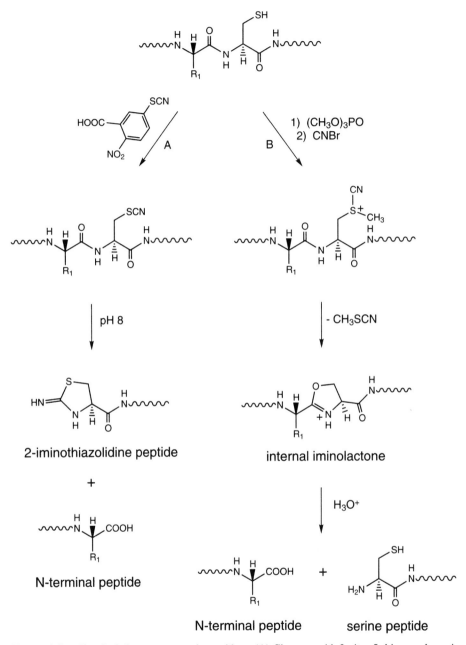

Figure 4-6. Chemical cleavage at cysteine residues. (A) Cleavage with 2-nitro-5-thiocyanobenzoic acid. (B) Cleavage of S-methylated cysteine with CNBr and acid.

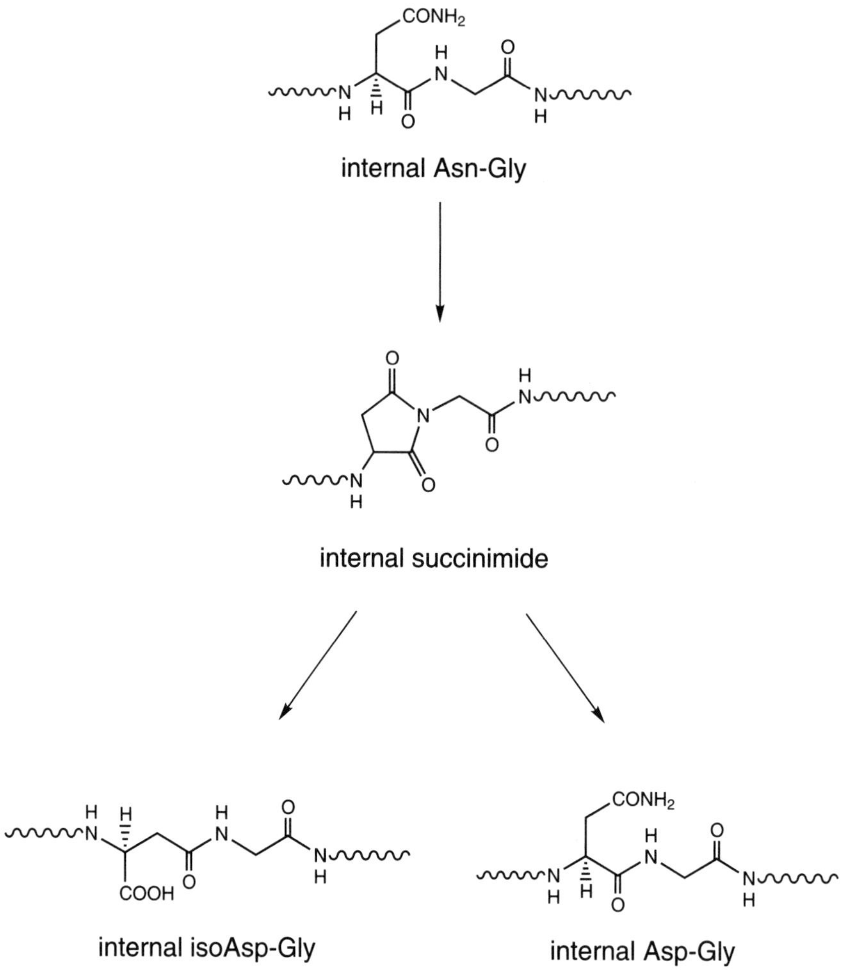

Figure 4-7. Mechanism of rearrangements at internal Asn-Gly bonds. All four possible products may be observed: Asn-Gly, Succ-Gly, Iso-Asp-Gly, and Asp-Gly.

bonds are prone to rearrangement, forming internal succinimides that may be stable or break down to a mixture of Asp and iso—Asp bonds (Fig. 4-7) (McFadden & Clarke, 1986). This reaction occurs to a certain extent in all proteins with these bonds and must be dealt with as a posttranslational modification. Iso—Asp bonds are resistant to Edman chemistry and, unless reconverted by methylation to Asp bonds, must be analyzed by MS methods. Peptides with Asp and iso—Asp bonds may either coelute or separate during peptide mapping.

HPLC Analysis

Peptide mapping is best performed by reversed phase HPLC on narrow bore (2.1 mm i.d.), microbore (1 mm i.d.), or capillary columns (<1 mm i.d.) packed with C-4, C-

8, or C-18 packing materials. The highest sensitivity is observed for packed capillary columns using methods described by Simpson et al. (1989a), Henzel et al. (1993), or Davis & Lee (1992). Capillary columns (0.18 or 0.32 mm inner diameter) can be purchased from LC Packings, Inc. or packed in the laboratory (Davis & Lee, 1992). Capillary columns may be run on conventional HPLCs such as the ABI 140A microgradient pump or the Michrom system using a flow splitter (e.g., the system can be run at 50 μL/min connected to a Valco tee with an output of 2 to 5 μL/min). Commercially available micro-flow cells (LC Packings, Shimadzu, and others) can be used, or conventional detectors (e.g., Kratos 759, Shimadzu SPD10A) at 190 to 200 nm can be modified to accept the capillary tubing as a flow cell (Davis & Lee, 1992). Capillary tubing should be used from the flow cell to minimize postdetector peak broadening and time delay during peak collection. Peaks should be hand-collected into 200 μL polypropylene tubes. The delay time from the detector to the end of the collection tubing can be measured by the formula D (sec) = 60 $\pi r^2 L/F$, where r is the internal radius of the capillary tubing in mm, L is the tubing length (mm), and F is the flow rate in μL/min. Gradient elution from solvent A (0.1% TFA) to solvent B (0.1:9.9:90 TFA-water-acetonitrile) over a period of 20 to 40 min should give a good peptide map. The percent TFA can be reduced to 0.05% to lower the UV absorbance of the solvent and improve electrospray performance if LC-MS is utilized. Water and TFA should be included in solvent B to maintain ion-pairing. If 100% acetonitrile is used as solvent B, peak broadening may occur midway through the gradient. It is usually a good idea to wash the column with solvent A after sample injection and wait for a return to baseline before starting a gradient. This ensures that peptides will not be lost in the "breakthrough" peak.

Peptide maps may not give adequate resolution of all peptides. Analysis of each collected peak by capillary electrophoresis (CE), or better, by mass analysis, may reveal the presence of multiple peptides in a single peak. In this case, it may be necessary to rechromatograph a collected peak. Rechromatography should be performed on a column with a different matrix (e.g., phenyl, polystyrene, ion exchange, hydrophilic, etc.) or on the same column but with a different solvent (e.g., 10 mM sodium phosphate or 10 mM ammonium acetate at pH 4-7). The sample should be diluted 1:1 or 2 before injection to assure binding to the column. Very small (1 to 3 amino acids) or very hydrophilic (e.g., phosphopeptides) peptides may not bind to reversed phased packings and may elute in the void volume. It may be useful to collect the void volume and perform rechromatography on this peak using hydrophilic or ion exchange chromatography. Hydrophilic packings such as polyhydroxyethylaspartamide (Poly LC, Inc.) are especially useful for the separation of glycopeptides (Swiderek et al., 1993). Collected peptides may be stored at 4°C for several weeks before sequence analysis, or spotted directly onto glass fiber filters or PVDF membranes. Alternatively, the peaks may be collected directly onto PVDF membranes as they elute from the HPLC (Murata et al., 1993). This method has the advantage that the peptides cannot be lost on the walls of the collection tube, but if rechromatography, CE, or mass analysis is desired, it will be necessary to elute the peptide from the PVDF membrane. The peptides bound to PVDF may be eluted with DMSO or TFA-acetonitrile.

A good review of peptide mapping and mass analysis of the resulting peptides is given in several chapters in volume 193 in *Methods of Enzymology* (Lee & Shively, 1990).

Mass Spectrometric Methods

Ionization methods suitable for peptides include secondary ionization mass spectrometry (SIMS), electrospray ionization (ESI) and matrix assisted laser desorption ionization (MALDI). These three ionization techniques are illustrated in Fig. 4-8. ESI and MALDI are also suitable for proteins. Each of the techniques has its advantages; they are described in reviews such as volume 198 of *Methods in Enzymology*. Typical protocols are described below. It should be noted that the mass spectra obtained by these techniques will differ depending on the resolution of the instrument, and the presence of multiply charged species. The mass for a given peptide may be calculated in two ways. The so-called exact or monoisotopic mass is the mass calculated assuming that all the carbon is C-12, all the nitrogen is N-14, all of the hydrogen is H-1. In reality, 1.1% of the carbon is C-13, 0.37% of the nitrogen is N-15, and 0.015% of the hydrogen is H-2 (negligible for most mass spectrometers). Thus, one may also calculate the average mass for a peptide. A low resolution instrument, such as most time of flight or quadrupole instruments, will not resolve the peaks corresponding to the isotopic abundances, and thus, average mass calculations are appropriate. For high-resolution instruments, such as magnetic sector or FT-MS, it is usually possible to resolve the monoisotopic peak, and with tandem instruments, select it for further analysis.

Mass Measurements of Peptides

Fractionated peptides may be analyzed by all three techniques. A sample aliquot (0.5 to 1.0 µL) containing 1 to 10 pmoles is a good starting amount, but with some skill it

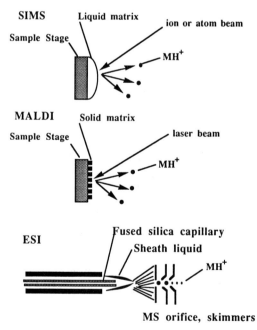

Figure 4-8. Common ionization techniques for peptides in mass spectrometry: SIMS, MALDI, and ESI.

may be possible to analyze as little as 0.1 picomole. SIMS or FAB-MS is the oldest of the three methods and was first described by Barber et al. (1981) for the analysis of peptides. SIMS is a general term referring to secondary ionization processes, including the use of ion or atom beams for the ionization of peptides and other molecules from a liquid matrix. FAB refers to the exclusive use of neutral atom beams such as xenon, while SIMS includes FAB and ion beams such as cesium ions. In the case of SIMS, the sample is mixed with an equal amount of matrix such as glycerol, thioglycerol, or a mixture of DTT and DTE on the sample stage. Results may be obtained over a mass range of 500 to 5000, depending on the instrument. Peptides less than 500 m/z may be lost in the matrix background, depending on sample amount and polarity. The peptide must bear a charge to be analyzed and detected; this is usually picked up from the matrix medium, which generally includes a small amount of acid. Most peptides are detected as their protonated molecular ions (MH^+), but, depending on sample amount, peptide fragments may also be detected. A typical SIMS spectrum for a peptide is shown in Fig. 4-9A. The most intense peak late in the spectrum is the protonated molecular ion, $MH^+ = 1818.1$, but matrix peaks may dominate the spectrum at the low mass end ($m/z < 400$). A small peak corresponding to the doubly charged species is seen, $MH_2^{2+} = 909.2$, a peak which occurs at one-half the mass of the MH^+ peak. This latter peak has a characteristic mass spacing that distinguishes it from singly charged species. Since mass spectrometers separate ions according to m/z; this is the expected result. SIMS produces relatively large amounts of singly charged peptides compared to doubly charged (although there are exceptions) and almost negligible amounts of triply charged species.

Peptide signals may be suppressed by the presence of salts, detergents, or other peptides. In order to give a good signal, the peptide must be soluble and able to migrate to the surface of the liquid matrix where the ionization process takes place. Small, polar peptides tend to give poor signals, perhaps due to the surface phenomenon, and must be derivatized (e.g., by esterification with 1-hexanol) to overcome this problem (Falick & Maltby, 1989). Since salts ionize more efficiently than peptides, and detergents compete for the surface better than peptides, it can be seen why these substances should be avoided. Salts also complex to the peptide and give peaks such as MNa^+, MK^+, and so forth. The number of salt ions per peptide depends roughly on the number of carboxyl and other anionic groups present and the exposure of the sample to these salts. Since each salt peak arises at the expense of the MH^+ peak, signal intensity is diminished, often to the point of being lost in the background. In most cases, salts are removed during peptide mapping on reversed phase HPLC where TFA is used as the ion pairing agent. It should be remembered that exposure of a sample to a glass tube is all that is necessary to add sodium to that sample. For this reason, all peptides from a map should be collected and stored in polethylene or polypropylene tubes. The collection tubes should be checked for the presence of other contaminating peaks such as polymerization agents. Proper tubes should add no background peaks to the mass spectrum, nor contribute to the phenomenon of sample suppression.

Mixtures of peptides are also problematic, since one or two may dominate the spectrum to the exclusion of the others, even though the most intense peak may be a minor species in the mixture. This is a reminder that this technique is qualitative, rather than quantitative. Signal intensity correlates with ion yield, a process that depends on the ability to migrate to and be ejected from the matrix surface, and to ionize (neglecting limits imposed by the mass spectrometer). As mass increases, signal intensity

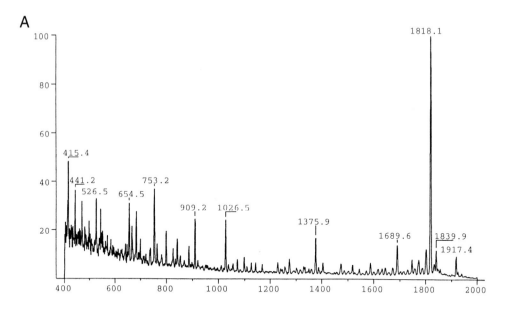
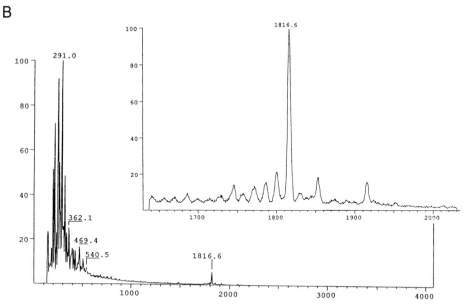

Figure 4-9. Mass analysis of a peptide by SIMS. The sequence is ALFHGRVSWAMFPNGK. The sample (30 pmoles) was analyzed in a 1 μL mixture of DTE and DTT on a Finnigan TSQ700 mass spectrometer. (A) Positive ion mode, mass scanned from 400 to 2000. (B) Negative ion mode, mass scanned from 120 to 4000; inset from 1600 to 2100.

will decrease, since it is harder to eject the sample from the liquid matrix without causing it to fragment. Thus, the useful mass range is less than 4000 m/z. Most peptides are run in the positive ion mode, since this mode is usually more sensitive than the negative ion mode. However, it is often instructive to run samples in both modes.

It is an obvious check on the molecular weight, and may reveal important information about the presence of charged groups or fragment ions. Peptides bearing strongly anionic groups such as sulfate or phosphate will show up strongly in the negative ion mode; in the case of sulfate, the protonated molecular ion peak may be very weak in the positive ion mode. An example of a peptide in the negative ion mode is shown in Figure 4-9B. It should be noted that sulfate and phosphate have the same average mass as protonated molecular ions and thus cannot be distinguished by mass measurements alone.

MALDI

The use of MALDI for peptides and proteins was developed mainly by Karas & Hillenkamp (1988). The sample (0.1 to 10 pmoles) is mixed with a matrix such as nicotinic acid (Karas & Hillenkamp, 1988), sinapinic acid (3,5-dimethoxy-4-hydroxycinnamic acid) (Beavis & Chait, 1989), α-cyano-4-hydroxycinnamic acid (Beavis et al., 1992), or 2,5-dihydroxybenzoic acid (Billeci & Stults, 1993), and allowed to crystallize on a sample stage. The sample is ionized by laser desorption, usually with a nitrogen laser (337 nm). The UV absorbing matrix vaporizes and ionizes, carrying the protein sample along with it. If laser power is properly controlled, the major signal observed is for the protonated molecular ion (or MH$^-$ in the negative ion mode). MALDI is ideally suited for time of flight (TOF) mass spectrometers, since the laser firings can be used as timed events (i.e., can start the clock for mass measurements). TOF mass spectrometers have very high sensitivity due to their high ion transmission rates, and it is not unusual to detect fmole amounts of peptide (or protein) in MALDI-TOF instruments. On the one hand, the technique is simple and the instruments are easy to operate, but on the other hand, the parameters that govern good ion yields are at the early stages of understanding. Good results depend on the use of the right matrix, the right laser power, and careful attention to mass calibration. A MALDI spectrum for a peptide mixture is shown in Figure 4-10. This is a practical example in which an immunoglobulin was reduced and alkylated, run on an SDS gel, transferred to nitrocellulose and stained, the heavy chain digested with trypsin, and peptide mapped on an 0.5 mm i.d. reversed phase HPLC column. The peptide peak T24 was mixed with two internal standards and analyzed by MALDI-TOF. The internal standards were used to allow a more accurate calibration than can be obtained with external standards (Henzel et al., 1993). The peptide shows an intense peak for its protonated molecular ion, and corresponds exactly to the predicted peptide shown in the figure legend. Several other peptides were also detected in this spectrum, demonstrating the efficiency of the method for peptide mixtures. The mass accuracy for the sample in this experiment was ± 1 Da (0.05%).

A general protocol for running samples by MALDI is to first calibrate the instrument with several known peptides that have been run using several dilutions, laser powers, and matrices. From these data, the best parameters are chosen for a general unknown. In order to obtain the most accurate mass measurement for an unknown, the sample should be run with and without an internal calibrant. In this way, most peptides can be correctly analyzed within 1 to 2 mass units. MALDI will give good results for peptide mixtures and is relatively insensitive to salts. Nonetheless, peak suppression and salt adducts have been observed. Like SIMS, the method is not quantitative, and the most intense peak may not be the most abundant peptide in a mixture. A typical procedure is given here.

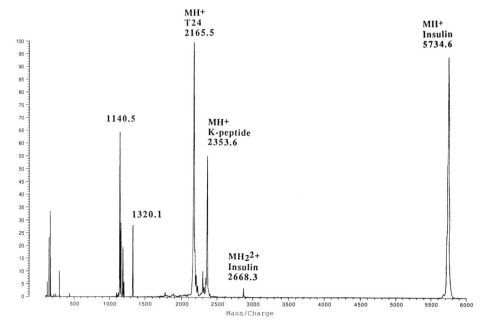

Figure 4-10. Mass analysis of a peptide mixture by MALDI-TOF MS. A tryptic peptide (about 1 pmole) from the heavy chain of an antibody isolated from a blot digest was mixed with two standard peptides (0.1 pmole of bovine insulin and K-peptide; 0.5 μL was spotted onto a sample stage with 0.5 μL of α-cyano-4-hydroxycinnamic acid and analyzed on a Kratos Kompact MALDI III mass spectrometer. The mass spectrum was calibrated with the two internal standards (MH^+ = 5134.6 and 2356.6). The mass of the tryptic peptide was 2165.5 corresponding to the sequence LGED-NINVVEGNEQFISASK. Several other tryptic peptides (1140.5 and 1320.1) are also seen.

1. Apply 0.2 to 0.5 μL of sample (1 to 2 pmoles) to the sample stage.
2. Add 0.2 to 0.5 μL of matrix (saturated α-cyano-4-hydroxycinnamic acid in 0.1:70:30 TFA-water-acetonitrile) containing 1 pmole/μL of internal standard (e.g., bovine insulin for a high mass, 5734, and gramacidin S for low mass, 1142), and then dry.
3. Gradually raise the laser power (and select a "good" spot) until the peak intensities of the internal standards are seen; if a new peak corresponding to the unknown is not seen, continue raising the laser power.

ESI

ESI was first described for biomolecules by Fenn and coworkers (Whitehouse et al., 1985). The sample is infused in a liquid medium (e.g., 1 to 20 μL/min of water-acetonitrile containing 0.05 to 0.1% acetic acid or TFA; the purpose of the acid is to encourage protonation of the sample) through a glass capillary surrounded by a stainless steel capillary set at 4 to 6 kV. The resulting liquid spray (at atmospheric pressure) is directed into an orifice in the mass spectrometer source. Since the electrospray may be unstable with respect to flow and solvent conditions, researchers have added a pneumatic nebulizer ion spray (Bruins et al., 1987) or sheath flow (Smith et al., 1988) to assist in the process. A popular sheath flow solvent is 2-methoxyethanol. The highly charged droplets are subjected to a counterflow of nitrogen gas as they enter the orifice and, in addition, may be driven through a heated capillary to help drive off solvent. As the charge to drop size ratio reaches a critical value, the drop explodes; this further reduces its size, until naked ions are exposed, carrying a variety of charges. In the case of positive ion ESI, a peptide may be detected as a mixture of singly, doubly, triply, and so on, charged ions, depending on its ability to carry positive charges. An example of a positive ion ESI mass spectrum of a peptide is shown in Figure 4-11A. In this example both the MH^+ and MH_2^{2+} ions are observed. The relative abundance of each charge state is a function of source and instrument parameters and the ability of the structure to hold charge. The maximum charge state usually corresponds to the total number of basic groups in a peptide (it is assumed that all carboxyl groups are fully protonated in the acidic medium). Since the higher charge states focus according to m/z, they are seen at lower masses on the mass scale. This has an obvious advantage for larger peptides and proteins, which may otherwise be out of the mass range for a given analyzer. Since all of the peaks for a given sample are related by one charge unit between peaks, the mass spectrum may be simplified by "deconvolution," that is, combining all of the charge states for a single molecule into a single peak. Most data systems perform this task automatically, and it is common now for most researchers to show only the deconvoluted mass spectrum. The sensitivity of ESI mass spectrometers are often measured by the timed infusion of a sample; for example, if a sample of concentration 6 pmole/μL were infused for 1 sec at a flow rate of 10 μL/min, then 1 pmole of sample would be consumed during the analysis. Negative ion spectra may also be recorded, but require special conditions. The sample is either infused in the presence of a volatile basic counter ion such as aqueous ammonia, or if analyzed by on-line LC, a make-up solution of aqueous ammonia is added to the sample via a coaxial sheath flow. In addition, a glass flow of air or oxygen is required to suppress coronal discharge. An example of a negative ion ESI mass spectrum is shown in Fig. 4-11B. Peaks corrsponding to the MH^- and MH_2^{2-} species are observed together with sodium and potassium adducts. Negative ion spectra may be especially useful for glycopeptides and sulfated or phosphorylated peptides (see the following).

Samples containing salts and detergents have major problems in direct infusion ESI, since the salts will carry the charge in preference to the peptide. Thus, the sample must be desalted, preferably by reverse phase HPLC, prior to ESI analysis. The simplest solution to this problem is to introduce the sample to the ESI source directly from an HPLC, hence LC-MS. Since the sensitivity of ESI is directly related to the concentration of the sample (pmoles/μL) and samples elute in a more concentrated form from capillary LC, capillary LC is an ideal interface to ESI mass spectrometers. While some ESI sources can accept high flow rates (mL/min), they have no advantage for sensitivity. Thus, the trend is towards interfacing capillary LCs with ESI sources. The main advantage of LC-MS is that an entire peptide map can be mass analyzed in as little as 20 minutes. Coeluting peptides are not a problem, as long as they do not have identical masses, which is a rather rare event. If LC-MS is used to complement Edman sequencing, a potential problem exists in relating the two peptide maps produced, one for the mass spectrometer, one for the Edman laboratory. In this case, it is important that the columns and HPLC systems be matched, and that UV detection be added to the LC-MS run. Tandem UV and ESI detection is usually not a problem if dead volume between the two detectors is kept to a minimum. A typical system has been described by Davis & Lee (1992).

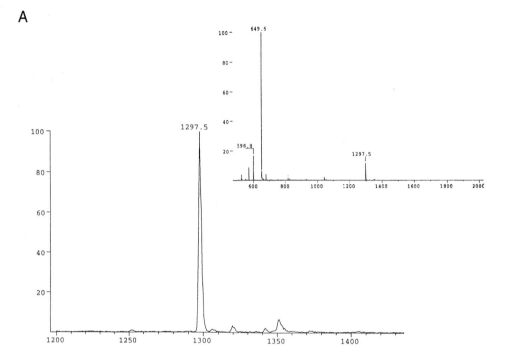
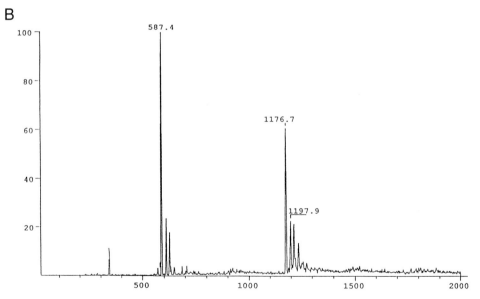

Figure 4-11. Mass analysis of peptides by ESI-MS. (A) Positive ion analysis of the peptide TG-PNLHGLFGRK. The sample (10 pmoles) was part of a cytochrome C endo Lys C digest analyzed by an on-line LC on a Finnigan TSQ700 triple quadrupole mass spectrometer. The inset shows the mass scan from 500 to 2000. (B) Negative ion analysis of the peptide DSDPR (10 pmoles in 1% aqueous ammonia). The mass scan was from 50 to 2000. $MH^- = 1176.7$, $MH_2^{2-} = 587.4$. The 1197.9 peak corresponds to the MNa^- ion.

ESI sources are usually found on quadrupole mass spectrometers, but are also becoming popular on magnetic sector and other instruments. The main advantage of the quadrupole instruments is that they are easy to couple to an ESI source, and that tandem quadrupole instruments are more compact and less expensive than their magnetic sector counterparts. ESI sources on magnetic sector instruments tend to give charge state distributions skewed towards the low charge end, but have the advantage of giving higher resolution for a given charge state than a quadrupole instrument.

Sequence Analysis of Peptides by MS-MS

Peptide bonds may be broken in the mass spectrometer by either high or low energy collision processes (collision induced dissociation, or CID). In magnetic sector instruments, the ions are accelerated at high potentials (10 to 20 kV) and thus fragment extensively when exposed to a collision gas. Fragments occur on both sides of the peptide bond and often at bonds in the amino acid side chains. In quadrupole instruments, ions are accelerated at a potential of 2 to 6 kV and fragment mainly at peptide bonds. In order to get complete sequence information, CID should produce fragments on both sides of the peptide bond for every amino acid in the peptide. While this process is more complete for high-energy CID, one can also obtain good sequence information from low-energy CID. A partial nomenclature for fragment ions is shown in Figure 4-12. The B ions arise from peptide bond cleavage and transfer of the positive charge to the N-terminal side of an amino acid (the acyl group carries the positive charge), resulting in a fragment ion series starting from the C-terminus, while Y ions arise from peptide bond cleavage and transfer of the positive charge to the N-terminal side of an amino acid (the amino group carries the positive charge), resulting in a fragment ion series starting from the C-terminus. In theory, a complete set of B or Y ions contains all of the necessary sequence information, but in practice, complete sets of either ion series are rarely observed. Details of the nomenclatures are described by Biemann (1990) and Roepstorff & Fohlman (1984). A typical MS-MS spectrum for a triple quadrupole instrument is shown in Figure 4-13.

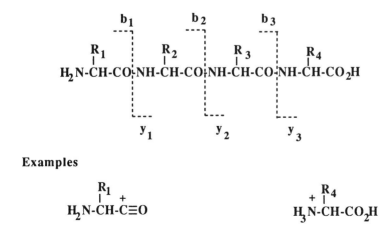

Figure 4-12. Nomenclature for ion fragment series. Examples are shown for the b_1 and y_3 positive ions.

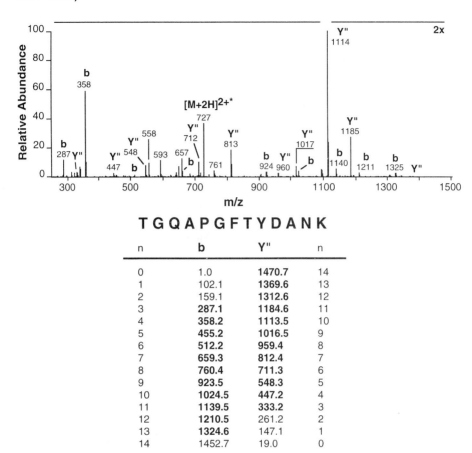

Figure 4-13. MS-MS analysis of a peptide. The sequence is TGQAPGFTYDANK, an Endo Lys C peptide from cytochrome C. The sample (10 pmoles) was analyzed by LC-MS-MS on a Finnigan TSQ700 triple quadrupole mass spectrometer. The MH_2^{2+} peak (727) was selected for MS-MS. The predicted positive fragment ions are shown below with those identified shown in bold.

In this example, mostly B and Y ions are produced, covering the majority of the fragments expected from the peptide. However, it should be noted that there are gaps in each of the ion series, and it is not a given that for each peptide analyzed there will be sufficient information for the complete structure of a peptide to be established.

Several ambiguities occur in MS-MS analysis of peptides. The first is the fact that two pairs of amino acids have the same mass, namely, Leu and Ile and Lys and Gln. In order to distinguish Leu and Ile, one must rely on fragments within the side chain, fragments which are only seen in high-energy CID. Lys and Gln can be distinguished by analyzing the sample before and after acetylation, resulting in a mass shift for Lys residues and the N-terminal amino acid of 42 mass units (CH_3CO group). This step is easily performed on the probe tip for FAB-MS, or can be done in solution for peptides analyzed by ESI. A typical protocol is to add 1 μL of pyridine and 1 μL of acetic anhydride to 1 μL of the sample, wait 1 to 10 minutes and reanalyze. Alternatively, the sample can be dried, dissolved in 10 μL of acetic acid plus 1 μL of acetic anhydride, heated at 50°C for 10 min, dried, and redissolved in 10 μL of 20% acetonitrile. If one is analyzing a tryptic map, it may be concluded that Lys (if present) is at the C-terminus. However, this is not always true. Trypsin will not cleave a Lys-Pro bond, and will occasionally cleave at anomalous positions. The majority of anomalous bond cleavages are due to contaminating chymotryptic activity in trypsin, a problem which is easily overcome by purifying the trypsin by HPLC (Titani et al., 1982). A final problem that is sometimes observed for sequence analysis on the quadrupole instruments is distinguishing N-terminus fragments from C-terminus fragments. If the protein is acetylated, the N-terminal amino acid fragments will increase in mass by 42. However, if the C-terminus is Lys, the C-terminal fragments will also increase by 42. Another approach is to treat the sample with methanolic HCl, a procedure that converts peptide carboxyl groups to their methyl esters (an increase in 14 mass units per carboxyl group). This procedure not only allows one to "count" carboxyl groups, but may also help to locate the C-terminal series (Hunt et al., 1986). The sample is dried, redissolved in 100 μL of 1 M methanolic HCl, tightly capped in a React-Vial, and heated at 40°C for 30 min, then dried and redissolved in 10 μL of 20% acetonitrile (Falick & Maltby, 1989).

Mass Measurements of Proteins

Mass measurements of proteins may be performed by either ESI or MALDI. ESI measurements usually will give a mass accuracy of 0.01 to 0.05%. ESI analysis for cytochrome C is shown in Figure 4-14. The mass error is ±0.016%. Mass accuracy of

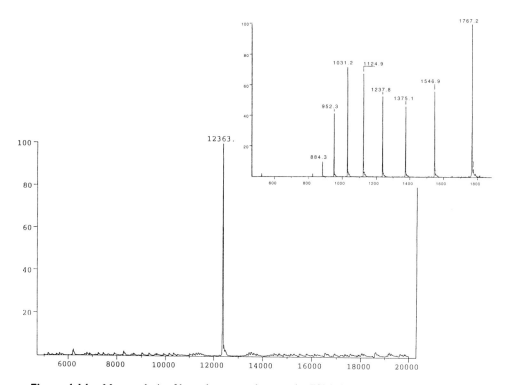

Figure 4-14. Mass analysis of horse heart cytochrome c by ESI-MS. The sample (10 pmoles) was analyzed by LC-MS on a Finnigan TSQ700 triple quadrupole mass spectrometer. The expected mass is 12,361.1. The inset shows the distribution of the multiply charged species from 500 to 2000.

this sort should allow one to conclude if the mass of the determined sequence agrees with the native protein. If the protein is heterogeneous, a common problem with glycoproteins, the ESI spectrum may be too complicated to deconvolute, and it may be necessary to remove some (e.g., silaic acid with neuraminidase) or all (e.g., with endoglycosidases) of the carbohydrate before obtaining an accurate mass. Proteins more than 50 kDa may not give the same mass accuracy as smaller proteins (for a variety of reasons), or may reach the size where the mass error is greater than the weight of most amino acid weight differences. Nonetheless, this information may go far toward convincing the researcher that the entire sequence is accounted for. Finally, the mass resolution required for a given instrument must be greater than ±0.5 unit for a protein of size 100 kDa; otherwise the various charge states will not be resolved.

Mass measurements of proteins by MALDI-TOF are usually accurate to ±0.05 to 1.0%. The ease of the technique and relative insensitivity to impurities often makes this the method of choice. The MALDI-TOF spectrum of a protein mixture is shown in Figure 4-15. The mass spectrum was calibrated with the internal standard horse heart myoglobin. The mass accuracy for cytochrome C was ±0.016% and for the single chain antibody ±0.75%. The poorer mass accuracy for the single chain antibody reflects the greater broadness of the peak, which may be due to microheterogeneity or matrix adducts. MALDI-TOF is especially useful for samples which are otherwise difficult to analyze by ESI. For example, glycoproteins give broad, but distinct peaks

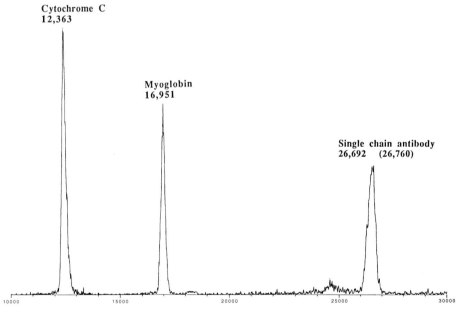

Figure 4-15. Mass analysis of a protein mixture by MALDI-TOF MS. The sample (1 pmole each in 0.5 μL of horse heart myoglobin and cytochrome c, and a single chain antibody) was spotted onto a sample stage with 0.5 μL of *a*-cyano-4-hydroxycinnamic acid and analyzed on a Kratos Kompact MALDI III mass spectrometer. The mass spectrum was calibrated to the myoglobin peak (16,951). The experimental molecular masses are shown together with the expected mass for the single chain antibody in parentheses.

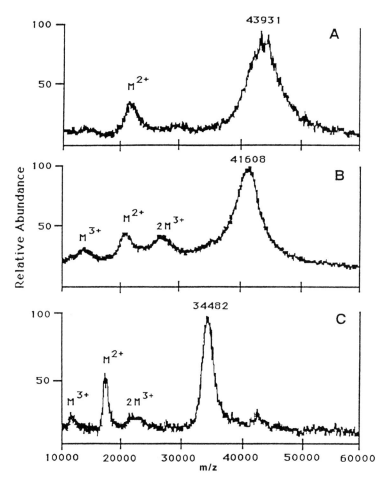

Figure 4-16. Mass analysis of a glycoprotein. The glycoprotein nonspecific crossreacting antigen (NCA, also known as CD66) was analyzed by MALDI-TOF on a Shimadzu LAMS 50KS instrument in a sinapinic acid matrix. (A) The intact glycoprotein (10 pmoles). (B) After treatment with endoglycosidase H. (C) After treatment with TFMSA.

when analyzed by MALDI-TOF (Fig. 4-16A). Similarly, proteins containing salts and other contaminants may not analyze by ESI, but will give nice spectra on MALDI. The example shown in Figure 4-16 illustrates how the peak becomes sharper after removal of some (Fig. 4-16B) or all (Fig. 4-16C) of the carbohydrate. A discussion of the deglycosylation procedure is given in the following.

Database Searches

Mass information on whole proteins or their peptide fragments may uniquely identify a known protein. Database searches have shown that mass information for as little as 2 or 3 peptides may allow identification of a known protein (Pappin et al., 1993). Programs for searching databases have been described (Henzel et al., 1993; Mann et al.,

1993; Pappin et al., 1993). Since many proteins that are sequenced as unknowns are later found to be known, database searches by mass are important and save the step of having to obtain actual sequence information, which is a much more time consuming task. Thus, one may perform mass measurements on an unfractionated digest by a technique such as MALDI, and even though not all of the masses for a given protein will be observed, sufficient data will be obtained to perform a mass scan of the database. If no hits are found, then it may be concluded that the protein of interest is truly new. If a hit is found, then one can compare the mass of the intact protein and other of its properties to the known protein in the database, allowing one to further judge its authenticity. In some cases, proteins may be identified from open reading frames for which no known protein has ever been isolated. A current goal of many laboratories is to perform database searches of unfractionated peptide maps directly from in-gel or on-blot samples from 2D gels. Data searches using unedited MS-MS data have been described by Eng et al. (1990).

Analysis of Posttranslational Modifications

Glycosylation

Many proteins have either N- or O-linked glycans. The glycan units may be heterogeneous in terms of their distribution on the protein and their individual structures. The structures of common O- and N-linked glycans and their analysis are described in a separate volume in this series. Glycoproteins can be treated (singly or sequentially) with specific glycosidases (neuraminidase, galactosidase, fucosidase, hexosaminidase, or endoglycosidases) and mass analyzed. If the situation is not too complicated, considerable information may be obtained by this approach. A general approach to mapping attachment sites is to compare the peptide map of the completely deglycosylated sample to the fully glycosylated sample. This strategy will reveal all glycopeptides by retention time and mass shifts. Complete deglycosylation can be performed by enzymatic or chemical methods.

O-Linked Glycans

The sample should be first subjected to carbohydrate compositional analysis to ascertain the type and amount of monosaccharides present. The most common glycans such as NaNa—GalNAc can be removed by sequential treatment with neuraminidase and α-galactosaminidase. Some may also contain galactose and fucose which can be removed by their specific glycosidases. A chemical method to specifically remove O-glycans (without removal of N-glycans) is alkaline elimination under reducing conditions. Typical conditions are 0.05 M NaOH-1.0 M $NaBH_4$ at 45°C for 16 to 45 hr. (Karlsson et al., 1989). More forcing conditions are 1 M NaOH-1 M $NaBH_4$ at 100°C for 6 hr. (Lee & Scocca, 1972). The sample is acidified with 2 M acetic acid to destroy excess borohydride and either desalted or digested with protease after the pH is adjusted. Ser and Thr residues are converted to their anhydro derivatives. The released oligosaccharides may also be analyzed as described in a separate volume in this series. Glycopeptides identified by their shifts (HPLC and mass) can be analyzed by MS-MS to identify the order and arrangement of the glycosyl units. The glycopeptides can also be permethylated and reanalyzed. The permethylated residues give linkage specific cleavages that aid in structural analysis. The sample is dried in a silylated Reacti-Vial, dissolved in 100 μL of anhydrous DMSO containing 8 mg of NaOH, purged with argon, and stirred with 25 μL of methyl iodide at 25°C for 10 to 30 min (Dell, 1990). The reaction is stopped with 200 μL of 30% acetic acid and either desalted by reversed phase HPLC or extracted with chloroform, dried, and redissolved in 70% acetonitrile (if the peptide is soluble in chloroform). This protocol will also methylate functional groups on amino acids.

N-Glycans

Enzyme digestion should be performed on reduced and alkylated glycoproteins to ensure complete deglycosylation (or performed on SDS-treated proteins on PVDF blots as previously described). The sample is dissolved in 50 μL of 0.1 M ammonium bicarbonate, pH 8.5, containing 0.005% EDTA and treated with PNGase F (4 μL of 0.25 units/mL) for 24 hr at 37°C (Carr et al., 1990). This protease removes the entire glycan, simultaneously converting the Asn residue at the attachment site to Asp (hence, an amidohydrolase). Other enzymes such as Endo H and Endo F are specific for biantennary and high mannose glycans, respectively, but usually do not give complete deglycosylation. In addition, these enzymes leave a single GlcNAc residue still attached to the Asn attachment site. This pendant GlcNAc may be identified by both Edman chemistry (as the PTH-Asn-GlcNAc derivative) and mass spectrometry (Paxton et al., 1987). Chemical deglycosylation may be performed with trifluoromethanesulfonic acid (TFMSA). The sample (0.1 to 1.0 mg) is dried in a Reacti-Vial under argon, dissolved in 100 μL of 2:1 TFMSA-thioanisole at 25°C, cooled to 0°C for 1 to 3 hr, and precipitated with 5 mL of 10:90 hexane-diethylether at −40°C. Ten μL of pyridine may be added to coprecipitate the sample with the pyridinium salt of TFMSA. The precipitate is resuspended and washed with cold 95% ethanol, dried, redissolved in 6 M GuHCl, and further desalted by reversed phase HPLC. This procedure removes all *O*-glycans and the *N*-glycans up to the pendant GlcNAc attached to the Asn residue (Edge et al., 1981).

Analysis of the isolated glycopeptides is performed as described above and in another volume of this series. Positive ion SIMS analysis of a glycopeptide is shown in Figure 4-17. Many fragment ions are observed, including the intact peptide and several of the expected losses of glycosyl units from the intact glycopeptide. This information can be used to confirm or suggest structures. Additional information can be gained from MS-MS analysis of glycopeptides. Negative ion MS-MS of a glycopeptide with a high mannose-type *N*-glycan is shown in Figure 4-18. The fragments are consistent with the proposed structure, but do not prove it. Further analysis, including methylation analysis and anion exchange chromatography of the released oligosaccharide would be required for proof of structure. Masses for the anhydro sugars are shown in Table 4-5. Isolated glycopeptides can also be analyzed by either SIMS or ESI without resorting to MS-MS, since they fragment easily between the glycosyl bonds, yielding a series of fragments that include all possible single bond cleavages, while the peptide bonds remain intact. Glycans attached to peptides are detected with greater sensitivity than free, underivatized oligosaccharides. The sensitivity can be further enhanced by methylation as described above or sometimes successfully performed

148 / Shively

directly on the SIMS probe tip. Glycans may also be sulfated, a modification more readily detected by negative ion MS (Gibson & Cohen, 1990).

Phosphorylation and Sulfation

A common indication of these modifications is observation of a given peptide at two positions in the HPLC peptide map and differing in mass by 80 units. This reflects the fact that it is not unusual to find that peptides are heterogeneous in their degree of phosphorylation or sulfation. Sulfated peptides are especially prone to loss of sulfate as protonated molecular ions in the mass spectrometer, and may only be detected in the negative ion mode (Gibson & Cohen, 1990). These modifications can be confirmed by treatment with alkaline phosphatase or aryl sulfatase as already described, and can be easily pinpointed by MS-MS as described by Gibson & Cohen (1990). Sulfate can also be distinguished from phosphate on tyrosine residues by acid hydrolysis (1 M TFA, 100°C, 5 min), conditions which completely hydrolyze sulfate, but not phosphate. Sulfate also has a diagnostic loss of HSO_3^- on negative ion MS, a weak peak

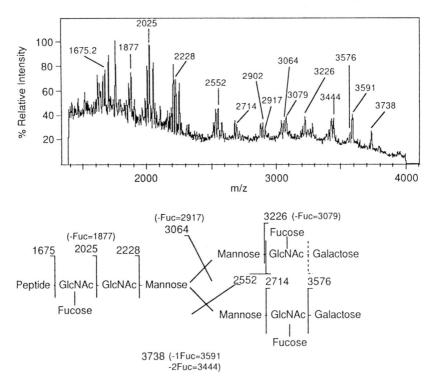

Figure 4-17. Mass analysis of a glycopeptide. A glycopeptide (about 50 pmoles) isolated from a chymotryptic digest of carcinoembryonic antigen (CEA) was analyzed by positive ion SIMS on a Finnigan TSQ700 triple quadrupole mass spectrometer (only the region from 1400 to 4000 is shown). The expected mass of the peptide (without carbohydrate) is 1675. The actual mass of the glycopeptide is 3738, which, together with the fragment ion series, corresponds to the glycosyl unit proposed below the mass spectrum.

Structural Analysis of Proteins / 149

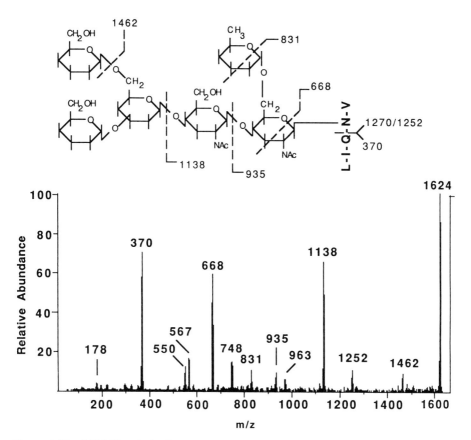

Figure 4-18. MS-MS analysis of a glycopeptide. A glycopeptide (about 50 pmoles) from the CD66 antigen related to CEA was isolated and analyzed by negative ion SIMS-MS on a Finnigan TSQ700 triple quadrupole mass spectrometer. The mass selected base peak (1624) corresponds to the molecular mass of the intact glycopeptide, which, together with the observed fragments, corresponds to the proposed glycosyl unit shown above the mass spectrum.

Table 4-5 Anhydro Masses for Common Sugars in Glycoproteins[a]

Sugar	Abbreviation	Monoisotopic Mass	Average Mass
Fucose	Fuc	146.0579	146.144
Hexose	Hex	162.0528	162.143
Hexosamine	HexN	161.0688	161.158
N-Acetyl hexosamine	HexNAc	203.0794	203.196
Sialic acid	NeuNAc	291.0954	291.259

[a]From Carr et al. (1990)

for phosphate (HPO_3^{2-}). Sulfate and phosphate can be detected on intact proteins if analyzed by ESI before and after enzymatic removal. The degree of heterogeneity can be assessed in intact proteins by this method.

N-Terminal Groups

N-terminal modifications such as formyl, acetyl, myristal, and so on are easily analyzed by SIMS and ESI MS-MS. A typical example of the MS-MS analysis of an *N*-acetylated peptide using lower energy CID is shown in Figure 4-19. In this example, the fragment ions containing the acetyl group are well represented. More diagnostic fragment ions can be generated in high energy CID experiments such as described by Biemann & Scoble (1987).

C-Terminal Groups

A common modification of small, naturally occurring peptides is a C-terminal amide. The mass difference of 1 unit from a free carboxyl is easily detected by MS. Proteins may have C-terminal glycans such as GPI (Doering et al., 1990) or GAG (Enghild et al., 1993). Examples of the structural analysis of the C-terminal GPI (glycosylphosphatidylinositol)-linked proteins of Leishmania major are given by McConville et al. (1990) and for human erythrocyte acetylcholinesterase by Deeg et al. (1992). This gly-

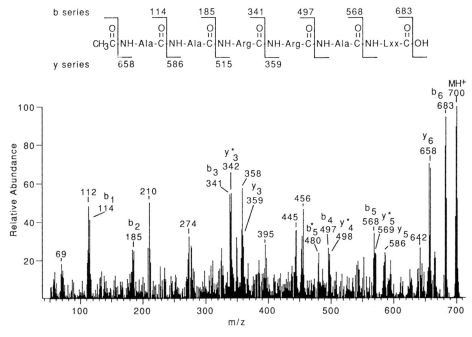

Figure 4-19. MS-MS analysis of an *N*-acetylated peptide. A chymotryptic peptide (about 10 pmoles) of murine DT diaphorase was analyzed by positive ion SIMS-MS on a Finnigan TSQ700 triple quadrupole mass spectrometer. The selected base peak corresponds to the intact peptide, which, together with the indicated b and y fragment ions corresponds to the structure indicated above the mass spectrum.

Structural Analysis of Proteins / 151

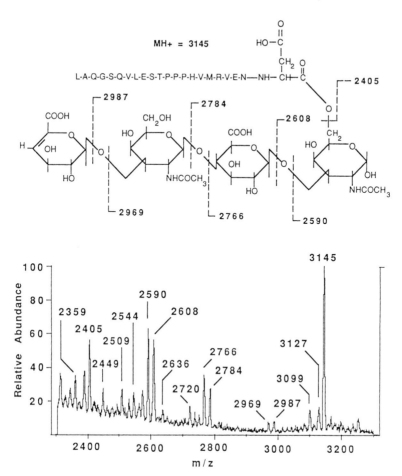

Figure 4-20. Mass analysis of a peptide-glycosaminoglycan (GAG) fragment. A bikunin containing plasma protein (HC2/bikunin) was digested with thermolysin and chondroitinase ABC, and chromatographed to produce the truncated froms of their GAG crosslinked chains. Positive SIMS analysis over the mass range 2300 to 3300 was performed on a Finnigan TSQ700 triple quadrupole mass spectrometer. The protonated molecular (3145) and fragment ions are consistent with the structure shown above the mass spectrum.

can is completely removed by TFMSA treatment except for the ethanolamine attached to the C-terminal amino acid (Paxton et al., 1987). The attachment of GAG (glycosaminoglycan) is a newly discovered posttranslational modification shown to link two polypeptide chains together. An example of the SIMS analysis of the peptide fragment from a bikunin protease inhibitor treated with chondroitinase ABC is shown in Figure 4-20. In order to obtain peptide, the protein was digested with both thermolysin and chrondroitinase. The GAG was shown to link the polypeptide chains through an O-glycosidic link to Ser-10 of bikunin (light chain) and through an ester bond to the C-terminal Asp-648 of the heavy chain (HC2). The peptide sequences were confirmed by Edman chemistry and the size of the glycosidic moiety by mass difference on SIMS.

The fragments observed were consistent with the presence of the expected tetrasaccharide from chondroitinase digestion. This is an excellent example of the power of the method for deduction or confirmation of posttranslational modifications.

Summary

Modern protein structural analysis combines the power of microisolation methods with extremely sensitive structural methods, including microsequence and mass spectral analyses. In order to be successful, the protein chemist must be aware of the strengths and weaknesses of each method, especially in choosing the correct combinations to solve the problem at hand. The many examples given in this chapter should provide a good introduction to the basic approaches. For more details, the reader is encouraged to refer to the many references cited.

Acknowledgments

I am indebted to Terry Lee, Stanley Hefta, Jerome Bailey, Kristine Swiderek, Mike Ronk, and Mike Davis for a critical reading of this chapter and many suggestions for examples of protein structural analysis.

5

Protein Structure
Charles W. Carter, Jr.

Structural molecular biology has experienced unprecedented growth in the past two decades, and there is ample evidence that the exponential growth in new structural data is spawning a similar growth in insight, largely because these data are avidly consumed by a corps of "molecular anatomists," physical biochemists, enzymologists, and molecular biologists armed with appropriate questions and techniques to answer them. Thus the subject of protein structure has assumed awesome dimensions. Yet the task of reviewing these developments has some compensations. We are on the verge of making rather precise statements of some of the most tantalizing questions ever posed, and the techniques to answer them may be coming into view.

Excellent sources exist for many aspects of protein structure. The Richardsons' two reviews (Richardson, 1981; Richardson & Richardson, 1990) are authoritative, comprehensive, well written and eminently readable for specialist and student alike. Textbooks (Brändén & Tooze, 1991; Creighton, 1995; Kyte, 1995) and a monograph (Perutz, 1992) are also recommended.

Structural Studies

An important reason why structural molecular biology has enjoyed such success in the past decade is that both X-ray crystallography and NMR spectroscopy have enjoyed similar growth and development. Both techniques now provide unprecedented atomic detail with surprisingly modest investment. The following discussion of the two methods introduces the experimental basis for, and the reliability of the atomic coordinates and their variance parameters, from which our understanding of protein structure derives.

X-ray Crystallography and NMR Spectroscopy Provide Complementary Kinds of Experimental Data

X-ray crystallography and NMR spectroscopy are neither exclusive nor competing techniques. An X-ray crystal structure is an atomic model based on the interpretation of a high resolution image, and an NMR structure is an atomic model based on the interpretation of a set of interatomic distances. Curiously, the raw experimental amplitude data provided by X-ray diffraction measurements are also limited to the interatomic vectors in the structure. However, in contrast to NMR, which provides *identifiable* interatomic distances, the maps of interatomic vectors provided by X-ray data are essentially impossible to deconvolute without an intermediate step, called phase determination. Phase determination constitutes the imaging step and completes a close, underlying connection between crystallography and microscopy. Thus, in a sense, crystallography is somewhat more complete with respect to the quality of information it provides. NMR measurements, on the other hand, provide crucial evidence regarding behavior in solution. In principle, both techniques can provide structural information on rapid time scales, and hence for kinetic processes.

X-ray crystallography can be used for samples of any molecular size. Size records of two different kinds were set twice in 1994, first with beta galactosidase, which is the largest asymmetric unit to be solved to date (M_r 1,861,648/asymmetric unit; Jacobson et al., 1994). Soon afterwards, solution of the F1-ATPase structure (M_r 371,000/unique structure; Abrahams et al., 1994) set the record for the largest chemically distinct mass. Problems as large and complicated as the ribosome are being studied, and the pace of technical advances raises the expectation that even this huge and asymmetric structure will be imaged in the near future (Yonath, 1992).

Multidimensional NMR spectroscopy gained substantial new power with the advent of multi-isotope or heteronuclear techniques, which facilitate the resolution of spectra that are tightly packed and impossible to resolve in two dimensions. Using two different isotopes (generally ^{14}C and ^{15}N) it is now possible to obtain a sufficient number of constraints to define the structures of proteins as large as 15 to 20 kDa (Clore & Gronenborn, 1994). Back transformation of model coordinates to NMR spectral data is problematic, but progress has been substantial even in this regard (Brunger et al., 1993; Clore & Gronenborn, 1991b; Clore et al., 1993). In favorable cases, distance constraints from NMR solution measurements define structures to the precision obtained with 2.0 Å X-ray diffraction measurements (Clore & Gronenborn, 1991a; Wagner, 1992).

Refinement Techniques Lead to More Reliable Models. X-ray crystallography is blessed with a strong mathematical formalism relating diffraction directly to the atomic coordinates, so one can evaluate gradient directions for coordinate shifts that minimize the sum of the squared differences between observed and calculated amplitudes. For this reason, X-ray diffraction also provides, via refinement techniques, the most accurate atomic models available. Indeed, recent advances in least squares refinement procedures (Brunger & Rice, 1996; Brunger et al., 1993; Lamzin & Wilson, 1993, 1996; Sheldrick & Schneider, 1996) have provided coordinate sets with estimated errors in the range of about 0.03 to 0.05 Å in favorable cases.

"Favorable cases" refers to situations where amplitudes can be measured to very high resolution. The reason for this is the fundamental dependence of the reliability of a parameter estimate on the ratio between its contribution to the model and the estimate for its uncertainty. This uncertainty, in turn, depends on the number of observations recorded, relative to the number of adjustable parameters in the model. This, "overdetermination ratio" is very modest for macromolecular X-ray crystal structures solved to moderate resolution (2.0 to 3.0 Å). As a consequence, structures refined to such data show substantial deviations from expected stereochemistry. Imposing idealized bond lengths and angles (Diamond, 1966) jointly with refinement to the observed amplitudes resulted in a better conditioned refinement (Carter et al., 1974; Freer et al., 1975). The stereochemical restraints impose correlations between the atomic parameters, thereby reducing the effective number of parameters to be refined and improving the overdetermination ratio. It is customary now to impose these restraints within the refinement algorithm (Konnert, 1976).

The weak link in forming an image from X-ray data is the way in which phases are assigned to the diffracted amplitudes. Phases fall into two broad categories, according to whether they derive from additional experimental measurements or have been transferred from the known structure of a related protein by a process called "molecular replacement" (Fitzgerald, 1991). The latter method is being used increasingly to solve new protein structures because it can reduce the number of experimental data sets required. However, it is vulnerable to unavoidable errors arising from "model bias" (Read, 1997). Such errors are often quite subtle and unless the overdetermination ratio is quite high (~8 to 10:1), they cannot be removed completely by the available refinement algorithms.

The reliability of the structures deposited in the Brookhaven Protein Databank therefore varies rather widely, according to the resolution limit and accuracy of the X-ray data, the nature and quality of the phases used for the initial interpretation of the map, the course of the refinement and the ultimate overdetermination ratio. The overdetermination of a particular structure is not related in a simple way to the resolution limit of the experimental data, because it depends on incidental particularities of crystal packing. Virus structures with 60 copies of a single subunit in a crystallographic asymmetric unit whose solvent content is high can be considerably better conditioned at 3.5 Å resolution than are many typical structures at 2.0 Å resolution.

Some guidelines are essential for prospecting the structural database (Table 5-1). The accuracy of the diffraction measurements, the number and quality of heavy atom derivatives and the resolution limits of the experimental amplitudes and of the phases should be scrutinized carefully to assure oneself of the reliability of an X-ray structure. In addition, one should be aware of whether and how additional constraints, such as non-crystallographic symmetry and solvent flatness, may have been imposed during the generation of an electron density map in order to improve phases. Burling et al. set a standard in 1996 against which refinement methods can be assessed by determining very accurate experimental phases using multiwavelength anomalous diffraction (Burling et al., 1996).

A procedure called "cross-validation" (Brunger, 1992) provides a quantitative estimate for the likely errors in a refined model by withholding some of the data from the refinement process. A random sample of the X-ray data (~1000 reflections) is reserved as a reference or "test" set and these reflections are not used to constrain the model. The crystallographic R-factor for the test set forms the "free R-factor," R_{free}, which should improve in parallel with the R-factor for the data being used to constrain the model. Structures for which R_{free} is more than about 20% higher than the quoted R-factor almost certainly contain errors or artifacts due to overfitting of the experimental data (Dodson et al., 1996). Model-independent phase (Xiang et al., 1993; Zhang et al., 1997) and structure refinement (Lamzin & Wilson, 1993) techniques recently introduced may in time reduce the uncertainty associated with the initial map interpretation.

Few structures reported at resolutions lower than about 3.5 Å are free of ambiguity regarding side chain placements or identity. Structures reported at resolutions between 2.9 Å and 3.5 Å probably have the correct path for the main chain, but required considerable guesswork in the placement of side chain groups. At 2.5 Å resolution, maps begin to reveal correct side chain orientations. At 2.0 Å one can often sequence much of the protein from the side chain identifications in the map. Isotropic, Debye-Waller thermal parameters begin to have significance at this resolution. At 1.5 Å resolution, five-membered rings have well-resolved holes, alternate conformations can be confidently identified for side chains, and anisotropic thermal parameters become significant. Beyond about 1.2 Å one can begin to locate hydrogen atoms in electron density maps.

Although these guidelines describe most structures, the temptation to publish an im-

Table 5-1 Reliability Characteristics of Atomic Coordinate Files

Resolution Limits	(>50% of measured I > $2\sigma(I)$)	Well-defined Features
	6.0–4.5Å	Placement of secondary structures
	3.0Å	Chain tracing
	2.5Å	Side-chain orientation, isotropic thermal parameters
	2.0Å	Side chain, bound water identification
	1.8Å	Alternate side chain orientations
	1.5Å	Anisotropic thermal parameters
	1.2Å	Hydrogen atoms
Source of phases	Multiple isomorphous replacement	Free of model bias; but noisy due to lack of isomorphism
	MAD (multiwavelength anomalous diffraction)	In general, the most reliable source of phases; isomorphism is nearly perfect
	Molecular replacement	Widely used, probably misused; errors due to model bias are variable and difficult to detect, correct
Model-independent map refinement	(Physical constraints on electron density)	
	Noncrystallographic symmetry (NCS)	Widely observed and enforced. Violations should be scrutinized
	Solvent flatness (SF)	Strength proportional to solvent volume-fraction
	Histogram matching (HM)	Complementary to NCS and SF
Refinement	$R_{free} < \sim 1.2 \times R_{cryst}$	Cross-validation statistic; indispensible to avoid model bias due to overfitting
Stereochemistry	<1–2% of residues with atypical (ϕ,ψ) angles	
	Bondlength +/− 0.005Å, Bond angles +/− 2°	

portant new structure has drawn premature reports of seriously wrong structures into wide circulation, occasionally with significant consequences (Brandén & Jones, 1990). Fortunately, incorrect structures usually contain clues to their own demise, and most practicing structural biologists can spot the trouble signs.

Among the factors limiting the accuracy of NMR structures, the most important is the degree of redundancy in the interproton distance measurements. These distance measurements are inherently divided into two different classes, those between unique protons and those for which stereochemical sources of ambiguity exist. Obviously, one hopes to have a larger proportion of the former type, relative to the latter. For an accuracy approaching a 2.0 Å X-ray crystal structure, one would like to see approximately 15 NOE restraints per residue of protein of which at least 60% are between uniquely defined proton positions (Clore et al., 1993).

Resonance Spectra and Diffraction Data can be Combined. This can produce better structures than those obtained from either of them alone. Images and interatomic distances are neither redundant nor mutually exclusive. In fact, it has been shown that if a model is forced to agree with both sets of data simultaneously, it can agree better with both sets of data than do models constrained by only one kind of information. This means that in cases where no structural changes arise from crystal packing during crystallization, NMR and X-ray crystal structures are highly complementary at a detailed level. Constraints provided by NMR and X-ray data were applied jointly to the coordinate refinement for interleukin-1β, and the joint refinement produced a model that agreed better with both sets of data than did the best models produced by refining to each set of data individually (Shaanan et al., 1992).

The importance of this remarkable result should not be underestimated. It means not only that the two kinds of data are complementary, but that in many cases the solution and crystal structures of proteins are the same within the experimental limitations of either technique. While exceptions most certainly do exist, this existence proof of one protein for which the best structure is that constrained by both solution and crystallographic measurements means that considerable confidence can be placed in the best atomic models produced by either technique.

Proteins Are Dynamic Structures

Proteins undergo motion on a very wide range of time scales. There are many techniques for studying this kinetic behavior: relaxation of most spectra, including NMR spectra (Campbell & Sykes, 1993; Lane & Lefèvre, 1994; London, 1989; Peng & Wagner, 1994); molecular dynamics simulations (Daggett & Levitt, 1993; van Gunsteren et al., 1994), and Laue diffraction (Bolduc et al., 1995; Hajdu & Andersson, 1993; Hadju et al., 1987). Structural characterization of protein mobility is technically difficult, however. Often, the roles of such conformational changes are inferred from comparison of static structures obtained from samples trapped in one or another conformation using specific ligands (Lolis & Petsko, 1990).

Motion in proteins was placed in an experimental context for the first time with the study of crystallographic "B-factors" obtained by refining protein crystal structures to the $|F_{obs}|$ (Watenpaugh et al., 1973). "B-factor" refers to an exponentially decreasing multiplier that reduces the amplitude of calculated structure factors, $\{F_{calc}\}$, at higher resolution, $\sin^2 \theta/\lambda^2$. Substantially improved agreement can be obtained between observed amplitudes and those calculated from the atomic model by allowing individual atoms to move within spheres or ellipsoids whose dimensions become adjustable parameters. The isotropic B-factor corresponds to convolution of the atomic position with a spherical Gaussian probability distribution proportional to $\exp(-4\pi^2\|x\|^2/B)$, where $\|x\|$ is the instantaneous displacement of the atom from its mean position in real space. Fourier transformation leads to the expression of the multiplier, $\exp(-B \sin^2 \theta/\lambda^2)$.

A refined isotropic $B = 8\pi^2\|u\|^2$ is a measure of the r.m.s. displacement, $\|u\|$, of the atom from its mean position. It really measures the displacements due to all sources, including thermal motion and lattice disorder. However, temperature-dependent measurements show that thermal motion is a major contributor to these displacements (Ringe & Petsko, 1986). The isotropic B values are usually the most readily accessible experimental data regarding atomic motions. However, since the atoms in proteins are bonded, their motions are highly correlated, and isotropic B values are a very crude guide to the thermal motions that actually occur, and which are of obvious relevance to biological processes. Extracting better models for motion from experimental data is an active research area (Kidera & Go, 1992; Kidera et al., 1992; Moss et al., 1996).

One of the more important observations regarding thermal motion in proteins has been the repeated demonstration of correlations between segmental mobility and the locations of antigenic determinants in proteins whose principal epitopes and crystal structures have both been worked out. Continuous regions of the tobacco mosaic virus coat protein, myoglobin, and lysozyme structures which have high B values corresponding to highly active antigenic determinants. This correlation is actually better than that between the hydrophilicity of the segment and its antigenicity. (Tainer et al., 1984; Westhof et al., 1984). Antibodies to the mobile regions react strongly to the native protein, whereas antibodies to the well-ordered regions do not (Tainer et al., 1984).

Doing Kinetics with X-ray Crystallography. There is no fundamental reason why we cannot take moving pictures of proteins in crystals. This staggering task would mean solving a separate crystal structure for each frame. This dream has recently been realized, in principle, by a new implementation of one of the oldest types of X-ray photography—the Laue method—adapted to exceedingly intense synchrotron radiation sources, which are on the order of a hundred to a thousand times brighter than laboratory X-ray sources.

X-ray intensity data are normally collected at a single wavelength by moving the crystal throughout a range between 30 and 180 degrees. The polychromatic spectra available at synchrotron sources facilitates very rapid data collection of the X-ray reflections from a crystal. Bragg's law, $\lambda = 2d \sin \theta$, relates the angle at which a reflection can be observed, θ, to the wavelength of the X-rays, λ, and the spacing, d, between planes in the crystal. With polychromatic radiation, Bragg's law holds simultaneously for reflections with many different spacings, without moving the crystal. An appreciable fraction of a complete data set can be recorded from a stationary crystal by illuminating it with polychromatic radiation. Exposures of tiny fractions of a second are sufficient with contemporary synchrotron sources. Recently, the ultimate limit was probed and usable photographs were obtained in fractions of a nanosecond. More typically, exposures accessible today have been in the millisecond range.

A requirement for doing X-ray kinetics is that one must be able to synchronize all the molecules in the crystal, so that they are doing essentially the same thing at the same time. The best way to do this is with a photoactivated substrate, such as a "caged" nucleotide (Schlichting et al., 1990). A photolabile protecting group renders the molecule inert to hydrolysis until it receives a flash of light. The result of this study was an unprecedented look at the reaction mechanism of a GTPase, in which the initial steps of hydrolysis were seen to involve a rearrangement of magnesium coordination.

Modeling the Forces Between Atoms: Molecular Dynamics. The energies experienced by atoms in molecules have been evaluated using approximate representations for the potential energies of stretching bonds, rotating torsion angles, bending bond angles, encountering the van der Waals fields of other atoms, and so on. These all depend on the positions of the atoms in the structure, and they evolve with time. Since these expressions are differentiable, the forces on the atoms can also be calculated from their atomic positions. Computer programs that simulate the motions of atoms in accordance with Newton's laws of motion are called molecular dynamics programs (Hermans, 1993; Karplus & Petsko, 1990; Levitt, 1981).

Generally, each atom is given an initial velocity, and new positions and forces are calculated after a given time interval, usually on the order of 1 to 2 fsec. At intervals of some multiple of this fundamental step the structure can be analyzed by writing out the coordinates to a computer file and statistics can be accumulated for the structural properties of the model.

There is a close correlation between the length of a hydrogen bond in the crystal structure and the tendency of that hydrogen bond to fluctuate narrowly about a well-defined hydrogen bonding distance (Levitt, 1981). This correlation is supported by a correlation between the strength of hydrogen bonds and the extent to which they are observed to be rapidly exchanging by proton NMR experiments. Simulations cannot, however, predict the observed exchange rates themselves to within more than a few orders of magnitude.

Motions accessible to molecular dynamics studies are limited to some tens of picoseconds, and hence are about three orders of magnitude faster than most biochemically interesting events, because encounter frequencies are in the neighborhood of 10^{-8} seconds. Large scale "breathing" motions in proteins probably occur on the scale of 10^{-6} seconds. Many important conformational changes occur with rates slower than this.

Presenting Protein Structure

Even the smallest proteins contain several hundred nonhydrogen atoms. Presenting salient features of protein structures thus poses a number of practical problems, which have been solved by abstracting and representing only selected properties from the ensemble of atoms. Different representations convey different aspects of structural organization. An important distinction should be made between *schematic* diagrams, which are interpretations, and plots of atomic coordinates, which actually represent data. Questions dealing with active site geometry, for example, can often be illustrated by actually representing a subset of the atomic positions. For other kinds of questions, some degree of abstraction is essential. Three kinds of abstractions are important: the path of the chain, volumes, and surfaces.

Block diagrams in projection and in perspective are useful for representing volumes occupied by different parts of a structure. Simply representing α-helices in a protein by circles and β-strands by squares (Levitt & Chothia, 1976) conveys underlying relationships in a globular protein structure; for example, the existence of pseudosymmetry relationships between different parts of the structure (Fig. 5-1).

Ribbon representations were introduced to faithfully chart the path of the polypeptide and were originally drawn manually by carefully tracing a plot of atomic coordinates (Carter et al., 1974). They quickly became and have remained a standard (Carson & Bugg, 1992; Kraulis, 1991; Richardson, 1985) because they provide a good compromise between the demands of clarity and ready access, on the one hand, and faithfulness to the data on the other hand. Computer programs to generate ribbon diagrams analytically, by spline fits to a space curve representation of the polypeptide backbone (Carson & Bugg, 1992; Kraulis, 1991), now mean that ribbon drawings carry nearly the authority originally conveyed by plots made from atomic coordinates.

Space-filling diagrams (Carson & Bugg, 1992; Kraulis, 1991) and more specialized abstractions of surface properties (Nicholls et al., 1991) can also be produced analytically. These, together with the use of color, are essential to the presentation of functionally significant features of, for example, the electrostatic channeling of substrates into enzyme active sites (Getzoff et al., 1983).

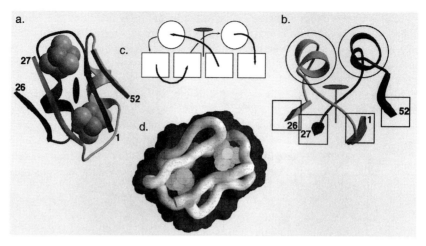

Figure 5-1. Representing protein structures. The small bacterial ferredoxin is represented as a "ribbon" drawing (a) and (b), as a simple block diagram (c) (Levitt, 1976), and as a composite showing its surface and the embedded trace of the backbone chain (d). The block diagram in (c) is obtained as suggested in (b) using the squares and circles to represent β-strands and α-helices.

Interactive computer graphic tools extend the resources for learning about protein structure far beyond what can be conveyed in even the most carefully prepared static images. Atomic coordinates are available for most, but not all, three-dimensional structures, and can be accessed from the World Wide Web. Although this area is changing rapidly, two different computer programs are so widely used that no student should be without them. MAGE (Richardson & Richardson, 1992) was developed for incorporating interactive graphic illustrations called "kinemages" into publications as pioneered in the journal *Protein Science*. RASMOL is a complementary program that offers more interactive control over the display format. It accepts files directly from the Brookhaven Databank, whereas coordinate files must be preprocessed to make kinemages. Both programs are easy to learn. They run on a variety of workstation and PC computers and incorporate numerous features similar to those in more sophisticated graphics programs. Important URL addresses include:

Brookhaven Protein Databank: (http://www.pdb.bnl.gov/cgibin.browse),

MAGE: (http://www.prosci.uci.edu/kinemages/kinpage.html)

RASMOL: (http://klaatu.oit.umass.edu/microbio/rasmol/).

Insight from Multiple Comparisons

Proteins exhibit regularities at several levels, which can be identified only by the fact that they recur approximately, but frequently, in different contexts. Much insight into the structures of proteins is therefore statistical in nature, coming from the comparison of large numbers of different proteins (Chothia, 1973, 1974; Chothia et al., 1977; Chou, 1995; Holm et al., 1992; Holm & Sander, 1991, 1992b, 1993a, b; Janin et al., 1978; Ouzounis et al., 1993; Richardson, 1973, 1977; Rooman & Wodak, 1988; Rost & Sander, 1993; Sander & Schneider, 1991, 1993; Sander et al., 1992).

Protein Structure / 161

Building Blocks

The Peptide Bond

The unique properties of the peptide bond, the repeating chemical linkage in proteins, underlie all of protein structure and function. Polypeptides are formed when the elements of water are removed from the carboxylate of one amino acid and the amino group of another to form peptide bonds (Fig. 5-2a). The distance between residues along a polypeptide chain is a very useful quantity that is seldom quoted. Reasons for this may include some variations in value according to changes in the constituent bond angles, which are less constant than the individual bond lengths. The value of 3.81 Å is characteristic and has a very small standard deviation (0.038 Å in a sample by the author).

The peptide bond has several important properties. It is close to being planar, with a barrier of 19 kcal/mole for interconversion between the *cis* and *trans* conformations. The isomerization rate at 313 K is $\sim 0.15 s^{-1}$ for model compounds; this rate implies an equilibrium constant for activation of 2.5×10^{-14}, and an activation free energy of ~ 18.8 kcal/mole (Radzicka et al., 1992). The equilibrium favors the *trans* conformation by a factor of 10^3, except in the case where the nitrogen is part of the five-membered ring of proline, in which case it is favored only by a factor of 4. Thus, peptide bonds in proteins are planar and *trans* in a large majority of cases. Exceptions, espe-

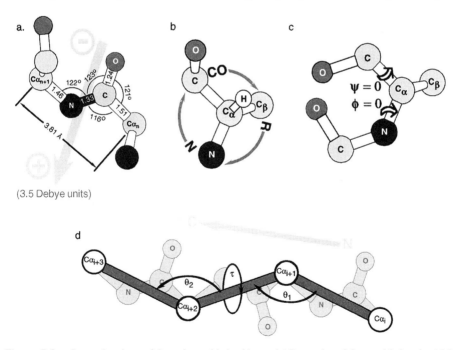

Figure 5-2. Stereochemistry of the polypeptide backbone. (a) Properties of the peptide bond, which constitutes the covalent backbone structure and is a dominant influence on formation of protein secondary structures. (b) The configuration of nonhydrogen atoms around the alpha carbon of an L-amino acid and the clockwise spelling of the word "CORN." (c) Definition of the Ramachandran conformational dihedral angles, $\phi = 0$ and $\psi = 0$. (d) An alternative description of backbone chain conformations in terms of "pseudobonds" between C^α atoms.

cially those involving proline in the C-terminal position, are observed (Stewart et al., 1990).

A second important property of the peptide bond is its dipole moment of ~3.5 Debye units. This value is more than threefold larger than that for an HCl molecule, and is therefore of considerable significance. This moment, the associated hydrogen bonding properties of the carbonyl oxygen and peptide nitrogen atoms, and the consequent impact on their solvation drive the formation of secondary structures and have other fundamental consequences for the structure of proteins and the function of enzyme active sites. Four such dipoles occur without compensation at the termini of α-helices, creating strong electrostatic fields (Hol et al., 1978).

Finally, the stereochemistry of the peptide unit generates much of the hierarchic organization in protein structure. Proteins are polymers comprised exclusively of L-amino acids. These can be identified by the sense of rotation of the three heavy atoms around the central, α-carbon (Fig. 5-2b). When the alpha carbon is viewed from the hydrogen atom, rotation from C=O to R to N is clockwise for L-amino acids, giving rise to the mnemonic, "CORN." The two dihedral angles associated with the polymer, ϕ and ψ, are quite useful in defining the path of the main chain atoms (Ramachandran et al., 1963). By convention, these rotation angles vary over the range $+/- \pi$ with a value of (0,0) when the two carbonyl oxygen atoms are eclipsed (Fig. 5-2c).

An alternative convention, useful in analyzing the geometry of the backbone chain, entails the description of pseudobonds between successive C^α atoms (Oldfield & Hubbard, 1994). Because it forms dihedral angles involving four consecutive C^α atoms, this formalism (Fig. 5-2d) captures concisely more characteristics of the space-curve traced by the main chain than do the Ramachandran angles. Thus, it has become a standard representation in studies of protein folding (Skolnick, 1997).

The Twenty Amino Acids

The twenty side chains normally found in proteins are illustrated in Figure 5-3. Each can be represented by an obvious three-letter code derived from its name. For purposes of efficiency, it is often useful to abbreviate their names still further, and a single letter code was devised for this purpose. All three designations are shown in Figure 5-3. Much about the behavior of proteins derives from the staggering range of side chain affinities for water, which are indicated by the vertical scale. Among the more surprising aspects of this representation is the singular location of the arginine side chain, which is hydrated four orders of magnitude more readily than the next most highly hydrated side chain, aspartic acid. The potential for hydrating the peptide bond is also uniquely high among neutral organic species. The competing tendencies of the peptide bond, neighboring, and interacting side chains to interact with water are dominant in determining the properties of proteins.

Protein Architecture: Linderstrøm-Lang's Hierarchy

Well before the first three-dimensional image of a protein was obtained by X-ray crystallography, Linderstrøm-Lang, then director of the Carlesberg Laboratory in Copenhagen and the universally acknowledged father of protein chemistry, described an analytical framework for understanding protein structure (Linderstrøm-Lang, 1952). The sequence of amino acids in a protein is referred to as its "primary" structure; local folding along

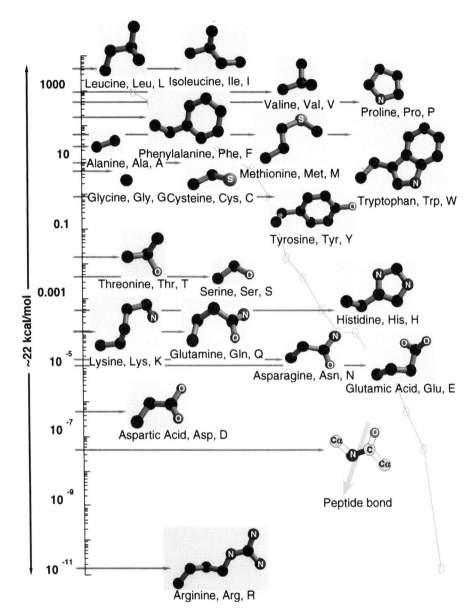

Figure 5-3. Amino acid side chain structures and properties. Side chains are distributed on a vertical scale according to their distribution coefficients for transfer from neutral aqueous buffer to cyclohexane, indicating relative affinities for water (adapted from Wolfenden, 1983). Nonhydrogen atoms, including C^α, are shown: carbon is black, other atom types are indicated with white letters. The peptide bond, with its strong dipole moment, is highly solvated, and shown for comparison. The vertical scale spans approximately 22 kcal/mol in free energies of transfer.

the chain as its "secondary" structure; and association of secondary structures via longer range contacts as its "tertiary" structure. Once the subunit structures of oligomeric proteins became understood, it was natural to include properties associated with intersubunit contacts, which are its "quaternary" structure. Many proteins appear to be assembled from smaller pieces with integral tertiary structure; these semi-autonomous folding units are "domains." That term applies equally to modules connected by muliple peptide linkers and larger domains connected by a single linker peptide (Fig. 5-25). Despite this lack of precision, the Linderstrøm-Lang hierarchy has endured for more than forty years because it captures a fundamental reality, ascribing natural levels of structural organization to proteins. It will be illustrated here by a series of case studies.

Primary Structure: Ultimate Source of Three-Dimensional Structure and Function

Despite a growing focus on proteins whose folding is assisted in a kinetic sense by accessory proteins called chaperonins (Gething & Sambrook, 1992), there remains general agreement that equilibrium three-dimensional structures of proteins are determined by their amino acid sequences. Proteins or parts of proteins that contain exclusively α-helical segments provide an opportunity for the most direct detection of relationships between sequence, structure, and function. The α-helical coiled-coil, exemplified by the protein tropomyosin, provides an excellent case study of such relationships.

Tropomyosin: a Linear Molecule Rich in Functionalities. Tropomyosin from skeletal muscle is composed of two identical, 284-amino acid polypeptides, each of which forms a single α-helix. Two molecules of tropomyosin form complementary packing of nonpolar amino acid side chains by twisting around one another to form a parallel

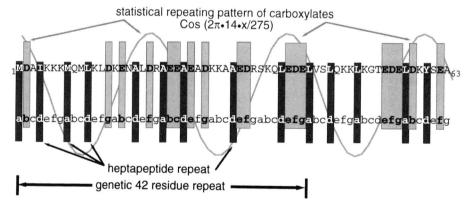

Figure 5-4. Repetitive structure in the first 63 amino acids from the primary structure of rabbit α-tropomyosin. Three different levels of repetition have been identified in the intact sequences of 284 amino acids. The heptapeptide repeat characteristic of α-helical structure is indicated by letters a to g below the sequence. Nonpolar components in positions a and d of this repeat are emphasized by dark shaded bars. Acidic side chains, Asp and Glu, generally occur in positions b, f, and g. However, they are clustered nonrandomly along the sequence, as emphasized by lightly shaded bars. Fourier analysis detects a very strong fourteenth order term (McLachlan & Stewart, 1976), suggested by the cosine function in light gray. This statistical repeat is implicated in complementarity to actin in muscle thin filaments. A 42-residue genetic repeat comprising 6 heptapeptides becomes evident when the entire sequence is aligned (McLachlan et al., 1975).

Protein Structure / 165

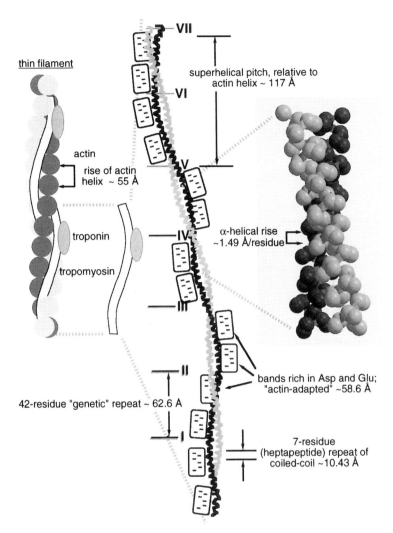

Figure 5-5. Structure-function relationships in tropomyosin and muscle thin filaments. Tropomyosin interactions with the filamentous actin to form thin filaments derive directly from the three levels of repetition in the amino acid sequence and their relationship to the coiled-coil (Fig. 5-4). Representative physical distances are given, as estimated from X-ray diffraction studies of fibers. Patches of negatively charged carboxylate side chains identified by Fourier analysis recur at intervals that more closely match the actin periodicity than does the underlying 42-residue genetic repeat. Only C^α and C^β atoms are drawn in the section of coiled-coil on the right.

coiled-coil in a manner first suggested by Francis Crick (Crick, 1953). Crick also described a triple helical interaction of this type, and such an arrangement has been proposed for the fibrous shaft in adenovirus (Stouten et al., 1992). The structure and function of tropomyosin were revealed largely through analysis of its amino acid sequence (McLachlan & Stewart, 1975, 1976; McLachlan et al., 1975; Stewart & McLachlan, 1975), a portion of which is illustrated in Figure 5-4. Because it has a linear, helical structure, the underlying structure function relationships can be attributed in considerable detail to repeating primary structure patterns (Fig. 5-5).

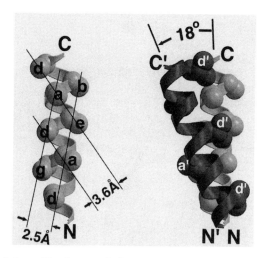

Figure 5-6. Side chain packing between helices in a coiled-coil and the origin of helix packing rules. Positions along the heptapeptide repeat are labeled a to g. Two sets of ridges and grooves are identified in the alignment of side chain C^β atoms along an α-helix. By convention, positions a and d are taken to be those that point toward the opposite strand, forming a nonpolar interface. The two-stranded coiled-coil is obtained from two parallel helices by rotating one of them by 180° about the vertical axis and then tilting it counterclockwise by approximately 18°. Side chains in positions d' and a' of the second, rotated helix pack into cavities between positions a, d, and e of the first. Other helix packing rules result from alternative alignments of the ridges and grooves (Chothia et al., 1977).

Consider an alpha helix of polyalanine (Fig. 5-6). Side chains, represented by the β-carbons, systematically face backward along the helix axis. In a coiled-coil the β-carbon atoms of side chains seven residues apart, for example, a and a', will be directly aligned along the helix axis, while halfway in between a and a' the line connecting them will pass midway between residues d and e of the heptad. The main chain in a helix ascends at an angle of approximately 67° ($\sin^{-1}(1.5 \text{ Å}/3.81 \text{ Å}) = 23°$) to the helix axis. Two sets of ridges connect successive residues, one involving positions e, a, and d; the other involving positions g, d, and a. Grooves halfway between these ridges accommodate the ridges from partner helices, leading to quantization of the observed crossing angles, which are related in an obvious way to the inclination of the ridges and grooves to the helix axis.

Complementary side chain packing between helices works best if they are tilted relative to one another (Fig. 5-6). A tilt angle of $\sim +18°$ (counterclockwise viewed from the helix that rotates) is the smallest crossing angle of those suggested by Crick. In fibrous or cytoskeletal proteins this crossing angle becomes continuous—in any local segment, the two helices are tilted relative to each other by this amount. To maintain this crossing angle over longer distances, the helices must coil round one another. As they coil, they unwind slightly, so that rather than 3.6 residues/turn, there are closer to 3.5 (Crick, 1953). Two turns comprise an integral, 7-residue helical repeat, $(abcdefg)_n$. This *heptapeptide* repeat is particularly characteristic of hydrocarbon side chains, that is, alanine, leucine, isoleucine, and valine, which occupy positions a and d in the typical registration. The structural reason underlying this periodicity is that these residues form a nonpolar surface complementary to that of the partner helix. This pattern has become a reliable diagnostic for the presence of coiled α-helices.

The 7-residue repeat of nonpolar side chains occurs 41 successive times throughout the length of tropomyosin. These repeats are grouped together approximately six at a time, giving tropomyosin an unusual repeat of 42 residues ($42 = 6 \times 7$) that repeats seven times in the intact molecule. This number is also somewhat mystical, because each tropomyosin dimer interacts with seven actins. The most probable origin for these 42-residue repeats is a combination of gene duplications (McLachlan et al., 1975).

A third periodicity relates the acidic side chains, aspartic acid and glutamic acid, which bear a negative charge at neutral pH. These residues recur primarily in positions b, f, and g of the heptapeptide repeat. However, they are conspicuously absent from some regions and concentrated in others. This periodicity is not genetic, but statistical in origin. It probably reflects the evolutionary adaptation of tropomyosin to the repeat length of the actin helix with which it interacts. For this reason, it eluded detection until Fourier analysis of a binary representation of the sequence revealed an unusually strong fourteenth Fourier coefficient indicating that clusters of Asp or Glu residues repeated almost exactly every 19.6 residues along the sequence (Figs. 5-4, 5-5).

The periodic repetition of the heptad in long coiled-coils affords insight into deeper sequence-structure-function relationships. Tropomyosin, together with the troponin regulatory complex, mediates the regulation of muscle contraction by calcium. A dimer of tropomyosin, together with a complex of calcium binding (C), tropomyosin binding (T), and inhibitory (I) subunits of troponin binds to seven actin monomers in skeletal muscle thin filaments. This arrangement has two important functional consequences.

The tropomyosin dimer rides in the grooves of the actin double helix, where it assumes one of two positions. Normally, at low calcium concentrations the muscle is resting, and the tropomyosin "rope" is anchored near the edge of the groove, where it inhibits the approach of myosin heads from the thick filaments. When the calcium concentration increases, the troponin C subunit binds Ca^{++} and this event triggers a movement of tropomyosin toward the center of the groove, leaving the myosin binding site open, so that contraction can proceed. Coordination of actin monomers by tropomyosin insures that they will behave cooperatively, approaching two-state, on-off behavior.

A very strong principle of biological architecture is that things that fit together should have complementary features. The functional and biological significance of the repeating intervals of negative charge also has to do with the interaction with actin. It occurs fourteen times along the sequence, which covers seven actin molecules, and hence is likely to correspond to two different sets of binding sites ("on" and "off") for actin monomers in thin filaments. The thin filament has a crystallographic repeat distance of 770 Å for 14 actin monomers. Actin monomers are thus about $770 \text{ Å}/14 = 55$ Å apart along the thin filament. In contrast, the 42-residue genetic repeat in tropomyosin spans a physical distance of $42 \times 1.49 \text{ Å} = 62$ Å. Negative patches exactly coincident with the genetic repeat would quickly disengage from the actin monomers along the actin helix. A 19.67×2 residue repeat recurs at physical intervals of ~58 Å, which provides a closer match to the actin helix. For this reason, it is referred to as "actin adapted" (McLachlan & Stewart, 1975).

In this way, the structure of tropomyosin, its interactions with thin filaments, its evolutionary history, and its functionality were deduced almost entirely from its primary sequence, together with a few physical repeat distances! The linear arrangement of information "coding" for the interhelix binding sites (nonpolar amino acids at positions a and d of the heptapeptide repeat) and for interactions with actin (the statistical repeat of Asp and Glu) is a paradigm for the much more complicated encoding of three-dimensional information in the sequences of globular proteins.

Coiled-coils as Subdomains in Globular Proteins. Bundles of α-helices in globular proteins assume geometries reminiscent of Crick's "knobs-in-holes" structures. They are invariably tilted, relative to one another, and can contain different numbers of helices. A second preferred way to bring two or more α-helices together involves a different tilt angle of $-50°$ (Chothia et al., 1977). Although the observed crossing angles are close to those identified by Crick, the underlying side chain packing rules identified by Chothia, Levitt and Richardson differ from those of Crick.

The coiled-coil motif provides quaternary contact surfaces involving otherwise globular domains in proteins and ribonucleoprotein complexes (Fig. 5-7). An important example is the "leucine zipper" motif, so called because residues at position d of the heptad are almost invariably leucine in the leucine-zipper motif (Fig. 5-7a). This motif involves an interaction between two parallel α-helices that mediate the dimerization of various transcription factors, such as the yeast GAL4 transcriptional activator (O'Shea et al., 1991). The sense of the component helices in such an interaction is unimportant, and both parallel and antiparallel coiled-coils (Fig. 5-7b) have been observed. Antiparallel coiled-coils occur in the amino terminal tRNA binding sites of seryl (Biou et al.,

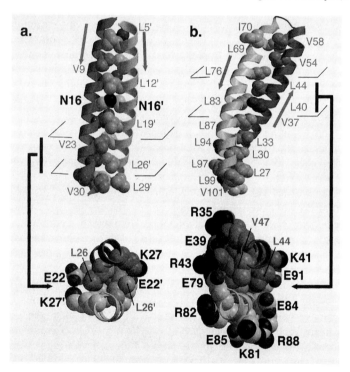

Figure 5-7. Coiled-coil motifs in otherwise globular proteins provide high resolution details. (a) The *parallel* "leucine-zipper" from the yeast GCN4 transcriptional regulator (O'Shea et al., 1991). (b) The *antiparallel* "leucine ladder" from seryl-tRNA synthetase (Cusack et al., 1990). The two different helices in each coiled-coil are shaded differently. Cross-sections of each coiled-coil are shown below to illustrate the complementary packing between nonpolar residues in positions a and d of the helices, and their reinforcement by complementary ion-pairing interactions between charged side chains in positions b, f, and g. Ion pairing also provides a means to surround the complex with groups having a high hydration potential, ensuring solubility. Gray labels indicate the inward pointing nonpolar side chains; black labels indicate charged residues that form salt bridges on the surface.

1994; Cusack et al., 1990) and phenylalanyl-tRNA synthetases (Goldgur et al., 1997). In another antiparallel variant, called the "Alacoil" (Gernert et al., 1995), positions characteristically occupied by leucine are actually alanine (as it frequently is in tropomyosin, Figure 5-4) and the two helices pack together unusually tightly. In these examples the details of the coiled-coil interaction can be examined much more precisely than they can be in the case of tropomyosin (Phillips et al., 1979), and they reveal delicate modulations of the rudimentary knobs-in-holes model first proposed by Crick.

Polar, and especially charged side chains recur at positions e, f, and g of the heptad. Both parallel and antiparallel coiled-coils make use of oppositely charged side chains in these positions to form both intra- and intermolecular ion pairs (lower illustrations in Fig. 5-7). The detailed patterns in such interactions are characteristic of the sense of the interacting helices (Cusack et al., 1990; O'Shea et al., 1989).

Coiled-coils and Protein Design. Coiled-coils represent the simplest level of assembly related to the formation of higher order protein structures. Specifically, the side chain packing rules are uncomplicated by the multiple conformations of beta and loop structures. Thus, the influence of packing can be scrutinized and tested in a straightforward fashion. It is, therefore, not surprising that the most substantial progress in protein design has come from the generation of model proteins based on the principles outlined here for tropomyosin (Lovejoy et al., 1992). Indeed, it appears that one such study has produced the first bona fide success in the design of a protein having a biological range of variation in rates of hydrogen exchange rates (Harbury et al., 1993, 1994, 1995; Lumb & Kim, 1994, 1995a, b; Lumb et al., 1994).

Secondary Structure: the Architectural Repertoire

The Linderstrøm-Lang hierarchy was formulated within a year of the description in 1951 by Pauling, Corey, and Branson (Pauling et al., 1951) of the secondary structures found in proteins, namely α-helices, and antiparallel and parallel β-sheets (Fig. 5-8). These helical structures implied the hierarchy by providing repeating sets of interactions between functional groups along the polypeptide backbone chain that created, in turn, irregularly shaped surfaces of projecting amino acid side chains.

The helical secondary structures in proteins arise from repeating patterns of similar peptide dihedral angles (ϕ,ψ; Fig. 5-2c) for successive residues. Reducing the conformation of each residue to two numbers facilitates their representation on a two-dimensional graph. Contours representing the stereochemically accessible regions of such plots were first compiled by Ramachandran, using a hard sphere potential treating atoms as billiard balls (Ramachandran et al., 1963). They have been revised using many successive levels of sophistication, initially involving potential energy calculation and most recently free energies obtained by thermodynamic integration of molecular dynamics simulations (Anderson & Hermans, 1988; Scully & Hermans, 1994).

These contours imply a thermodynamic tendency for successive residues along a chain to collect at the locations of the minima. Refined proteins in the database behave in the expected fashion; strings of successive residues often lie in the same potential energy well (Fig. 5-9). The resulting distribution means that proteins contain uninterrupted segments in which the residues have nearly the same (ϕ,ψ) angles, and are thus either α-helices or β-strands.

Residues in α-helices cluster near (ϕ,ψ) = ($-60, -45$), while those in β-sheets fall in the upper left-hand quadrant of the Ramachandran plot. Figure 5-10 (Anderson & Hermans, 1988) shows free energies of (ϕ,ψ) space as normalized probabilities for the alanine

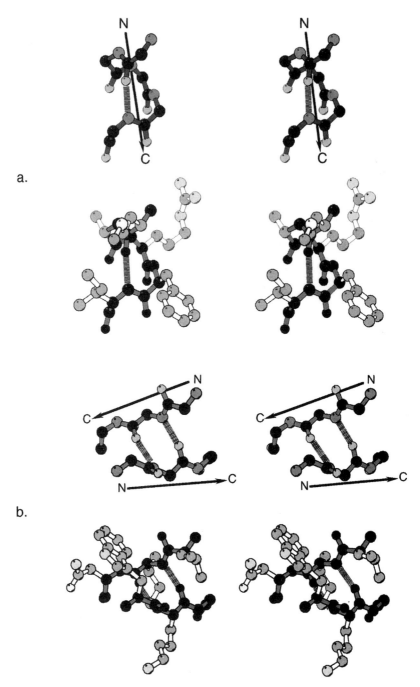

Figure 5-8. Secondary structures in proteins. All six images are stereo pairs that can be viewed with crossed eyes. (a) One turn of α-helix, with and without side chain atoms. (b) a stretch of antiparallel β-structure, without and with side chain atoms. (c) a stretch of parallel β-structure, with and without side chain atoms. Polar hydrogen bonding partners of the backbone chain are innermost, while the side chain atoms constitute the surface of all three secondary structures. Note that the hydrogen bonding directions in (b) are parallel whereas they are not in (c).

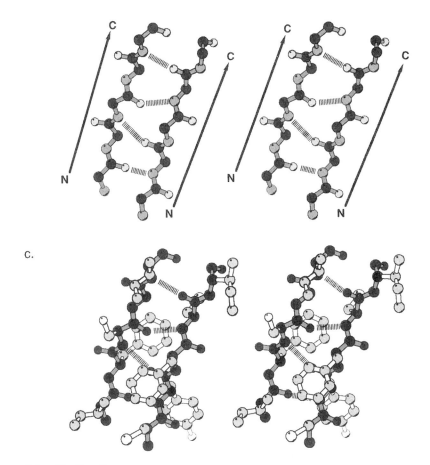

Figure 5-8. (Continued)

dipeptide; extended β-conformations have a deeper free energy minimum than the α-helical conformation, and are stabilized by conformational energy as well as by entropy. This plot demonstrates that conformational angles belong almost without exception to one of the two secondary structures. There is little detectable population of the transition regions.

The intermediate chain folding represented by secondary structures has two important consequences: it buries a considerable number of backbone carbonyl and amide groups, joining them in hydrogen bonding configurations (Fig. 5-9 second from left), and exposes a highly textured surface composed almost entirely of side chains (Fig. 5-9 far right). Thus, the innermost interactions in a folded protein find suitable intramolecular partners for a major proportion of the highly solvated peptide functional groups in a protein without forcing ϕ or ψ outside the favored regions for α and β secondary structure. Molecular dynamics simulations also indicate that the interactions between side chains are a major influence even on the formation of very short fragments of secondary structure (Zhang & Hermans, 1994). The textured exterior surfaces ensure strong van der Waals interactions with those of complementary secondary structures elsewhere in the sequence.

Complementary packing surfaces are the key link to the next hierarchical level, ter-

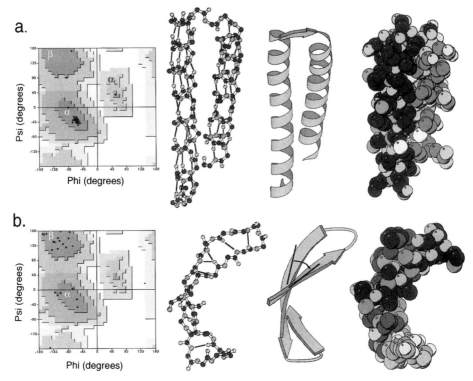

Figure 5-9. (a) Alpha and (b) beta secondary structures as "building blocks" for tertiary structure. The (ϕ,ψ) signature of each type of structure is illustrated in the Ramachandran plots to the left. Backbone atom ball-and-stick figures, ribbon drawings, and space-filling representations are presented from left to right. The helical structure in (a) is the ROP protein, a repressor (Banner et al., 1987). The fragment in (b) is from α-chymotrypsin (Blivens & Tulinsky, 1985). Note the high degree of curvature afforded by the antiparallel β-structure.

tiary structure. In secondary structural elements of the size normally found in proteins (Fig. 5-9), projecting side chain surfaces behave like velcro, binding secondary structures from distant parts of the primary structure via the interdigitation of side chains and generating considerable stabilization through van der Waals interactions. Thus, the Linderstrøm-Lang hierarchy arises from the distinction between forces orienting backbone atoms within secondary structures—mainly preferred dihedral angles and hydrogen bonds—and the hydrophobic and van der Waals forces between their resulting surfaces.

Changing Direction. It is common to use the word "fold" to describe the course of a polypeptide chain in a protein (Flöckner et al., 1995; Jones & Thornton, 1993; Jones et al., 1992; Shortle, 1995; Srinivasan & Rose, 1995). Its meaning, "to double over," implies a fundamental property without which proteins would not assume compact, globular shapes and would, therefore, remain fibrous. In this sense, a critical element of protein structure is the ability to change direction abruptly. Most direction reversals in proteins result from a limited repertoire of discrete structures in which the chain direction is changed within the span of several residues. These structures are termed "hairpin," "reverse," or simply "β-" turns (Crawford et al., 1973; Lewis et al., 1973).

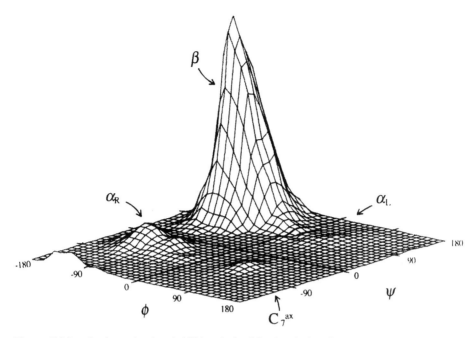

Figure 5-10. Conformational probabilities obtained for the alanine dipeptide by free energy perturbation molecular dynamics simulations. Reprinted from Anderson & Hermans, 1988, with permission.

Hairpin or β-turns are specialized secondary structures that effect chain reversal without violating the conformational probabilities shown in Figure 5-10. They are characterized by the (ϕ,ψ) angles of two successive residues, referred to here as $i + 1$ and $i + 2$. The two residues can be represented as a vector on a Ramachandran plot. They fall into three principal categories (Fig. 5-11), called simply Types I, II, and III. Types I and III are

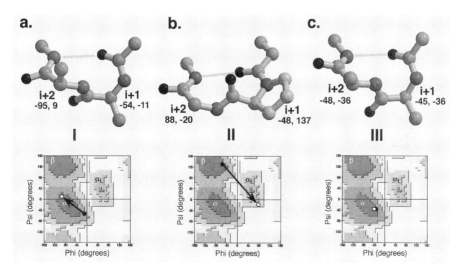

Figure 5-11. The common hairpin turn conformations. The vectors specified by the initial and final (ϕ,ψ) values for each turn are shown below each illustration.

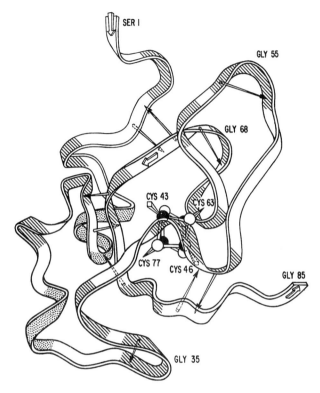

Figure 5-12. A small (8.5 kDa) iron-sulfur protein constructed extensively from hairpin turns. Approximately 40% of the main chain conformations are involved in residues $\{i + 1, i + 2\}$ of hairpin turns. Reprinted from Carter et al., 1974, with permission.

closely related; from their dihedral angles it can be seen that the Type I turn begins roughly in the region of right-hand α-helix and ends close to the region occupied by a β-strand, whereas Type III is actually a helical configuration; both residues falling close to the α-helical minimum. From the number of residues and atoms (including hydrogen) contained in the hydrogen-bonded loop closed by one turn, that helix is designated as a 3_{10} helix. By comparison, the α-helix is a 4_{14} helix. For this reason Type III turns are sometimes excluded from this classification (Hutchinson & Thornton, 1994). The properties of the 3_{10} helix make it more rigid and hence generally less stable thermodynamically than the α-helix (Zhang & Hermans, 1994). Observed 3_{10} helices in proteins are rarely longer than one turn (Karpen et al., 1992), and they occur most often at the ends of α-helices where they serve to change direction. They are grouped here with the conventional hairpin turns.

A large variance in the observed conformational angles for turns has led to considerable confusion in the literature regarding their proper designation. Molecular dynamics simulations have clarified the situation substantially. The precise geometry of turns in proteins is determined by a number of factors. Among these is the fact that the conformational energy minima for the two residues are somewhat different from those leading to good hydrogen bonds between $C{=}O_{i+1}$ and N_{i+2}, which close the ten-atom loop in each case. Thus, the variance in turn conformations is correlated with the extent to which these two groups can also hydrogen bond to water and hence with the proximity of the turn to the surface of the protein. When they are sufficiently protected from water, the ψ_{i+2} of approximately 0° improves the energy of the hydrogen bond (Scully & Hermans, 1994). This phenomenon accounts for separation of the observed type I turns into two subclasses, one of which has a ψ_{i+2} value close to 0°, while for the other (Type VIII, Hutchinson & Thornton, 1994) it is about 120°. If these two types of turns are considered to be generically Type I, Types I, II, and III (3_{10} helical) turns account for 92% of nearly 4000 turns analyzed in the most comprehensive compilation to date (Hutchinson & Thornton, 1994). The remaining 8% are formed with conformational angles that are approximately the negative of those for Types I (Type I′) and II (Type II′). The Type I′ turn has an important distinction: it is the only turn that naturally has the correct out-of-plane disposition of the four successive C^α atoms to accommodate the right-handed twist of an antiparallel β-loop (Richardson & Richardson, 1987; Sibanda & Thornton, 1985).

Turns can become a dominant element in small proteins, especially those that bind metals. Turns serve dual functions of creating internal volumes for the metal ions or clusters and, alternatively, providing NH—S hydrogen bonds to cysteine S_γ atoms that bind the metals (Carter et al., 1972). An extreme example is the high potential iron protein, which uses 17 turns in a sequence of 85 residues to bind a cluster of four iron and four inorganic sulfur atoms (Fig. 5-12).

A small number of significant correlations exist between specific amino acids and positions $i + 1$ and $i + 2$ of specific turns (Hutchinson & Thornton, 1994). Proline is acceptable in either position of Type I and Type III turns, but only in position $i + 1$ of Type II turns, as illustrated in Figure 5-11b. The Type II turn is distinctive because the dihedral angles of the second residue, $i + 2$, are "forbidden" for any residue having a β-carbon atom. For this reason, Type II turns almost invariably contain glycine in this position, and are actually stabilized by having a D-alanine in this position, the D configuration forcing position $n + 2$ to adopt the ϕ angle of a Type II turn.

A final unusual, but potentially important, means of changing the direction of a polypeptide chain is the left-handed α-helical conformation. This conformation has an only slightly less favorable energy than that for right-handed helices. However, the rarely observed occupants of this region (about 1% of all residues) are invariably single residues (Fig. 5-13). These α_L residues occur at the carboxy termini either of α-helices or β-strands, abruptly terminating these secondary structures and introducing substantial changes in direction. The other frequent location of α_L conformations is in position 1 of a G1 β-bulge (Chan et al., 1993; Richardson et al., 1978).

Curvature. The β-structures, antiparallel and parallel, display a new, higher-level folding characteristic: their right-handed twist (Figs. 5-8b, c and 5-9b) (Chothia, 1973). The preferred crossing angle for two interacting β-strands is nearly always clockwise when viewed from the strand that rotates (Fig. 5-9b). Thus, β-structure actually coils in much the same way as was described for α-helical coiled-coils, the clockwise crossing angles generating a right-handed twist.

The right-handed twist has a dominant impact on the architecture of proteins containing β-structure. It derives from thermodynamic properties of the extended polypeptide conformation. Chothia advanced an insightful, but qualitative explanation in terms of an excess in the numbers of states available to right-, relative to left-handed twists. The locus of points on the Ramachandran diagram for untwisted helices with $n = 2$ residues/turn is a line running approximately down the diagonal from upper left $(\phi,\psi) = (-180,180)$ to lower right through the origin (0,0). As this line passes through the region of β-structures, it divides the β-conformations into those that curve to the right (above the line $n = 2$) and those that curve to the left (below it). Since more accessible conformations lie

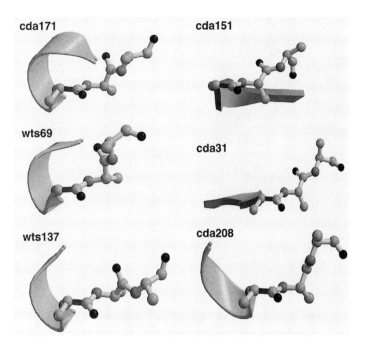

Figure 5-13. A selection of left-handed alpha helical configurations, showing their role in changing direction. The C$^\alpha$-C bond of each has been rotated to the same orientation in each case, and is black. The examples are taken from cytidine deaminase (cda) (Betts et al., 1994) and tryptophanyl-tRNA synthetase (wts) (Doublié, 1993) and are indicated by their residue numbers as cda n or wts n.

above the line, β-sheets nearly always twist to the right, because this conformation has greater entropy for comparable enthalpy (Chothia, 1973). Recently, Wang and Hermans updated this argument, showing by molecular dynamics calculations that the *enthalpy* also favors right-handed over left-handed β-conformations (Wang et al., 1996).

Various mechanisms augment the curvature in naturally occurring β-structures. The most important of these entails the inclusion of two residues between two β-type hydrogen bonds on one strand of a sheet opposite a strand that has a single residue between the same hydrogen bonds. This feature is called, felicitously, a "β-bulge" (Richardson et al., 1978). There are two predominant configurations, and they are surprisingly widespread in proteins with β-structure (Chan et al., 1993). The outcropping from the additional residue increases the natural right-handed twist (Chan et al., 1993); β-bulges are, therefore, nearly always found in highly twisted β-structures. An unusual cooperative array of stacked bulges actually leads to a reversal in the normal sense of the twist in thymidylate synthase (Matthews et al., 1989). A more dramatic way to increase the curvature of a sheet is to insert multiple turns in the chain reversals [see the turns associated with Gly55, Cys63, and Gly68 in Fig. 5-12].

Parallel β-Structures. These are actually a special case of secondary structure because they cannot form without intervening secondary structures, which usually take the form of an alpha helix. The most common such structure was identified by Richardson (1973), who called it a "$\beta\alpha\beta$ crossover connection" (Fig. 5-14).

a.

b.

c.   d.

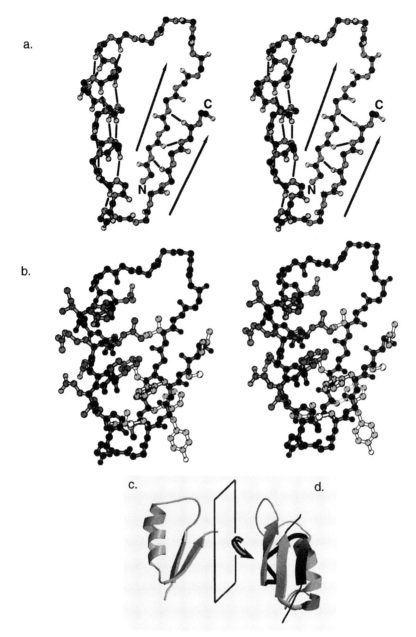

Figure 5-14. The β-α-β crossover connection is an important building block intermediate between secondary and tertiary structures. Formation of parallel β-structure invariably requires a crossover. Usually this crossover assumes an α-helical conformation, as in the N-terminal crossover in tryptophanyl-tRNA synthetase (Doublié, 1993). Side chains are omitted in (a) to emphasize the course of the main chain. Side chains from the two β-strands and the α-helix are included in (b) and the mainchain atoms are all black. A ribbon drawing of this crossover (c) is compared to the one existing case of a left-handed crossover (d). The main secondary structure elements in the lightly shaded portion in (d) are related to those in (c) by reflection in the suggested mirror. As discussed in the text, adjacent peptide segments in dark shading suggest that the underlying connection is composed of the adjacent darkly and lightly-shaded β-strands, and is actually right-handed. The left-handed connection may result from folding down an extended loop connecting the first and second β-strands, as indicated by the curved arrow.

178 / Carter

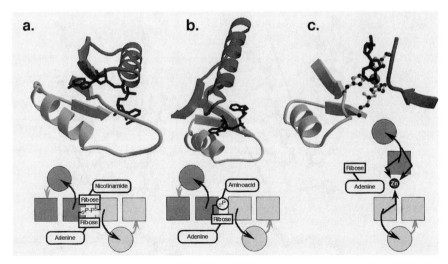

Figure 5-15. The Rossmann dinucleotide-binding fold and related enzyme active sites. (a) The Rossmann fold originally identified in lactate dehydrogenase (Buehner et al., 1973; Rossmann et al., 1974). (b) The Rossmann fold from tryptophanyl-tRNA synthetase (Doublié, 1993). (c) The active site of adenosine deaminase (Wilson et al., 1991). In each case crucial active site interactions originate from the C-terminal loop of the first β-strand of β-α-β crossover connections. The block diagrams and ribbon drawing are shaded consistently with the light and dark shading of the component crossovers.

This crossover connection is such an integral part of so many proteins that it deserves further description. It comes in several varieties, depending on the number of additional strands that intervene between the leading and trailing β-strands (Richardson, 1977). By far the most common crossover connection is one where these strands are adjacent. The earliest suggestion of the importance of this motif was the recognition by Rossmann that an underlying tertiary folding pattern in lactate dehydrogenase was nearly superimposable on that of an unrelated protein, flavodoxin (Buehner et al., 1973; Rossmann & Argos, 1977). It was described, appropriately, as a "supersecondary structure," capturing the higher level organization while preserving the underlying hierarchy. The common structure in this case was a pair of $\beta\alpha\beta$ crossover connections oriented approximately 180° apart such that the four beta strands formed a single parallel β-sheet (Fig. 5-15).

At about the same time, Jane Richardson observed that just as the strands of β-structure were always twisted in a right-handed sense, so the common $\beta\alpha\beta$ crossover nearly always makes a right-handed spiral gyre, and suggested that the two twist senses were related. In fact, this is probably the case. Only one left-handed crossover connection has ever been identified, from the protein subtilisin. That structure is compared to the right-handed configuration in Figure 5-14c and d. It is, in fact, nearly "knotted." The N-terminal flanking segment, consisting of a β-strand and a coil (darkly shaded) passes between the two parallel β-strands from the crossover connection itself. It could be argued consistently that this is actually a right-handed crossover connection. An obvious way to assemble this structure would be first to form the *right-handed* $\beta\alpha\beta$ crossover connection involving the darkly shaded β-strand and the middle strand. The long intervening loop in an extended configuration could subsequently fold back, forming the third strand of the β-sheet, as suggested by the arrow. The helix itself is noteworthy in the subtilisin substructure. It is interrupted by a long looping segment that returns for one additional turn (medium shading), before initiating the final β-strand.

The Rossmann Nucleotide Binding Fold and Supersecondary Structure. The common right-handed βαβ crossover connection is the fundamental active site building block for a statistically very significant fraction of all enzymes (Fig. 5-15). Representative enzymes include examples from intermediary metabolism (Farber & Petsko, 1990; Neidhart et al., 1990), amino acid biosynthesis (Hyde et al., 1988; Wilmanns et al., 1991), translation (Carter, 1993), energy transduction (Abrahams et al., 1994; Rayment et al., 1993), and signal transduction (Milburn et al., 1990; Pai et al., 1990). The appearance of proteins containing this motif must, therefore, be considered an extraordinary and fundamental evolutionary achievement.

The Rossmann nucleotide binding fold (Figure 5-15 a, b) (Buehner et al., 1973; Rossman et al., 1974) was first observed in the structures of flavodoxin, where it is involved in flavin binding, and in lactate dehydrogenase, where it mediates coenzyme (NAD) binding. It is almost invariably and almost exclusively found in proteins that bind to purine nucleotide coenzymes. This one-to-one correspondence between the presence of this fold and the binding of purine nucleotide coenzymes led to examination of the evolutionary relationships between different nucleotide binding proteins.

In dehydrogenases, the dinucleotide NAD binds so as to conform to this approximate symmetry, with the adenine and nicotinamide moieties bound to pockets more or less on opposite sides of the sheet from the crossover helices. In such cases, the loop between a central β-strand and helix invariably forms a binding site for one of the phosphates connecting the two nucleotide moieties. A similar arrangement occurs in unrelated proteins containing the Rossmann fold, such as the class I aminoacyl-tRNA synthetases (Carter, 1993). In tryptophanyl-tRNA synthetase, the Rossmann fold binds the activated amino acid in very similar fashion, the amino acid occupying the location occupied by the nicotinamide in the dehydrogenases (Fig. 5-15b).

Apparently, many of the proteins containing the dinucleotide-binding fold [many dehydrogenases and ten of the aminoacyl-tRNA synthetases (Carter, 1993)] descended from common ancestors. The consistent association of this fold with nucleotide binding is intriguing because of the widespread use of nucleotide triphosphates in energy metabolism. Relationships observed in such complexes may represent important evolutionary solutions to problems of storing and utilizing chemical free energy, which must have been one of the earliest "biological" properties acquired by macromolecules.

β-Helices. The obvious alternative way to form parallel β-structure is to form a continuous helix, allowing the strands to align along the helix axis. An exceptional example of such a structure (Fig. 5-16) was identified only recently, and termed the parallel β-helix (Yoder et al., 1993b). It consists of cylindrical arrangements of two or three successive β-strands which, upon repetition of the motif, form opposing faces, each of which is a parallel β-sheet. It is a very specialized structure, as evidenced by the fact that the hydrophobic core within the three-stranded cylinder shows highly regular vertical packing of identical side chains that repeat in the primary sequence (Jurnak et al., 1994; Yoder et al., 1993a, b). As is evident from a view down the cylinder (Fig. 5-16b), the three classes of β-strands are quite inequivalent. The two longer classes (light and medium shading) form a sandwich, whose inward-facing side chains can interact. The third class consists of dipeptides and forms an asymmetric "corner."

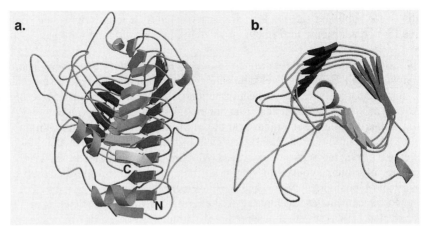

Figure 5-16. Parallel β-helix substructure in pectate lyase (Yoder et al., 1993b). Parallel strands recur in groups of three, which are shaded differently. (a) The entire protein showing the ascending cylinder of parallel β-structure. (b) An orthogonal view of the central section of the cylinder. Note that the environments of the three classes of β-strands are distinctly different, and that two sets of the connecting segments are quite uniform whereas the third is quite idiosyncratic.

Stereochemistry. Refinements of many proteins at high resolution have shown that very few residues have unfavorable conformational energies. One of the earliest (Freer et al., 1975) and most consistent (Morris et al., 1992) observations resulting directly from the refinement of protein crystal structures is that the more rigorously the X-ray and chemical bonding constraints are satisfied, the more closely the torsion angles of the polypeptide backbone conform to local potential energy minima characteristic of well-defined secondary structures. Subsequent analysis soon extended this conclusion to amino acid side chain conformations; energetically preferred conformations were substantially favored over less stable ones in the folded conformations of a large database of highly refined coordinates (Janin et al., 1978).

Structure refinement adjusts the coordinates of the model to minimize $\Sigma(|F_{obs}| - |F_{calc}|)^2$ while preserving standard bond lengths and angles. Since it is determined only by the X-ray intensity data and the stereochemical restraints, it introduces no information concerning the backbone dihedral angles, and is, therefore, an objective indication that protein conformations lie almost entirely in low energy regions of the Ramachandran plot (Fig. 5-17).

Exceptional residues with disallowed conformations occur only in situations where there is some compensating advantage. Generally, such residues are in turns, where a ψ angle close to 0° occurs frequently to improve the geometry of the hydrogen bond closing the turn (Scully & Hermans, 1994). Another specific example occurs in the active site of *E. coli* cytidine deaminase, where a loop of one subunit crosses the dimer interface to form a hydrogen bond to an active site histidine on the opposite monomer (Fig. 5-18) (Betts et al., 1994). The ψ angle of Phe233' is obviously strained in this configuration, but the resulting hydrogen bond orients an active site histidine. Moreover, the Phe233' side chain is in position to interact with part of the substrate. Deviations from the canonical dihedral angles are, therefore, not random, but are correlated with relevant structural variations whose function can be identified.

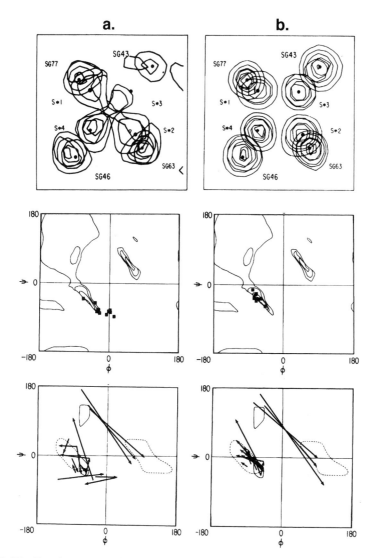

Figure 5-17. Protein structure refinement improves polypeptide chain stereochemistry (Freer et al., 1975). Three properties are shown (a) before and (b) after crystallographic refinement. The first row shows the electron density for eight sulfur atoms in the active site and demonstrates that the phases calculated from the refined model are superior to those based on the initial model. The second row shows α-helical residues, several of which lie outside the preferred regions of the Ramachandran plot in (a) and move into the preferred region after refinement. The third row shows the vectors representing the hairpin turns. Type I and type III turns occupy the left cluster, type II turns the upper right cluster. Before refinement, a number of these are actually pointing nearly opposite to where they should. The tight clustering after refinement indicates that the local geometry of the main chain has experienced *correlated* improvements in stereochemistry, bringing these dipeptides into canonical turn stereochemistry.

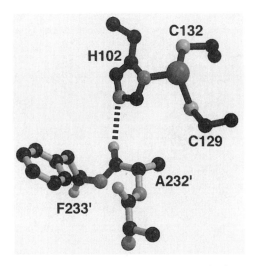

Figure 5-18. A legitimate conformational outlier. Residue Phe233' in cytidine deaminase has $(\phi,\psi) = (68°, -75°)$, which lies in a forbidden region of the Ramachandran plot. Contributions to the active site of this enzyme arise from two different subunits across the dimer interface, as indicated by light and dark shading and the use of primed residue numbers to indicate the second monomer. The disallowed ϕ angle allows the carbonyl oxygen of the preceding residue Ala232 (dark shading) to form a hydrogen bond with the active site Zn^{++} ligand, His102 from the other subunit of the dimeric enzyme.

A second generalization from protein refinements is that there is very little evidence in refined protein structures for what is frequently described inappropriately as "random coil" structure. Not only do α- and β-structures converge toward favored regions of the Ramachandran plot, pairs of residues within hairpin turns also assume characteristic (ϕ, ψ) angles of such turns upon refinement (Fig. 5-17). The first protein structure to be refined using combined X-ray, and bond length and angle constraints was the high potential iron protein, HiPIP (Freer et al., 1975). Analysis of that structure revealed that despite the fact that the 85-residue polypeptide had to accommodate a large internal cavity to bind a Fe_4S_4 cluster, all but two of the residues participated either in secondary structures (two helices and five short β-strands) or one or more of 17 hairpin turns (Fig. 5-12). The canonical building blocks accounted for nearly the entire length of the chain (Carter et al., 1974; Freer et al., 1975). The original observation has recently found an elegant echo in a detailed study of the "statistical coil" properties of polypeptide chain conformations in a large subset of the protein structures refined up to 1995 (O'Connell et al., 1997). As originally proposed, once the canonical α, β, and turn structures are deleted from the database of main chain conformations, very little is left that could be described as "statistical coil."

The important conclusion is that a small number of canonical secondary structures provide the repertoire for essentially all of the known protein structures. Since there are a number of different β-turn geometries, and since residues at the junction between different secondary structures can participate in either or both of the two in somewhat different ways, the manifold of possible structures built from this repertoire of secondary structural building blocks is very large. Specialized structures based on integration of the three have been identified at the beginnings and ends of helices (Presta & Rose, 1988; Richardson & Richardson, 1988) and in turns associated with certain recurring β-structures (Hemmingsen et al., 1994).

***Cis* Peptide Bonds.** The energy difference between *cis* and *trans* peptide conformations (about 2.6 kcal/mole) is of about the same magnitude as that provided by hydrogen bonds, and is even less for peptide bonds arising from imides of the form X-Pro. This fact, plus the existence of enzymes whose function is believed to involve isomerization of such imide peptide bonds, strengthens the suspicion that *cis* peptide bonds involving proline are important in some way for protein folding or protein structure.

A statistical study by Stewart et al. (1990) shows that the vast majority of *cis* peptide bonds in proteins (85%) are found in X-pro sequences. Moreover, the greater the frequency of their occurrence, the higher the resolution and refinement of the structures. Since protein models are built with the assumption of *trans* peptide bonds, and since higher resolution implies that the structures are also more accurate, this suggests that the true frequency of *cis* X-Pro peptide bonds in proteins is really closer to the higher value of about 8%.

Another important number to bear in mind is the energy barrier preventing isomerization, estimated to be about 20 kcal/mole, which implies a half-time for the reaction of about seven seconds. One possible role for *cis* X-Pro sequences is to introduce slow steps into protein folding pathways. This is consistent with the fact that X-Pro sequences nearly always occurred in external loops. This observation would be even more provocative if it were the case that *cis* imide peptide bonds occurred frequently in loops connecting independently folded modules in proteins. Apparently, no one has yet done a systematic study to test this possibility.

Bound Solvent Water Molecules. Associated with the refinement of protein models, per se, is the introduction of a number of fixed solvent molecules, assumed to be water, which interact with specific surface functional groups. Since they generally comprise the first sphere of hydration, they complement main chain-main chain hydrogen bonding interactions within the structure. One can, therefore, assess how fully satisfied these requirements are (McDonald & Thornton, 1994). For high resolution structures, there are three times as many unsatisfied backbone NH groups (5.8%) as CO groups (2.1%). Surprisingly, perhaps, less than 2% of all main chain groups are sequestered from water without some type of electrostatic compensation in the form of neighboring hydrogen bonding partners.

Intramolecular and Intermolecular Hydrogen Bonding. As with other thermodynamic properties of native proteins estimated from crystal structures, the number of polar groups without hydrogen bonding partners decreases as the resolution of the structures increases. However, although this is a statistically valid observation, its interpretation is unclear. On the one hand, one is tempted to conclude that the deficiencies of lower resolution data sets are masking effects that would be seen at higher resolution. That is, the presumption is that what is observed in higher resolution crystal structures can be interpreted as implying that the structures at lower resolution would show the same features, if one could measure the higher resolution data. However, an alternative interpretation, that the structural properties of proteins that crystallize in forms with high resolution diffraction patterns are somehow different from those of crystal forms that diffract only to lower resolution limits, cannot be excluded.

A qualitatively different conclusion has been drawn from the spatial distribution of water molecules surrounding proteins in crystals (Kuhn et al., 1992). The geometric requirements for binding water at protein surfaces are well defined enough that one can frame a predictive model, based on analysis of the configurations of exposed main and side chain hydrogen bonding groups. This model predicts correctly the positions of waters identified in high resolution (1.2 Å) crystal structures to better than the crystallographic resolution of the structures (Roe & Teeter, 1993). At a higher level of integration, the distribution of water molecules on protein surfaces is also decidedly nonrandom. Protein surfaces are often highly convoluted and characterized by deep grooves or crevices. These crevices, which include most active sites, are shaped closely to the dimensions of water molecules, and the number of bound water molecules within the grooves is proportionately greater (2 to 3 times greater) than on convex portions of the surfaces (Kuhn et al., 1992).

Tertiary Structure: Stability and Uniqueness

Formation of unique and stable three-dimensional structure by a polypeptide requires a new kind of bonding interaction that is scarcely even prefigured in the discussion of secondary structure. Uniqueness and stability are hardly obvious companions; as the number of possible configurations for a polymer grows, the entropic cost of sustaining only one of many increases rapidly (Dill et al., 1995). The new bonds that pay this rapidly increasing cost are called "tertiary" interactions, and they arise because distant parts of a protein interact across what might be called an "intramolecular interface." Tertiary interactions probably play a dominant role in stabilizing the small number of native conformations, relative to the overwhelming excess of denatured and other nonnative conformations. Although it appears that no completely satisfactory description of tertiary structure has been assembled, there is a consensus that tertiary structure entails at least two new concepts.

1. The main chain adopts a characteristic "fold" with some degree of imprecision, both with respect to primary structure encoding (Flöckner et al., 1995; Jones et al., 1992; Orengo & Thornton, 1993; Orengo et al., 1993; Shortle, 1995; Srinivasan & Rose, 1995) and to the mechanism of renaturation (Dill & Chan, 1997).
2. "Packing" or stereochemical complementarity (Richards, 1991; Richards & Lim, 1993) is an important source of the new bonding interactions required to stabilize unique native structures.

These two ideas seem to be contradictory at some level. It has been proposed on the one hand that (1) implies (2) (Behe et al., 1991) and hence that (2) is a secondary consequence of as yet unspecified forces governing (1). On the other hand, requiring too much precision in packing interactions appears incompatible with (1). The interplay between these two statements, therefore, probably lies close to the heart of the outstanding problems in protein structure.

Stereochemical Complementarity and Packing. Qualitatively, tertiary interactions include the same types of interactions that stabilize secondary structure: intramolecular hydrogen bonds and intramolecular hydrophobic bonds and, to a variable extent, electrostatic interactions between charged groups. Folding an extended chain into the component secondary structures of a protein involves a significant reduction in accessible surface area for side chains, but the surfaces that remain are all "outside" surfaces. Formation of tertiary structure converts a subset of these surfaces into "inside" sufaces. The macroscopic velcro analogy was developed in the previous section on secondary structure. Another appropriate analogy is that of a three-dimensional jigsaw puzzle. Indeed, the term "puzzle pieces" has been used to describe specific and recurring patterns of tertiary interaction (Gernert, 1994).

Proteins do achieve an extraordinary degree of complementarity between distant parts. The backbone chain hydrogen bonding functionalities (Fig. 5-19a) and the textured surfaces formed by the side chains (Fig. 5-19b) are almost perfectly matched across the intramolecular interface of the high potential iron protein (Carter et al., 1974). It is a source of wonderment that the two very different types of interactions simultaneously assume nearly ideal configurations.

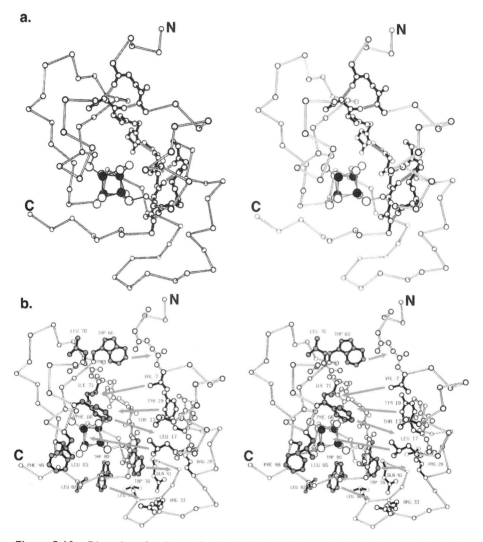

Figure 5-19. Dissection of an intramolecular interface. (a) Intramolecular hydrogen bonds (gray) between the first and second halves of the sequence. (b) Intramolecular hydrophobic bonds arising from nonpolar side chain packing across the interface. The C-terminal half is internally stabilized by four covalent bonds from cysteine thiolate atoms to the iron-sulfur cluster. Reprinted from Carter et al., 1974, with permission.

In general, "tertiary" interactions tend to involve a higher proportion of nonpolar side chain packing, relative to the number of intramolecular hydrogen bonds between main chains. The stabilizing effects of these nonpolar contacts are defined by opposing attractive and repulsive components of the van der Waals interaction potential. The former leads to relatively high packing densities; the latter to the avoidance of steric overlap (Richards & Lim, 1993). Imperfections in stereochemical complementarity can and are compensated by including buried water molecules wherever there are a sufficient number of accessible hydrogen-bonding sites (Radzicka & Wolfenden, 1994; Williams et al., 1994; Zhang & Hermans, 1996). Moreover, across the database there are numerous examples of side chain-side chain and side chain-main chain hydrogen bonds.

Stereochemical complementarity is apparently required in order to achieve the high side chain packing density observed in proteins. Various model systems (Chan & Dill, 1990; Dill et al., 1995; Richards & Lim, 1993; Yee et al., 1994) show that high packing density is also requisite for maintaining a specific structure; the number of different structures accessible to a polymer rises sharply as the packing density decreases below a threshold near the limit of closest packing. Condensed structures with excessive empty cavities are therefore disfavored, relative to those that fill space efficiently. Moreover, evidence from a large database of small molecule crystals indicates that close packing is the dominant determinant of crystal structures actually observed (Kitaigordosky, 1973). Thus the intuitive notion that the close side chain packing observed in proteins is at least in some sense a requisite for producing unique folded structures has a strong empirical basis (Richards & Lim, 1993).

The actual contributions of stereochemical complementarity to protein stability in general and to the determination of a particular fold are hard to assess because they intrinsically involve interactions between multiple groups. As a result, the implications available from the known protein structures have become amenable to quantitative analysis only very recently. The reasons have to do with the size of the structural database itself. Each part of a structure can be said to participate in a large number of different interactions, involving contributions from all neighboring parts. These interactions must be enumerated before they can be characterized. One way to do this is to consider an ascending hierarchy of interactions, grouping together first all the pairwise interactions, and then the higher order interactions between three, four, and more participating groups. Then interactions from the database of known structures can be sorted, either according to the distance between participating atoms along the polypeptide chain or to their composition, for the accumulation of probability distributions. The resulting probabilities provide a wealth of insight about packing interactions. However, this procedure leads rapidly to a very large number of different combinations. This combinatorial explosion means that the significance of specific patterns of nearest neighbor tertiary contacts cannot be assessed without a large number of observations.

An early attempt to discern significant patterns that might implicate stereochemical complementarity examined deviations in the number of contacts actually observed between different pairs of amino acids from those expected on the basis of random packing (Behe et al., 1991). Although enough accurate structures were present to provide reasonable statistics for that study, it failed to identify significant patterns. Properly speaking, however, three-dimensional tertiary interactions require a minimum of four interacting components; two such interactions lie on a line and three on a plane. Nearest-neighbor interactions in three dimensions are, therefore, uniquely defined by quadruplets (Singh et al., 1996; Tropsha et al., 1996). Thus, it becomes meaningful to speak of tertiary *patterns* only for those interactions involving four components.

Recently the database has reached a sufficient size to begin such an assessment, by using Delaunay tessellation of protein structures (Singh et al., 1996; Tropsha et al., 1995, 1996). This procedure, illustrated in two dimensions in Figure 5-20, is related to the construction of Voronoi polyhedra, used by Richards to represent the space-filling properties of side chains in protein interiors, and to analyze packing (Richards, 1974, 1977, 1985; Richards & Lim, 1993). For a collection of points, for example, the C^α positions in a protein structure, the Voronoi polyhedron surrounding each point is the collection of all points closer to that point than to any other members of the set. The polyhedra themselves can have irregular shapes, and different numbers of vertices. However, the number of such polyhedra that intersect at each vertex is constant, and constitute a *simplex* called the Delaunay simplex (a triangle in a plane and a tetrahedron in three dimensions).

The uniquely defined Delaunay simplices have three properties of interest for applications to protein structure: their volumes, shapes, and compositions. For a partic-

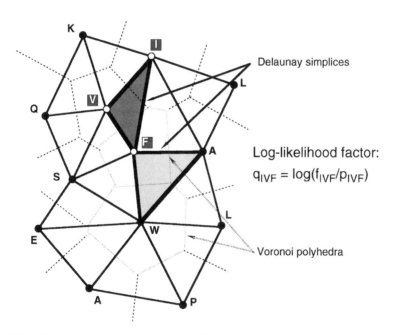

Figure 5-20. Delaunay tessellation and the identification of nearest neighbors demonstrated in two dimensions. Voronoi polyhedra are developed around each point in a set of points, here the C^α atom of a residue or the centroid of its side chain, by expanding the neighborhood of points in space that lie closer to a particular point than to any of the other points in the set. Whenever Voronoi polyhedra meet each other, the intersection invariably involves the same number of points, three in two dimensions, four in three dimensions. The three or four contributors to each intersection form a unique simplex, called the Delaunay simplex. The log-likelihood factor can be used to estimate a quasipotential of mean force. Properties of high-scoring tetrahedra in proteins are illustrated in Figure 5-26.

ular protein, all three properties can be compared with the statistical properties of identical simplices in the database. In one subset of high resolution structures, 94% of the possible quadruplets are observed at least once, and the total number of quadruplets is sufficient to contain roughly thirteen instances of each one (Tropsha et al., 1996). Thus, the statistical dispersion of observed frequencies, properly normalized for the length of the protein and the frequency of occurence of each amino acid, provides meaningful evidence regarding which quadruplets occur more or less often than expected in the absence of specific neighbor preferences. The frequency observed in the database for a quadruplet with a particular composition can be normalized to the frequency expected for a random distribution, and the ratio provides a likelihood function whose logarithm is an index of how unusually frequent or infrequent is the occurence of the quadruplet.

This likelihood function provides a useful figure of merit for the compatibility of a protein sequence with a particular structure (Tropsha & Carter, unpublished). Moreover, the distribution of quadruplets with high log-likelihood in any given protein is distinctly nonrandom (Fig. 5-26, vide infra). This and similar statistics regarding quadruplet volumes and shapes will provide increasingly useful evidence regarding the role of three-dimensional packing interactions and stereochemical complementarity in protein folding.

Quantitation of the importance of packing interactions has also begun to accumulate from analysis of the distributions of interatomic distances between corresponding pairs of atoms or residues in a large number of refined protein structures (Subramanian et al., 1996). These probability distribution functions, or PDFs, turn out to be rather sharply defined, and to be characteristic of the participating components. Moreover, potentials derived from these PDFs have proven useful in refining protein structures for which there are only moderately high resolution data. Enforcing the PDFs during the refinement leads to better agreement of the structures with the experimental diffraction data.

Mutational analysis provides a complementary window, showing that simplified attempts to relate stereochemical complementarity to protein stability are naive. Proteins respond to mutations of core residues with a variety of mechanisms which limit deleterious effects introduced by mutation (Erickkson et al., 1992; Lim & Sauer, 1989, 1991; Lim et al., 1992; Matthews, 1993; Pakula & Sauer, 1989). An important factor complicating the relation between packing complementarity and stability is the opposing effects of Van der Waals (dispersion) forces, which stabilize as complementarity and packing density increase, and the consequent decrease in mobility, which destabilize the system as a whole (Richards & Lim, 1993).

Mutation of internal residues can increase the mobility of the surrounding residues, with the increased entropy compensating the deleterious effect. Moreover, packing in the protein can rearrange, which also compensates by restoring some of the original packing density. An elegant analysis of these effects comes from the study of the structures and thermodynamic stabilities of mutant phage T4 lysozyme molecules. The mutants replaced leucine or phenylalanine by alanine in core packing combinations. The structural rearrangements refilled the resulting cavities by different extents. The resulting destabilization was related to the increase in cavity volume, ΔV, determined from the crystal structures by a linear relation, $\Delta\Delta G = a + b \cdot \Delta V$ (Erickkson et al., 1992; Matthews, 1993). The constant term, -1.9 kcal/mol, presumably represents the differential cost of removing leucine and alanine from water, and the proportionality to ΔV, -0.024 kcal/mol, reflects the increase in surface area in the new cavity.

Systematics and Fold Detection. Topologically, most proteins are unknotted, linear polypeptide chains. Recent identification of an exceptional, truly knotted protein (Takasugawa & Kamitori, 1996) will doubtless provoke a reexamination of this generalization. With nearly 1000 entries in the Brookhaven Protein Databank, it has become clear that the multiplicity of unique structures has grown rather slowly as more structures have been solved. Many new structures are quite similar to others already in the databank, and families of related structures can easily be identified (Holm & Sander, 1992a; Jones & Thornton, 1993). Indeed, one folding pattern, the eight-stranded parallel β-barrel, now accounts for more than 10% of the known protein structures! Statistical procedures for comparing different structures and identifying "clusters" of similar structures are under intense development and have involved some subtlety (Holm & Sander, 1997). They will remain a fertile source of new observations for some time to come.

The criterion for attributing a particular family name to a new structure is a property of protein tertiary structure that invites systematization and has come to be known as the "fold." Defining the idea of a fold quantitatively is a treacherous undertaking, requiring a consistent metric for structural similarity and criteria for identifying clusters when all structures are characterized by this metric. The distance functions are usually based on a root mean square distance for equivalent points after optimal superposition of related structures. Although several procedures have been suggested, the ultimate arbiter seems still to be a subjective recognition of the "sameness" of two protein backbone tracings.

One objective and sensible procedure uses, as a reference distance, the comparison of a structure with its mirror image (Maiorov & Crippen, 1994, 1995). This apparently arbitrary procedure leads to distance classifications in reasonable accord with the subjective assessment. Using this criterion together with an analytical procedure for generating all self-avoiding chain traces of 170 residues or less suggested that only about 130 distinct motifs (other than β-barrels) are possible, of which about 100 have been found in nature (Crippen & Maiorov, 1995). This consistency suggests a solid conceptual basis for defining "folds."

The concept of a fold is useful in several respects. First, it allows grouping of related structures for descriptive purposes and potentially for phylogenetic studies. Second, many protein scientists feel that it simplifies problems related to the encoding of structure in primary sequences. To those researchers it is axiomatic that a fold is encoded by a simpler set of rules than those that determine the precise tertiary structure (Bowie et al., 1991; Kamtekar et al., 1993; Ouzounis et al., 1993).

Four Tertiary-Structural Classes. Proteins with related folds fall into one of four broad groupings, depending on the constituent secondary structures and how they are combined (Levitt & Chothia, 1976). These are designated α, β, α/β, and $\alpha + \beta$, to indicate proteins with all α, all β, interleaved α and β, and segregated α and β structures. The designations also apply to parts of larger proteins built from domains that fit one of the four preceding categories. Proteins in the first two categories are relatively easy to identify because they consist of a single secondary structural element. The remaining categories are less clearly defined, so there is a need to provide criteria on which to classify them. There is some variation in how these criteria are formulated. The scheme used here is proposed by Chou (Chou, 1995; Chou & Zhang, 1995), as described in Table 5-2.

Recent statistical studies have greatly strengthened the foundations of this broad

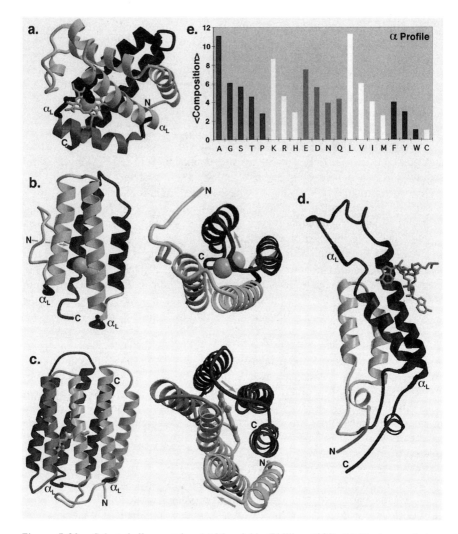

Figure 5-21. Selected all-α proteins. (a) Myoglobin (Phillips, 1980). (b) Myohemerythrin, another oxygen transport protein utilizing a binuclear iron cluster (Sheriff et al., 1987). Two orthogonal views are shown for this protein and the next. (c) Bacteriorhodopsin, the light-dependent proton pump from the membrane of photoauxotrophic bacteria (Henderson et al., 1990). This is a transmembrane protein. (d) Tobacco mosaic virus coat protein subunit (Namba et al., 1989). A four-helix bundle similar to that in (b), but with a more pronounced crossing angle between the pairs of helices. Proteins in (a) and (b) were oriented to emphasize the pronounced tendency to fold with approximate two-fold rotation symmetry. The two related segments are drawn in light (N-terminus) and dark (C-terminus) gray. Left-handed alpha helical turns are labeled and colored black. (e) Profile of average compositions for all α proteins, sorted according to a reduced, six-letter code that groups together side chains that interconvert frequently by mutation (Goldstein et al., 1992). Profiles here and in Figures 5-23 to 5-25 were adapted from Chou (1995) and Chou & Zhang (1995).

classification. Every protein can be represented as a twenty-dimensional vector, whose components consist of the compositions for each amino acid. One would like to show that these vectors cluster consistently into separate clusters, each with a similar amino acid composition profile corresponding to one of the four classes. Previous attempts to identify such profiles failed more than 30% of the time. However, when proper account is taken of correlations between the compositions of different amino acids in proteins—covariation of one amino acid composition with another—the discrimination for proteins excluded from the initial training sets is much higher (Chou, 1995; Chou & Zhang, 1995). This landmark result demonstrates not only that the class of a protein is determined statistically by a property as general as its overall composition, but also that the key to this encoding lies predominantly in correlated changes for different pairs of amino acids. Two aspects of this conclusion were challenged on technical grounds (Eisenhaber et al., 1996). First, Chou tested his predictions on a set of proteins that included homologues of those in his "training set." Second, the definition of structural classes (Table 5-2) was questioned because it does not permit unambiguous, automatic classification by computer algorithms (Michie et al., 1996). The former objection is valid; using homologues will certainly improve scores somewhat. The latter objection is more subtle, because successful class prediction can be used in a reciprocal sense, to validate the choice of structural classes. It is possible that computer programs should decide structural classes using "fuzzy" logic, where the criteria in Table 5-2 are weighted by soft paramters, rather than imposed rigorously. In any case, Chou's conclusions about classification and the importance of compositional covariance appear both to be correct and to point toward greater insight into folding rules (Bahar et al., 1997).

In Figures 5-21–5-24 these "compositional signatures" are illustrated as profiles in which similar amino acids are grouped together, reflecting statistical tendencies toward mutational interconversion. Protein encoding studies are often simplified by assigning a single "letter" to all members of such a group. The differently shaded groups in the profiles thus illustrate a reduced, six-letter alphabet (Goldstein, et al., 1992) to facilitate comparisons within and between groups. As an example, the α-profile employs fewer aminoacids from the first group (A, G, S, T, P; 6%) than does the β-profile (7%), but has an exceptionally high composition of A (11%).

Proteins built up entirely from α-helices include the globin and hemerythrin families, tobacco mosaic virus coat protein, and the purple membrane protein bacteriorhodopsin (Fig. 5-21). With few exceptions, these are transport proteins (the viral coat protein being a special case, binding and transporting the viral RNA). Bacteriorhodopsin is representative of a much larger class of membrane-bound proteins whose structures and activities have grown to include large external domains with more elaborate binding and catalytic activities. Despite their simplicity of design, many all-α proteins have sophisticated structural polymorphisms related to their function, and studies of one of them, hemoglobin, raise it to a status that could indeed be described as the "hydrogen atom" of protein chemistry (Baldwin & Chothia, 1979; Perutz, 1970, 1978, 1979).

All-β proteins (Fig. 5-22) invariably consist of a sandwich of two layers, each a sheet held together by a hydrophobic core formed by one side from each sheet. Often the sheet is rounded to form a "β-barrel." The β interactions in these proteins are usually antiparallel because there are no helices available to form the necessary crossover connections required for parallel β-sheets. The novel family of proteins with an all-parallel "β-helix" (Fig. 5-16), (Jurnak et al., 1994; Yoder et al., 1993a, b) is a remarkable exception.

Figure 5-22D shows an Fab fragment which is a domain from an intact IgG molecule. Proteins from this superfamily are constructed from twelve discrete domains,

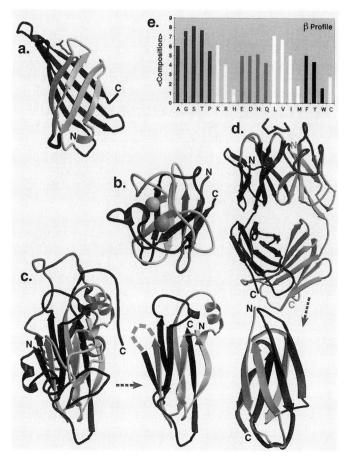

Figure 5-22. Selected all-β proteins. (a) The streptavidin-biotin complex (Weber et al., 1989). (b) Superoxide dismutase, an oxidoreductase using a binuclear copper-zinc site (Tainer et al., 1983). (c) Subunit 2 from the coat of polio virus, a "Swiss roll" protein (Filman et al., 1989). (d) Immunoglobulin Fab fragment with a bound peptide (dark gray trace at the top of the figure). Hypervariable regions at the top of the two chains are drawn with reversed shading. Two-fold symmetry is emphasized by differentiating related fragments by light and dark gray coloring. It is evident throughout the chains in (a) and (b). Subsets of the polio virus protein 2 and the Fab fragment showing twofold symmetry are shown separately. The dashed line connecting the light and dark gray portions of the subset in (c) represents a much longer inserted sequence evident in the lower foreground of the intact protein. The approximate intramolecular dyad is vertical in all cases except (b), where it is normal to the page. (e) Profile of average compositions for all β proteins.

probably arising from gene duplication. Each is folded in very much the same way as all the others. These are distributed among four separate polypeptide chains. Each of two heavy chains contain four of these modules, and two light chains each contain two. The domain structure is constructed from a β-barrel whose strands are organized into two opposite sheets, one containing four strands, the other containing three. The repeating pattern of such modules may play an important role in the transmission of signals from the Fab fragment to the base of the Fc fragment when antigens are bound. This would involve cooperative behavior of the successive domains.

Protein Structure / 193

The question of how such signals are transmitted is one of the present challenges in immunology.

Although there are several broad classes of immunoglobulins, they all contain the same basic fold. Most of the variation in structure is contained in the amino terminal 110 amino acids of both the light and the heavy chains, the so-called hypervariable regions. Hypervariable sequences typically make up less than 25% of the 110 amino acids in the variable domains of light and heavy chains. They provide the direct interface between anti-

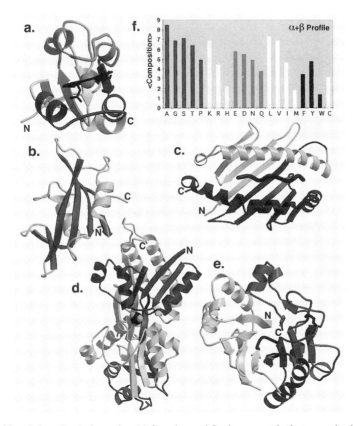

Figure 5-23. Selected $\alpha + \beta$ proteins. (a) Cytochrome b5, a heme protein that serves in electron transport (Mathews et al., 1972). (b) The SRC homology domain, a recurrent motif found in proteins involved in intracellular signal transduction (Waksman et al., 1992). (c) The peptide-binding domain of the class I major histocompatibility protein involved in presenting foreign peptides for recognition by the immune system proteins (Fremont et al., 1995). A bound peptide is indicated by the thick black line. (d) The bacterial periplasmic binding protein for phosphate (Kabena et al., 1986). The phosphate group is bound in a highly symmetrical site in this monomeric protein, where it lies in the positive electrostatic field provided by the N-termini of four α-helices. A striking two-fold rotation axis relates the entire first half of the sequence with the second half. Functionally, the protein opens by hinge-bending motion between the top and bottom halves. (e) Deoxyribonuclease I is a β-sandwich consisting of two layers of six β-strands each (Lahm & Sack, 1991). The connection between first and second halves of the sequence, residues 129–136, is emphasized by a thick, black coil, and a single disulfide bridge is shown in black. With the exception of (b), each protein displays extensive internal two-fold rotation symmetry perpendicular to the page, one half drawn in light, the other in dark, gray. The two-fold symmetry is intact throughout nearly the entire chains in (c), (d), and (e). (f) Profile of average compositions for $\alpha + \beta$ proteins.

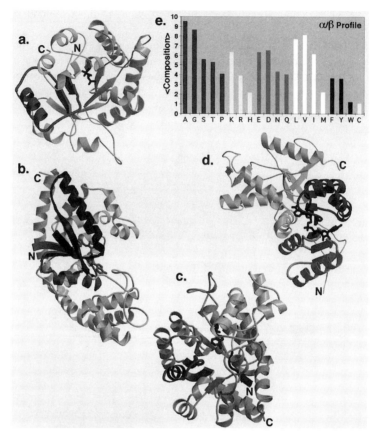

Figure 5-24. Selected α/β proteins. (a) Triosephosphate isomerase, the original 8-stranded β-barrel (Nobel et al., 1991). One β-α-β crossover connection is emphasized in dark gray. This motif recurs in an overlapping pattern around the barrel. (b) Tryptophanyl-tRNA synthetase (Doublié et al., 1995). The two β-α-β crossover connections of the Rossmann dinucleotide binding fold are colored medium (N-terminal) and dark (C-terminal) gray. The activated amino acid is shown in the active site (cf. Fig. 5-17). (c) Adenosine deaminase (Wilson et al., 1991), an eight-stranded β-barrel similar to that in (a). The active site is constructed from β-strands across the barrel (dark gray). In both (a) and (b), the N- and C-termini are in close proximity. (d) Lactate dehydrogenase, the protein in which the dinucleotide-binding fold was first identified (Abad-Zapatero et al., 1987). The view emphasizes the two-fold rotation axis that relates the two halves of this fold (medium and dark gray). (e) Profile of average compositions for α/β proteins.

body and antigen, while the rest of the antibody molecule serves as the scaffold presenting the repertoire of hypervariable sequences in a cavity capable of forming a binding site.

The third category (Fig. 5-23) contains proteins with distinct regions of all alpha or all beta structure, and are called α + β proteins. Within these regions, the tertiary structures are similar to one or another of the first two categories.

The most complex class are the α/β proteins. In these, segments of α-helix and β-sheet follow one another, and consequently each of the two classes of secondary structures is intimately involved in stabilization of the other (Fig. 5-24). For this reason, the β-struc-

Table 5-2 Classification of Protein Tertiary Structures

Class	Characteristics[a]
α proteins	$\alpha \geq 40\%, \beta \leq 5\%$
β proteins	$\alpha \leq 5\%, \beta > 40\%$
$\alpha + \beta$ proteins	$\alpha \geq 15\%, \beta \geq 15\%$; >60% of β strands antiparallel
α/β proteins	$\alpha \geq 15\%, \beta \geq 15\%$; >60% of β strands parallel
ζ proteins	$\alpha \leq 10\%, \beta \leq 10\%$

[a]Secondary structure definitions are those suggested by Kabsch and Sander (1983).

tures are often parallel, since the presence of helices provides the necessary crossover connections. There are two dominant motifs in this category: the Rossmann fold and more generally a single β sheet (Fig. 5-24b,d), and the parallel β-barrel (Fig. 5-24a,c).

The ζ-class of proteins (Table 5-2) is an ill-defined catch-all (Chou, 1995). It includes a number of large peptides with no globular structure; proteins whose structure has been determined by NMR methods only, and which have apparent similarities to all-α and all-β structures; and several very large viral coat proteins with large domains of all-β structure, together with long, apparently soluble segments.

Changing Course. Richardson examined the pattern of connecting loops and the progression of strands along a β-sheet (Richardson, 1977). This progression changes direction when a connecting loop skips one or more strands. This phenomenon was first identified in the Cu, Zn superoxide dismutase (Fig. 5-22b) and the constant region of the immunoglobulins (Fig. 5-22d). It can equally well be described as a long, single antiparallel loop that has wrapped around itself. This motif was dubbed by Richardson the "Greek key." The IGG barrels contain a Greek key motif (three strands of the four-stranded sheet and two of the three-stranded sheet). It also has a widespread occurrence in viral coat proteins, where it has elaborated into what has been called a "Swiss roll" (Fig. 5-22c).

When the Greek key structure is wrapped around a barrel, there is a difference between the two possible directions of the swirl because it has been combined with the curvature of the sheet around the barrel. Only one has been observed. The long single loop always folds backward, never forward under these circumstances (Brändén & Tooze, 1991).

Another important example of changes in direction arises in proteins with a single, parallel β-sheet (Fig. 5-24b). It is important to protect both sides of the sheet from dissolving into water, and to accomplish this the direction must change at least once. Otherwise all of the crossover connections would be made on only one side of the sheet. This is probably the reason why the α/β proteins have so many structural similarities and so few distinct families. Either the helices lie all on the same side, in which case the β-strands are protected by forming the widely observed $(\beta\alpha)_8$ TIM barrel, or else the helices lie on opposite sides and the protein has a single sheet.

Internal Symmetry. An unexpectedly large number of the proteins in each class exhibit strong internal symmetry, whereby repeated structural elements are arranged in nearly symmetrical fashion, usually related by two-fold rotations. This phenomenon was first observed in the eight-iron bacterial ferredoxins (Rossmann & Argos, 1976) (Fig. 5-1). However, the orientations and shading in Figures 5-21–5-24 illustrate that

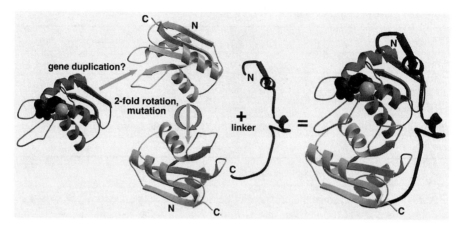

Figure 5-25. Evolution of complexity by duplication and variation. This process appears to have been at work in a large number of cases. An example is the dimeric cytidine deaminase from *E. coli*, ECCDA. The monomer has a nearly exact repetition of tertiary structure in two of its three domains. The dimeric enzyme has approximate 222 symmetry. This symmetry is similar to that of the tetrameric cytidine deaminases, which have subunits with a similar molecular weight to that of the repeated domain. Gene duplication and fusion appear to have been accompanied by the insertion of a long "linker" in the case of ECCDA.

this phenomenon is quite commonplace, and sometimes startling (Fig. 5-23c, d, e). Holm & Sander (1997) discuss the possible evolutionary processes that might give rise to this widespread evidence of internal redundancy.

An unusually fully developed example of internal two-fold symmetry was observed in the subunit of the dimeric *E. coli* cytidine deaminase (Fig. 5-25; Betts et al., 1994). Little trace of the likely sequence homology remains in the sequences of the two domains in the monomer, which are related to each other by nearly exact two-fold rotation symmetry. This internal symmetry is particularly interesting because the human and *B. subtilis* cytidine deaminases are tetramers, whose subunits have a similar molecular weight to that of the repeated domain in the *E. coli* enzyme monomer. The dimeric *E. coli* enzyme apparently shares with the mRNA editing enzyme APOBEC-1 cytidine deaminase the fact that its active site is composite, built from contributions from different domains and across the dimer interface (Navaratnam et al., 1993, 1997). The tetrameric cytidine deaminases apparently have discrete and independent active sites (Carlow et al., unpublished).

Despite the absence of obvious sequence homology between the two halves of the cytidine deaminase sequence, they clearly share similar folding instructions. Delaunay tessellation and likelihood scoring identified a small number of tetrahedra with high likelihood scores. These tetrahedra, which involve primarily leucine and isoleucine side chains, occur in almost exactly the same locations in the two domains (Fig. 5-26).

Disulfide Bonding and Exceptional Tertiary Structures. In extracellular proteins the oxidizing environment leads to formation of disulfide bonds whenever two free cysteine side chains can assume the appropriate configuration. This covalent bond provides a new source of stability by reducing the configurational entropy of the denatured state, relative to the native state. Not surprisingly, the range of tertiary structures in proteins with disulfide bonds is more varied than that normally observed for intracellular proteins, as illustrated by the following two examples.

The human chorionic gonadotropin, HCG (Figs. 5-27a, b), is a heterodimer in which one subunit grabs hold of the other by means of a structure called, imaginatively, the "seat belt" (Lapthorne et al., 1994; Wu et al., 1994). Both subunits assume nearly the same fold, and the dimer possesses approximate two-fold symmetry. The fold is apparently an all-β structure. However, the hydrogen bonding between strands and the backbone chain conformations are atypical, either because of the constraints imposed by the intricate pattern of disulfide bonds (Fig. 5-27b) or because those bonds permit the protein to assume a noncanonical conformation. HCG is a member of a group of

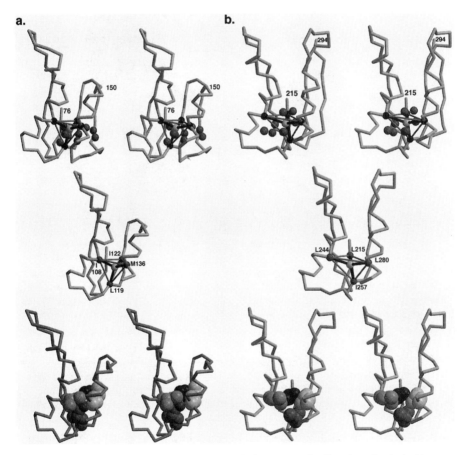

Figure 5-26. The highest-scoring Delaunay tetrahedra are not distributed randomly inside a protein. Those in the two domains of cytidine deaminase, (a) and (b), shown in stereo, are used to illustrate the use of Delaunay tessellation and likelihood scoring to filter the most significant packing contacts. The first row shows all of the tetrahedra that exceed a threshold of 0.95, that is, those that are more than approximately 10 times as frequent in known protein structures as would be expected by chance. The second row highlights one of these tetrahedra, formed in almost exactly the same location in both domains, but from a different set of side chains. The third row shows the actual packing of the four side chains that make up the highlighted tetrahedron.

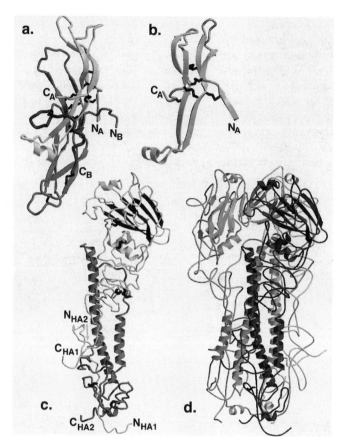

Figure 5-27. Exceptional tertiary structures. (a) The human chorionic gonadotropin hormone is typical of proteins characterized by the "disulfide" or "cystine knot." HCG is a heterodimeric protein whose subunits are each stabilized by several disulfide bonds. The larger subunit (B) is shaded in medium-dark gray, the smaller (A) in light grey. Both subunits have essentially the same fold, and the intact molecule has approximate two-fold rotation symmetry. The "seat belt" portion of the larger subunit is drawn as a larger tube, and is shaded dark gray. (b) The "disulfide knot" in subunit A, showing how the otherwise marginal folding is stabilized by covalent "tertiary" contacts. (c) Influenza haemagglutinin monomer. HA1 is shaded light gray, HA2 medium gray. The eight strands of the "Swiss roll" fold are dark gray. (d) The haemagglutinin trimer in the same orientation as in (c). The subunit shown in (c) is dark gray. In (a), (b), and (c) disulfide bonds are indicated by black linkages.

similar proteins called the "cystine knot" superfamily (Sun & Davies, 1995) whose members have the same fold and are stabilized by a similar pattern of disulfide bonds.

The influenza virus haemagglutinin structure (Wilson et al., 1981) (Fig. 5-27c, d) is a dramatic violation of any expectation that secondary structures close to one another in sequences should show a neighborhood correlation in tertiary structure. A trimer of identical polypeptide chains is the naturally occurring form in the virus. The polypeptide chain of the intact protein begins and ends inside the viral membrane. Its entire ~90 Å excursion from the membrane can be considered to be a single, grand loop (Fig. 5-27a). Proteolytic release of this fragment by bromelain digestion was necessary for isolation and crystallization.

In each monomer, a globular "head" contains the host-receptor binding site and the antigenic determinants. The head contains an eight-stranded, antiparallel β-sheet, with the "Swiss roll" motif found in most virus coat proteins. (Fig 5-22c). It lies at the top of a long, fibrous, coiled-coil structure built from two alpha helices from each of three subunits. The conformation of the fibrous coiled-coil seems to be stabilized only by contacts with partners in the trimer, as it does not make substantial interactions within the subunit itself. The extreme lack of regular secondary structure in the long meander of HA1 in this region may reflect both the stabilization arising from trimer formation as well as from disulfide bond formation.

Haemagglutinin is an important case for the study of contemporary evolution. Antigenic regions have been identified in looped-out segments between the β-strands. Amino acid substitutions alter these antigenic determinants of the molecule, and these changes are responsible for recurring influenza epidemics (Wiley et al., 1981).

During assembly, the monomer is cleaved twice, with the removal of Arg329, giving two fragments, HA1 and HA2. A substantial conformational rearrangement shifts the N-terminus of HA2 about 20 Å away from the C-terminus of HA1 after cleavage of Arg329. In the newly synthesized protein, these groups must be continuous. A "fusion-activation" peptide at the amino terminus of HA2 promotes fusion of the viral and host cell membranes. It is located at the bottom of the coiled-coil, where the haemagglutinin enters the viral membrane. Fusion requires an extraordinary, pH-dependent conformational change, described below, to project this peptide into the vicinity of the host cell membrane (Bullough et al., 1994).

Quaternary Structure: Symmetry, Control, and Cooperativity

Quaternary structure describes the formation of oligomeric proteins from monomeric subunits. The thermodynamic properties of the interfaces between subunits are generally similar to those associated with tertiary structure, with the extra complication that complex formation involves a reduction in the rotational and translational entropy of the participating subunits. This entropy is a destabilizing factor that must be overcome by forming sufficient "bonds" between subunits. A crude, but effective description of these bonds is that they arise from the burial of solvent accessible surface area between subunits, and semiquantitative estimates can be made using the proportionality 25 cal/mole/Å^2 of buried surface area (Chothia, 1974; Chothia et al., 1976).

The specialized questions associated with quaternary structure have to do with the symmetry of association. Subunits of an oligomeric protein associate with one another in regular patterns, which are described succinctly by the nomenclature provided by a particular type of symmetry. Different types of symmetry are appropriate for proteins localized in different environments. In assemblies with intermediate numbers of protomers, the symmetry is often reduced by the inclusion of genetic variants, in order to restrict the tendency to polymerize. Finally, symmetry tends to be preserved when subunits change conformation, giving rise to the phenomenon of cooperativity, which is often crucial to function.

Rotation Symmetry. Pure rotation symmetry allows the polar structures of oligomeric membrane protein subunits to persist in their quaternary structures (Fig. 5-28a). Membrane-bound proteins function at the boundary between two very different environments—the membrane itself is nonpolar while the exterior environment is aqueous.

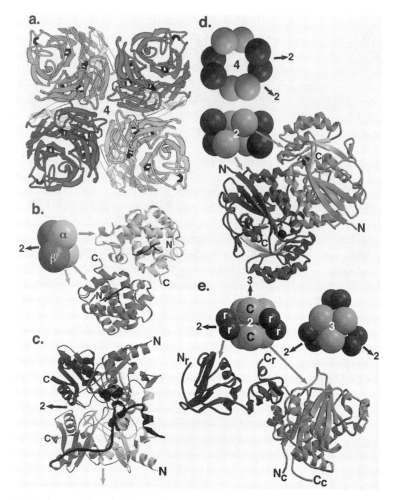

Figure 5-28. Symmetries of oligomeric proteins are correlated with their environments. (a) Tetrameric influenza virus neuraminidase. Pure rotation symmetry allows the polar structures of oligomeric membrane protein subunits to persist in their quaternary structures. See also Figure 5-27d. (b–d) Oligomeric soluble proteins usually associate with dihedral symmetry, in which two-fold axes normal to the principal rotation axis override the polarity of individual subunits. Thus, an essentially isotropic polar surface is exposed to the aqueous environment. (b) The hemoglobin tetramer has 2 symmetry and approximately 222 symmetry (Liddington et al., 1992). (c) The cytidine deaminase dimer has 2 symmetry with approximate 222 symmetry. (d) mandelate racemase, an octamer with 42 symmetry (Neidhart et al., 1991). (e) aspartate carbamoyl transferase is an $\alpha_6\beta_6$ dodeaamer with 32 symmetry (Stevens et al., 1990).

They consequently require a polar structure that projects two different surfaces, one toward each environment. Cyclic symmetry matches this requirement.

The influenza neuraminidase structure (Fig. 5-28a) has an exquisite hierarchy of rotational symmetries both internally and in its quaternary structure (Colman et al., 1983). The monomer consists of six nearly identical β-sheets arranged with near six-fold rotation symmetry; hence the fact that the tetramer can be generated conceptually from

24 copies of this β-sheet by first generating the (approximately) symmetry-related copies of the sheet within the monomer, and then by generating the tetramer by four-fold rotation symmetry. The quaternary structure of the enzyme shows four-fold rotation symmetry and, as with the haemagglutinin, that is its only symmetry. The analogy of this high degree of symmetry to a flower has been widely noted. The neuraminidase structure determination completed a structural inventory of two components through which the flu virus interacts with human cell surfaces.

Other membrane-bound proteins with cyclic symmetry include the ryanodine receptor or calcium release channel of sarcoplasmic reticulum (Orlova et al., 1996; Wagenknecht et al., 1996), the photoreaction center (Deisenhofer & Michel, 1991; Deisenhofer et al., 1984) and the F1 ATPase (Abrahams et al., 1994). The former example is extraordinary for the sheer size of the relevant polypeptide, which has approximately 5600 amino acids and represents challenges at every level of analysis. The latter two examples are high resolution X-ray crystal structures, in which the symmetry was shown to be violated in functionally significant ways. The ATPase contains five principal subunits ($\alpha \sim 51$ kDa, $\beta \sim 48$ kDa, $\gamma \sim 27$ kDa, $\delta \sim 15$ kDa, and $\varepsilon \sim 5$ kDa), two of which (α and β) are present in three copies. These six subunits alternate like the segments of an orange around an approximate three-fold axis. These subunits are the ones that bind nucleotides, and the β subunits are exclusively responsible for the actual binding of ADP and phosphate and the synthesis of ATP. The six subunits sit atop a stalk, formed by the γ subunit. This stalk resembles the stalk of the haemagglutinin (Fig. 5-27c) in having a 90 Å α-helix which forms, in its lower regions, a coiled-coil with a second α-helix.

This second helix serves to break the rotation symmetry relating the three pairs of $\alpha\beta$ dimers. Each $\alpha\beta$ dimer assumes a slightly different conformation with respect to the binding of nucleotide. One is empty, or open; one is occupied by ADP and is somewhat closed; and the other binds ATP and is closed. ATP is synthesized by this enzyme in response to a protein gradient that drives the rotation of the "orange" around the stalk. Each third of a rotation drives the subunit with tightly bound ATP past the second helix, where it must open to allow a helical extension from the β subunit to pass. This transition from "closed" to "open" conformation is necessary and sufficient to make ATP because inside the active site of the β subunit the equilibrium constant for hydrolysis is close to 1.0. The structural asymmetry is thus an important clue to the mechanism of energy transduction by the ATPase.

The photosynthetic reaction center complex structure (Deisenhofer & Michel, 1991; Deisenhofer et al., 1984) is actually not symmetric. It is a heteromeric complex of four proteins: a cytochrome (~ 34 kDa), and subunits M (~ 32 kDa), L(~ 27 kDa) and H(~ 26 kDa). However, the L and M subunits have essentially the same tertiary structures, including the appropriate electron transport cofactors, and they associate about a two-fold rotation symmetry axis normal to the membrane. This provides essentially equivalent electron transport pathways from cytochrome. For reasons that are only partially understood, only the pathway through the L subunit is used.

Dihedral Symmetry. Oligomeric soluble proteins usually associate with dihedral symmetry, in which two-fold axes normal to the principal rotation axis override the polarity of individual subunits (Fig. 5-28b–d). Thus, an essentially isotropic polar surface is presented to the aqueous environment.

The classic oligomeric protein, hemoglobin, is a dimer of dimers. Its alpha and beta subunits associate tightly to form dimers, which then associate to form a tetramer. Although there is only one true symmetry axis, the approximate symmetry of the dimer lends approximate 222 symmetry to the tetramer. The association constant for tetramer formation is highly dependent on the ligation state, being approximately 100 times stronger in the deoxy state. This differential affinity is coupled to the configuration of the heme and the iron ligands, and regulates its affinity for oxygen.

Other rotation orders are observed. Mandelate racemase (Fig. 5-27c) is an octamer with 42 symmetry, and aspartate carbamoyltransferase has six catalytic and six regulatory subunits that are related by 32 symmetry (Kantrowitz & Lipscomb, 1988).

Regardless of whether or not there are additional two-fold axes, the primary symmetry axis may have an arbitrary order, but is usually between 2 and 5.

Helical Symmetry and The Cytoskeleton. The polymeric proteins of filamentous viruses and the cytoskeleton possess helical symmetry, in which subunits are related by a translational, as well as a rotational component. Actin, myosin, tubulin, and various other filamentous and fibrous proteins all interact with helical symmetry, which is often also called "screw" symmetry. Screw symmetry combines a translation along the helix axis with the rotation.

The tobacco mosaic virus coat protein forms a 34 subunit subassembly, called the "disk," which assumes either a polar 17-fold rotation point group or a helical assembly known colloquially as the "lockwasher," depending on the solution pH. The mechanism for this transition involves competing forces of close packing, achieved in the lockwasher and virus, and electrostatic repulsion, which forces the subunits slightly apart in the disk (Butler & Klug, 1978). The repulsion arises because in the helical configuration carboxylate side chains are forced into proximity to one another. Physically, this can occur only if they can share a proton, so the lockwasher is stable only at pH values below 6.5, which is the approximate pKa of the "high affinity" proton-binding site formed by the 34 carboxyl-carboxylate pairs.

Actin forms a two-stranded helix of globular actin subunits. Important variations in the helix parameters occur, however (Egelman et al., 1982). The rise per subunit is relatively constant, but the twist or relative rotation around the helix axis is highly variable. This polymorphic tendency is probably important for the smooth functioning of muscle contraction, which involves considerable force generation.

Icosohedral Symmetry. The small genome size of viruses also implies that their coding information is at a premium, giving an advantage to systems that can use many copies of the same gene product. This was recognized first by Crick and Watson (1956) and elaborated into a persuasive analytical theory by Caspar and Klug (1962). The latter authors recognized in a fairly deep way the importance of symmetry in efficient use of identical subunits for constructing shells that approximated spheres.

The icosahedral 5:3:2 point group generates 60 "asymmetric units", the largest possible number for a point group. It has 12 (= 60/5) five-fold axes, 20 (= 60/3) three-fold axes, and 30 (= 60/2) two-fold axes. It is mathematically impossible to generate an object with any higher combination of symmetry elements that form a group; one cannot construct an object with 6:3:2 or 7:3:2 symmetry because these combinations of operations do not cover a sphere without discontinuities. All "spherical" viruses known have approximately this symmetry, which is also the symmetry of a European football.

Another way to describe this restriction makes use of a theorem by Gauss, which states that to enclose a volume by folding a flat surface one must introduce wedge disclinations totalling 720°. A wedge disclination is simply a wedge cut from the surface at a point followed by a joining of the two sides of the wedge to form a cone at the point. The highest-order regular polygon that tiles a flat surface is a hexagon. Introducing 12-wedge disclinations of 60° into a surface tiled by hexagons converts 12 hexagons into pentagons, leaving behind an integral number of equal wedges that total 720°.

Most viruses with icosahedral shells contain many more than 60 protein subunits. To make bigger shells, one must sacrifice the exact bonding interactions permitted in a strictly icosahedrally symmetric shell. The most common macromolecular solution to this problem is to give up as little of the bonding equivalence as possible, that is, to make all of the bonding patterns as similar as possible. This naturally leads to approximate symmetry or quasi equivalence. The basic assumption of quasi equivalence is that the shell is held together by the same types of bonds everywhere, except that the bonds are slightly deformed in different, nonsymmetric environments. This mechanism retains the local environment involved in intersubunit bonding, even though the subunits are not equivalent by global symmetry operations.

According to the theory of Caspar and Klug, augmented icosahedral shells should contain $60T$ subunits, 60 of which obey the icosahedral symmetry, in T distinct sets each of which is slightly different. The triangulation number, T, is restricted to having values $T = Pf^2$, where f is any integer and P is obtained from the series 1, 3, 7, 13 ... given by $(h^2 + hk + k^2)$ for all pairs of integers h and k having no common factor. The arrangements of quasiequivalent points on the shells with $T \geq 7$ are handed, and therefore exist as pairs related by mirror symmetry. Thus, there are $T = 7d$ and $T = 7l$ shells, both of which have $T = 7$. Viruses with triangulation numbers 1, 3, and 7 have been identified and studied by X-ray crystallography.

Tomato bushy stunt virus (Harrison, 1978; Harrison et al., 1978) incorporates 180 identical subunits into its shell, and so has a triangulation number $T = 3$. This means that there are three different "kinds" of subunits. They are denoted A, B, and C. The 60 C subunits are linked together by special structures involving their NH terminal tails, which interlock as a β-sheet around the true three-fold axes, forming a structure called a β-annulus. The corresponding N-termini of the A and B subunits are disordered and do not interact in any way, either with each other or with any other subunits. They may be involved in RNA binding.

Intrasubunit flexibility provides the key to implementing Caspar and Klug's quasi equivalence (Fig. 5-29). The TBSV subunit has two domains; a "radial" domain projects from the shell (the P domain) and a tangential domain forms the shell itself (the S domain). The hinge between domains assumes a different angle in the three, quasi-equivalent positions. The projecting domain contains a "Swiss roll" β-structure very similar to that observed in the flu virus haemagluttinin. This fold is often, if not always, observed in the projecting domains of icosahedral RNA viruses of diverse origins.

In contrast to the small plant RNA viruses, the small mammalian RNA viruses (picorna viruses) have used gene duplication and proteolysis to generate individually customized gene products. The alteration of bonding surfaces demanded by quasiequivalence and which is mediated entirely by segmental flexibility in TBSV is mediated by altered sequences in the VP1, VP2, and VP3 subunits of polio virus, which have built-

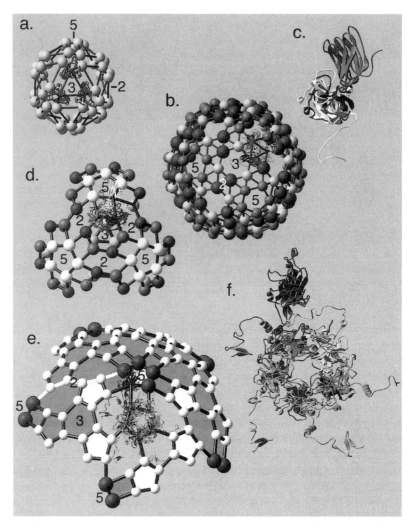

Figure 5-29. Icosahedral symmetry and quasiequivalence. a: A rare T=1 virus, satellite tobacco necrosis virus (Jones & Liljas, 1984). Three of the 60 subunits are represented as ribbon drawings. The remaining 57 subunits are represented by spheres at their centers of mass. The view is along one of the 20 three-fold axes. Sample five-fold and two-fold axes are as indicated. b: Inclusion of extra subunits via quasi-equivalence in the T = 3 virus, tomato bushy stunt virus, the first sperical virus solved to atomic resolution (Harrison et al., 1978). 180 identical subunits pack in three groups, A (medium gray, at five-fold axes), B (light gray), and C (dark gray, larger radius). Projecting domains of the C subunits interact with each other at the two-fold axis. Their extended N-termini form a "β-annulus" around the three-fold axis. c: Mediation of quasiequivalence by interdomain flexibility in TBSV. The A and C subunits have been superimposed in their projecting domains to illustrate the different conformations of their "shell domains", whose orientations differ by 22°. d: Made to measure quasi-equivalent subunits in polio virus (Filman et al., 1989). Approximately a quarter of the shell is illustrated. Three subunits (VPI white, at the five-fold; VP2, dark gray, alternating with VP3, medium gray, around the three-fold) occupy the three quasi-equivalent positions of a T = 3 lattice. A fourth subunit (VP4, light gray) ties the structure together in an "inner layer" along the three-fold axis. e: Disregarding the conservation of bonding surfaces in papilloma viruses. Approximately a quarter of the polyoma virus shell

in conformational and surface differences that allow each subunit to fit into place in a different position of the $T = 3$ lattice (Hogle et al., 1987). These proteins are synthesized, together with a fourth, VP4, as a single polyprotein. Specific proteolysis of this polyprotein is carried out by a viral protease during assembly. The interaction between VP4 and the N-terminus of VP3 at the five-fold axes probably acts as an internal clamp locking the structure together.

The final cleavage step releasing VP4 has a profound affect on the stability of polio virus. Before cleavage, the immature virus can be disrupted by modest concentrations of mild detergents. After cleavage it becomes one of the most stable macromolecular aggregates known, resisting disruption by 0.1% SDS! This association between VP4 and viral coat stability is reinforced by the fact that the infection process involves an inflation of the viral shell that leads to release of VP4.

A bound lipid molecule, tentatively identified as sphingosine, was found in a cleft inside the VP1 barrel. This hydrocarbon binding site is also the site of action of a series of potential antiviral drugs that have been developed against human rhinovirus.

Papilloma viruses do not obey quasi equivalence. The $T = 7d$ shell of polyoma (Rayment et al., 1982) and SV40 (Liddington et al., 1991) viruses revealed a profound violation of quasiequivalence. These viruses appear in the electron microscope to be constructed from 72 clumps, called capsomeres because they are the morphological units from which the capsid is constructed. Twelve of the capsomeres are surrounded (in electron micrographs) by five other capsomeres. The other sixty are surrounded by six other capsomeres. These two different classes of capsomeres are called, by an obvious convention, pentavalent and hexavalent, where valency refers to the intrinsic layout of molecular bonding partners.

Quasiequivalent bonding would require that the pentavalent capsomeres contain five subunits, while the hexavalent capsomeres would contain six subunits. That arrangement would allow every subunit in the shell to be surrounded by other subunits in nearly identical directions, and hence allow the valency of each subunit to be filled in approximately the same way. The X-ray structures show quite the contrary. All capsomeres, including the hexavalent ones, are constructed from only five subunits. The tendency of the subunit to associate in pentamers is apparently quite strong, revealing itself even in artifactual aggregates (Baker et al., 1983). The subunit is evidently designed to accommodate six different bonding environments, which it is able to do by varying the orientation and conformation of a long, C-terminal extension (Fig. 5-29d) (Liddington et al., 1991).

(Stehle & Harrison, 1996) is shown. The shell is constructed from 360 identical subunits, which pack in six different environments. The shell, however, is constructed entirely of pentamers. Consequently, there are pentamers (large, dark spheres) which have exact five-fold symmetry, and others (that with subunits drawn as ribbons) which are surrounded by six pentamers, including the strict pentamer (white pentagons). The T = 7d lattice has a possible handedness, determined by the path illustrated in white lines between subunits between pentamers at the five-fold axes.

Limiting Polymerization in Mesomolecular Assemblies. A small, but significant, class of molecules have quaternary structures involving 10 to 50 protomers. These "mesomolecular" assemblies confront and must solve a specialized problem arising from the thermodynamics of polymerizing identical building blocks.

Building an oligomeric complex from identical subunits is akin to a phase transition. It is governed by a substantial activation energy barrier to "nucleating" assembly. There is a large reduction in the entropy of two macromolecules when they join to form a dimer. Adding each additional molecule involves a comparable loss in the entropy of the new monomer, but the reduction in the entropy of all the other subunits is divided amongst them all. So the reduction in entropy per subunit goes down with each addition. The reduction in entropy creates a barrier to assembling a polymer. This barrier represents the transition state for assembly.

Once the initial barrier for nucleating a polymer is overcome, however (through strong specific interactions that can be propagated indefinitely), the most stable aggregate is actually the one containing the most subunits. The energy gained by making the polymer one larger is approximately the same for each new addition, so the overall free energy change becomes increasingly favorable as the size of the polymer increases. Once an assembly reaches a critical size, enough of its subunits are "internal", in some sense, that additional subunits "bury" others already in place without changing the configuration that attracts additional subunits.

Thus, having solved first the problem of nucleating their assembly, mesomolecular assemblies then must face the problem of restricting the forces so unleashed so as to retain a discrete size intermediate between those of most soluble oligomers with less than about 10 subunits and those of large, polymeric aggregates, which are usually filamentous. Often, this problem is solved by using slightly different subunits for the outside and inside of a mesomolecular assembly (Carter, 1978; Hendrickson & Ward, 1977; Klotz et al., 1976) This paradigm appears in a variety of contexts, two of which are described here (Fig. 5-30).

Arthropod Hemocyanin. The scorpion and horseshoe crab circulatory system, their hemolymph, has neither red cells nor structures analogous to capillaries to force it into close contact with target tissues that require oxygen. Dispersing oxygen binding equivalents of one hemocyanin molecule as tetramers would increase the number concentration by a factor of 6, thereby increasing both the osmolarity and the viscosity of the hemolymph. Both would be disadvantageous. Efficient oxygen delivery also requires a highly cooperative transporter; hemocyanins are extraordinarily cooperative, having Hill-plot slopes more than 20 (Gaykema et al., 1984; Van Holde & Miller, 1982).

Hemocyanin molecules from arthropods are large mesomolecular assemblies composed of different subunits whose amino acid sequences, though similar, are distinct. The scorpion, for example, requires eight slightly different sequences to make a single molecule containing 24 subunits (Lamy et al., 1981). The arthropod hemocyanin subunit (~75 kDa) contains more mass than the entire tetramer of hemoglobin (54 kDa), and instead of a heme bound iron, the prosthetic group is a binuclear copper cluster. The stoichiometry of different subunits is quite specific and nonuniform. The 24 subunits of the intact molecule include four each of four different subunits: 2, 4, 5A, and 6, and two copies each of four other subunits: 3A, 3B, 3C, and 5B.

The high genetic and metabolic cost of maintaining eight different genes for a single oxygen-binding activity reflects two conflicting biological demands: (1) binding as

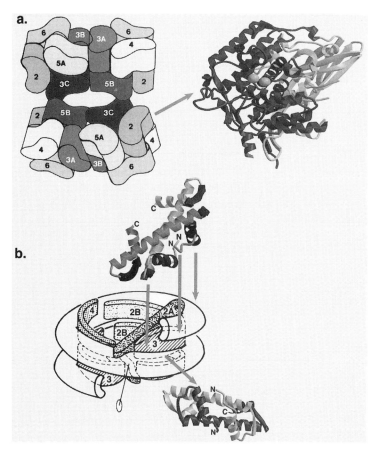

Figure 5-30. Using multiple genes to control mesomolecular assembly. (a) The arthropod hemocyanin from *Androctonus australis* (Reprinted from Lamy et al., 1981, with permission). The intact assembly contains 24 subunits, transcribed from eight different but closely related genes, indicated by their numbering and shading. The degree of exposure indicated by comparative line electrophoresis is encoded by progressive shading, the lighter shades being more exposed, and hence closer to the surface. Modifications within each discrete kind of subunit enable them to fit only into prescribed locations in the 24-mer, despite the fact that all subunits closely resemble the one illustrated to the right. (b) The histone octamer in the nucleosome core particle of chromatin. Four different genes are required to encode the ability to nucleate a DNA supercoil without propagating. The stoichiometry of the octamer is $(H3)_2(H4)_2(H2B)_2(H2A)_2$ and the locations within the supercoil are indicated schematically. The X-ray structure of the octamer (Arents et al., 1991) revealed that the fundamental organization involved heterodimers, of very similar structure to the example shown here, which is taken from the NMR structure of an archebacterial histone B (Starich et al., 1997). The monomers (in this case identical) in the dimer are shaded light and dark gray to distinguish them. In the eukaryotic core particle, the dimers are heterodimers, but H3H4 and H2BH2A dimers are essentially similar to the example shown except for N- or C-terminal extensions not seen in the crystal structure. Two orientations are shown, one along the intermolecular two-fold axis, showing the "handshake motif," the other from the top of the particle, showing the regular repetition of DNA binding motifs, the "paired helices motif" created at the two-fold axis from the amino termini and the "β-bridge" motifs at either end of the extended dimer. The H2A molecules occupy a unique position at the beginning and end of the nucleosome, as indicated by the schematic diagram, owing to the extension of their carboxy termini toward the dyad axis where the dimers in the (H3H4)2 tetramer bind. Adapted from Carter et al., 1980, with permission.

many individual oxygen carriers together as possible to reduce osmotic pressure and ensure cooperative oxygen release and, (2) doing so without forming a crystal containing essentially all of the hemocyanin subunits. Such aggregates are indeed observed when mixtures of only subunits 3C and 5B are allowed to polymerize. Without the protection of the "external" subunits, these two "internal" subunits, 3C and 5B (Fig. 5-30a), self-assemble to form very large, sheet-like aggregates. To effect a compromise between this Scylla and Charybdis, the scorpion maintains eight different genes, each of which contributes in nearly identical ways to the cooperative interactions between subunits in the intact molecule. However, each subunit plays also a role in maintaining a discreet degree of polymerization. The "external" subunits are designed to discourage further polymerization onto the same surfaces used by the internal ones to hold the 24-mer together!

Fab fragments prepared from antisera to specific subunits were shown using a line electrophoresis procedure to divide the eight subunits into two groups: those with a high degree of accessibility in the intact molecule (subunits 2, 4, 5A, and 6), and those with little accessibility (subunits 3A, 3B, 3C, and 5B). Electron microscopy and immunochemical staining with specific Fab fragments localized individual subunits within the 24-mer. High resolution crystallographic analysis of a related hemocyanin confirmed this interpretation (Gaykema et al., 1984): the intact enzyme consists of four hexamers, organized as a dimer of two dodecamers, each of which consists of two hexamers. It contains many approximate symmetries (each hexamer has approximate 32 symmetry, and the 12-subunit substructures have in addition approximate 222 symmetry). However, the overall symmetry of the 24-mer is only two-fold! Much of the potential symmetry of the 24-mer is discarded to ensure formation of a discrete polymer size. This loss of symmetry does not extend to the functional linkages between subunits, as the 24-mer remains highly cooperative.

The Histone Octamer: Discreet Coiling of DNA. The histone core particle of chromatin illustrates the same idea in a different context (Carter, 1978; Carter et al., 1980). The function of this octameric mesoassembly is to serve as a reel on which chromosomal DNA can wind up. The path of the DNA is a nearly regular supercoil (Finch et al., 1981; Richmond et al., 1984). It would be catastrophic in several respects, however, if that helical path were not limited to a discrete length. An indefinitely long superhelix would restrict access to the information contained in the DNA as well as making it hard to introduce higher-order packing. This fascinating battle juxtaposes the conflicting demands of the DNA superhelical screw symmetry and the two-fold point group symmetry of the histone octamer. In fact, the octamer produces internal protein-protein contacts along a helical path to nucleate one coil of superhelix without allowing similar contacts to develop between histones in successive core particles. It does its best to impose pseudo point group symmetry on the DNA, thereby effectively breaking the screw symmetry of the supercoil (Carter, 1978).

As with the hemocyanin, multiple genes are used to mediate this broken symmetry. The octamer is built from two copies each of four slightly different histones, H3, H4, H2B, and H2A, which associate in pairs. In the 3.1 Å crystal structure of the octamer itself, only one copy of each protein is present in the crystallographic asymmetric unit (Arents et al., 1991), which, therefore, contains two different heterodimers, H3-H4 and H2B-H2A. Thus, the particle has strict, crystallographic two-fold symmetry. The complementary fragment of DNA presumably also has a nearly exact dyad axis (Fig. 5-30b).

The detailed realization of the pseudosymmetry (Arents & Moudrianakis, 1993) involves a novel "handshake" configuration, in which two different histone subunits, H3 with H4 and H2A with H2B, bind to each other via a long central α-helix that runs roughly parallel to the direction of the DNA (Fig. 5-30b). Both ends of the two helices have specialized helix-turn-strand-helix motifs, and in the heterodimers the strands from this motif form a short parallel β-sheet. The N-terminal α-helices from the first motif in each monomer face outward from the octamer, and have been termed a "paired-ends of helix" element. The exposed N-terminal amide groups are related by a pseudo-two-fold symmetry axis normal to the circumference of the octamer and create a region of considerable positive electrostatic charge (Hol et al., 1978).

The four heterodimers follow one another from one end of the nucleosomal DNA to the other in the sequence, H2A:H2B, H4:H3, H3:H4, H2B:H2A. Contacts between the two H3:H4 dimers are strong enough to sustain an independent tetrameric configuration when the nucleosome is dissociated, whereas the two slightly lysine-rich dimers dissociate as dimers. All of the histones are rich in basic side chains, and many of these lie along the helical path presumably followed by the DNA. Unlike the N-terminal segments of the other six histones, which emanate from within the successive turns of DNA supercoil, those of H2A lie *outside* the DNA strands on top and bottom of the core particle modeled by Arents, et al. (Arents & Moudrianakis, 1993), acting as "caps" demarking the beginning and end of the nucleosome and discouraging association between octamers from successive nucleosomes, thus delimiting the length of DNA allowed to supercoil (Carter et al., 1980).

Evolution of New Functions by Variation of Subunit Interfaces. An important source of evolutionary variation, first described by Brian Hartley (1970), involves altering the interfaces between subunits in an oligomeric enzyme. Hartley pursued this suggestion with an intriguing kind of Darwinian evolution, using the poor, but adequate ability of ribitol dehydrogenase to metabolize sorbitol. It also was soon verified by structural studies of the dehydrogenase family. The lactate dehydrogenase (LDH) tetramer has three intersecting, mutually perpendicular two-fold axes, labeled P, Q, and R, whose evolutionary origins are distinct (Buehner et al., 1973). The interaction between subunits across the Q axes is quite similar in all of the dehydrogenases, but interactions across the P and R axes of tetrameric dehydrogenases can be quite different. The Q axis dimer appears to persist from quite distant ancestors. Subunits in the dimeric dehydrogenases, like liver alcohol dehydrogenase and the cytoplasmic malate dehydrogenase, interact with each other in a similar manner to the association across the Q axis of LDH.

Significantly, the nucleotide binding cavities (Fig. 5-17) both face in roughly the same direction along the Q-axis of all dehydrogenases. Comparison of the tetrameric glyceraldehyde phosphate dehydrogenase (GPD) revealed that it also was constructed by the association of two Q-axis dimers. However, they were associated face-to-face, whereas the similar Q-axis dimers in LDH associated tail-to-tail. This contrast evidently has functional significance: the face-to-face association in GPD forces the nicotinamide portions of the NAD coenzyme to share the nucleotide binding fold across the interface between Q-axis dimers. Consequently only half (two of the four) active sites in GPD are active at any given time, whereas in LDH the four monomers are nearly independent and noncooperative.

A recent variation on the use of subunit interfaces was identified in two superoxide dismutases (Fig. 5-22b) (Bourne et al., 1996). Two different dismutases form dimers using the opposite sides of the active site cavity, which are related by approximate twofold symmetry.

A related phenomenon is what Edmundson (Edmundson et al., 1975) refers to as "rotational allomerism" in the immunoglobulin superfamily (Fig. 5-22d). The switch regions between each Greek key module in Fab fragments introduce a near 180° rotation of the domain structure. This rotation changes the interfaces between the light and heavy chains from the four β-strands on one side of the β-barrel in the constant domain to the three β-strands on the opposite side of the barrel in the interface between the variable domains. The more extensive interface between constant regions anchors the light and heavy chains together. The interface between variable regions is less extensive, permitting a cavity to develop at the distal tip of the interface. The result is akin to a jaw, anchored behind the molars and open at the mouth!

Binding, Plasticity, and Catalysis

The quintessential functional behaviors of proteins are related to their abilities to bind, and change shape, and to the coupling between binding and conformational change. The last section of this chapter is, therefore, devoted to a brief tour of principles underlying these properties.

"Topological Switch Points" and Active Sites

There is an uncanny association between the approximately symmetric arrangements of long stretches of polypeptide within a protein (Figs. 5-15, 5-21–5-25) and the location of binding sites. In such cases, two polypeptide segments enter or leave a region of space in opposite directions. This phenomenon was first recognized for β-structures, since discontinuities in β-structure invariably lead to places where the surface of a protein must open into a cavity. These regions have been called "topological switch points" (Brändén, 1980), and they are a reliable indicator of ligand binding sites (Figs. 5-15, 5-31, and 5-32).

The topological switch point evident in the Rossmann fold appears to be the quintessence of the ability of proteins to manage the free energy of hydrolysis of nucleotide triphosphates. Ten of the twenty aminoacyl-tRNA synthetases use it to bind ATP for amino acid activation (Carter, 1993). A close topological relative of the Rossmann fold, but with a somewhat different nucleotide binding mode, is a highly conserved motif in kinases, G-proteins, and motors (Smith & Rayment, 1996). This variant does not involve successive β-α-β crossovers, but retains only the two central β-strands, each of which is followed by an α-helix that veers away in the opposite direction (Fig. 5-31).

Related to the Rossmann dinucleotide fold in its secondary structural organization, this motif invariably utilizes a conserved loop between the central β-strand and the following helix. This loop, called the "P-loop," was identified by Walker (Walker et al., 1992) and is a "signature" of nearly all enzymes that utilize the γ-phosphate of ATP or GTP. The consensus sequence of this loop, Gly-X-X-X-X-Gly-Lys-Thr-Ser, is adapted to bind the α-phosphate of the nucleotide close to the amino terminus of the following helix. Across the switch point in these proteins, there is invariably a region (related by the approximate symmetry to the P-loop) called "switch II," which is cru-

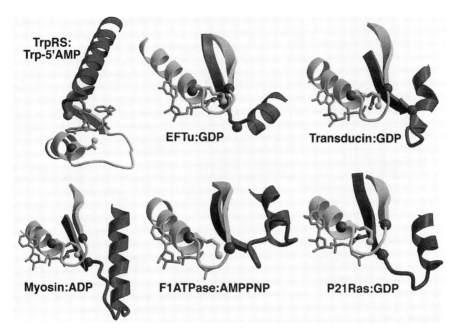

Figure 5-31. The topological switch points in G-proteins and motors. The topological switch point for tryptophanyl-tRNA synthetase is shown in the upper left-hand corner for comparison. The nucleotide complexes for five transducing nucleotide triphosphatases are as labeled. Different features are color coded. The amino terminal strand-helix segment is light gray, the carboxy terminal segment is dark gray. Adenine nucleotides are dark gray, guanine nucleotides are light gray. A glycine residue that is invariant in all such proteins and is also invariant in all class I aminoacyl-tRNA synthetases is illustrated as a spacefilling atom at the C^α site, and the following lysine residue is drawn in ball-and-stick representation. Protein coordinates all have been oriented to superimpose the amino terminal helix and, especially, the invariant glycine. The essential difference between those enzymes with a proper Rossmann fold (TrpRS) and other transducing proteins is that the turn segment following the invariant glycine loops (*top*) of the β-phosphate of the nucleotide, which sites to the lower left of the invariant glycine in the five nucleotide triphosphatases that cleave the γ-phosphate. In the synthetases, the adenine nucleotide sits above and to the right of the invariant glycine. The β-phosphate sits snugly into the strong positive electrostatic field of the unpaired amide groups at the helix terminus, and the invariant lysine side chain orients toward the γ-phosphate. In the synthetases, which release pyrophosphate, the positive electrostatic field of the carboxy-terminal helix faces the ribose hydroxyl group. Among the five transducing NTPases, two (EFTu and Transducin; top row) share the β-configuration of the synthetases and the second β-strand lies above the plane of the amino terminal helix and the first β-strand. In the remaining three (myosin, F1ATPase, and P21 Ras) (*bottom*), the second β-strand of the topological switch point lies below the plane.

cial to the coupling between nucleotide-triphosphate hydrolysis and communication of this event to and from other protein partners.

Two further examples, in which the active-site configurations are nearly identical despite profound differences in tertiary structure, are shown in Figure 5-32. The nucleoside deaminases for cytidine (CDA) and adenosine (ADA) illustrate the significance of topological switch points in a lively manner. They are both zinc hydrolases,

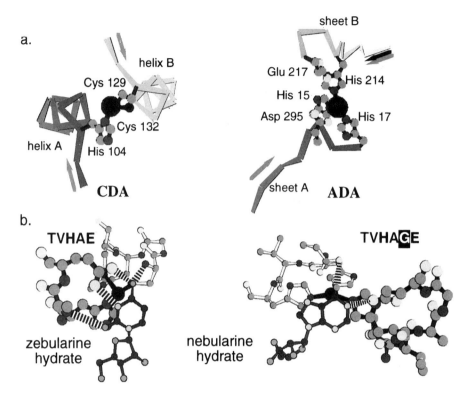

Figure 5-32. Topological switch points and active sites. (a) The nucleoside deaminases for cytidine and adenosine have utterly different tertiary structures, but are able to configure nearly identical active sites at two contrasting types of switch point, each indicated by the light gray arrows, one at the amino termini of two helices, the other at the carboxy termini of two β-strands from across a TIM barrel. The active site residues include three zinc ligands and a glutamate side chain which serves to stabilize the tetrahedral configuration of the transition state and shuttle protons to the leaving ammonia. (b) The active sites, together with bound homologous hydrated transition state analogue inhibitors (dark gray bonds). Diastereoisomerism of the two sites results from the fact that the helical turns containing one zinc ligand and the carboxylate (black bonds; black, bold-face letters in the signature sequences) are of opposite hand. That in CDA is right-handed; that in ADA is left-handed owing to the presence in the sequence of the extra glycine. Reprinted from Carter, 1995, with permission.

and at first glance there seem to be extensive similarities between their active sites (Betts et al., 1994; Wilson & Quiocho, 1993, 1994; Wilson et al., 1991; Xiang et al., 1995, 1996). The active-site residues originate from entirely different tertiary structures (Fig. 5-32a). All they have in common is the fact that the incoming chains come from opposite directions, lending approximate two-fold rotation symmetry to both zinc binding sites.

Detailed comparison of the active sites occupied by comparable transition state analogue inhibitors (Carter, 1995) shows that they are diastereoisomers. There are two chiral centers associated with the reaction, and CDA has the configuration C4(R) O4(S). ADA, on the other hand, has C6(S) O6(S). The diastereoisomerism is a curious correlate of the fact that the signature sequence TVHAGE in ADA (Yang et al., 1990) has an extra glycine that can assume the conformation of a left-handed α-helix, a functional mirror image of the right-handed α-helical conformation in CDA.

Stable Secondary Structures in Proteins Complement Double Helical Nucleic Acid Structures

Binding (and hence recognition) of polynucleotides by polypeptides was probably a necessary function of the earliest biological polymers. The size (diameter and typical lengths) of α-helices are almost perfectly suited to fit into the major grooves of B-form DNA (Anderson et al., 1981; Warrant & Kim, 1978). Model building studies (Carter & Kraut, 1974; Church et al., 1977) have demonstrated that the thermodynamically favored right-handed twist of extended β-polypeptide double helices is also potentially complementary to the grooves of both RNA and DNA double helices. The most prominent features in the protein-nucleic acid complexes observed to date involve complementarity at the level of secondary structure.

A large number of transcriptional regulators interact with DNA primarily via the insertion of an alpha helix into the major groove (Brennan & Matthews, 1989a, b; Harrison & Aggarwal, 1990). There also is growing evidence for the participation of β-structures in genetic regulation (Kim, 1992; Phillips, 1991; Rice et al., 1996). A rudimentary structural complementarity of polypeptide secondary structures to the helical structures of nucleic acids underlies many of the observed modes for protein-nucleic acid interaction.

Interaction Via α-Helices. The structure of the CRO repressor (Anderson et al., 1981) introduced the first example of a large and growing family of DNA binding proteins using a helix-turn-helix motif in their complexes with DNA (Fig. 5-33a). The proposed repressor-operator complex based on this structure illustrated fundamental rules for model building. Chemical modification studies indicated the general regions of contact between protein and DNA. The local symmetry of the tetramer suggested a complementarity to the approximate symmetry of the operator sequence. The distance between the two helical protuberances (34 Å) matched exactly the distance between two successive appearances of the major groove in the same face of the DNA, and the tilt of the helices along that direction matched the tilt of the major groove (32°).

The fact that the four subunits in the tetramer are not identical suggests that if the active molecule is actually a tetramer, then different monomers in the tetramer may have different conformations. The key feature of the DNA operator sequence relevant to this question is that the operator is only approximately two-fold symmetric itself. The existence and functional significance of this type of broken symmetry is a puzzling and elusive question in structural molecular biology.

The advent of rapid oligonucleotide synthesis has meant that many such regulatory protein structures have now been solved complexed to DNA targets. Two additional examples shown in Figure 5-32 are the tryptophan operon repressor (Otwinowski et al., 1988) (Fig. 5-32b), which shows the same helix-turn-helix motif, and a fragment of the *MYOD* basic helix-loop-helix motif (Ma et al., 1994). The Trp repressor structure presented an interesting sidelight: the interface between repressor and operator was mediated not by intimate contact, but by the intermediary participation of water molecules. This surprising development suggests that the overall shape properties of the two polymers actually do play important roles not only in binding, but also in specific recognition.

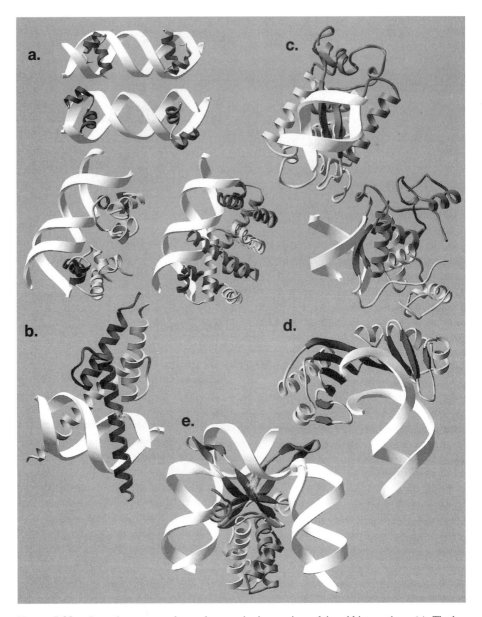

Figure 5-33. Secondary structural complementarity in protein:nucleic acid interactions. (a): The helix-turn helix motif in the CRO repressor (Anderson et al., 1981) and the tryptophan repressor (Otwinowski et al., 1988) in that order. The top two figures compare the helix-turn-helix motifs themselves; the bottom two compare intact, dimeric protein complexes. The CRO repressor binds DNA as a tetramer composed of the dimer shown and another. (b): The basic helix-loop-helix (Ma et al., 1994). (c): The methionine repressor: operator complex (Somers & Phillips, 1992). (d): The DNA binding region of a TFIIA/TATA box binding protein: DNA complex (Tan et al., 1996). (e): An IHF:DNA complex (Rice et al., 1996). Elements of secondary structural complementarity are highlighted in dark gray shading. Examples a and b illustrate the use of α-helices in binding DNA; examples c-e illustrate the use of β-sheets. Examples d and e show the power of β-sheet structures to bend DNA considerably.

Interaction via β-Strands. The methionine operon repressor (Somers & Phillips, 1992) is a dimer that is induced to associate with the Met operon by sufficiently high concentrations of *S*-adenosyl methionine. The dimer involves an intermolecular antiparallel β-sheet that fits into the major groove of the operator (Fig. 5-32c). The TATA box binding protein (Kim et al., 1993) and the associated IHF protein-DNA complex (Rice et al., 1996) both demonstrate the power of the antiparallel β-loop motif to reshape the DNA helix by bending it. Curiously, these "bending" interactions occur in the minor groove, which was originally judged to be too narrow to accommodate antiparallel β-structure (Carter & Kraut, 1974). Binding in this manner may, therefore, induce bending of the DNA in order to widen the narrow groove.

Plasticity (Polymorphism)

Although it has not been proven, it is very likely that a correlation exists between the affinity of a bound complex and the degree to which a protein must rearrange to bind and release ligands. In order to develop and then relinquish the strong forces required for high affinity, binding sites must be equipped to close over their ligands and then reverse the process to release them. The choreography and integration of cell physiology thus depend crucially on the interplay of structural polymorphism and ligand binding, whether it consists of microfolding reactions explored by a polypeptide as it accumulates tertiary structure on the way to its folded state, allosteric regulation of binding affinity in transport proteins, variation of binding affinity as an enzyme guides its substrate through the conversion to product, or, especially, the elegant sequence of conformers encountered along energy- and information-transduction pathways.

Polymorphism: A Sleeping Giant. The enduring challenge of structural molecular biology is to document the manifold different structures accessible to a particular protein and to rationalize the interactions between these different structures and its various ligands. Structural variation comes in a variety of forms: genetic variation among different species provides valuable clues to the relative importance of specific regions and amino acids for function; conformational changes of a specific protein map out the repertoire involved in function.

The study of variation is one of the most general tools available to biological chemistry. Fortunately, both NMR and X-ray crystallography have provided extensive evidence concerning the degree of structural polymorphism for some systems. Solution structures can be manipulated by varying the kinds and concentrations of ligands, and the consequences monitored by NMR. Crystal growth polymorphisms are occasionally observed in the course of initial screening for crystal growth conditions (Carter, 1979), and the structures afforded by such accidental crystallogenesis have led to considerable insight. However, in the future it will likely become important to understand relationships between polymorphism and the physical chemistry of crystallogenesis (Ries-Kautt & Ducruix, 1997) and to seek out crystalline representatives of expected states different from those obtained in the random screening by a more "directed crystallogenesis" (Bullough et al., 1994; Carter et al., 1994, 1971).

The kinds of structural variation available to proteins have begun to be examined systematically (Gerstein et al., 1994). Functionally important conformational changes range from discrete changes of one or a few amino acid side chain rotamers (Matthews et al., 1977), to movement of a single loop over an active site (Fitzgerald & Springer, 1991), to progressively more involved motions including both "hinge-bending" (Gerstein et al., 1993). and "shear" movements, as well as more complicated changes. An accessible directory for examples of the different types of changes is available from the Protein Motion Database:

http://cb-iris.stanford.edu/~mbg/ProtMotDB

Loop Closure. A very simple example of a functionally significant conformational change involves the active site serine in serine proteases. This side chain has one adjustable bond, and in the ligand-free enzyme it is apparently not oriented to participate in nucleophilic attack on substrates until the substrates bind (Matthews et al., 1977). A celebrated example of a slightly more involved conformation change involves the HIV protease. This enzyme is a small (~10 kDa) all beta protein (Fig. 5-34) that forms a homodimer (Wlodawer et al., 1989). The active site lies at the interface between monomers, and is covered by a loop consisting of residues 46–55 that closes by ~7.0 Å in inhibited complexes (Fitzgerald et al., 1990; Wlodawer et al., 1989) (Fig. 5-34a).

Hinge-Bending: The Arabinose-Binding Protein (Quiocho & Vyas, 1984; Quiocho et al., 1989). A number of different binding proteins (sulfate-, phosphate-, leucine-isoleucine-valine-, D-galactose, and arabinose-binding proteins) are all composed of two domains of roughly equal size and similar structure connected by a flexible hinge (Fig. 5-23d). The structures of the two domains are extraordinarily similar and are disposed toward each other with approximate two-fold symmetry, such that each side of the cleft contains nearly the same mass and structure.

The hinge connecting these two domains has been shown to close upon binding of substrate: low-angle diffraction measurements showed that the radius of gyration of the L-arabinose binding protein decreases upon binding of ligand (Newcomer et al., 1981). This finding was consistent with the observation that the initial structure contained a bound arabinose molecule that was inaccessible to solvent. Structures of both free and bound forms of the closely related iron-binding protein, transferrin, showed that a hinge-bending motion between domains accompanied ligand binding (Anderson et al., 1990).

The resulting picture of ligand binding and its interaction with the conformational change that closed the hinge is unmistakably similar to a "venus fly trap," and this model has been put forward explicitly as a metaphor for the functionality of this protein, whose job is to bind arabinose, transport it into the cell, and signal its presence to the chemotaxis response elements. The conformational change that accompanies ligand binding is therefore a signaling mechanism as well, requiring an outward manifestation of the change in its status when the ligand is bound.

The elements of this mechanism involve hinge-bending energies on the one hand, which tend to stabilize the closed conformation, and eight charged side chains projecting into the cleft from both domains. In the closed configuration, these charged side chains need to be "solvated" or to interact with watery partners. Thus, in the absence of ligand, they drive the mouth open, so that they can interact with water. When a sugar is bound, its hydroxyl groups tend to "solvate" these side-charged side chains from both sides of the cleft, reducing their demands to be in contact with water. As a result, the cleft closes onto the ligand once it is bound and makes good contacts with the charged groups.

Three water molecules are present in the active site in the arabinose complex. They are involved in a total of eleven hydrogen bonds interconnecting the polar side chains in the active site and the bound ligand. In addition, they undoubtedly are important for the van der Waals contacts they make in the active site. They fill otherwise void vol-

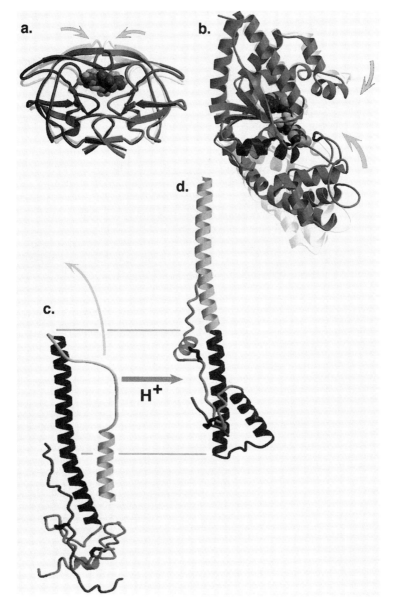

Figure 5-34. Functionally important conformational changes. (a) Closure of the "flaps" at the entrance to the active site of the HIV protease. (b) Movement of the helical domain containing the anticodon-binding domain of tryptophanyl-tRNA synthetase upon synthesis of the activated amino acid, tryptophanyl-5′AMP. (c) and (d) Comprehensive refolding of the "stalk" of the influenza virus haemagglutinin induced at low pH, and which is probably required for the activation of membrane fusion on infection.

umes and complement the hydrogen-bonding patterns of the ligand and protein. Finally, they provide a kind of structural redundancy, allowing the arabinose-binding protein to bind anomeric and other allomeric sugars with nearly the same affinity (Quiocho et al., 1989).

Bound water molecules have turned up in the active sites of all highly refined enzymes. Many times, the hydrogen-bonded pattern of bound waters strongly resembles substrate molecules (James et al., 1981). In some cases these active site waters have subsequently been shown not to be water, but to be various bits of unidentified organic material—peptides in the case of proteases, and so on. Nevertheless, there remain many of these "bound waters" which can only be water; they are, therefore, a ubiquitous feature of binding sites, making it hazardous to predict binding affinity from modeling studies (Weber et al., 1992; Xiang et al., 1995).

Combined Hinge and Shear Motions. A considerably more complicated conformational change mediates the activation of tryptophan and the acyl transfer to tRNA^{Trp} by tryptophanyl-tRNA synthetase (Fig. 5-17b). Interdomain movement is much more pronounced and integrated with catalysis and with recognition of the tRNA substrate. The dimeric enzyme has two potential active sites and two potential binding sites for tRNA. The monomer is subdivided into two domains, the Rossmann dinucleotide binding fold domain and a smaller, C-terminal domain consisting of an irregular four helix bundle (Doublié et al., 1995). Two different conformations are shown in Fig. 5-34b. It is unknown at present whether or not the 14° movement of the helical domain improves the fit between the anticodon and acceptor stem of the tRNA with their cognate binding sites on the enzyme.

Large Scale Rearrangements: Haemagglutinin. The curious tertiary structure of haemagglutinin is apparently connected with an extraordinary, pH-dependent conformational change that is involved in the infection process. In order to induce the host cell membrane to fuse with the viral membrane, this protein must convey the fusion activation peptide, located near the base of the stalk at the viral membrane surface, to the opposite end of the haemagglutinin, some 90 Å up the stalk (Fig. 5-34c, d). The mechanism by which this happens involves a complete remodeling of the structure of the stalk and its base in the HA2 peptide (Fig. 5-27d) (Bullough et al., 1994). The long connecting segment between the two helices of the stalk is apparently in a delicate balance between the extended conformation and an α-helix (Carr & Kim, 1993). When the pH is lowered, the helical form is preferred, and the long stalk rearranges, adding the short helix onto the top of the long helix, which is shortened at the other end by an abrupt foldback of the carboxy terminus. The fusion peptide at the amino terminus of the short helix is thereby transported all the way to the other end of the stalk, where it can promote fusion with the infected cell.

Cooperativity

An important functional property of multisubunit proteins arises from the fact that the bonds between subunits can be used to communicate information about ligand binding. When one subunit in a cooperative system changes its ligation state, that change is coupled to structural rearrangements that influence the binding affinity to the other subunits. This intersubunit communication is amplified by preserving the symmetry of the system, and may be considered in terms of the reciprocal linkage between binding and conformational equilibria (Wyman & Gill, 1990). Consider that an oligomer of n subunits exists in two *conformational* states, A and B, that each A-state monomer can bind a ligand, L, with dissociation constant, K_L, but that no appreciable binding occurs to state B. Then the binding equilibrium

$$A[L_i] + L \Leftrightarrow A[L_{i+1}], \quad K_L = \frac{A[L_i][L]}{A[L_{i+1}]}$$

and the conformational equilibrium

$$A \Leftrightarrow B, \quad K_0 = \frac{[A]}{[B]}$$

are said to be "linked", since the total concentration of A subunits, $[A_{total}]$, increases as the ligand concentration, $[L]$, increases. In terms of the concentration of *free* A subunits, $[A]$

$$[A_{total}] = [A]\left(1 + \frac{[L]}{K_L}\right)^n$$

The situation becomes interesting when the intermolecular bonds in the B state are stronger than those in the A state, so that at low $[L]$ the system remains in the low affinity B state. Then, as $[L]$ increases, ligand binding will shift the conformational equilibrium leading to the appearance of $A[L_i]$.

From the ligand concentration, $[L_{switch}]$, for which $[A_{total}] = [B]$

$$[B] = [A]\left(1 + \frac{[L_{switch}]}{K_L}\right)^n; \quad K_0 = \left(1 + \frac{[L_{switch}]}{K_L}\right)^{-n}.$$

In a reciprocal sense, the affinity of the system for L will also depend on K_0. Thus, preservation of symmetry can ensure that all or most subunits of a multimeric protein will do the same thing at the same time, thereby avoiding wasteful differences of opinion between subunits. Moreover, if other ligands also bind preferentially to one of the two conformational states, then similar linkage relationships give additional control over the conformational equilibrium and, hence, the functional state of oligomeric proteins through variation in ligand concentrations. The dependence of the conformational equilibrium on **n** shows why the degree of cooperativity increases with the number of participating subunits.

The requirement for cooperativity in oxygen binding is an important reason why oxygen transport proteins have multiple subunits. Reversible oxygen binding is a clever bit of engineering, since oxygen generally binds very tightly to metals, forming oxides. Three different types of molecules have evolved to carry out this function: hemoglobin, hemerythrin, and hemocyanin. Each uses a very different mechanism for oxygen binding and dissociation. The mechanisms are best known for hemoglobin, but structures are known now for all three different classes of protein. Hemoglobins and hemocyanins both use cooperativity to insure that oxygen is transported efficiently to appropriate target sites in the respective organism.

Structures are known for two different conformers, one (R) active, the other (T) inactive of several oligomeric enzymes, notably aspartate carbamoyltransferase (Kantrowitz & Lipscomb, 1988) and glycogen phosphorylase (Barford et al., 1991). As is also true for hemoglobin, both R and T structures have the same point group symmetry. Although the conservation of symmetry cannot be absolute because it must break during the transition between the two forms, it nevertheless accounts satisfactorily for all but the most exacting physical measurements: the activity of the enzyme is essentially either "on" or "off" because the conformational change cannot take place between one pair of subunits without communicating this change to the others.

Hemoglobin. Analysis of liganded and deoxy hemoglobins presented the first extensive look at the interplay between structural polymorphism and ligand binding phenomena (Baldwin & Chothia, 1979; Perutz, 1970, 1978, 1979). This system has been much discussed elsewhere, and it is entirely appropriate to consider it as analogous to the hydrogen atom in quantum chemistry.

In the presence of the allosteric effector diphosphoglycerate, hemoglobin has about 1000 times lower affinity for oxygen in peripheral tissues where [O_2] is low. This behavior cannot possibly be explained with four independent sites having different affinities for oxygen. Why not? The high affinity sites would fill up first at low [O_2], and the remaining unoccupied sites would, therefore, have low affinity. The only way to induce a higher binding affinity for the final oxygen bound is via an intersubunit communication called positive cooperativity.

The underlying quaternary polymorphism involves two different ways to assemble the hemoglobin tetramer from two $\alpha\beta$ dimers. The deoxy quaternary structure has a strong binding constant ($K_d = 4.5 \times 10\text{-}11$ M; $\Delta G = -14.3$ kcal/mole) and for this reason it is called the "T" or "tense" structure. The oxy quaternary structure has a much weaker binding constant ($K_d = 1.6 \times 10\text{-}6$ M; $\Delta G = -8.0$ kcal/mole) and is called the "R" or "relaxed" structure. The ratio of these two binding constants is related to the cooperative free energy. The structures of the $\alpha\beta$ dimers do not change very much in the allosteric transition between the two quaternary structures. Most of the free energy difference between the two states arises from the different characteristics of the interface between the dimers.

The $\alpha\beta$ dimer is a reference state in somewhat the same sense as the vapor phase is a reference state for analysis of interactions in water. In the absence of ligand, the reference state of two dimers is in equilibrium with both T and R states, with free energies and equilibrium constants given above. The cooperative free energy is $(-8 - (-14.3)) = 6.3$ kcal/mole. This corresponds to an equilibrium constant of about 35,000 in favor of the T state in the absence of oxygen.

The intermolecular interactions responsible for this cooperative free energy have been inferred from the crystal structures of oxy and deoxy hemoglobin structures. Since the isolated monomer chains and $\alpha\beta$ dimers have noncooperative, high affinity oxygen binding, constraints must come into play in the deoxy tetramer to reduce the oxygen affinity of the heme group. In the deoxy quaternary structure the carboxy terminal residues tether to residues with complementary charges and allow considerable surface area to be buried when these residues form ordered interactions across the tetramer interface in the T state. These "quaternary constraints" buckle the hemoglobin subunits in a tertiary conformation that has a low oxygen affinity. The binding of oxygen breaks open these buckles and the four subunits revert to the conformation they have as monomers, in which the oxygen affinity is high.

The R and T interfaces between $\alpha\beta$ dimers involve many alternative configurations with nearly the same stability, so that the dimers can fit together almost equally well in either T or R states over most of the subunit interface. One example of such alternative conformations is the fit of His97β, which fits alternately in between Thr38α and Thr 41α in the oxy state and between Thr41α and Pro44α facing outward from the C helix in the deoxy state.

The R state structure may be an artifact of crystallization. The fully liganded form of the Ypsilanti mutant hemoglobin has a very different quaternary structure than either R or T. In this new quaternary structure, called the Y state (Smith et al., 1991), the interactions between His97β and the C helix of the opposing α subunit are completely disengaged by screw displacements of the dimers by approximately 22° from the T state and by +12° from the R state about different axes approximately 50° apart. Its ligand-free form is evidently isomorphous to deoxy HbA. The functional significance, if any, of the Y quaternary structure is currently a mystery. However, it has recently been observed for a normal human carbon monoxy-hemoglobin crystallized under physiological conditions (Smith & Simmons, 1994). Recent speculation (Janin & Wodak, 1993; Srinivasan & Rose, 1994) is that this Y quaternary structure may actually be the physiologically important liganded state, a "super R state."

The potential importance of the Y quaternary state of hemoglobin should not be minimized. It is an extraordinary contrast to the canonical R state crystallized under high salt conditions. It raises the specter of other potentially artifactual results among published crystal structures, particularly with respect to quaternary interactions, and emphasizes the importance of careful characterization of the dependence of quaternary structure on crystal growth conditions.

Disproportionation and Broken Symmetry

An utterly remarkable observation was made regarding the first Bence-Jones protein structure (Edmundson et al., 1975) . This abnormal immunoglobulin is a dimer of light chains expressed in very high levels in patients with multiple myelomas. Despite the fact that it contains two identical light chains, it forms a complex that resembles an Fab fragment containing a light chain and a heavy chain fragment! The angle made in the hinge region is different for the two subunits, being about 70° in one of the two chains (the "light" behaving light chain) and 110° in the other (the "heavy" pretender).

The two identical light chains are both "looking" for a heavy chain. The missing interfacial bonds are powerful enough inducement that one of the two chains consents to assume the conformation of a heavy chain in order that the other can achieve happiness as a fully complemented light chain!

Enzyme Catalysis Involves Differential Binding of Ground States and the Transition State

Enzymes are simply ligand-binding proteins that have learned how to differentiate between the different transition states and intermediates along the reaction coordinate of a chemical reaction. In favorable cases where stable analogues of the transition state can be prepared (Wolfenden, 1976), the crystal structures of their complexes provide comprehensive catalogs of the interactions important for specific binding (Lolis & Petsko, 1990). The structural reaction profile of the *E. coli* cytidine deaminase has provided a detailed sequence of different complexes related to hydrolytic deamination catalyzed by a zinc-activated nucleophilic water molecule (Xiang et al., 1995, 1996, 1997). An unexpected conclusion of that series of structures is that binding sites on the enzyme for the products are set apart in opposite directions, so that the active site literally pulls the substrate apart during hydrolysis, and that this separation of the products is accompanied by the facilitated proton transfer from the hydrolytic water to the leaving ammonia group by rotation of a single bond in a crucial glutamic acid (Fig. 5-35) (Xiang et al., 1997).

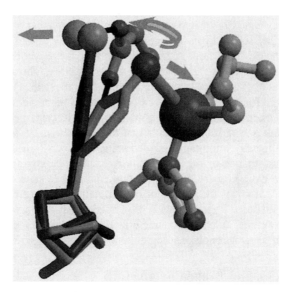

Figure 5-35. Substrate and product complexes of *E. coli* cytidine deaminase. Trajectory of the hydrolytic deamination of cytidine to uridine as catalyzed by the *E. coli* cytidine deaminase. Positions of a stable substrate analogue, 3-deazacytidine (dark gray), and the products, uridine and ammonia (light gray) were determined in separate experiments by X-ray crystallography. The pyrimidine moiety rotates by 30° to the right toward a zinc-activated water molecule (dark gray) through a tetrahedral transition state, while the ammonia moves 0.8 Å into a pocket to the left. A proton is liberated from the nucleophilic water molecule by the carboxylate of a neighboring glutamate (E104) which rotates in the opposite direction, transporting it to the leaving ammonia group. Reprinted from Xiang et al., 1996b, with permission.

Linked Equilibria and Transduction

A crucial set of linked ligand binding and conformational equilibria underlie all transduction systems. The nucleotide triphosphate hydrolysis changes the structure of the substrate in what is ultimately an essentially irreversible way. Transducing systems have evolved to bind the different forms of nucleotide di- and triphosphates with slightly different conformations, which endow the relevant NTPases with differential abilities to bind to other macromolecular ligands. The coupling mechanism described above for the F1 ATPase reverses the coupling of ATP hydrolysis by the actomyosin ATPase mechanism first worked out by Lymn and Taylor (1970, 1971).

The three-dimensional structures of myosin (Rayment, 1993; Rayment & Holden, 1994; Rayment et al., 1993) have recently provided a detailed picture of how this occurs in muscle contraction, and the structures of the kinesin motor subsequently revealed profound similarities in that structure (Rayment, 1996; Sablin et al., 1996). Kinetic analyses of the kinesin-mictotubule ATPase mechanism suggest that there are differences in detail regarding the rates at which the products of ATP hydrolysis are released from the motor-substrate-microtubule complex. A similar change in the free energy of tubulin-bound guanosine triphosphate hydrolysis inside tubulin has been demonstrated and linked to mechanisms of conformational coupling that operate in the assembly of microtubules from tubulin and their stabilization (Caplow et al., 1994;

Caplow & Shanks, 1996). It, therefore, appears to be a general property of transducing systems.

Summary

Protein structure has become an unwieldy field to review comprehensively, as the number of new structures provided by NMR and diffraction methods grows exponentially. Electrostatic properties of the peptide bond and the relative solubilities of amino acid side chains determine three-dimensional structures by hierarchical interactions that sequester first as many hydrogen bonding groups and then as many nonpolar side chains as possible. The Linderstrøm-Lang structural hierarchy captures this physicochemical hierarchy and still provides a strong conceptual framework. Ligand-driven conformational polymorphism provides the crucial link between structure and function, and is an area where much remains to be learned.

Acknowledgments

This chapter is dedicated to the memory of Mary Ellen Jones for her steadfast support of my teaching efforts. I wish to thank many years of Biochemistry 134 students for helping me to see the beauty and scope of protein structure in continually new lights. Many of my colleagues in the Department of Biochemistry and Biophysics, including Jan Hermans, Barry Lentz, Richard Wolfenden, Aziz Sancar, Gerhard Meissner, and Michael Caplow, have helped clarify points in this chapter. In particular, Jan Hermans provided the simple treatment of linked equilibria. I am indebted to Joël Janin for a critical reading of the chapter.

6

Protein Folding

Zhi-Ping Liu, Josep Rizo, and Lila M. Gierasch

The protein folding problem deals with the questions of how and why a protein adopts a specific native conformation. It has been shown that all the information required to direct the folding of a protein to its native conformation is contained in its amino acid sequence (Anfinsen, 1973). Thus, the ultimate goal of studying protein folding is to understand how the specific amino acid sequence of a protein encodes such information, and to predict its native structure using this information. The protein folding problem has been one of the most fundamental and challenging problems in the life sciences since the discovery of the reversibility of protein denaturation in the 1930s. Much has been learned in the past six decades, but many aspects of the protein folding problem still remain poorly understood. With the rapid advancement of modern molecular biology, there is growing urgency to find solutions to the folding problem. The nucleotide sequences of genes can be determined much faster than their protein products can be isolated and subsequently characterized. More gene sequences will be revealed through the efforts of the Human Genome Project. Elucidation of the general principles of protein folding will help us predict the structure and function of the protein product of a gene sequence or an open reading frame. Knowledge of how a protein folds will also help us to understand the causes of certain cancers and genetic diseases and thus to find corresponding therapeutic treatments. For example, wild-type p53 acts as a tumor suppressor by halting the abnormal growth in cells (Donehower & Bradley, 1993). Mutations in p53 can eliminate the activity of the protein and turn it into a tumor promoter. A temperature-sensitive mutant of p53 (p53Val135) (Michalovitz et al., 1990) was found to have a conformation similar to wild type at low temperature and a novel conformation at high temperature (Milner & Medcalf, 1991). This conformational alteration has been proposed to underlie the change in activity of p53 from a growth suppressor to a growth promoter (Milner, 1991). Similarly, several other genetic diseases, including cystic fibrosis (Thomas et al., 1992), Alzheimer's (Haass et al., 1992a,b), and scrapie (Xi et al., 1992), may arise from defects in the folding pathways of the corresponding mutant gene products.

Because of the rapidly growing literature in the protein folding field, it is impossible to provide a comprehensive review. Rather, we will first discuss how the facts and concepts in the field have been discovered and developed, and then we will summarize the current understanding of protein folding with some specific examples. We hope that this chapter will provide a starting point for readers and introduce the questions and how to approach them. We limit our coverage to the experimental aspects of *in vitro* folding of small, single-domain proteins. Much theoretical research on protein folding, which will not be discussed here, has been directed to methods to simulate the folding process (Abagyan, 1993; Daggett & Levitt, 1993; Godzik et al., 1993; Skolnick & Kolinski, 1990) and to algorithms to predict the secondary and tertiary structure of proteins from their amino acid sequences (Blundell et al., 1988; Fasman, 1989; Garnier, 1990; Greer, 1991; Ring & Cohen, 1993). Protein folding models will be described briefly in the historical introduction. Several books have chronicled developments in the field of protein folding (Creighton, 1992b,c; Ghelis & Yon, 1982; Gierasch & King, 1990; Jaenicke, 1980). A number of excellent reviews have appeared (Baker & Agard, 1994; Creighton, 1990; Dill, 1990; Dill & Shortle, 1991; Haynie & Freire, 1993; Kim & Baldwin, 1990; Matthews, 1993a,b; Rose & Wolfenden, 1993).

Historical Perspective

Protein Denaturation and Renaturation

The protein folding problem was first articulated in the early 1930s by Wu (1929, 1931), Northrop (1932a,b), and Anson & Mirsky (1934a,b), who studied the denaturation process of proteins. These early works demonstrated that proteins lose their biological activities upon heating, acidification, or dissolution in the presence of a high concentration of urea or guanidinium chloride (GdmHCl), a process referred to as *denaturation*. These researchers also found that the denatured proteins have different physical and chemical properties from the biologically active *native* proteins, for instance, their solubility, crystallizability, and viscosity. Contrary to the general belief that the process of denaturation of proteins was irreversible, it was found that, at least for a few proteins, their biological activities, as well as their physical and chemical properties, could be regained after a protein had been denatured through a process called *renaturation* (Northrop, 1932a,b; Anson & Mirsky, 1934a,b). This discovery established the significance of the denaturation reaction and led to the most intriguing question, posed by Max Perutz in 1940: "Can a boiled egg be made raw again? Can it, so to speak, be unboiled?"

The view that protein denaturation may correspond to the unfolding of a chain appeared during early protein folding studies (Wu, 1931). In the following 20 years, this view became clear with the progress of our understanding of the physical chemistry of proteins and the development of spectroscopic methods that allowed researchers to follow quantitatively the conformational change occurring during the denaturation process. A critical step was the development by Sanger of a method to determine the amino acid sequence of proteins, which was first demonstrated for insulin (Sanger et al., 1952). A few years later the sequence of ribonuclease was determined (Hirs et al., 1960). These results proved that the covalent structure of a protein consists of a linear peptide chain and that the amino acid sequence of the chain is unique for a given protein. Tanford and coworkers performed extensive studies on denaturation of proteins with various physicochemical perturbations in the 1960s. They concluded that denaturation of proteins can be defined as a major conformational change from the original native structure, without alteration of the amino acid sequence (Tanford, 1968). They also found that the extent of the conformational change upon denaturation depends on the perturbation method used; GdmHCl and/or urea induce complete unfolding of protein chains, whereas heat and pH-denatured proteins often contain residual structure that can be further unfolded by addition of GdmHCl or urea (Aune et al., 1967).

Although the reversibility of the process of protein denaturation was demonstrated in the 1930s, at the time this was not accepted to be the general rule. It was believed that the reversibility of the denaturation process was due to the presence of covalent disulfide bonds and cleavage of the disulfides would cause the denaturation to be irreversible (Kauzmann, 1954). The reversibility of protein denaturation, independent of the presence of disulfide bonds, was demonstrated by Anfinsen and coworkers with bovine pancreatic ribonuclease (RNase) (Anfinsen et al., 1961; Sela et al., 1957). This protein contains four disulfide bonds. Native RNase becomes fully denatured in the presence of 8 M urea and reducing agents such as β-mercaptoethanol. By allowing reoxidation of the reduced RNase in the presence of denaturant and then removing the denaturant, a mixture of RNase derivatives (called "scrambled" RNase) was produced, which contained many or all of the possible 105 isomeric disulfide-bonded forms. Anfinsen and coworkers demonstrated that this mixture of RNase derivatives could be readily converted to the native protein with full biological activity upon exposure to β-mercaptoethanol (Haber & Anfinsen, 1962).

Anfinsen's experiment established that all the information necessary to define the native three-dimensional structure of a protein, and therefore its function, is contained in the sequence of the protein (Anfinsen, 1973). This conclusion constitutes the basis for *in vitro* protein folding studies; these folding studies can be done in the test tube and require only that the conditions and properties of the solvent be properly adjusted.

In the last several years, it has been shown that folding of some proteins can be assisted by other proteins called molecular chaperones (Ellis, 1994) and by enzymes that catalyze specific isomerizations such as peptidyl prolyl *cis-trans* isomerase (PPI) and protein disulfide isomerase (PDI) (Gething & Sambrook, 1992; Schmid, 1993). However, it appears that the main function of these biological factors is to influence the kinetics of protein folding and to prevent off-pathway aggregation or degradation, thus increasing the yield of the intrinsic process of protein folding (Hendrick & Hartl, 1993). These biological factors do not carry specific information to direct the folding of a protein.

Anfinsen's observation further argues that the native state is the thermodynamically most stable one relative to all other accessible states and that the native state thus is a state function that does not depend on the process or initial conditions leading to that state (Anfinsen, 1973). Therefore, it is possible to study the folding problem thermodynamically: by determining experimentally the free energy, enthalpy, and entropy of the folded form relative to those of unfolded forms, one may be able to determine the forces that stabilize and perhaps direct the folding of a protein toward the native state.

Many studies of protein folding under equilibrium conditions have shown that for many small, single-domain proteins, unfolding is a reversible cooperative two-state process, and that the native protein is the most stable state under physiological conditions. Particularly compelling are the experiments performed by Wüthrich and coworkers, who used nuclear magnetic resonance (NMR) spectroscopy to study the conformation of the N-terminal DNA-binding domain of phage 434-repressor (residues 1–69) under various urea concentrations (Neri et al., 1992a,b). The relative concentration of the fully folded (in the absence of urea) and fully unfolded (in the presence of 7 M urea) states of the protein is a function of the concentration of urea. In the unfolding transition region (4.2 M urea), only fully folded and unfolded forms are present in equal concentration and with an exchange lifetime of 1 sec. The unfolded and native states exist in thermodynamic equilibrium within the unfolding transition region, and the native state in the transition region is the same as the one present under physiological conditions.

The Hydrophobic Effect

A concept that is crucial to understanding protein folding is that of hydrophobicity. Nonpolar substances have low solubility in water and tend to associate in aqueous media. Thermodynamically, this means that transfer of nonpolar substances from pure liquid phase (l) into water (w) requires energy ($\Delta G_{l \rightarrow w}$ is positive). Because the enthalpy of transfer of nonpolar substances ($\Delta H_{l \rightarrow w}$) is close to zero at room temperature and $\Delta G_{l \rightarrow w} = \Delta H_{l \rightarrow w} - T\Delta S_{l \rightarrow w}$, the large positive $\Delta G_{l \rightarrow w}$ of transfer of nonpolar substances into water must result from the negative entropy of transfer ($\Delta S_{l \rightarrow w} \ll 0$). Frank and Evans (1945) recognized the entropic origin of the low solubility of nonpolar substances in water. They proposed that the decrease of entropy upon transfer of nonpolar substances into water is due to the formation of so-called iceberg structure; the water molecules that are in the vicinity of the nonpolar molecules must be more ordered and have more persistent hydrogen bonding networks than in ordinary liquid water at the same temperature (Frank & Evans, 1945). Consistent with this is the observation that there is a linear relationship between the free energy of transfer of a nonpolar amino acid and its surface area (Richards, 1977).

Nearly all proteins contain a relatively high proportion of nonpolar amino acid side chains and the nonpolar side chains tend to exist in the interior of the proteins [ca. 40%, (McCaldon & Argos, 1988)], "out of the water." Based on this fact and the thermodynamic data characterizing the transfer of hydrocarbons and other nonpolar molecules from a nonpolar solvent to water, Kauzmann introduced the idea that the tendency of the nonpolar groups of proteins to cluster in aqueous environments so as to minimize their water accessible surface would be the dominant force in stabilizing the native protein. He referred to this tendency as *hydrophobic bonding* (Kauzmann, 1954, 1959). At the time, this idea contrasted with Pauling's view, which ascribed a crucial role for intramolecular hydrogen bonding in the formation of α-helices (Pauling et al., 1951) and β-sheets (Pauling & Corey, 1951), and, consequently, in the overall stabilization of protein structure.

Kauzmann's idea has now been supported by many lines of experimental data. (1) Experiments from differential scanning microcalorimetry indicate that the temperature-dependence of the free energy of the protein unfolding reaction resembles that of transfer of nonpolar model compounds from nonpolar media into water (Privalov & Gill, 1988); they both involve a large heat capacity change. (2) The three-dimensional structures of proteins obtained by X-ray crystallography and more recently by NMR spectroscopy show that the nonpolar residues are clustered inside the molecule, forming a core where they largely avoid contact with water (Miller et al., 1987; Richards, 1977). (3) The hydrophobicities of the amino acid residues in the cores of proteins appear to be highly conserved. Mutation of these core residues has a more severe effect on the protein structure than that of residues involved in other types of interactions (Bowie et al., 1990; Lim & Sauer, 1989). (4) Thermodynamic analysis of unfolding reactions using mutant proteins shows that the contribution of the hydrophobic interaction to the stability of native proteins can be estimated based on the value of free energy transfer of side chains of amino acids from water to an organic solvent (Matthews, 1993a). All these observations have led to the contemporary textbook view that hydrophobic interactions are the driving force for protein folding (Dill, 1990; Rose & Wolfenden, 1993). Interestingly enough, however, the concept that hydrogen bonds also have a substantial contribution to the stabilization of protein structures has recently been revived in work by Privalov and Gill (1988) and by Pace and coworkers (Shirley et al., 1992).

Pathways of Protein Folding

The importance of folding pathways was realized in the late 1960s with the so-called Levinthal paradox (Levinthal, 1969): how does the polypeptide chain search in a finite time through the extensive conformational space available to it to find its native state? The point is often illustrated in the literature by considering, for instance, that the total number of possible conformations for a polypeptide chain with 100 residues and two equally probable conformations per residue is on the order of 10^{30}, and that, even assuming interconversion times among conformations as short as 10^{-13} sec, it would take 10^9 years to sample all these conformations. Since single-domain proteins usually fold on the time scale of seconds, it is then concluded that specific pathways of protein folding must exist. However, it is important to realize that this argument is somewhat fallacious, since there is no reason *a priori* to think that proteins should behave differently from all other physical systems. Following general laws of statistical mechanics, a system consisting of unfolded protein molecules that is suddenly placed under conditions favoring the native state will evolve to populate, according to a Boltzmann distribution, the microstates (conformations) accessible to it in the new conditions. Many of the possible conformations may be ignored for practical purposes due to their very high energy or due to high energy barriers that make them inaccessible. During the evolution of the system with time, the populations of the accessible conformations will change. Some common structural features may exist at each time among the many different populated microstates, and the convergence of structural features will increase with time until the native state is predominantly populated. The description of protein folding pathways consists (or should consist) in the definition of what are the common structural features, how they are distributed among the microstates and how they change with time, keeping in mind this statistical view. It is only in this sense that we will use the term "pathway," which otherwise can be very misleading. Accordingly, "intermediate states" along the pathway correspond to transient ensembles of microstates with common structural features.

One of the first and clearest demonstrations of the existence of a specific folding pathway comes from folding studies of bovine pancreatic trypsin inhibitor (BPTI) (Creighton, 1974a,b). Native BPTI contains three disulfide bonds (5–55, 14–38, 30–51) and can be unfolded by simply reducing the disulfide bonds. BPTI refolds under conditions where disulfide bond formation is favored. Creighton devised strategies for trapping the intermediates with one, two, or three disulfide bonds, formed at various folding stages, by irreversibly blocking all thiol groups in the solution with iodoacetate. The trapped intermediates were isolated and characterized by nondenaturing polyacrylamide gel electrophoresis and ion-exchange chromatography. Because the formation of disulfide bonds is coupled to the formation of conformations that favor them, the trapped species containing disulfide bonds are also intermediate folding species.

Creighton's experiments revealed two features of the BPTI folding mechanism. First, the process is distinctly nonrandom: only a small fraction of the possible disulfide-bonded intermediates are detected during folding and unfolding. The folding process starts randomly with heterogeneous populations of unfolded protein conformations and rapidly converges to a limited number of intermediates, and then the proteins acquire the native conformation in the rate-limiting step (Goldenberg & Creighton, 1985). The second feature is that the trapped intermediates contain both native (those present in the final folded native protein) and "nonnative" disulfide bonds (i.e., those not present in the native protein). The non-native intermediates, as interpreted by Creighton, play a central role in guiding the refolding of BPTI.

The nonrandom folding pathway revealed in the BPTI folding study has now been accepted as a general feature of the folding mechanism of single-domain proteins with and without disulfide bonds. However, whether the non-native disulfide intermediates are involved in the folding pathway of BPTI is open to debate (Creighton, 1992a; Weissman & Kim, 1992b). Creighton's original studies identified three intermediates with one disulfide bond (30–51, 5–55, and **5–30**) (nonnative disulfide bonds are indicated in boldface), and five intermediates with two disulfide bonds [(30–51,14–38), (30–51, **5–14**), (30–51, **5–38**), (5–55, 14–38), and (30–51, 5–55)] (Fig. 6-1). Recent experiments by Weissman & Kim (1991, 1992a), using acid for trapping and reverse phase HPLC to isolate the disulfide intermediates, show that the nonnative disulfide

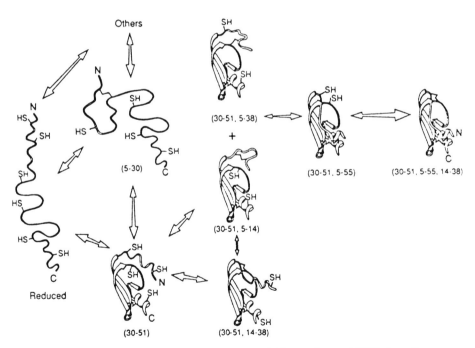

Figure 6-1. Schematic representation of the disulfide folding pathway of BPTI. Open lines represent the conformation of the polypeptide chain that is not regular or well defined, arrows represent β-strands, and coils represent α-helices. The intermediates are named by their disulfide bonds (solid crosslink) formed at each stage of the folding. The intermediates (5,55) and (5–55, 14–38) are not on the productive folding pathway, and they are not included here. The relative rates of the reactions are depicted semiquantitatively by the thickness of the arrow heads; the wider the arrowhead, the greater the rate in that direction. Reprinted from Creighton, 1990, with permission.

species are in much lower quantities than previously thought when they were trapped with iodoacetate. The authors argue from these data that only native disulfide intermediates are involved in the folding pathway of BPTI.

The nonnative one-disulfide intermediate (5–30) now appears not to play a role in the formation of subsequent intermediates. However, the role of the nonnative two-disulfide intermediates remains controversial. The central issue is the conversion from the two-disulfide intermediate (30–51,14–38) to (30–51, 5–55). Since in the intermediate (30–51, 14–38) the thiol groups of Cys5 and Cys55 are buried and nonreactive, formation of the disulfide bond 5–55 in the two-disulfide bond intermediate (30–51, 5–55) could happen either by partial unfolding of the protein so that Cys5 and Cys55 become accessible (Weissman & Kim, 1991) or by intramolecular disulfide rearrangement through the nonnative intermediates (30–51, **5–14**) and/or (30–51, **5–38**) (Creighton, 1992c). In Creighton's model, regions around nonnative disulfide bonds are not structured, and the only elements of well-defined structure present along the folding pathway are nativelike (Darby et al., 1992).

Metastable States of Proteins

The importance of kinetics in protein folding also raises the possibility that the observed folded conformation is not necessarily the one with the global minimum free energy but rather the one with lowest kinetically accessible free energy, that is, the native form may be a metastable conformation (Levinthal, 1969). Indeed, this has been demonstrated for plasminogen activator inhibitor-1 (PAI-1) (Carrell et al., 1991; Mottonen et al., 1992). PAI-1 folds initially to an active but relative unstable form. The active form then undergoes slow conversion with a half-life of 1 hr at 37°C to a form that is more stable but is inactive as an inhibitor (the latent form). The metastable active form of PAI-1 can be regenerated from the latent form by unfolding and then refolding the protein.

It has also been suggested that folding of α-lytic protease is under kinetic control (Baker et al., 1992; Baker & Agard, 1994). α-Lytic protease has been shown to have two conformational forms under native conditions, one active and another inactive (Baker et al., 1992). α-Lytic protease is synthesized and folds as a precursor containing a 166-amino acid Pro region. Mature, active protease is formed after cleavage of the Pro region. Agard and coworkers have found that the protease does not refold to its active form after being unfolded, but rather to a stable, inactive form (Baker et al., 1992). The conversion from the inactive form to the active form requires the Pro region, which presumably functions as a catalyst to bypass a high kinetic barrier separating the inactive and active forms (Baker et al., 1992). In the absence of the Pro region, the two forms do not interconvert. Because a thermodynamic equilibrium between the active and inactive forms cannot be established in the absence of the Pro region, the relative stability of the two forms cannot be assessed. Although it seems likely that the inactive form is a metastable state, it could also be that the active form is actually the kinetically trapped state, while the inactive form is the lowest energy state.

The Molten Globule Form of Proteins

As mentioned earlier, the reversible folding-unfolding of small single-domain proteins is usually a two-state cooperative process in which only fully folded native and fully unfolded molecules are populated at equilibrium. The existence of intermediate states of proteins is interesting because these states may help elucidate the conformational properties of the protein chains and provide clues about their folding mechanisms.

In the mid-1970s, Kuwajima and coworkers found that α-lactalbumin (α-LA) forms a stable conformation at low pH (originally called the A state for "acid state") that is different from both the native and the unfolded forms (Kuwajima et al., 1976). The A state of α-LA contains essentially an identical amount of secondary structure as the native protein, but lacks specific tertiary interactions (Kuwajima, 1977). The A state is more compact than unfolded α-LA and slightly less compact than the native form of the protein (Dolgikh et al., 1981). Because of these characteristics, it was given the name "molten globule" by Ohgushi and Wada (1983). Later, it was found that a variety of other proteins adopt conformations with similar properties under mild denaturing conditions. This led to the "molten globule" model for the description of a novel physical state of protein molecules by Ptitsyn (1987): "a compact globule with fluctuating tertiary structure." Analysis of molten globules that exist under equilibrium conditions and comparison with kinetic folding intermediates are currently among the most active areas in research on protein folding.

Protein Folding Models

During the last two decades, several models of how protein folding proceeds have been proposed. These models provide critical frameworks to design experimentation on protein folding and to interpret the experimental results. We finish this historical perspective by briefly summarizing some of these protein folding models and pointing out relationships among them.

In the diffusion-collision model (Karplus & Weaver, 1976), a protein is divided into several elements ("microdomains"). Two or more of these microdomains diffuse together in an effective collision to form a stable structural entity with native or near native conformation. In the framework model (Kim & Baldwin, 1990), the hydrogen-bonded secondary structure forms first to provide a framework for the formation of tertiary structure in the final folding process. The concepts underlying the diffusion-collision model and the framework model appear to be the same. A hierarchical process, in which simple structures form first and these interact to give more complex structures, is implied in both models. In contrast, the hydrophobic collapse model (Chan & Dill, 1991) emphasizes the role of hydrophobic interactions in protein folding. The model suggests that the "hydrophobic collapse" takes place before any secondary or tertiary structure formation. Secondary and tertiary structures are the consequence of the compactness of the intermediate and are formed in later steps through local reorganization. The molten globule model (Ptitsyn, 1987) has elements of both the framework model and the hydrophobic collapse model, and assumes that molten globule intermediates are formed universally in the folding pathways of proteins. The molten globule is formed by hydrophobic collapse of preformed pieces of native secondary structure. The native tertiary interactions are formed in later stages of folding.

Most protein folding models in the literature, including those just described, assume that folding of a protein proceeds through one or several preferred pathway(s) that can be described in terms of distinct structural intermediates. In contrast, the jigsaw puzzle model (Harrison & Durbin, 1985) suggests that proteins fold by a number of different, parallel pathways. The model postulates that the folding process of a protein is analogous to solving a jigsaw puzzle. There are no preferred pathways or definable structural intermediates. Each folding attempt will have a different path.

Conformational States of Proteins

The Native State

The native folded conformations of proteins are well known from X-ray crystallography and NMR spectroscopy studies. The Brookhaven Protein Databank (Bernstein et al., 1977) maintains a compendium of the known three-dimensional structures of proteins. The native states are highly specific and have varying degrees of flexibility (Williams, 1989). Conformational fluctuation occurs usually, but not always, at the surface or loop regions of proteins. Flexibility is least in the interior of proteins, although side chain rotation does occur, especially for the aromatic rings of tyrosine and phenylalanine. Nevertheless, all protein molecules of the same sequence (with very rare exceptions) take up the same native three-dimensional state. This homogenous nature accounts for the fact that the unfolding of proteins usually proceeds with a single kinetic phase without a lag period. All of the molecules unfold with the same probability (Creighton, 1990).

Although the structures of proteins are complex and irregular at atomic level, the folding patterns are usually simple (Richardson & Richardson, 1989; Richardson et al., 1992). Only a limited number of secondary structure elements are found in the context of widely different primary sequences: α-helix, β-sheet, reverse turns, and loops. In most proteins, α-helices and β-sheets are assembled in typical topological patterns (Orengo et al., 1993; Richardson & Richardson, 1989).

The native states of proteins are well packed, with an optimal reduction in solvent exposed hydrophobic surface area and all hydrogen bonding groups having either other protein groups or solvent as partners (Richards, 1977). Most hydrogen bonds are local or correspond to paired strands of β-sheet, and predominate within single elements of secondary structure (Stickle et al., 1992). Despite the high packing density in the folded state, proteins tolerate amino acid substitutions remarkably well (Bowie et al., 1990). Variability in the sequence is tolerated most readily at solvent-exposed or mobile sites in the folded proteins, whereas replacement of residues located in the interior of the protein often leads to destabilization. These properties result from a delicate balance of forces that stabilize or destabilize the native structures. These forces are discussed in the following.

The Unfolded State

Unlike the structures of native states, the structures of nonnative (or denatured) states of proteins are much less well understood due to their inherent properties and the lack of techniques to characterize them. It has long been known from measurements of the hydrodynamic and optical spectroscopic properties of protein solutions that proteins unfolded with 6 M GdmHCl or 8 M urea have characteristics of a random coil (Tanford, 1968, 1970). Although Tanford (1970) did show that structures somewhat less like "random coil" were obtained for proteins under some other unfolding conditions, the term "random coil" has been traditionally used to describe an unfolded protein. The view of the structure of unfolded proteins as a random coil is now changing, and the unfolded state of proteins is being examined with renewed interest. Part of this interest is because unfolded forms of proteins are the substrates of molecular chaperones, which are known to be involved in a number of biological processes such as *in vivo* protein folding and protein transport across biological membranes (Georgopoulos &

Welch, 1993; Hartl et al., 1994; Nelson et al., 1992; Neupert et al., 1990). Although studies with model peptides have shown that some structural elements such as α-helix and patches of hydrophobic surface are implicated in the recognition process (Landry et al., 1992, 1993), much still needs to be learned about protein-chaperone complexes. Understanding the unfolded states of proteins will help us elucidate the structural basis of the recognition process.

Interest in the unfolded forms of proteins has also been revived by the observation of well-defined conformational tendencies in some short model peptides, which are very useful to study the effect of local interactions on unfolded states of proteins (Dyson et al., 1988a,b; Dyson & Wright, 1991; Marqusee et al., 1989). The study of unfolded states is also being facilitated by the enormous progress in the development of NMR techniques to study the structure of polypeptides in solution (Clore & Gronenborn, 1991; Wüthrich, 1994). Particularly useful are heteronuclear NMR methods that allow assignment of proton spectra of proteins in the absence of stable structure (Broadhurst et al., 1991; Evans et al., 1991; Logan et al., 1993; Neri et al., 1992a).

Residual structure has been observed in unfolded forms of several proteins, including the thermally unfolded states of ribonuclease A (Labhardt, 1982; Robertson & Baldwin, 1991) and hen lysozyme (Aune et al., 1967), and acid-unfolded states of cytochrome c (Jeng et al., 1990) and cellular retinoic acid binding protein (CRABP) (Liu et al., 1994). The term "premolten globule" has been proposed for these unfolded forms, which are expanded and nonglobular (Alonso et al., 1991; Baldwin, 1991), in contrast to other unfolded forms of proteins that are compact and loosely packed (i.e., the "molten globule" defined above and described in more detail in the next section). Experimentally, the premolten globule forms of proteins have significant amounts of circular dichroism (CD) signal around 220 nm (indicative of formation of secondary structure), but low protection against amide proton exchange.

It appears that the presence of residual structure in unfolded states of proteins is rather general (for recent reviews see Dill & Shortle, 1991; Dobson, 1992; Shortle, 1993). The nature of the unfolded states is dependent on the physical influence that is used to create them. Additionally, their properties are determined by the sequence of amino acids, like those of the native state. The ensemble of states created under denaturing conditions likely represents a dynamically varying, interconverting distribution of sequence-biased conformations. The implication of the residual structure present in the unfolded states of proteins for the mechanisms of protein folding is difficult to assess, partly because information on the residual structure is still limited, due to its transient and fluctuating nature. As a driving force (e.g., cooling or removal of a denaturant) is applied to an unfolded state, it is likely that its conformational biases will govern the approach to the native state. Indeed, this idea is consistent with the framework model of protein folding (Kim & Baldwin, 1990), which posits a hierarchical acquisition of native-like secondary structure, supersecondary structure, subdomain structure, and so forth, en route to the native state. Structural studies on unfolded states of proteins and comparison with kinetic studies of protein folding will shed light on these issues.

The Molten Globule State

The molten globule state has attracted great attention in recent years because it has distinct structural and energetic characteristics that differ from those of both native and unfolded states. In addition, it has been proposed that equilibrium molten globule forms of proteins resemble kinetic intermediates in protein folding pathways (Barrick & Baldwin, 1993; Christensen & Pain, 1991; Kuwajima, 1989; Ptitsyn, 1987). Therefore, one can obtain information on the kinetic intermediates in protein folding by studying partially folded forms that can be obtained and examined directly under equilibrium conditions. Extensive structural or energetic studies of molten globules have been carried out for several proteins, including α-lactalbumin, carbonic anhydrase B, cytochrome c, apomyoglobin, and T4 lysozyme. Several reviews have summarized the results obtained (Haynie & Freire, 1993; Kuwajima, 1989; Ptitsyn, 1992). The general consensus on the structural properties of the molten globule state is that it is compact and has a sizable hydrophobic core (hence the term "globule"); it has high content of secondary structure and native-like tertiary fold; and yet it has significant structural flexibility with mostly nonspecific tertiary interactions (hence the term "molten"). It is worth emphasizing that the compactness is a key property of the molten globule forms of proteins. The molten globule form of proteins is also often referred to as "compact intermediate" (Kim & Baldwin, 1990) and "collapsed form" (Creighton, 1990) to reflect the fact that portions of the hydrophobic interior of the molten globule state, as revealed by NMR spectroscopy, are relatively stable and have well-ordered three-dimensional conformations with amide protons highly protected from solvent exchange (Baum et al., 1989).

Thermodynamics of Protein Folding

Studies of the thermodynamics of protein folding provide a quantitative measure of the conformational stability of proteins in their native states. The noncovalent physical forces that stabilize the native conformation of proteins were mostly identified in the 1950s, and include van der Waals interactions, electrostatic interactions, hydrogen bonds, and the hydrophobic effect (Dill, 1990; Kauzmann, 1959). The questions that remain elusive are: What are the contributions of each of these forces to the stability of a protein? How do these forces integrate to produce the observed thermodynamic stability? Which residues are essential for stability and which direct protein folding? Several reviews have been written recently on these subjects (Dill, 1990; Matthews, 1993a; Rose & Wolfenden, 1993). Before summarizing what is currently known about the above questions, we first describe how to determine the conformational stability of a protein experimentally and how to estimate the relative contribution of a specific interaction or amino acid residue to the stability.

Methods of Thermodynamic Analysis of Protein Folding

The conformational stability of a protein is defined as the difference between the Gibbs free energy of the folded (N) and the unfolded (U) states:

$$\Delta G_{N \to U} = G_U - G_N.$$

For most small single-domain proteins, the reversible unfolding reaction is a two-state cooperative process. Only fully folded native and unfolded states are populated significantly at the unfolding transition. Intermediate states (I) are rarely detected at equilibrium. Therefore, the conformational stability of a single-domain protein can be studied from the unfolding reaction using a two-state model of thermodynamic analysis.

The two-state thermodynamic model of protein unfolding has been discussed in several excellent papers (Becktel & Schellman, 1987; Pace et al., 1989; Privalov, 1992). Pace et al. (1989) gave a detailed account of the practical aspects of the determination of thermodynamic parameters using spectroscopic methods. Experimental details on the study of heat-induced unfolding of proteins using differential scanning microcalorimetry can be found in papers by Privalov and Potekhin (1986) and by Sturtevant (1987). Assuming the unfolding reaction is a two-state cooperative process, numerical values of the enthalpy change $\Delta H_{N \to U}$, the entropy change $\Delta S_{N \to U}$, and the free energy change $\Delta G_{N \to U}$ between the folded and unfolded states of a protein at a temperature T can be calculated from thermal unfolding experiments with the following equations:

$$\Delta H_{N \to U} = \Delta H_m + \Delta C_p(T - T_m) \quad (6\text{-}1)$$

$$\Delta S_{N \to U} = \Delta S_m + \Delta C_p \ln(T/T_m) \quad (6\text{-}2)$$

$$\Delta G_{N \to U} = \Delta H_{N \to U} - T\Delta S_{N \to U} = \Delta H_m - T\Delta S_m + \Delta C_p[T - T_m - T\ln(T/T_m)] \quad (6\text{-}3)$$

where T_m is any reference temperature (often set to the melting temperature), ΔH_m and ΔS_m are the enthalpy and entropy differences between the unfolded and folded states at T_m, and ΔC_p is the change in heat capacity that accompanies protein unfolding at constant pressure.

Equation 6-3 provides a measure of the stability of proteins as a function of temperature. ΔH_m can be determined by calorimetry (see Fig. 6-2) or from the slope of van't Hoff plots (natural logarithm of the equilibrium constant of the unfolding reaction, K_U, as a function of inverse absolute temperature, $\ln K_U$ versus $1/T$). ΔC_p can be determined by calorimetry (see Fig. 6-2) or by studying ΔH_m as a function of pH (Privalov, 1979; Schellman, 1987).

A large and positive value of ΔC_p is often observed in heat-induced unfolding of native folded proteins. The ΔC_p between the folded and unfolded states is constant or nearly constant in the temperature range 0 to 80°C (Privalov, 1979). ΔC_p is observed to be directly proportional to the nonpolar surface area that is buried in the interior of the folded state and becomes exposed in the unfolded state (Haynie & Freire, 1993; Privalov, 1992; Spolar et al., 1992). Therefore, ΔC_p provides a quantitative way to measure the importance of hydrophobic interactions to the stability of a protein.

A special feature of this model is that, as a consequence of the nonzero ΔC_p, ΔH and ΔS are temperature dependent. At certain low temperatures, the model predicts a cold denaturation (Privalov & Griko, 1986). Unlike the high temperature-induced unfolding, which results in large heat absorption with an increase in enthalpy and entropy, low temperature-induced unfolding results in heat release with loss of entropy (Privalov, 1992). Experimental evidence of cold denaturation was found for several proteins including apomyoglobin (Privalov & Griko, 1986), staphylococcal nuclease (Griko et al., 1988a), and T4 phage lysozyme (Chen & Schellman, 1989; Chen et al., 1989).

Temperature-induced unfolding of proteins often yields a sharp peak of heat absorption which allows one to follow the unfolding reaction by differential scanning microcalorimetry (DSC) (Fig. 6-2). All of the thermodynamic parameters associated with the unfolding reaction can be obtained by analyzing the heat capacity curve as a function of temperature (Privalov & Potekhin, 1986; Sturtevant, 1987). DSC is often used to check the reliability of the two-state approximation for protein unfolding. The identity of the calorimetric enthalpy ΔH_{cal} and the effective enthalpy (van't Hoff) ΔH_{vH}, supports an effective two-state behavior. When significantly populated states occur be-

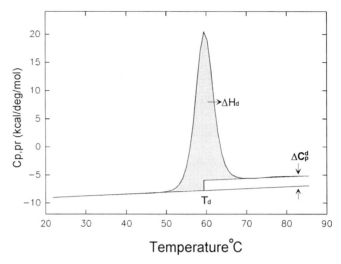

Figure 6-2. A differential scanning calorimetric curve of thermal unfolding of the Arg111 → Gln mutant of CRABP. The curve corresponds to the fitting of the experimental data to a two-state, nonzero ΔC_p model. The differential scanning calorimeter measures directly the difference in the amount of heat required to raise the temperature of a protein solution by 1°C relative to the solvent alone, at constant pressure (i.e., the specific heat capacity $C_{p,pr}$). The difference between the measured heat capacity of the pre- and post-transition region gives the ΔC_p of unfolding. The heat absorbed during unfolding, when integrated over the entire transition region (the shaded area) gives the specific enthalpy change upon unfolding ΔH_d, also known as ΔH_m (unpublished results from the authors).

tween the initial and final states, $\Delta H_{vH} < \Delta H_{cal}$. A broad asymmetric heat absorption peak will be seen. When $\Delta H_{vH} > \Delta H_{cal}$, it may be concluded that the process involves intermolecular interactions.

The unfolding reaction can also be induced by increasing the concentration of a denaturant, and by decreasing or increasing the pH (Tanford, 1968). The methods commonly used to monitor these types of unfolding reactions include CD and fluorescence spectroscopy, which take advantage of the high sensitivity of the spectroscopic properties of proteins to their conformation.

When denaturation is induced by a denaturant such as urea or GdmHCl, or when a thermal denaturation is monitored by spectroscopic methods rather than scanning microcalorimetry (Pace et al., 1989), the stability of a protein (ΔG) can be estimated by using the thermodynamic formula $\Delta G = -RT \ln K_U$, where K_U is the equilibrium constant of the unfolding reaction at the transition region of the unfolding curves. Assuming that Y is the observable parameter monitored during unfolding (such as intrinsic fluorescence or circular dichroism of a protein) and that Y_N and Y_U are the values of this parameter for the folded and unfolded protein, respectively, ΔG at the transition region can be calculated with the following equation:

$$\Delta G = RT \ln K_U = RT \ln(Y_N - Y)/(Y - Y_U) \qquad (6\text{-}4)$$

Values of Y_N and Y_U in the transition region are obtained by extrapolating from the pre- and post-transition regions (Pace et al., 1989). The conformational stability of the native state at zero denaturant concentration ($\Delta G°$) is often obtained by linear extrapolation,

$$\Delta G = \Delta G° - m \times [\text{denaturant}] \tag{6-5}$$

where m is a measure of the dependence of ΔG on the denaturant concentration. This method assumes that the linear relationship in Equation 6-5 holds true in the entire urea concentration range, which may not necessarily be correct.

One of the most dramatic changes in a protein upon unfolding is the increase of its hydrodynamic radius. This can be visualized by transverse urea gradient gel electrophoresis, where protein samples are applied across the top of a slab gel containing a urea concentration gradient perpendicular to the direction of electrophoresis (Goldenberg, 1989). Figure 6-3 shows the refolding and unfolding of the predominantly β-sheet protein that we have studied, CRABP, followed by transverse urea gradient gel electrophoresis. At low urea concentration (<3 M), the protein is folded and migrates rapidly due to its small hydrodynamic volume. At high urea concentration (>4 M), the protein is unfolded and migrates more slowly. At intermediate urea concentration (3 to 4 M urea), only folded and unfolded conformations are present in a significant amount, and the mobility of the mixture is determined by the weighted average of the mobilities of the folded and unfolded forms (Liu et al., 1994). The folding-unfolding of CRABP is reversible: the electrophoretic patterns obtained when folded protein (Fig. 6-3A) or unfolded protein (Fig. 6-3B) are applied to the gel are almost superimposable, and both display a continuous transition between the folded and unfolded bands with a single inflection point at 3.2 M urea concentration.

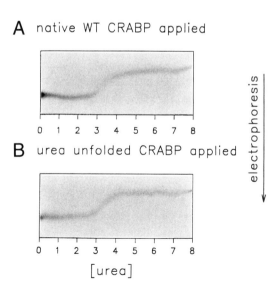

Figure 6-3. Transverse urea gradient gel electrophoresis of CRABP. The gels contained a linear gradient of 0 to 8 M urea, from left to right, perpendicular to the direction of electrophoresis. (A) Folded and (B) urea-unfolded protein was applied across the top of the gel and electrophoresed toward the cathode (indicated by the arrow) at room temperature, pH 7.5. Reprinted from Liu et al., 1994, with permission.

Thermodynamic measurements of protein stability only provide information on the sum of all interactions in the system. Mutational analysis is usually performed to estimate the relative contribution of a specific interaction or amino acid residue to the stability. A specific interaction is removed by amino acid substitution with genetic methods and the magnitude of the specific interaction is estimated from the difference in the $\Delta G_{N \rightarrow U}$ between the wild-type protein and its variant, $\Delta\Delta G$ (Alber, 1989a; Matthews, 1993a). One should keep in mind that there are two critical factors that can affect the interpretation of such analyses. (1) The mutant proteins usually retain an overall structure that is very similar to the wild-type protein. However, local structural changes at the mutation site do occur sometimes even for a simple mutation. Therefore, knowledge of the structure of both the wild-type protein and its variants is important. (2) The conformational stability of the native protein could be enhanced by either decreasing the free energy of the native state or by increasing the free energy of the unfolded state, or by a combination of both, since it has been observed that mutations could affect the unfolded state (Dill & Shortle, 1991). Thus, knowledge of the unfolded state is critical to understanding protein stability as well. An additional issue to consider in the analysis of protein stability by mutagenesis is to what extent the effects of multiple mutations are additive.

Extensive studies using this approach have been carried out for a large number of model protein systems, including bacteriophage T4 lysozyme (Matthews, 1993a), barnase (Fersht, 1993), staphylococcal nuclease (Green & Shortle, 1993), lambda repressor (Sauer et al., 1990), ribonuclease T1 (Pace, 1990), and several other proteins. These studies have allowed examination, in a systematic way, of the role and contribution of the different forces that stabilize proteins. Mutational analysis also provides information concerning the structure and energetics of kinetic intermediate states and of transition states in the folding pathway (Fersht et al., 1992; Matouschek et al., 1989; Matthews, 1993b; Matthews & Hurle, 1987).

Forces That Stabilize Native Proteins

Native folded proteins are marginally stable considering the large number of interactions involved in the stabilization and destabilization of the structure. Common values for ΔG at physiological conditions ($\Delta G°$) are found to be between 5 and 20 kcal/mol (Pace, 1990; Privalov & Gill, 1988), which are equivalent to only a few hydrogen bonds, ion pairs, or hydrophobic interactions. The lower size limit for a stable folding domain has been postulated to be about 50 amino acid residues (Privalov, 1979). This is based on the observation that the free energy of folding for single-domain proteins rarely exceeds about 120 cal/mol per residue. For a protein structure to be stable, the $\Delta G°$ of unfolding needs to be about 10 times greater than the available thermal energy (kT, 0.6 kcal/mol at 25°C); hence the minimum of about 50 residues would be required.

There seems to be little doubt that the *hydrophobic interaction* is the dominant factor in stabilizing the folded structure, as originally proposed by Kauzmann (1959). Still, the debate on quantitative measurement of the hydrophobic effect continues (Dill, 1990; Matthews, 1993a; Pace, 1992; Rose & Wolfenden, 1993; Sharp et al., 1991). The hydrophobic contribution of individual amino acid residues to the stability of a protein is usually quantitated based on the free energies of transferring the corresponding side chains from water to an organic solvent (Tanford, 1980). There are many published hydrophobicity scales, and they do not agree in several respects (Cornette et al., 1987). Nonetheless, there is consensus that the hydrophobicity of a nonpolar amino acid depends linearly on its surface area (Richards, 1977). Richards (1977) proposed that a scalar factor of 22 to 25 cal/mol per Å² of buried hydrophobic surface area is a good estimate of the hydrophobic effect. However, calculations performed by Honig and coworkers (Sharp et al., 1991) including entropic effects due to the distinct molecular volumes of the solute and water led to a substantially larger value (46 cal/mol per Å² of buried hydrophobic surface area). This subject is still under debate (Sun et al., 1992).

Mutagenesis approaches have also been used to estimate the magnitude of the hydrophobic effect (Eriksson et al., 1992, 1993; Kellis et al., 1989; Matsumura et al., 1988; Shortle et al., 1990). In this method, an internal hydrophobic residue is mutated to another of smaller size, and the change in the stability of a protein ($\Delta\Delta G$) is measured to give an estimate of the difference between the hydrophobic stabilization by the two amino acid residues. Results from such mutational analyses show that the change of the stability of a protein ($\Delta\Delta G$) upon amino acid substitution varies widely and is in general significantly larger than those expected from the amino acid transfer experiments ($\Delta\Delta G_{tr}$) (Pace, 1992).

Matthews and coworkers have recently performed a series of experiments in which leucine residues within the core of T4 lysozyme were replaced by the smallest nonpolar residue (alanine) (Eriksson et al., 1992, 1993). The change in protein stability associated with each mutation ($\Delta\Delta G$) was measured and the crystal structure of each mutant was determined. They found that the change in stability of T4 lysozyme due to a Leu → Ala mutation ($\Delta\Delta G_{Leu \to Ala}$) could be approximated by the relationship

$$\Delta\Delta G_{Leu \to Ala} = \Delta\Delta G_{tr(Leu \to Ala)} + \Delta\Delta G_{cav(Leu \to Ala)}$$

where $\Delta\Delta G_{tr(Leu \to Ala)}$ is the difference between the free energies of transferring Leu and Ala from water to an organic solvent and $\Delta\Delta G_{cav(Leu \to Ala)}$ is a term that is dependent on the size of the cavity created by the mutation. When the cavity size created by mutation is zero (the local structure of the protein readjusts to fill the whole), the decrease of energy stabilization due to the mutation ($\Delta\Delta G_{Leu \to Ala}$) is 1.9 kcal/mol, or 25 cal/mol/Å², which agrees with the difference in hydrophobicity between leucine and alanine estimated by transfer from water to octanol ($\Delta\Delta G_{tr(Leu \to Ala)}$) (Fauchere & Pliska, 1983) and with the value proposed by Richards (1977). This suggests that it may be valid to estimate the contribution of the hydrophobic interaction to the stability of native proteins using ΔG_{tr} values based on n-octanol (Matthews, 1993a; Pace, 1993). The cavity-dependent variable term ($\Delta\Delta G_{cav(Leu \to Ala)}$) is about 24 cal/mol/Å³. This may be attributed to a loss of favorable van der Waals contacts due to the mutation (Eriksson et al., 1992; Pace, 1993).

Hydrogen bonds and van der Waals interactions are often treated together when considering the forces that stabilize protein structures because of their similar nature. Van der Waals forces exist between any two atoms that are close to each other, and originate in interactions involving instantaneous, induced or permanent dipoles. Hydrogen bonds can be viewed as arising mainly from an electrostatic interaction involving a hydrogen atom between two electronegative atoms in a linear arrangement. It is not clear at present if van der Waals interactions contribute to stabilize or destabilize folded protein conformations. These interactions depend very strongly on interatomic distances. Since protein unfolding does not result in a substantial change of the volume of the system (thus in the average interatomic distance), one may argue that van der Waals forces can only have a small contribution to the free energy of unfolding. However, the distribution of interatomic distances certainly varies upon unfolding, since the interior of folded proteins is as densely packed as crystals of small organic molecules (Richards, 1977). Such dense packing could provide a substantial stabilization energy, but the formation of optimum van der Waals contacts is hindered by the irregularities in the packing due to the covalent structure of the protein and by the entropic forces opposing rigidity in the molecule. Despite the lack of a good quantitative estimate of the van der Waals contribution to the overall protein stability, it is clear that these interactions influence the specificity of folding, as only sequences that are able to form a densely packed core adopt stable folded conformations.

The contribution of hydrogen bonding to the stability of native proteins has been controversial and remains so (Dill, 1990; Privalov & Gill, 1988; Rose & Wolfenden, 1993). Hydrogen bonds have traditionally been assumed to provide no net, or at best marginal contribution to the stability of folded proteins, based on studies with model compounds (for references see Dill, 1990; Kauzmann, 1959). However, results from protein mutagenesis (Fersht et al., 1985; Shirley et al., 1992) and from surveying the crystal structures of proteins (Baker & Hubbard, 1984; Jeffrey & Saenger, 1991; Stickle et al., 1992) suggest that hydrogen bonds could contribute significantly to the stability of proteins. To prove that hydrogen bonding would have a dominant contribution to the stability of folded proteins, one must demonstrate that intramolecular hydrogen bonds in folded proteins are energetically more favorable than intermolecular hydrogen bonds between unfolded proteins and water molecules. Although it has been difficult to obtain quantitative measurements, several arguments support the view that hydrogen bonds could play a major role in folding and stabilization of native proteins. (1) Intramolecular hydrogen bonding within the folded state is more favorable than intermolecular hydrogen bonding between unfolded molecules and water because of the smaller loss of translational and rotational entropy in the former case (Page & Jencks, 1971; Stahl & Jencks, 1986). (2) Most hydrogen bonds within water and between protein and water are usually present only a fraction of the time, whereas those within folded proteins are present essentially all the time; as a consequence, hydrogen bonds in native proteins will have more negative enthalpy (note that, although this argument is given often, it is somewhat in contradiction with argument (1) since a hydrogen bond present only a fraction of the time will involve a smaller loss of translational and rotational entropy than the same hydrogen bond present all the time). (3) Hydrogen bonding plays a dominant role in stabilizing secondary structures such as α-helix and β-sheet (Pauling & Corey, 1951; Pauling et al., 1951). (4) In native proteins, nearly half of the polar groups are buried in the interior of proteins and they are almost invariably hydrogen bonded (Stickle et al., 1992). Helix capping could play a directing role in protein folding (Rose & Wolfenden, 1993).

The contributions of *electrostatic interactions* to the stability of a protein are pH dependent. At low or high pH, proteins are unstable due to the increasing charge repulsion. At neutral pH or at a pH near the isoelectric point of a protein, electrostatic interactions are expected to contribute to the free energy of stabilization, but they do not play a dominant role (Dill, 1990; Matthews, 1993a). Electrostatic interactions on the surface of proteins are offset by the entropy loss of immobilizing the interacting partners (Daopin et al., 1991), and thus contribute very little, overall, to protein stability (Matthews, 1993a). In some cases, mutation of amino acid residues that affect ion pairing has changed the stability of proteins by about 1 to 3 kcal/mol per ion pair (Fersht, 1972; Pace & Grimsley, 1988), or even 3 to 5 kcal/mol (Anderson et al., 1990). However, the number of ion pairs in proteins is small (about 5 ion pairs per 150 residues)

(Barlow & Thornton, 1983) and, therefore, the total contribution of ion pairing to the free energy of stabilization is still much smaller than that of hydrophobic interactions.

Another factor to be considered is that of *conformational entropy*. In general, unfolded states of proteins consist of an ensemble of many different conformations, none of which is highly populated. Thus, adoption of a unique folded conformation involves the loss of many conformational degrees of freedom. This loss of configurational entropy is the main force opposing protein folding. Unfortunately, the magnitude of this force remains perhaps the most difficult to assess (Dill, 1990). The importance of the conformational entropy term in the free energy of unfolding has been illustrated by experiments where covalent links compatible with the native structure of a protein are introduced to stabilize the structure. Stabilization occurs by reduction of the number of accessible conformations in the unfolded state. Single crosslinks introduced in hen lysozyme and ribonuclease A increase the temperature of denaturation at pH 2.0 by 29°C and 25°C, respectively, and thermodynamic analysis indicates that the crosslink in ribonuclease A provides about 5 kcal/mol of stabilization energy, mostly of entropic origin, at 53°C (Alber, 1989b).

Thermodynamics of Unfolded and Molten Globule States

Privalov and coworkers investigated the energetics of denaturation of hen lysozyme by GdmHCl, by pH, and by temperature (Pfeil & Privalov, 1976a,b,c). They found that there is no difference in the enthalpy change and the heat capacity increments associated with unfolding either by GdmHCl, by pH, or by heat. Thus, they concluded that the residual structure remaining in the unfolded state of proteins denatured in different ways is not significant from the energetic point of view, and all the denatured forms of proteins can be treated as the same macroscopic state (Privalov, 1992). Yutani et al. (1992) concluded that the molten globule state is enthalpically equivalent to the unfolded state based on the evidence that no measurable thermal transition was observed in a calorimetric measurement on a molten globule form of apo-α-LA at low ionic strength and neutral pH.

On the contrary, Freire and coworkers argue that the absence of a thermal denaturation transition for the molten globule state of apo-α-LA may be the result of the specific energetic balance in this special case, rather than a general property of molten globule states (Haynie & Freire, 1993). Using a statistical thermodynamic model and thermodynamic parameters derived from scanning microcalorimetry, Haynie and Freire (1993) did a series of computer simulations of the population of the molten globule state of apo-α-LA as a function of the temperature and transition entropy, and as a function of temperature and GdmHCl activity; they also calculated the heat capacity as a function of pH and temperature. They concluded that the molten globule state and the unfolded states of proteins are different thermodynamically. It can be expected that calorimetric experiments on other proteins forming molten globules will clarify this issue in the near future.

Kinetics of Unfolding and Refolding of Single-Domain Proteins

It is now generally accepted that folding of a single-domain protein *in vitro* occurs along some pathway(s) that can be described in terms of intermediate, partially folded states (Creighton, 1990; Dobson et al., 1994; Kim & Baldwin, 1990; Matthews, 1993b).

Thus, the ultimate goal of studying mechanisms of protein folding is to identify and characterize these intermediate states in terms of their structures and energetics, and to place them in correct order along the pathway. In practice, study of the kinetics of protein folding is hindered by three main characteristics of the folding reaction: high cooperativity, rapidity, and the conformational heterogeneity of the unfolded state. As a consequence of the high cooperativity, only a limited number of steps can be observed in the folding kinetics. Intermediates are usually formed fast, on the time scale of milliseconds to seconds, and are not stable enough to be highly populated. Therefore, rapid kinetic methods using stopped-flow (SF) instrumentation are required in order to detect these intermediates. Structural information at the level of individual residues is not obtainable by using conventional spectroscopic methods such as SF-CD and SF-fluorescence. It was not until recently that quenched-flow techniques combined with 2-D NMR spectroscopy have provided the opportunity to study the structures of intermediates in detail (Baldwin, 1993; Englander & Mayne, 1992). The heterogeneity of the unfolded state often complicates the kinetic analysis. Kinetic phases observed in the folding reactions may not represent the formation of structural intermediates but rather arise from interconversion between different conformers of the unfolded state [e.g., prolyl peptide bond *cis-trans* isomerization (Brandts et al., 1975)]. Thus, it is necessary to study the folding reaction under various conditions and, using various probes, to differentiate one kinetic mechanism from another.

The concept of a two-state equilibrium must not be confused with the concept of a kinetic two-state mechanism. A two-state equilibrium implies that only fully folded and fully unfolded molecules are populated and detectable at equilibrium. This does not exclude the possibility of transient accumulation of intermediates during a folding reaction. These intermediates may be unstable and not populated at equilibrium, but may be detectable by kinetic analytical techniques. Indeed, structural intermediates are often detected under conditions that favor the formation of the native state.

To analyze a protein folding mechanism from the kinetic point of view, one must first obtain the number of kinetic phases ($i = 1,..., n$) as well as the time constants t_i and the amplitudes A_i of each phase. For unfolding studies, the solution containing the native protein is diluted into a buffer solution containing a high denaturant concentration. The procedure for a refolding reaction is the same except that the protein is first unfolded, equilibrated in a high denaturant concentration, and then diluted into a refolding buffer. If the reaction is monomolecular, the number of kinetic phases (n), the time constant t_i, and the amplitude A_i of each phase can be obtained by fitting the reaction curve with the following equation:

$$A(t) = A(\infty) - \Sigma A_i * \exp(-t/\tau_i) \tag{6-6}$$

where $A(t)$ is the signal at time t, and $A(\infty)$ is the value of A at the end of the reaction. Detailed accounts of the practical aspects of this type of studies can be found in Utiyama and Baldwin (1986) and Schmid (1992).

Prolyl Peptide Bond Isomerization

Kinetic analysis of protein folding is often complicated by the presence of prolyl peptide bond *cis-trans* isomerization in the unfolded state, which gives multiple unfolded forms with different rates of refolding. When a protein is folded, its prolyl peptide bonds are usually in a defined unique conformation, either *cis* or *trans*, and it is rare to observe equilibria between these two forms. However, when the protein is unfolded, the prolyl peptide bonds are free to isomerize due to the lack of structural constraints, thus leading to an equilibrium mixture of unfolded molecules with different prolyl peptide bond conformations. *Trans* conformations are usually favored by a ratio of about 4:1 over *cis* conformations. Prolyl peptide bond isomerization is intrinsically slow with time constants around 10 to 100 sec at 25°C and activation energies of about 25 kcal/mol (Schmid, 1992).

If all the prolyl peptide bonds must be in the correct form prior to folding, molecules with correct sets of prolyl peptide bond conformations (U_F) will fold faster than those containing one or more incorrect prolyl peptide bond conformations (U_S). Fast and slow refolding phases will be seen during the course of refolding: $U_S \rightarrow U_F \rightarrow N$. The relative magnitude of the slow refolding phases depends on the fraction of U_S molecules. After a protein is unfolded with a denaturant, U_S will depend on the time the protein has been unfolded because the $U_F \rightarrow U_S$ equilibrium occurs only in the unfolded state. This is the basis of the "double jump" technique to test whether a slow kinetic phase is due to slow *cis-trans* isomerization in the unfolded state (Brandts et al., 1975; Schmid, 1992). In this method, aliquots of unfolded proteins are taken at various time intervals after unfolding has been initiated, and are transferred into refolding buffer. The relative amplitude of the slow refolding phase compared to the unfolding time interval is then determined. An increase in the relative amplitude with increasing time interval suggests that a conversion from U_F to U_S in the unfolded state causes the slow refolding phase. This technique also allows one to find appropriate conditions in which only U_F molecules are significantly populated and refolding proceeds in the virtual absence of U_S molecules (Fink et al., 1988). Isomerization can be also probed by peptidyl prolyl *cis-trans* isomerase (PPI) (Schmid, 1993) or by mutagenesis (Nall et al., 1989). Catalysis of the folding reaction by PPI or disappearance of the slow folding phase in mutant proteins with replaced proline residues are indicative of prolyl peptide bond isomerization.

It is also possible that prolyl peptide bond reisomerization occurs coupled with folding or after molecules refold to nativelike structures with an incorrect conformation in one or more prolyl peptide bonds. The rate of reisomerization could be increased or decreased by the conformation of the protein. In either case, the kinetic analysis will be complicated. A good example, which has been discussed extensively in the literature, is provided by ribonuclease T1 (Schmid, 1992). Not all the proline residues in a protein necessarily affect the kinetics of folding. "Nonessential" proline residues do not interfere with folding, or sometimes the extent of the isomerization in the unfolded state may be so small that its effect on the kinetics of refolding can be practically ignored. Ubiquitin, which has three proline residues, offers an example of this behavior (Briggs & Roder, 1992). A slow phase with an amplitude of 10% is observed in the kinetics of overall folding of ubiquitin. This slow phase can be ascribed to the presence in the unfolded state of 10% population of isomers with Pro19 preceded by a *cis* peptide bond. On the other hand, the populations of isomers with *cis* peptide bonds preceding Pro37 and Pro38 in the unfolded ubiquitin were estimated to be 14% and 19%, respectively; the presence of these isomers does not affect the overall folding kinetics, and only has a local effect on the kinetics of protection of neighboring amide protons.

Detection of Structural Intermediates in the Folding Pathway

Evidence of structural intermediates in the absence of prolyl peptide bond isomerization can be obtained by studying the folding reactions with different probes and under various final conditions. In the absence of partially folded intermediates, the relative amplitude of the different kinetic phases only depends on the initial distribution of molecules in different unfolded states and should be the same in the folding reactions studied with different probes and at various final conditions (Utiyama & Baldwin, 1986). The rate constants of the folding phases that originate from the actual folding reaction have a characteristic dependence on the denaturant concentration. The logarithm of the rate constant shows a V-shaped dependence on the denaturant concentration, with a minimum near the midpoint of the respective equilibrium unfolding transition (Matthews & Hurle, 1987). Partially folded intermediates have low stability and are normally observed only at low denaturant concentration, where folded native structure is favored. In this case, deviation of the observed rate constant from values predicted by a kinetic two-state model provides positive indication for the presence of folding intermediates.

In general, kinetic studies of protein folding by conventional spectroscopic methods using real time stopped-flow (SF) instruments give only overall structural information along the folding pathway. Recently, hydrogen-deuterium (H-D) exchange pulse labeling combined with subsequent 2D NMR analysis (Baldwin, 1993; Englander & Mayne, 1992) and electrospray ionization mass spectrometry (ESI-MS) (Miranker et al., 1993) have emerged as very powerful methods to obtain detailed structural information on kinetic folding intermediates. The general scheme for the H-D exchange pulse labeling experiment, also referred to as quenched-flow H-D exchange, is illustrated in Fig. 6-4. In this experiment, a protein dissolved in **D₂O** is unfolded with a high concentration of a denaturant so that all the amide NHs are exchanged to NDs (amide deuterons and protons are depicted by the letters D and H, respectively, in Fig. 6-4). Refolding is initiated by rapid dilution of the unfolded protein solution into refolding buffer. After a folding interval (t_f), a labeling pulse is then introduced by a second step of mixing with a high pH **H₂O** solution so that ND-to-NH exchange is favored. During the labeling period, only solvent-exposed amide deuterons will exchange with H₂O. Structural intermediates may have formed during the time t_f and thus a proportion of exchangeable deuterons may have become protected from the solvent (for example, those depicted by the representative deuterons in the α-helix and the turn of the intermediate (I) in Fig. 6-4). The labeling pulse is terminated by rapid mixing with acidic buffer, in conditions where the hydrogen exchange rate is at a minimum yet the folding of the protein can still proceed. The final solution containing folded protein molecules is concentrated and the buffer exchanged with D₂O by lyophilization-redis-

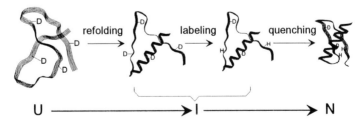

Figure 6-4. Schematic illustration of the H-D exchange pulse labeling experiment.

solution or repeated dilution and ultrafiltration. The extent of labeling at different residues of the protein after different refolding times is analyzed by recording 2D NMR spectra [e.g., heteronuclear single quantum coherence (HSQC) spectroscopy if the protein is uniformly ^{15}N-labeled; correlation spectroscopy (COSY) or total correlation spectroscopy (TOCSY) if the protein is not labeled]. In this case, the sites that were protected in the structures formed before introducing the label will have little or no proton incorporation, while the other amides, which were protected after t_f, will have significant intensities in the NMR spectra. By varying the folding time t_f, one can obtain plots of proton occupancy as a function of t_f for each amide proton that can be monitored (those that are protected in the native form). From such plots, the kinetics of proton protection can be calculated, which yield residue specific information on the sequence of events that leads to the native conformation.

The refolding buffer in the first mixing step of the quenched-flow experiment can be made in either H$_2$O or D$_2$O. If during the refolding time t_f the ND-to-NH exchange can be neglected ($t_f <<$ the time constant of NH-to-ND exchange at the pH of the refolding buffer), then refolding in H$_2$O is preferred because in this case large dilution with the pulse labeling buffer in the second step of mixing is no longer necessary. The pH can be simply changed by addition of a small volume of high pH labeling buffer, thus minimizing the final volume.

Pulsed H-D exchange experiments yield information only on amide protons that are protected from exchange in the native proteins, as any information concerning amide protons that are exposed in the native structure will be lost after folding is complete, even if these protons were protected at some point along the folding pathway. However, this does not mean that these experiments cannot yield information on the formation of nonnative secondary structure. For instance, observation that all the amide protons of a central strand of a β-sheet in a given protein are protected quickly in the folding pathway, with no early protection observed for the contiguous strands, would suggest that, in the initial stages of folding, the amide groups of the central strand are hydrogen bonded to unknown partners, probably forming nonnative α-helix or β-sheet conformations.

Analysis of pulsed H-D exchange results by NMR only monitors the average proton occupancy at individual sites and cannot tell whether these hydrogens are labeled altogether in the same molecule or independently in different molecules. This limitation can be overcome by application of electrospray ionization mass spectrometry (ESI-MS) (Katta & Chait, 1993; Miranker et al., 1993). Consider, for instance, two cases: in one case, 50% of the molecules have all the amide groups exchanged and the other 50% of the molecules did not exchange at all (Fig. 6-5A); in the other case, all the amides in a molecule have 50% probability of being deuterated (Fig. 6-5B). The NMR spectra will have the same pattern in both cases, but ESI-MS gives two distinct mass spectra.

Dobson and coworkers employed both NMR spectroscopy and ESI-MS to analyze the results of pulsed amide hydrogen labeling experiments during refolding of hen egg white lysozyme (Miranker et al., 1991, 1993). These studies, which are described in more detail below, provide an exquisite illustration of the power and the complementarity of NMR spectroscopy and mass spectrometry in the study of protein folding. NMR spectroscopy, but not mass spectrometry, can identify the precise locations of hydrogens in the folded protein that are or are not exchanged. Mass spectrometry, but not NMR spectroscopy, can tell us about the molecular distribution of H-D labeling. In a broad sense, when two structural elements fold on the same time scale, NMR spectroscopy can help us to identify the structural elements, whereas mass spectrometry

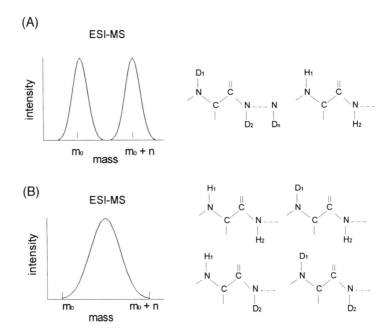

Figure 6-5. Diagrammatic representation of the ESI-MS spectra of protein molecules with n exchangeable amides after 50% H-D exchange. (A) 50% of the molecules have all the amide groups exchanged and the other 50% of the molecules did not exchange at all; (B) all the amides in a molecule have 50% probability of being deuterated. Both cases produce identical 2D ^{1}H NMR spectra but different ESI-MS spectra. Reprinted from Miranker et al., 1993, with permission.

will tell us whether they fold together in the same molecule or independently in different molecules.

Kinetic Folding Intermediates and Transition States

Most kinetic folding studies described in the literature were performed on proteins containing substantial amounts of α-helix. As indicated below, the folding kinetics of mostly β-sheet proteins may be significantly different from those of α- and α/β proteins. We will summarize here kinetic folding results obtained for α- and α/β proteins; studies on β-sheet proteins are discussed later.

The earliest intermediates in protein folding pathways are those formed within the dead time (burst phase) of stopped-flow instruments, on the time scale of 10 milliseconds or less (for examples see Baldwin, 1993; Matthews, 1993b). A significant amount of secondary structure is formed in this time range, as suggested by the far-UV CD and by protection of amide protons against solvent exchange in H-D exchange pulse labeling experiments. Formation of both α-helix and β-sheet is observed in the burst phase for mixed α/β proteins. H-D exchange pulse labeling experiments show that the secondary structure formed in the burst phase is native in almost all the proteins studied so far. However, the possibility that nonnative secondary structure may be formed transiently can not be excluded (Chaffotte et al., 1991; Kuwajima et al., 1987). The early kinetic intermediates may have a hydrophobic core, at least for some proteins (for references, see Baldwin, 1993). Such intermediates have been shown to bind a hydrophobic dye, 8-anilino-1-naphthalene sulfonate (ANS), resulting in enhancement of the fluorescence of the dye. Binding of ANS was proposed to be due to the nonpolar surfaces developed in these intermediates (Semisotnov et al., 1991). The early folding intermediates do not have absorption in the near-UV CD, which indicates absence of asymmetric packing environments for aromatic side chains. Hence, the tertiary structure formed in these intermediates is most likely nonspecific. The burst phase intermediates resemble equilibrium molten globule states in several aspects, including high content of secondary structure and nonspecific tertiary interactions. Thus, structural details of the kinetic intermediates may be elucidated from analysis of molten globules, which can be characterized by conventional spectroscopic techniques under equilibrium conditions.

Intermediate folding steps, after the burst phase and before the final formation of native structures, are observed on the time scale of hundreds of milliseconds in the folding of several proteins (for reference, see Matthews, 1993b). Further enhancement of ANS binding and protection of amide protons against exchange are usually observed, concomitant with changes in fluorescence of Trp and near-UV CD, suggesting that compact and specific tertiary interactions are formed. Semisotnov et al. studied the ANS binding of folding intermediates of carbonic anhydrase (Semisotnov et al., 1991). Although ANS binding was found in both the burst phase and subsequent intermediate phase of the folding of carbonic anhydrase, Semisotnov et al. suggested that the burst phase intermediate is not compact (thus not molten globulelike), whereas the intermediate formed in the subsequent phase is compact and molten globulelike.

Information on transition states of protein folding pathways is available from studies on hen egg lysozyme (Segawa & Sugihara, 1984a, b), α-LA (Kuwajima, 1989), and on several other proteins studied by mutagenesis, including barnase (Serrano et al., 1992), a subunit of tryptophan synthase (Beasty et al., 1986; Tsuji et al., 1993), dihydrofolate reductase (Jennings et al., 1993), and T4 lysozyme (Chen et al., 1992). The general consensus is that the rate-limiting step of folding, excluding the prolyl peptide bond isomerization and disulfide bond formation, involves the formation of the specific native tertiary structure, including acquisition of native side chain packing, hydrogen bond formation throughout the entire polypeptide chain, and exclusion of water molecules from the interior of proteins (Matthews, 1993b).

Case Studies

We summarize folding studies of three proteins, apomyoglobin, hen lysozyme, and cellular retinoic acid binding protein. These represent three different classes of proteins: α-helix, mixed α-helix and β-sheet, and predominantly β-sheet proteins, respectively.

Apomyoglobin, an α-Helical Protein

Myoglobin (Mb) (MW 17,000) is a small single-domain protein containing eight α-helices (A to H) (Fig. 6-6A). The crystal structure of apomyoglobin (apoMb), that is, myoglobin without the heme group, is not known. However, it is believed to be similar to the holoprotein except for the region around the heme binding site (helix F) (Hughson et al., 1990). The folding mechanism of apoMb has been studied extensively. ApoMb is one of the few proteins that shows an equilibrium intermediate. At low temperature, the acid-induced unfolding transition agrees well with a three-state model (N↔I↔U) (Griko et al., 1988b). The intermediate (I) formed at pH 4 is a molten globule (Barrick & Baldwin, 1993): I is compact, as judged by intrinsic viscosity mea-

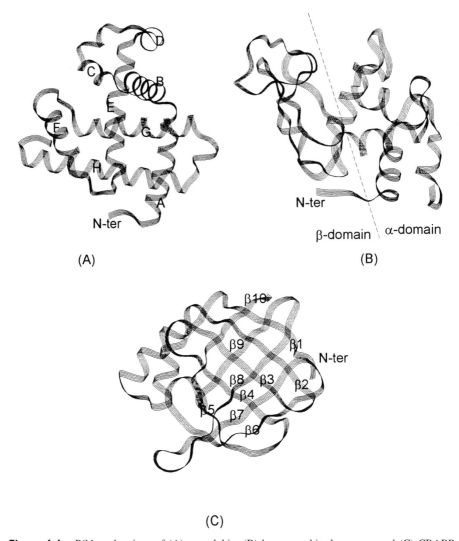

Figure 6-6. Ribbon drawings of (A) myoglobin, (B) hen egg white lysozyme, and (C) CRABP. The coordinates of myoglobin and lysozyme are from the protein data bank (Bernstein et al., 1977). Those for CRABP correspond to a structure modeled by homology to the crystal structure of myelin P2 protein (Zhang et al., 1992). The ribbons were drawn using the program InsightII (Biosym Technologies Inc.).

surements, and its CD spectrum indicates that I contains 35% α-helix, compared to 55% α-helix in the native state. The helicity corresponds mostly to helices A, G, and H, as determined by H-D exchange followed by ^{1}H NMR spectroscopy at pH 4.2 (Hughson et al., 1990). Results from mutagenesis experiments suggest that the subdomain formed by the A, G, and H helices in the molten globule lacks native tertiary interactions between helices (Hughson et al., 1991).

The kinetics of folding of apomyoglobin at pH 6.1 has been studied by stopped-flow CD and H-D exchange pulse labeling (Jennings & Wright, 1993). An early intermediate with complete protection of protons in the A, G, and H helices and part of helix B was detected in the burst phase (6.1 msec). The burst phase intermediate has an apparent stability of about 2.5 kcal/mol relative to the unfolded state. Because they share similar H-D protection properties, the equilibrium molten globule of apoMb at pH 4.2 and the kinetic intermediate at pH 6.1 were proposed to be the same (Jennings & Wright, 1993).

A simple sequential folding pathway has been proposed for apoMb:

$$U \to A \cdot G \cdot H \to A \cdot B \cdot G \cdot H \to A \cdot B \cdot C \cdot CD \cdot E \cdot G \cdot H \to N.$$

In this mechanism, a compact molten globule intermediate involving helices A, G, and H is formed in less than 5 msec. Studies of peptide fragments suggest that the G-H helix pair may be formed transiently, prior to the formation of the A·G·H molten globule (Shin et al., 1993). The B helix, C helix, CD loop, and E helix appear to be formed later in a sequential order and assembled onto the A·G·H molten globule (Jennings & Wright, 1993). The studies on apomyoglobin provide perhaps the best experimental evidence supporting the hypothesis that molten globule states are structurally analogous to kinetic intermediates in protein folding pathways (Ptitsyn, 1987).

Hen Lysozyme, a Mixed α-Helix and β-Sheet Protein

The crystal structure of hen egg white lysozyme (MW 14,000) comprises two subdomains (α- and β-domains) connected by a short region of double stranded antiparallel β-sheet (Fig. 6-6B). The α-domain contains four α-helices made from both amino and carboxy terminal sequences of the protein, and one 3_{10} helix; the β-domain contains a triple-stranded antiparallel β-sheet and one 3_{10} helix.

Hen lysozyme has two prolines. Both prolyl peptide bonds are *trans* in the native protein and most unfolded lysozyme molecules (90%) contain correct prolyl peptide bond conformations (Kato et al., 1982). Therefore, the amount of slow *cis-trans* isomerization in the folding studies of lysozyme can be practically ignored. There are four disulfide bridges in the native hen lysozyme. Most of the equilibrium and kinetic folding studies of hen lysozyme were done with the oxidized form and the four disulfide bridges were maintained intact in the folding studies reviewed here. Unfolding of hen lysozyme is reversible under a variety of conditions. The thermodynamics of the folding-unfolding transition fits well to a two-state model (Dobson & Evans, 1984; Privalov & Khechinashvili, 1974; Tanford, 1970). No equilibrium molten globule has been described for hen lysozyme so far. However, an equilibrium molten globule was found for a homologous protein, α-lactalbumin (Kuwajima, 1989).

The kinetic mechanism of folding of lysozyme has been studied with a wide variety of techniques, including SF fluorescence, SF CD in the near and far UV regions, and H-D exchange pulse labeling combined with NMR spectroscopy and electrospray ionization mass spectrometry (Dobson et al., 1994; Miranker et al., 1991, 1993). The NMR spectral analysis shows that there are two distinct folding domains that coincide with the two structural lobes of lysozyme (the α- and β-domains). The kinetics of protection against hydrogen labeling are biphasic for both the α- and the β-domain. In the fast phase, with time constant of 5 to 10 msec, 40% of the molecules have their α-domain quickly protected and 25% of the molecules have their β-domain protected. The fast phase kinetics can be interpreted with two different folding models (Miranker et al., 1993). One is that the α- and β-domains fold independently, 40% of the molecules having only their α-domain quickly folded and another 25% of the molecules

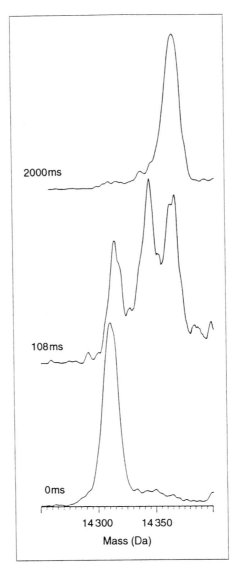

Figure 6-7. The mass spectra of hydrogen-labeled hen lysozyme samples at folding times $t_f = 0$, 108, and 2000 msec from the H-D exchange pulse labeling experiments. Reprinted from Dobson et. al., 1994, with permission. P0, P28, and P50 denote the peaks in order of increasing mass.

having only their β-domain quickly folded. In the second model, 40% of the molecules have their α-domain folded and within this 40%, 25% of the molecules have their β-domain folded as well. The NMR method cannot distinguish between these two possibilities, whereas ESI-MS can. The mass spectra of hydrogen-labeled samples at folding times between 20 ms and 2000 ms show three peaks (Fig. 6-7): one with light mass at 14,313 Da (denoted P0) corresponds to the unprotected molecules; the second peak, with intermediate mass (14,341 Da, P28), corresponds to molecules having only α-domain protected; and the third peak with largest mass (14,363 Da, P50) corresponds to the molecules having both α- and β-domains protected. The kinetics of formation of peak P28 is consistent with the second model, which predicts that about 15% of the molecules would have mass peak P28 on the time scale of 5 to 10 msec.

A folding mechanism for hen lysozyme (Fig. 6-8) has been proposed, based on the

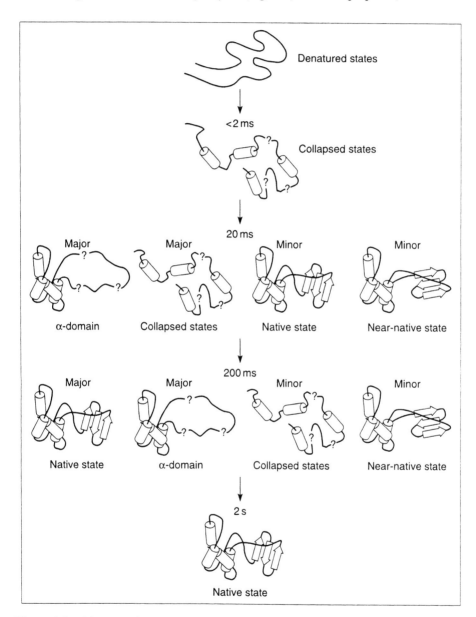

Figure 6-8. Diagrammatic representation of the folding pathway of hen lysozyme. The polypeptide chain of the protein is shown by cylinders for α-helix, arrows for β-strand, and question marks for irregular or undefined conformation. Major species are those present at more than 30% while minor species are about 10% of the total population at each refolding stage shown here. Reprinted from Dobson, 1994, with permission.

results summarized above and a wealth of other experimental evidence (Dobson et al., 1994). This mechanism is more complicated than the simple sequential pathway shown for apomyoglobin. Hen lysozyme folds via multiple parallel paths, as evidenced by the heterogenous structural intermediates at various stages of folding. In a majority of hen lysozyme molecules, the α-domain folds earlier than the β-domain. Simultaneous folding of both domains was also found in a minor percentage of molecules. The second feature of the lysozyme folding mechanism is that, unlike apomyoglobin, secondary structure formation in hen lysozyme is observed before the protection of amide hydrogens from exchange. In hen lysozyme, formation of α-helix, as detected by CD, occurs in the burst phase of folding (2 to 5 msec) (Chaffotte et al., 1992; Kuwajima, 1989). However, the protection of amide protons from solvent exchange is not developed until approximately 10 msec. For the majority of hen lysozyme molecules, protection of amides in the α-domain is complete within 200 msec, whereas the complete protection of amides in the β-domain requires more than 1 sec.

CRABP, a Predominantly β-Sheet Protein

Cellular retinoic acid binding protein (CRABP) (MW 15,500) is a 136-residue long single-domain protein. Both a preliminary model built from our NMR data (Rizo et al., 1994), and structural modeling based on the crystal structure of the homologous protein myelin P2 (50% sequence homology) (Zhang et al., 1992), reveal that the structure of CRABP resembles a clam shell and consists of two nearly orthogonal β-sheets, formed by five antiparallel β-strands each, and a short region of α-helix-turn-helix near the N-terminal end (Fig. 6-6C). This model has been confirmed recently by X-ray crystallography (Kleywegt et al., 1994).

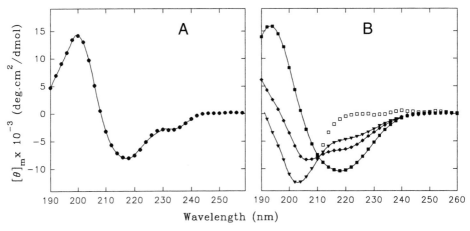

Figure 6-9. CD spectra of CRABP (ca. 3 mM). (A) In H_2O at pH 7.0 (●); (B) in H_2O with 5 mM HCl pH 2.6 (▼), 5 mM $HClO_4$ pH 2.6 (◆), 5 mM HCl/2 mM Na_2SO_4 (■), and 5 M urea/14 mM HCl (□). The symbols represent experimental data and are plotted every 2 nm. The lines are best fits from the CONTIN program (Provencher & Glöckner, 1981). Reprinted from Liu et al., 1994, with permission.

Protein Folding / 253

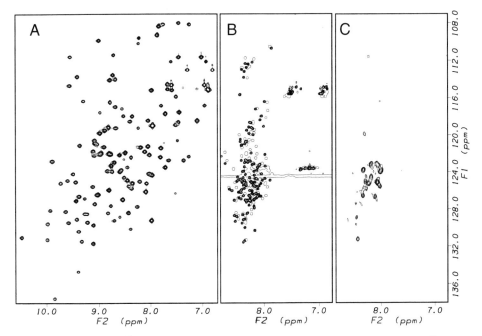

Figure 6-10. (A) ^{1}H-^{15}N heteronuclear single quantum coherence (HSQC) spectrum of ^{15}N uniformly labeled CRABP in 12 mM Tris-d_{11} at pH 7.5; (B) superposition of the ^{1}H-^{15}N HSQC spectra of CRABP in 14 mM HCl and in 8 M urea with 14 mM HCl; only one contour has been plotted for the former, while 12 contours were plotted for the latter; (C) ^{1}H-^{15}N HSQC spectrum of CRABP in 20 mM DCl/D$_2$O at 5°C after 3.5 hr of H-D exchange. In ^{1}H-^{15}N HSQC spectra, a cross peak is observed for every hydrogen directly bonded to a nitrogen atom. The distribution of cross peaks for the acid-unfolded protein is much more like that of urea-unfolded CRABP than the distribution observed for the native form. Despite the overall similarity in the distribution of cross peaks for the acid- and urea-unfolded CRABP, there appears to be a somewhat larger dispersion of chemical shifts in the HSQC spectrum of acid-unfolded CRABP, consistent with the presence of residual structure indicated by CD. There is also a clear tendency for both the proton and the ^{15}N resonances of the acid-unfolded state to be shifted upfield with respect to the urea-unfolded state, which is consistent with the formation of α-helical structure. Reprinted from Liu et al., 1994, with permission.

Unfolding of CRABP by either urea (see Fig. 6-3) or acid is reversible and the unfolding transitions fit well to a two-state model (Liu et al., 1994). However, the unfolded states in urea and in acid are different, as shown by CD (Fig. 6-9) and heteronuclear NMR (Fig. 6-10) (Liu et al., 1994). CRABP unfolded by acid contains a substantially larger amount of α-helix than does native CRABP. The residual structure in acid-unfolded CRABP can be further unfolded by addition of urea, which results in a CD spectrum characteristic of a random coil (Fig. 6-9). The nonnative α-helical structure in the acid-unfolded form of CRABP becomes more prominent when HClO$_4$ rather than HCl is used (Fig. 6-9). In addition, CRABP adopts up to 75% α-helix in solutions containing a high percentage of 2,2,2-trifluoroethanol, which suggests that the sequence of CRABP has the tendency to adopt α-helix in the absence of specific tertiary interactions (Liu et al., 1994). The residual structure in the acid-unfolded form of CRABP is not very stable and the protection against amide proton exchange is low (Fig. 6-10C)

(Liu et al., 1994). The acid-unfolded state of CRABP may be a premolten globule similar to the thermo-unfolded forms of ribonuclease A (Labhardt, 1982; Robertson & Baldwin, 1991) and hen lysozyme (Aune et al., 1967), and to acid-unfolded cytochrome c (Jeng et al., 1990).

Experiments based on kinetic competition between refolding and H-D exchange of amide hydrogens suggest that there is at least one early intermediate present during refolding of CRABP (our unpublished results). In these experiments, denatured ^{15}N-labeled CRABP in a solution of 8 M urea in H_2O was diluted into D_2O at pD ranging from 7.0 to 8.3, which initiated refolding and H-D exchange of amide protons simultaneously. In control experiments, the native protein in H_2O was diluted into the same buffer with the same amount of final urea concentration (0.16 M). The percentage of protons remaining in the folded protein depends both on the intrinsic H-D exchange rate (k_{HD}) of NHs and on the rate of protection due to folding (k_F). The larger k_F with respect to k_{HD}, the more protons remain in the folded protein, and thus the larger the intensities of the cross peaks in the HSQC spectra.

Figure 6-11 shows the results of the competition experiment at pD 7.5. In the HSQC spectrum of a control experiment done under the same conditions, but without unfolding the protein (Fig. 6-11A), there are about 70 amide protons that remain protected from H/D exchange. Of the 70 amide protons, 30 amide protons are quickly protected in the competition experiment (Fig. 6-11B). The strongest intensities correspond mainly to the regions L19 to V24 and E117 to V124, that is, the helix αI and the β-strand β9 in the native protein (see Fig. 6-6C). Interestingly, the amide protons in the strands β8 and β10 that form hydrogen bonds with the carboxyl groups of the strand β9 were not protected or were only weakly protected (Y108, T110, E112, C129, R131, and Y133), which suggests that nonnative secondary structure (perhaps α-helix) involving residues in strand β9 may be formed in the early folding stages of CRABP.

Preliminary SF-CD experiments showed that the native ellipticity at 220 nm appears in less than 5 msec. This observation suggests that most of the native secondary structure is formed very fast, but does not exclude the possibility that some nonnative secondary structure participates in the early stages of folding. On the other hand, H-D exchange pulse labeling experiments indicate that most of the protons in the β-sheets become protected against amide exchange on the time scale of about 1 sec, and SF fluorescence shows that formation of the native tertiary structure also occurs on the time scale of 1 sec (our unpublished data). Thus, it appears that the folding mechanism of CRABP involves the formation of an early intermediate with a secondary structure content similar to that of the native protein, where only the amide protons of helix αI and part of the strand β9 are protected from exchange. Consolidation of the tertiary structure and protection of most of the amide protons of the protein occur more slowly and in a highly cooperative manner. Early formation of secondary structure and late protection of amide protons was also observed for the β-sheet protein interleukin-1β, where some of the protons required more than 25 seconds to acquire protection against solvent exchange in H/D exchange pulse labeling experiments (Varley et al., 1993).

As the formation of stable β-sheets involves numerous interstrand interactions and α-helices can be formed autonomously in a single sequence of consecutive residues, it is not surprising that α-helices form faster than β-sheets. It is likely that formation of the correct native β-sheets is slower in mostly β-sheet proteins than in mixed α/β proteins because, in the latter, early formation of α-helices may constrain and induce the adoption of β-sheet structure in the appropriate residues. More folding studies with

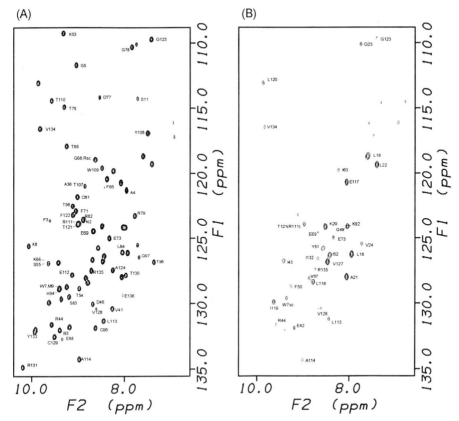

Figure 6-11. Detection of an early intermediate of folding of CRABP by using the refolding and H-D exchange competition method. The refolding was done by diluting 15 mg of ^{15}N-CRABP unfolded in 1 ml of 8 M urea/H_2O to 40 ml of 10 mM Tris-d_{11} in D_2O at pD 7.5 and 5°C. After 2 min of refolding, retinoic acid was added to a final molar ratio of 1.5:1 (RA/protein). The exchange was quenched by addition of 10 ml of 200 mM deuterated acetic acid to pD 3.8. The protein was concentrated and the HSQC spectrum was taken at 25°C (B). In a control experiment (100% protection) (A), the same process was applied except that native protein dissolved in water was diluted into 10 mM Tris-d_{11} in D_2O, containing 0.2 M urea. (Unpublished data from the authors). RA was added so that the protein would be stable at pH 3.8 and 25°C, conditions under which the complete assignment of the HSQC cross peaks is available (Rizo et al., 1994).

β-sheet proteins must be done before general conclusions can be drawn on the folding mechanisms of such proteins.

Summary

Where do we stand today? There is little doubt that the information required to direct the folding of a protein to its native conformation is contained in its amino acid sequence. The native structure of a protein is energetically the most stable among the accessible conformations, although exceptions may exist. The stabilization of native pro-

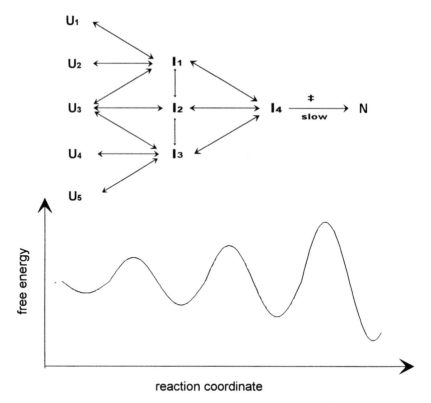

Figure 6-12. A general schematic diagram of pathways of protein folding. Folding of a protein likely follows a sequential mechanism with ensembles of rapidly interconverting conformers (I) present at various stages of the process. The number of conformers in each folding stage decreases and the stability of the intermediates increases as higher order structure develops, which finally converges to a single form, the native structure (N). The rate-limiting step ($\ddagger$) likely occurs in the late stages of the pathway, perhaps right before the final formation of the native state. (Adapted in part from Creighton, 1990.)

teins is maintained through the integration and delicate balance of hydrophobic effects, hydrogen bonding, van der Waals interactions, electrostatic interactions, and opposing forces arising from the loss of conformational entropy. Proteins tolerate amino acid substitutions remarkably well, in particular at solvent-exposed or mobile sites.

Folding of single-domain proteins is rapid and highly cooperative. At equilibrium, the process can be described by a two-state model for most proteins. The pathway of folding can be described in terms of intermediate partially folded states. Figure 6-12 illustrates the kinetic and thermodynamic aspects of a folding model that is consistent with current experimental data on *in vitro* protein folding (Creighton, 1990; Matthews, 1993b). Folding of a protein likely proceeds through a series of sequential stages with ensembles of rapidly interconverting conformers present at each stage of the process. Folding from one stage to the next does not necessarily follow a single pathway but rather multiple parallel pathways. The number of conformers in each folding stage decreases as higher order structure develops, which finally converges to a single form, the native structure. There appear to be three common stages during folding: rapid collapse of unfolded polypeptide chains to more compact intermediates, further structural development and stabilization of early intermediates, and final formation of native states. The driving force in early stages of protein folding is likely to be hydrophobic interactions: the necessity to sequester the nonpolar amino acid side chains from the aqueous solvent (hydrophobic collapse model). The secondary structure formed in the early intermediates may be either due to intrinsic propensities of the amino acid sequence, or as a consequence of the hydrophobic collapse, or both. Hydrogen bonds stabilizing the secondary structure may be formed concomitantly to the hydrophobic collapse due to the necessity of pairing the polar amide groups when they are buried into an hydrophobic environment. The distribution of conformations in unfolded states is likely nonrandom, with biases similar to those displayed in the final native state. The residual structure in unfolded states may also contain nonnative secondary structure.

The progress of the field has been critically dependent on the development of techniques. The advent of directed mutagenesis has made it possible to alter the sequence of a protein almost at will and to study the energetics and kinetics of intermediate and transition states of folding. H-D exchange pulse labeling combined with NMR analysis allows us to follow the structural build-up of intermediates during the folding at residue level. Recent application of electrospray ionization mass spectrometry has opened a new avenue so that multiple pathways of protein folding can be explored.

The protein folding problem has been studied for about six decades, but we are still far from being able to predict the native three-dimensional structure of a protein from its primary sequence. The so-called second half of the genetic code still remains a mystery. Where do we go from here? Nonrandom distribution of unfolded states has been shown for many proteins under a variety of chaotropic conditions. Interest in denatured states of proteins is growing rapidly for the various reasons mentioned earlier. With the advance of NMR spectroscopy, more detailed structural studies on unfolded states will certainly appear. An essential difference between unfolded states and native states is the heterogeneity and conformational flexibility of the former. As a consequence, new ways of describing and quantitating unfolded structures will be required. In the past, most studies on pathways of protein folding have focused on identification and characterization of structural intermediates as a function of time. Is there any other way to describe a folding pathway? Shortle proposed that the folding pathway of a protein may be described in terms of the free energy distance of the denatured states from the native state (Shortle, 1993). Such an "energetic pathway" of protein folding could consist of energetic terms of the various interactions that stabilize the polypeptide chain as higher order of structure develops, and may provide insights into the interactions responsible for the specific structure of the native state. Lastly, most of our knowledge about protein folding has come from studies with proteins containing mixed α-helix and β-sheet proteins. Only a few studies have been reported for β-sheet proteins. More folding studies with β-sheet proteins need to be performed before general schemes of protein folding can be drawn.

7

Nucleic Acid Interactive Protein Domains That Require Zinc

Michael A. Massiah, Paul R. Blake, and Michael F. Summers

Since 1985, a wealth of information has been generated regarding the structure and function of novel, miniglobular nucleic acid interactive domains (NAIDs) that require zinc for proper folding and function. These domains have been referred to historically as "zinc fingers" due to the original proposal by Klug and coworkers in 1985 that short repeating stretches of amino acids in transcription factor IIIA (TFIIIA) from *Xenopus* oocytes that contain Cys and His residues form "zinc-binding fingers" (Miller et al., 1985). As of the writing of their review, a total of nine unique motifs have been identified in proteins from organisms ranging from retroviruses to humans that require zinc for proper folding and nucleic acid-binding.

Nuclear magnetic resonance (NMR) spectroscopy has played a leading role in developing knowledge of the structure and function of zinc-containing NAIDs. In fact, three-dimensional structures for each new class of Zn-NAID were first determined by NMR methods. In the next section of this chapter, we present brief overviews of the Zn-NAIDs that have been structurally characterized to date. In the following section, details are presented on how NMR methods have been employed to gain insights into the determinants of stability in miniglobular Zn-NAIDs. In this latter section, emphasis is placed on experiments of our own laboratory that employed the iron-sulfur protein rubredoxin from the hyperthermophile, *Pyrococcus furiosus*, as a model for miniglobular zinc finger domains.

Zinc-Containing Nucleic Acid Interactive Domains

As will be discussed, all of the known Zn-NAIDs that have been structurally characterized thus far contain NH—S hydrogen bonds between specific backbone amide protons and zinc-coordinated cysteine sulfurs. These local structural elements are often very similar to those observed in the iron binding domain of rubredoxin (Rd). As such, this section begins with a review of the structural elements of the Rd metal binding site.

All rubredoxins contain two copies of a conserved iron-binding sequence, X-X-**Cys**-X-X-**Cys**-**Gly**-X-X-X (X = variable amino acid), and high resolution X-ray and NMR structures of proteins from *Clostridium pasteurianum*, *Desulfovibrio gigas*, *Desulfovibrio vulgaris*, *Desulfovibrio desulfuricans*, and *Pyrococcus furiosus*, reveal that the four Cys residues are coordinated to the single iron atom. The backbone conformations of the N- and C-terminal X-X-**Cys**-X-X-**Cys**-**Gly**-X-X-X residues are essentially identical. Adman and coworkers discovered that three of the backbone amide protons within each of the **Cys**$_{(i)}$-X$_{(i+1)}$-X$_{(i+2)}$-**Cys**$_{(i+3)}$-**Gly**$_{(i+4)}$-X$_{(i+5)}$ loops are involved in amide-to-sulfur hydrogen bonding (Adman et al., 1975). The backbone amide protons of residues X$_{(i+2)}$ and Cys$_{(i+3)}$ are oriented in the direction of the Cys$_{(i)}$ side chain sulfur atom, forming what are referred to as Types I and III tight turns, respectively (Fig. 7-1). In addition, the backbone amide proton of residue X$_{(i+5)}$ is oriented toward the side chain sulfur atom of Cys$_{(i+3)}$, forming a Type II NH—S tight turn. Rubredoxins also contain a conserved Gly residue at position $(i + 4)$, and it was proposed that this residue is conserved in order to stabilize the Type II NH—S tight turn (Adman et al., 1975). The glycine NH is positioned to hydrogen bond with the carbonyl of Cys$_{(i)}$, forming a Type II NH—O bond. Throughout this review, we will refer to this conserved folding pattern as a "Rd-knuckle" (Fig. 7-1).

Classical CCHH Zinc Fingers

As indicated in the introduction, the study of "zinc fingers" began in 1985 with the proposal by Klug and coworkers that each of the nine-tandem repeats of the sequence, C-X$_4$-C-X$_{12}$-H-X$_3$-H (CCHH), that are present in transcription factor IIIA (TFIIIA) from *Xenopus laevis* binds an atom of zinc (Miller et al., 1985). TFIIIA serves as both a positive transcription factor for the synthesis of *Xenopus* 5S RNA genes and a storage protein for 5S RNA prior to ribosome assembly during the late stages of oogenesis (Clemens et al., 1993).

To date, more than a dozen classical CCHH zinc finger domains have been structurally characterized by NMR and X-ray crystallographic methods. Except for a few cases, the experimentally derived CCHH zinc finger structures are consistent with a structure predicted on the basis of sequence comparisons with other metalloproteins, including rubredoxin (Green & Berg, 1990). Klevit and coworkers (Klevit et al., 1990)

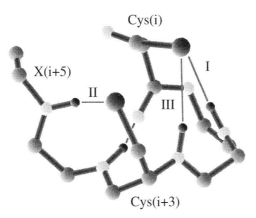

Figure 7-1. Backbone folding of a Rd-knuckle [**Cys**$_{(i)}$-X$_{(i+1)}$-X$_{(i+2)}$-**Cys**$_{(i+3)}$-**Gly**$_{(i+4)}$-X$_{(i+5)}$] showing backbone NH—S and NH—O hydrogen bonds. The coordinates were obtained from the NMR structure of *P. furiosus* Zn(Rd). Roman numerals denote the three types of NH—S tight turns common to all rubredoxins. Reprinted from Blake & Summers, 1994, with permission.

Figure 7-2. Molscript (Kraulis, 1991) representation of a 34-amino acid residue peptide corresponding to the CCHH zinc finger, ADR-1b (Klevit et al., 1990).

were the first to provide experimental NMR evidence in support of the predicted model, and shortly thereafter atomic coordinates for a peptide with the sequence of the thirty-first zinc finger of the *Xenopus* protein, Xfin, were published by Wright and coworkers (Lee et al., 1989). A representation of the ADR-1b zinc finger structure determined by Klevit and coworkers (Klevit et al., 1990) is shown in Figure 7-2.

The CCHH zinc finger consists of a three-turn helix and a short antiparallel β-sheet. The strands of the β-sheet are linked by the CXXCXX metal-binding turn. The CXXCXX turns for five representative CCHH zinc finger structures are shown in Figure 7-3. For comparison, relevant residues of the N-terminal CXXCXX turn of *C. pasteurianum* Rd, and the associated NH—S and NH—O hydrogen bonds, are included in this figure. These residues within zinc fingers Xfin-31, ADR-1 and Zif-268-2,3 are folded in a manner similar to the folding of analogous residues in rubredoxins. All of these structures exhibit the Types I and III NH—S hydrogen bonding found in rubredoxins. However, only the zinc fingers of ADR-1b and Zif-268 exhibit the Type II NH—S turn after the second Cys. The Cys($i + 3$)-X-X residues of the Xfin-31 structure were disordered, precluding classification of the potential NH—S hydrogen bonding for the X($i + 5$) amide proton.

The CXXCXX loops of the Zif-268 and ADR-1 zinc fingers contain Type II NH—S turns after the second cysteine, whereas finger-3 of Zif-268 contains a Gly residue at position ($i + 4$). Finger-2 of Zif-268 and the ADR-1 zinc finger contain Met and Thr residues, respectively, at position ($i + 4$). Type II folding results in a positive ϕ value for the residue at position X($i + 4$), and as such, it was proposed early on that Type II NH—S folding should be stabilized by a glycine residue at position X($i + 4$). This provided a rationale for the conservation of glycine residues at position X($i + 4$) in the CXXCXX loops of all known rubredoxins (Adman et al., 1975). As shown in Figure 7-1, Type II NH—S folding leads to hydrogen bonds between the X($i + 4$) amide proton and the Cys(i) carbonyl and between the X($i + 5$) amide proton and the Cys($i + 3$) sulfur. Thus, the unfavorable steric interactions associated with the positive ϕ-value may be offset by the additional hydrogen bond formed between the X($i + 4$) amide and the Cys(i) carbonyl (Adman et al., 1975). A positive ϕ-value has also been observed for the Type-II NH—S turn of residues Cys-Ala-Gly in the iron site of ferredoxin, where the Ala-methyl group eclipses the Cys carbonyl (Adman et al., 1975).

The amide proton of the X($i + 5$) residue of the ZFY-even finger and an even-finger mutant are directed toward the β-carbon of the Cys(i) residue and not toward the sulfur atoms of residue Cys($i + 3$) (Kochoyan et al., 1991a, b). In addition, the Cys($i + 3$)-to-X($i + 4$) peptide bond is oriented in a manner that would preclude hydrogen bonding to the Cys(i) carbonyl (Fig. 7-3). Thus, although the folding observed for residues Cys-X-X-Cys of the ZFY even finger domains are consistent with the rubredoxin fold, the conformation of the Cys($i + 3$)-X-X residues differs from that observed in rubredoxins. The folding observed for the zinc finger domains of the human en-

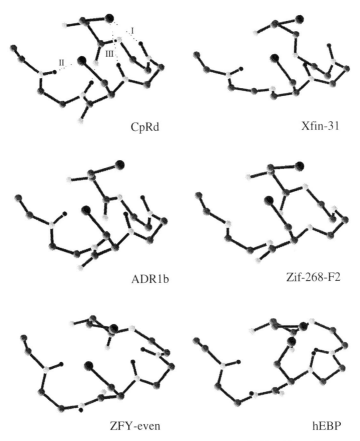

Figure 7-3. Comparison of the folding determined for the CXXCXX residues of representative, classical-type CCHH zinc fingers with related residues of *C. pasteurianum* Rd (Watenpaugh et al., 1973). The folding of these residues in the Xfin-31 (Lee et al., 1989) ADR1b (Klevit et al., 1990) and Zif-268-2 (Pavletich & Pabo, 1991) zinc fingers is essentially identical to that observed in the X-ray structure of *C. pasteurianum* Rd. The ZFY-even finger domain, (Kochoyan et al., 1991a) is similar except that residues Cys(i + 3)-X(i + 4)-X(i + 5) form a Type-I NH—S turn. The zinc fingers of the human Enhancer DNA binding domain (Omichinski et al., 1990) are unusual in that (1) the X1-value of the Cys(i + 3) residue precludes the formation of an NH—S hydrogen bond between the amide proton of X(i + 5) and the Cys(i + 3) sulfur, and (2) the amide proton of X(i + 2) does not appear to form a hydrogen bond to the Cys(i) sulfur. Reprinted from Blake & Summers, 1994, with permission.

hancer protein DNA-binding domain (hEBP) (Omichinski et al., 1990, 1992) are unusual in that the side chain orientation of the Cys($i + 3$) residues precludes formation of a Type II NH—S turn, and amide orientations are generally inconsistent with NH—S hydrogen bonding.

Thus, with the exception of the human enhancer protein DNA-binding domain, all of the classical CCHH zinc finger structures exhibit Types I and III NH—S hydrogen bonds within the Cys-X-X-Cys knuckles. In addition, all CCHH zinc fingers except those of the hEBP zinc fingers exhibit an NH—S hydrogen bond from the X($i + 5$) backbone amide to the Cys($i + 3$) sulfur, forming either a Type I or Type II NH—S turn.

Retroviral CCHC Zinc Fingers

The proposal that the conserved CCHH residues in TFIIIA bind zinc stimulated studies of cysteine and histidine-rich sequences found in other nucleic acid interactive proteins. Our laboratory has focused on "CCHC-type" NAIDs observed originally in retroviral nucleocapsid proteins. The conserved sequence, **C**-X_2-**C**-X_4-**H**-X_4-**C**, has recently been found in proteins ranging from plant viruses to frogs to humans, and thus may represent a common motif for binding single-stranded nucleic acids (Summers, 1991). For example, a nine-tandem repeat of the conserved sequence is found in a protein encoded by the *Leishmania major* gene of the protozoan parasite *Leishmania* (Webb & McMaster, 1993); a seven-tandem repeat is found in human cellular nucleic acid-binding protein (hCNBP) that binds to the sterol regulatory element (Rajavashisth et al., 1989); and a single motif is present in a protein encoded by the *Xenopus-Posterior* (*Xpo*) gene of frog embryos (Sato & Sargent, 1991).

NMR structures have been determined for peptides with sequences of several CCHC zinc finger domains, as well as for the intact NC (nucleocapsid) protein from HIV-1 (Morellet et al., 1992; Omichinski et al., 1991; South et al., 1990; Summers et al., 1990, 1992). Like TFIIIA, overall folding of the NC protein consists of the independent CCHC zinc finger domains separated by a labile linker region. The global folding CCHC Zn-NAIDs consists of a short antiparallel β-sheet with an Rd-turn, followed by a short β-like stretch that precedes a 3_{10}-helix (Fig. 7-4). Like CCHH zinc fingers and

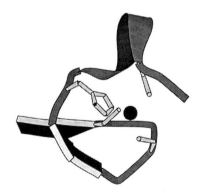

Figure 7-4. Molscript (Kraulis, 1991) representation of the N-terminal CCHC zinc finger from the nucleocapsid protein of human immunodeficiency virus type 1 (HIV1-F1) (South et al., 1990; Summers et al., 1992).

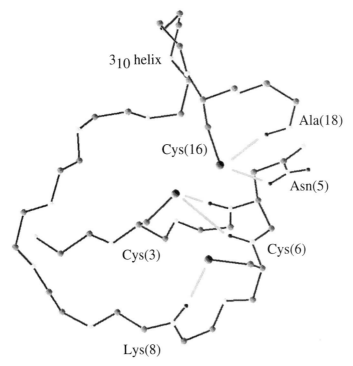

Figure 7-5. Representation of the backbone and selected side-chain atoms of a representative Zn(HIV1-F1) DG/SA model. The shaded lines denote the five hydrogen bonds to the metal-coordinated cysteine sulfur atoms. Reprinted from Blake & Summers, 1994, with permission.

rubredoxin, the metal-binding site of CCHC NAIDs fold with a number of NH—S hydrogen bonds, including a hydrogen bond between the sidechain NH of Asn($i + 2$) and the Cys($i + 3$) sulfur, and between the backbone NH protein of residue X($i + 15$) and the Cys($i + 13$) sulfur (Fig. 7-5). In addition to these five NH—S hydrogen bonds, there are a total of six NH—O hydrogen bonds that further stabilize the CCHC zinc finger structure (Fig. 7-6) (Blake & Summers, 1994).

Steroid Hormone Receptor DNA-Binding Domain

Steroid hormone receptor (SHR) superfamily is a class of transcription regulatory proteins whose activity is controlled by the binding of specific hormones. This class is so far comprised of more than a dozen proteins, including the glucocorticoid receptor, retinoic acid receptor-β, progesterone receptor, estrogen receptor, and retinoid X receptor-α proteins. Typically, members of the SHR family bind as dimers to specific DNA sequences called hormone response elements (HREs) (Lee et al., 1993). The SHR protein binds to the HRE, forming a hormone-receptor complex that activates transcription (Schwabe et al., 1990).

The global folding observed for the DNA-binding domains of steroid hormone receptor (SHR) proteins differs significantly from that of classical and retroviral NAIDs. These structures contain two independent zinc modules, with each of the zinc atoms

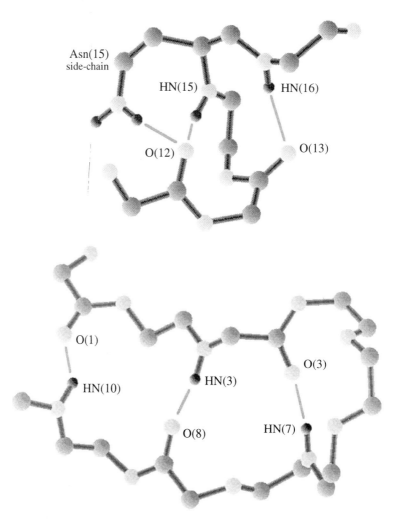

Figure 7-6. Diagram showing portions of Zn(HIV1-F1) that comprise a miniantiparallel β-sheet and a 3_{10}-helical turn. Hydrogen bonds are denoted by shaded lines. The side chain amide proton of Asn(15) forms a hydrogen bond to the backbone carbonyl of Ile(12), which appears to lend stability to the helical corner. Reprinted from Blake & Summers, 1994, with permission.

coordinated via four cysteine residues (Fig. 7-7) (Hard et al., 1990; Knegtel et al., 1993; Luisi et al., 1991; Schwabe et al., 1990). Unlike classical CCHH zinc fingers, the two zinc atoms are incorporated into a single domain, with several contacts between residues that comprise the independent zinc sites.

The N-terminal zinc coordination site of the SHR proteins contains two **CXXC** arrays, and the C-terminal zinc site contains one **CXXC** array. Luisi and coworkers indicated that several backbone amide protons are involved in NH—S hydrogen bonding to the metal-coordinating sulfurs (Luisi et al., 1991). Comparison of the three **CXXC** segments of the glucocorticoid receptor DBD reveals a variation in $\chi 1$-values of the cysteine side chains, with only the N-terminal knuckle exhibiting a Cys(i) side chain orientation similar to that of rubredoxins (Fig. 7-8). The backbone folding within each of the CXXC knuckles is generally consistent with the folding observed in rubredoxins, with backbone amide protons participating in Types I and III NH—S hydrogen bonding. The Cys(i + 3)-X-X residues of the N-terminal CXXC knuckle form a Type II NH—S turn, with the X(i + 4) amide forming a hydrogen bond to the Cys(i) backbone carbonyl. The Cys(i + 3)-X-X residues of the two subsequent CXXC segments do not form NH—S turns because these residues reside within an α-helix (Fig. 7-8).

In summary, the N-terminal CXXCXX array exhibits Types I, II and III NH—S turns and exists in a conformation very similar to that observed in rubredoxins (Fig. 7-8). The two subsequent arrays are similar in that they contain the Types I and III NH—S turns within the CXXC segment, but differ in that (1) the χ-values of the Cys(i + 3) side chains are significantly different, and (2) the Cys(i + 3)-X-X residues do not form an NH—S turn.

GAL-4 DNA Binding Domain

GAL4 is a yeast transcription activator that binds to an upstream activating sequence (UAS$_G$) for the genes (GAL1 and GAL10) encoding galactose-metabolizing enzymes (Marmorstein et al., 1992). The 881-amino acid residue GAL4 protein binds as a dimer to the UAS$_G$-DNA site (Gardner et al., 1991). In the absence of galactose, GAL4 remains on the DNA but is inactivated by a complex formed with another protein, GAL80 (Johnston, 1987; Ma & Ptashne, 1987). In the presence of galactose, a metabolite derived from galactose binds to and dissociates GAL80, thus activating GAL4 (Gardner et al., 1991). The DNA-binding domain of GAL4 is a 62-amino acid sequence located on the N-terminus of the protein (Kraulis et al., 1992). This sequence contains a cys-

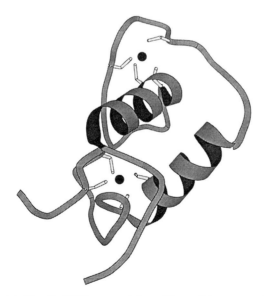

Figure 7-7. Molscript, (Kraulis, 1991) representation of the peptide corresponding to the DNA binding domain of glucocorticoid receptor protein (Luisi et al., 1991).

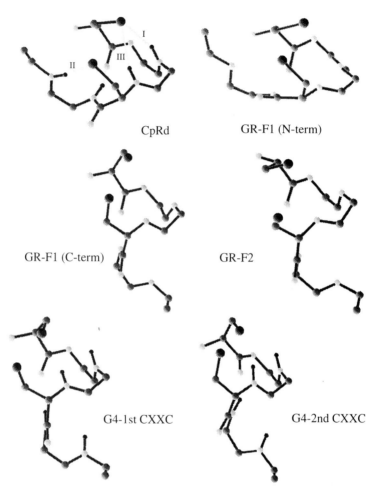

Figure 7-8. Comparison of the folding determined for the CXXCXX residues of the glucocorticoid (GR, *top*) (Luisi et al., 1991) and GAL4 (*bottom*) (Marmorstein et al., 1992) DNA-binding domains with related residues of *C. pasteurianum* Rd (Watenpaugh et al., 1973). The GR protein contains three CXXCXX arrays. The folding of the N-terminal array [GR-F1(N-term)] is highly similar to that observed in rubredoxins. The remaining arrays contain Types I and III NH—S turns within the CXXC knuckles common to rubredoxins; however, the Cys(i + 3)-X-X residues do not from an NH—S turn due to their involvement in α-helical conformations. Interestingly, the amide proton of residue X(i + 4) participates in a hydrogen bond to the Cys(i) carbonyl, an interaction that also occurs in rubredoxins. The folding of the two CXXCXX arrays in GAL-4 (bottom) matches closely the folding exhibited by the latter two arrays of the GR protein. Reprinted from Blake & Summers, 1994, with permission.

teine-rich array, $C\text{-}X_2\text{-}C\text{-}X_6\text{-}C\text{-}X_6\text{-}C\text{-}X_2\text{-}C\text{-}X_6\text{-}C$, which is conserved among 13 known fungal transcription factors (Kraulis et al., 1992). Unlike other Zn-NAIDs, the two zinc atoms of GAL4 form a binuclear cluster that contains two bridging cysteines (Fig. 7-9) (Kraulis et al., 1992; Pan & Coleman, 1991).

Structures of the GAL4 NAIDs have been determined by NMR (Baleja et al., 1992; Kraulis et al., 1992;) and X-ray crystallographic (Marmorstein et al., 1992) methods. The GAL4 DBD contains two helix-turn-strand motifs that are packed around the $Zn(II)_2Cys_6$ binuclear cluster (Fig.7-9) (Kraulis et al., 1992). The structure has a two-fold pseudosymmetry axis that passes between the two helices and the center of the $Zn(II)_2Cys_6$ binuclear cluster. The GAL4 zinc-binding domain contains two **CXXC** arrays that participate in metal coordination. Both the N- and C-terminal **CXXC** arrays exhibit backbone folding that is generally consistent with the folding observed in rubredoxins. Amide protons of residues $X(i + 2)$ and $Cys(i + 3)$ are oriented in a manner consistent with Types I and III NH—S hydrogen bonding to the $Cys(i)$ sulfur (Fig. 7-8). However, the $\chi 1$-values of the Cys side chains differ from those of Rd by almost 180°. The $Cys(i + 3)$-X-X residues comprise the N-terminal portions of α-helical stretches, and the associated backbone conformations preclude the formation of Type II NH—S turns. The folding of these residues is similar to that observed for analogous residues in the steroid hormone receptor proteins where an α-helical conformation exists (Fig.7-8).

GATA-1 DNA Binding Domain

The erythroid-specific transcription factor GATA-1 contains a DNA-binding domain that has two conserved zinc-binding sequences of the form $C-X_2-C-X_{17}-C-X_2-C$. These motifs are separated by 29 amino acid residues. Named for the (T/A)GATA(A/G) DNA sequence it binds to, GATA-1 regulates the transcription of erythroid-expressed genes and facilitates the generation of the erythroid lineage (Omichinski et al., 1993). Mutational studies indicated that the N-terminal motif may be unimportant for specific DNA-binding (Martin & Orkin, 1990; Yang & Evans, 1992), although more recent studies indicate that sequence specificity is a function of both domains (Whyatt et al., 1993).

The solution structure of the C-terminal GATA- DBD (residues 158–223) was determined recently by NMR methods (Omichinski et al., 1993). The global folding of the peptide consists of a core (residues 2–51) that contains the zinc binding region, and

Figure 7-9. Molscript representation (Kraulis, 1991) of the DNA binding domain of GAL4 (Kraulis et al., 1992).

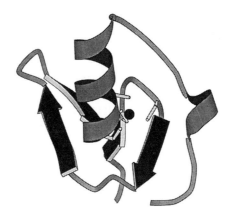

Figure 7-10. Molscript representation (Kraulis, 1991) of DNA binding domain of GATA-1 (Omichinski et al., 1993).

an extended C-terminal tail (residues 52–59) (Omichinski et al., 1993) (Fig. 7-10). The folding of the majority of the core (residues 3–39) is very similar to the folding exhibited by the N-terminal CCCC zinc module of glucocorticoid receptor protein (Omichinski et al., 1993). Interestingly, both the GATA-1 domain and the N-terminal CCCC module of GR participate directly in protein-nucleic acid interactions.

The N-terminal CXXCXX array of GATA-1 forms a classical Rd-knuckle that contains Types I, II, and III NH—S turns. The Type II turn is present despite the fact that the array lacks a Gly residue at position $X(i + 4)$. In addition, the backbone NH of residue $Cys(i + 21)$ forms a hydrogen bond to the $Cys(i)$ sulfur, and the sulfur of $Cys(i + 21)$ accepts a hydrogen bond from the backbone amide proton of $Cys(i + 24)$.

Methionyl-tRNA Synthetase NAID

Aminoacyl-tRNA synthetases are responsible for the esterification of tRNAs with the cognate amino acid during protein translation (Fourmy et al., 1993a). A number of aminoacyl-tRNA synthetases have been analyzed and shown to display a zinc finger domain of the type $C-X_2-C-X_9-C-X_2-C$ (Fourmy et al., 1993b). Among these enzymes are the methionyl-tRNA, alanyl-tRNA, and isoleucyl-tRNA synthetases, all of which have been shown to contain zinc (Mayaux & Blanquet, 1981; Miller et al., 1991; Posorske et al., 1979). On the other hand, tyrosyl- and phenylalanyl-tRNA synthetases, which do not contain zinc finger domains, apparently do not contain zinc (Mayaux & Blanquet, 1981).

Dardel and coworkers (Fourmy et al., 1993a) have recently determined the three-dimensional structure of a peptide corresponding to the zinc coordination site of methionyl-tRNA synthetase by NMR methods. The structure differs significantly from that determined by X-ray crystallography for the intact protein. The zinc-binding site is located within residues 138–163 of the protein, and contains a single zinc atom coordinated via four cysteine residues. Unlike other Zn-NAIDs, the zinc-binding region of methionyl-tRNA synthetase does not contain any distinguishable helices or extended β-sheet structure. Instead, the domain contains a series of four consecutive tight turns separated by short loops (Fig. 7-11). Turns 2 and 4 form Rd-knuckles. Superposition of the backbone and Cys-side chain atoms of the XCXXCX residues of the methionyl-tRNA synthetase metallopeptide with analogous residues of *C pasteurianum* Rd and Zn(HIV1-F1) afforded RMSD values of 0.65 to 0.90 Å (Fourmy et al., 1993a). The backbone amide protons of residues X(i + 2) and Cys(i + 3) are oriented in a manner consistent with Types I and III NH—S hydrogen bonding. The N- and C-terminal Rd-knuckles contain Gly and Lys residues, respectively, at position (i + 4). As expected, the backbone amide of residue X(i + 5) from the **CXXCGX** array forms a hydrogen bond to the Cys(i + 3) sulfur, with the backbone amide proton of Gly(i + 4) forming a hydrogen bond to the Cys(i) carbonyl. Thus, the Cys(i + 3)-Gly(i + 4)-X(i + 5) residues exist in a Type II NH—S turn common to rubredoxins.

The backbone amide proton of the X(i + 5) residue from the N-terminal **CXXCXX** array also appears to form a hydrogen bond with the Cys(i + 3) sulfur. In addition, the backbone amide proton of the lysine at position (i + 4) is oriented in the general direction of the Cys(i) backbone carbonyl. As such, the lysine at position (i + 4) possesses a positive ϕ value characteristic of Type II NH—S tight turns. Thus, as observed in the zinc finger of Zif-268-F2 and ADR-1b, the presence of a nonglycine residue at position X(i + 4) does not preclude the formation of a Type II NH—S turn.

Transcription Elongation Factor TFIIS

Transcription elongation factor TFIIS is a protein that binds to the largest subunit of RNA-polymerase II (RNAP-II) in a paused elongation complex (Archambault et al., 1992; Pappaport et al., 1988) and allows elongation through DNA pause and termination sites (Reines & Mote Jr., 1993; Siva-Raman et al., 1990). It also facilitates transcription of sites blocked by sequence-specific DNA-binding proteins (Reines & Mote, Jr., 1993). The 50-amino acid zinc-binding domain contains a C-X_2-C-X_{25}-C-X_2-C motif. The structure of a peptide with the sequence of the TFIIS zinc-binding site (residues 235–280) was determined recently by NMR methods (Qian et al., 1993). Unlike other Zn-DBDs, the domain does not contain an α-helical structure but consists of a three-stranded antiparallel β-sheet (Fig. 7-12). Two of the sheets are located between the C(15)-X_{25}-C(40) region, while the third exists after the second C(40)-X_2-C(43) array. TFIIS also contains three β-turns (residues 14–16, 30–32, and 42–44) two of which contain the zinc-binding cysteines (Qian et al., 1993). The zinc atom is chelated via two Rd-knuckles, both of which contain Types I, II and III NH—S turns (residues **Cys**(12)-X_2-**Cys**(15)-X-X and **Cys**(40)-X_2-**Cys**(43)-**Gly**-Asp(45)). The N-terminal Rd-knuckle contains a Type II NH—S turn even though the sequence lacks a glycine at residue 16.

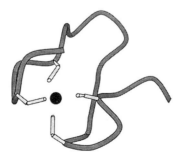

Figure 7-11. Molscript representation (Kraulis, 1991) of a peptide corresponding to the RNA binding domain (residues 168 to 163) of methionyl-tRNA synthetase (Fourmy et al., 1993a).

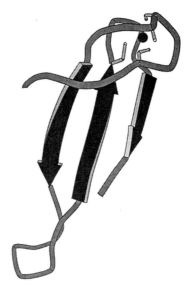

Figure 7-12. Molscript representation (Kraulis, 1991) of a 50-amino acid residue peptide corresponding to the DNA binding of transcription elongation factor TFIIS (Qian et al., 1993).

Ring (C3HC4) Fingers

Over the past several years, more than 40 proteins have been identified from organisms including plants, fungi, viruses, and vertebrates that contain a "C_3HC_4" or "RING finger" motif (C-X_2-C-X_{12}-C-X-H-X_2-C-X_2-C-X_{10}-C-X_2-C) (Everett et al., 1993; Freemont, 1993). Proteins that contain a C_3HC_4 motif are involved in numerous processes including transcription, site-specific recombination, differentiation, and oncogenesis (Everett et al., 1993).

The three-dimensional solution structure of a peptide of the sequence encompassing the C_3HC_4 domain from a herpes virus protein was recently solved by Barlow and coworkers (Barlow et al., 1994). The two zinc atoms are bound in separate sites, with cysteines 8 and 11 and cysteines 29 and 32 coordinating to one zinc, and cysteines 29, 43 and 46 and histidine 26 coordinating to the second zinc atom. The peptide binds two zinc atoms and adopts a $\beta\beta\alpha\beta$ structure (Fig. 7-13). Barlow and coworkers showed that the amphipathic α-helix is essential for *trans*-activation of gene expression (Barlow et al., 1994). Interestingly, the N-terminal $\beta\beta\alpha$ portion of the protein is structurally very similar to that of classical CCHH zinc fingers.

NMR Studies of *P. Furiosus Rubredoxin*, a Hyperthermostable Model for Miniglobular Zinc Finger Domains

Based on the findings that the metal binding sites of zinc finger proteins and rubredoxins contain similar structural elements, our laboratory has carried out detailed NMR studies of the rubredoxin from the hyperthermophile, *Pyrococcus furiosus* (PfRd), an organism that grows optimally at 100°C (Adams, 1990; Stetter et al., 1990). Native PfRd contains one iron atom in either the +2 or +3 oxidation states. The protein is paramagnetic in both states, and the line broadening that results from the paramagnetic-induced relaxation precluded the application of correlation experiments essential for high resolution NMR studies (Blake et al., 1991). However, the structural similarities of the metal-binding sites in CCHC zinc fingers and in rubredoxins suggested to us that the iron may be substituted by zinc. Indeed, we found that the native iron atom could be substituted by the divalent cations of the entire zinc triad (Blake et al., 1991, 1992a, c, 1994)! The results obtained from NMR studies of metal-substituted pfRd have provided new insights into the molecular level determinants of stability of small globular zinc finger minidomains. As such, this hyperthermostable protein has served as a useful "model" for the smaller zinc finger structures.

The solution structure of the zinc-substituted Zn-pfRd was determined by NMR spectroscopy (Blake et al., 1992a, d). The global folding consists of a three-stranded antiparallel β-sheet formed by residues A1-C5, Y10-E14, and F48-L51. Two additional short stretches of antiparallel β-sheetlike structure involving residues G17-P19 paired with I23-S24 and residues W36-C38 paired with P44-K45 also exist. Residues C5-Y10 and C38-A43 form a series of Types I, II, and III NH—S tight turns, and residues D13-G17, D18-G22, K28-L32, and P44-F48 are folded to form four 3_{10} helical corners. In addition, the side chain of D13 appears to be hydrogen bonded to the amide of D15, forming an Asx turn (Abbadi et al., 1991; Rees et al., 1983; Richardson, 1981) and residues P25-T27 and P33-W36 form Types I and II β turns, respectively. The hydrophobic side chains of residues W3, Y10, Y12, I23, F29, L32, W36, and F48 contribute to the hydrophobic core of the protein, with apparent stacking of the six-membered aromatic rings of Trp 3 and Phe 29. These structural elements are very similar to those observed in the high-resolution X-ray structure of native *C. pasteurianum* Rd (CpRd). Superposition of the backbone C, Cα, N atoms of residues A1-L51 of Zn-pfRd onto relevant atoms of residues K2-V52 of CpRd resulted in an average pairwise RMSD value of 0.77 ± 0.06 Å. The conformations of the aromatic groups that comprise the hydrophobic cores are also very similar. Thus, the enhanced thermal

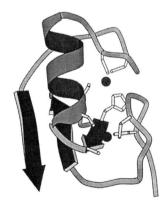

Figure 7-13. Molscript representation (Kraulis, 1991) of the N-terminal region of the RING-1 protein. The nucleic acid interacting C_3HC_4-RING domain encompasses about 60 amino acid residues (Barlow et al., 1994).

stability of PfRd relative to CpRd is not attributable to differences in global tertiary structure (Blake et al., 1992a).

In addition, superposition of the C, C_α, N, C_β, and S_γ atoms of the metal coordination site residues C5-Y10 and C38-A43 of Zn-PfRd onto relevant residues C6-Y11 and C39-V44 of native CpRd afforded pairwise RMSD values of 0.33 ± 0.02 Å (Blake et al., 1992a) demonstrating that zinc substitution did not lead to a significant structural perturbation at the metal binding site. In separate studies, Rees and coworkers (Day et al., 1992) determined the structures of the oxidized and reduced forms of native *P. furiosus* (Pf) Rd by high resolution X-ray crystallography. The X-ray and NMR structures were found to be fully compatible (Blake et al., 1992a); superposition of the mean backbone C, $C\alpha$, N coordinates for residues 1:51 of Zn-PfRd onto related atoms of oxidized and reduced, native PfRd yielded pairwise RMSD values of 0.62 and 0.63 Å, respectively. Most importantly, the folding of those residues that comprise the metal-binding sites of both proteins were also essentially identical. Superposition of the mean backbone atoms for residues 5–10 and 38–43 of Zn(Pf Rd) onto the same atoms of oxidized and reduced, native Pf Rd gave pairwise RMSD values of 0.29 and 0.31 Å, respectively (Blake et al., 1992a). This observation shows that the conformation of the metal center appears to be insensitive to the oxidation state of the iron and also to the nature of the bound metal (at least for Zn^{+2}, Fe^{+2}, and Fe^{+3}).

HMQC and HSED Experiments for Detection of NH—S Hydrogen Bonds in CXXCGXX Rd-Knuckles

Due to apparent structural lability of a ^{113}Cd-substituted CCHC peptide, our attempts to observe long-range scalar connectivities between the metal and exchangeable protons were not successful. However, since the folding of the **CXXCGX** residues and associated NH—S hydrogen bonds of the CCHC zinc fingers are essentially identical to those observed in rubredoxins, some of the focus in our laboratory shifted to heteronuclear NMR studies of the metal-substituted PfRd.

Heteronuclear spin-echo difference (HSED) and heteronuclear multiple quantum correlation (HMQC) experiments were used to probe the metal binding site of the ^{113}Cd- and ^{199}Hg-substituted PfRd. In ^{1}H-X HSED experiments (X = magnetic nucleus such as ^{113}Cd or ^{199}Hg), only signals of protons that are scalar coupled to the X nucleus are detected. Since scalar coupling is mediated by electron density of bonding orbitals, the presence of ^{1}H-X HSED signals provides direct evidence for covalent character in the intervening bonding orbitals between the proton and the X nucleus.

Based on similar principles, HMQC experiments reveal the environment of the X nucleus in the protein by virtue of the chemical shift, and allow identification of protons that are scalar coupled to the metal. For ^{113}Cd, the metal chemical shift can span a range of over 1,000 ppm, and ^{113}Cd has proven to be extremely useful in probing the metal coordination site in many metalloproteins (Summers, 1988). The chemical shift of ^{199}Hg can exceed 3,000 ppm (Brevard & Granger, 1981), and this nucleus may, therefore, serve as a useful metallobioprobe, in addition to ^{113}Cd. Also, novel mercury-binding regulatory proteins such as Mer-R (Helmann et al., 1990; Wright et al., 1990) have been identified, and knowledge of the ^{199}Hg NMR parameters may be useful for characterization of these proteins. Chemical shift anisotropy (CSA), which is influenced by the symmetry of the mercury site, is likely to be severe for mercury sites with reduced symmetry, and may exceed 1200 ppm for trigonal sites (Santos et al., 1991) such as those expected for MerR (Helmann et al., 1990; Wright et al., 1990). This could limit the utility of such experiments.

The ^{1}H-^{113}Cd HSED spectrum obtained for ^{113}Cd-PfRd exhibits four backbone NH signals corresponding to the Ile(7), Cys(8), Ile(40) and Cys(41) backbone NH protons (Blake et al., 1992c) of which Ile(40) and Cys(41) are part of the metal binding CXXCGX loop (C38-X-I40-C41-G42-A43). The observation of scalar coupling between ^{113}Cd and NH protons Ile(40) and Cys(41) could not occur through the peptide backbone since the NH protons are eight and five bonds removed from the metal, respectively. The observed amide HSED signals are thus attributed to two-bond $^2J(^1$H-^{113}Cd) scalar coupling mediated by NH—S(Cys) hydrogen bonds. These novel findings provided evidence that Types I and III NH—S tight turns exist in PfRd, and further demonstrate that these NH—S hydrogen bonds contain significant covalent character (Blake et al., 1992c).

The X($i + 5$) (Ala43) NH proton proposed to form Type II NH—S tight turns did not give rise to detectable HSED signals in ^{113}Cd substituted PfRd, despite the fact that this proton exhibited the lowest deuterium exchange rate (Blake et al., 1992c). The lack of observable HSED signals was attributed to the presence of weaker scalar coupling constants. However, since scalar coupling is influenced by multiple factors (such as bond length, bond angle, electronegativity, etc.), the lack of observed HSED signals for the X($i + 5$) NH protons did not indicate that the proposed Type-II NH—S hydrogen bonds are weaker or that they contain less covalent character.

^{1}H-^{199}Hg HSED NMR experiments were then performed for the ^{199}Hg adduct to evaluate the influence of this larger metal on the nature of the NH—S hydrogen bonds (Blake et al., 1992b). A portion of the ^{1}H-^{199}Hg HSED spectrum is shown in Fig. 7-14. As observed for the ^{113}Cd adduct, HSED signals were observed for the Ile(7,40) and Cys(8, 41) backbone NH protons (Blake et al., 1992c), confirming the presence of the Types I and III NH—S turns. Interestingly, the magnitudes of the scalar couplings, determined using a novel pulse sequence that affords pure phase HSED spectra (Blake et al., 1992b), were approximately twofold greater than the coupling constants determined for ^{113}Cd-PfRd. This finding confirms that the HSED signals result from scalar coupling rather than from a dipolar cross-correlation mechanism. In addition, a weak ^{1}H-^{199}Hg HSED signal ($^2J_{H-Hg} = 0.4$ Hz) was detected for the Tyr(10) backbone amide proton. Although a similar signal for the analogous Ala(43) amide could not be distinguished due to signal overlap, the observation of a Tyr(10) NH HSED signal confirmed the existence of the Type II NH—S turn. Thus, for the ^{199}Hg adduct, where the magnitudes of the scalar couplings are greatest, HSED spectroscopy provided confirmation for the existence of all three types of NH—S tight turns (Fig. 7-14).

Additional ^{1}H-^{113}Cd and ^{1}H-^{199}Hg scalar coupling was observed to the Ala(43) methyl group (Blake et al., 1992b, c) (Fig. 7-14). This scalar coupling is probably due to direct orbital overlap between the methyl protons of Ala(43) and the Cys(42) sulfur atom, since 6- and 11-bond ^{1}H-^{113}Cd scalar coupling mediated by the Ala(43) NH—S hydrogen bond or by the covalent bonds, respectively, is implausible. This finding has important implications for understanding intraprotein electron transfer, since $CH_3\cdots S(Cys)$ overlap provides a direct electron-transfer pathway between the metal center and the hydrophobic core of the protein.

Two-dimensional ^{1}H-^{113}Cd and ^{1}H-^{199}Hg HMQC spectra were obtained for the ^{113}Cd- and ^{199}Hg-substituted proteins to evaluate the metal chemical shifts and to identify protons of amino acids that exhibit scalar coupling to the metals. The ^{113}Cd chem-

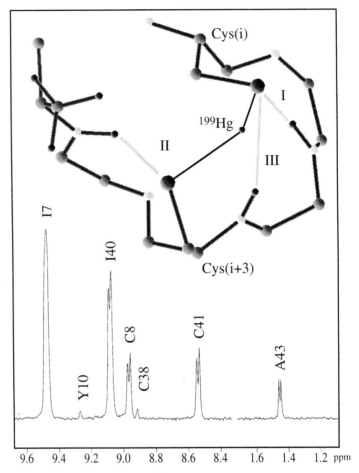

Figure 7-14. Selected regions of the ^{1}H-^{199}Hg heteronuclear spin echo difference (HSED) spectrum obtained for P. furiosus ^{199}Hg(Rd). Signals observed for the I7/I40, C8/C41, and the Y10 NH protons, which result from NH—S hydrogen bond mediated scalar coupling, confirm that these backbone amide protons participate in Types I, III, and II NH—S tight turns, respectively. The C(38) backbone amide proton exhibits weak, five-bond scalar coupling via the side chain covalent bonds. "Through-space" scalar coupling to the Ala(43) methyl protons results from the close proximity of the methyl group to the Cys(41) sulfur. Reprinted from Blake & Summers, 1994, with permission.

ical shift of 750 ppm observed for ^{113}Cd-PfRd is identical to the shift observed for ^{113}Cd in the metal site of horse liver alcohol dehydrogenase (ADH) (Bobsein & Myers, 1981). The chemical shift of the ^{113}Cd resonance is consistence with ^{113}Cd coordinated by cysteines. The metal binding region of horse ADH utilizes four cysteine-sulfur atoms as ligands to the metal. The ^{199}Hg chemical shift of −241 ppm (relative to an external dimethylmercury shift of 0.0 ppm, calculated from the ratio of magnetogyric ratios: $\gamma H/\gamma^{199}Hg$ = 1.0/0.179107) (Brevard & Granger, 1981) is reasonably consistent with the shift of −433 ppm observed using cross polarization-magic angle spinning (CP-MAS) solid-state ^{113}Cd NMR methods for the mercury tetrathiolate complex, [Et$_4$N][Hg(S-2-PhC$_6$H$_4$)$_4$] (Santos et al., 1991).

The upfield region of the ^{1}H-^{199}Hg HMQC spectrum is shown in Fig. 7-15. Included in this figure is the upfield region of a high-sensitivity ^{1}H-^{199}Hg HSED spectrum obtained with J evolution and refocusing delay periods of 60 msec. Numerous ^{1}H signals observed in the spectrum indicate that the associated protons are scalar coupled to mercury. First, intense correlation signals are observed for the cysteine H$_\beta$ protons that are three bonds removed from the metal. The intensities of these signals are disproportionately smaller than the intensities of other signals in the spectrum due to the long J-evolution-refocusing periods employed ($t = 60$ msec). As indicated below, the coupling constants for these protons can be as large as 116 Hz; as such, the optimal J-evolution and J-refocusing periods for detecting these Cys H$_\beta$ protons would be on the order of 5 msec.

Weak correlation signals are also observed for all of the cysteine Hα protons due to long-range, four-bond scalar coupling. It is particularly interesting that protons of several aliphatic side chain groups from amino acids which comprise the metal binding CXXCGX loops exhibit scalar coupling to the metal (Fig. 7-15). For example, HSED signals are observed for the Ala(43) methyl protons and for the Hβ protons of residues Ile(7) and Ile(40) (Blake et al., 1994). These protons are separated from the metal by a minimum of six covalent and/or hydrogen bonds. The most plausible explanation for such coupling is that the interaction occurs via a so-called "through space"

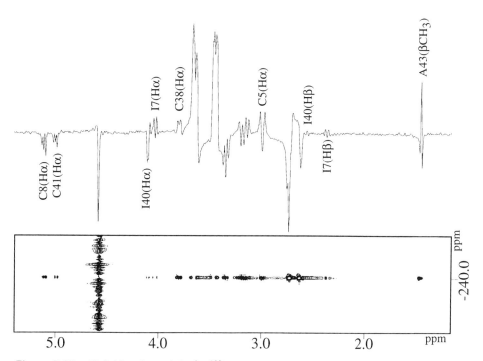

Figure 7-15. Upfield regions of the ^{1}H-^{199}Hg HSED (top) and 2D HMQC (*bottom*) spectra obtained for ^{199}Hg(*Pf* Rd). Correlation signals are labeled for all protons except the Hβ protons of the cysteines. At the long τ values employed (60 msec), signals of weakly coupled protons are enhanced relative to those of the strongly coupled protons. Scalar coupling to the Hβ protons of Ala(43), Ile(7), and Ile(40) appears to result from a "through-space" coupling mechanism (see text). Reprinted from Blake et al., 1994, with permission.

mechanism. Weak five-bond scalar coupling is also observed for the Cys(5) and Cys(38) backbone amide protons, where the coupling occurs via five covalent intervening bonds (Fig. 7-15).

Structural Information from Cysteine Hβ-Metal Coupling Constants

Three-bond H_β-C_β-S-^{199}Hg coupling constants were measured from resolution-enhanced 2D homonuclear correlated spectroscopy. Values measured for each of the Cys-H_β protons for ^{199}Hg-PfRd correlated extremely well with the H_β-C_β-S-^{199}Hg torsion angles (Blake et al., 1994). The magnitudes of the coupling constants range from 13 Hz to 116 Hz. The largest couplings (116 Hz) are observed for the proS-H_β protons of residues Cys(8) and Cys(41) and the H_β-C_β-S-^{199}Hg torsion angles associated with these protons are 144° and 151°, respectively. A smaller coupling constant of 59 Hz was observed for the ProR-H_β protons of residues Cys(8) and Cys(41). The H_β-C_β-S-^{199}Hg dihedral angles associated with these protons range from 27° to 35°. Coupling constants in the range of 21 to 22 Hz and 13Hz were observed for the proR- and proS-Hb protons, respectively, of residues Cys(5) and Cys(38). The magnitudes of the coupling constants follow a Karplus-type relationship, where torsion angles close to 90° lead to weak three-bond coupling constants and torsion angles that approach 0° or 180° give rise to the largest three-bond couplings (Karplus, 1959). Similar findings have been obtained for ^{113}Cd-substituted *Desulfovibrio gigas* Rd and metallothionein (Helmann et al., 1990; Zerbe et al., 1994). These findings are significant since it now appears that ^{1}H-metal scalar couplings may be exploited as restraints in distance geometry-based structure refinement strategies.

"Through-Space" Mediated Scalar Coupling: Implications Regarding Electron Transfer in Redox Metalloproteins

As indicated above, protons as far as 11 covalent bonds away from the metal exhibited scalar coupling to the ^{199}Hg in ^{199}Hg-PfRd. Specifically, the Hβ protons of Ala(43), Ile(7) and Ile(40), which are 11, 8, and 8 covalent bonds away, respectively, gave rise to signals in HSED spectra. Since long-range coupling via the covalent bonds of the protein backbone and side chain groups is implausible, the most reasonable mechanism for the observation of scalar coupling to these protons is the so-called through space mechanism. Unfortunately, the "through space" terminology is inappropriate since scalar coupling is actually mediated by intervening electron density and not by dipolar (true through space) interactions. The observation of scalar coupling indicates the existence of continuous molecular orbitals between the metal center and the hydrophobic side chains of residues Ala(43), Ile(7) and Ile(40). As shown in Figure 7-16, the Ala(43) methyl protons are actually in van der Waals contact with the Cys(41) sulfur atom. In addition, the Ile(7) and Ile(40) Hβ protons are also in van der Waals contact with the Cys(5) and Cys(38) sulfur atoms, respectively. Thus, overlap of orbitals associated with these protons and sulfur atoms is most likely responsible for the observed scalar interactions.

These heteronuclear NMR studies of the ^{113}Cd- and ^{199}Hg-substituted pfRd have not only established the presence of NH—S hydrogen bonds within the CXXCGX Rd knuckle, but have important implications regarding potential mechanisms of intrapro-

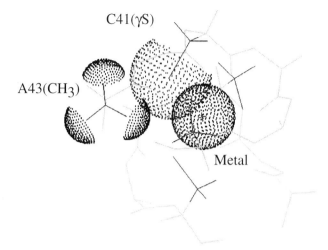

Figure 7-16. Representation of selected atoms in the metal binding site of *Pf* Rd. The Ala(43) protons, Cys(41) sulfur, and metal are represented as van der Waals spheres. In this static model, the Ala(43) methyl group is in van der Waals contact with the cysteine sulfur atom. Orbital overlap between the Ala(43) methyl protons and the Cys sulfur account for the observed "through-space" scalar coupling to ^{113}Cd and ^{199}Hg. Reprinted from Blake et al., 1994, with permission.

tein electron transfer in rubredoxins and other electron transport proteins. Our studies indicate that "electron jumps" from the metal-coordinated sulfurs directly to hydrophobic side chains may not necessarily require a true "through-space" jump, since continuous orbitals between specific hydrophobic side chains and the metal-coordinated sulfur atoms are apparent in *P. furiosus* rubredoxin. The overlap of the sulfur and hydrophobic side chain orbitals could indeed provide a low-energy pathway for electron transfer from metal centers directly to the hydrophobic side chains of the core or surface amino acids.

Determinants of Stability in Miniglobular Zinc NAIDs

All of the zinc finger structures reviewed in this article contain at least one copy of a **CXXC** sequence at the metal binding site. As indicated above, many of these segments adopt similar backbone conformations when bound to zinc, with the NH protons of residues $X(i + 2)$ and $Cys(i + 3)$ oriented in a manner consistent with hydrogen bonding to the $Cys(i)$ sulfur. In addition, many of the structures exhibit Type-II NH—S folding for the subsequent $Cys(i + 3)$-$X(i + 4)$-$X(i + 5)$ residues. In fact, the folding of the entire **CXXCXX** segments of rubredoxins, of the CCHC zinc fingers, of the CCHH zinc fingers from Zif-268 and ADR-1, of the N-terminal **CXXCXX** array of glucocorticoid receptor DNA-binding domain all adopt highly similar backbone and Cys-side chain conformations (Figs. 7-3 and 7-8). Clearly, the formation of a Type II NH—S turn by residues $Cys(i + 3)$-$X(i + 4)$-$X(i + 5)$ does not depend on the presence of a Gly at position $X(i + 4)$, even though the ϕ-value for the $X(i + 4)$ residue in a Type II NH—S turn is positive. The NH—O hydrogen bond between the $X(i +$

4) amide and the Cys(i) backbone carbonyl apparently compensates for the unfavorable steric interactions between the X($i + 4$) side chain and the Cys($i + 3$) carbonyl (35).

Heteronuclear ^{1}H-metal correlated NMR studies of ^{113}Cd- and ^{199}Hg-substituted forms of *P. furiosus* rubredoxin have provided the most detailed insights into the nature of the NH—S hydrogen bonds in the metal center. Protons that participate in Types I and III NH—S hydrogen bonding exhibited hydrogen bond mediated scalar coupling to both ^{113}Cd and ^{199}Hg, providing direct evidence that these protons are indeed H-bonded to the cysteine sulfurs and demonstrating that the NH—S hydrogen bonds contain significant covalent character. The ^{1}H-metal scalar coupling is substantially weaker for the protons involved in Type II NH—S hydrogen bonding, despite the fact that these protons exhibit the slowest chemical exchange with solvent deuterons.

Very recently, NH—S hydrogen bond mediated scalar coupling has been observed in our laboratory for a ^{113}Cd-substituted LIM domain (unpublished) and by Gardner, Coleman and coworkers for ^{113}Cd-GAL4 (personal communication). Such coupling is likely to be detected only via application of a 1D or 2D heteronuclear-correlated NMR experiment since the small magnitude of the couplings will most likely preclude the observation of splitting of normal 1D ^{1}H signals. Of course, the detection of such coupling can only be achieved for metal-substituted forms of the proteins since the zinc nucleus does not possess a stable isotope with a nuclear spin quantum number of 1/2. Future studies with other classes of zinc domains are likely to lead to similar findings for what appear to be important determinants of stability in miniglobular Zn-NAIDs.

Summary

Nuclear magnetic resonance spectroscopy has played a major role in developing an understanding of the structures and nucleic acid interactive properties of zinc finger domains. The zinc finger motif appears to be nature's choice for providing a highly stable miniglobular domain using a minimal number of amino acid residues. Structures within a given classification, such as those of the classical CCHH zinc fingers, adopt a common backbone fold, and the specificity and function of the domain is determined by residues at key surface positions. The remarkable stability of these small domains appears to be attributable, at least in part, to the extensive network of NH—S hydrogen bonds. Indeed, the heteronuclear magnetic resonance experiments reviewed in this chapter indicate that these NH—S hydrogen bonds contain significant covalent character. Such hydrogen bonds were observed originally in the iron binding site of rubredoxins, and have now been observed in every class of zinc finger motif. The conserved Cys-X-X-Cys-Gly sequence could actually serve as a predictive tool for identifying potential metal-binding sites in proteins of unknown structure.

8

Understanding the Mechanisms and Rates of Enzyme-Catalyzed Proton Transfer Reactions to and from Carbon

John A. Gerlt

Some readers might think it oversimplified to begin a chapter on the mechanisms of enzyme-catalyzed reactions by noting that enzymes, be they proteins or nucleic acids, are extremely efficient (rapid) and specific (stereospecific) catalysts. Furthermore, some readers might also think that it is obvious that both the efficiency and the specificity of biological catalysts can be explained by the interactions of functional groups in the active site with the substrate, product, reaction intermediates, and the transition states through which these species are interconverted. However, even for perhaps the "simplest" enzyme-catalyzed reactions, acid-base reactions in which protons are transferred heterolytically to and from carbon atoms, a *quantitative* understanding of the relationship between active site structures and both the rates and mechanisms of the reactions has not been available. In this chapter, unifying principles are described for understanding the rates and mechanisms of enzyme-catalyzed reactions, with primary emphasis placed on proton transfer reactions to and from carbon. These principles may be extended to certain other classes of enzyme-catalyzed reactions that also involve heterolytic bond cleavage (Gerlt & Gassman, 1993a).

The mechanisms of many reactions (either enzyme catalyzed or nonenzymatic) can be written either as (1) concerted processes in which bound substrate is converted directly to bound product *without* the formation of any transiently stable intermediate, or (2) stepwise processes in which bound substrate is converted *first* to a transiently stable intermediate and *then* in a subsequent reaction to bound product.

In the biochemical literature and even in textbooks of biochemistry, the distinction between concerted and stepwise mechanisms is often blurred. In fact, the hybrid term "transition state-intermediate" is commonly encountered in descriptions of the mechanisms of various enzyme-catalyzed reactions. However, a clear distinction between a concerted mechanism with a single transition state and no intermediates and a stepwise mechanism with a single intermediate and two transition states, one for formation and one for decomposition of the intermediate, is important. In stepwise mechanisms that proceed via a transiently stable intermediate, functional groups in the active site would be able to "grasp" the intermediate by the formation of hydrogen bonds or electrostatic interactions, thereby stabilizing the intermediate. By the Hammond postulate (Hammond, 1955), this stabilization will lower the energies of the transition states for formation and decomposition of the intermediate, thereby increasing the rate of the

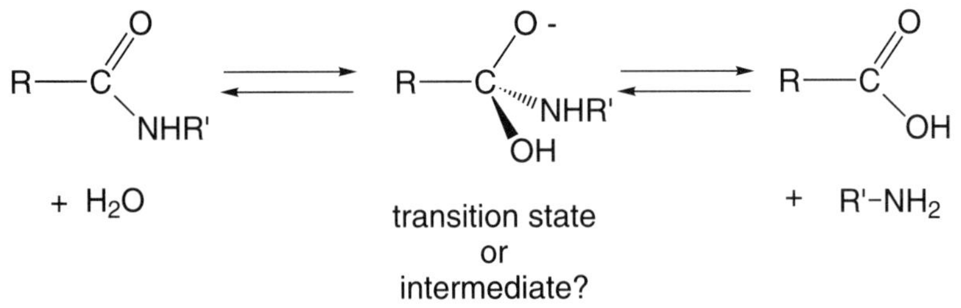

$$R-C(=O)-NHR' + H_2O \rightleftharpoons R-C(O^-)(NHR')(OH) \rightleftharpoons R-C(=O)-OH + R'-NH_2$$

transition state
or
intermediate?

Figure 8-1. The general mechanism for acyl transfer reactions in which the a tetrahedral species may be either a transition state or an intermediate.

reaction relative to the rate in the absence of the enzyme. In contrast, in a concerted mechanism, the functional groups in the active site must stabilize the transition state even though it has, at best, a fleeting lifetime. Structural explanations for the differential stabilization of a transition state relative to a bound substrate are not as obvious as explanations for the stabilization of an intermediate relative to a bound substrate. Despite these differences and the accompanying difficulties in understanding the possible magnitudes of rate accelerations for concerted enzyme-catalyzed reactions, concerted mechanisms have been proposed for enzyme-catalyzed reactions since they appear to avoid the formation of intermediates that are too unstable to exist in aqueous solution.

The Transition State-Intermediate Dilemma

Three related examples of this mechanistic dilemma can be readily identified (Gerlt & Gassman, 1993a).

1. The hydrolysis of a peptide bond catalyzed by a protease can be envisioned to proceed either via a tetrahedral transition state in which a bond from the acyl carbon is being made to the nucleophilic water molecule as the bond to the amine leaving group is being broken, or a tetrahedral intermediate in which both the nucleophile and the leaving group are stably bonded to the acyl carbon atom (Fig. 8-1). Although the existence of a tetrahedral intermediate is well documented for nonenzymatic reactions of carboxylic esters and amides (Guthrie, 1974), the evidence for the formation and stabilization of a tetrahedral intermediate in enzyme active sites is less direct.

2. The hydrolysis of a phosphodiester bond catalyzed by a nuclease can analogously proceed either via a pentacoordinate transition state in which a bond from the phosphorus atom to the nucleophilic water is made as the bond to the hydroxyl leaving group is broken, or a phosphorane intermediate in which both the nucleophile and leaving group are stably bonded to the phosphorus atom (Fig. 8-2). While recent nonenzymatic studies have pointed to the importance of monoanionic phosphorane intermediates in imidazole-catalyzed hydrolysis and transesterification reactions of ribonucleic acids (Breslow, 1991), no definitive evidence for the formation and stabilization of phosphorane intermediates in the active sites of nucleases is yet available.

3. The racemization of a chiral molecule catalyzed by a racemase can proceed via either a transition state in which two simultaneous proton transfers occur from and to the chiral atom, or an electron rich enolic intermediate in which the proton is first abstracted from the substrate to form an intermediate that is then protonated on the opposite face to form

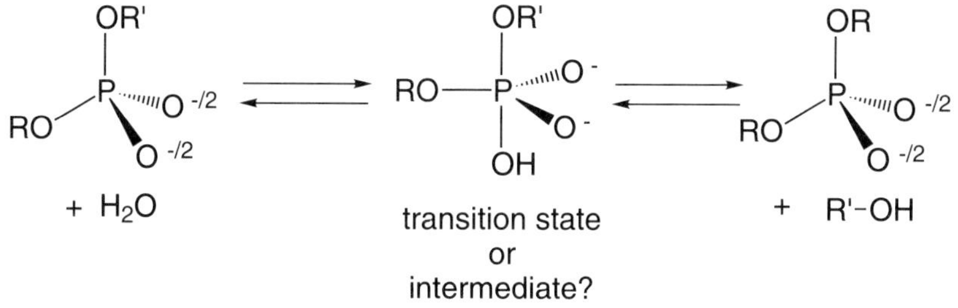

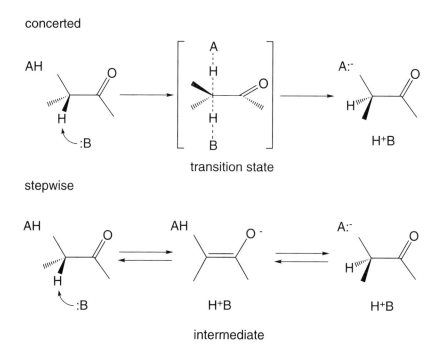

Figure 8-2. The general mechanism for displacement reactions of phosphodiesters in which a pentacoordinate species may be either a transition state or an intermediate.

product (Fig. 8-3). In contrast to the stabilities of tetrahedral and phosphorane intermediates in at least some nonenzymatic reactions, the enolate anions (also called carbanions) that might occur as intermediates can be too unstable to exist in aqueous solution and, as a result, the mechanisms of at least some enzyme-catalyzed proton abstraction reactions from carbon are thought to proceed via concerted pathways (Thibblin & Jencks, 1979).

The actual mechanism of an enzyme-catalyzed reaction will be determined by the relative activation energy barriers for the concerted and stepwise pathways (Jencks, 1981). As suggested previously, these, in turn, will be determined by the strengths of the interactions of the functional groups in the active site with the substrate and either the transition state or intermediate.

Figure 8-3. General mechanisms for proton transfer reactions from and to carbon atoms in which the two proton transfer reactions are either concerted (*top*) or stepwise (*bottom*).

Prior to the availability of high resolution structures of enzymes, mechanistic enzymologists rarely worried about the distinction between concerted and stepwise mechanisms, since they had little idea as to the identities and positions of the functional groups present in active sites. Therefore, they could neither analyze the relative energetics of the two pathways nor devise experiments that would unequivocally determine the actual mechanistic pathway. Presumably, it is for this reason that the mechanistically confusing term "transition state-intermediate" noted previously crept into biochemical thinking. However, the number of structures available for enzymes is increasing at a dramatic rate, and site-directed mutagenesis has become a routine experimental tool in mechanistic enzymology laboratories. Thus, mechanistic enzymologists are now able both to determine the identities and positions of functional groups in active sites and to design and construct structurally conservative replacements that differ from the wild-type functional groups in, for example, acid-base properties. It is now both possible for, and incumbent upon, enzymologists to make the distinction between concerted and stepwise mechanisms and, as a corollary, to understand the absolute rates of (i.e., the activation energy barriers or, equivalently, the energies of the transition states in) enzyme-catalyzed reactions.

The mechanisms of acyl transfer reactions (Fig. 8-1), displacement reactions of phosphodiesters (Fig. 8-2), and proton transfers from and to carbon atoms (Fig. 8-3) are similar in one respect that is likely to be important in determining the mechanisms of enzyme-catalyzed reactions. In each of these reactions, the anionic intermediate that would be formed in a stepwise reaction is significantly more basic than the substrate. In acyl transfer reactions (Fig. 8-1), the pK_a of the conjugate acid of the carbonyl oxygen of the substrate is ~ -4 as compared to ~ 13.4 for the neutral tetrahedral intermediate (Guthrie, 1974). In displacement reactions of phosphodiesters (Fig. 8-2), the pK_as of the conjugate acids of the substrate phosphodiester anion are ~ -3.5 and ~ 1.6 as compared to 7.2 and 12 for the neutral phosphorane intermediate (Guthrie, 1977). In proton transfer reactions, which always occur via abstraction of a proton adjacent to a carbonyl or carboxylic acid group (α-proton of a carbon acid) (Fig. 8-3), the pK_a of the conjugate acid of the carbonyl oxygen ranges from ~ -2 (aldehydes and ketones) to ~ -8 (carboxylic acids) as compared to a range of ~ 6.5 to 10 for neutral enols (Chiang & Kresge, 1991; Gerlt & Gassman, 1993b). The increased basicity of the potential reaction intermediate relative to the substrate can provide the "handle" by which the enzyme active site can grasp the intermediate and channel the mechanism to a stepwise pathway that is lower in energy than either the stepwise pathway in the absence of the enzyme active site (i.e., in aqueous solution, the uncatalyzed reaction) or the concerted pathway in which the formation of any intermediate is avoided. In the opinion of the author, this mechanistic attribute serves as a very useful focal point in deducing unifying mechanistic principles for a variety of enzyme-catalyzed reactions.

In the remainder of this chapter, the rates and mechanisms of enzyme-catalyzed proton transfer to and from carbon acids will be examined.

Types of Enzyme-Catalyzed Proton Transfer Reactions to and from Carbon

Many types of enzyme-catalyzed reactions involve heterolytic abstraction of the α-proton of a carbon acid. These include 1,1-, 1,2-, and 1,3-migrations of protons, β-elimination reactions, and Claisen condensations (Fig. 8-4).

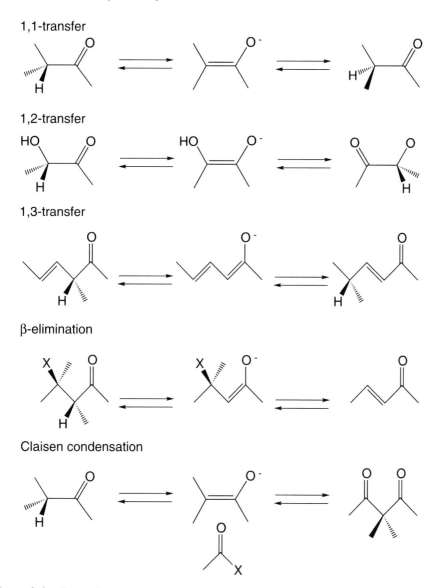

Figure 8-4. Types of enzyme-catalyzed proton transfer reactions.

Each of these reactions is initiated by the abstraction of the α-proton of the carbon acid substrate. The usual rationale for the ability of an active site base to accomplish the abstraction of the α-proton is that the pK_a of the α-proton of a carbon acid (18 to 20 for aldehydes, ketones, and thioesters, 22 to 25 for carboxylic acids, and 29 to 32 for carboxylate anions; Gerlt et al., 1991) is much less than that for the pK_a of an unactivated carbon bound proton (≥50 for alkanes). The lower pK_as of the "activated" α-protons of carbon acids can be explained by the fact that the product of proton abstraction is a resonance-stabilized enolate anion rather than a highly unstable "naked" carbanion (Fig. 8-3).

The 1,1-proton transfer reactions catalyzed by racemases appear to be the simplest proton transfer reactions, since a proton is removed from one face of the substrate and a proton (either the same substrate-derived proton or a new solvent-derived proton) is delivered to the same carbon atom but on the opposite face. In fact, one might imagine that these could be catalyzed by a *concerted* mechanism in which proton abstraction from carbon occurs in the same *transition state* as proton delivery to carbon (Fig. 8-3). However, this concerted mechanism does not take advantage of the activation provided by the carbonyl-carboxylic acid group: in a *truly* concerted mechanism, *no* negative charge develops on the carbonyl-carboxylic acid oxygen. Instead, the developing negative charge is located along the reaction coordinate for the proton transfer reaction. As a result, the pK_a of the proton being abstracted is not that of the α-proton of a carbon acid but should approach that of an alkane. For this reason, these "simple" reactions presumably proceed via mechanisms in which an intermediate is on the reaction coordinate. The reactions catalyzed by racemases do offer the distinct experimental advantage that the equilibrium constant for the reaction *in solution* is necessarily unity. This allows the reaction to be studied readily in either direction (e.g., R to S or S to R).

In 1,2 and 1,3-proton migration reactions, the influence of the carbonyl-carboxylic acid group on the pK_a of the α-proton is more readily apparent. In these reactions, abstraction of the α-proton can be envisaged to result in the formation of an enolate anion (Fig. 8-4). In 1,2-proton migration reactions, the anion that is generated is a enediolate anion, since a hydroxyl group is also located on the α-carbon of the substrate carbon acid (for example, the reaction catalyzed by triose phosphate isomerase). In 1,3-proton migration reactions, the anion that is generated is a dienolate anion, since a double bond is adjacent to the α-carbon of the substrate carbon acid (for example, the reaction catalyzed by Δ^5-3-ketosteroid isomerase). Subsequent protonation of the enolate anion on carbon results in the 1,2- or 1,3-proton migration. In the enzyme-catalyzed reactions that have been studied, 1,2- and 1,3-proton migrations *always* occur via a *syn* stereochemical course (Schwab & Henderson, 1990; Creighton & Murthy, 1992). Furthermore, the same proton that is abstracted from the α-carbon is transferred, at least in part, to the product, implying the formation of a transiently stable enolate intermediate.

A *priori*, β-elimination reactions can occur via either a concerted mechanism (an E2 mechanism) in which abstraction of the α-proton and departure of the β-leaving group occur in the same transition state or a stepwise mechanism (e.g., an E1cb mechanism) in which abstraction of the α-proton results in the formation of an enolate anion. The stereochemical courses of enzyme-catalyzed β-elimination reactions are observed to be *syn* when the substrate is an aldehyde, ketone, imine, or thioester and *anti* when the substrate is a carboxylate anion (Creighton & Murthy, 1992; Gerlt & Gassman, 1992). However, neither stereochemical outcome places any restriction on the timing of the abstraction of the α-proton and departure of the β-leaving group, that is, the reactions can be either concerted or stepwise. In analogy to the discussion of 1,1-proton transfer reactions, the E2 mechanism does not appear to take full advantage of the activation provided by the carbonyl group. Thus, a stepwise mechanism may be expected in which an enolate anion intermediate is formed as the result of abstraction of the α-proton and prior to vinylogous elimination of the β-leaving group.

In those cases for which stereochemical information is available, Claisen condensations can also be envisaged to occur via a concerted mechanism in which proton abstraction from carbon occurs in the same *transition state* as carbon-carbon bond formation. For example, in the reaction catalyzed by citrate synthase, the condensation of acetyl CoA with oxalacetate to form citrate proceeds with inversion of configuration at the methyl carbon (Kluger, 1990; Creighton & Murthy, 1992). Again, a truly concerted mechanism cannot take advantage of the activation provided by the carbonyl group of the thioester since no negative charge develops on this oxygen. Thus, a stepwise mechanism can be expected in which an enolate anion intermediate is formed between abstraction of the α-proton and formation of the carbon-carbon bond.

Enolate Anions Cannot Be Intermediates in Stepwise Proton Transfer Reactions

Despite the considerable acidifying effect of a carbonyl or carboxylic acid group on the α-proton ($\sim$20–30 pK_a units) and the conclusion that the various types of enzyme-catalyzed proton transfer reactions are expected to proceed via intermediate enolate anions as described in the previous section, the reduction in pK_a of the α-proton by formation of an enolate anion is insufficient to allow both the enolate anion to be an intermediate *and* the k_{cat}s of the enzyme-catalyzed reactions to be as rapid as actually observed (Gerlt & Gassman, 1993a, b; Thibblin & Jencks, 1979). The reason for this very important conclusion is as follows. While the pK_as of the α-protons of carbon acids are reduced relative to alkanes (18 to 20 for aldehydes, ketones, and thioesters, 22 to 25 for carboxylic acids, and 29 to 32 for carboxylate anions), they are still considerably greater than the pK_as of the conjugate acids of the general base catalysts in active sites. In virtually every case known to this author, the pK_a of the conjugate acid of the general base residue is less than $\sim$7. From a teleological point of view, a pK_a less than $\sim$7 allows the functional group to be in the proper state of protonation so that it can function as a general base at neutral pH. However, for most carbon acid substrates the difference in pK_a between the carbon acid and the conjugate acid of the general base catalyst (ΔpK_a) is too large to allow a sufficient concentration of an enolate anion intermediate to accumulate so that the observed k_{cat} can be explained, that is, the enolate anion is not kinetically competent. This conclusion follows from the fact that the values for the k_{cat}s of most enzyme-catalyzed proton abstraction reactions occur in the range from 10 sec^{-1} to 10^4 sec^{-1}. From transition state theory and the assumption that the rate of formation of an enolate anion intermediate is determined only by the thermodynamics for its formation (Fig. 8-5), this range of values for k_{cat} requires that the ΔpK_a range from 9 to 12 (i.e., the activation energies for the enzyme-catalyzed reactions fall in the range 13 to 17 kcal/mol).

For example, for triose phosphate isomerase (TIM) the k_{cat} is 600 sec^{-1} when dihydroxyacetone phosphate (DHAP) is substrate and 8,300 sec^{-1} when glyceraldehyde 3-phosphate (G3P) is substrate (Blacklow & Knowles, 1990). The pK_a of the α-proton of G3P is estimated as $\sim$18, that of DHAP is estimated as $\sim$20, and the pK_a of the active site general base catalyst, Glu165 (Coulson et al., 1970; Waley et al., 1970; Hartman, 1971), is measured as $\sim$6 (Plaut & Knowles, 1972; Hartman & Ratrie, 1977). The measured ΔpK_as are $\sim$12 (G3P) and $\sim$14 (DHAP); those predicted from the k_{cat}s are $\sim$9 (G3P) and $\sim$10 (DHAP). In this case, the k_{cat}s of the reactions catalyzed by TIM are *only* 10^3–10^4 greater than that predicted from the ΔpK_a.

However, the difference between the ΔpK_as calculated from the value for the k_{cat} and pK_as of the substrate and the conjugate acid of the active site general base cata-

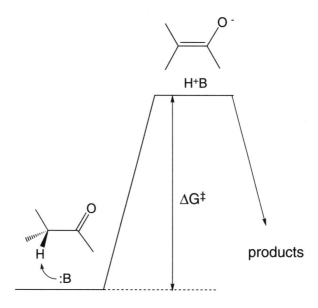

$$\Delta G^{\ddagger} = \Delta G^{\circ} = -2.303 \, RT \, \Delta pK_a$$

Figure 8-5. A free energy diagram for a reaction involving the formation of an enolate anion as an intermediate assuming that the $\Delta G^{\ddagger}$ is determined only by the thermodynamic barrier (ΔG°) associated with transferring a proton from the substrate carbon acid to the general base catalyst.

lyst can be much greater. Using mandelate racemase (MR) as an example, the k_{cat} is ~500 sec^{-1} when either enantiomer of mandelate is substrate (Whitman et al., 1985; Powers et al., 1991). The pK_a of the α-proton of mandelate anion is estimated as ~29 (Gerlt et al., 1991), and the pK_as of Lys166 and His297, the active site general base catalysts, are ~6 (Landro et al., 1991; Kallarakal et al., 1994). Thus, the measured ΔpK_a is ~23, but that predicted from the k_{cat} is ~10. In this case, the k_{cat} of the reactions catalyzed by MR is *10^{13} greater* than that predicted from the ΔpK_a.

From this analysis, enolate anions cannot be intermediates in stepwise mechanisms for enzyme-catalyzed proton abstraction. However, this simplified analysis does not provide any insight into whether the mechanisms of the enzyme-catalyzed reactions are concerted, thereby avoiding the formation of enolate anions which are too unstable to be intermediates, or stepwise with kinetically competent intermediates that are stabilized by interactions with active site functional groups.

In the following sections of this chapter, the reactions catalyzed by TIM and MR will be considered in detail. In the next section, the experiments that allowed construction of the free energy diagram for the 1,2-proton transfer reaction catalyzed by TIM will described and the implications of that diagram will be discussed. Then the structures of the active sites of both TIM and MR will be discussed, since these provide the rationale for the site-directed mutagenesis experiments that have allowed the general base and electrophilic (general acid) catalysts to be identified. Finally, a proposal for understanding the rapid rates of the reactions catalyzed by TIM and MR in the context of their active site structures will be described.

Although this chapter focuses on the reactions catalyzed by TIM and MR, mechanistic and structural data are available for a number of other enzyme-catalyzed reactions that involve abstraction of the α-protons of carbon acids, for example, citrate synthase and ketosteroid isomerase. TIM and MR were chosen for detailed discussion in this chapter since (1) TIM has been considered a structural and mechanistic paradigm for enzymes that catalyze abstraction of the α-protons of carbon acids, and (2) MR is studied in the author's laboratory.

Triose Phosphate Isomerase: The Free Energy Diagram for an Enzyme-Catalyzed Proton Transfer Reaction

TIM catalyzes a 1,2-proton transfer reaction, the interconversion of dihydroxyacetone phosphate (DHAP) and glyceraldehyde 3-phosphate (G3P).

$$\text{DHAP} \rightleftharpoons \text{G3P}$$

DHAP and G3P are the products of the aldolase-catalyzed cleavage of fructose 1,6-bisphosphate. Since the catabolism of G3P to pyruvate (or acetyl CoA or CO_2) provides a major source of metabolic energy (in the form of ATP and NADH), nature undoubtedly has been subject to considerable selective pressure to isomerize DHAP to G3P efficiently (Albery & Knowles, 1976b).

The stepwise nature of the 1,2-proton transfer reaction catalyzed by TIM was first established by Rose (Rieder & Rose, 1959) and later characterized in considerable detail by Knowles and Albery (summarized in Albery & Knowles, 1976a). The structure of TIM was first elucidated in Phillips' laboratory (Banner et al., 1975), and, more recently, high resolution structures of wild-type TIM from various sources as well as site-directed mutants have been solved in Petsko's laboratory (Davenport et al., 1991; Komives et al., 1991). This combination of mechanistic and structural information has provided much insight into the mechanism of the proton transfer reactions catalyzed by TIM.

Early Evidence for a Stepwise Proton Transfer Mechanism

A stepwise proton transfer mechanism for the reaction catalyzed by TIM was established by Rieder and Rose (1959). Three key observations were made that implicate the necessary participation of a transiently stable intermediate.

1. When DHAP is incubated in $^3HO^1H$ the presence of TIM, the equilibrated mixture of DHAP and G3P contains 1 g-atom of ^3H per mole of trioses (96% DHAP and 4% G3P). The ^3H in the DHAP cannot be exchanged with solvent hydrogen by aldolase, thereby establishing that the aldolase and TIM-catalyzed reactions are stereospecific, but with the opposite stereospecificity at the 1-carbon.

2. When DHAP that is tritiated by TIM is used as substrate for TIM under irre-

versible conditions, that is, the G3P so produced is oxidized to 3-phosphoglycerate by G3P dehydrogenase, the G3P that is formed contains almost no ^{3}H.

3. As a control, when DHAP that is tritiated by aldolase is used as substrate for TIM under irreversible conditions, the G3P that is formed contains 1 g-atom of ^{3}H. This control demonstrates that the exchange observed in the presence of TIM must be enzyme-catalyzed.

The nearly complete exchange of ^{3}H from DHAP catalyzed by TIM implicates the formation of an intermediate (presumably a *cis*-enediolate) that is sufficiently stable that the conjugate acid of the active site general base catalyst can exchange with solvent, thereby accounting for the lack of transfer of ^{3}H from substrate DHAP to product G3P.

The Free Energy Diagram

Knowles and Albery and their coworkers extended these studies by carefully quantitating (1) the amount of ^{3}H remaining in unreacted labeled substrate (DHAP) or appearing in product (G3P) and (2) the amount of ^{3}H incorporated from solvent into unreacted substrate or appearing in product. From these studies, a (nearly complete) free energy diagram was constructed for the reaction catalyzed by TIM (Albery & Knowles, 1976a); this was the first such diagram for an enzyme-catalyzed reaction.

As noted in the previous section, Rose and Rieder observed that essentially no ^{3}H from DHAP labeled with TIM was found in the G3P product. Knowles and Albery and their coworkers (Herlihy et al., 1976) quantitated the amount of this intramolecular transfer more carefully and found that between 3 and 6% of the ^{3}H from DHAP was transferred to G3P, with greater transfer being observed at greater extents of reaction. This experiment demonstrates that the exchange of the conjugate acid of the general base catalyst that abstracts the proton-triton from DHAP competes effectively but not completely with transfer of the proton-triton to the transiently stable intermediate to form G3P. Thus, the relative rates (or transition state energies) for the exchange and intramolecular transfer reactions can be calculated.

Knowles and Albery and their coworkers also studied the incorporation of solvent ^{3}H into both substrate and product. Tritium from solvent can be incorporated into substrate and product only by exchange of the conjugate acid of the general base catalyst with solvent. Since this exchange reaction is possible only when the transiently stable enediolate intermediate is present in the active site, the rate of appearance of ^{3}H in substrate can be used to measure the relative rates at which the enediolate intermediate is converted to substrate and product (Maister et al., 1976). If, for example, ^{3}H is "slowly" incorporated into unreacted substrate as the substrate is consumed and product is formed, the intermediate partitions to product with incorporation of ^{1}H from the solvent much more frequently than it partitions to substrate as revealed by incorporation of ^{3}H (Fig. 8-6, curve a). (Since ^{3}H is used as a tracer, most of the solvent hydrogen is ^{1}H and the vast majority of the substrate is necessarily converted to product by incorporation of ^{1}H.) In contrast, if ^{3}H is "rapidly" incorporated into unreacted substrate as the substrate is consumed and the product is formed, the intermediate partitions to substrate with incorporation of ^{3}H much more frequently than it partitions to product with incorporation of ^{1}H (Fig. 8-6, curve b). In addition, the amount of isotopic discrimination against ^{3}H in formation of the product can be used to determine whether proton transfer occurs in the rate-limiting transition state in conversion of the intermediate to product: no discrimination (i.e., $k_H/k_T = 1$) implies that a proton transfer reaction is not rate-limiting, whereas discrimination (i.e., $k_H/k_T > 1$) implies that a pro-

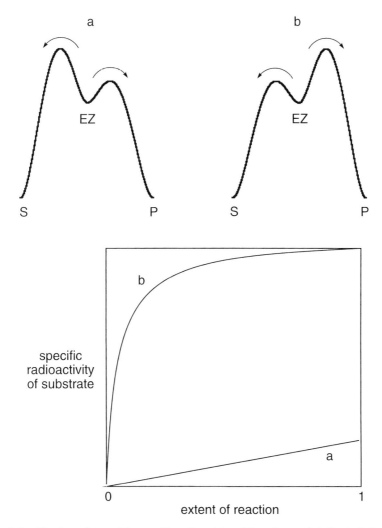

Figure 8-6. The dependence of the specific radioactivity of the substrate (relative to that of the solvent) in the reaction catalyzed by TIM assuming that in (a) the intermediate partitions to substrate with incorporation of ^3H more slowly that it partitions to product with incorporation of ^1H; and in (b) the intermediate partitions to substrate with incorporation of ^3H more rapidly that it partitions to product with incorporation of ^1H. Adapted from Maister et al. (1976).

ton transfer reaction is at least partially rate-limiting. Taken together, these experiments allow the relative energies of the transition states in the free energy diagrams for incorporation of ^1H and ^3H into the intermediate to be compared, thereby allowing the free energy diagrams for incorporation of either ^1H or ^3H to be deduced.

When DHAP is used as substrate, ^3H is incorporated into the intermediate to reform DHAP more slowly than ^1H is incorporated into the intermediate to form G3P (Maister et al., 1976), that is, their data qualitatively resembled curve a in Fig. 8-6. Quantitative analysis of the data revealed that the intermediate partitions to product with incorporation of ^1H at three times the rate it partitions to substrate with incorporation of

^{3}H. Furthermore, the specific radioactivity of the product G3P was approximately equal to the specific radioactivity of the solvent. Since ^{1}H and ^{3}H are incorporated into the intermediate at equal rates to form product, the rate limiting step cannot be the proton transfer reaction in which the intermediate is converted to G3P bound to the enzyme since this would be subject to a kinetic deuterium isotope effect of ~9; instead, it must be diffusion of the G3P from the active site of the enzyme.

These data can be summarized with the following description of the relative rates, where the intermediate (EZ or EZ′) can accept ^{1}H or ^{3}H from the conjugate acid of the active site base to form DHAP (S or S′) or G3P (P or P′), where the unprimed species S, EZ, and P represent protiated species and the primed species S′, EZ′, and P′ represent tritiated species. The enzyme bound enediolate intermediates EZ and EZ′ are in rapid equilibrium, with this reaction representing the exchange of the conjugate acid of the active site base with solvent hydrogen.

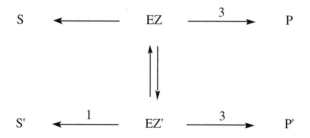

When G3P is used as substrate, ^{3}H is incorporated into the intermediate to reform G3P more slowly than ^{1}H is incorporated into the intermediate to form DHAP (Fletcher et al., 1976), that is, their data again qualitatively resembled curve a in Figure 8-6. In this case, quantitative analysis of the data again revealed that the intermediate partitions to product (now DHAP) with incorporation of ^{1}H at three times the rate it partitions to substrate (now G3P) with incorporation of ^{3}H. In contrast to the situation when DHAP was substrate, the specific radioactivity of the product DHAP was approximately one-ninth the specific radioactivity of the solvent. This suggests that the rate-limiting step is the proton transfer reaction by which the intermediate is converted to DHAP bound in the active site and that dissociation of DHAP from the enzyme must be a faster process.

These data can be summarized with the following description of the relative rates.

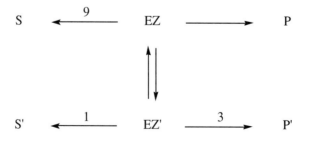

Taken together, these comparisons can be combined to describe the relative rates at which the intermediate (EZ or EZ′) can accept ^{1}H or ^{3}H from the conjugate acid of the active site base to form DHAP (S or S′) or G3P (P or P′).

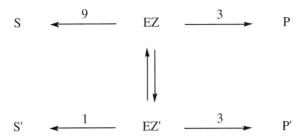

When these data were combined with other kinetic data, the free energy diagram for the 1,2-proton transfer reaction catalyzed by TIM shown in Figure 8-7 was obtained (Albery & Knowles, 1976a). Two noteworthy features are apparent from inspection of this free energy diagram. First, the rate determining transition step is dissociation of G3P (P) from the active site when DHAP (S) is substrate or binding of G3P to the active site when G3P is substrate. The important implication of this result is that the k_{cat} of the reaction catalyzed by TIM is determined by a physical process, that is, the diffusion of G3P to and from the active site. Since enzymes can do little to influence the rate of a diffusion controlled association-dissociation process (other than to provide favorable electrostatic interactions that might direct the substrate to the active site), the k_{cat} cannot be increased by selective pressure, that is, TIM is a "perfect" enzyme that has reached its catalytic limit. Perfection is expected for an enzyme whose function is

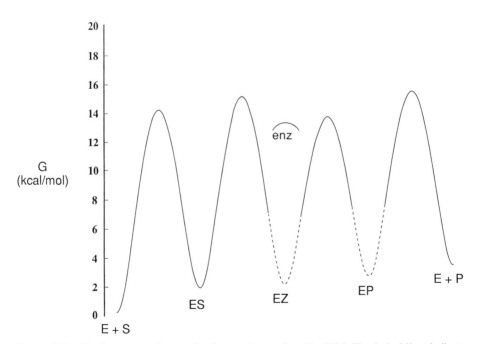

Figure 8-7. The free energy diagram for the reaction catalyzed by TIM. The dashed lines indicate uncertainties in the concentrations of bound enediolate intermediate (EZ) and bound G3P product (EP). The barrier labeled "enz" is that for the exchange of the conjugate acid of Glu165 with solvent hydrogen. Adapted from Albery and Knowles (1976a).

to maximize the flux of a metabolic intermediate. Since the function of TIM is to ensure that DHAP as well as G3P is channeled to pyruvate in glycolysis, the flux of DHAP to G3P should be maximized by natural selection.

Second, the equilibrium constant for the interconversion of DHAP and G3P bound in the active site is closer to unity than the equilibrium constant in solution. Although the concentration of the enediolate intermediate was not measured by Albery and Knowles and is still unknown (dotted line for EZ in Fig. 8-7), Albery and Knowles proposed that the equilibrium constant for formation of the intermediate from bound DHAP or G3P should also be approximately unity. The author notes that this represents a considerable stabilization of the enediolate intermediate by the active site, since, as noted earlier in this chapter, the pK_a of DHAP is ~20, the pK_a of G3P is ~18, and the pK_a of the conjugate acid of the general base catalyst (Glu165, see later) is ~6. If the mechanism of the interconversion of DHAP and G3P catalyzed by TIM only involves general base catalysis by the carboxylate group of Glu165, the equilibrium constant should be ~10^{-14} for formation of the enediolate intermediate from DHAP and ~10^{-12} for formation of the enediolate intermediate from G3P.

From this diagram Albery and Knowles (1976b) proposed that the evolution of enzyme function and the development of catalytic efficiency followed three sequential steps.

1. Uniform binding of the substrate, product, and reaction intermediates. By this process, a decrease in the energy of the rate-limiting transition state is expected, that is, catalysis can be achieved, since substrate, product, and reaction intermediates are stabilized equally by binding to the active site. Uniform binding does not alter the equilibrium constants relating bound substrate, product, and reaction intermediates.

2. Differential binding of the substrate, product, and reaction intermediates. As the result of differential binding, the equilibrium constants relating bound substrate, product, and reaction intermediates are made to approach unity. As a result of differential binding, the transition states relating substrate, product, and reaction intermediates will be decreased in energy, thereby achieving catalysis. This effect is a consequence of the Hammond postulate. In the case of the reaction catalyzed by TIM, differential binding will make the equilibrium constant relating the concentration of bound substrate and product approach unity. Since the equilibrium constant for the interconversion of DHAP and G3P is not much larger than unity in solution, differential binding of DHAP and G3P may be easy to accomplish. However, since the equilibrium constants for interconversion of the enediolate intermediate with both DHAP and G3P is much less than unity (see previous discussion), the enzyme must provide significant differential binding (stabilization) of the enediolate intermediate. A structural explanation for how differential binding of unstable intermediates (enediolates) is accomplished was not provided by Albery and Knowles.

3. Catalysis of elementary steps or, equivalently, differential binding of transition states is the last step. By differential binding of transition states of chemical steps, for example, the proton transfer reactions in the case of TIM, these transition states can be sufficiently stabilized that the transition states for one of the physical steps (substrate binding-product dissociation) is rate limiting. When this situation is achieved, the k_{cat} of the enzyme-catalyzed reaction cannot be increased. The free energy diagram that Albery and Knowles (1976a) established for the TIM-catalyzed reaction is an example of such a reaction.

In a following section dealing with mandelate racemase, an alternate strategy by which the rates of enzyme-catalyzed proton transfer reactions can be optimized will be presented.

Triose Phosphate Isomerase: Active Site Structure and the General Base and General Acid (Electrophilic) Functional Groups

While the free energy diagram for the reaction catalyzed by TIM was being investigated by Albery and Knowles and their coworkers, the three-dimensional structure of TIM from chicken muscle was being solved by Phillips and his coworkers (Banner et al., 1975). Both the dependence of k_{cat} on pH ($pK_a \leq 6$; Plaut & Knowles, 1972) and chemical modification studies (e.g., inactivation by the affinity labels such as 1-halohydroxyacetone phosphate; Hartman, 1971) implicated a carboxylate group as being essential for catalysis and, presumably, the general base catalyst. However, the high resolution structural study of TIM from chicken muscle and subsequent high resolution structural studies of TIM from yeast provided descriptions of the active site of TIM so that the interactions of bound substrate, product, and a stable analogue of the reaction intermediate with active site functional groups can be specified. These structures confirmed not only the identity of the general base catalyst, Glu165, but also revealed the identities of at least one general acid catalyst, His95, and perhaps a second general acid catalyst, Lys12.

Active Site Structure of TIM

The structure of the active site of TIM from yeast complexed with an analogue of the enediolate derived from either DHAP or G3P, phosphoglycolo-hydroxamate (PGH), is shown in Figure 8-8 (Davenport et al., 1991). The structural similarity of PGH and the

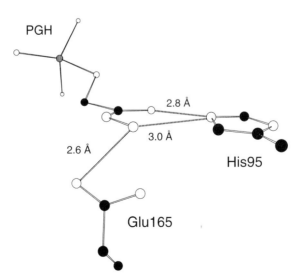

Figure 8-8. The active site of TIM showing the interaction of PGH with Glu165 and His95.

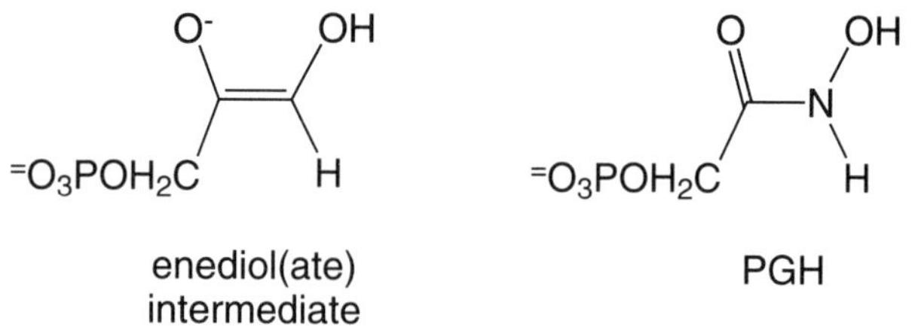

enediol(ate) intermediate

PGH

Figure 8-9. A comparison of the structures of the enediol(ate) intermediate in the reaction catalyzed by TIM and of the intermediate analogue PGH.

enediolate is shown in Figure 8-9; in these structures, the enediolate and hydroxamate are planar. This structure and those of site-directed mutants that are discussed in this section were determined by Petsko and his coworkers.

The carboxylate group of Glu165 is located appropriately to act as the general base catalyst that abstracts the α-proton from either DHAP or G3P, thereby generating the conjugate acid of the general base, the neutral carboxylic acid, and the enediolate. Either rotation of the carboxylic acid group or a prototropic shift will permit the proton to be transferred to the adjacent carbon, thereby accomplishing the 1,2-proton transfer reaction. The exchange of the substrate-derived proton with solvent that is observed to compete with the isomerization reaction involves the neutral carboxylic acid of Glu165; Albery and Knowles have established that the rate-limiting step in this exchange reaction does not involve a transition state in which a proton transfer reaction occurs, that is, the same extent of exchange is observed with ^2H and ^3H (Fisher et al., 1976).

The carbonyl and hydroxamate oxygens of PGH are hydrogen bonded to His95. Based on these interactions, His95 is located appropriately to hydrogen bond to the carbonyl and hydroxyl oxygens of both DHAP and G3P as well as to both oxygens of the enediolate. Thus, His95 has been termed an "electrophilic" catalyst since it has been demonstrated to polarize the carbonyl group of the substrate (either DHAP or G3P) and proposed to stabilize the enediolate intermediate by hydrogen bonding (Komives et al., 1991). Interestingly, the structure of the PGH complex also demonstrates that the imidazole functional group is uncharged since it is also hydrogen bonded to the N—H group of a backbone peptide bond. This conclusion has important implications for understanding both the mechanism and rate of the isomerization reaction catalyzed by TIM.

These proposed functions for both Glu165 and His95 have been investigated in Knowles' laboratory by constructing, and both structurally and mechanistically characterizing, the E165D and H95Q site-directed mutants. The E165D mutant retains the general base carboxylate group but "retracts" it ~ 1 Å from the substrate. The H95Q mutant retains the ability to hydrogen bond to the oxygen atoms of both the substrate and enediolate intermediate derived by abstraction of the α-proton from the substrate.

Glu165, the General Base Catalyst

The high resolution X-ray structure of the E165D mutant revealed that this substitution did not perturb the overall structure of the enzyme: the only change was the desired retraction of the carboxylate group from the binding site for the substrate. The k_{cat} of the E165D mutant is reduced by $\sim 10^3$-fold, demonstrating the importance of the carboxylate group of Glu165 as a general base catalyst (Strauss et al., 1985).

His95, the General Acid (Electrophilic) Catalyst

The high resolution X-ray structure of the H95Q revealed that this substitution also did not perturb the overall structure of the enzyme, but the glutamine side chain was observed to move away from the binding site of the substrate and participate in a hydrogen bond that is not found in the structure of wild-type TIM complexed with PGH (Komives et al., 1991). The k_{cat} of the H95Q mutant is reduced by $\sim 10^2$-fold, suggesting the importance of the imidazole group of His95 as an electrophilic catalyst (Nickbarg et al., 1988). However, careful mechanistic analysis of the reaction catalyzed by H95Q revealed that the mechanism of the reaction has been altered by the substitution, since *no* intramolecular proton transfer from substrate to product can be detected (in contrast to the 3 to 6% transfer observed for wild-type TIM). Thus, the decrease in k_{cat} cannot be used directly to quantitate the importance of the uncharged imidazole group of His95 as an electrophilic catalyst.

The role of the imidazole group of His95 in catalysis has been further characterized by measuring its pK_a by ^{15}N NMR spectroscopy (Lodi & Knowles, 1991). As noted previously, the X-ray studies suggested that it hydrogen bonds to PGH as the neutral imidazole. This conclusion was verified in Knowles' laboratory by determining that His95 cannot be titrated over the accessible pH range in which TIM is stable (5 to 10). The ^{15}N chemical shifts of the imidazole nitrogens of His95 were those of neutral histidine. Since the pK_a of imidazolium ion (to yield neutral imidazole, $pK_a = 7$) and of neutral imidazole (to yield imidazolate anion, $pK_a = 14$) are similarly effected by electron donating or withdrawing substituents, the pK_a of protonated His95 must be more than or equal to 4 and the pK_a of neutral His95 must be less than or equal to 11. Knowles therefore concluded that neutral His95 and not protonated His95 must be the electrophilic catalyst in the active site of TIM. This conclusion was met with surprise by both Knowles and the enzymological community, since electrophilic catalysts are also considered to be proton donors, that is, general acid catalysts. As such, the high pK_a of the functional (neutral) form of His95 was unexpected.

The pKas of the Enediol Intermediate and His95

The pK_a of the *neutral* vicinal enediol derived by tautomerization of either DHAP or G3P is expected to be ~ 11 (Chiang & Kresge, 1991). Knowles proposed that the similar pK_as of the electrophilic catalyst and the neutral enediol would allow rapid proton transfer (Lodi & Knowles, 1991), thereby facilitating the prototropic shift between the oxygens of the enediolate intermediate that is necessary for the 1,2-proton transfer reaction. An alternate explanation for the similar pK_as of the electrophilic catalyst and the enediol intermediate in the active site of TIM is presented below.

Mandelate Racemase: Active Site Structure and the General Base and General Acid (Electrophilic) Functional Groups

As noted above, the 1,1-proton transfer reactions catalyzed by racemases superficially are the simplest enzyme-catalyzed proton transfer reactions, since a proton is removed from one face of the substrate and a proton is delivered to the same carbon atom on the opposite face.

At least three distinct classes of racemases are known in biochemistry (Gerlt et al., 1992).

1. The most familiar racemases are those that utilize pyridoxal phosphate (PLP) as cofactor. These enzymes catalyze the equilibration of the enantiomers of α-amino acids. As an example, alanine racemase is important for microbial metabolism since cell walls contain R-alanine that is derived from S-alanine by the action of alanine racemase. Although an alanine racemase from *Bacillus stearothermophilus* has been crystallized (Neidhart et al., 1987), no high resolution structure is yet available.

2. Various microorganisms contain PLP-independent racemases that also equilibrate the enantiomers of some α-amino acids. Examples of these enzymes include proline racemase, glutamate racemase, and aspartate racemase. These enzymes have no known cofactor requirement and are also thought to be metal ion independent. The active sites of these enzyme contain two general acid-base catalysts that abstract the α-proton from the substrate and deliver a solvent-derived proton to generate the product. In each of these enzymes, these catalysts are thought to be the thiol-thiolate groups of cysteine residues (Cardinale & Abeles, 1968; Gallo & Knowles, 1993; Tanner & Knowles, 1993).

3. Strains of Pseudomonads are able to utilize both enantiomers of mandelate (2-hydroxy-2-phenylacetate) as sole carbon and energy sources. Only the S-enantiomer can be oxidatively degraded to acetyl CoA and succinate; the R-enantiomer must be racemized so that it can also be degraded. This racemase, mandelate racemase (MR), is metal ion dependent, with Mg^{2+}, Co^{2+}, Ni^{2+}, and Mn^{2+} being able to satisfy the metal ion requirement (Kenyon & Hegeman, 1979). The X-ray structure of MR has been determined at high resolution in the presence of a substrate analogue-competitive inhibitor (Landro et al., 1994), so that the interactions of bound substrate enantiomer, product enantiomer, and the presumed reaction intermediate with active site functional groups can be specified. My laboratory has studied structure-function relationships in the active site of mandelate racemase in collaboration with Professors George L. Kenyon, University of California, San Francisco, John W. Kozarich, University of Maryland, and Gregory A. Petsko, Brandeis University. MR is the focus of the remainder of this section.

MR catalyzes the Mg^{2+}-dependent racemization of both enantiomers of mandelate (as it must, since catalysts cannot alter the equilibrium position for a reaction).

R-mandelate ⇌ S-mandelate

Although the Haldane relationship requires only that the values for k_{cat}/K_m be identical for both enantiomers of mandelate, the values for k_{cat} and K_m for racemization of R and S-mandelates are approximately identical, ~ 500 sec^{-1} and 0.25 mM, respectively (Whitman et al., 1985).

Active Site Structure of MR

The overall structure of the active site of MR from *Pseudomonas putida* ATCC 12633 complexed with *S*-atrolactate (α-methylmandelate), a competitive inhibitor and a substrate analogue, is shown in Figure 8-10 (Landro et al., 1994). This high resolution structure (2.1 Å) and those of site-directed mutants that are discussed in this section were determined by Petsko and his coworkers.

S-Atrolactate is coordinated to the essential divalent metal ion (Mg^{2+}) by the α-hydroxyl group and one of the carboxylate oxygens (Fig. 8-11). The same carboxylate oxygen is hydrogen bonded to the ϵ-ammonium group of Lys164. The interactions with the carboxylate oxygen with these two cationic groups are thought to at least partially neutralize the negative charge of the mandelate anion, thereby allowing it to behave electronically more like mandelic acid. Since the pK_a of the α-proton of mandelate anion has been estimated as ~29 (Gerlt et al., 1991) and the pK_a of the α-proton of mandelic acid has been measured to be 22.0 (Chiang et al., 1990), these electrostatic interactions should help catalyze abstraction of the α-proton by an active site general base catalyst. The reasonable assumption is that the same coordination geometry describes the complex of *S*-mandelate with the active site, since the K_i for *S*-atrolactate and the K_m for *S*-mandelate are similar.

This structure also reveals the presence of two general base catalysts that can abstract the α-proton from the substrate enantiomer and, as their conjugate acids, deliver a solvent-derived proton to generate the product enantiomer (Fig. 8-10). The α-methyl group of *S*-atrolactate is pointed toward the side chain of Lys166. The N^ϵ of the imidazole functional group of His297 is on the opposite face of the bound inhibitor. On

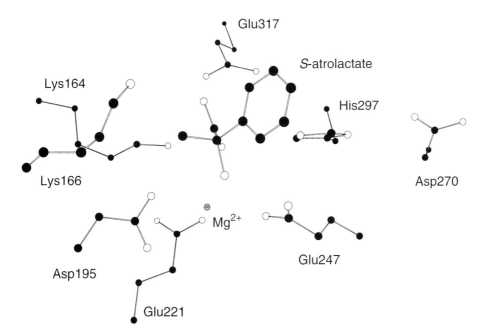

Figure 8-10. The active site of MR complexed with *S*-atrolactate.

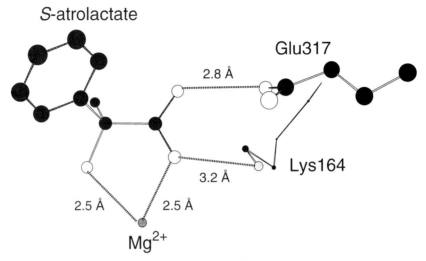

Figure 8-11. The interaction of S-atrolactate with active site functional groups in the active site of MR.

the basis of these geometric relationships and since S-atrolactate and S-mandelate have the same relative configurations, the active site general base catalyst that abstracts the α-proton from S-mandelate can be assigned to the ε-amino group of Lys166 (the S-specific base) and the active site general base catalyst that abstracts the α-proton from R-mandelate can be assigned to N^ϵ of the imidazole group of His297 (the R-specific base).

These assignments are consistent with results that were obtained when the racemization of R and S-mandelates were studied separately in ^2H_2O (Powers et al., 1991).

1. When either R or S-mandelate is substrate, the product enantiomer was found to contain only ^2H in the α-position, that is, the α-hydrogen that enters the active site with substrate does not depart the active site with the product. Instead, the product contains a solvent-derived hydrogen. The simplest mechanism to explain this result is that the active site has two general acid-general base catalysts. One functions as a general base catalyst to abstract the α-proton from the substrate enantiomer, and the second is protonated at the start of the reaction cycle so that it can deliver a solvent-derived proton to generate the product enantiomer. This mechanism, often called a "two-base" mechanism, can be contrasted with the labeling pattern expected for a "one-base" mechanism in which a single general acid-general base catalyst is present in the active site. In the one-base mechanism, the single catalyst first functions as a general base to abstract the α-proton from the substrate enantiomer. Then, after either movement of the conjugate acid of the catalyst to the opposite face of the transiently stable intermediate derived by proton abstraction or flipping of the intermediate, the protonated catalyst delivers the same proton to the intermediate to generate the product enantiomer. In this mechanism, the α-hydrogen that enters the active site with substrate is expected to depart the active site with the product [unless, of course, the conjugate acid of the catalyst can exchange with solvent in competition with protonation of the intermediate to form product as was observed in the reaction catalyzed by TIM (Fisher et al., 1976)].

2. When S-mandelate is substrate, solvent ^{2}H is incorporated into the remaining pool of S-mandelate at a rate comparable to the rate of racemization. When R-mandelate is substrate, no solvent ^{2}H is incorporated into the remaining pool of R-mandelate. This pattern of ^{2}H incorporation provides independent support for the two-base mechanism described in the previous paragraph, since if the one-base mechanism were operative, solvent-derived hydrogens would be incorporated at equal rates into the remaining pools of both R and S-mandelates. However, this pattern of ^{2}H incorporation also provides important information about the chemical constitution of the general base catalyst that abstracts the α-proton from the substrate enantiomer. The lack of incorporation of solvent ^{2}H into the R-enantiomer of mandelate indicates that the conjugate acid of the R-specific general base catalyst is monoprotic; this is consistent with the assignment of His297 as the R-specific base. The rapid incorporation of solvent ^{2}H into the S-enantiomer of mandelate indicates that the conjugate acid of the S-specific general base catalyst is polyprotic; this is consistent with the assignment of Lys166 as the S-specific base.

Lys166 and His297, the General Basic Catalysts

The assignments of Lys166 as the S-specific general base catalyst and His297 as the R-specific general base catalyst have been examined by constructing and both structurally and mechanistically characterizing the K166R and H297N mutants, as well as by studying the inactivation of MR by the enantiomers of the oxirane-containing irreversible inhibitor α-phenylglycidate.

In the H297N mutant of MR, the imidazole functional group is replaced by the weakly acidic-basic carboxamide functional group (Landro et al., 1991). The high resolution X-ray structure of the H297N mutant revealed that the functional group substitution did not perturb the overall structure of the enzyme. The only change was the replacement of the imidazole ring of histidine with the carboxamide group of asparagine, with the carboxamide group in the mutant oriented in the same plane as the imidazole group in the wild-type. The H297N mutant has no detectable racemase activity with either enantiomer of mandelate, thereby supporting the importance of the imidazole functional group and its conjugate acid in catalysis.

However, further assessment of the role of His297 was possible since the H297N mutant catalyzed the rapid abstraction of the α-proton of S-mandelate. The exchange of the α-proton with solvent ^{2}H was studied by NMR spectroscopy. Wild-type MR is obviously able to catalyze the incorporation of solvent ^{2}H into the product enantiomer as the racemization reaction proceeds, so the resonance associated with the α-proton of mandelate decreases in intensity after enzyme is added. When H297N was studied by the same procedure, no change in the intensity of the resonance associated with the α-proton of R-mandelate was observed even after days of incubation. In contrast, the resonance associated with the α-proton of S-mandelate decreased in intensity at one-third the rate at which wild-type MR catalyzed the exchange of the α-proton of S-mandelate. This suggests that the ϵ-amino group of Lys166 remains catalytically competent for abstraction of the α-proton of S-mandelate in the active site of H297N and, importantly, that a transiently stable intermediate is necessarily generated (Fig. 8-12). The lifetime-stability of this intermediate must be sufficient for the ϵ-ammonium group to rotate about the C$^\epsilon$—N$^\epsilon$ bond so that ^{2}H rather than ^{1}H can be delivered to the intermediate from the ^{1}H^{2}H$_2$-ammonium group, thereby exchanging the α-proton with solvent ^{2}H without racemization. The rate of the exchange reaction catalyzed by H297N is consistent with

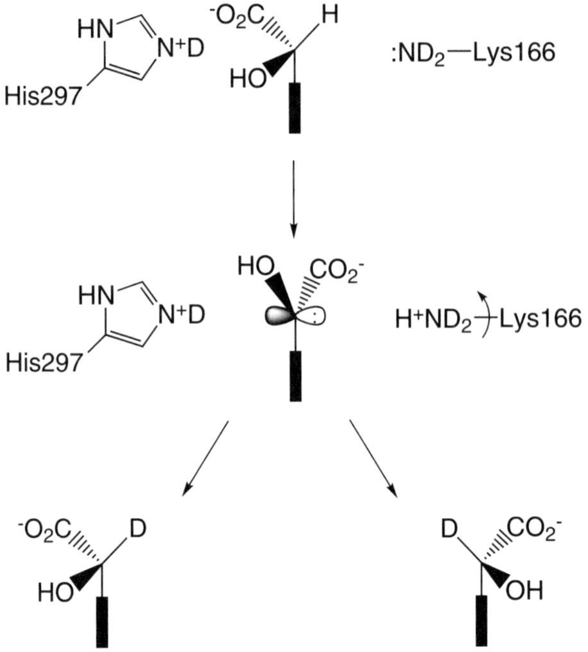

Figure 8-12. The roles of Lys166 and His297 in the reaction catalyzed by MR, showing the involvement of a transiently stable intermediate.

the α-proton of S-mandelate having a $pK_a \approx 15.4$ in the transition state for the proton abstraction by Lys166. Since, in solution, the pK_a of the α-proton of mandelate anion has been estimated as ~29 (Gerlt et al., 1991) and the pK_a of the α-proton of mandelic acid is 22.0 (Chiang et al., 1990), the active site must be able to decrease the pK_a of the bound mandelate by an amount greater than that which might be expected from neutralization of the bound mandelate anion by its interactions with the essential Mg^{2+} and the ε-ammonium group of Lys164. A structural explanation for the decreased pK_a of the α-proton in the transition state for proton transfer will be discussed in the section below.

The H297N mutant was studied further by examining the rate and stereospecificity of the 1,6-elimination of bromide ion from a racemic mixture of p-(bromomethyl)mandelates. Wild-type MR catalyzes the elimination of bromide ion from both enantiomers of p-(bromomethyl)mandelate at a rate approximately 0.005% that of the rate of racemization of the enantiomers of mandelate (Lin et al., 1988; Fig. 8-13). This elimination reaction has been attributed to abstraction of the α-proton to generate an electron rich (enolic) intermediate which can partition between racemization or elimination of bromide ion. H297N catalyzes elimination of bromide ion from only the S-enantiomer of p-(bromomethyl)mandelate at a rate comparable to that measured for the elimination reaction catalyzed by wild-type MR (Landro et al., 1991; Fig. 8-13). [The inert R-enantiomer is solvolyzed to p-(hydroxymethyl)mandelate, and the configuration of this material is used to deduce the configuration of the reactive enantiomer.] This finding is in accord with the interpretation of the experiments described in the previous paragraph, that is, the substitution of histidine by asparagine has "inactivated" the R-specific acid-base catalyst and the ε-amino functional group of Lys166 remains catalytically competent as the S-specific acid-base catalyst.

Analogous experiments have been performed with the K166R mutant of MR, in which the ϵ-amino functional group of lysine is replaced by the δ-guanidino functional group of arginine (Kallarakal et al., 1995). The high resolution X-ray structure of the K166R mutant revealed that the functional group substitution did not perturb the overall structure of the enzyme. The only change was the replacement of the side chain of lysine with the longer side chain of arginine. This difference in length necessarily places the guanidine-guanidinium functional group in suboptimal positions for abstraction of the α-proton from S-mandelate and delivery of a proton to the stabilized enolic intermediate to generate S-mandelate.

In contrast to H297N that has no detectable racemase activity, K166R catalyzes the racemization of both enantiomers of mandelate but with the values for k_{cat} reduced 5 × 10^3-fold in the R to S-direction and 1 × 10^3-fold in the S to R direction. Presumably, the decreased values for k_{cat} reflect the suboptimal positioning of the guanidine-guanidinium functional group of Arg166 for proton transfer reactions. K166R is unable to catalyze the rapid exchange of the α-proton of either substrate enantiomer with solvent ^{2}H. However, this lack of activity is mechanistically uninformative concerning the ability of the imidazole functional group of His297 to rapidly abstract the α-proton of R-mandelate, since the conjugate acid of this general base functional group is monoprotic, that is, N^ϵ is protonated with the α-proton that would be abstracted from R-mandelate. What is more informative is the observation that K166R catalyzes elimination of bromide ion from the R but not the S-enantiomer of p-(bromomethyl)mandelate at a rate that is comparable to that measured for the elimination reactions catalyzed by both wild-type MR and H297N (Fig. 8-13). Thus, the substitution of lysine by arginine has reduced but not totally eliminated the activity of the S-specific acid-base catalyst and the imidazole functional group of His297 remains catalytically competent as the R-specific acid-base catalyst.

Additional evidence that supports the assignment of the ϵ-amino functional group of Lys166 as the S-specific acid-base catalyst was obtained by characterizing the na-

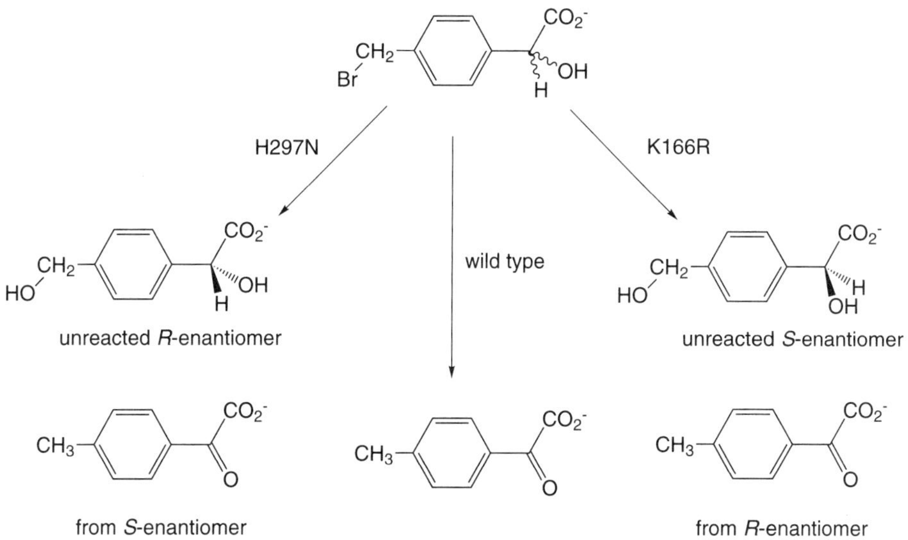

Figure 8-13. The products of bromide ion elimination from a racemic mixture of p-(bromomethyl)mandelates in the presence of wild-type MR, H297N, and K166R.

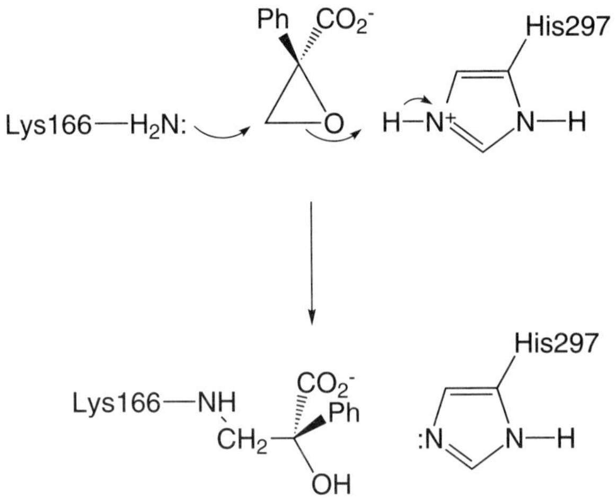

Figure 8-14. A possible mechanism for the inactivation of MR by R-α-phenylglycidate.

ture of the irreversible inactivation of MR by the enantiomers of α-phenylglycidate (Landro et al., 1994; Fig. 8-14). Kenyon and his coworkers first characterized the irreversible inactivation of MR by α-phenylglycidate (Fee et al., 1974). However, from this work the chemical mechanism of the inactivation was unclear, although the inactivation was active site-directed, since it required Mg^{2+} and was competitively inhibited by the enantiomers of mandelate. The enantiomers of α-phenylglycidate have now been stereoselectively synthesized in highly enriched (~80% ee) form (Landro et al., 1994). This enrichment was sufficient to demonstrate that R but not S-α-phenylglycidate is the irreversible inactivator of MR. R-α-Phenylglycidate has the same relative configuration as S-mandelate, with the methylene carbon of the oxirane ring having the same spatial disposition as the α-proton of S-mandelate (Fig. 8-15). The high resolution X-ray structure of enzyme inactivated by R-α-phenylglycidate allows a description of the mechanism of inactivation. MR is inactivated by alkylation of the ε-amino group of Lys166, with the amino group reacting with the less hindered methylene carbon of the oxirane ring (Fig. 8-14). That R-α-phenylglycidate and S-mandelate have the same relative configurations suggests that the ε-amino group of Lys166 should also be oriented properly in the active site to be the S-specific acid-base catalyst.

The pK_as of the conjugate acids of Lys166 (the ε-ammonium group) and of His297 (the imidazolium group) have been estimated by measuring the pH dependence of k_{cat} for both enantiomers. In both the R to S and the S to R directions, the dependence of k_{cat} on pH for wild-type MR is bell-shaped (Landro et al., 1991). These plots are virtually superimposable, with the value of the pK_a of the ascending limb being ~6 and that of the descending limb being ~10. This behavior might suggest the requirement for a general base catalyst (the ascending limb) and a general acid catalyst (the descending limb). However, when the racemization reactions catalyzed by K166R are similarly studied (Kallarakal et al., 1995), in the S to R direction, the pK_as are ~8 and ~10; in the R to S direction, the pK_as are ~6 and ~10. The Lys to Arg substitution alters only the pK_a of the ascending limb when Lys166-Arg166 is the S-specific base by an amount (~2 pK_a units) that is consistent with the difference in the pK_as of the conjugate acids of Lys and Arg in solution. Thus, the simplest, but not necessarily correct, interpretation of these data is that the pK$_a$s deduced from the ascending limbs are those associated with the conjugate acid of the base that abstracts the α-proton of the substrate enantiomer, that is, Lys166 in the S to R direction and His297 in the R to S direction. Since these pK$_a$s have been determined by kinetic methods rather than direct measurement, they are subject to the usual caveats of interpretation (Bruice & Schmir, 1959; Knowles, 1976; Schmidt & Westheimer, 1971).

Glu317, the General Acid (Electrophilic) Catalyst

The structure of the active site of wild-type MR complexed with S-atrolactate (Figs. 8-10 and 8-11) also reveals the potential presence of a third electrophilic catalyst in addition to the Mg^{2+} and the ε-ammonium group of Lys164 that are coordinated to one of the carboxylate oxygens of the bound inhibitor. (Recall that the function of these electrophilic catalysts is to at least partially neutralize the negative charge of the mandelate anion, thereby allowing it to behave electronically more like mandelic acid and reducing the pK$_a$ of the α-proton.) The other carboxylate oxygen of the bound inhibitor is hydrogen bonded to the carboxylic acid functional group of Glu317, since the O-O distance between the inhibitor carboxylate oxygen and the enzyme carboxylate oxygen is 2.8 Å. Is the carboxylic acid group of Glu317 an electrophilic (general acid) catalyst, or is it merely a "spectator" that is involved in substrate binding but not in facilitating the abstraction of the α-proton by Lys166-His297?

The importance of the carboxylic acid group of Glu317 in catalysis has been investigated by constructing and characterizing, both structurally and mechanistically, the E317Q site-directed mutant (Mitra et al., 1995). The high resolution structure of E317Q complexed with S-atrolactate revealed that this substitution did not perturb either the overall structure of the enzyme or any feature of the active site geometry, that is, this structure was indistinguishable from that of wild-type MR complexed with S-atrolactate. E317Q catalyzes the racemization of both enantiomers of mandelate but with the values for k_{cat} reduced 4.5 × 10^3-fold in the R to S direction and 2.9 × 10^4-fold in the S to R direction. These reductions in k_{cat} strongly suggest that the carboxylic acid group of Glu317 participates in catalysis as a general acid that transfers a proton toward the neutralized carboxylate group of bound substrate. This suggestion is supported by the additional observation that E317Q does not catalyze detectable elimination of bromide ion from either enantiomer of p-(bromomethyl)mandelate but is irreversibly inactivated by α-phenylglycidate at a rate comparable to that measured for wild-type MR. Since the elimination of bromide ion requires reduction in the pK$_a$ of the α-proton so that it can be rapidly abstracted, but inactivation by α-phenylglycidate requires only the proper orientation between the ε-amino group of Lys166 and the methylene group of the oxirane ring, these mechanistic attributes further support the involvement of the carboxylic acid group of Glu317 as an electrophilic catalyst.

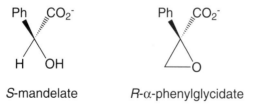

Figure 8-15. A comparison of the structures of S-mandelate and R-α-phenylglycidate.

The pK$_a$s of the Enediol Intermediate and Glu317

The pK$_a$ of the *neutral* geminal enediol (actually an enetriol) derived by tautomerization of mandelic acid has been measured as 6.6 (Chiang et al., 1990). The pK$_a$ of the enediolate monoanion has been estimated to be greater than or equal to 10 (Gerlt et al., 1991). Although the pK$_a$ of Glu317 has not yet been measured directly, a likely value is ≈6. [NMR methods such as those used by Knowles in determining the pK$_a$ of the electrophilic catalyst in the active site of TIM (Lodi & Knowles, 1991) are not expected to be useful in determining the pK$_a$ of Glu317 since MR is an octamer of identical polypeptides with M_r = 38,570 Da.] Although the similar pK$_a$s of Glu317 and the neutral geminal enediol of mandelic acid might allow rapid proton transfer reactions, in analogy with the explanation provided by Knowles for the similar pK$_a$s of the neutral vicinal enediol derived from DHAP/G3P and the neutral imidazole functional group of His95, an alternative explanation is provided in the next section.

Understanding the Relationship between the Mechanisms and Rates of Enzyme-Catalyzed Proton Transfer Reactions: The Quantitative Importance of Electrophilic Catalysis

The active sites of both TIM and MR contain general base catalysts that abstract the α-proton from the substrate to generate an intermediate and deliver a proton to the intermediate to generate the product [the same catalytic group in TIM (Glu165) and two catalytic groups in MR (Lys166 and His297)] *and* general acid (electrophilic) catalysts (His95 in TIM and Glu317 in MR) that are necessary for high catalytic activity. In view of the kinetic and thermodynamic problems associated with enolate anions as intermediates (see discussion earlier in chapter), how can both the rapid rates and chemical mechanisms of these reactions be rationalized? As described in this section, the rates and mechanisms and their interrelationship can be understood when the quantitative importance of the electrophilic (general acid) catalyst is fully appreciated.

Marcus Formalism for Describing Reaction Coordinates

The author and the late Paul G. Gassman analyzed the kinetic factors that determine the rates of proton abstraction from carbon acids (Gerlt & Gassman, 1993a, b). This analysis was based on a Marcus formalism that describes the free energy diagram for a reaction with the equation for an inverted parabola. The shape of the parabola (the free energy, G, as a function of the reaction coordinate, x) is determined by two independent parameters, $\Delta G°$, the thermodynamic barrier to the reaction and $\Delta G^{\ddagger}_{int}$, the intrinsic kinetic barrier to the reaction

$$G = -4\Delta G^{\ddagger}_{int}(x - 0.5)^2 + \Delta G°(x - 0.5) \qquad (8\text{-}1)$$

where the value of x can range from 0 (substrate) to 1 (product). $\Delta G°$ is the instability of the product relative to the substrate. $\Delta G^{\ddagger}_{int}$ is the barrier to reaction that remains when $\Delta G° = 0$. In the case of understanding the rates of enzyme-catalyzed abstraction of the α-protons of carbon acids, the substrate is the bound keto tautomer of the carbon acid and the product is the bound enolic tautomer of the carbon acid.

From this equation, the activation energy barrier to the reaction, $\Delta G^{\ddagger}$, is given by the equation

$$\Delta G^{\ddagger} = \Delta G^{\ddagger}_{int}(1 + \Delta G^{\circ}/4\Delta G^{\ddagger}_{int})^2 \qquad (8\text{-}2)$$

According to this equation, $\Delta G^{\ddagger}$ is larger than ΔG° by an amount that is related to the magnitude of $\Delta G^{\ddagger}_{int}$: the larger the value for ΔG°, the smaller the contribution of $\Delta G^{\ddagger}_{int}$ to ΔG°. The free energy diagrams in Figure 8-16 illustrate relationships among $\Delta G^{\ddagger}$, ΔG°, and $\Delta G^{\ddagger}_{int}$. These diagrams were calculated with equation (8-1), assuming that in (a) $\Delta G^{\circ} = 0$ kcal/mol and $\Delta G^{\ddagger}_{int} = 10$ kcal/mol (the conditions that define $\Delta G^{\ddagger}_{int}$); and in (b) $\Delta G^{\circ} = 10$ kcal/mol and $\Delta G^{\ddagger}_{int} = 10$ kcal/mol. In diagram (a), $\Delta G^{\ddagger} = 10$ kcal/mol [$(\Delta G^{\ddagger} - \Delta G^{\circ}) = 10$ kcal/mol]; in diagram (b), $\Delta G^{\ddagger} = 15.63$ kcal/mol [$(\Delta G^{\ddagger} - \Delta G^{\circ}) = 5.63$ kcal/mol]. The contribution of $\Delta G^{\ddagger}_{int}$ to $\Delta G^{\ddagger}$ is less in diagram (b) than in diagram (a).

Since $\Delta G^{\ddagger}$ is a function of both ΔG° and $\Delta G^{\ddagger}_{int}$, the rate of the enzyme-catalyzed reaction can be increased if the values of ΔG° or $\Delta G^{\ddagger}_{int}$ are decreased. From transition state theory and the measured k_{cat}s for the reactions catalyzed by TIM and MR, the $\Delta G^{\ddagger}$s for these enzyme-catalyzed reactions are ~ 14 kcal/mol.

The Magnitudes of ΔG° and $\Delta G^{\ddagger}_{int}$

How are the values for ΔG° and $\Delta G^{\ddagger}_{int}$ reduced so that the $\Delta G^{\ddagger}$s can be reduced to ~ 14 kcal/mol? The active sites of both TIM and MR contain a general base that abstracts the α-proton from the substrate as well as a general acid (electrophilic) catalyst

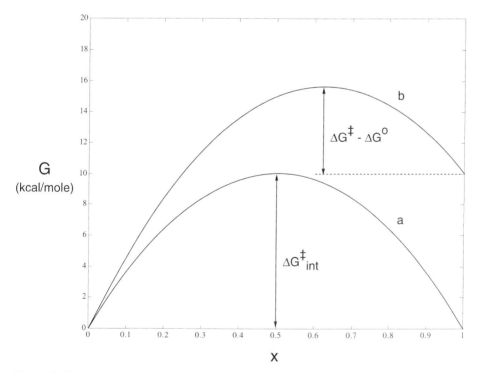

Figure 8-16. Reaction coordinates calculated from equation (1) assuming that $\Delta G^{\circ} = 0$ kcal/mol and $\Delta G^{\ddagger}_{int} = 10$ kcal/mol; and that $\Delta G^{\circ} = 10$ kcal/mol and $\Delta G^{\ddagger}_{int} = 10$ kcal/mol.

that is positioned to transfer its proton to the developing negative charge on the carbonyl oxygen of the substrate, thereby converting it to an enolic intermediate. Thus, the reactions catalyzed by these enzymes can be represented by the equation:

$$B: \quad H-\overset{O}{\underset{}{\bigtriangleup}} \quad \overset{HA}{\quad} \quad \underset{BH^+}{\overset{K}{\rightleftharpoons}} \quad \overset{OH \quad :A^-}{\underset{}{\bigtriangleup}}$$

The $\Delta G°$ for this reaction that involves the participation of both a general acid and a general base catalyst can be calculated from the equation:

$$\Delta G° = 2.303 \, RT \, [pK_E + \{pK_a(HA) - pK_a(BH^+)\}] \tag{8-3}$$

where pK_E is the negative logarithm of the equilibrium constant for interconversion of the keto and enol tautomers of the carbon acid, $pK_a(HA)$ is the pK_a of the general acid catalyst, and $pK_a(BH^+)$ is the pK_a of the conjugate acid of the general base catalyst. [This expression for $\Delta G°$ assumes that there is no special stabilization available to the enolic intermediate by virtue of hydrogen bonding interactions with $:A^-$; this assumption is actually incorrect (see the following).]

For DHAP (a ketone) and G3P (an aldehyde), $pK_E = 6 - 8$ (Chiang & Kresge, 1991). For mandelic acid, $pK_E = 15.4$ (Chiang et al., 1990). [For aliphatic carboxylic acids, $pK_E \geq 15.4$ (Gerlt et al., 1991).] For TIM, since $pK_a(HA) \geq 11$ and $pK_a(BH^+) \leq 6$ (see above), $\Delta G° \geq 18$ kcal/mol. For MR, since $pK_a(HA) \approx 6$ and $pK_a(BH^+) \approx 6$ (see above), $\Delta G° \approx 22$ kcal/mol. Both of these values for $\Delta G°$ are greater than the $\Delta G^\ddagger$s calculated for the enzyme-catalyzed reactions, demonstrating that the enzyme active site [and, in particular, the general acid (electrophilic) catalyst] must be able to reduce $\Delta G°$ for formation of the enolic intermediate.

The rate of abstraction of the α-proton of a carbon acid is slower than the rate of abstraction of a proton from a heteroatom or normal acid (HX, where X = O, N, or S) of equal acidity (Bernasconi, 1992). As implied earlier in this section in the discussion of equation (8-2), the proton transfer reaction from a carbon acid is, therefore, described by a larger value for $\Delta G^\ddagger_{int}$ than that which describes the proton transfer reaction from a normal acid. The $\Delta G^\ddagger_{int}$ for abstraction of the α-proton of a carbon acid is ~12 kcal/mol; the $\Delta G^\ddagger_{int}$ for abstraction of the proton from a normal acid is less than or equal to 3 kcal/mol. The larger value for $\Delta G^\ddagger_{int}$ for proton transfer from a carbon acid is thought to result from changes in the orientations of solvent dipoles as the negative charge develops on the carbonyl oxygen as the α-proton is abstracted. Thus, the $\Delta G^\ddagger_{int}$ for carbon acids is dominated by a negative entropic contribution associated with solvent ordering that is unimportant for normal acids (i.e., $\Delta G^\ddagger_{int} = \Delta H^\ddagger_{int} - T\Delta S^\ddagger_{int}$, where $\Delta S^\ddagger_{int}$ is negative for carbon acids).

This author and the late Paul G. Gassman proposed that general acid (electrophilic) catalysts located in the active sites of both TIM and MR are responsible for the reduction in the values for both $\Delta G°$ and $\Delta G^\ddagger_{int}$ (Gerlt & Gassman, 1993a, b). The reductions that are possible for both $\Delta G°$ and $\Delta G^\ddagger_{int}$ are sufficient to reduce $\Delta G^\ddagger$ to ~14 kcal/mol, the value that describes the reactions catalyzed by both TIM and MR.

Reduction in $\Delta G°$: Short, Strong Hydrogen Bonds

The reduction in $\Delta G°$ is attributed to the formation of a very strong hydrogen bond

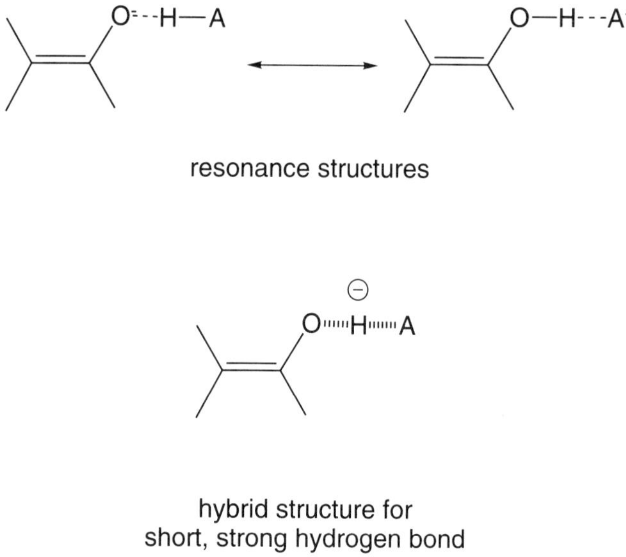

Figure 8-17. The resonance and hybrid structures for a short, strong hydrogen bond.

between the general acid catalyst and the enolic intermediate that is not present when the substrate is hydrogen bonded to the general acid catalyst (Hibbert & Emsley, 1990). The formation of this hydrogen bond provides differential binding of the enolic intermediate, using the terminology put forth by Albery and Knowles (1976b) to describe one of the steps in the evolution of catalytic efficiency. As noted earlier, although Albery and Knowles proposed differential binding of reaction intermediates, they did not indicate a specific mechanism by which differential binding might be accomplished.

The very strong hydrogen bond between the general acid catalyst and the enolic intermediate is a consequence of the matched pK_as of the general acid catalyst and the neutral enol intermediate that would be formed if a proton were quantitatively transferred to the carbonyl group of the carbon acid as the α-proton is abstracted. As noted previously, in the active sites of both TIM and MR, the pK_as of the general acid catalyst and the neutral enol intermediate are closely matched. In the active sites of TIM and MR which are separated from bulk aqueous solvent by a mobile flap that closes when substrate is bound, the matched pK_as of the general acid catalyst and the neutral enol intermediate are expected to result in the formation of a "short, strong hydrogen bond." In short, strong hydrogen bonds, the hydrogen bond lengths can be as short as 2.45 Å (as compared to hydrogen bond lengths greater than or equal to 2.6 Å between the general acid catalyst and the substrate) and as strong as *20 kcal/mol* (as compared to hydrogen bond strengths of less than or equal to 5 kcal/mol between the general acid catalyst and the substrate).

The short, strong hydrogen bond between the enolic intermediate and the general acid catalyst results from a sharing of the proton initially associated with the general acid catalyst. In these hydrogen bonds, the proton is located midway between the general acid catalyst and the intermediate rather than being closer to the general acid catalyst one-half the time and closer to the intermediate the other one-half of the time. Each of these latter two configurations may be viewed as resonance structures for the short, strong hydrogen bond, with the actual structure of the bond having the proton midway between the general acid catalyst and the enolic intermediate (Fig. 8-17).

According to this proposal, the intermediates in the reactions catalyzed by both TIM and MR are neither enolate anions nor neutral enols. The possibly ambiguous term "enolic intermediate" is used to describe the structures and charges associated with the intermediates in these reactions (Fig. 8-17), since the sharing of the proton *between* the general acid catalyst and the intermediate is crucial to understanding the rates and mechanisms of these reactions (by providing the necessary reduction in $\Delta G°$).

An important consequence of forming a stabilized enolic intermediate in which the proton from the general acid catalyst is shared by the intermediate and the catalyst is that in the transition state for proton transfer the pK_a of the α-proton will be significantly reduced from the value that describes the substrate carbon acid. Protonation of the carbonyl group of a carbon acid reduces the pK_a of the α-proton by ~15 pK_a units, irrespective of the nature of the carbon acid (Gerlt et al., 1991). If formation of the short, strong hydrogen bond is nearly complete in the transition state for abstraction of the α-proton [which requires that $\Delta G^{\ddagger}_{int}$ be reduced (see the following)], the carbonyl group will be approximately one-half protonated in the transition state, and the pK_a of the α-proton will be reduced by ~7.5 pK_a units.

Reduction in $\Delta G^{\ddagger}_{int}$

The reduction in $\Delta G^{\ddagger}_{int}$ by the general acid catalyst is attributed to its close approximation to the carbonyl group of the substrate carbon acid so that as the α-proton is abstracted by the general base catalyst, the negative charge that develops on the carbonyl group can be stabilized (negated) by the partial transfer of a proton from the general acid catalyst to the carbonyl group. Since, as noted previously, the large value of $\Delta G^{\ddagger}_{int}$ for carbon acids is attributed to solvent ordering that is necessary for stabilization of the developing negative charge on the carbonyl oxygen, the prepositioning of the general acid catalyst adjacent to the carbonyl oxygen reduces $\Delta G^{\ddagger}_{int}$ by allowing the $\Delta S^{\ddagger}_{int}$

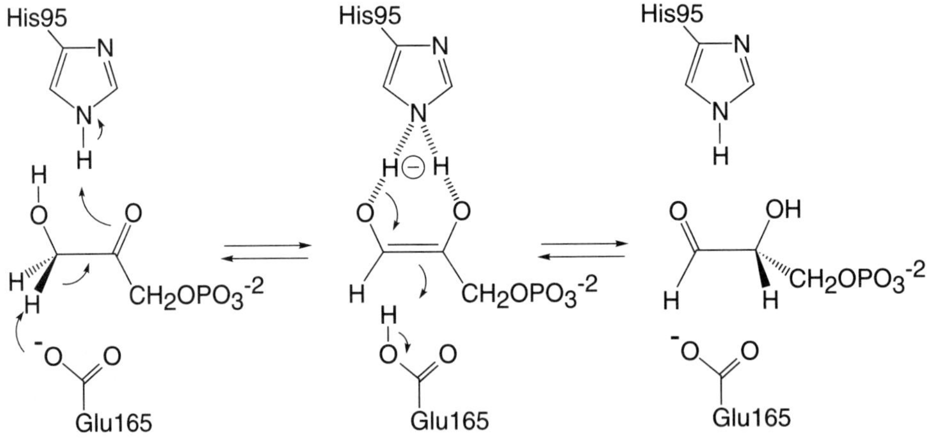

Figure 8-18. The mechanism of the reaction catalyzed by TIM involving formation of an enolic intermediate that is stabilized by short, strong hydrogen bonds (hash marks) with His95. The negative charge (⊖) is dispersed in the hydrogen bond and not localized on the heteroatoms or the bridging protons.

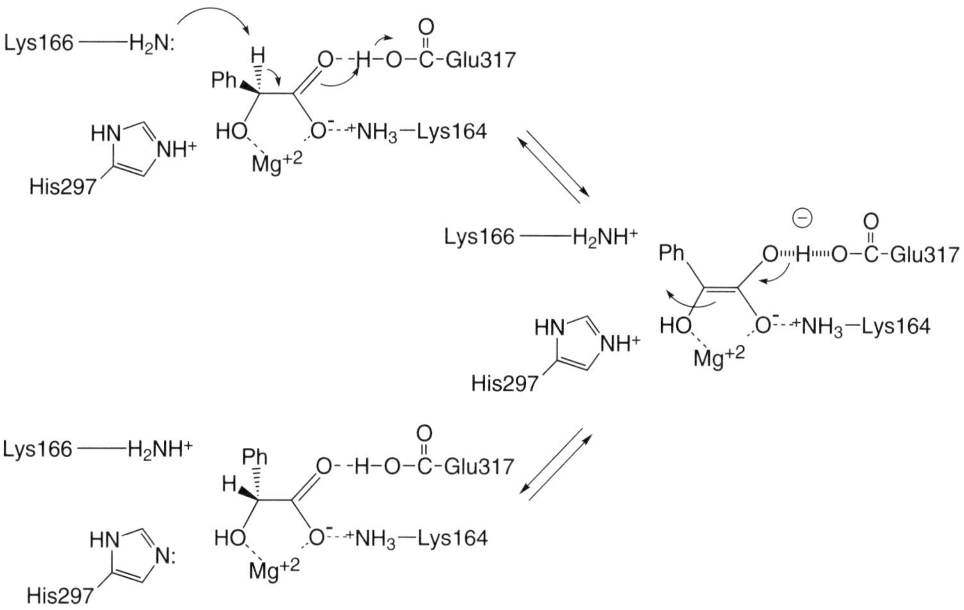

Figure 8-19. The mechanism of the reaction catalyzed by MR involving formation of an enolic intermediate that is stabilized by a short, strong hydrogen bond (hash marks) with Glu317. The negative charge (⊖) is dispersed in the hydrogen bond and not localized on the heteroatoms or the bridging proton.

component to be less negative. Although the data from physical organic chemistry on the magnitude of the reduction in $\Delta G^{\ddagger}_{int}$ that is possible from this effect is limited (Gerlt & Gassman, 1993b), the value for $\Delta G^{\ddagger}_{int}$ may be reduced sufficiently so that from equation (2) $\Delta G^{\ddagger}$ can be determined almost entirely by $\Delta G°$.

If $\Delta G^{\ddagger}_{int}$ can be reduced, the known strengths of short, strong hydrogen bonds are more than sufficient to reduce $\Delta G°$ and $\Delta G^{\ddagger}$ to ~14 kcal/mol, the value that describes the reactions catalyzed by both TIM and MR. Earlier in this section, equation (8-3) was used to calculate that $\Delta G°$ must be reduced by only ~4 kcal/mol for the reaction catalyzed by TIM and by ~8 kcal/mol for the reaction catalyzed by MR to achieve the observed $\Delta G^{\ddagger}$s of ~14 kcal/mol. (This calculation assumes that the contribution of $\Delta G^{\ddagger}_{int}$ to $\Delta G^{\ddagger}$ is negligible, i.e., $\Delta G^{\ddagger}_{int}$ is reduced by the general acid catalyst.) Thus, the absolute amount of the differential binding of the enolic intermediates need not be very large.

Mechanisms of the Reactions Catalyzed by TIM and MR

According to these proposals for understanding the rates of the reactions catalyzed by TIM and MR, an enolic intermediate stabilized by a short, strong hydrogen bond is the reaction intermediate. The proton in the short, strong hydrogen bond is derived from the general acid catalyst that is positioned adjacent to the carbonyl group of the substrate carbon acid. The intermediate is not an enolate anion that is too unstable to be kinetically competent.

The mechanisms proposed for the reactions catalyzed by TIM and MR are sum-

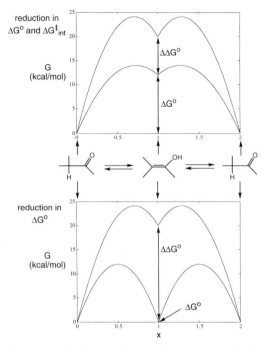

Figure 8-20. The dependence of G on reaction coordinate, x, for enolization and ketonization reactions of a carbon acid bound in an enzyme active site. The keto tautomer of the bound substrate carbon acid is at $x = 0$; the bound enolic intermediate is at $x = 1$; and the bound keto tautomer of the product carbon acid is at $x = 2$. The free energy difference between the enolic intermediate and the substrate carbon acid is labeled as $\Delta G°$. In both panels, the upper reaction coordinates are those calculated for nonenzymatic enolization and ketonization of a carbon acid ($\Delta G^{\ddagger}_{int} = 12$ kcal/mol and $\Delta G° = 20$ kcal/mol). In the top panel, the lower reaction coordinate was calculated assuming reductions in both $\Delta G^{\ddagger}_{int}$ and $\Delta G°$ ($\Delta G^{\ddagger}_{int} = 6.65$ kcal/mol and $\Delta G° = 12$ kcal/mol). In the bottom panel, the lower reaction was calculated assuming only a reduction in $\Delta G°$ ($\Delta G^{\ddagger}_{int} = 12$ kcal/mol and $\Delta G° = 0$ kcal/mol). The free energy difference, $\Delta\Delta G°$, between the enolic intermediates in the upper and lower reaction coordinates in each panel is the stabilization of the enol intermediate required to produce the observed rates of the enzyme-catalyzed reactions.

marized in Figures 8-18 and 8-19, respectively. These mechanisms involve concerted general acid-general base catalyzed formation of an enolic intermediate that is facilitated by the formation of a short, strong hydrogen bond (represented by hash marks).

Evolution of Catalytic Efficiency

A result of this analysis of the rates and mechanisms of the reactions catalyzed by TIM and MR is that $\Delta G^{\ddagger}$ is numerically dominated by $\Delta G°$ and not by $\Delta G^{\ddagger}_{int}$ (Fig. 8-20, top panel). This is a consequence of the reduction in $\Delta G^{\ddagger}_{int}$.

This conclusion contrasts with that proposed by Albery and Knowles (1976b) for understanding the evolution of catalytic efficiency. According to that proposal, $\Delta G°$ for formation of the enolic intermediate from the bound substrate approaches 0 (Fig. 8-20, bottom panel). This decrease in $\Delta G°$ is the result of differential binding of substrate, product, and reaction intermediates so that the equilibrium constants for interconversion of all bound species approach unity. In this way, the energies of the transition states for formation and decomposition of the reaction intermediate are reduced and are approximately equal. Therefore, according to the Marcus formalism described in this section, the $\Delta G^\ddagger$ is associated solely with $\Delta G^\ddagger_{int}$. The role of the general acid catalyst is simply to reduce $\Delta G°$; the general acid catalyst does not reduce $\Delta G^\ddagger_{int}$ since the observed $\Delta G^\ddagger$ (~14 kcal/mol) is approximately equal to the $\Delta G^\ddagger_{int}$ for proton abstraction from a carbon acid (~12 kcal/mol). The reduction of $\Delta G°$ that is necessary to achieve $\Delta G° = 0$ kcal/mol would require that the strengths of the short, strong hydrogen bonds that stabilize the enolic intermediates approach the maximum possible values. Thus, while the proposal made by Albery and Knowles based on the Hammond postulate and the proposal described in this section based on Marcus formalism both allow the observed values for the $\Delta G^\ddagger$s to be explained (~14 kcal/mol), the manner in which the catalytic efficiency is accomplished is quite different.

These conflicting proposals for explaining the rates and mechanisms of the reactions catalyzed by TIM and MR can be resolved by measuring the concentrations of the bound enolic intermediates. However, at present, the concentrations of the enolic intermediates in the reactions catalyzed by TIM and MR (or in the active site of any enzyme catalyzing abstraction of the α-proton of a carbon acid) have not yet been quantitated.

Summary

The interrelationship between the rates and mechanisms of proton transfer reactions to and from the α-carbons of carbon acids has been a long standing problem in mechanistic enzymology. While mechanisms have been proposed for a variety of proton transfer reactions that involve the formation of enolate anion intermediates, such intermediates are too unstable to be kinetically competent. This problem can be resolved, not only for the reactions catalyzed by TIM and MR but also for other enzymes including citrate synthase and ketosteroid isomerase, by recognizing the presence of a general acid (electrophilic) catalyst adjacent to the carbonyl group of the substrate carbon acid and realizing its importance in determining both the rates and mechanisms of these reactions.

9

Site-Directed Mutagenesis

Paul J. Loida, Ronald A. Hernan, and Stephen G. Sligar

Site-directed mutagenesis is an extraordinarily powerful tool for the elucidation of structure-function correlations in a wide variety of protein systems. Indeed, it is difficult to pick up any modern journal in the biochemical, biophysical, or molecular biology areas that does not contain numerous articles that utilize the technique of site-directed mutagenesis to aid in understanding mechanisms. The techniques of site-directed mutagenesis have matured drastically from the advent of recombinant DNA technology in the late 1970s. In those early times only a very few laboratories in the world had on hand the needed reagents to manipulate DNA at the molecular level. Following this, through the mid-1980s, numerous companies were founded to provide the necessary reagents, enzymes, and techniques for a variety of recombinant DNA methodologies. During this period, however, one needed to know which company provided the most reliable and active enzymes of particular type and which did not. In addition, the various procedures for manipulating recombinant DNA molecules were somewhat tenative and often labeled as "black art." During the last five years or so, the competitive edge of the free market has brought the uniform quality and availability of recombinant enzymes and reagents necessary to elicit the directed mutagenesis of specific amino acids in a protein. Numerous courses now exist to train even the uninitiated to a level such that all fields of science can make use of recombinant methodologies to probe the details of structure-function correlations in enzyme, protein, and nucleic acid systems. It is interesting also to note the tremendous advances in structure elucidation that have occurred during the past decade. It is now possible for a single laboratory to not only mutate a given gene, express the protein in a host environment, purify the protein, and characterize its function at the molecular level but also to obtain diffraction quality crystals such that a high resolution structure can be obtained. Several cycles through the mutation, chemical characterization and structure determination processes are now possible, such that a detailed understanding of the effects of mutations on global protein parameters as well as on the exact residues responsible for molecular recognition and catalytic processing can be defined.

Strategies for Site Directed Mutagenesis

Over the years many methodologies have emerged that define specific procedures for the manipulation of DNA to elicit replacement of the amino acid side chain with another at a given location in the gene and derived protein. The earliest methods involved a means for generating single-stranded DNA molecules, which could then be hybridized with *in vitro* synthesized oligonucleotide probes carrying the desired mutation. DNA polymerase is used to replicate the remaining complementary DNA and generate a double-stranded duplex molecule. This breakthrough technology was acknowledged in the 1993 Nobel Prize in Chemistry. A variety of methods were developed for screening the resultant duplex DNA such that the progeny descending from the mutant single strand could be amplified (Kunkel et al., 1987; Olsen et al., 1993). Current practice, however, tends to focus primarily on the application of two major classes of mutagenesis: cassette and polymerase chain reaction. In the first, two complementary strands of DNA, corresponding to coding sequences between two restriction endonuclease recognition sites are synthesized to generate a cassette that contains the desired alteration in the primary protein sequence. The second major method has largely replaced single strand mutagenesis techniques, and uses oligonucleotides to direct the desired replacement of the amino acid side chain and a polymerase chain reaction to selectively amplify the mutant DNA strand. As these two methods represent state-of-the-art mutagenesis techniques, we will restrict our discussion to their applications in this chapter. The beauty of recombinant DNA methodology is that "you only need one to succeed." Here "one" is a single colony or clone containing the desired alteration. The various technologies, therefore, are all equivalent ways of accomplishing the same thing. We also give several examples which illustrate the power of site-directed mutagenesis to understanding structure-function correlations in protein systems. We have chosen examples primarily from our own research endeavors.

Recombinant DNA methodologies can indeed be taught to the uninitiated biochemist in an extremely short period of time. A one-semester course currently offered at the University of Illinois at Urbana-Champaign focuses on site directed mutagenesis of a totally synthetic gene and the detailed biophysical characterization of the resultant purified proteins. Thus, recombinant DNA technology, and site-directed mutagenesis as an example, has become so pervasive in the biochemical community that it is no longer a rate-limiting step in generating new and novel structures. The rapid and detailed characterizations of structure and molecular function remain the tasks to close the loop in mechanistic interpretation. We will begin our description then with the methodologies involved in eliciting mutants by cassette methods followed by a discussion of techniques utilizing the current polymerase chain reaction.

Cassette Mutagenesis

Cassette mutagenesis is a powerful technique that has been applied effectively in the total synthesis of genes as well as to the generation of site directed mutants. The technique involves the insertion of duplex DNA between two restriction endonuclease recognition sites on a piece of plasmid DNA, and gives one the ability to insert, delete or change amino acids conveniently (Fig. 9-1). Cassette mutagenesis circumvents some of the fidelity problems associated with the use of polymerase-based reactions (discussed in the second part of this chapter) and single-stranded vector techniques. The technique has been used to study structure-function relationships of proteins and peptides, protein folding, macromolecular interactions, molecular recognition events, and enzymatic catalysis.

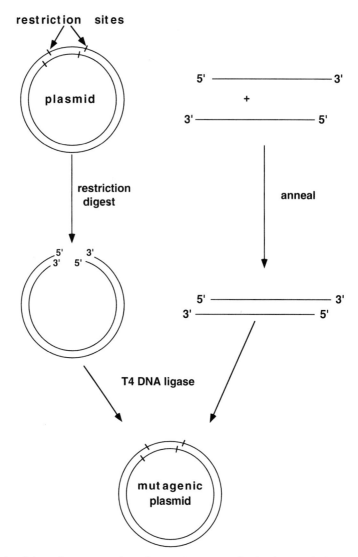

Figure 9-1. Schematic representation of cassette mutagenesis showing restriction digestion, annealing of oligonucleotides, and ligation of a new cassette into the digested plasmid.

Oligonucleotide Design, Construction and Purification

Design of the oligonucleotides for cassette mutagenesis is a critical first step in order to perform mutagenesis easily and screen effectively for potential mutants. Important characteristics that must be considered in the design of an oligonucleotide are the overall oligonucleotide length, the incorporation of silent mutations, which utilize the degenerate amino acid code to change the DNA but leave the amino acid sequence unaltered at points that are unaffected by the mutation, introduction or destruction of restriction endonuclease recognition sites for use in later screening steps, complementary base pairing within the heteroduplex, and, of course, complementarity between restriction endonuclease site overhang on the plasmid and the mutagenic DNA. Introduction of new or destruction of existing restriction endonuclease recognition sites are ideally incorporated using silent mutations to screen large numbers of potential clones rapidly, thereby eliminating the time-consuming need to sequence a significant number of possible mutants.

With the advent of automated DNA synthesizers, it is now possible to synthesize segments of single-stranded DNA up to a few hundred base pairs in length in a reasonably short period of time. Solid phase synthesis produces yields of typically over 98% per cycle, making large amounts of material for single-stranded DNA of 50 bases in length easily obtainable. If lengths of DNA greater than 50 oligonucleotides are required, the yield of desired material diminishes significantly. Insertion of large fragments of DNA or synthetic genes of hundreds of base pairs between restriction endonuclease sites is, therefore, usually accomplished by synthesizing many smaller segments of DNA and then assembling them together to produce a large heteroduplex; this can then be inserted between the restriction endonuclease sites of the digested plasmid. Using this method of assembling many fragments of oligonucleotides together, it is convenient to insert large heteroduplexes corresponding to entire synthetic genes into plasmid DNA for expression of genetically engineered protein.

Oligonucleotides can be synthesized with the dimethoxytrityl protecting group attached or removed depending upon the choice of purification. The major problem in the synthesis is the production of deletion sequences ranging in size from one base to lengths one base shorter than the desired product. Purification of oligonucleotides to remove unwanted by-products can be accomplished using a variety of techniques. Among these are paper, thin layer, and polyacrylamide gel electrophoresis, ion exchange chromatography, and high performance reversed phase, normal phase, and ion exchange chromatography.

Polyacrylamide gel electrophoresis is probably the most economical method for purification of a crude mixture of synthetic oligonucleotides. Good results are generally obtained using a 20% acrylamide gel and 8 M urea under denaturing conditions. In addition, the percentage of acrylamide can be varied depending upon the length of the oligonucleotide. Migration of the oligonucleotide in polyacrylamide is a function of the log of the chain length. Ultraviolet shadowing is probably the most commonly used technique for visualization of the oligonucleotides when carrying out purifications using gel electrophoresis; however, the use of stains has been successfully employed. The band is excised from the gel, crushed, and eluted with buffer. Relatively large quantities of crude material are used, and the detection of minor components is sometimes difficult. In addition, the procedures are time consuming, laborious, and the recovery of pure product is low, about 70% for a 20-mer, and dependent upon the length of the oligonucleotide.

Reversed phase HPLC provides a useful and convenient method for separating deletion products and unwanted by-products of DNA synthesis. In this procedure the hydrophobic dimethoxytrityl group is left attached to the 5'-position of the synthetic oligonucleotide after completion of synthesis. Failure or deletion sequences generated during the automated synthesis are capped with a 5'-O-acetyl cap and hydrolyzed during the workup of the oligonucleotide. Since the final product will have the hydrophobic dimethoxytrityl group and the deletion or failure sequences will have a 5'-hydroxyl group, reversed phase separation involves separation of the 5'-dimethoxytritylated

DNA from the 5′-hydroxylated DNA (Sonveaux, 1986). Many methods exist for the purification of crude oligonucleotide products using reversed phase HPLC. Most use a gradient consisting of 0.1 M triethylamine-acetate, pH 7.0, and acetonitrile. The course of the chromatography is monitored at 260 nm and the major peak is collected. The putative product is dried under vacuum and resuspended in 80% acetic acid, hydrolyzing the dimethoxytrityl group. The solvent is then removed under vacuum and resuspended in water, followed by ether extraction and ethanol precipitation. A method that works well in our laboratory uses a C-4 reversed phase HPLC column and incorporates an acid hydrolysis step with trifluoroacetic acid during the chromatography. The major peak is collected, dried under vacuum, resuspended in water, extracted with ether, and precipitated with ethanol to yield a clean oligonucleotide.

The use of cassette mutagenesis in many cases allows for the introduction of a restriction site within the mutagenic heteroduplex, or the destruction of the terminal recognition site at the end of a mutagenic duplex, without changing the amino acid sequence. This process allows one to screen large numbers of potential mutagenic plasmid more quickly and effectively, with less ambiguity than one is able to achieve with colony hybridization. The removal of terminal endonuclease restriction sites usually involves the changing of the first base pair in the palindromic sequence of the recognition site, thereby eliminating the recognition site. Introduction of restriction endonuclease sites within the mutagenic cassette involves the analysis of the DNA using computer programs to compile a list of endonuclease sites which may be introduced within a fragment of mutagenic DNA. Both insertion and deletion of endonuclease restriction sites are useful for screening large numbers of potential clones; however, the destruction of a restriction endonuclease site is more beneficial, since this can be used to enrich mutagenic plasmid, as described later.

The introduction of silent mutations into the design of the mutagenic oligonucleotide is accomplished by taking advantage of the degeneracy of the genetic code. Silent mutations facilitate the use of colony hybridization and make the identification of mutagenic DNA by dideoxy sequencing less difficult. For small fragments of DNA, 12 to 20 bases in length, the introduction of one silent mutation will destabilize the heteroduplex formed between the mutagenic and wild-type DNA. For fragments of DNA larger than 20, many more silent mutations need to be introduced into the oligonucleotide in order to destabilize the duplex formed between the mutagenic and wild-type DNA. Introduction of many silent mutations will make the discrimination between wild-type and mutagenic DNA less ambiguous during colony hybridization, thereby increasing the probability of isolating mutants.

Construction of Mutants: Plasmid Preparation

Plasmids are prepared from bacterial cultures on a small or large scale depending upon the intended use of the plasmid in subsequent steps. The bacteria are harvested by centrifugation and prepared for lysis. Lysis can be accomplished by a variety of different methods including the use of ionic and nonionic detergents, organic solvents, alkali, or heat (Sambrook et al., 1989). The method used depends upon the size of the plasmid, the strain of *E. coli*, and the subsequent techniques to be used in the experiment. Generally for plasmids less than 15 kb, either the boiling lysis or the alkaline lysis method is preferred. Both alkaline lysis and boiling lysis methods can be used for cultures ranging in size from 1 milliliter to 1 liter. These methods are simple, reliable, reproducible, and produce sufficient material in good quality to be used for cassette mutagenesis. Commercial kits are available for small cultures; they afford similar results but at a slightly higher cost.

The plasmid is prepared for cassette mutagenesis by digesting the plasmid with a restriction endonuclease (Fig. 9-1). Typically, an aliquot of plasmid preparation, usually one microgram, is digested with the appropriate restriction endonuclease whose sites flank the region where the cassette will be inserted. The digested plasmid may be used without further purification for mutagenesis if the piece of DNA released is small. However, further purification may be necessary to reduce the amount of self-ligation if the insert is large or if screening will be difficult. The digested plasmid can be purified using agarose gel electrophoresis, with the assistance of DNA extraction kits that are commercially available. Alkaline phosphatase, an enzyme that removes the 5'-phosphate from DNA, can also be used on the plasmid to reduce self-ligation.

Construction of Mutants: Preparation and Insertion of Cassette

Generation of a cassette that is to be ligated into the digested vector requires the addition of a 5'-phosphate to the oligonucleotide; this is accomplished using ATP + T4 polynucleotide kinase. This enzyme catalyzes the addition of phosphate from ATP to the 5'-end of DNA. After phosphorylation, the mutagenic heteroduplex is generated by heating the two complementary oligonucleotides together at 95°C and slowly cooling to 30°C.

Using T4 DNA ligase, which catalyzes the formation of a phosphodiester bond between 3'-OH and 5'-phosphate termini in double-stranded DNA, the mutagenic insert is ligated into the digested plasmid. The success of a ligation reaction depends upon the nature of the termini, whether the ends are compatible, and the optimal concentration of DNA for the correct ligation products. Ligation reactions with compatible protruding ends are typically run at 16°C for four hours with insert-to-plasmid ratios between 1:1 and 5:1. Ligation of termini with blunt ends are less efficient than those with protruding termini. Hexamine cobalt or polyethylene glycol are added to the ligation reaction to accelerate the rate by 1 to 3 orders of magnitude and suppress intramolecular ligation. Blunt-end ligations require higher concentrations of ligase, DNA, and plasmid, as well as incubation at 4°C.

In some cases it is possible to enrich the transformation of the mutagenic plasmid while suppressing the transformation efficiency of wild-type plasmid. This technique works when insertion of a mutagenic heteroduplex destroys a restriction endonuclease site. The ligation mixture is heated to 70°C for five minutes to denature T4 DNA ligase. The ligation mixture is digested with the restriction endonuclease whose site was destroyed during insertion of the mutagenic cassette. During the enrichment the wild-type plasmid will be digested, while the mutagenic plasmid will remain circularized. Transformation of linearized plasmid will occur approximately 10 times less efficiently than that of circular plasmid, thereby increasing the probability of finding a mutant. The transformed competent cells are grown on plates and prepared for screening.

Screening of Potential Mutants

There are several useful methods for screening large numbers of potential mutants. If a restriction site has been removed, restriction enzyme mapping of a plasmid preparation is a convenient method. In this case a plasmid preparation is digested with the appropriate restriction endonuclease along with the wild-type plasmid, and the digestion mixture is then analyzed by agarose gel electrophoresis.

Hybridization techniques provide a way to screen hundreds of bacterial colonies at once. Colonies are randomly picked and grown on nitrocellulose filter paper. The cells are then lysed *in situ* and rinsed to remove cell debris, followed by baking the DNA to the nitrocellulose. Usually one of the mutagenic oligonucleotides is radioactively labeled with ^{32}P using [γ-^{32}P]ATP + T4 polynucleotide kinase. The probe is then added to the filter and incubated, typically at 65°C. After incubation, the nitrocellulose filter is washed close to the heteroduplex melting point, T_m. The filters are exposed to film and potential mutant colonies appear as dark spots on an autoradiograph. Wild-type colonies may also appear on the autoradiograph but with a lower intensity than a mutant colony. Potential mutagenic plasmids are usually verified by DNA sequencing using the Sanger dideoxy method.

Examples of the Application of Cassette Mutagenisis

Cytochrome b_5

The use of site directed mutagenesis of rat liver cytochrome b_5 and cytochrome c for the study of macromolecular association and the electrostatic contributions to the overall Gibbs free energy of protein-protein complex assembly has been investigated. Site-directed mutagenesis coupled with high pressure techniques were used to directly measure and identify the residues involved in formation of the protein-protein complex (Rodgers & Sligar, 1991b). These studies defined the area of the protein surface involved in molecular recognition and the thermodynamic parameters involved in electrostatic interaction of macromolecular assembly. Amino acids which may be involved in the protein-protein complex were identified using a computer-assisted docked model complex. Using a synthetic gene (Beck von Bodman et al., 1987), point mutations were generated via cassette mutagenesis. The glutamic acid residues at positions 43, 44, and 48 of cytochrome b_5 were mutated to glutamine. Carboxylic acid and amide groups are isosteric with a change in the volume of 5%. Single point mutations showed a decrease in affinity of 0.5 kcal/mole and a volume change of 38 mL/mole. Double and triple mutants were also used to map the interaction domain. Multiple mutations destabilized the complex further, but the differences were not additive. Electrostatic interactions involving charged groups on the cytochrome b_5 surface only account for 14% of the total free energy for the cytochrome b_5-cytochrome c interaction (Fig. 9-2). A recent theoretical treatment of the diprotein complex confirms the overall scaling of association free energy with charge mutation. The use of mutants demonstrated that solvent electrostriction accounts for two-thirds of the total volume change for complex dissociation. Because electrostatics and surface charges play a role in the affinity and selectivity of the recognition process, the use of site-directed mutagenesis in conjunction with electrochemical techniques has been used to determine the electrostatic potential boundaries of cytochrome b_5 (Rodgers & Sligar, 1991b). Potentials of the prosthetic group and ionization of amino acids are affected by the surface charge residues, so that mutagenesis of the glutamic acid residues at positions 43, 44, and 48 are ideal for this study. Negatively charged amino acids near the heme group were reversed, or else these amino acids were made neutral. These changes in polarization of the glu-

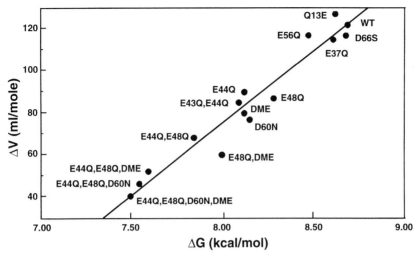

Figure 9-2. Thermodynamic values of ΔG and ΔV for the cytochrome b_5-cytochrome c complex of different mutants of cytochrome b_5. The nomenclature for the modified proteins is as follows: the first letter is the one-letter abbreviation for the amino acid of the native protein at the position specified by the number. The last letter is the abbreviation for the amino acid introduced at this position.

tamic acids permitted the determination that individual surface charges play only a minor role in governing the redox potential of the proteins (Rodgers & Sligar, 1991a).

Sperm Whale Myoglobin

Site-directed mutagenesis of sperm whale myoglobin was used to investigate the role of the distal pocket mutants at position 64 (E7). Site-directed mutagenesis in conjunction with stopped flow and flash photolysis studies were used to probe the functional role of histidine E7. These studies assessed the role of size, charge, polarity, and stereochemistry of the amino acid at this position. In the original complete synthesis of the sperm whale myoglobin gene the degeneracy of the amino acid code was used to introduce specific restriction sites that bracketed the distal pocket regions. For example, using cassette mutagenesis the histidine at E7 was replaced with Gly, Val, Leu, Met, Phe, Gln, Arg, and Asp (Rohlfs et al., 1990). Oxygen and carbon monoxide association rate constants increased 5- to 10-fold and showed little dependence upon the size of the amino acid at position E7. The values for the Gln substitution at E7 show only a moderate increase, suggesting that the polar interactions at position E7 provide a kinetic barrier to oxygen and carbon monoxide binding (Table 9-1). In contrast to small ligands, large ligands such as isocyanides produced a trend in the association rate of Gly >> Val ≈ Leu > Met > Phe ≈ His ≈ Gln indicating that the kinetic barrier for isocyanide binding is primarily through the interaction with the amino acid at position E7.

Valine E11 in sperm whale myoglobin is a conserved residue in many myoglobins and is of interest since it is in van der Waals contact with the bound ligand and the distal histidine, E7, restricting the size of the binding site of the protein. Flash photolysis and stopped flow techniques were used to investigate the binding kinetics of myoglobin mutants substituted at this position. Valine E11 was replaced with Ala, Ile, and Phe using

Table 9-1 Rate and Equilibrium Constants for Ligand Binding to His E7 Mutants of Sperm Whale Myoglobin

Ligand	E7 Residue	k′ × 10^{-6} M^{-1} s^{-1}	k s^{-1}	K × 10^{-6} M^{-1} s^{-1}
O_2	Gly	140	1,600	0.088
	Val	250	23,000	0.011
	Leu	98	4,100	0.023
	Met	75	1,700	0.045
	Phe	75	10,000	0.0074
	Gln	24	130	0.18
	Arg			
	Asp	79	880	0.09
	His	14	12	1.2
CO	Gly	5.8	0.038	150
	Val	7.0	0.048	150
	Leu	26.0	0.024	1,100
	Met	4.6	0.023	200
	Phe	4.5	0.054	83
	Gln	0.98	0.012	82
	Arg	5.7	0.014	400
	Asp	4.4	0.052	85
	His	0.51	0.019	27
MNC	Gly	10	6.3	1.6
	Val	0.71	12	0.059
	Leu	1.80	2.1	0.86
	Met	0.44	4.0	0.11
	Phe	0.18	2.4	0.075
	Gln	0.20	5.6	0.037
	Arg	1.2	5.6	0.21
	Asp	2.9		
	His	0.12	4.3	0.028

cassette mutagenesis. The role of side chain volume on ligand association was investigated utilizing flash photolysis and stopped flow techniques to measure the kinetics of ligand binding (Egeberg et al., 1990). Association rate constants for ligands such as oxygen, carbon monoxide, and isocyanides decreased in the order Ala > Val > Ile (Table 9-2). This order implies that the residue at position E11 is also involved in the overall kinetic barrier to ligand binding. This extensive set of distal pocket mutants can also be used to probe the steric and polarity effects that control the detailed structure of the bound ligand.

Human Hemoglobin

Cassette mutagenesis techniques have recently been applied to the area of total gene synthesis for the expression of heterologous proteins in bacterial hosts. Overexpression of heterologous proteins using cDNA clones can produce gene products that accumulate in amounts that constitute a large proportion of total cellular protein; however, in many cases there is little overexpression of heterologous proteins. Expression of tetrameric hemoglobin in *E. coli* using cDNA clones proved difficult, but the use of synthetic genes for expression produced tetrameric hemoglobin that accumulated to 10% of the total cellular protein (Hernan et al., 1992). The overexpression of hemoglobin with synthetic genes is attributed to the use of codons optimal for *E. coli*. Total gene synthesis in some cases allows for the optimization of protein expression machinery and incorporation of convenient restriction sites for cassette mutagenesis.

PCR for Site-Directed Mutagenesis

The second major technique for introducing mutations at defined positions in an amino acid sequence makes use of temperature stable polymerases to extend replication from a given primer. The polymerase chain reaction (PCR) is a method in which a sequence of template DNA between two flanking primers is amplified by repeating a series of

Table 9-2 Rate and Equilibrium Constants for Ligand Binding to Val E11 Mutants of Sperm Whale Myoglobin

Ligand	E11 Residue	k' $\times 10^{-6}\, M^{-1}\, s^{-1}$	k s^{-1}	K $\times 10^{-6}\, M^{-1}\, s^{-1}$
O_2	Ala	22	18	1.2
	Val (native)	14	12	1.2
	Ile	3.2	14	0.22
	Phe	1.2	2.5	0.48
CO	Ala	1.2	0.021	56
	Val	0.51	0.019	27
	Ile	0.05	0.024	2.1
	Phe	0.25	0.018	14
MNC	Ala	0.38	0.76	0.50
	Val	0.12	4.3	0.028
	Ile	0.050	21	0.0024
	Phe	0.013	0.030	0.43
ENC	Ala	0.18	0.070	2.6
	Val	0.069	0.30	0.23
	Ile	0.047	3.4	0.014
	Phe	0.061	0.0035	1.7

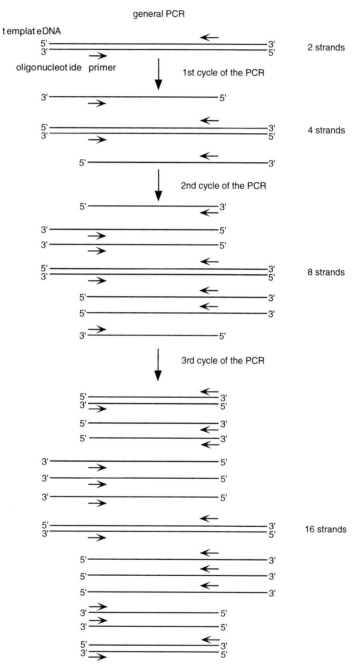

Figure 9-3. Schematic representation of PCR-mediated exponential amplification of a DNA sequence spanning two flanking primers. At the end of the fourth cycle, 22 out of 32 strands are the desired reaction product. By the fifteenth cycle, the increase is 10^4.

three steps: (1) denaturation of the template DNA strands, (2) annealing of primers and template to form duplex DNA, and (3) extension of the primers from their 3'-ends by a DNA polymerase. The availability of a thermostable DNA polymerase isolated from *Thermus aquaticus* (*Taq*), enables the cycle to be repeated multiple times without interruption, utilizing an automated heating block (thermal cycler). The number of DNA strands is doubled during each cycle, thus the amplification is exponential (Fig. 9-3). PCR is a rapid, efficient method for site-directed mutagenesis compared to other techniques such as single strand M13-based methods. The time required to apply this technique is short, few reagents and enzymes are required, and there is no need to culture phage in the laboratory. A point mutation is incorporated into the DNA sequence, and subsequently into translated protein by designing one of the primers (mutagenic primer, Fig. 9-4) to overlap the codon corresponding to the targeted residue and to contain a nucleotide mismatch such that the sequence codes for the desired amino acid at that position. The three most common approaches to utilizing PCR for site-directed mutagenesis are described herein; innumerable variations on these methods are present in the literature.

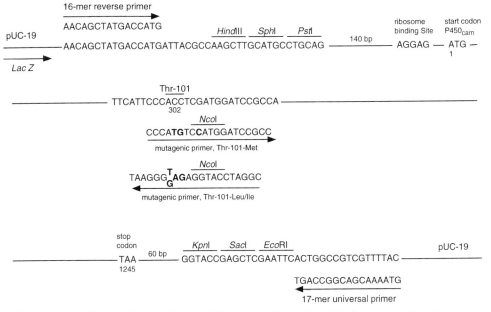

Figure 9-4. The gene for cytochrome P450$_{cam}$ inserted into the polylinker region of the plasmid pUC-19 at the *Pst*I and *Kpn*I restriction sites. The remaining unique restriction sites of the divided polylinker, the reverse and universal primers which flank the polylinker region, and the orientation of the *Lac Z* promoter are indicated. Nucleotide base mismatches are shown in bold lettering. The mutagenic primer coding for the Thr-101-Met amino acid substitution was used in the one-primer secondary PCR mutagenesis method (see Fig. 9-6) to incorporate the methionine codon, ATG, and nucleotide mismatches for the *Nco*I restriction site. The 17-mer universal primer was used in the primary reaction and the 16-mer reverse primer was used in the secondary reaction. The *Nco*I restriction site and 16-mer reverse primer were utilized in subsequent mutagenesis of the Thr-101 residue employing the cassette with PCR method (see Fig. 9-5). Mutagenic primers for sites Thr-185 and Val-247 are also shown.

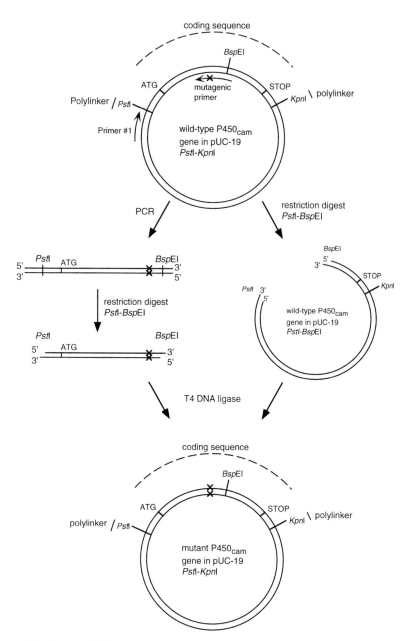

Figure 9-5. Schematic illustration of the cassette mutagenesis with PCR method for site-directed mutagenesis. A single PCR is required to produce the modified DNA fragment for subcloning into a vector which has had the corresponding gene segment excised. The general conditions for a 100 μL PCR are: 100 μM primers, 20 ng template DNA (P450 gene in pUC-19, 4000 base pairs), 200 μM of each dNTP, 1.5 mM $MgCl_2$ and a 10-fold dilution of the commercially supplied reaction buffer, resulting in final concentrations of 50 mM KCl, 10 mM Tris-HCl, pH 9.0, at 25°C containing 1% Triton X-100. The reaction mixture (99 μL) is incubated for 5 min at 95°C to thoroughly denature the template, and then 1 unit of *Taq* DNA polymerase is added. The temperature cycle of annealing at 55°C for 1.0 min, extension at 72°C for 1.5 min and denaturation at 95°C for 30 sec is repeated 15 times and the final polymerization period is lengthened to 5 min.

Primary PCR Method

In many instances, conveniently located restriction sites for use with classical cassette mutagenesis are not available. For the case in which there is one unique site close to the target codon, PCR can be utilized to essentially "fill in" the remainder of the DNA cassette by designing the mutagenic primer to overlap the unique restriction site. The other primer flanks a unique restriction site in the gene or in the polylinker region of the construction (Fig. 9-5). The product of the PCR is treated as a cassette and ligated with a vector in which the corresponding gene segment has been removed. The undisputed advantage of this approach is that the product of the primary PCR is subcloned directly without the need for a secondary PCR, which is typically more difficult to optimize (Kadowaki et al., 1989).

The One-Primer Secondary PCR Method

If there is no restriction site near the codon targeted for mutagenesis the product of the primary PCR can be used as the mutagenic primer in a second PCR (Sligar et al., 1991). The initial PCR is performed under typical conditions with a mutagenic primer and a flanking primer and the product is gel purified. In the secondary PCR, the DNA segment is extended in the opposite direction of the primary PCR with the relatively long mutagenic primer and a second oligonucleotide primer which flanks the corresponding end of the gene. The amplified sequence is designed to contain unique restriction sites on either side of the mutation (Fig. 9-6). The product of the secondary PCR is isolated and subcloned into a suitable plasmid vector. A common variation is to use the primary PCR product without purification (Zhao et al., 1993). In order to avoid the amplification of the wild-type template by the flanking primers, one of which is carried over from the primary reaction, the first 5 to 10 cycles of the secondary PCR are run without additional oligonucleotide primer. In this initial phase the mutagenic primer is essentially extended on the wild-type template and the residual flanking primer is consumed. For the secondary reaction to be successful, it is often necessary to linearize the template. Presumably this is due to the inefficient annealing of plasmid DNA to the relatively long primer represented by the product of the primary reaction. It is standard procedure to vary the amount of primary reaction product (10 to 100%) used in the secondary reaction in order to optimize the yield of the PCR.

The Two-Primer Secondary PCR Method

Incorporating a mutation in a codon distant from convenient restriction sites can also be performed by generating two overlapping DNA segments in separate primary PCRs. The primary PCRs extend in opposite directions from the site of the mutation and span restriction sites in the polylinker or unique sites internal to the gene (Fig. 9-7). The segments are used to "mega prime" each other in the secondary PCR to generate the full length gene sequence with the desired base substitutions (Higuchi et al., 1988). In order to maximize the yield of the secondary PCR an equimolar ratio of the gel purified primary PCR products are used in varying concentrations. This method can also be used to generate insertions and deletions in the sequence by designing 5' overhangs that are self-complementary (Ho et al., 1989).

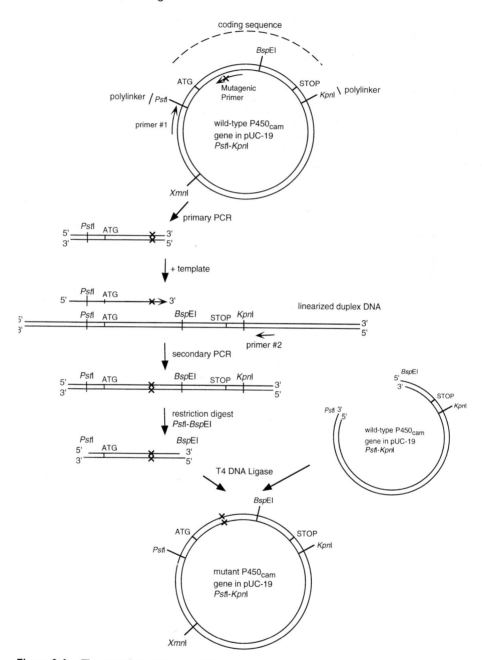

Figure 9-6. The one-primer secondary PCR method is applicable when there are no restriction sites near to the mismatched nucleotide(s). The product of the primary reaction is used as the mutagenic primer in the secondary PCR to extend the sequence in the opposite direction of the initial reaction such that unique restriction sites are available on both sides of the altered codon.

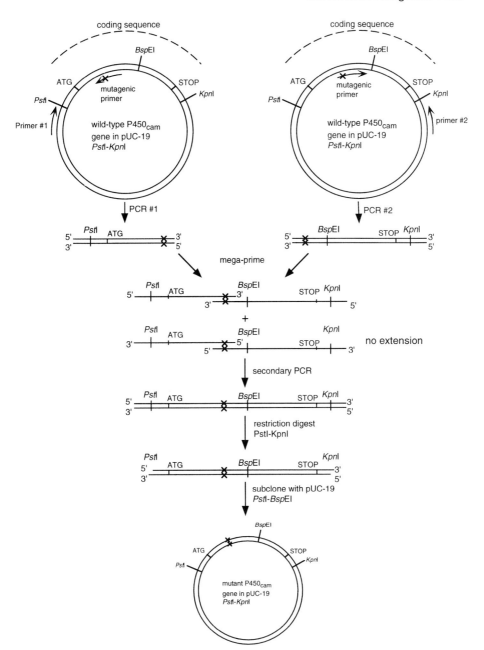

Figure 9-7. In the two-primer secondary PCR method the 3′-ends of one pair of primary PCR products are complementary and thus can anneal and subsequently megaprime off of one another to fill in the remaining portion of the duplex DNA. The other pair of DNA fragments anneal but cannot be extended in the 3′ → 5′ direction by the DNA polymerase.

PCR Conditions and Considerations

Since the *Taq* DNA polymerase does not have a $3' \rightarrow 5'$ exonuclease activity, random errors within the PCR-amplified region can occur. The dNTP and free Mg^{2+} concentrations (20 to 200 μM and 0.5 to 2.5 mM, respectively), the temperature of the annealing phase of the reaction ($>50°C$) and the complementarity of the template and 3'-terminus of the mutagenic primer are important for minimizing the misincorporation of nucleotides. Vent DNA polymerase is thermostable and offers the advantage of an inherent proofreading activity, but is more sensitive to reaction conditions such as Mg^{2+} concentration, and annealing and extension temperatures. Regardless which enzyme is used, the error rate is such that the accuracy of the polymerase must be checked by determining the sequence of the entire fragment produced in the PCR.

Standard denaturation conditions range from 90°C for 2 min to 97°C for 15 sec with *Taq* polymerase. The enhanced thermal stability of Deep Vent DNA polymerase (New England Biolabs) allows higher denaturation temperatures without appreciable loss of enzyme activity, but the conditions for efficient primer annealing and DNA polymerization may require adjustment in order to maximize product yield. The optimal annealing temperature is approximately 5°C below the T_m of the primer. In cases where the primer is very long (>200 nucleotides), such as in a secondary PCR, the annealing stage of the cycle can often be omitted and the extension phase lengthened (denaturation at 95°C for 0.5 min with an annealing-extension phase at 72°C for 2 min). Mispriming of the template and the formation of primer-dimers is promoted by too low annealing temperatures. The extension phase of the cycle is typically conducted at 72°C as the activity of *Taq* is maximal between 70 and 80°C.

Mutagenic primers should be at least 15 nucleotides in length and extend 3 to 4 nucleotides beyond the last mismatched base pair. A 4 to 8 nucleotide extension is preferred if a restriction site located near the terminus of the resulting PCR product is to be used for subsequent subcloning and the corresponding restriction enzyme is active at temperatures greater than 37°C (*Bst*EII, $T_{optimum}$ 60°C). The primer should not be self-complementary, especially at the 3'-end, in order to avoid secondary structure and the consumption of primer, dNTPs, and enzyme by contaminating primer-dimer artifacts (Innis et al., 1990). Runs of three or more Gs and Cs positioned near the 3'-end of the primer can promote mispriming. *Taq* DNA polymerase has been found to add an extra adenine on the 3' terminus of the transcribed DNA strand (Hemsley et al., 1989). This terminal transferase activity is not a problem when the end of the PCR product is to be cleaved by a restriction enzyme. If the DNA fragment is to be used in a second PCR, then the adenine overhang may result in an unwanted base pair mismatch in the final product of the secondary PCR. The primer should be designed with the extra A added in the third position of an appropriate codon (wobble nucleotide) such that the base pair change does not alter the amino acid sequence (Fig. 9-4). Introducing restriction sites into the sequence along with the desired mutation can be useful for screening potential mutant clones and future site-directed mutagenesis. If a potential site is found near the target codon, then a variation on the PCR-aided cassette method can be used to facilitate the mutagenesis by avoiding the secondary PCR (Fig. 9-8). Two primers with base pair mismatches, both of which contain the sequence for the new restriction site, are used to amplify the DNA sequence in separate PCRs, which extend upstream and downstream of the introduced restriction site. One of the primers also contains the nucleotide mismatches coding for the altered amino acid. The re-

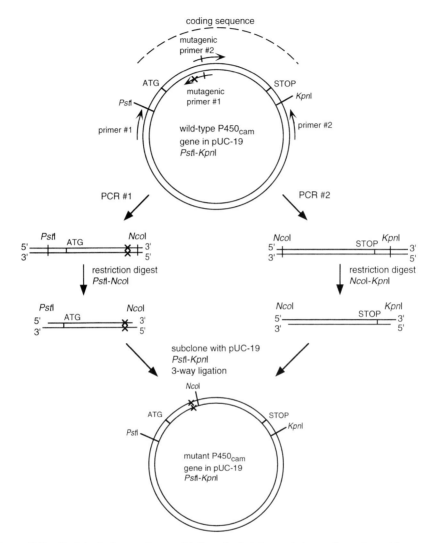

Figure 9-8. By introducing a unique restriction site that does not change the amino acid sequence of the translated protein (silent substitution) a derivative of the cassette-by-PCR method can be used to introduce the mutation without the need for a secondary PCR. The primary PCR products are ligated with a suitable vector using the new restriction site and other unique sites of the polylinker or within the gene.

striction site common to both of the fragments produced and unique sites flanking the gene are then used in a three-way ligation with the PCR products and suitable vector to construct plasmid DNA containing the mutant gene sequence.

Problems of low yields in PCR are often due to regions of extensive secondary structure in the template that can cause inefficient denaturation and subsequent annealing with primers. Incorporation of 7-deaza-2'-deoxyguanosine is often used to reduce the effect of these so-called compressions and increase the yield of the PCR (Innis et al.,

1990). The fidelity of the polymerase is not affected by this modified nucleotide and the product can be subcloned efficiently with pUC vectors and transformed into E. coli.

Examples of Applications of PCR Mutagenesis

Active Site Mutants of Cytochrome $P450_{cam}$

Cytochrome P450 is a monooxygenase that catalyzes dioxygen bond cleavage, resulting in the reduction of one oxygen atom to water and insertion of the other oxygen atom into the substrate. The superfamily of cytochromes P450 is important in the synthesis of steroids and fatty acids and in the oxidation of xenobiotics. The isozymes activate a wide range of substrates from ethylene to cyclosporin A with a variety of chemical reactions, including hydroxylation of carbons and heteroatoms, dealkylation of amines and ethers, epoxidation of double bonds, and reductive dehalogenation. Atmospheric oxygen and reducing equivalents derived from nicotinamide adenine dinucleotide (NADH) are used to activate a tertiary, secondary, or even primary CH bond. Cytochrome $P450_{cam}$, isolated from *Pseudomonas putida*, catalyzes the 5-*exo* hydroxylation of *d*-camphor as the first step in a metabolic pathway that allows the parent organism to live with camphor as its sole source of carbon.

$$\text{d-camphor} + O_2 + 2H^+ + NADH \xrightarrow{P\text{-}450_{cam}} \text{5-exo-hydroxycamphor} + H_2O + NAD^+$$

d-camphor

Cytochrome $P450_{cam}$ has served as a prototype, yielding enormous biophysical, chemical and mechanistic information regarding this class of enzymes (Sligar & Murray, 1986); the gene for $P450_{cam}$ has been cloned and functional protein isolated from *E. coli* in high yield. By utilizing site-directed mutagenesis, several structure-function relationships of cytochrome $P450_{cam}$ have been addressed. These include substrate access, protein-protein association, and electron transfer.

One area of active interest is an effort to control the regiospecificity, stereospecificity, and coupling of the hydroxylation reaction by altering specific substrate-protein contacts that determine the position and orientation of the substrate in the active site. The X-ray structure of cytochrome $P450_{cam}$ reveals that the active site is a buried cleft largely isolated from solvent and the C5 of the camphor molecule is adjacent to the heme iron. Twelve amino acid residues from six different pieces of secondary structure along with the protoporphyrin IX form the inside surface of the pocket. Extensive hydrophobic contacts are made with camphor, as the shape of the active site is highly complementary to the three-dimensional structure of this molecule. As a result, camphor has very little rotational mobility when bound to $P450_{cam}$ and the classic hand-in-glove fit of the native substrate camphor into the binding site of the enzyme is responsible for the observed regiospecificity. Initial investigations have defined the importance of various substituents in the binding site of the protein. In particular, site-directed mutagenesis experiments have shown that F87, T101, T185, V247, and V295 all contribute to the orientation of camphor and other substrates in the active site through steric interactions (Atkins & Sligar, 1989; Loida & Sligar, 1993a).

The design of new substrate-protein contacts for controlling regiospecificity and reaction efficiency has been the focus of extensive mutagenesis experiments. Camphor analogues and substrates bearing no structural resemblance to camphor are oxidized with reduced specificity. Hydroxylation occurs at three substrate carbons and the coupling of NADH to hydroxylated product is reduced tenfold. These molecules lack the methyl groups protruding from the bicyclic rings of camphor and, therefore, make fewer specific contacts with the protein active site. Site-directed mutagenesis was used to change the topology of the active site in an attempt to make it more structurally complementary to the substrate analogs. Molecular modeling based on the P450$_{cam}$ X-ray structure resulted in the design of amino acid substitutions that cause predictable changes in the regiochemistry and efficiency of the reaction with the camphor analogs norcamphor and 1-CH$_3$ norcamphor (Atkins & Sligar, 1989). The specificity of the hydroxylation reaction is increased by hydrophobic residues which pack the substrates more tightly into the active site and position C5 of the substrate closer to the heme iron.

Site-directed mutagenesis has also been used to elucidate the aspects of a substrate-P450$_{cam}$ complex that affect the individual branch points in the catalytic cycle (Loida & Sligar, 1993b). In the cytochrome P450 reaction cycle there are three branch points for the uncoupling of NADH-derived reducing equivalents whereby reduced dioxygen side products are liberated without activation of the substrate (Fig. 9-9). At the first branch point, the oxygen bound reduced form of the enzyme (oxy-P450) either disproportionates to superoxide anion and ferric P450$_{cam}$, or undergoes second electron reduction in the native reaction coordinate. The peroxy-P450 species produced marks the second branch point at which dissociation of hydrogen peroxide or dioxygen bond cleavage can occur. At the third branch point, the putative iron-oxo species is partitioned between hydrogen abstraction and subsequent substrate hydroxylation or further reduction to water in a classical oxidase stoichiometry (2 NADH:1 dioxygen). A set of mutants were designed to alter the coupling of ethylbenzene hydroxylation by in-

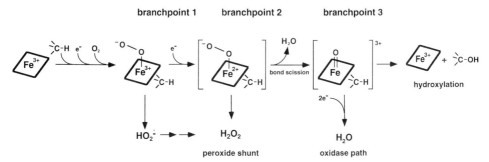

Figure 9-9. Schematic of the cytochrome P450 reaction illustrates the unproductive pathways that give rise to the side products hydrogen peroxide and water. The one electron reduced oxy-P450 intermediate marks the first branch point at which autoxidation or second electron reduction occurs. The two-electron reduced peroxy-P450 species either undergoes dioxygen bond cleavage or peroxide release at the second branch point. At the third fork the putative iron-oxo intermediate is partitioned between substrate hydroxylation and additional reduction via the oxidase pathway to form water.

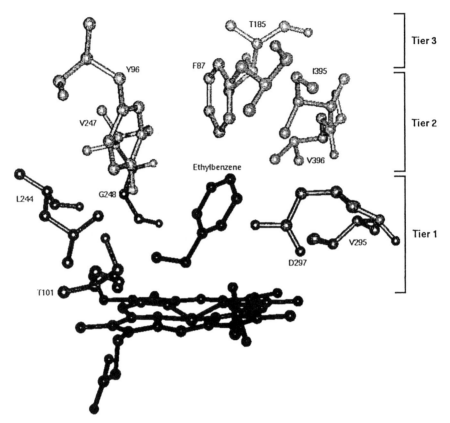

Figure 9-10. Side-on view of the cytochrome P450$_{cam}$ active site with a model of ethylbenzene oriented for benzylic hydroxylation. Residues in Tier 1, shown in dark gray, are positioned near to the heme and include T101, L244, V295, G248, and D297. Tiers 2 and 3, shaded light gray, form the upper region of the binding pocket and are comprised of residues F87, Y96, T185, V247, I395, and V396.

troducing steric bulk, in the form of large and hydrophobic amino acids, at various positions around the perimeter of the active site. The substrate was modeled into the active site of wild-type P450$_{cam}$ in a rigid mode and oriented to optimize hydrogen abstraction at the benzylic carbon. Residues T101, T185, V247, and V295 made extensive contacts with the substrate in the static complex and were, therefore, chosen for site-directed mutagenesis (Fig. 9-10). The NADH-derived reducing equivalents recovered in hydroxylated substrate, hydrogen peroxide and water were measured by standard biochemical techniques for eight single mutants and one triple mutant (Loida & Sligar, 1993b).

With respect to reaction efficiency, the direct release of superoxide at the first branch point never competes with second electron input for the wild-type and mutant enzymes. At the second and third branch points the reaction specificity is affected by alterations in the topology of the binding pocket (Fig. 9-11A). Increased commitment to catalysis at the second branch point is observed for all mutants and suggests that active site hydration is important in the uncoupling to form hydrogen peroxide. The increased steric bulk of the leucine, isoleucine, methionine, and phenylalanine side chains makes the active site more hydrophobic by filling volume otherwise occupied by water. A more nonpolar environment disfavors charge separation at the heme and dissociation of peroxide. At the third branch point, a strong correlation is observed between water production and the location and size of the amino acid side chain substitution (Fig. 9-11B). Larger hydrophobic side chains introduced in the upper regions of the binding

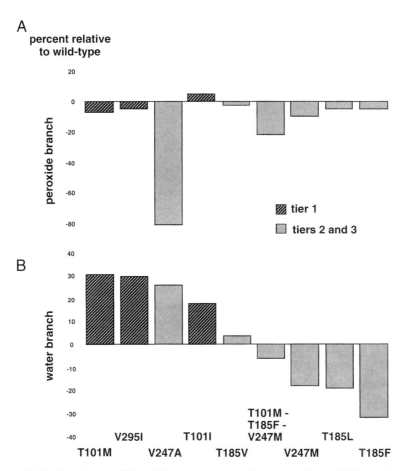

Figure 9-11. Reaction specificity differences between wild-type cytochrome $P450_{cam}$ and a series of active site mutants in the hydroxylation of ethylbenzene. The large hydrophobic side chains of Ile, Leu, Met, and Phe were substituted for Val and Thr at four positions in the binding pocket. Residues in Tier 1 are located near the heme and are shown in hatched bars, and residues in Tiers 2 and 3 are distant from the heme and are indicated by shaded bars. (A) Each bar represents the difference in the partitioning of oxygen to peroxide at the second branch point for the mutants minus wild-type $P450_{cam}$. The hydrogen peroxide is expressed as a percentage of the total oxygen consumed in the reaction. (B) Each bar represents the difference in water formed at the oxidase pathway relative to wild-type $P450_{cam}$. The water is expressed as a percentage of the total oxygen equivalents passing through the third reaction branch point. For active site residues in Tiers 2 and 3, there is a negative correlation between side chain packing volume and the production of water. This relationship is reversed for mutations in Tier 1.

pocket (Tiers 2 and 3, Fig. 9-10) increase the ratio of hydroxylated product to water production two- to fourfold, relative to wild type. Similar substitutions in residues near the heme plane (Tier 1) result in diminished 1-phenylethanol production. The partitioning between oxidase and hydroxylation activities at the iron-oxo species can be controlled to favor the production of either water or alcohol by reengineering the active site via site-directed mutagenesis. The results show that access of ethylbenzene to the iron is the key determinant in tight coupling at the third reaction branch point and is rationally modulated by changing the position and steric packing volume of nonpolar amino acid substitutions in the binding pocket. These mutagenesis studies provide insights into the molecular recognition of substrates by cytochrome P450$_{cam}$ and serve as a general basis for continuing investigations of structure-function relationships in protein-ligand interactions and attempts at rational redesign of enzyme active sites.

Summary

In summary, two easy methodologies for introducing site-directed mutations into a protein structure have been presented. Each of these methodologies can be readily learned in a very short period of time and, when coupled with detailed physical biochemical functional studies and the structural elucidation of mutant proteins, can provide powerful closure in the loop of structure-function elucidation. Although the completely uninitiated may require several months to become adept at handling the minute quantities of material that characterize recombinant DNA technology, it is clear that future use of this technology opens vast arenas for understanding the basis and function of cellular processes.

10

The Structural Basis of Antibody Catalysis

Donald Hilvert, Gavin MacBeath, and Jumi A. Shin

Complementarity is the basis for many chemical and biological interactions. The tremendous efficiency displayed by enzymic catalysts is in large part due to the shape and charge complementarity provided by enzyme binding sites to specific reaction transition states (Pauling, 1948). Complementarity is also the basis for the precise specificity displayed by receptors for their cognate ligands and by antibodies for particular antigens.

A revolution in molecular biology, which began two decades ago, has provided researchers abundant use of natural enzymes and is changing the face of molecular science. Recent advances in immunology have similarly enabled the exploitation of antibodies. Immunization with an appropriate antigen has been successfully used in countless cases to generate immunoglobulins with desired characteristics, such as high-affinity binding to a specific molecule or class of molecules.

The similarity of enzymes and immunoglobulins has also inspired researchers to contemplate how the exquisite fidelity of the immune system might be exploited to create antibodies with tailored catalytic properties (Lerner et al., 1991). Such agents would be especially valuable for those chemical transformations commonly used in the laboratory for which there exists no natural enzymic counterpart. Recently, these theoretical ambitions have become practical reality.

The Antibody-Antigen Complex

Antibodies are produced in all vertebrates during the B-cell mediated humoral immune response to antigenic challenge by foreign substances (Kabat, 1976). Under appropriate conditions, virtually any foreign molecule can elicit an immune response. It is estimated that the primary repertoire of the immune system contains on the order of 10^8 different antibodies (Alt et al., 1987), and this number may be further expanded by several orders of magnitude through somatic mutations which accumulate as the immune response to a particular antigen is refined (Rajewsky et al., 1987). The remarkable diversity and specificity that can be achieved by antibodies is now understood in structural terms (Davies et al., 1990; Wilson & Stanfield, 1993).

Antibody Structure

Antibodies are large modular proteins comprised of two heavy (H) and two light (L) chains, each made up of several homologous domains approximately 110 amino acids in length (Davies et al., 1990). The heavy chain has four such domains, and the light chain two. In the intact immunoglobulin, the heavy and the light chains are intimately paired by disulfide bonds and by packing of complementary domains to produce a characteristic Y-shaped scaffold (Fig. 10-1).

Antigen recognition is mediated by the variable F_v domain formed by the N-terminal regions of the light and heavy chains (Fig. 10-2). The combining site is constructed from six peptide loops that extend from a highly conserved structural framework of antiparallel β-barrels. Three loops (H1, H2, and H3) are contributed by the heavy chain and three loops (L1, L2, and L3) by the light chain. These loops, also called complementarity determining regions (CDRs), display a high degree of variability with respect to length and amino acid sequence. It is this hypervariability that provides the basis for both the specificity and diversity displayed by antibody molecules.

Structural studies of numerous antibodies and their complexes have shown that the immunoglobulin binding site can vary extensively in size, shape, and charge distribution (Davies et al., 1990; Wilson & Stanfield, 1993). Depending on the antigen, deep pockets, long clefts, or flatter, more undulating surfaces can be induced. In all cases,

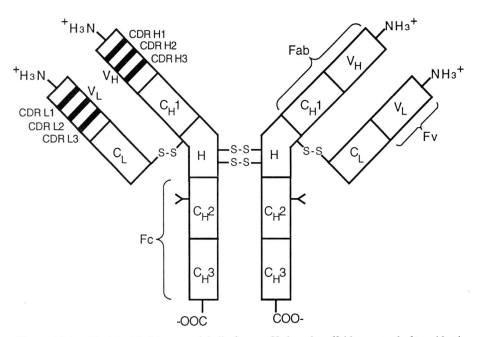

Figure 10-1. The intact IgG immunoglobulin forms a Y-shaped scaffold composed of two identical heavy chains and two identical light chains. Each heavy chain is composed of four domains (V_H, C_H1, C_H2, and C_H3), as well as a hinge region (H) connecting C_H1 and C_H2. The light chains are each composed of two domains (V_L and C_L). The two immunoglobulin binding sites are each defined primarily by the six complementarity determining regions (CDRs), shown in black for one of the variable domains.

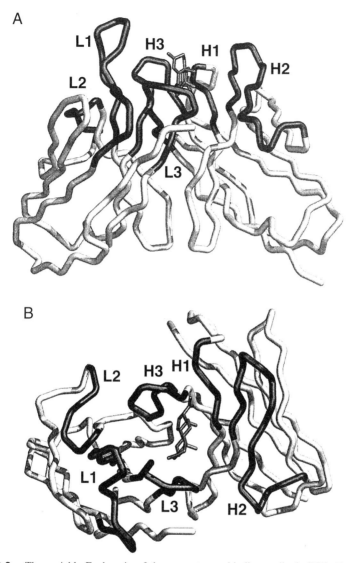

Figure 10-2. The variable F_v domain of the progesterone-binding antibody DB3 (Arevalo et al., 1993a) is illustrated, viewed with the light chain on the left and the heavy chain on the right. The six CDR loops of the antibody are shown in black, labeled L1, L2, and L3 for the light chain and H1, H2, and H3 for the heavy chain. The progesterone ligand is shown in gray. (A) Side view. (B) Top view, looking into the active site. The figure was generated with the program MIDAS (Ferrin et al., 1988).

a high degree of shape and chemical complementarity has been observed at the interacting surfaces of antibody and antigen, although recent investigations have indicated that water molecules may be essential for achieving an optimal fit in some cases (Mariuzza & Poljak, 1993). The amount of antibody surface area available for molecular recognition ranges from about 200 to 900 $Å^2$, while the percentage of the buried surface area of the antigen can vary from 6 to 95%. Thus, small ligands can be almost completely enveloped in cavitylike pockets, whereas larger protein antigens tend to interact with more extended binding surfaces.

The versatility of the immune response in recognizing the universe of possible antigens is a consequence of the combinatorial association of multiple CDR loops. All six loops are not necessarily required for molecular recognition, however, even for large antigens. In general, the heavy chain has been found to contribute slightly more interactions with the antigen than the light chain, and the H3 loop in particular appears to dominate (Fig. 10-2). In experiments involving shuffling of heavy and light chains, for example, certain heavy chains were capable of retaining antigen specificity upon recombination with multiple light chains, but the converse was not found (Collet et al., 1992). Nevertheless, heavy chain dominance is not completely generalizable: for the antidinitrophenol antibody AN02, light chain interactions predominate (Brunger et al., 1991).

Contributions to Antibody-Antigen Stability

The free energy change that occurs upon ligand binding to an antibody represents a balance of enthalpic and entropic factors (Mariuzza & Poljak, 1993; Novotny et al., 1989). In the cases studied to date, the principal driving force for association appears to be enthalpic in origin. The main contributions to complex stability thus include van der Waals and electrostatic interactions and hydrogen bonding. Entropic effects, on the other hand, can be either positive or negative. Release of ordered water molecules from the binding pocket and/or solvent shell surrounding the ligand leads to a favorable increase in entropy, but this is offset by the loss of conformational and translational entropy of the antigen and antibody upon complex formation. Recruitment of solvent molecules from bulk water to improve the complementarity of the interacting surfaces extracts a further entropic cost.

An unusual feature of the antibody combining site is the abundance of aromatic amino acids, particularly tyrosine (Padlan, 1990). Structural studies show that such residues play key roles in the recognition of a wide variety of ligands, including small charged haptens, peptides, and large proteins. Aromatic side chains present large surface areas for interaction with antigen and, because they are relatively rigid (compared to apolar aliphatic amino acids), lose little conformational entropy upon complex formation. Tyrosine can also interact with polar functionalities on an antigen via hydrogen bonding through its phenolic hydroxyl group. In addition, the aromatic ring itself may serve as a hydrogen bond acceptor. In energetic terms, the latter interaction has been estimated to be worth roughly half a normal hydrogen bond (Levitt & Perutz, 1988). It is consequently no surprise that the immune system has evolved to favor tyrosines over apolar aliphatic residues in the antibody response to foreign antigens.

Two other amino acids, histidine and asparagine, are also more likely to be found in the complementarity determining regions than in the immunoglobulin framework (Padlan, 1990). Asparagines probably serve a predominantly structural role, stabilizing and perhaps rigidifying the binding site through hydrogen bonding to the side chains of other residues and to the protein backbone. Histidine, on the other hand, is a particularly versatile residue. In addition to possible structural roles, it can contribute to the recognition of a wide range of antigens as a consequence of its aromaticity and physiologically accessible pK_a. These same properties are exploited in a variety of enzymes where histidines can function as nucleophiles or as acids-bases to facilitate a wide variety of chemical transformations (Fersht, 1985).

Conformational Changes

For many years, our view of immune recognition was informed by Fischer's classical "lock-and-key" model in which preformed antibody combining sites are essentially unaltered upon antigen binding. In support of this model, the first structures of antibody-protein complexes exhibited relatively small conformational changes, restricted to alterations in the peptide backbone of 1 to 2 Å. More recently, comparative studies of unliganded antibodies and the corresponding complexes with antigen have suggested a more complex situation (Arevalo et al., 1993a; Rini et al., 1992). A wide variety of conformational changes associated with antigen binding have been detected, including alterations in side chain rotamers, segmental movements of hypervariable loops, and changes in the relative disposition of the V_H and V_L domains (Wilson & Stanfield, 1993). On the basis of this apparent flexibility, an "induced-fit" model has been proposed as a more apt description of the antibody-antigen recognition process (Wilson & Stanfield, 1993).

Conformational flexibility on the part of the antibody presumably maximizes the ability of the immune system to recognize diverse populations of antigens. Indeed, the immune response is both heterogeneous, such that different antibodies can respond to a single antigen, and degenerate, in that the same antibody may respond to different antigens (Colman, 1988). Precise complementarity is probably not necessary for achieving an initial immune response, allowing a variety of molecules to initiate the microevolutionary processes leading to high affinity antibody-antigen complexes. Once an initial match has been made, somatic mutation and selection can be exploited to optimize recognition on a relatively short time scale. The inherent flexibility of antibodies also makes cross-reactivity with antigens of similar structure possible (Arevalo et al., 1993b).

Similarity of Antibodies and Enzymes

Antibodies share many common features with enzymes (Pauling, 1948). Enzymes, like antibodies, possess pockets capable of highly specific molecular recognition. Despite great variability in tertiary structure, their binding pockets are comparable in shape and size and tend to be constructed from peptide loops that link elements of secondary structure rather than from secondary structural elements themselves. The active sites of triose phosphate isomerase and chymotrypsin are cases in point (Branden & Tooze, 1991). Moreover, enzymes and antibodies exploit the same palette of amino acids and the same set of molecular interactions for their recognition tasks, employing hydrogen bonds, hydrophobic contacts, and van der Waals and dispersion interactions to achieve the size, shape, and charge complementarity necessary to bind ligands with high specificity and affinity. The ability to modulate interactions with ligands through conformational changes is another shared attribute.

Enzymes differ from antibodies in their ability to chemically transform the ligands they bind. Pauling (1948) explained this difference by noting that the active sites of enzymes are complementary to the transition states of the reactions they promote, whereas antibody binding pockets are configured for recognition of ground state molecules. That is, enzymes bind ephemeral transition states more tightly than ground

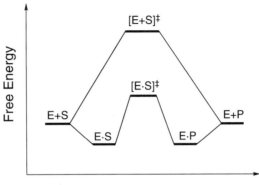

Figure 10-3. Reaction coordinate for uncatalyzed versus catalyzed reactions where E is enzyme, S is substrate, P is product, and ES and EP are Michaelis complexes of enzyme with substrate and product, respectively. Enzyme catalysis is achieved by preferential stabilization of the transition state of the reaction relative to the ground state of the substrate.

states (Fig. 10-3), while the opposite is true for antibodies. Of course, without preferential stabilization of transition states relative to ground states, no catalysis can result. Given this, it should be theoretically possible to create highly selective antibody catalysts by redirecting binding energy to the recognition of transition state structures.

Catalytic Antibodies

Theory and Practice

The notion that antibodies could be induced to act as catalysts originated with Jencks in 1969. He proposed that stable compounds that mimic the stereoelectronic properties of transition states might elicit antibodies capable of accelerating the corresponding chemical transformations (Jencks, 1969). Although this suggestion is simple and elegant, several factors precluded its direct implementation.

One practical obstacle to harnessing the catalytic potential of the immune system lay in identifying suitable transition state analogues. Much progress has been made in this regard over the past 25 years (Wolfenden, 1976). Chemists have created many potent enzyme inhibitors through transition state analogy, and many of these compounds have served as successful haptens for creating catalytic antibodies. Because small molecules, or haptens, are not immunogenic, generation of a potent immune response also requires coupling the transition state analogue to a suitable carrier protein. General principles regarding the length and composition of the linker have been established empirically (Erlanger, 1980), but viable coupling strategies must still be developed for each hapten on an individual basis.

Another difficulty lay in the polyclonal response of the immune system to invading substances. The immune system fashions a large number of antibodies capable of binding a given antigen, but these antibodies may vary significantly in structure and affinity. The polyclonal response, while good for the host and potentially useful for the rapid evaluation of new transition state analogues (Gallacher et al., 1993; Wilmore

& Iverson, 1994), dilutes the best catalysts with less effective agents and noncatalytic hapten binders. The use of polyclonal antibodies also precludes detailed study of the kinetic properties, specificity and mechanism of individual catalysts. Consequently, most experiments have relied on homogeneous and chemically well-defined monoclonal antibodies (Lerner et al., 1991). Monoclonals are prepared by standard methods involving fusion of antibody-producing B-cells with a myeloma cell line followed by subcloning of the resulting hybridomas (Goding, 1983; Harlow & Lane, 1988). This technology revolutionized the field of immunology in the 1970s and has also enabled rapid progress in the field of catalytic antibodies. Once identified, individual immunoglobulins can be obtained reproducibly in large quantity in tissue culture or by propagation of their parent hybridomas in mouse ascites. Moreover, purification is straightforward, so that each catalyst can be characterized by the same methods used to analyze naturally occurring enzymes.

Effective screening of the immune response poses additional challenges. Because the immune system manufactures countless antibodies that may bind hapten, linker, or carrier protein in various combinations, and because all haptens are necessarily imperfect transition state mimics, the fraction of antibodies that are catalytically active is often very small. It follows that the chances of finding an active catalyst increase with the number of antibodies screened. Typically, antibodies that bind the haptenic transition state analogue are first identified by an enzyme-linked immunosorbent assay (ELISA) (Clark & Engvall, 1980); individual binders are then tested for catalytic activity using conventional spectroscopic or HPLC assays. However, the expense and labor required to maintain large numbers of hybridomas generally limits the number of antibodies that can be evaluated in this way (typically to 10 to 50 candidates). Implementation of methodologies that allow larger numbers of hybridoma supernatants to be screened directly for catalytic activity would greatly facilitate the identification of active clones. The use of chromogenic substrates (Lewis et al., 1991; Thorn et al., 1995), tagging protocols (Lane et al., 1993), novel immunoassays (catELISA) (MacBeath & Hilvert, 1994; Tawfik et al., 1993), and biological assays (Lesley et al., 1993; Smiley & Benkovic, 1994; Tang et al., 1991) are among the most promising techniques being developed for this purpose.

Design of Transition State Analogues

The first attempts to create catalytic antibodies focused largely on proper mimicry of the shape and electrostatic properties of the transition state of the target reaction (Lerner et al., 1991). For instance, negatively charged phosphonate esters and phosphonamidates (**4**, Fig. 10-4), which are effective inhibitors of hydrolytic enzymes (Bartlett & Marlow, 1983), resemble the anionic tetrahedral transition state arising during the hydrolysis of esters and amides. These molecules have been employed extensively to induce antibody catalysts for simple hydrolytic reactions. For example, antiphosphonate and antiphosphonamidate antibodies have been shown to hydrolyze structurally analogous esters and carbonates, and in one case an amide, with rate accelerations in the range of 10^2 to 10^6 over background (Lerner et al., 1991; Stewart et al., 1993). The antibodies displayed classical Michaelis-Menten kinetics, exhibited substantial substrate specificity and, in some instances, high enantioselectivity. The observation that the transition state analogue is generally bound considerably more tightly than substrate is consistent with the notion of transition state stabilization. However, some catalytic anti-

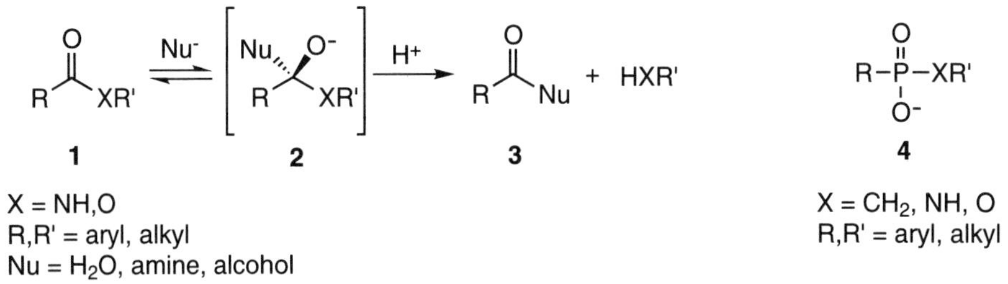

Figure 10-4. Acyl transfer reactions of ester or amide substrates (**1**) proceed through a high energy intermediate (**2**) that resembles the flanking tetrahedral, anionic transition states. The geometric and electronic properties of these transient species have been effectively mimicked for hydrolytic reactions (Nu = H_2O) by the corresponding phosphonate esters and phosphonamidates (**4**) (Bartlett & Marlow, 1983).

bodies achieve high rates by exploiting factors in addition to those programmed by their transition state analogue (see following), underscoring the potential versatility of the immune system in the manufacture of immunoglobulin combining sites.

In a second approach, bisubstrate analogues and conformationally constrained compounds have been developed to elicit antibodies capable of catalyzing reactions with unfavorable entropies of activation. By utilizing binding energy to freeze out rotational and translational degrees of freedom necessary to reach the transition state, antibodies can act as "entropy traps" (Westheimer, 1962). Bimolecular reactions are particularly susceptible to proximity effects, with large rate accelerations expected simply from increasing the effective concentrations of the two substrates at the active site of an antibody. This is illustrated by antibody catalysis of the bimolecular Diels-Alder reaction

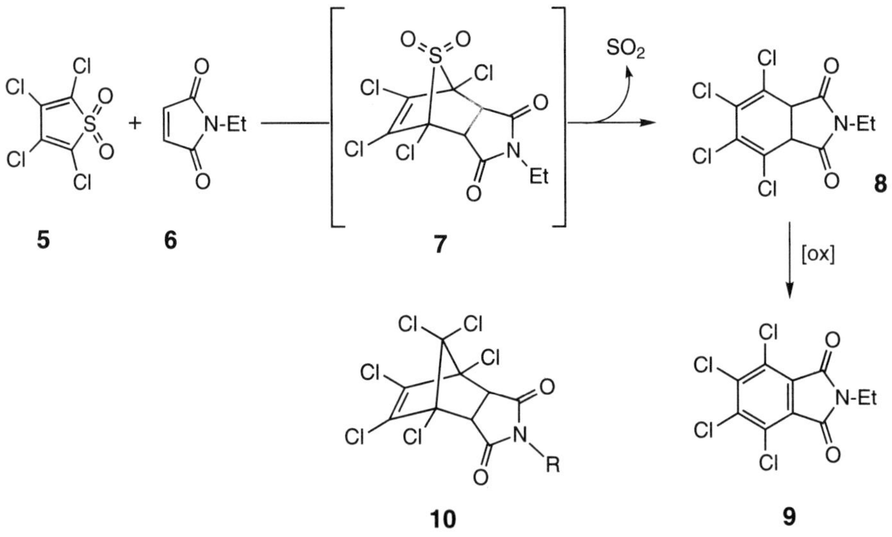

Figure 10-5. Antibodies raised against the hexachloronorbornene hapten **10** catalyze the Diels-Alder reaction between **5** and **6** by increasing the effective concentration of the substrates in the immunoglobulin binding site (Hilvert et al., 1989). Rapid elimination of SO_2 from product **7** and subsequent oxidation of **8** minimizes complications due to product inhibition.

(Fig. 10-5). The Diels-Alder reaction between dienes and alkenes to form cyclohexenes is among the most powerful transformations in synthetic organic chemistry. The unfavorable entropies of activation associated with the cycloaddition (typically in the range -30 to -40 cal K^{-1} mol^{-1}) reflect the fact that two substrates must be brought together to form a highly ordered boat-like transition state (Sauer & Sustman, 1980). Antibodies raised against bicyclic compounds that mimic this transition state structure and contain elements of both substrate molecules have achieved significant catalytic effects, including multiple turnovers, effective molarities in excess of 200 M, and exacting control of the *exo* and *endo* reaction pathways (Braisted & Schultz, 1990; Gouverneur et al., 1993; Hilvert et al., 1989). Antibodies have also been used as entropy traps for unimolecular reactions. Preorganization of a flexible substrate molecule into a reactive conformation has been realized, for example, in an intramolecular lactonization (Napper et al., 1987) and for model Cope and Claisen rearrangements (Braisted & Schultz, 1994; Hilvert et al., 1988; Jackson et al., 1988). The conversion of chorismate into prephenate, a biologically relevant [3,3]-sigmatropic rearrangement, is perhaps the best studied example (Hilvert et al., 1988; Jackson et al., 1988); it is discussed in depth in the following section of this chapter.

Strain and distortion have also been utilized to achieve large rate accelerations with antibodies. In this strategy, antibody binding energy is used to force a substrate into a destabilizing microenvironment. If the induced destabilization is relieved at the transition state, significant rate increases can be achieved (Jencks, 1975). In one example, geometric strain was exploited to generate mimics of the enzyme ferrochelatase (Cochran & Schultz, 1990a). Antibodies were raised against an *N*-methylated porphyrin (**13**, Fig. 10-6), which has a nonplanar structure corresponding to a distorted substrate conformation, and were shown to facilitate the metallation of mesoporphyrins **11** by Zn^{2+}. Antibody-catalyzed decarboxylations provide another example of this approach. The large rate accelerations observed in enzymatic decarboxylations are believed to arise, in part, from partitioning of the negatively charged carboxylate into a relatively apolar binding site where it is destabilized relative to the charge-delocalized transition state. Consistent with this view, hydrophobic haptens have yielded antibodies with decarboxylase activity in the range 10^4 to 10^7 over background (Ashley et al., 1993; Lewis et al., 1991; Smiley & Benkovic, 1994). Detailed investigation of one antibody that

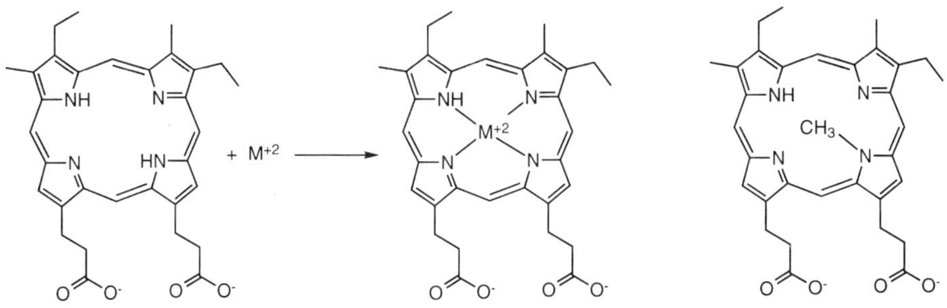

Figure 10-6. Metallation of the planar porphyrin substrate **11** is promoted by an antibody raised against the nonplanar hapten **13** (Cochran & Schultz, 1990a). The antibody utilizes binding energy to induce geometric strain in the substrate, which is partially relieved at the nonplanar transition state.

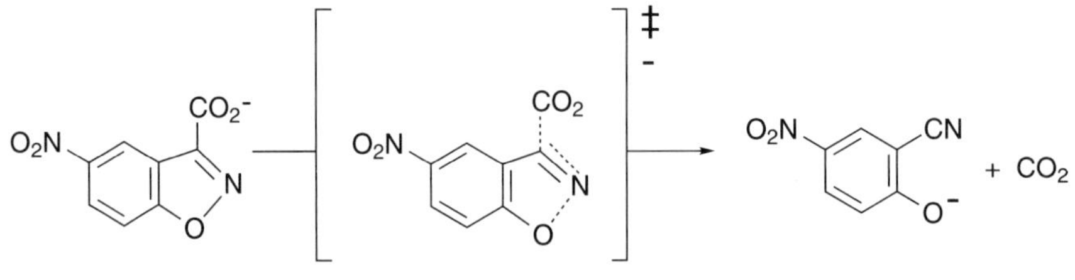

Figure 10-7. Antibodies elicited by hapten **17** accelerate the decarboxylation of 3-carboxybenzisoxazoles (e.g., **14**) via medium effects (Lewis et al., 1991). The anionic substrate is destabilized relative to the charge-delocalized transition state **15** in the nonpolar microenvironment of the antibody binding site.

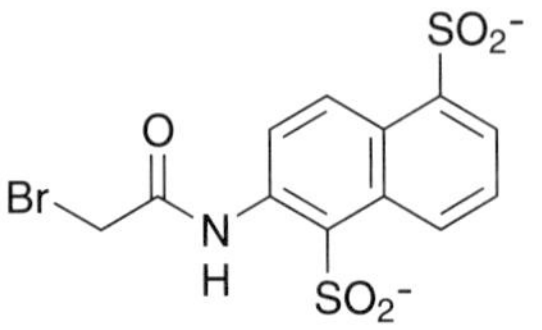

accelerates the decarboxylation of 3-carboxybenzisoxazoles to give salicylonitriles (Fig. 10-7) showed that medium effects can fully account for the more than 10^4-fold rate enhancements (Lewis et al., 1991, 1993; Tarasow et al., 1994). In essence, the antibody provides a tailored and preorganized solvent microenvironment for the reaction. Because many important chemical transformations, including aldol condensations, S_N2 substitutions, E2 eliminations, and even some hydrolytic reactions, are sensitive to solvent microenvironment, this strategy is likely to be increasingly exploited in the development of a wide variety of catalytic antibodies. In principle, the catalytic advantage to be derived from substrate destabilization is limited only by the amount of binding energy available to force the substrate into the destabilizing environment (Jencks, 1975).

Catalytic Groups and Cofactors Provide Alternative Pathways to Product

Many enzymes exploit multistep pathways to convert substrate into product. Serine proteases are among the best-studied examples. Rather than catalyzing the direct attack of hydroxide on the scissile carbonyl of the amide substrate, these enzymes utilize a two-step mechanism in which an active site serine is acylated transiently (Fersht, 1985). In addition to the nucleophilic serine, several other residues contribute to the efficiency of the catalyst, including a histidine that functions as a general base, an aspartate that orients the histidine, and side chain and backbone hydrogen bonds that help stabilize the tetrahedral, anionic intermediates and their flanking transition states. Removal of any one of these residues results in drastically decreased activity (Carter & Wells, 1988).

The difficulty in exploiting catalytic groups in antibodies is that of correct positioning within the binding pocket. Charge complementarity between hapten and antibody has been used successfully in a number of cases to elicit acids and bases in the immunoglobulin active site. For instance, positively charged ammonium salts have been used as haptens to produce antibodies containing complementary negatively charged carboxylate groups. Depending on the system, the induced carboxylate has served as a general base capable of effecting β-eliminations (Cravatt et al., 1994; Shokat et al., 1989; Thorn et al., 1995), as a general acid (in its protonated form) in a variety of hydrolytic reactions (Reymond et al., 1991, 1992, 1993; Uno & Schultz, 1992), and as a nucleophile that facilitates the interconversion of the geometrical isomers of an α,β-unsaturated ketone (Jackson & Schultz, 1991). By optimizing hapten design and efficiently screening the immune response, highly efficient catalysts can be produced. This was recently demonstrated for the base-promoted decomposition of benzisoxazoles, a concerted E2 elimination (Fig. 10-8) (Thorn et al., 1995). The best antibodies exhibited more than 10^3 turnovers and rate accelerations of greater than 10^8. Simultaneous induction of acid-base pairs capable of bifunctional catalysis may provide even larger effects. Heterologous immunization with two different but structurally related haptens, each containing a different charged moiety, may provide a simple means of eliciting arrays of multiple catalytic groups without requiring complex hapten synthesis (Suga et al., 1994). Direct selection of specific catalytic residues through the use of mechanism-based inhibitors as haptens has great promise as well (Barbas et al., 1993; Janda et al., 1997; Wagner et al., 1995).

Catalytic functionality can also be introduced into existing antibody combining sites through semisynthesis (Pollack et al., 1988; Pollack & Schultz, 1989) or by site-directed mutagenesis (Baldwin & Schultz, 1989; Stewart et al., 1994b). For example, thiol and imidazole moieties were engineered into the binding pocket of immunoglobulin MOPC315 using cleavable affinity reagents. The resulting semisynthetic antibodies substantially accelerated the cleavage of activated esters (Pollack & Schultz,

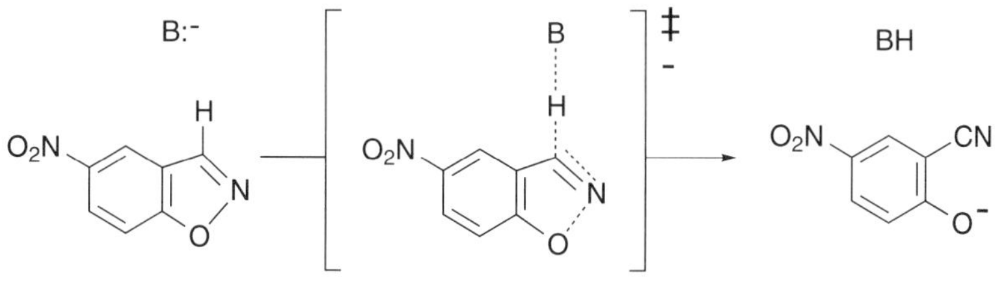

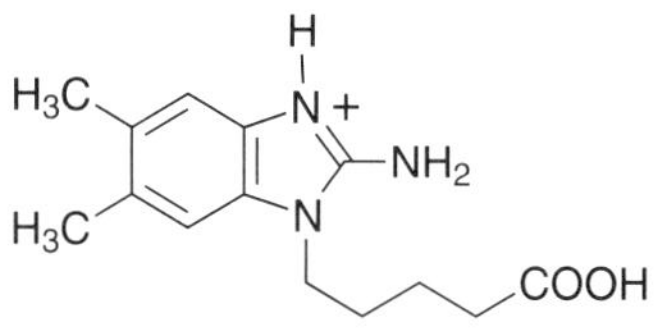

Figure 10-8. The base-promoted decomposition of **18** is greatly accelerated by an antibody raised against hapten **21** (Thorn et al., 1995). The hapten was designed to induce a general base in the antibody binding site, appropriately oriented to abstract the proton from the 3 position of the benzisoxazole substrate.

1989). Similarly, site-directed mutagenesis has been used to produce a hybrid Fv fragment of MOPC315 in which Tyr34 in the V_L domain was substituted with histidine (Baldwin & Schultz, 1989). This engineered antibody accelerated the hydrolysis of the 7-hydroxycoumarin ester of 5-(2,4-dinitrophenyl)aminopentanoic acid by 90,000-fold over the reaction catalyzed by 4-methylimidazole. Future studies to redesign existing antibodies will benefit greatly from forthcoming structural information.

The scope of antibody catalysis can be expanded further through the use of nonprotein catalytic cofactors. Enzymes frequently exploit metal ions, vitamins and other prosthetic groups to catalyze a wider spectrum of reactions than can be achieved with the 20 natural amino acids. Low molecular weight coenzymes are especially important for redox chemistry and functional group transformations. Metal ions (Iverson et al., 1990; Iverson & Lerner, 1989) and cofactors such as flavins (Shokat et al., 1988) and porphyrins (Cochran & Schultz, 1990b; Schwabacher et al., 1989) have already been used successfully in antibody catalysis, and this approach, which couples the intrinsic chemical reactivity of a well-defined catalytic agent with the specificity of the inducible antigen-binding site, is likely to become increasingly important. Moreover, because the development of catalytic antibodies is not limited by nature's choices, stable and inexpensive cofactors can be selected for specific tasks. This could prove especially important for preparative work, as practical applications of many enzymes are limited by the need to recycle expensive cofactors.

Structural Studies of Catalytic Antibodies

The field of catalytic antibodies has developed rapidly in the last ten years, but much remains to be learned. The scope and limitations of this technology still remain largely undefined, prompting ongoing efforts to expand the repertoire of reactions amenable to catalysis. Transformations that cannot be achieved efficiently or selectively with available chemical methods are particularly attractive targets for investigation (Schultz & Lerner, 1993). Integral to these studies, and of increasing importance, is the elucidation of the mechanistic and structural basis of antibody catalysis. Improved design of transition state analogues and optimization of the catalytic efficiency of first generation catalytic antibodies via protein engineering will require detailed understanding of the structure-function relationships that govern catalysis. Structural studies are also important for determining what if any limitation antibody structure places on the nature and efficiency of the reactions that can be carried out within the combining site. As outlined below, the first structures of catalytic antibodies complexed with their haptens are providing exciting insights into these issues and raising provocative new questions of their own.

Antibodies as Entropy Traps

Chorismate Mutase

The rearrangement of chorismate into prephenate (Fig. 10-9) was one of the first nonhydrolytic reactions to be catalyzed by antibodies (Hilvert et al., 1988; Jackson et al., 1988). This transformation is formally a Claisen rearrangement and, as noted previously in the chapter, a rare example of a biologically important pericyclic reaction. In bacteria, fungi, and higher plants, it is the committed step in the biosynthesis of the aromatic amino acids tyrosine and phenylalanine (Weiss & Edwards, 1980). The enzyme chorismate mutase accelerates the reaction by a factor of two millionfold over

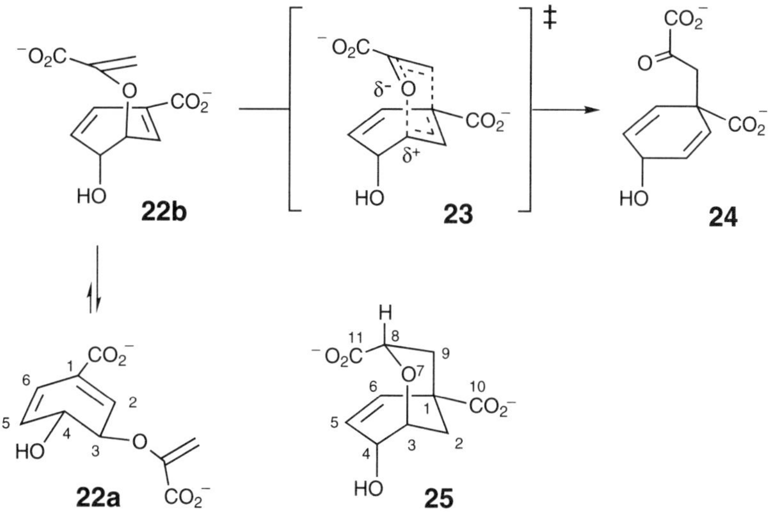

Figure 10-9. The conversion of chorismate (**22**) to prephenate (**24**) is catalyzed by antibodies raised against the bicyclic transition state analogue **25** (Hilvert et al., 1988; Jackson et al., 1988). Chorismate prefers the pseudodiequatorial conformation **22a** in solution, but must adopt the disfavored pseudodiaxial conformation **22b** to reach the pericyclic transition state **23**.

the uncatalyzed thermal process (Andrews et al., 1973), but how this level of activity is achieved has been a matter of debate for some time.

Stereochemical studies have shown that the enzymic and nonenzymic rearrangements proceed through the same chairlike transition state **23** (Copley & Knowles, 1985; Sogo et al., 1984). Solvent and kinetic isotope effects have established that the reaction proceeds asynchronously with carbon-oxygen bond cleavage preceding carbon-carbon bond formation (Addadi et al., 1983; Gajewski et al., 1987; Guilford et al., 1987). Consequently, a significant portion of the observed rate acceleration achieved by chorismate mutase may result from stabilization of the dipolar transition state through specific hydrogen bonding or electrostatic interactions (Severence & Jorgensen, 1992).

Entropic contributions must also be significant, since the flexible chorismate molecule preferentially adopts an extended pseudodiequatorial conformation **22a** in aqueous solvent that is not disposed for reaction (Copley & Knowles, 1987). To reach the pericyclic transition state, chorismate must undergo an energetically unfavorable conversion to the pseudodiaxial conformer **22b**, in which the enol pyruvate side chain achieves the proper orientation for rearrangement. Chorismate mutase could thus function as an "entropy trap" (Westheimer, 1962), utilizing binding energy to freeze out rotational degrees of freedom in the transition state. The favorable entropy of activation observed for the enzyme-catalyzed reaction compared to the spontaneous thermal rearrangement ($\Delta\Delta S^{\ddagger} = 13$ cal mol^{-1} K^{-1}) supports this view (Görisch, 1978). The importance of entropy is also indicated by the fact that conformationally locked transition state analogues, like **25** (Fig. 10-9), are the most potent inhibitors of chorismate mutase known (Bartlett et al., 1988).

Antibody-Catalyzed Rearrangement of Chorismate

The shape-selective character of the chorismate mutase reaction makes it an ideal candidate for antibody catalysis. Participation by general acids-bases and nucleophiles is

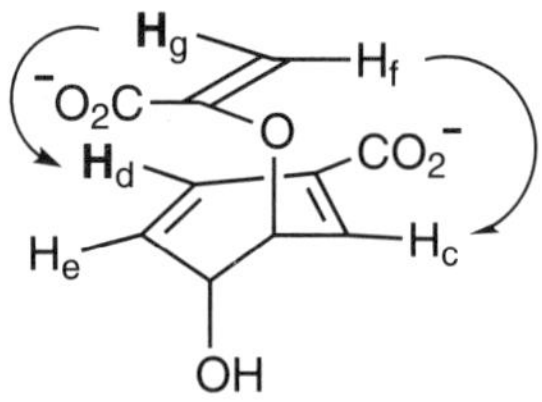

Figure 10-10. Transferred nuclear Overhauser effects (TRNOEs) from H_f to H_c and H_g to H_d are observed for chorismate **26** bound in the active site of the catalytic antibody 1F7 (Campbell et al., 1993).

not required, and the topological differences between substrate, product, and transition state minimize potential problems with product inhibition. Furthermore, because the catalyzed and uncatalyzed reactions are both unimolecular, direct comparisons of the two are expected to shed light on the basic mechanisms by which transition state stabilization is achieved.

Induction of a binding pocket capable of preorganizing the flexible substrate molecule into the reactive pseudodiaxial conformation represents the major challenge in the creation of chorismate mutase mimics. This problem was solved independently by two groups using conformationally restricted **25** as the template for antibody production (Hilvert et al., 1988; Jackson et al., 1988). For the purposes of immunization, **25** was coupled to carrier proteins via its C4 hydroxyl group. This coupling strategy makes the two carboxylates of the hapten the primary determinants of immune recognition and was expected to induce interactions capable of locking the flexible chorismate molecule into its reactive conformation.

This approach successfully yielded two antibodies with significant chorismate mutase activity. The better catalyst (11F1-2E9) is enantioselective and only two orders of magnitude less efficient than chorismate mutase itself (Jackson et al., 1988). Moreover, like its natural counterpart, 11F1-2E9 has a very favorable entropy of activation compared to the uncatalyzed process ($\Delta \Delta S^{\ddagger} = 12$ cal mol^{-1} K^{-1}) (Jackson et al., 1992). A second antibody (1F7), also highly enantioselective (Hilvert & Nared, 1988), appears to function in a fundamentally different way. Its 200-fold rate acceleration is achieved entirely by lowering the enthalpy of activation; its entropy of activation is actually 10 cal mol^{-1} K^{-1} less favorable than that of the uncatalyzed thermal rearrangement (Hilvert et al., 1988). Although activation parameters are difficult to interpret under the best of circumstances, the differences exhibited by 1F7 and 11F1-2E9 illustrate the capacity of the immune system to provide multiple solutions to a specific chemical problem.

The unfavorable $\Delta S^{\ddagger}$ observed for 1F7 also raises interesting questions regarding the extent to which the antibody stabilizes the diaxial conformer **22b**. Diagnostic transferred nuclear Overhauser effects (TRNOEs) between protons on the enol pyruvate side chain and protons on the cyclohexadiene ring (Fig. 10-10) demonstrate that a substantial fraction of bound chorismate *does* adopt the reactive geometry "programmed" by the transition state analogue (Campbell et al., 1993). Although chorismate may retain some flexibility at the active site, as evidenced by smaller nonspecific TRNOE signals, the NMR results rule out substrate conformational changes as the major contributor to the high entropy barrier for the 1F7-catalyzed reaction. Some combination of protein conformational changes and solvent reorganization must be involved.

The Structural Basis of Antibody Catalysis / **349**

Crystallographic Studies of 1F7

The three-dimensional structure of the Fab' fragment of 1F7 complexed with **25**, determined to 3.0 Å resolution (Haynes et al., 1994), has yielded an informative "snapshot" of an evolutionarily young catalyst. As shown in Fig. 10-11A, the Fab' closely resembles other known antibody structures. The transition state analogue binds at the apex of the

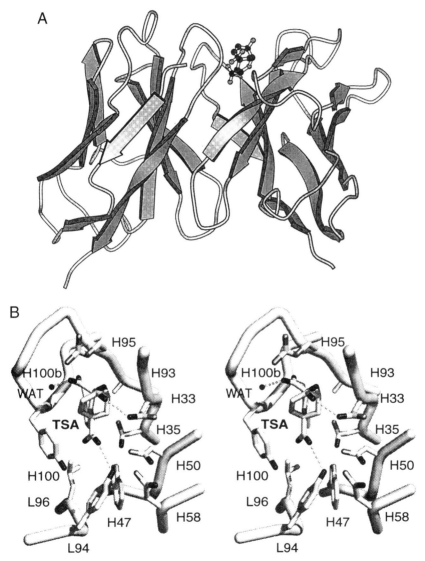

Figure 10-11. (A) Ribbon drawing of the Fab' fragment of 1F7 generated with the MOLSCRIPT program (Kraulis, 1991). β-Strands are shown as arrows directed toward their C-termini. The transition state analogue **25** is shown in ball-and-stick representation, with carbon atoms black and oxygen atoms gray. (B) Stereoview of the 1F7 active site with bound transition state analogue **25** (Haynes et al., 1994). Only the side chains of residues that make contacts with the ligand are displayed, with carbon atoms white, oxygens black, and nitrogens gray.

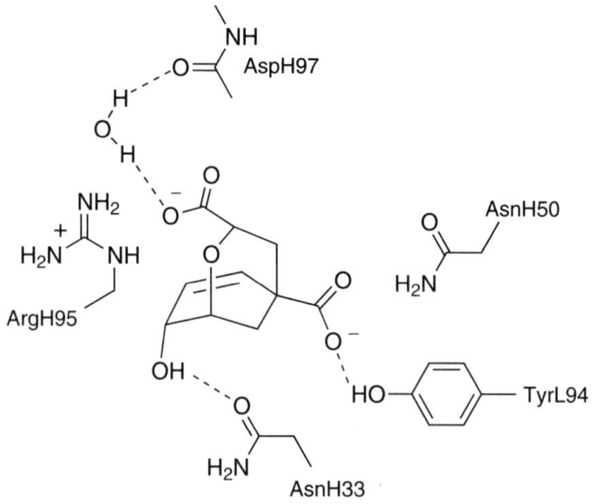

Figure 10-12. Schematic representation of the polar interactions between amino acid residues in 1F7 and the transition state analogue **25** bound in the active site (Haynes et al., 1994).

variable domain and, as is often found, interacts primarily with the heavy chain. More than 60% of the important protein interactions are contributed by CDR H3 alone.

The combining site of the antibody exhibits excellent shape and chemical complementarity to **25** (Fig. 10-11B). The hapten is roughly 90% buried upon complex formation, and its orientation within the binding pocket reflects the coupling strategy adopted for immunization. The C4 hydroxyl group is located at the mouth of the pocket, while the two carboxylates are more deeply buried. These polar functional groups are fixed in place by a combination of electrostatic, hydrogen bonding, and hydrophobic interactions (Figs. 10-11B, 10-12). The structural properties of the hapten are thus imprinted accurately on the antibody binding pocket.

The structural data suggest convincingly that the antibody-catalyzed rearrangement occurs via the same concerted pericyclic mechanism deduced for the uncatalyzed reaction. There are no acids, bases, or nucleophiles available for catalysis at the active site. This is consistent with the pH-independence of the antibody-catalyzed reaction and rules out a number of plausible mechanistic alternatives (Shin & Hilvert, 1994). The high degree of shape complementarity observed in the antibody-hapten complex presumably favors binding of the correct substrate enantiomer in a conformation appropriate for reaction. Protein conformational changes, detected in crystallographic studies of unliganded 1F7 and its complex with prephenate (M. Haynes, unpublished work), may modulate reactivity somewhat and could account for the observed unfavorable entropy of activation.

Crystallographic Studies of Natural Chorismate Mutases

The mechanism shared by the uncatalyzed and 1F7-promoted rearrangement of chorismate also appears to be used by the natural enzyme (Gray et al., 1990). Structural information has recently become available for chorismate mutases from several organisms (Lee et al., 1995b), and comparison of the enzymes and the antibody offers a rational basis for evaluating alternative strategies for accelerating this pericyclic reaction.

Despite similar steady-state kinetic profiles and comparable levels of inhibition by the transition state analogue **25**, chorismate mutases from different organisms possess little sequence similarity. This dissimilarity extends to their secondary structures and tertiary folds (Fig. 10-13). The monofunctional chorismate mutase from *Bacillus subtilis* (BsCM) is a symmetric homotrimer, packed as a pseudo-α/β-barrel, with three identical active sites formed at the interface of adjacent subunits (Chook et al., 1993).

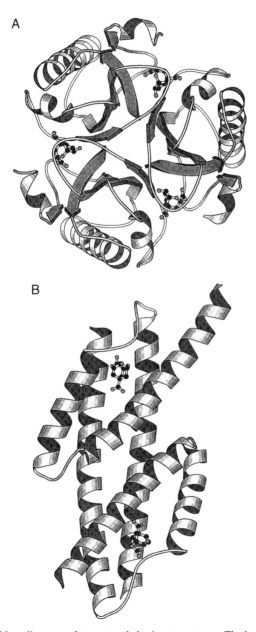

Figure 10-13. Ribbon diagrams of two natural chorismate mutases. The bound transition state analogue **25** is shown in ball-and-stick representation, with carbon atoms black and oxygen atoms gray. (A) Monofunctional chorismate mutase from *B. subtilis* (Chook et al., 1993). (B) Chorismate mutase domain of the bifunctional chorismate mutase-prephenate dehydratase from *E. coli* (Lee et al., 1995a).

In contrast, an allosteric chorismate mutase from the yeast *Saccharomyces cerevisiae* (ScCM) is an all α-helical homodimer (Xue et al., 1994), the core of which is related to the fold of the chorismate mutase domain from the bifunctional chorismate mutase-prephenate dehydratase from *Escherichia coli* (EcCM) (Lee et al., 1995a).

The active sites of BsCM and EcCM were identified with the transition state analogue **25** (Chook et al., 1993; Lee et al., 1995a). Like 1F7, both proteins provide environments tailored to the topology of the conformationally restricted inhibitor. However, they utilize many more strong polar interactions to bind ligand (Fig. 10-14). Given their disparate tertiary structures, the similar choice and placement of functional groups in the two enzymes represents a remarkable example of convergent evolution. In each active site, a glutamate residue is used to form a hydrogen bond with the hydroxyl group of **25**, and numerous positively charged residues are available for orienting the two carboxylates of the inhibitor. Also notable is the placement of a positively charged residue (Arg90 in BsCM and Lys39 in EcCM) within hydrogen bonding distance of the ether oxygen of **25**.

The functional groups in the enzyme active sites are thus strategically arranged to exert exacting conformational control over the flexible chorismate molecule. Hydrogen bonding and electrostatic interactions with the migrating enolpyruvate moiety probably stabilize the incipient oxyanion in the charge-separated transition state, and the anionic glutamate residue could stabilize the partial positive charge that develops on C5 of the cyclohexadiene ring. Experiments in which Glu78 and Arg90 in BsCM were mutagenized support these conclusions (Kast et al., 1996a,b). More elaborate mechanisms involving active participation of the C4 hydroxyl group have been ruled out in the case of EcCM by studies with substrate analogues (Pawlak et al., 1989), and dissociative pathways, including a nucleophile-assisted heterolytic process (Guilford et al., 1987), are unlikely in the absence of reactive active site residues.

The structural data also explain the observation that chorismate mutase is limited by product dissociation under saturation conditions, rather than by a chemical step (Gray et al., 1990). The active site of EcCM is completely shielded from solvent (Lee

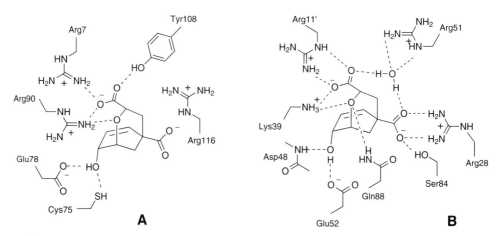

Figure 10-14. Schematic illustration of the polar interactions between amino acid residues and the transition state analogue **25** in the active sites of natural chorismate mutases from *B. subtilis* (**A**) and *E. coli* (**B**).

et al., 1995a), and substrate binding and product release will entail significant protein conformational changes. Although the active site of BsCM appears more open than that of EcCM in the crystal structure, ordering of the C-terminus has been proposed to sequester ligand from bulk solvent (Chook et al., 1994; Gray et al., 1990).

Why is 1F7 Less Active than Natural Chorismate Mutases?

Although chorismate mutases and 1F7 have active site pockets of similar shape and size, the antibody is about 10^4-fold less active than the enzymes. Comparison of the three crystal structures suggests that its low chemical efficiency is a consequence of fewer conformational constraints and poorer electrostatic complementarity to the transition state.

The relative paucity of functional groups in the antibody active site should make it a poorer "entropy trap" than either enzyme. The interactions available for orienting the migrating enolpyruvate with respect to the cyclohexadiene ring are illustrative. Both enzymes exploit multiple hydrogen bonding and electrostatic interactions to the C11 carboxylate to accomplish this task (Fig. 10-14), but in the antibody, a single water molecule is used (ArgH95 in 1F7 is probably important for charge neutralization, but it is too distant to donate a hydrogen bond to either of the carboxylate oxygens) (Fig. 10-12). A water molecule is unlikely to be effective in completely freezing out rotational degrees of freedom in the transition state; its extraction from bulk solvent and fixation at the active site will also be entropically costly. Furthermore, conformational changes detected in the unliganded antibody (M. Haynes, unpublished work) indicate that 1F7 is substantially less preorganized than BsCM, which requires comparatively little change in structure to bind ligand (Chook et al., 1994).

Interestingly, although 1F7 employs fewer strong polar contacts, it actually binds **25** more tightly (~5-fold) than either BsCM or EcCM (Haynes et al., 1994). The lower affinity of the enzymes for this ground state molecule may reflect unfavorable desolvation of the many charged active site residues upon ligand binding. Indeed, replacement of the ether oxygen of **25** with a methylene group decreases the stability of the EcCM-inhibitor complex by a factor of 250 (Bartlett et al., 1988), as expected for juxtaposition of a hydrophobic moiety with the ammonium group of Lys39.

While the lower charge density in the antibody active site may facilitate binding of ground state molecules, it is less suitable for stabilizing charge separation in the transition state. The side chain of ArgH95 occupies a similar position in the antibody as Arg90 in BsCM and Lys39 in EcCM, but because it must neutralize the full negative charge of the enolpyruvate carboxylate, it will be much less effective in accommodating additional negative charge on the ether oxygen at the transition state. In addition, 1F7 has no counterpart to Glu78 in BsCM or Glu52 in EcCM. The lack of a dipolar binding site is not surprising: hapten **25** effectively mimics the conformationally restricted geometry but not the polarized character of transition state **23**.

Improved electrostatic complementarity at the active site could conceivably be achieved by replacing the water molecule that contacts the secondary carboxylate of the hapten with an exogenously added cation. However, no effect on the rate of reaction is detected upon addition of either ammonium or guanidinium ions (Shin & Hilvert, 1994). It is possible that the antibody active site is unable to accommodate these cations, even though the relatively small ammonium ion might be expected to replace the bound water molecule with relative facility. Desolvation of the cation may be too costly entropically. Alternatively, bound ammonium and guanidinium ions may be no better than water at restricting the conformational flexibility of the substrate at the active site.

Next Generation of Experiments on 1F7

Given the ability of the completely unrelated tertiary folds of EcCM and BsCM to achieve comparable levels of catalysis, it seems unlikely that immunoglobulin architecture is incompatible with high chorismate mutase activity. The low efficiency of 1F7 is more plausibly a consequence of imperfect hapten design and short evolutionary history. Rational modifications in the basic strategy for generating catalytic antibodies might, therefore, yield better catalysts.

More extensive screening of the immune response to **25** is perhaps the simplest refinement that can be implemented. This approach has already produced a second antibody, 11F1-1E9, with substantially greater activity than 1F7 (Jackson et al., 1988). Nevertheless, only a tiny fraction of the immune response to **25** ($\sim$50 antibodies) has been tested for catalytic activity and even better catalysts might be found with more efficient assay protocols. Recently developed methods for displaying antibody fragments on phage and expression in microorganisms (Garrard & Zhukovsky, 1992), in conjunction with powerful biological screening and selection techniques (Lesley et al., 1993; Smiley & Benkovic, 1994; Tang et al., 1991), will facilitate this effort.

Refinements in hapten design could substantially increase the fraction of highly active catalysts present in the population of induced antibodies. Improved knowledge of enzyme structure and mechanism will be invaluable in making suitable alterations in hapten structure. In the present instance, derivatives of **25** containing additional charged moieties—an ammonium ion in place of the C4 alcohol or a phosphonate in place of the C11 carboxylate—could be used to induce binding pockets that more closely mimic the electrostatic environment of the BsCM or EcCM active sites.

Redesign of the first generation catalyst through site-directed mutagenesis or random mutagenesis coupled with classical genetic selection may be possible as well. The availability of structural data (Haynes et al., 1994) enables rapid identification of sites suitable for modification. For example, it might be possible to redesign CDR H3 of 1F7 to improve complementarity to the C11 carboxylate of the hapten. Replacement of the water molecule with properly oriented backbone or side chain interactions would be expected to greatly improve the conformational control of the antibody over the flexible substrate molecule. A biological selection assay has been developed which is based on the ability of the antibody to confer a growth advantage to a chorismate mutase-deficient yeast strain under auxotrophic conditions (Fig. 10-15) (Tang et al., 1991). Complementation of the chorismate mutase deficiency can now be exploited to select the most active antibodies from a large pool of variants generated by random mutagenesis. Direct selection for catalysis, rather than the conventional, more limited approach of screening for hapten binding, provides a rapid and efficient path to obtaining the most active catalysts.

Although the chorismate mutase rearrangement is one of the simplest reactions in mechanistic terms to be catalyzed by an enzyme, generating antibodies with the same level of activity as their natural counterparts represents a considerable challenge for the field of catalyst design. A combined chemical and biological approach to this problem may well have the greatest promise of success.

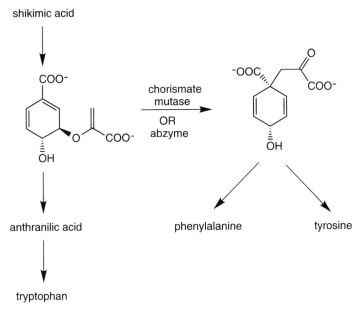

Figure 10-15. A catalytic antibody with modest chorismate mutase activity is able to complement a chorismate mutase-deficient yeast strain that is auxotrophic for phenylalanine and tyrosine (Tang et al., 1991).

Electrostatic and Nucleophilic Catalysis in Antibody Binding Pockets

Acyl Transfer Reactions

Given the potential utility of tailor-made peptidases and proteases in biology, the production of antibodies that cleave amide bonds site specifically has been an important goal in the field of catalytic antibodies since its inception (Lerner et al., 1991). While this objective remains elusive, much has been learned over the last several years about inducing antibodies capable of promoting other group transfer reactions, especially acyl transfer from esters and carbonates.

Ester hydrolysis (i.e., acyl transfer to water) has been investigated in particular detail (Lerner et al., 1991). This reaction is more facile kinetically than amide cleavage, and its solution mechanism has been studied extensively. In addition, a variety of excellent transition state analogues have been described. Of these, phosphonate and phosphonamidate haptens have proved particularly useful for generating hydrolytic catalysts: more than 50 antiphosphonate (phosphonamidate) antibodies with esterolytic activity have been reported to date. Structural and mechanistic studies of three catalysts that promote increasingly difficult acyl transfer reactions (CNJ206, 17E8, and 43C9) are contributing to our understanding of how phosphonate (phosphonamidate) recognition translates into catalysis (MacBeath & Hilvert, 1996).

Esterase CNJ206 was generated with phosphonate hapten **32** and selected for its high affinity for the short transition state analogue **31** (Fig. 10-16) (Tawfik et al., 1990). The selection strategy was designed to maximize the relative contribution of transition

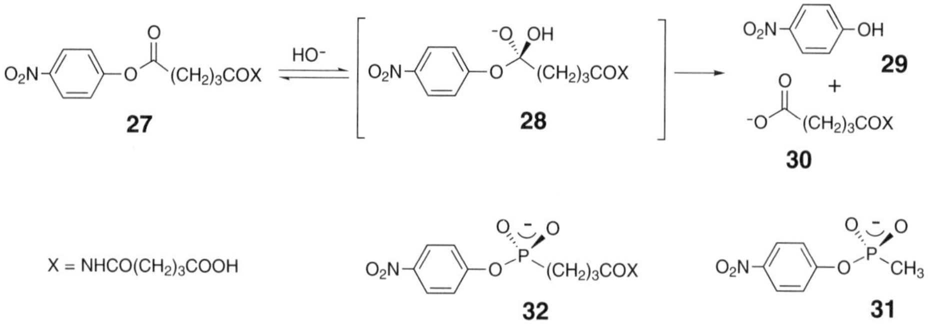

Figure 10-16. Antibody CNJ206 elicited by phosphonate hapten **32** and selected for high affinity to phosphonate **31** catalyzes the hydrolysis of *p*-nitrophenyl ester **27** (Tawfik et al., 1990; Zemel et al., 1994).

state structural elements to binding and yielded a panel of antibodies that catalyze the cleavage of activated *p*-nitrophenyl ester **27** with rate enhancements (k_{cat}/k_{un}) of 10^3 to 10^4 over background. Detailed mechanistic studies have yet to be reported, but CNJ206 accelerates substrate hydrolysis by a factor of 1,600, achieves multiple turnovers, and exhibits substrate specificity consonant with the structure of the hapten.

An analogous strategy (Fig. 10-17) yielded antibody 17E8 which promotes the enantioselective hydrolysis of unactivated phenyl esters of *N*-formyl-L-norleucine (**33**) and *N*-formyl-L-methionine (**34**) with rate enhancements of $\sim 10^4$ over the uncatalyzed reaction (Guo et al., 1994). The pH-rate profile for the antibody-catalyzed reaction is bell-shaped, suggesting the possible participation of two ionizable groups at the active site. Although the phosphonate transition state analogue implies a simple mechanism in which hydroxide attacks the scissile carbonyl group of the ester substrate, a two-

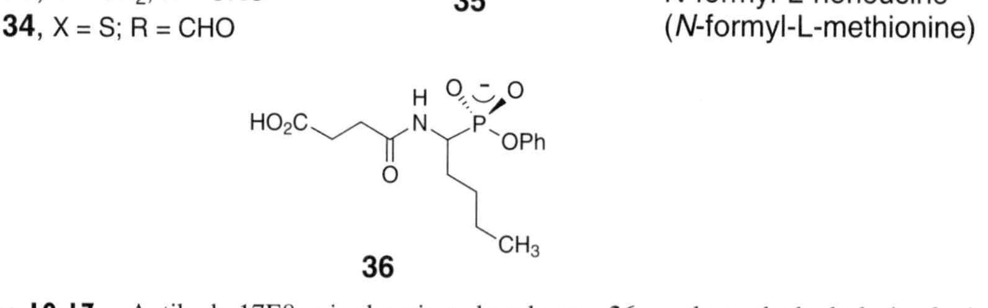

Figure 10-17. Antibody 17E8, raised against phosphonate **36**, catalyzes the hydrolysis of substrates **33** and **34** (Guo et al., 1994).

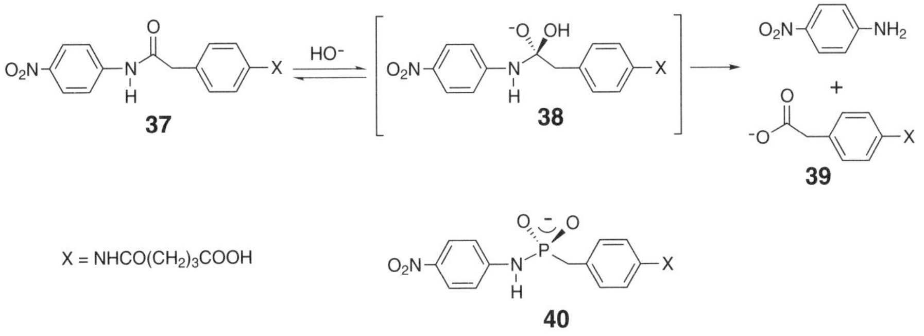

Figure 10-18. Antibody 43C9 was elicited by phosphonamidate hapten **40** and catalyzes the hydrolysis of the structurally analogous *p*-nitrophenylanilide **37** (Janda et al., 1988).

step mechanism with transient acylation of the antibody by substrate has been proposed for 17E8 on the basis of hydroxylamine partitioning data. This proposal warrants some skepticism, however, given inconsistencies between the hydroxylamine-dependent rates of phenol release and quantitation of the hydroxamic acid product (Guo et al., 1994).

Much stronger evidence for the formation of a covalent acyl-antibody intermediate has been obtained for the antiphosphonamidate antibody 43C9 (Fig. 10-18) (Stewart et al., 1994a). This catalyst is one of the most active hydrolytic antibodies known and a rare example of an amidase. At pH 9.0, 43C9 accelerates the hydrolysis of *p*-nitroanilide **37** by a factor of 2.5×10^5 (Janda et al., 1988); it also accepts a variety of substituted phenyl esters as substrates (Gibbs et al., 1992a), albeit with somewhat diminished efficiency relative to the corresponding uncatalyzed reactions. Mutagenesis experiments (Roberts et al., 1994; Stewart et al., 1994b) and extensive steady-state and pre-steady-state kinetic data (Benkovic et al., 1990) support a multi-step mechanism involving attack of a histidine side chain on substrate to form an acyl-intermediate which subsequently decomposes with hydroxide. The acylated antibody does not accumulate under most conditions, but it could be detected by electrospray mass spectrometry at pH 5.9 where it comprises roughly 10% of the antibody sample (Krebs et al., 1995). The mechanistic sophistication of 43C9 is thus reminiscent of more highly evolved natural enzymes and reveals the chemical capabilities inherent within the immunological repertoire.

Crystallographic Studies of Esterolytic Antibodies

The structures of CNJ206 (Charbonnier et al., 1995) and 17E8 (Zhou et al., 1994) complexed with their phosphonate hapten were recently determined at 3.2 and 2.5 Å resolution, respectively, affording a valuable opportunity to compare and contrast the active sites of two independently generated esterolytic antibodies. Moreover, the structure of uncomplexed CNJ206 is available (Golinelli-Pimpaneau et al., 1994), so that ligand-induced conformational changes can be assessed as well.

In the absence of hapten, the active site of CNJ206 is a long, shallow groove (Golinelli-Pimpaneau et al., 1994). The binding pocket contains a large proportion of aromatic residues (four tyrosines, two tryptophans and one phenylalanine), as well as a histidine residue and an arginine, in accord with the results of earlier chemical modification studies. The side chain of TyrH97 (Kabat numbering system) from the CDR H3 loop binds within the cavity and prevents formation of the deep pocket that is generally found in antibodies specific for small molecule haptens. Upon complex formation, however, a significant rearrangement of the CNJ206 Fv domain is observed (Charbonnier et al., 1995). The V_H domain not only moves relative to V_L but the CDR H3 loop undergoes a large conformational change to create a well-defined pocket that almost completely (95%) buries the transition state analogue. Although this structural reorganization may reflect specific packing constraints in the different crystals and have little relevance to catalysis in solution, it illustrates the intrinsic flexibility of antibody molecules and raises the possibility that catalytic antibodies may be able to adjust their shape in response to subtle changes along the reaction coordinate.

Hydrogen bonding and hydrophobic interactions dominate the binding of transition state analogue **31** by CNJ206 (Fig. 10-19). In the complex (Charbonnier et al., 1995), the *p*-nitrophenyl group sits in a deep hydrophobic pocket, while the phosphonate is located closer to the entrance of the cavity where it forms hydrogen bonds with HisH35 and the peptide NH groups of two consecutive residues of CDR H3. These observations suggest that CNJ206 functions according to the mechanism programmed by its hapten. It appears to be a relatively simple catalyst that facilitates direct hydroxide attack on the scissile carbonyl of *p*-nitrophenyl esters by stabilizing the resulting oxyanion through specific hydrogen bonding interactions. The magnitude of the rate acceleration ($\sim 10^3$-fold) corresponds roughly to the contribution of an oxyanion binding residue in the serine protease subtilisin (Bryan et al., 1986; Wells et al., 1986) and is compatible with this conclusion. Conceivably, HisH35 could function as a general base to assist in the delivery of the water molecule, but titratable groups have not (yet) been reported for this antibody.

Antibody 17E8, which catalyzes the hydrolysis of unsubstituted aryl esters (Guo et al., 1994), shares only modest sequence identity with CNJ206 (48% for V_H and 59% for V_L) (Guo et al., 1995; Zemel et al., 1994). Nevertheless, their active sites are surprisingly similar. As in CNJ206, hapten **36** binds to 17E8 so that its aryl moiety is embedded in a deep hydrophobic pocket. Remarkably, the contact residues that line this pocket are highly conserved between the two antibodies (Table 10-1; Fig. 10-20). The

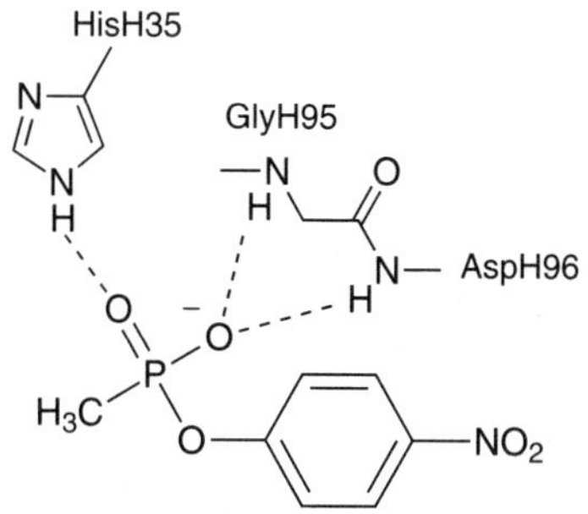

Figure 10-19. Schematic representation of the polar interactions between short transition state analogue **31** and active site residues of the catalytic antibody CNJ206 (Charbonnier et al., 1995). Residues are numbered according to Kabat et al. (1987).

Table 10-1 Comparison of Residues That Line the Binding Sites of Three Different Hydrolytic Antibodies

a. Residues of 17E8 in contact with the aryl phosphonate moiety of hapten **36** (Fig. 10-17) and the corresponding residues in CNJ206 and 43C9[a]

Residue[b]	Location[c]	CNJ206	17E8	43C9
L36	FR2	L	Y	Y
L89	CDR3	L	L	Q
L96	CDR3	Y	R	R
L98	FR4	F	F	F
H35	CDR1	H	H	H
H37	FR2	V	V	V
H47	FR2	W	W	W
H93	FR3	A	K	V
H95	CDR3	G	S	Y
H103	FR4	W	W	W

b. Residues of 17E8 in contact with the alkyl side chain moiety of hapten **36** (Fig. 10-17) and the corresponding residues in CNJ206 and 43C9[a]

Position[b]	Location[c]	CNJ206	17E8	43C9
L34	CDR1	S	G	A
L46	FR2	R	L	L
L49	FR2	Y	H	Y
L91	CDR3	Y	Y	H
H96	CDR3	D	Y	G
H97	CDR3	Y	Y	Y
H98	CDR3	Y	G	G

[a]Residues in 17E8 that make contact with hapten **36** were identified using the program CONTACSYM (Sheriff et al., 1987). The corresponding amino acids in CNJ206 and 43C9 were identified by primary sequence alignment and three-dimensional structure comparison.
[b]Residue positions are labeled according to the Kabat numbering system (Kabat et al., 1987) here and throughout the text. Note that this convention differs from the sequential numbering system previously adopted for 17E8 (Zhou et al., 1994) and CNJ206 (Charbonnier et al., 1995; Golinelli-Pinpaneau et al., 1994).
[c]Abbreviations: FR, framework region; CDR, complementarity determining region.

combining sites exhibit greater individuality in those regions that contact components of the hapten that are not common to both. Antibody 17E8 also provides more extensive interactions with the polar phosphonate group of the hapten than CNJ206. In addition to the hydrogen bond from the backbone NH of residue H96 (which is present in both antibodies), the side chains of two cationic residues, LysH93 and ArgL96, provide complementary electrostatic-hydrogen bonding interactions with the pro-*S* and

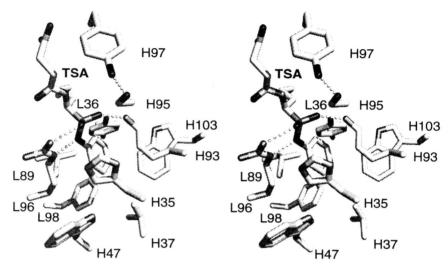

Figure 10-20. Stereoview depicting the active site of the esterolytic antibody 17E8 (Zhou et al., 1994). With the exception of TyrH97, only residues in contact with the phosphonate and aryl moieties of the transition state analogue **36** are shown. Residues are numbered according to Kabat et al. (1987).

pro-R phosphonate oxygens, respectively (Figs. 10-20 and 10-21). HisH35, which is an important functional group in CNJ206, appears to be inappropriately oriented in 17E8 and too distant (3.2 Å) to contibute a stabilizing hydrogen bond to the phosphonate (Fig. 10-20).

The simplest interpretation of the structural similarities between 17E8 and CNJ206 is that both antibodies exploit the same hydrolytic mechanism: direct hydroxide attack on the ester substrate. In this view, the greater number of stabilizing interactions that 17E8 provides to the negatively charged tetrahedral transition state accounts for its enhanced activity relative to CNJ206. Cationic residues are known to be very effective in

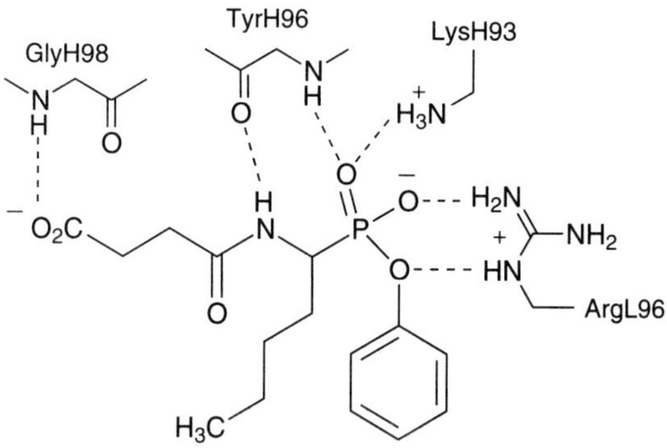

Figure 10-21. Schematic representation of the polar interactions between transition state analogue **36** and active site residues of the esterolytic antibody 17E8 (Zhou et al., 1994). Residues are numbered according to Kabat et al. (1987).

the stabilization of oxyanions. Mutagenesis studies on the metalloprotease carboxypeptidase A showed that a single arginine residue stabilizes the transition state for ester hydrolysis by as much as 6.0 kcal/mol (Phillips et al., 1990, 1992). The availability of multiple cationic residues may also explain the ability of 17E8 to hydrolyze unsubstituted aryl esters. The side chain of ArgL96, in particular, appears poised to complement the incipient phenolate anion electrostatically. In CNJ206 this arginine is replaced by a tyrosine which will be less able to accommodate negative charge on the leaving group.

As noted above, an alternative mechanism involving rate-limiting formation of a covalent acyl-antibody intermediate has also been considered for 17E8 (Guo et al., 1994). Modeling studies based on the structure of the antibody-hapten complex led to the hypothesis that SerH95 is the active site residue that becomes acylated during catalysis (Zhou et al., 1994). Because the serine side chain is hydrogen bonded to TyrH97 and oriented away from the active site, this mechanism requires a novel pH-sensitive switch. Ionization of the tyrosine phenol at high pH would allow the serine side chain to undergo a 180° rotation that would place its hydroxyl group within hydrogen-bonding distance of HisH35 and orient it for attack on the scissile carbonyl of the substrate. Although an appealing analogy can be made between this Ser-His dyad and the Ser-His-Asp catalytic triad of serine proteases, convincing evidence in support of the two-step mechanism is lacking. Indeed, counterevidence is provided by a closely related antibody (29G11), obtained in the same fusion that yielded 17E8 (Guo et al., 1995). Antibody 29G11 has a glycine residue in place of the "nucleophilic" serine but no other active site changes. Despite this substitution, the SerH95Gly variant retains 26% of the hydrolytic activity of 17E8 and exhibits essentially the same pH-rate profile. These observations are inconsistent with critical catalytic roles for SerH95 and TyrH97. For comparison, mutation of the nucleophilic serine in subtilisin causes a greater than 10^6-fold drop in activity, and the serine of the Ser-His dyad in a subtilisin variant lacking the triad aspartate contributes $\sim 10^2$-fold to catalytic efficiency (Carter & Wells, 1988). Additional kinetic and mutagenesis studies on 17E8 are clearly needed to clarify the extent to which SerH95, TyrH97, HisH35, and the other amino acids that line the binding pocket participate in ester cleavage.

Homology Modeling of 43C9

Analysis of antibody-antigen interactions is essential for a complete understanding of catalytic mechanism. Unfortunately, the structures for only a tiny fraction of all antibodies can ever be elucidated by X-ray crystallography or NMR spectroscopy. In the absence of high resolution data, homology modeling represents a potentially valuable approach for establishing preliminary structure-function relationships and for proposing mechanistic hypotheses that can be tested by mutagenesis. Homology modeling is possible because antibodies of divergent binding specificity share a common tertiary structure, differing mainly in the length and sequence of the six complementarity determining regions that mediate antigen recognition. This approach has been used to gain structural insight into the amidase 43C9 (Roberts et al., 1994).

A three-dimensional model of 43C9 was constructed in a stepwise fashion utilizing an antibody structural database (Charbonnier et al., 1995). Sequence comparisons were first used to select appropriate framework regions. The CDRs were then modeled using structural templates from antibodies with closely related sequences. The conformation of five of the six loops could be predicted with a high degree of confidence, but successful modeling of H3 was made difficult by the lack of highly homologous sequences. After pairing the V_H and V_L domains and docking the hapten, the H3 loop was optimized with the rest of the model in a final step involving manual adjustment and energy minimization of side chain positions.

The resulting computer model bears many similarities to CNJ206 and 17E8 (neither of which was available when the model was constructed). Indeed, the groups that line the binding pocket are highly conserved in each of the antibodies (Table 10-1; Fig. 10-20). Specifically, seven of the ten residues that contact the aryl phosphonate in 17E8 are unchanged in 43C9 and another is mutated conservatively (LeuL89Gln). Replacement of LysH93 with valine represents a substantive change, but CNJ206 has an alanine at this position, so a significant alteration in active site geometry is unlikely. Incorporation of a histidine at position L91 (in place of tyrosine) is unique to 43C9 and may account for the special properties of this catalyst (see the following).

Despite these shared features the 43C9 model differs greatly from the crystallographically determined 17E8 and CNJ206 complexes with respect to the purported conformation of CDR H3 and the orientation of bound hapten. In the model, the hydrophobic pocket that binds the aryl leaving group in the CNJ206 and 17E8 structures is occupied by the side chains of PheH100b and TyrH95. Hapten **40** was consequently docked in a rather shallow depression on the antibody surface to allow formation of a salt bridge between its phosphonamidate moiety and the guanidinium group of ArgL96. While a distinctive binding mode for this ligand cannot be ruled out, it seems more plausible that the modeled conformation of the H3 loop is incorrect and that 43C9 recognizes the aryl phosphonamidate moiety in much the same way as the two phosphonate-binding antibodies. This conclusion is consistent with the otherwise remarkable similarities of the three binding pockets and rationalizes the finding that *p*-nitrophenol release limits the rate of ester hydrolysis at high pH (Benkovic et al., 1990). If correct, it also illustrates the limitations of homology modeling in the absence of appropriate template molecules. Even assuming such templates are available, the inherent flexibility of the H3 loop, exemplified by the large ligand-induced conformation changes in CNJ206 (Charbonnier et al., 1995), may ultimately limit the utility of this approach for establishing accurate binding site geometries.

The ambiguities of the 43C9 model must eventually be resolved crystallographically, but mutagenesis experiments have already established the importance of three active site residues. The computer model implicated ArgL96 as the amino acid probably involved in stabilization of the oxyanionic intermediates and transition states (Roberts et al., 1994). In accord with this hypothesis, substitution of ArgL96 by glutamine produced a mutant antibody with no measurable esterase activity but essentially unaltered affinity for the reaction products (Roberts et al., 1994). Mutagenesis of two histidine residues identified by the model similarly yielded inactive 43C9 variants (Stewart et al., 1994b). In addition to their deleterious consequences for catalysis, substitutions for HisH35 produced mutant proteins with substantially weaker affinity for hapten and products. In contrast, replacement of HisL91 by glutamine decreased catalytic efficacy at least 50-fold but had little effect on ligand binding. These results were interpreted in favor of a structural role for HisH35 and a catalytic role for HisL91 as the active site nucleophile (Stewart et al., 1994b). Failure to detect a covalent intermediate by electrospray mass spectrometry when the HisL91Gln mutant was incubated with the ester analogue of **37** has been cited as further support for transient acylation of HisL91 during catalysis (Stewart et al., 1994a).

A mechanism compatible with the experimental results is summarized in Fig. 10-22 (Stewart et al., 1994a). In the first step, binding of the leaving group in the hydrophobic

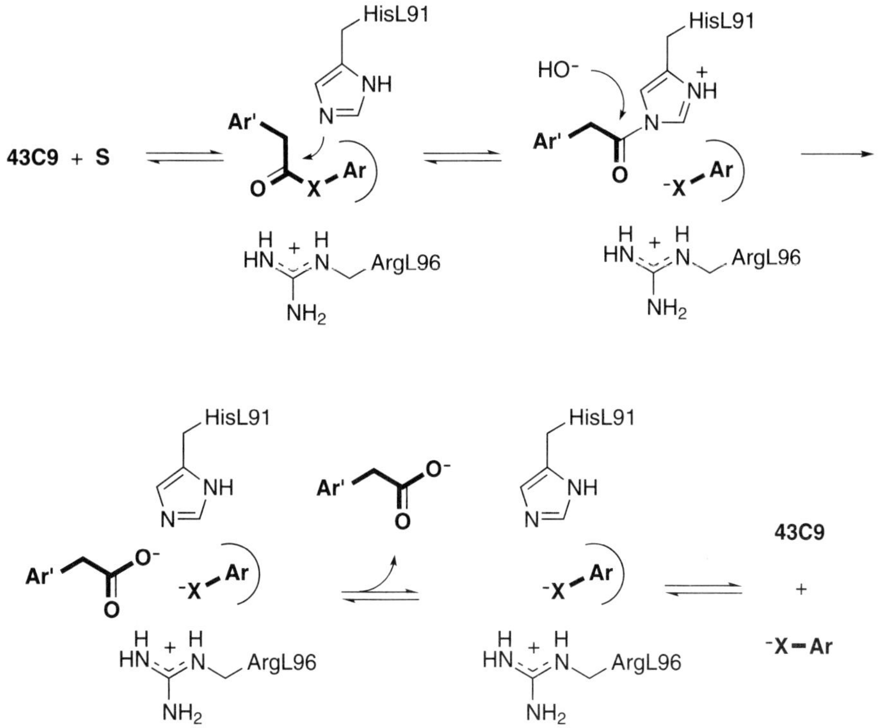

Figure 10-22. Proposed two-step mechanism for the hydrolysis of ester and amide substrates by the catalytic antibody 43C9. This differs from the published mechanism (Stewart et al., 1994a) in the roles of HisH35 and ArgL96.

pocket orients substrate for attack by the proximal imidazole group of HisL91. In a second step, the acylated imidazole is hydrolyzed by hydroxide to afford products and regenerate the catalyst. In analogy with 17E8, ArgL96 will electrostatically complement the oxyanionic species that occur along the reaction coordinate and may facilitate breakdown of the tetrahedral intermediate and expulsion of the anionic leaving group.

Nucleophilic catalysis may explain why 43C9, but not 17E8, is capable of cleaving an acyl derivative featuring the relatively poor p-nitroanilide leaving group. This strategy, which has also been adopted by serine proteases (Fersht, 1985), subdivides a difficult chemical reaction into a series of more tractable steps. In the case of 17E8, however, the proposed serine ester intermediate would be nominally *less* reactive than the aryl ester substrate. The histidine nucleophile is unique to 43C9, and it would be interesting in this context to substitute TyrL91 in 17E8 with histidine to determine whether the properties of the mutated antibody approximate those of the amidase. The complementary experiment in which ValH93 in 43C9 is replaced by lysine, as in 17E8, might provide valuable insight into alternative mechanisms for stabilizing oxyanionic intermediates.

Mechanistic and Evolutionary Considerations

The similarity of CNJ206, 17E8, and 43C9 is perhaps the most startling observation emerging from the comparative analysis. These antibodies were generated in separate experiments with different haptens, have low overall sequence identity, yet evolved closely related binding pockets. The immune system has apparently devised one basic solution to the problem of high affinity recognition of aryl phosphonates (phosphonamidates). This may mean that there are relatively few ways to bind such molecules using the antibody scaffold. In fact, a large proportion of the amino acids used to contruct the arylphosphonate binding pocket are contributed by framework region residues (Table 10-1). Alternatively, it is possible that only a small fraction of the immunological repertoire was selected for affinity maturation during the initial encounter with antigen. The Balb/c and 129GIX$^+$ mice used to produce these antibodies are only two highly inbred strains, and immunization of additional strains of mice might increase the chances of finding radically different binding pocket configurations. Mice prone to autoimmunity have been shown to yield unusually large numbers of esterolytic antibodies and may prove useful in expanding the repertoire of catalytic clones elicited by a single transition state analogue (Tawfik et al., 1995).

The many parallels between the hydrolytic antibodies and natural enzymes are also striking. The use of directed hydrogen bonds and electrostatic effects to stabilize oxyanionic intermediates and transition states is well precedented in the mechanisms of serine proteases and metalloproteases, and the increasing efficacy of CNJ206, 17E8, and 43C9 seems to correlate with the provision of more elaborate chemical functionality in response to the anionic phosphonate moiety. Although the precise energetic contribution of oxyanion stabilization in the antibody systems has yet to be quantified, there is little doubt that it will account for much of the observed rate accelerations. Mimicry of the multistep reaction pathways that characterize serine proteases is even more remarkable. Nothing in the design of the phosphonamidate hapten of 43C9 programs for nucleophilic catalysis or an ability to stabilize consecutive transition states. Because catalytic activity was not the basis for immunological selection, these properties must reflect a serendipitous coincidence in the requirements for high affinity recognition and catalytic efficacy. The probability of covalent chemistry may, however, increase with the affinity of the antibody for the transition state analogue, provided that nucleophilic residues are induced to satisfy the electrostatic and hydrogen bonding needs of the hapten (Stewart et al., 1994a).

Despite their sophistication, even the most efficient hydrolytic antibodies are still primitive compared to highly evolved proteases like subtilisin. For example, nitroanilide hydrolysis by 43C9 ($k_{cat} = 1.1 \times 10^{-4}$ s^{-1} at pH 8.5, 37°C; Janda et al., 1988) is more than 400,000 times less efficient than the subtilisin-catalyzed cleavage of succinyl-Ala-Ala-Pro-Phe-p-nitroanilide ($k_{cat} = 44$ s^{-1}, pH 8.0, 25°C; Bryan et al., 1986). Subtilisin also hydrolyzes unactivated alkyl amides under the same mild conditions, whereas 43C9 is restricted to substrates having relatively stable leaving groups (p$K_a < 12$). The high efficiency of the enzyme is a consequence of a multiresidue network of chemical groups that facilitates proton transfers and stabilizes the amide leaving group (Wells & Estell, 1988). Unlike subtilisin, 43C9 does not appear to exploit general acid-base chemistry.

The low probability of eliciting complex constellations of catalytic groups may explain why it has proved so difficult to create antibodies that function as site-specific peptidases and proteases. To address the challenge posed by alkyl amide hydrolysis, new strategies for generating catalytic functionality in antibody pockets will have to be developed. This might be accomplished through more clever hapten design or by site-specific mutagenesis. In the case of 43C9, preliminary efforts to introduce a general acid into the active site by mutation of TyrL32 to histidine were unsuccessful in augmenting catalytic efficiency (Stewart et al., 1994b), but high resolution structural data may allow more effective targeting of other sites in the future. Substrate-assisted catalysis (Carter & Wells, 1987), as recently demonstrated with antibodies that promote the rearrangement of an asparaginyl-glycine peptide bond (Gibbs et al., 1992b), may prove to be another broadly useful strategy for catalyzing energetically demanding chemical transformations. Incorporation of external cofactors, such as metal ions (Iverson & Lerner, 1989; Iverson et al., 1990), should also extend the capabilities of antibody catalysts.

Structural Limitations to Antibody Catalysis

Does antibody structure limit catalysis in a fundamental way? While a definitive answer to this question is not at hand, the structures of the chorismate mutase and esterolytic antibodies suggest that low activity more likely results from imperfect hapten design and the method by which the catalysts are generated than from architectural constraints. Nevertheless, some structure-related problems are apparent and will need to be addressed if we hope to create antibodies with enzyme-like efficiency.

First, the repertoire of binding pockets available for recognition of certain classes of hapten may be more restricted than previously thought. The primary immune response in mice is believed to contain more than 10^8 distinct antibodies (Alt et al., 1987; Rajewsky et al., 1987), but the structural similarities noted for CNJ206, 17E8, and 43C9 suggest that structural diversity is far more limited. It will be important to establish through additional research whether fundamentally different binding pockets can give rise to catalysts with equal or greater activity than those already identified.

Second, the number and placement of potentially useful functional groups in a generic binding pocket can vary significantly from antibody to antibody. In the case of the esterolytic antibodies CNJ206, 17E8, and 43C9, small refinements of the same basic active site yielded substantial increases in catalytic efficiency. This observation underscores the need for broad screening of the immune response with catalytic activity, rather than hapten binding, as the criterion for propagation of an individual clone. It also implies that first generation catalysts may prove amenable to incremental improvement. The upper limit on activity that can be achieved by modifying an existing antibody is unknown, however, and may depend on the starting point. For example, increasing the activity of a relatively primitive, unadorned active site (such as 1F7 or CNJ206) might be much easier than improving a more active agent (like 43C9). Furthermore, augmentation of catalytic efficiency may be complicated by the need to alter residues that are not directly involved in ligand recognition and hence more difficult to identify by inspection of crystal structures.

Third, despite the ability of the immune system to fashion tailored antibody pockets for individual haptens, affinities for ligand are generally much lower than for enzymic complexes of transition states (Stewart & Benkovic, 1995). Dissociation constants for antibody-hapten complexes are typically in the range 10^{-6} to 10^{-9} M, whereas enzymes bind transition states with affinities as high as 10^{-24} M (Radzicka & Wolfenden, 1995). A correlation between the observed rate accelerations and the ratio of equilibrium binding constants of the reaction substrate and transition state analogue shows that the efficiency of antibody catalysis is determined primarily by transition state stabilization (Stewart & Benkovic, 1995). Consequently, transition state affinities will have to be increased greatly to achieve enzyme-like rate accelerations. This poses a daunting challenge, since concomitant stabilization of the ground state must be avoided in order to minimize problems with product inhibition, which already limits the turnover of many catalytic antibodies.

Finally, the dynamics of the antibody-antigen interaction may need to be optimized for catalysis. Enzymes like chorismate mutase are conformationally labile and their flexibility may be essential for function. In addition to providing a means of excluding water from the active site, protein conformational changes may enable the catalyst to adjust to changes in substrate as the reaction coordinate is traversed. In contrast, the conformational changes observed in 1F7 and CNJ206 are probably deleterious, since they result in collapse of the active site in the absence of substrate. Although useful conformational changes are conceivable in antibody catalysis, it is difficult to imagine how they can be controlled *a priori* through antigen design.

Nature has solved many of these problems in enzymes through natural selection. Catalytic antibodies, in contrast, have not been subjected to a function-based selection. The application of mutagenesis and biological selection methods may ultimately yield much higher catalytic activities and, in so doing, provide a direct answer to the general question of structural limitations to antibody catalysis posed at the beginning of this section. In lieu of biological selection methods, the production of increasingly active catalysts will benefit from more sophisticated transition state analogues and the implementation of new methods for generating and screening large immunoglobulin libraries.

Summary

Catalytic antibodies combine programmable design with the powerful selective forces of biology. This synergy has allowed the field to progress in a relatively short period of time from simple reactions with well-studied mechanisms to processes difficult to achieve via existing chemical methods. Importantly, the first structures of catalytic antibodies confirm the superb shape and chemical complementarity between the binding pocket and the haptenic transition state analogue. Insofar as an antigen accurately reflects the transition state for a given reaction, the chances are excellent that an active site will be generated that can envelop the transition state and provide appropriate stabilizing interactions. Moreover, although catalytic antibodies have more primitive active sites than enzymes, they emulate their more highly evolved counterparts in many important respects. The structural work has highlighted the need to develop better haptens for immunization and more creative selection systems for catalytic antibodies. Efforts in this direction are likely to enhance the degree to which antibody binding energy can be exploited to control reactivity and mechanism. Knowledge of the complex interplay between structure and function as deduced by continued structural investigation will lay a solid foundation for the next generation of experiments in antibody catalysis.

Acknowledgments

The authors gratefully acknowledge the Natural Science and Engineering Research Council of Canada for a Centennial Postgraduate Scholarship to G.M., the American Cancer Society for a Postdoctoral Fellowship PF-3761 to J.A.S., and financial support from the National Institutes of Health, National Science Foundation, American Cancer Society and the Army Research Office.

11

Peptide Hormones

Arno F. Spatola

Classically, hormones have been defined as chemical entities released by one organ and having effects on cells at sites remote from the point of origin; most hormones were found to be steroids, peptides, or diverse structures such as prostaglandins. A prototypical example is insulin's action on cells throughout the body. More recently, these definitions have been broadened to include entities whose targets are not far removed, but where the hormone receptors might be located on adjacent cells or even the same cell in a particular organ or tissue.

Figure 11-1 demonstrates the four major types of hormone action, endocrine, paracrine, autocrine, and neurocrine. Stimulation of a nearby cell in the same tissue is defined as a paracrine action. Autocrine control refers to the location of a target receptor on the same cell which releases the hormone. Neurocrine peptides traverse the synaptic space between a nerve cell and their targets. In contrast, the definition of a hormone where the target receptor is far removed and peptides are delivered by the blood (circulatory system) is traditionally described by the term "endocrine action."

Production and Identification of Peptide Hormones

Peptides and peptide hormones are synthesized by all cells, but most peptide hormones have been discovered in four major sources: mammalian brains, gastrointestinal organs, animal venoms, and the skins of amphibians, an unexpectedly rich source of peptides. Some of these latter nonmammalian products such as the marine snail conotoxin GI (Gray et al., 1981) and α-*bungarotoxin* are toxins (Grant & Chiappinelli, 1985), while frogs have also provided important antibacterial agents such as *brevinins* (Morikawa, 1992) and the *magainins* (Zasloff, 1987). Frog skins have also provided analogues of opioid peptides including *dermorphins* and *deltorphins* (Erspamer & Melchiorri, 1980; Erspamer et al., 1989) as well as *sauvagine* (Montecucchi & Henschen, 1981; Vale et al., 1972) a peptide with considerable homology to human corticotropin releasing factor.

Table 11-1 provides a listing of representative peptides and peptide hormones, their sources, number of amino acids, and presence of disulfide bridges. A number of useful reviews have been published that augment this listing by providing full structures, additional modes of action, and potential biological and clinical uses for many of these hormones (Jones, 1993). The table does not include such growth factors as epidermal growth factor (EGF), transforming growth factor alpha or beta (TGF-α or TGF-β) or

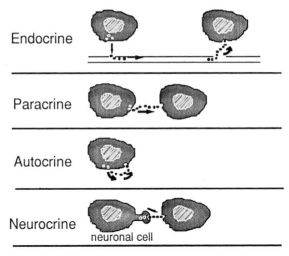

Figure 11-1. Representation of four major types of hormone action.

Table 11-1 Selected Peptide Hormones and Their Sources

Name	Total No. Amino Acids	Disulfide Bridges	Isolated from or Major Organ Source
Thyroid stimulating hormone (TRH)	3	0	Hypothalamus
Leucine enkephalin	5	0	Adrenal medulla
Bradykinin	9	0	Brain; liver; kidney
Delta sleep-inducing peptide	9	0	Brain
Angiotensin I, II	10, 8	0	Blood; kidney
Neuromedin B	10	0	Brain; spinal cord
LH-RH (GnRH)	10	0	Hypothalamus
Neurokinin A (substance K)	10	0	Spinal cord
Neuromedin K	10	0	Spinal cord
Substance P	11	0	Brain
Neurotensin	13	0	Brain
α-MSH (α-melanotropin)	13	0	Pituitary
Neurotensin	13	0	Pituitary
Pancreastatin	17	0	Pancreas
Dynorphin	17	0	Submucous plexus
β-MSH (β-melanotropin)	18–22	0	Pituitary
Motilin	22	0	Ileal mucosa
Histatin	24	0	Parotid
Secretin	27	0	Duodenum
Gastrin releasing peptide	27	0	Gastric tissue

(continued)

Table 11-1 (Continued)

Name	Total No. Amino Acids	Disulfide Bridges	Isolated from or Major Organ Source
Vasoactive intestinal peptide	28	0	Small intestine
Galanin	29	0	Small intestine
Glucagon	29	0	Pancreas
β-Endorphin	31	0	Brain
Calcitonin	32	0	Thyroid
Cholecystokinin-33 (CCK-33)	33 (also 58, 39, 22, 8, 4)	0	Brain; intestine; pancreas
Gastrin	34 (also 17, 14, 6, 4)	0	Duodenum
Neuropeptide Y	36	0	Brain
Pituitary adenylate cyclase activatory polypeptide	38	0	Pituitary
Adrenocorticotropic hormone (ACTH)	39	0	Plasma
Corticotropin-releasing hormone	41	0	Hypothalamus
Gastrin inhibitory peptide	42	0	Duodenum
Growth hormone-releasing hormone	44	0	Hypothalamus
Parathyroid hormone	84	0	Bone
Oxytocin	9	1	Posterior pituitary
Arginine vasopressin (ADH)	9	1	Posterior pituitary
Somatostatin-14 (SRIF)	14	1	Hypothalamus
C-type natriuretic peptide	22	1	Brain
Atrial natriuretic peptide	28	1	Heart
Brain natriuretic peptide	32	1	Brain
Amylin	37	1	Pancreas
Adrenomedullin	52	1	Adrenal medulla
Guanylin	15	2	Intestine
Endothelin-1	21	2	Endothelium
Defensin	30	3	Neutrophils
Insulin	51	3	Pancreas
Sodium potassium ATPase inhibitor	49	4	Duodenum
Elafin	57	4	Skin

immunologically active peptides such as interferon or the interleukins. Additional details regarding their biological functions may be found in references by Houben and Denef (1994) or Yamada and Owyang (1991). An excellent overview on peptides in medicinal chemistry is found in Hirschmann (1991).

Three major strategies have characterized the isolation of peptide hormones. The first is identification from bioassay; most hypothalamic-releasing factors were identified in this way. For example, the groups of Guillemin and Schally used the hypothalamic glands from hundreds of thousands of sheep or cows to isolate TRH (thyrotropin-releasing hormone), somatostatin, and LH-RH (luteinizing hormone-releasing hormone), guided by the release (or inhibition of release) of thyrotropin, growth hormone (also known as growth hormone-releasing inhibitory hormone), and luteinizing hormone. By contrast, Victor Mutt and his colleagues have laid claim to over a dozen new peptides found in gastrointestinal organs, in most cases without a definitive bioassay (Mutt, 1993).

Nearly all of these latter compounds were discovered by virtue of their characteristic C-terminal amidation. Thus, this represents a second major path for isolation; namely, by means of a common structural element, but where no bioassay guides the isolation. In almost all cases, the C-terminal NH_2 is derived from a glycine residue. The C-terminal region contains a characteristic basic extension such as Gly-Arg-Arg-Lys or Gly-Lys-Arg-Lys and the glycine is cleaved by the enzyme peptidyl glycine hydroxylase giving the peptide amide and glyoxalic acid. In several cases, the actual biological functions of these peptide amides have yet to be determined adequately. In general, these compounds are linear peptides with 20–50 residues and many share common structural motifs with the long-known GI secretogogues, secretin and glucagon. Intriguingly, several of these compounds are characterized by aromatic residues (Tyr, Phe, or His) at positions 1, 6, and 10, a trait also shared with the hypothalamic peptide, growth hormone-releasing factor (Table 11-2).

A third major source of new peptide hormones is emerging from the sequencing work on the human genome and the isolation of RNA sequences from a variety of species. Most peptides are synthesized as part of larger precursors and are embedded

Table 11-2 Structures of Several Human Gastrointestinal Peptide Hormones

Glucagon (29)	H-*His*-Ser-Gln-Gly-Thr-*Phe*-Thr-Ser-Asp-*Tyr*-Ser-Lys-Tyr-Leu-Asp-Ser-Arg-Arg-Ala-Gln-Asp-Phe-Val-Gln-Trp-Leu-Met-Asn-Thr-OH
Growth hormone-realeasing factor (44)	H-*Tyr*-Ala-Asp-Ala-Ile-*Phe*-Thr-Asn-Ser-*Tyr*-Arg-Lys-Val-Leu-Gly-Gln-Leu-Ser-Ala-Arg-Lys-Leu-Leu-Gln-Asp-Ile-Met-Ser-Arg-Gln-Gln-Gly-Glu-Ser-Asn-Gln-Glu-Arg-Gly-Ala-Arg-Ala-Arg-Leu-NH_2
Gastric inhibitory polypeptide (42)	H-*Tyr*-Ala-Glu-Gly-Thr-*Phe*-Ile-Ser-Asp-*Tyr*-Ser-Ile-Ala-Met-Asp-Lys-Ile-His-Gln-Gln-Asp-Phe-Val-Asn-Trp-Leu-Leu-Ala-Gln-Lys-Gly-Lys-Lys-Asn-Asp-Trp-Lys-His-Asn-Ile-Thr-Gln-OH
Secretin (27)	H-*His*-Ser-Asp-Gly-Thr-*Phe*-Thr-Ser-Glu-**Leu**-Ser-Arg-Leu-Arg-Glu-Gly-Ala-Arg-Leu-Gln-Arg-Leu-Leu-Gln-Gly-Leu-Val-NH_2
VIP (1-28)	H-*His*-Ser-Asp-Ala-Val-*Phe*-Thr-Asp-Asn-*Tyr*-Thr-Arg-Leu-Arg-Lys-Gln-Met-Ala-Val-Lys-Lys-Tyr-Leu-Asn-Ser-Ile-Leu-Asn-NH_2

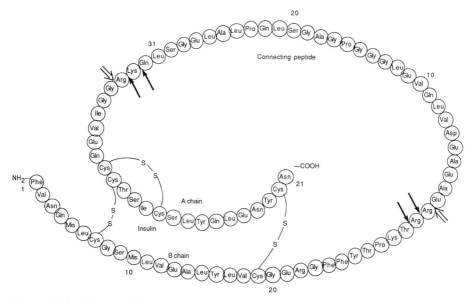

Figure 11-2. Structure of human proinsulin, insulin and C-peptide molecules are connected at 2 sites by dipeptide links. An initial cleavage by a trypsin-line enzyme (⇒) followed by several cleavages by a carboxypeptidase-like enzyme (→) results in the production of the heterodimeric (AB) insulin molecule and the C-peptide. Reprinted from Granner, 1993, with permission.

between pairs of dibasic amino acids as shown in Figure 11-2 using the proinsulin to insulin conversion as an example. It is apparent that as new nucleic acid fragments are decoded, the presence of these characteristic cleavage sites will lead to identification of potentially new peptide hormones. In some cases, the biological actions may be apparent from sequence homology. This was the case with several new enkephalin structures which were found to have an N-terminal tyrosine, as well as being located with numerous other opioid peptides, in a large precursor molecule (Noda et al., 1982). In other cases, it may be anticipated that, as with some of the GI peptides, the structure will *predate* the recognition of their actual biological response.

It is becoming apparent that, as with many early dogmas in science, the classic picture of endocrine hormones is faulty in another respect. It was initially considered that the endocrine organs such as the pituitary, adrenal cortex, and ovaries were the only sources of hormones. But with the recent discovery of atrial natriuretic peptide (synthesized and released from the heart) (Kangawa & Matsuo, 1984) and the large group of endothelins (secreted by endothelial cells lining all blood vessels) (Yanagisawa et al., 1988), it is now apparent that virtually all cells in the human body are known sources of hormones, many of which are peptides. Thus the more limited view of cell differentiation has had to be revised such that virtually all organs and tissues may be considered as endocrine (or at least autocrine or paracrine) glands. As the partial list of structures in Table 11-1 implies, peptide hormones are found, and influence organs, throughout the body. Medicinal chemists seek to produce new drugs based on naturally occurring structures but usually hope to devise analogues that are more potent and long lasting, and which have improved selectivity. The problem of bioavailability

Table 11-3 Examples of Peptide Hormones and Their Second Messengers

Peptide Hormone	Second Messenger
ACTH; angiotensin II; LH, FSH; opioids; MSH; glucagon; somatostatin	cAMP
Angiotensin II; oxytocin; LH-RH; substance P; TRH; vasopressin	Ca^{++} or phosphatidylinositides
Atrial natriuetic factor	cGMP
Insulin; EGF; growth hormone	Unknown second messenger

is also of critical importance (Humphrey & Ringrose, 1986) and efforts to improve transport and achieve oral activity have stimulated a variety of solutions (Bai & Amidon, 1992).

An alternative way of defining peptide hormones is through the nature of their receptors and the second messengers involved in delivering their message to the cell. When peptide hormones bind to cell surface receptors, the latter undergo a presumed conformational change that activates an enzyme or chemical within the cell (signal transduction event). Most hormones act through one of two second messengers: adenosine 3′,5′-monophosphate (cyclic AMP or cAMP) through activation of the enzyme *adenylate cyclase*, or calcium (Ca^{++}) or phosphoinositides (phosphorylated carbohydrates); each of these paths serves to amplify the peptide message. But other second messengers are known or suspected. The peptide atrial natriuretic factor functions through cyclic GMP as a second messenger, while insulin, selected growth factors, and other hormones have yet to have their second messengers fully identified. Selected examples of peptides and their presumed second messengers are summarized in Table 11-3.

Considerable research on the cAMP mechanism has begun to unravel the complex stimulatory or inhibitory controls involving the G-proteins (Gilman, 1987; Rodbell, 1985). These are large multiunit components that amplify the hormone signal through the cell membrane and, ultimately, affect cAMP levels. G-proteins were so named for their ability to bind GTP and act as the intermediate step between hormone-receptor binding and activation of adenylate cyclase with formation of cAMP. In turn, cAMP activates protein kinases that again function in diverse paths but often phosphorylate serine or threonine residues in key proteins at consensus phosphorylation sites such as Arg-Arg-X-Ser.

In summary, peptide hormones, unlike steroid hormones, tend to act on surface membrane receptors, and need not enter the cell to elicit their responses. Increases in cAMP or Ca^{++} levels act within the cell interior cycloplasm largely through kinase actions to amplify the response, with additional controls present to modulate or terminate their actions when necessary. Hormones such as insulin and EGF exert their biological actions via receptors through phosphorylation of tyrosine since their receptors are tyrosine kinases, but a true second messenger has not yet been established.

Later in this chapter, we will focus on four important examples of peptide hormones. In each case, their isolation, structure-function relationships, or analogue design reveal important lessons that come from the studies of these fascinating molecules.

In general, improved peptide analogues can be devised by a series of gradual replacements of backbone and side chain elements intended to afford compounds (peptide analogues) with qualities of improved potency, selectivity, or bioavailability; this gradual transition involving more and more peptide structural replacements may be viewed as a continuum between peptides and nonpeptide mimetics (Spatola, 1983). Alternatively, it has been argued (Farmer, 1980) and convincingly demonstrated by numerous groups (Olson, 1993; Hirschmann, 1991; Hirschmann et al., 1993) that peptide analogues can often be discovered or designed based on the appropriate positioning of active elements on a nonpeptide scaffold, even one as simple as a cyclohexane platform (Olson, 1993). In the sections following, the first approach will be emphasized in order to convey some basic lessons concerning analogue design. But clearly both approaches have had many successes as the studies following will confirm.

Oxytocin-Vasopressin

Much of the development of the field of peptide hormones has been punctuated by the award of Nobel prizes. This includes prizes to Vincent du Vigneaud in 1955 for the synthesis of oxytocin and vasopressin; and to Andrew Schally and Roger Guillemin in 1977 for their discovery of the hypothalamic peptides LH-RH, TRH, and somatostatin. (Rosalyn Yalow also shared this prize for developing the vital technique of radioimmunoassay). More recently a Nobel prize was awarded to R. Bruce Merrifield in 1984 for inventing solid phase peptide synthesis, which made it possible to rapidly synthesize so many of these peptides and their analogues.

The chemical synthesis of oxytocin in 1953 by du Vigneaud and coworkers stands as a milestone of structural analysis and synthetic achievement (du Vigneaud et al., 1953a, b). Although most short peptide hormones are linear, oxytocin, vasopressin, and somatostatin share a disulfide bridge and short extended tails at the C- or N-terminus (Fig. 11-3). A typical solution or solid phase strategy involves synthesis of the linear peptide with the cysteine pair protected by benzyl or methylbenzyl functions, followed

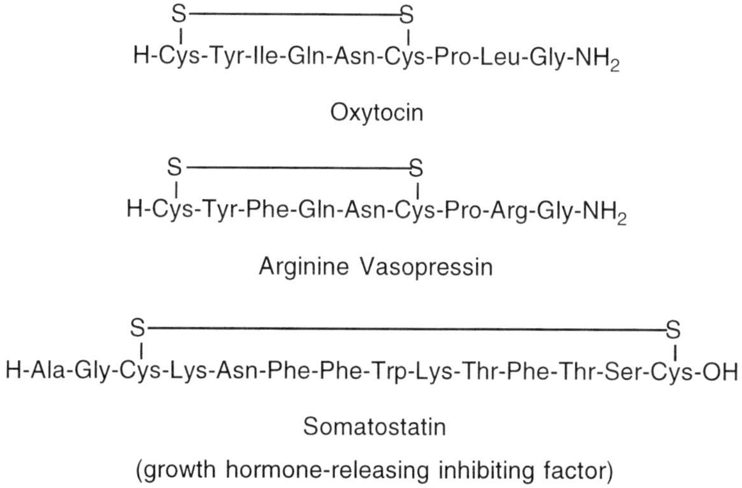

```
        S─────────S
        |         |
H-Cys-Tyr-Ile-Gln-Asn-Cys-Pro-Leu-Gly-NH₂
```
Oxytocin

```
        S─────────S
        |         |
H-Cys-Tyr-Phe-Gln-Asn-Cys-Pro-Arg-Gly-NH₂
```
Arginine Vasopressin

```
        S─────────────────────────────S
        |                             |
H-Ala-Gly-Cys-Lys-Asn-Phe-Phe-Trp-Lys-Thr-Phe-Thr-Ser-Cys-OH
```
Somatostatin
(growth hormone-releasing inhibiting factor)

Figure 11-3. Structures of oxytocin, arginine vasopressin, and somatostatin using amino acid nomenclature.

Table 11-4 Comparison of Biological Activities of Several Neurohypophyseal Hormone Analogues

		Biological Activities (international units per micromole)			
Name	Peptide Hormone Structure	Rat Uterus	Mammary Gland	Anti-Diuretic	Rat Pressor
Oxytocin	H-Cys-Tyr-Ile-Gln-Asn-Cys-Pro-Leu-Gly-NH$_2$ (S—S)	450	450	5	5
Arginine vasopressin	H-Cys-Tyr-Phe-Gln-Asn-Cys-Pro-Arg-Gly-NH$_2$ (S—S)	17	69	465	412
Arginine vasotocin	H-Cys-Tyr-Ile-Gln-Asn-Cys-Pro-Arg-Gly-NH$_2$ (S—S)	120	220	260	255

by hydrogenolytic removal of the sulfur protecting groups by sodium in liquid ammonia, and conversion into the disulfide ring form using air oxidation or potassium ferricyanide ($K_3Fe(CN)_6$).

This approach has led to the synthesis of thousands of analogues of these hormones and to the subtle distinctions of how minor side chain alterations between two closely related hormones (oxytocin and arginine vasopressin) can lead to major bioactivity changes. Thus, oxytocin is a potent agent for uterine contraction or milk ejection, while arginine vasopressin (AVP) is known as the antidiuretic hormone (promotes water retention by the kidney) and also is a pressor agent (increases blood pressure). An evolutionary pattern of other closely related peptides, also containing a six-residue disulfide bridge and a three-amino acid "tail" extends throughout the vertebrates and even the invertebrates, with at least 16 known examples discovered thus far. Of special interest is the hybrid structure, arginine vasotocin (Table 11-4) which is found in fish, amphibians, and reptiles. It possesses both oxytocic and vasopressin-like activities in relevant human or mouse assays. It is likely that arginine vasotocin played a key role in maintaining water balance through the intricate transformations that led fish to evolve through amphibians and reptiles on the path toward mammals. Perhaps significantly, arginine vasotocin is an antidiuretic in amphibians and reptiles but is a diuretic in freshwater fishes including lungfish (Acher, 1993).

From numerous structure-activity studies, it became possible to distinguish, at least partially, between those residues essential for biological function ("active" or "message" elements) and those leading to higher affinity for the relevant receptor ("binding" or "address" elements) (Schwyzer, 1977; Walter et al., 1977). In the case of oxytocin, Walter and coworkers focussed on tyrosine-2 and asparagine-5 as being indispensable for the transduction step (Walter, 1977); these residues thus make it possible for the interaction of oxytocin in its membrane-bound receptor (functioning through a Gp/1 phospholinase C complex: Jard et al., 1988) to lead to the ultimate biological event, namely smooth muscle contraction. Other residues, such as Pro^7, found in the corner of one of the reverse turns in the Urry-Walter oxytocin conformation

(Urry-Walter, 1971), may contribute to binding but are apparently not essential for hormonal action (Walter et al., 1972). Schwyzer has argued that the location of the receptor within the membrane compartment is also an important contributor to hormone/receptor binding and selectivity (Schwyzer, 1986).

Oxytocin is also instructive as a model for antagonist design. In theory, the elimination of an "active" element with retention of a binding element should lead to a competitive antagonist (Fig. 11-4). But inhibitor design has rarely been this straightforward. In reality, many antagonists also appear to bind differently than agonists (Hruby, 1992). As will be seen with LH-RH antagonists, highly potent inhibitors have been prepared, but these have often been very dissimilar to their parent hormone in both primary structure and, often, in their binding modes.

Oxytocin and vasopressin antagonists have often developed from common modifications. Thus, conversion of Tyr^2 to $Tyr(OMe)^2$ tends to be effective in both series as does incorporation of a hindered 1-position residue such as penicillamine (Pen) or cyclo-pentamethylene β-mercaptopropionic acid (Pmp) in place of cysteine. In fact, a very potent antagonist of oxytocin action is the compound [1-β-mercapto-β,β-cyclopentamethylene propionic acid^1,Tyr(Me)^2,Orn^8]oxytocin, while [β-mercapto-β,β-cyclopentamethylene propionic acid^1,Tyr(Me^2)]AVP is a potent antagonist of vasopressin (Hruby & Smith, 1987).

To what extent are structural changes additive? If two particular modifications increase the activity of a peptide hormone independently, can they be combined to yield an even more potent analogue? And how often do such common changes transfer between two closely related peptides? Clearly the answer appears to be that this general principle *is* operative among oxytocin and vasopressin antagonists, although many exceptions do exist.

This "principal of additivity" also appears to extend to related agonists. Thus, replacement of glutamine-4 by threonine-4 leads to an increase in activity in both oxytocin and vasopressin analogues (Hruby & Smith, 1987; Jost et al., 1987).

Both oxytocin and vasopressin have been studied intensively in terms of synthesis, conformational analysis, and biological actions. Among literally thousands of ana-

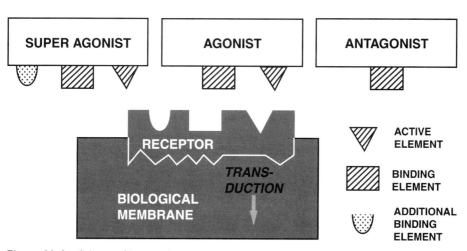

Figure 11-4. Scheme of hypothetical superagonist design and function.

```
        ┌─────────────────────────────────┐
Mpa–Tyr–Ile–Glu–Asn–Cys–Pro–Lys–Gly–NH₂
                    └─────────────────────┘
```

Figure 11-5. A bycyclic analogue of deamino oxytocin (Mpa = mercaptopropionic acid) with potent oxytocic antagonist activity. Reprinted from Hill et al., 1990, with permission.

logues, several may prove particularly instructive. Hill and coworkers prepared an interesting bicyclic analogue in which positions 4 and 8 were covalently linked, while position 1 contained a deamino penicillamine analogue known to confer antagonist activity (Fig. 11-5) (Hill et al., 1990). The resulting structure showed 50% inhibition at $10^{-8} - 10^{-9}$ M in an oxytocic assay but had limited antipressor activity, thus effectively "freezing out" the undesired action. A related monocyclic precursor to the bicyclic antagonist was actually a weak oxytocin agonist; this reemphasizes the role that conformational constraints may play in antagonist design.

These analogues are also of interest in view of the diverse locations in which these peptides are found. Oxytocin is found in male and female sex organs as well as in the brain-hypothalamus-posterior pituitary. These diverse sources also correlate with newly recognized functions for the hormone, including behavioral effects such as influence of maternal instincts in women and penile erection in men. Manning and coworkers have been particularly successful in preparing both agonists and antagonists of vasopressin using similar approaches (Manning & Sawyer, 1993; Manning et al., 1993). Subtle structural changes have led to analogues able to discriminate between several vasopressin receptor classes known as V_1 and V_2 receptors (along with subclasses known as V_{1a} and V_{1b} receptors). In addition to its pressor and ADH actions, vasopressin has long been suspected of affecting memory processes (DeWied, 1971). Some analogues devoid of the standard AVP activities also show these memory enhancement properties, thereby allowing separation of physiological from behavioral actions.

In contrast to their conformationally constrained analogues, Manning and coworkers have also found an open chain analogue of vasopressin showing surprisingly high biological potency. An example is the potent antagonist to both antidiuretic (V_2) and vasopressor (V_1) receptors for vasopressin (Manning et al., 1987) shown in Figure 11-6. Much recent work has focused on the exciting discovery that many nonpeptide structures possess significant agonist and antagonist activity. Many of these are quite far removed from a peptide structure such as a derivative of camphor sulfonamide (Figure 11-7) that is a potent antioxytocic compound with nanomolar affinity in various *in vivo* models (Williams et al., 1994). Oxytocin antagonists might well be useful clinical agents to prevent undesired uterine contractions leading to premature birth.

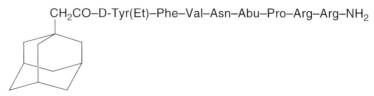

Figure 11-6. Structure of a linear vasopressin antagonist with good selectivity and potency for the V_2 receptor. Reprinted from Manning et al., 1987, with permission.

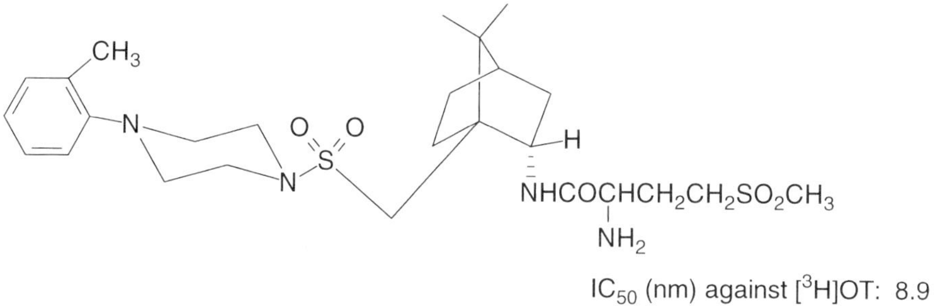

IC$_{50}$ (nm) against [^{3}H]OT: 8.9

Figure 11-7. Structure of L-368,899, a nonpeptide oxytocin antagonist based on a camphor sulfonamide: a dipeptide mimic orally bioavailable in rat, dog, and chimpanzee. Reprinted from Williams et al., 1994, with permission.

LH-RH Agonists and Antagonists

LH-RH or GnRH, otherwise known as luteinizing hormone-(gonadotropin)-releasing hormone, was first isolated by Andrew Schally and coworkers (Schally et al., 1971). This decapeptide, pGlu-His-Trp-Ser-Tyr-Gly-Leu-Arg-Pro-Gly-NH$_2$, has been the subject of one of the most intense series of structure-function studies among peptides in history, with literally thousands of analogues of both agonists and antagonists having been synthesized and tested biologically. The result has been a variety of superagonists used clinically for such disorders as prostate cancer or endometriosis, as well as many promising antagonists tested as both male and female contraceptive agents (Karten & Rivier, 1986; Nestor & Vickery, 1988; Zatuchni, 1981).

LH-RH is synthesized in the human hypothalamus and controls the release of LH and FSH (follicle-stimulating hormone) from the anterior pituitary. Careful synthetic modification has provided several analogues of LH-RH that are 10-fold or more as potent as the parent hormone, i.e. that are "superagonists." A key finding is that superagonists of LH-RH can actually reduce LH release, and thus suppress ovulation or spermatogenesis; they consequently function as contraceptive agents in a process sometimes termed the "paradoxical" effect. In fact, these superagonists do so by a type of feedback inhibition that results in decrease in content (down-regulation) of LH-RH receptors. The superagonists and LH-RH antagonists can affect both men and women and, therefore, provide a confusing choice of four possible new methods of birth control (Table 11-5). To date, none of these analogues is used clinically for birth control due to a number of major or minor side effects. What are the structures of these compounds

Table 11-5 Approaches to Contraception Using LH-RH (GnRH) Analogues

	LH-RH super-Agonists	LH-RH Antagonists
Male	Inhibit spermatogenesis (down regulation)	Block LH-RH receptors
Female	Decrease LH-RH receptors (down regulation)	Block receptors; prevent LH surge

and how were they devised? And, more importantly do they provide any lessons regarding future analogue development?

Superagonists

Several clinical agents based on LH-RH analogues include leuprolide (Abbott), nafarelin (Syntex) and buserelin (Hoechst). These peptides are generally characterized by a hydrophobic D-amino acid at position 6 and a glycine replacement at the C-terminus [proline ethylamide; also known as the "Fujino" modification (Fujino et al., 1972)] (Fig. 11-8). The resulting compounds are superagonists since they are reported to be 50-, 200-, and 40-fold more potent, respectively, than LH-RH (Dutta & Furr, 1985).

These compounds were discovered after a lengthy series of structure-activity studies. Typical strategies in such studies include an "alanine scan" (where each individual amino acid is serially substituted by an L-alanine residue) and a "D-amino acid scan" (the absolute configuration, where possible, is individually reversed). Other common replacements involve Trp → Nal, Tyr → Phe, Arg → Lys, or Ser → Thr. These strategies are designed to help distinguish binding elements from active elements as previously described.

The acceptance of a D-amino acid is often interpreted as evidence of a reverse turn, which may be stabilized by moving a quasi-axial (L-residue) substituent to a quasi-equatorial (D-residue) position. But D-amino acids are also known to retard or prevent proteolytic degradation. Various experiments are available to help separate these factors.

Early studies (Marks et al., 1974) on LH-RH established that the molecule was most susceptible to degradation by endopeptidases at the 6–7 and 9–10 amide bonds (the 3–4 and 5–6 positions are also likely chymotrypsin cleavage sites):

$$\text{pGlu—His—Trp—Ser—Tyr—Gly—Leu—Arg—Pro—Gly—NH}_2$$
(with arrows ↓ above Ser, Tyr, Leu, Pro)

It has been shown that at least part of the enhancement in potency for the superagonists is based on their enhanced biostability. This affects not only potency but also results in an increase in half-lives ($t_{1/2}$), ordinarily a weak point for peptide hormones as drugs.

In retrospect, the LH-RH findings suggest a preferred initial strategy for analogue design. First, find the weak links in the peptide primary structure, through qualitative

pGlu–His–Trp–Ser–Tyr–Gly–Leu–Arg–Pro–Gly–NH$_2$	LH-RH
pGlu–His–Trp–Ser–Tyr–<u>D-Leu</u>–Leu–Arg–Pro–<u>NHCH$_2$CH$_3$</u>	leuprolide (Abbott)
pGlu–His–Trp–Ser–Tyr–<u>D-Nal</u>–Leu–Arg–Pro–Gly–NH$_2$	Nafarelin (Syntex)
pGlu–His–Trp–Ser–Tyr–<u>D-Ser(tBu)</u>–Leu–Arg–<u>Pro–NHCH$_2$CH$_3$</u>	Buserelin (Hoechst)

Figure 11-8. Structure of potent LH-RH superagonists.

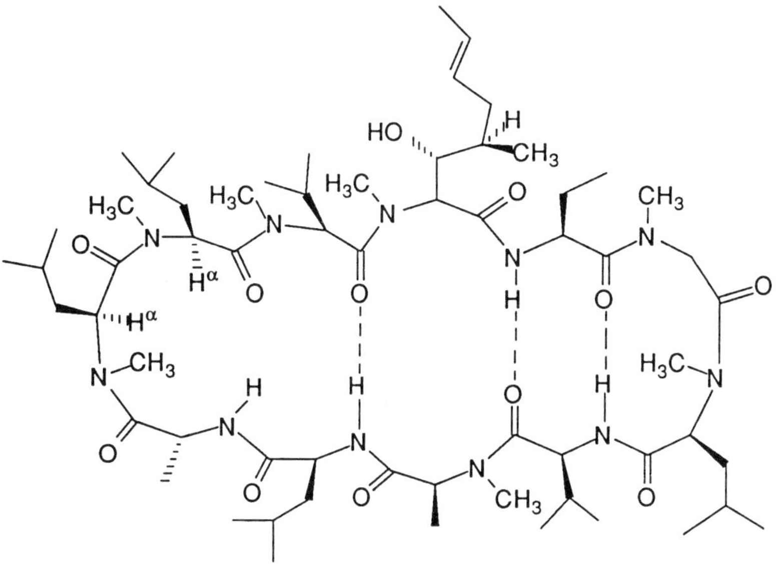

Figure 11-9. Cyclosporin structure.

degradation assays using blood serum, purified enzymes or enzyme mixtures, or organ homogenates (especially those most likely to be target organs). Once the most vulnerable amide bonds have been identified, additional structural modification [N-methylation; D-amino acids; α-methylation, or amide bond replacements (Spatola et al., 1980; Spatola, 1983; Aubry & Marraud, 1989)] may be introduced at these sites to enhance enzyme stability before entering a second round of degradation studies for further structure-activity optimization.

This approach was used in the design of a series of highly potent appetite suppressants, based on the naturally occurring peptide "satiety" agent, CCK-8. CCK-8 has the structure Asp-Tyr(SO_3H)-Met-Gly-Trp-Met-Asp-Phe-NH_2. Because of its unusual sulfated tyrosine moiety at the 2-position, the molecule was found to be unusually stable in serum. After many largely unsuccessful structure-activity studies, a new approach based on discovering the weak links was initiated. This combination "synthesis-degradation" strategy was far more successful and led to a modified analogue with 100-fold greater potency in a feeding inhibition bioassay (Rosamond et al., 1988). Among the more important structural changes was an N-methyl substitution at the C-terminal phenylalanine that protected against proteolytic action (Rosamond et al., 1988).

LH-RH superagonist analogues have also been used as hosts for the study of N-methylation. This simple amide modification both provides stability against proteolysis and also removes a potential or real H-bond donor function. The latter effect has been blamed as a potential barrier to effective membrane translocation; conversely, the nearly global N-methylation found in the peptide immune suppressant cyclosporin is believed to be responsible for its oral bioavailability (Fig. 11-9).

In any case, when Haviv and coworkers (Haviv et al., 1993) carried out a serial N-methylation study on the LH-RH superagonist leuprolide, and several clinical competitors, they found that few were as active as the parent, although N-MeSer[4] and N-MeTyr[5] were able to protect completely the 3–4 bond from chymotrypsin cleavage.

Interestingly, several N-methyl analogues exhibited antagonist properties. When the same modifications were examined within true antagonists, a 5-position substitution (N-MeTyr^5) enhanced potency, reduced 4–5 amide cleavage, and resulted in dramatically enhanced water solubility. The analogue Ac-D-Nal^1-D-pClPhe^2-D-Pal^3-Ser^4-NMeTyr^5-D-Lys(Nic)^6-Leu^7-Lys(Isp)^8-Pro^9-D-Ala^{10}-NH$_2$ (A-75998) is a potent suppressor of testosterone and is in clinical trials for the treatment of prostate cancer (Haviv et al., 1993). At a dose of 0.1 mg/kg over a period of 10 days, the compound was effective at completely suppressing testosterone release in male Synomologus monkeys.

LH-RH Antagonists

The first LH-RH antagonists involved position 2 replacements since the normal His^2 residue in the parent compound was required for agonist activity (Vale et al., 1972). This began a great synthetic effort largely focused on the critical N-terminal portion of the molecule, with the recognition that a concentration of hydrophobic groups at this position and elimination of His^2 led to quite potent inhibitors. Thus, in place of the normal tripeptide N-terminus, "pGlu-His-Trp," were incorporated such replacements as:

Ac-D-pClPhe^{1,2}, D-Trp^3; or

Ac-D-Nal(2)^1,D-pClPhe^2,D-Trp^3; or

Ac-Δ^3Pro^1,D-pFPhe^2,D-Nal(2)^3.

The aromatic residues and their D-configuration tended to prevent proteolysis and eliminate residual agonist activity. These analogues also contained a hydrophobic residue (D-configuration) at the 6-position as previously used in the superagonists. Thus a typical early potent LH-RH antagonist was [Ac-D-pClPhe^1,D-pClPhe^2,D-Trp^{3,6}]LH-RH. But these compounds proved to be so hydrophobic that a later generation of compounds was synthesized that replaced the usual D-Trp^6 or D-Nal^6 with the highly hydrophilic residue D-Arg^6, presumably to help solubilize the compound by introducing an additional positive charge at a remote location (Coy et al., 1982). This interpretation was further supported by the surprising activity that had been found for a rather simple replacement at the 1-position when glycine was used to replace the bulky Nal or pClPhe residues. When Ac-Gly^1 was combined with hydrophobic residues at positions 2 and 3, this combination led to the new analogue [Ac-Gly^1,D-pClPhe^2,D-Trp^{3,6}]LH-RH, which showed significant inhibitory activity (2/10 rats ovulating) at a dose of 10 μg/rat (Spatola & Agarwal, 1980). This seemed to argue in favor of a proper balance between local versus global hydrophobicity and apparently afforded better pharmacokinetics through optimization of these factors.

Interestingly, the D-Arg^6 analogues that were prepared initially to enhance stability generally retained substantial N-terminal hydrophobicity and led to LH-RH antagonists that had potency better than 1 μg/rat for antiovulatory activity. Initially, these appeared very promising as clinical agents. Nevertheless, problems developed with a disturbing side effect (anaphylactic reactions) observed in rats when testing the D-Arg^6 analogues. These effects were attributed, at least partially, to the presence of two Arg residues in close proximity. By replacing one of the two Arg residues (LH-RH itself has an Arg^8 moiety) with another hydrophilic moiety such as pyridylalanine^6 (Pal^6) or even a hindered homoarginine derivative (homo-Arg(Et)$_2$), the edema side effect could be sup-

Figure 11-10. Representation of stabilized turn structures in linear peptides by introduction of a D-amino acid at the $n/2 + 1$ position.

pressed while retaining or enhancing antiovulatory potency. It has also been suggested that the charged residues enhance initial binding to the negative phospholipid membrane components, thus facilitating receptor interaction (Nestor et al., 1988). In any case, the most potent linear LH-RH antagonists currently available are active in the antiovulatory assay (AOA) at submicrogram doses per rat. Nevertheless, their possible clinical utility is still somewhat controversial in view of the apparent requirement for a "totally safe" contraceptive agent.

Cyclic Gn-RH Analogues

A key theme in recent peptide analogue design is the conversion of potent peptide analogues to more biologically stable analogues or "peptide mimetics" (see Chapter 12). In the case of LH-RH, these efforts have followed the example of highly potent "min-isomatostatin" analogues by first defining the most probable conformation(s) of the parent form and then stabilizing this conformation through backbone or side chain cyclization without interference with critical *active* elements.

Early attempts to produce biologically active cyclic LH-RH analogues through head-to-tail amide formation failed, probably due to an inadequate base of conformational underpinnings (Rizo & Gierasch, 1992). Rivier and colleagues have argued that linear peptides may be stabilized by introduction of a D-residue at or next to position $n/2 + 1$, where n is the number of amino acids in the peptide or in its bioactive core (Fig. 11-10) (Rivier et al., 1992). This presumably enhances various types of β-turn structures and antiparallel hydrogen bonding along the chain halves and suggests models for new cyclic structures for LH-RH as it did for somatostatin, another hypothalamic peptide hormone with significant clinical potential.

An analogous approach that did prove successful involved formation of a small ring through a lactam "bridge" connecting residues 6 and 7, thus simulating a β-turn structure formed in the linear bioactive form (Fig. 11-11). This compound proved to be 8

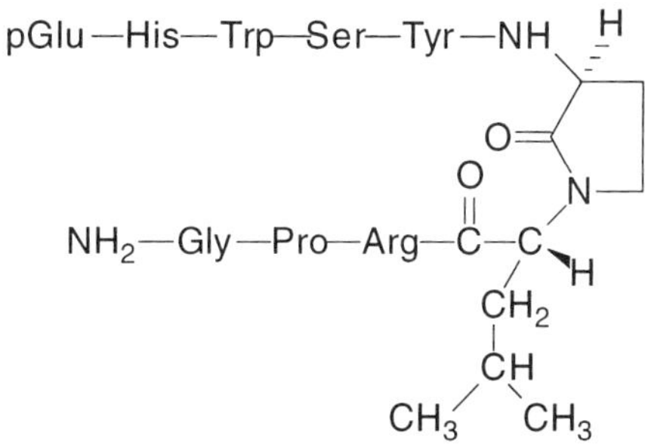

Figure 11-11. Structure of active linear LH-RH analogue with lactam bridge simulating a reverse turn. Reprinted from Freidinger et al., 1980, with permission.

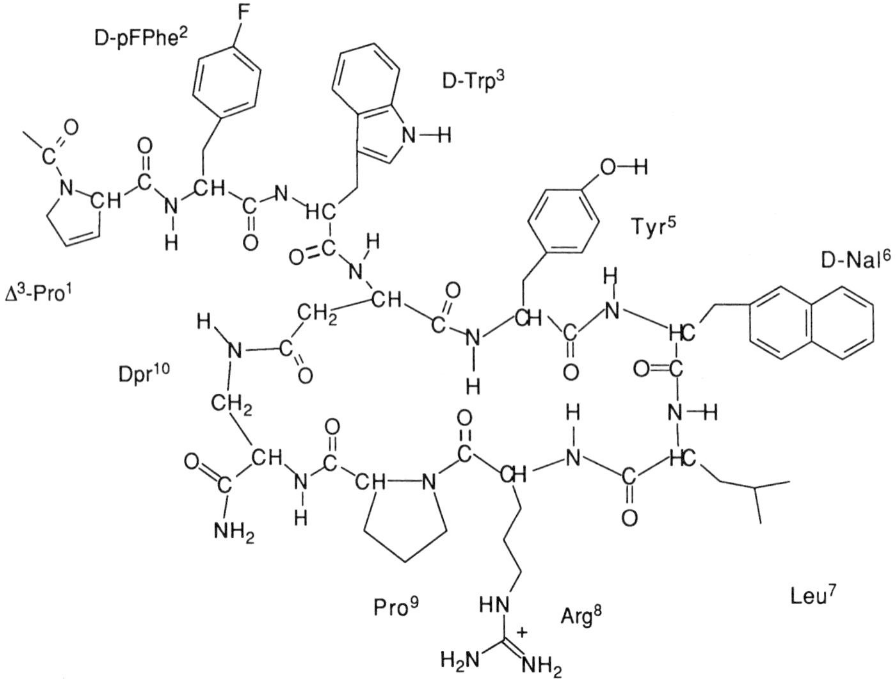

Figure 11-12. Representation of the cyclic LH-RH antagonist AcΔ^3-Pro-D-pFPhe-D-Trp-c(Asp-Tyr-D-Nal-Leu-Arg-Pro-Dpr)-NH$_2$.

to 9 times as potent *in vitro* and 24 times as potent in rats as LH-RH (Freidinger et al., 1980).

Head-to-tail cyclization of a potent linear LH-RH antagonist led to cyclo (Δ^3-Pro^1-D-pClPhe^2-D-Trp^3-Ser^4-Tyr^5-D-Trp^6-N-MeLeu^7-Arg^8-Pro^9-β-Ala^{10}) which showed only weak binding activity. Presumably, involving the critical N-terminal portion as part of a ring was deleterious to antagonist potency. But by replacing a nonessential Ser^4 residue with Asp^4, and by forming a lactam bridge with a diamino propionic acid at the 10 position, a modified 4–10 bridged cyclic analogue with a "free" N-terminus tripeptide tail was produced (Fig. 11-12) which was considerably more active, showing 80% inhibitory activity at 10 μg in the rat AOA test (Struthers et al., 1990). Detailed NMR and molecular dynamics analysis (Rizo et al., 1992a, b) of [Ac-Δ^3-Pro^1-D-pFPhe^2-D-Trp^3-c(Asp^4-Tyr^5-D-Nal(2)^6-Leu^7-Arg^8-Pro^9-Dpr^{10})NH$_2$ led to the conclusion that the molecule adopts a Type II' β-turn about residues 6 and 7, and promotes a second hydrogen bond between the Arg^8 CO and Tyr^5 NH. The N-terminus appears to lie above the ring in this model.

Introducing further constraints within peptide hormone analogues is a logical approach for designing peptide mimetics. Since the available evidence suggested that the 4–10 cyclic analogue retained considerable flexibility, a bicyclic derivative was prepared with a second amide bridge between a Glu^5 and a Lys^8. By using optimized side chain replacements, an analogue with 80% activity at 5 μg in the rat antiovulatory AOA test was obtained (Struthers et al., 1990). A similar derivative lacking the potent D-Arg^6 residue and several other substitutions was less active (Table 11-6).

Table 11-6 Comparison of Two Bicyclic LH-RH Antagonists in the Rat Antiovulatory Assay (AOA)

Peptide Analogue	Dose (mg)	In vivo Rat Assay	Percent Inhibition
Bicyclo (4/10, 5/8) [Ac-D-Nal^1-D-pClPhe^2-D-Pal^3-Asp^4-Glu^5-D-Nal^6-Leu^{7}Lys^{8}Pro^{9}Dpr^{10}]	50	2/8	75%
Bicyclo (4/10, 5/8) [Ac-D-Nal^1-D-pClPhe^2-D-Trp^3-Asp^4-Glu^5-D-Arg^6-Leu^{7}Lys^{8}Pro^{9}Dpr^{10}]	5	2/10	80%

Abbreviations: Pal = 3-pyridylalanine; Nal = 2-naphthylalanine

Thus, LH-RH represents an effective case study of an approach being taken to produce peptides with reduced flexibility and with built-in barriers to enzyme degradation. Constrained, *active* analogues are often the best approach to discover the receptor-bound conformation in the absence of a three-dimensional receptor binding site (Hruby, 1982; Hruby et al., 1990; Toniolo, 1989, 1990). It may be anticipated that future studies with LH-RH analogues will focus on reducing the flexibility of the N-terminal "tail," either through additional cyclic constraints or by the incorporation of α-substituted nonnatural amino acids that discourage free rotation about the ϕ, ψ, and χ torsional angles as shown in Figure 11-13. The importance of these analogues will be examined in later sections.

α-Melanophore Stimulating Hormone

A molecule known as β-lipotropin (β-LH) has proven to be the source of a number of fascinating peptide hormones. As shown in Figure 11-14, within the larger β-LH structure may be found β-endorphin, the potent opioid peptide, and β-melanotropin, also known as β-MSH or β-melanophore stimulating hormone.

β-LH is in turn part of a larger precursor known as pre-pro-opiomelanocortin, which in the cow consists of 265 amino acids. Within the large peptide may be found additional peptide hormones including gamma-MSH and adrenocorticotropic hormone

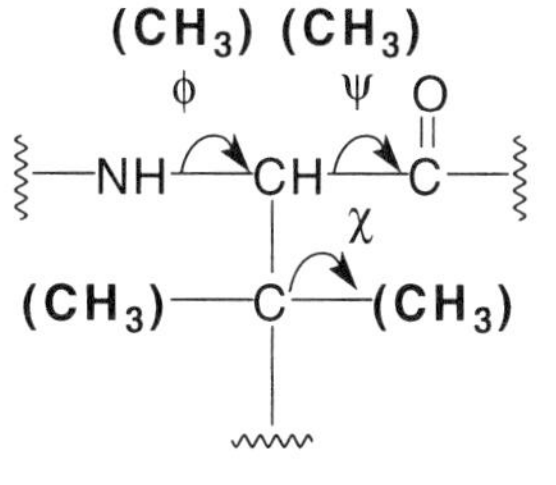

Figure 11-13. Methylation at amino nitrogen, α-carbon, or β-carbon positions can introduce constraints against free rotation at adjacent ϕ, ψ and χ bonds.

384 / Spatola

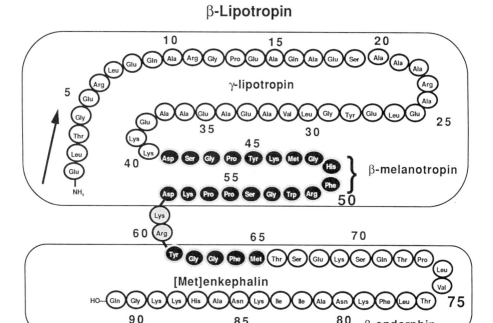

Figure 11-14. Structure of ovine-β-lipotropin showing relationships of several hormones (β-MSH, endorphin and Met-enkephalin) within common precursor. Note dibasic sequences flanking two hormone segments. See Li, 1981 and Yamashiro & Li, 1984.

(ACTH) (Fig. 11-15). Just as the bovine pre-pro-opiomelanocortin N-terminus of β-endorphin consists of the pentapeptide opioid, methionine enkephalin, the ACTH 39-peptide contains within its N-terminus a 13-amino acid peptide known as α-MSH. The peptide hormones are released by sequence specific proteases that cleave the dibasic residues Arg-Arg, Arg-Lys, Lys-Arg, or Lys-Lys, that flank the hormone sequence within the precursor molecule.

The peptide α-MSH has important peripheral bioactivities, including control of pigmentation, and central effects on memory and thermal regulation. The peptide has some proteolytic protection from exopeptidases by virtue of N-terminal acetylation and C-terminal amidation. Frogs and certain lizards respond to a dark-colored background by releasing MSH from the pars intermedia of the pituitary, resulting in the dispersion of melatin granules, and thereby causing the skin to darken accordingly (Hadley, 1989). When a solution of α-MSH is applied to an *in vitro* assay consisting of frog (*Rana pipiens*) or lizard (*Anolis carolinensis*) skin, the color change may be monitored in a dose-dependent manner via photometric reflectance measurements. With a potent α-MSH analogue, a light green frog will typically turn dark green and remain so for several hours or longer.

Evaluation of α-MSH activities of various analogues has shown that the lizard assay may be more diagnostic of potency expected in humans and other mammals than the frog assay (Hadley et al., 1993). The melanoma tyrosinase assay involving a melanoma cell line has also been used and may provide insights of potential utility for controlling cell growth. For this reason and others, developing potent and stable analogues of α-MSH has been an important objective of peptide scientists.

It had been noted in the 1960s that a heat-alkali treatment of α-MSH resulted in a prolonged skin-darkening activity. Lande and Lerner (1971) speculated that racemization of one or more amino acids in the 13-residue chain was responsible. Hruby and coworkers (Engel et al., 1981) explored this proposal by quantifying the extent of racemization of each amino acid using high resolution gas chromatography. While Ser and Met were found to suffer the greatest degree of racemization, Phe seemed to be the more likely candidate as it was also partially epimerized during alkali treatment of the peptide. Indeed, when a stabilized analogue 4-norleucine-7-D-phenylalanine-α-MSH was synthesized, it was found to be a superagonist with important melanotropic activity: a 26-fold increase in its *in vitro* potency was confirmed, along with a remarkable (>40 days) skin-darkening effect in frogs (Sawyer et al., 1980). This raises a number of questions regarding the ultimate fate of this analogue *in vivo* and the nature of the hormone-receptor complex that appears to "lock in" this response so well.

As with LH-RH, cyclization of α-MSH was attempted and was quite successful. A potent cyclic analogue was prepared that linked position 4 (Met) with position 10 (Gly) to yield Ac-Ser-Tyr-Ser- c(Cys-Glu-His-Phe-Arg-Trp-Cys)Lys-Pro-Val-NH$_2$, through a "pseudoisosteric" replacement (Fig. 11-16). This compound [Cys^4,Cys^{10}]α-MSH is up to 1000-fold more potent than the parent hormone, again depending on the assay used (Sawyer et al., 1982). But disulfide bridged molecules are not always very stable. More recently, Hruby and coworkers have prepared a truncated, D-Phe^7 analogue with a (shortened) lactam bridge between Glu^5 and Lys^{10} to produce a highly potent analogue. This species was 90-fold more potent in the ligand skin assay than α-MSH and had prolonged activity in both frog and skin assays (Al-Obeidi et al., 1989).

Some of these compounds have proven to be not only very potent but, as described above, very long lasting. Most peptides, and in fact many drugs, are rarely active for more than a few hours. In order to retain activity for weeks, the receptor response seems

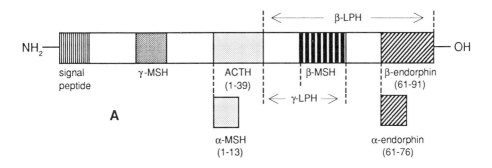

Figure 11-15. (A) Schematic representation of bovine pre-pro-opiomelanocortin showing the presence of various hormones and the C-terminal portion, β-lipotropin. (B) Structure of α-MSH.

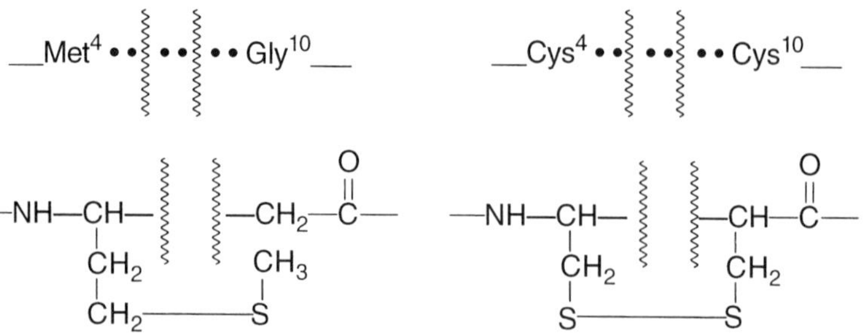

Figure 11-16. The cyclic α-MSH analogue with a Cys^4, Cys^{10} substitution (*right*) is "isosteric" with the parent Met^4, Gly^{10} residues (*left*).

to be permanently locked into the "on position" which raises the question of whether the peptide agonist is still resident within the active site of the receptor.

The results of these studies are especially interesting because of the prospect of designing other peptide hormone analogues with prolonged action. This behavior does not seem to correlate with enzymatic stability, and prolongations are also receptor specific, that is, analogues differ in their respective behaviors toward frog compared to lizard receptors (Hruby et al., 1984). Perhaps most interesting would be the prospect of measuring k_{on} versus k_{off} rates to establish kinetic parameters that might quantify this effect. Unfortunately, this is difficult when receptor concentrations are highly variable and when the k_{off} rates are so low. Nevertheless, there is considerable evidence that suggests that the peptide analogues bind tightly to the receptor, that both entities are internalized within the cell, and that the activity persists through several cell divisions (Hadley, 1992). Most importantly, these results suggest that peptide analogues may be synthesized with conformational and topographic features crucial to the receptor-bound state, regardless of their effects on biostability or other factors not directly related to the transduction event.

From a large series of structure activity studies, it has been established that a central core of α-MSH comprising residues His-Phe-Arg-Trp is the absolute minimum substructure required for full bioactivity. Longer regions (4-11 or 4-12) are required for full activity in the frog and lizard skin bioassays (Hruby et al., 1993). Through experimental determination of association constants with various lipids, a correlation was confirmed between high affinity to lipids and high biological activity. From fluorescence decay measurements, it has been concluded that strong binding requires formation of a β-turn structure along with penetration of a critical tryptophan residue into the lipid component (Ito et al., 1993). Thus the various linear and cyclic α-MSH analogues provide further indications of the importance of cyclization or other conformational constraints for producing peptide analogues with enhanced potencies and with the potential for dramatic enhancement of biological half-lives or prolongation of the biological response.

The potent α-MSH analogues themselves could prove clinically useful both as tanning agents or as potential therapeutic agents for melanoma (skin cancer) or other melanocyte related skin disorders. This application suggests a requirement for both prolonged stability and the ability to penetrate the skin following topical application. In fact, various researchers have shown that transdermal delivery of such analogues as [Nle^4,D-Phe^7]α-MSH is possible (Hadley et al., 1993) and is more effective than subcutaneous injection. In human studies, using Melano-Tan I ((([Nle^4,D-Phe^7])α-MSH) or Melano-Tan II (the cyclic analogue

$$\text{Ac-[Nle}^4\text{, }\overline{\text{Asp}^5\text{,D-Phe}^7\text{,Lys}^{10}\text{]}}\alpha\text{-MSH}_{4\text{-}10}\text{-NH}_2)$$

both caused skin darkening, especially in the face and neck following subcutaneous injection during phase I clinical trials. The authors suggest that ozone depletion may accelerate the need for effective sun protection agents, especially those that may be administered either orally or through other routes.

Endothelins and Endothelin Antagonists

The endothelins are a group of recently identified peptide hormones that promote vasoconstriction and thus increase blood pressure (Yanagisawa et al., 1988). Thus these peptides, along with angiotensin II and arginine vasopressin, are part of a diverse set of hormones that, in concert with vasodilators such as atrial natriuretic peptide (from heart muscle), bradykinin, and substance P, act to maintain homeostasis.

In humans, three closely related endothelins (Fig. 11-17) have been isolated, each of which has 21 amino acids and a conserved pair of disulfide bridges. These are, in turn, derived from larger precursor molecules known as the big endothelins I, II, and III, with 38, 37, and 41 amino acids, respectively (reviewed by Doherty, 1992). Cleavage in each case involves scission of a Trp-Val(Ile) bond by a phosphoramidon-sensitive enzyme, possibly [EC 3.4.24.11], a neutral endopeptidase (metalloprotease) referred to as enkephalinase or, more recently, *neprilysin*.

The endothelins possess a wide variety of physiological actions. Thus analogues designed to probe their role, either as selective agonists or antagonists, to the two classes of receptors (ET$_A$ and ET$_B$) (Sakurai, et al., 1992) are desired. In the endothelial cells lining blood vessels may be found the endothelins, together with the recently described endothelium-derived relaxing factor; the latter has been shown to be the simple gaseous molecule, nitric oxide. Although their role in renal failure, atherosclerosis, wound healing, gastrointestinal disorders, and asthma are only some of their reported pathophysiological effects (Doherty, 1992), most pharmaceutical interest appears to lie in blood pressure control.

Before considering general strategies for endothelin-based drug design, the issue of multiple receptors should be addressed. Many peptide hormones appear to have a series of closely related receptors that often exist in diverse locations of the body. Schwyzer has argued that receptors often differ in their location within the membrane compartment (Schwyzer, 1986; 1992), thus rationalizing some of the site preferences of hydrophilic ligands (interacting preferentially with receptor sites exposed on the aqueous compartment or cell surface) versus the hydrophobic ligands (which may penetrate within the lipid to buried, nonpolar receptor sites). In the latter case, polar peptide "tails" may remain closer to the surface aqueous compartment, not unlike a diver whose head has penetrated to the lower reaches of the cellular pool, while his or her feet remain near the air-water surface. In the case of the endothelin receptors, the two known classes, ET$_A$ and ET$_B$, differ both in their predominant locations and functions. The ET$_A$ receptors are found primarily in peripheral tissues such as the heart, lung,

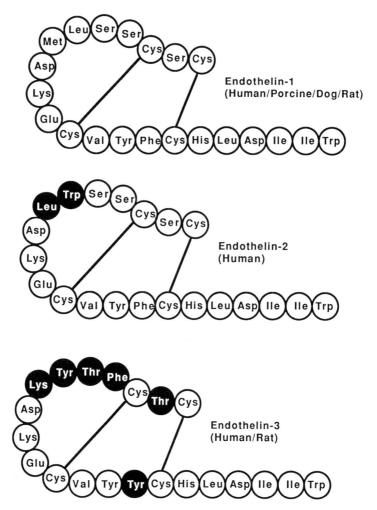

Figure 11-17. Structures of endothelins 1, 2, and 3 as found in humans and related species

and aorta (Sakurai et al., 1992). Here they mediate vasoconstriction. In contrast, ET_B subtypes are found in the central nervous system and in endothelial cells; they appear to be linked to vasodilation. As will be discussed below, the availability of endothelin antagonists with selected binding affinity to ET_A or ET_B receptors would be extremely useful in further defining the roles of these peptides and their receptors. (Spinella et al., 1991).

Since the endothelins are potent vasoconstrictors and the big endothelins are not, devising antagonists to the endothelins or preventing the degradation of the big endothelins represent potential routes to new antihypertensive agents. The endothelin family presents a nearly ideal case study since most of the approaches that may be envisaged for discovering and controlling peptide hormone action have been reasonably successful.

As regards enzyme inhibition, phosphoramidon blocks the action of an enzyme that converts the big endothelins to endothelins (Fukuroda et al., 1994; Matsumura et al., 1990; Okada et al., 1990). Peptide hormone structural modifications have also been succesful in that modifications of the hormone structure to remove "active" elements have provided analogues that function as selective agonists or which have antagonist activity. Naturally occurring antagonists include a potent cyclic pentapeptide antagonist found in an extract of *Streptomyces* by random screening. A modified analogue of the microbial inhibitor BQ-123, has been found to be selective for the ET_A receptor with an IC_{50} value of 22 nM (Fig. 11-18) (Ishikawa et al., 1992). A linear modification, BQ-610, is even more potent, having an IC_{50} of 0.7 nM (Fig. 11-18).

Inhibition of Enzyme Processing—Phosphoramidon

The nature of the enzyme (endothelin-converting enzyme or ECE) that converts big endothelins to endothelins is not precisely known, but is believed to be similar to EC.24.11 (enkephalinase). As with other metalloproteases, a reasonable inhibitory strategy would be to find compounds capable of coordinating to the active site metal in place of the scissile amide bond. Phosphoramidon (Fig. 11-19) is an established protease inhibitor with a mode of action that derives functionally from its structure, in which a PO_2NH moiety is present in lieu of the CONH peptide linkage. This functionality is stable to proteolysis and thus serves to prevent the cleavage of the 1-39 precursor of endothelin (1-21); in so doing, it blocks the pressor action (McMahon et al., 1991). It may be anticipated that the search for more potent and selective agents with

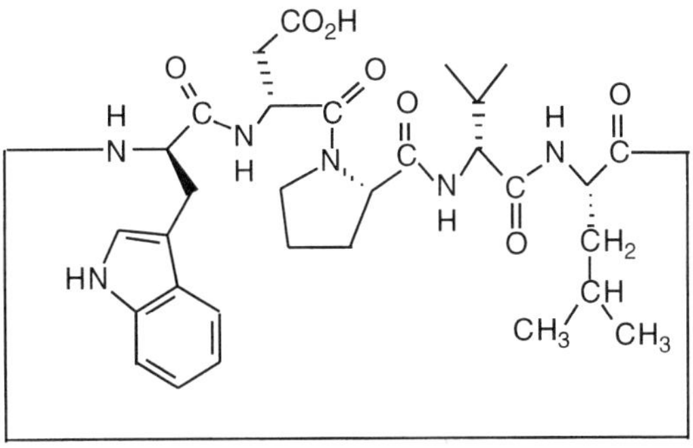

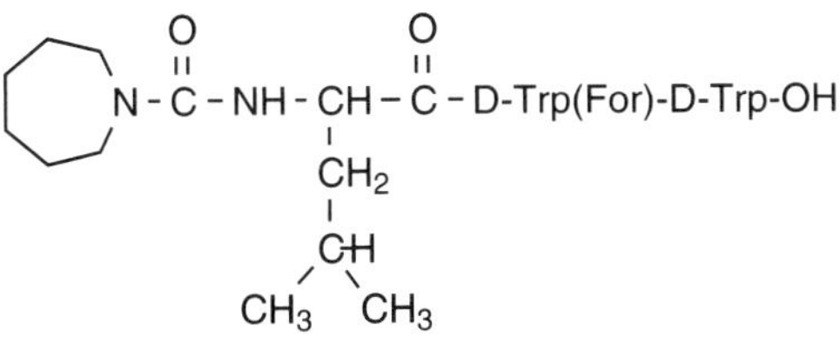

Figure 11-18. Structure of cyclic (BQ-123) and linear (BQ-610) endothelin receptor antagonists.

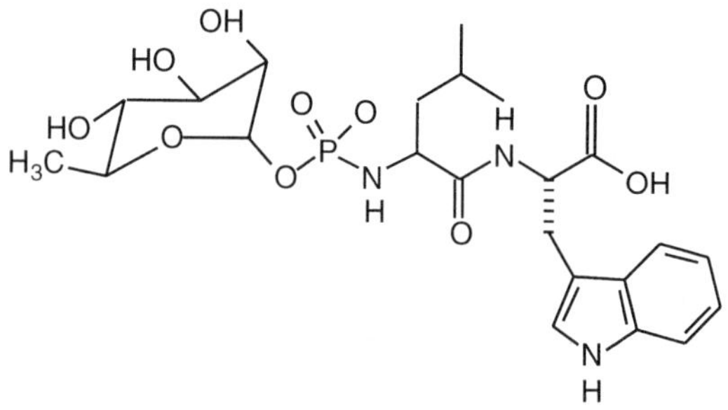

Figure 11-19. Structure of the metalloprotease inhibitor phosphoramidon.

similar mechanisms would provide a useful rationale for the design of new types of antihypertensive agents. In this regard, a report that replacement of the rhamnose sugar group in phosphoramidon by an alkyl moiety with retention of some potency (~60-fold less active; IC_{50} = 109 nM) suggests that analogues with other amide bond replacements might prove at least equally effective (Bertenshaw et al., 1993).

Endothelin Analogues

A second antagonist design approach involves the concept discussed earlier of retaining binding elements while replacing active elements. Endothelins 1–3 contain three conserved acidic residues at positions 8, 10, and 18, and basic residues at positions 9 (Lys) and 16 (His). Kikuchi et al. (1993) reported that Asp^{18} was essential for expression of ET_A receptor antagonist activity, but replacement with hydrophilic amino acids (Thr, Ser, or Asn) caused the analogues to retain agonist activity. The authors suggested that this behavior meant that activation of the ET_A receptor required formation of a hydrogen bond between the 18-position side chain and some corresponding group on the receptor. In fact, a Thr^{18} analogue had affinity (IC_{50} = 0.19 nM) similar to the Asp^{18} analogue (IC_{50} = 0.15 nM) although it was a *partial agonist*, meaning that maximal vasoconstriction could not be reached with the Thr^{18} analogue, even at higher concentrations.

To probe the significance of the position 18 active element further, Kikuchi and coworkers varied the amino acids at position 19, normally an Ile in ET-1. Surprisingly, very subtle structural changes at position 19 were able to convert a peptide agonist into an antagonist. As shown in Table 11-7, the key feature appears to be the presence or absence of a single methyl group at the γ-position of the residue-19 side chain, since Leu^{19} is an antagonist while the Ile^{19}, Val^{19}, and Nva^{19} variants are reasonably effective agonists. The behavior of some of these peptides as ET_B antagonists was also reasonably dramatic, although several of the analogues were shown to be noncompetitive antagonists toward the ET_B receptor subtypes.

The preceding analysis provides a clear demonstration of the need for careful analysis of peptide hormone actions to distinguish between full and partial agonists; binding (receptor occupancy) versus activation (receptor transduction event); and competitive versus noncompetitive antagonism. But it is not always necessary to use the full hormone structure to discover such important structure activity relationships. Researchers at Parke Davis (Doherty et al., 1993) recognized the presence of strong binding elements within the C-terminal "tail" of the endothelin, and the importance of His^{16}

for agonist activity. A series of C-terminal hexapeptide analogues were prepared based on the parent Ac-His-Leu-Asp-Ile-Ile-Trp, which had modest (mid-micromolar) binding affinity. Substitution of His^{16} by either D-His^{16} or D-Phe^{16} led to increases in binding affinity and to antagonist activity at both ET$_A$ and ET$_B$ receptors.

By introducing a larger, hydrophobic aromatic group in place of His^{16}, a fairly potent, though nonselective, ET$_A$ and ET$_B$ antagonist was produced having the structure Ac-D-Dip-Leu-Asp-Ile-Ile-Trp (Table 11-8), where Dip = 3,3-diphenylalanine (Cody et al., 1992). These short peptide antagonists appear to have significant structural similarities to endothelin antagonists isolated from natural sources, as discussed below.

Natural Product Leads

In 1991, Banyu (Japan) scientists reported that a cyclic pentapeptide isolated from a cultured broth of *Streptomyces misakiensis* was a modestly potent endothelin antagonist, having selectivity for the ET$_A$ receptor subtype (Ihara, 1991). The lead compound, cyclo(D-Trp-D-Glu-Ala-D-Val-Leu) was subsequently modified to afford a new synthetic analogue, BQ-123 (Fig. 11-18A) having greater than two orders of magnitude improvement in inhibitory potency (Table 11-8) (Ishikawa et al., 1992).

The major changes in BQ-123 were replacement of D-Glu by D-Asp and replacement of L-Ala by L-Pro. The first change was designed to reduce hydrophobicity; the resulting compounds, especially in the form of their Asp sodium salts, showed greatly improved water solubility. The second change was based primarily on an analysis of the predicted conformations and was designed to stabilize intramolecular hydrogen bonding, according to the following rationale.

Banyu scientists realized that their lead structure possessed a DDLDL pattern of side chain configurations within the cyclic pentapeptide framework. Previous work from several laboratories, using model systems, had established that many cyclic pentapeptides, and especially those with a DDLDL pattern, formed stable conformational isomers with two *intra*-molecular hydrogen bonds involving 4 of the 5 amide bonds as shown

Table 11-7 ET$_A$ Antagonist Activity of Thr^{18} Analogues of ET-1 Correlates with Structure of Position 19 Substituents

ET Analogues	Position 19 Side Chain	Agonist or Antagonist	Antagonist pA^2
ET-1	-CH(CH$_3$)CH$_2$CH$_3$	Agonist	—
[Leu^{18}]ET-1	-CH(CH$_3$)CH$_2$CH$_3$	Agonist	—
[Thr^{18}]ET-1	-CH(CH$_3$)CH$_2$CH$_3$	Agonist	—
[Leu^{19}]ET-1	-CH$_2$CH(CH$_3$)CH$_3$	Agonist	—
[Thr^{18},Leu^{19}]ET-1	-CH$_2$CH(CH$_3$)CH$_3$	Antagonist	7.7
[Thr^{18},Cha^{19}]ET-1	-CH$_2$C$_6$H$_{11}$	Antagonist	7.7
[Thr^{18},MeLeu^{19}]ET-1	-CH$_2$C(CH$_3$)$_3$	Antagonist	7.4
[Thr^{18},Val^{19}]ET-1	-CH(CH$_3$)$_2$	Agonist	—
Thr^{18},Nva^{19}]ET-1	-CH$_2$CH$_2$CH$_2$CH$_3$	Agonist	—

Table 11-8 Comparison of Relative Potencies and Selectivities of Cyclic and Linear Endothelin Antagonists against ET_A and ET_B Receptors

Peptide Antagonists	$ET_A{}^a$ IC_{50} (μM)	$ET_B{}^b$ IC_{50} (μM)	Ref.
Cyclic			
cyclo(D-Trp^1-D-Glu^2-Ala^3-D-Val^4-Leu^5)	3.0 ± 0.46	>100	Ishikawa et al., 1992
cyclo(D-Trp-D-Glu-Ala-D-alloIle-Leu)	1.4 ± 0.19	>100	Ishikawa et al., 1992
cyclo(D-Trp-D-Asp-Ala-D-Val-Leu)	0.11 ± 0.012	>100	Ishikawa et al., 1992
cyclo(D-Trp-D-Asp-Ala-D-Val-Pro)	8.5 ± 0.87	>100	Ishikawa et al., 1992
cyclo(D-Trp-D-Asp-Pro-D-Val-Leu) (BQ-123)	0.022 ± 0.0017	18 ± 22	Ishikawa et al., 1992
Linear			
Hexahydroazepinylcarbonyl-D-Leu-D-Trp (For)-D-Trp (BQ-610)	0.0007	24	Ishikawa et al., 1993
N-cis-2,6-dimethylpiperidinocarbonyl-L-γ-Me-Leu-D-Trp(COOMe)-D-Nle-OH (BQ-788)	0.28 1.3^c	900 0.0012^d	Fukuroda et al., 1994
Ac-D-Dip-Leu-Asp-Ile-Ile-Trp	0.015	0.015	Cody et al., 1992
Ac-D-Bhg-Leu-Asp-Ile-Ile-Trp	0.0026	0.019	Cody et al., 1992

Abbreviation: Bhg = 5H-dibenzyl(a,d)cycloheptene-10,11-dihydroglycine
^{a}Porcine aortic smooth muscle cells (rich in ET_A receptors)
^{b}Porcine cerebellar membranes (ET_B receptors)
^{c}Human neuroblastoma (SK-M-MC) cells (ET_A receptors)
^{d}Human Girardi heart cells = (ET_B receptors)

in Figure 11-20. Two major turn features commonly found in peptides and proteins, a β-turn and a γ-turn, were usually present in such structures, and the hydrogen bonds could be confirmed using both proton NMR (solution studies) and X-ray crystallography (solid state) for several models (Pease (Gierasch) & Watson, 1978; Karle, 1978). Model compounds have shown clearly that a proline residue in the middle of a γ-turn or at the $i + 1$ position of a β-turn, tend to favor and stabilize these important structural features (Smith & Pease, 1980).

When this hypothesis was applied to the cyclic pentapeptide endothelin antagonist lead, the results were somewhat surprising. While a Pro^3 analogue (in the putative γ-turn region) led to a five-fold improvement in antagonist potency, a Pro^5 substitution (in place of L-Leu) led to a large *decrease* in potency. This suggests that whatever conformational advantage that *may* have accrued from stabilizing a β-turn was more than

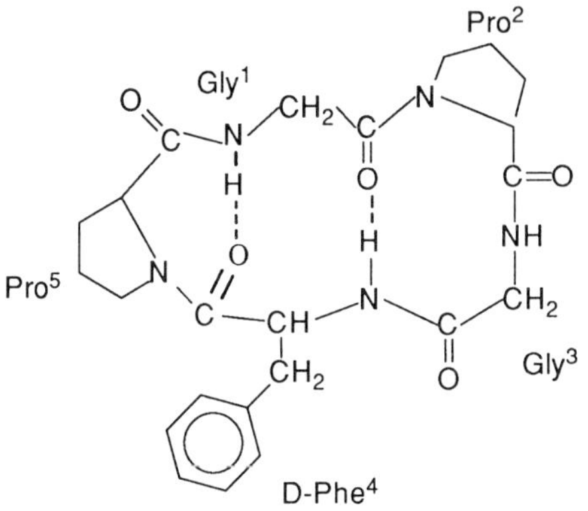

Peptide Hormones / 393

Figure 11-20. Structure of a hypothetical cyclic pentapeptide showing proline residues at position 2 and at position 5, thereby stabilizing the 10-atom β-turn hydrogen bond (proline in the preferred $i + 1$ position) and the 7-atom γ-turn, respectively. The NMR and X-ray parameters supporting this generalized structure have been published (Karle, 1978; Pease & Watson, 1978).

offset by the negative loss of a presumably critical L-Leu side chain binding interaction. These results were later confirmed by more detailed structural analysis of BQ-123 using proton and carbon NMR analysis (Ishikawa et al., 1992).

More recently, Japanese scientists have been able to prepare linear analogues with Leu-Trp or modified Leu-Trp-Trp structures that represent new types of ET_A selective antagonists (Table 11-8 and Fig. 11-18B). By further modification of the N-terminus, a linear analogue was converted into a selective endothelin antagonist that is effective against the ET_B class of receptors (Fig. 11-21). As shown in Table 11-8, this compound was tested and found to be selective toward both porcine and human tissue types known to be dominated by only a single receptor class.

In view of the important role of the endothelins in physiology, it is apparent that, as with virtually all peptide hormones, there will be many more analogues, structure-activity relationships and related studies in the years ahead. The future appears bright indeed for chemists and biologists fascinated with this important group of compounds.

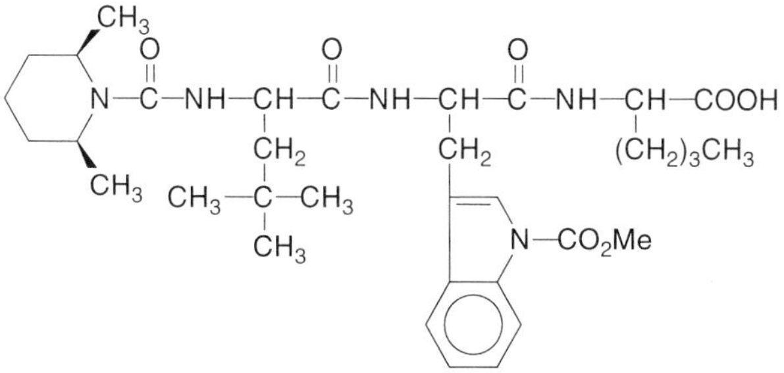

Figure 11-21. Structure of the potent ET_B-selective endothelin antagonist (N-*cis*-2,6-dimethylpiperidinocarbonyl-L-γ-MeLeu-Trp(COOMe)-D-Nle-OH. Reprinted from Ishikawa et al., 1994, with permission.

Summary

Peptide hormones remain attractive targets for rational drug design because of their ubiquitous nature and physiological significance. Problems of bioavailability and stability are being solved through judicious structure modification or structural simplification and mimicry using nonamide components, as the four examples given involving analogues of oxytocin, LH-RH, α-MSH, and the endothelins have demonstrated.

The following are some of the more important questions necessary for a more complete understanding of peptide hormones and their functions: (1) Are amide bonds essential for activity? Problems of desolvation have been cited as reasons for replacing amide bonds for more effective transport. But to what extent are peptide backbone elements, as distinct from the side chains, important for activity? (2) How do hormones interact with their receptors and how is receptor selectivity achieved? As hormone receptors are cloned and compared, these answers should be more evident. (3) How do antagonists work? A simplified picture of antagonists as merely hormone agonists without the "activating element" is not correct, at least not in all cases. A rational approach to inhibitor design will follow once we have a better understanding of where and how some of our most potent peptide hormone antagonists actually interact with their targets. (4) What steps follow hormone binding? Signal transduction is becoming one of the most exciting areas of current research, primarily due to our insufficient understanding of these critical events.

It is apparent that peptide hormones will continue to be at the center of chemistry and molecular biology for the years ahead and represent fertile areas for both research and future drug development.

12

Peptide Mimetics

Hiroshi Nakanishi and Michael Kahn

Peptides and proteins control all biological processes at transcriptional, translational or posttranslational levels. Yet, at the molecular level, our understanding of the relationship between structure and function remains rudimentary. The problems involved in clarifying these issues are somewhat different for peptides and proteins. The dissection of multidomain proteins into small, synthetic, conformationally restricted components is an important step in the design of low molecular weight nonpeptides that mimic the activity of the native protein. Mimetics of critical functional domains might possess beneficial properties in comparison to the intact proteinaceous species with regard to specificity and therapeutic potential, and are valuable probes for the study of molecular recognition events (Chen et al., 1992). On the other hand, peptides are characteristically highly flexible molecules whose structure is strongly influenced by their environment (Marshall et al., 1978). Their conformational mobility in solution complicates their use to determine their receptor bound or bioactive structures (Fauchère, 1987; Hruby, 1987). Conformational constraints can significantly aid this determination (Hruby et al., 1990a). Peptide mimetics are powerful tools for the study of molecular recognition and are providing a unique opportunity to dissect and investigate structure-function relationships in peptides and complex proteins. This chapter briefly describes the very broad and rapidly expanding area of peptidominetic research which, according to one's personal definition, can be limited to *N*-methylated amino acids or include species as diverse as steroids.

Modified Peptides

Native peptides are generally rapidly inactivated via proteolytic degradation by enzymes. Additionally, peptides possess poor bioavailability, particularly in passing through lipophilic barriers such as intestinal mucosa or the blood-brain barrier. In an attempt to overcome these detrimental properties, extensive modifications of peptides have been undertaken. Although these analogues are often referred to as peptidomimetics, it may be more logical to classify them as modified peptides. Nevertheless, the information garnered from investigations with modified peptides is important for the design of peptidomimetics.

N-methylated Amino Acids

N-Methylated amino acids are commonly found in naturally occurring peptide antibiotics (e.g., cyclosporin). *N*-Methylation has a number of significant effects. It elimin = 0,1 (proline), 2

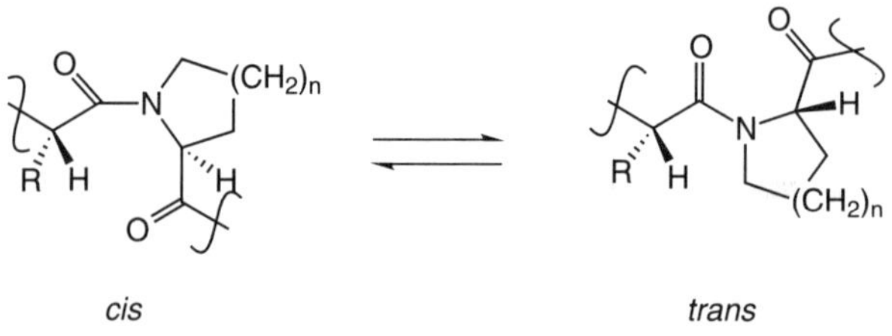

Figure 12-1. *cis*- and *trans*-isomers of N^α-C^α cyclized amino acids.

nates the hydrogen atom on the nitrogen of the amide and, perforce, alters the hydrogen bonding pattern of the peptide. N-Methylated amides often exist in equilibrium between a *cis* and *trans* geometry (Fig. 12-1). Spectroscopic and computational investigations of the dipeptide Sar-Sar (**1**) (Sar = sarcosine or N-methylglycine) found the *cis* isomer to be less than 1 kcal/mol higher in energy (Toniolo, 1990). N-Methylation restricts the allowed conformational space of the preceding amino acid residue, in a manner very similar to the effect of proline residues (Howard et al., 1973) (Yamazaki et al., 1993). N-methylated amino acids have been incorporated into bioactive analogues of opioid peptides (Manavalan & Momany, 1980; Morley, 1980), bradykinin (Kawai et al., 1990), TRH (Filatova et al., 1986), angiotensin II (Bovy et al., 1989), and CCK (Hruby et al., 1990b).

α-Alkyl Amino Acids

α-Alkyl amino acids have a structure in which the α-hydrogen atom is replaced by an alkyl group. The effect of this modification is that rotation around the N—$C_\alpha(\phi)$ and C_α—C(O)(ψ) bonds is severely restricted. Approximately 70% of the conformational space available to glycine is precluded by the addition of one methyl group (Ala). The addition of a second methyl group (Aib, α-amino isobutyric acid) (**2**) eliminates an additional 20% of the conformational space available to glycine (Degrado, 1988; Paterson et al., 1981). Aib (α-aminoisobutyric acid or α-methylalanine), a naturally occurring amino acid observed in channel-forming peptides, is the most extensively investigated of the α-alkyl amino acids. The conformational space favored by Aib residues includes regions of both the left- and right-handed α and 3_{10} helices (Karle & Balaram, 1990; Toniolo & Benedetti, 1991). This residue has been incorporated into numerous bioactive peptides including enkephalin (Balaram & Sudha, 1983; Nagaraj & Balaram, 1978), angiotensin (Samanen et al., 1991), and bradykinin (London et al., 1990) in an attempt to obtain potent analogues and define their bioactive conformations.

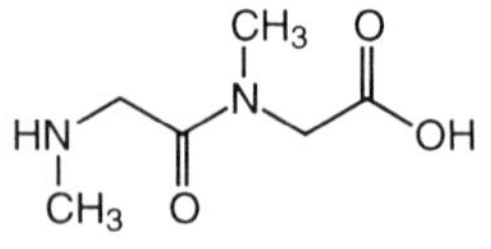

Sar–Sar (**1**)

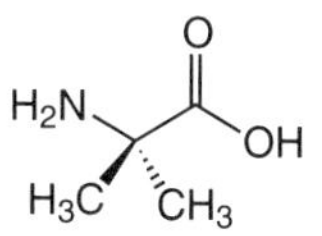

Aib (α-amino isobutyric acid)

2

Toniolo has recently reviewed the conformational preferences for several additional α-methylated amino acids including isovaline (L-α-ethylalanine) (**3**), α-methylvaline (**4**), α-methylleucine (**5**), and α-methylphenylalanine (**6**) (Toniolo et al., 1993). These preferences have been determined experimentally (either by X-ray crystallography or ^{1}H-NMR spectroscopy) and via computational energy calculations. The consensus is that peptides (tripeptides and longer) adopt Type I or Type III (3_{10} helical) β-turn conformations, and rarely adopt extended ($\phi = 180°$, $\psi = 180°$) structures. Similarly, α-amino cycloalkanecarboxylic acids (**7**) generally prefer to adopt a Type III β-turn (3_{10} helix) (Di Blasio et al., 1992). Interestingly, diethylglycine (**8**) and dipropylglycine (**9**) exhibit fully extended structures where ϕ and ψ angles both approach 180° (Benedetti et al., 1984; Marshall et al., 1988). A similar minimum energy conformation is predicted for diphenylglycine (Crisma et al., 1990).

N$^\alpha$—C$^\alpha$ Cyclized Amino Acids

The N$_\alpha$—C$_\alpha$ cyclized amino acids are essentially analogues of proline. The tertiary amide bond in these compounds leads to facile *cis-trans* amide bond rotation. This is because the difference in energy between these conformers is significantly less (approximately 2 kcal/mol) than a normal peptide bond (approximately 12 kcal/mol)

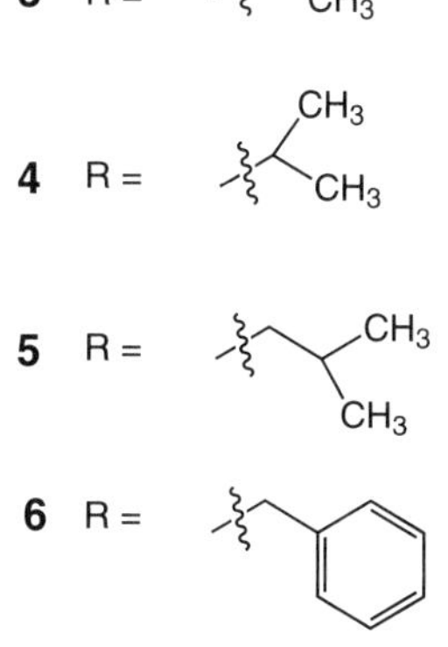

3 R = —CH$_3$

4 R = —CH(CH$_3$)$_2$

5 R = —CH$_2$CH(CH$_3$)$_2$

6 R = —CH$_2$C$_6$H$_5$

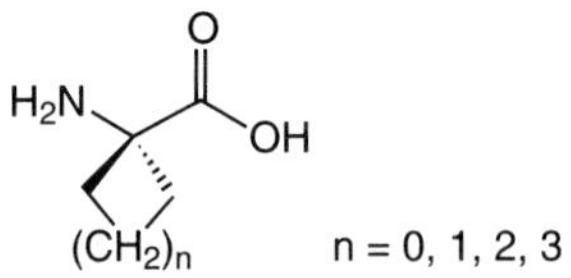

H₂N—C(—(CH₂)ₙ)—C(=O)OH   n = 0, 1, 2, 3

7

H₂N—C(R)(R)—C(=O)OH

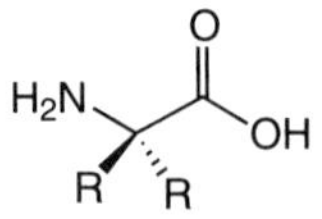

8 R = CH$_2$CH$_3$

9 R = CH$_2$CH$_2$CH$_3$

(Creighton, 1984) (Fig. 12-1). Like proline, these derivatives adopt a highly restricted set of ϕ and ψ angles.

Other Short Range Cyclizations (Residue i to i − 1 or i + 1)

Short range cyclization of the type N↔C_α^1, and C_α^1↔C_α can significantly reduce the conformational space accessible to the peptide segment in which they are incorporated. The range of examples of this type of conformational restriction is beyond the scope of this chapter (For a recent review, see Toniolo, 1990).

Peptoids

The many undesirable properties of peptides have prompted the development of unnatural biopolymers. Oligomers of N-substituted glycines, termed peptoids (**10**), have recently been described (Simon et al., 1992). Nowick and coworkers developed an oligourea scaffold that is stabilized by internal hydrogen bonding (**11**) (Nowick et al., 1992). Additionally, oligocarbamates (**12**) (Cho et al., 1993), polypyrrolinones (**13**) (Smith et al., 1992) and vinylogous polypeptides (**14**) (Hagihara et al., 1992) have been suggested as alternatives to polyamide backbones.

Peptide Bond Isosteres

Many amide bond isosteres have been devised. These analogues resemble the peptide bond to varying degrees; however, they are more resistant to proteolytic cleavage. This generally comes at the expense of modifications in the geometric or topochemical structure, electronic and hydrogen bonding interactions, and general hydrophilicity. Examples of this type of modification include retro-inverso [NH—C(O)] (Chorev & Goodman, 1993), reduced amide [CH$_2$—NH] (El Masdouri et al., 1988), thiomethylene [CH$_2$—S] (Spatola & Edwards, 1986), oxomethylene [CH$_2$—O] (TenBrink, 1987), ethylene [CH$_2$—CH$_2$] (Rodriguez et al., 1990), thioamide [C(S)—NH] (Jones et al., 1973), *trans*-olefin and *trans*-fluoroolefin [CH=CH and CF=CH] (Cox et al., 1980; Felder et al., 1992; Spaltenstein et al., 1987), ketomethylene (Jennings-White & Almquist,

1982; Vara Prasad & Rich, 1990), and fluoroketomethylene [C(O)—CFR, R = H or F] (Damon & Hoover, 1990; Thaisrivongs et al., 1991) analogues.

Nonpeptide Mimetics

The isolation and identification in 1975 of the endogenous opioid pentapeptides, methionine and leucine enkephalin (Hughes et al., 1975) represents the intellectual groundbreaking in the field of peptidomimetics. The work described demonstrated that despite their highly disparate structures, these linear pentapeptides and the condensed

15   **16**   **17**

heterocyclic species morphine, all elicit their biological response, analgesia, by binding to the opiate receptor. However, despite intensive efforts to understand this relationship at a structural level, it is fair to say that it remains far from clear some twenty years later.

Screening

To date, unquestionably the most successful approach for developing nonpeptide leads for peptide ligands has involved screening. A wide array of ligands, often bearing little, if any, structural resemblance to the endogenous peptide ligand that they mimic, have been uncovered through receptor-based screening programs. These include CCK-A and CCK-B mimetics of the benzodiazepine (**15**) (Bock et al., 1993), β-carboline (**16**) (Evans et al., 1993) and diphenylpyrazolidinone (**17**) (Howbert et al., 1993) classes of molecules, angiotensin II antagonists incorporating a broad range of heterocyclic structures [e.g., losartan (**18**) and PD-123,319 (**19**)] (Mantlo et al., 1994, and references therein), and an array of neurokinin antagonists (**20, 21**) (Advenier et al., 1992; Lowe et al., 1993; Watling, 1992). Additional nonpeptide ligands have been discovered for an array of peptide ligands including neuropeptide Y (Doughty et al., 1992), C5a (Lanza et al., 1992), glucagon (Collins et al., 1992), melatonin (Yous et al., 1992), oxytocin (Evans et al., 1992; Salituro et al., 1993), vasopressin (Otsubo et al., 1993), bombesin (Valentine et al., 1992), growth hormone-releasing peptide (Smith et al., 1993a) and neurotensin (Snider et al., 1992).

Despite the diversity of structural classes that have been uncovered, a number of generalities have emerged. The nonpeptide ligands generally bear little, if any, structural resemblance to the peptide they are mimicking. Although not a drawback inherently, this precludes the use of already generally available peptide structure activ-

losartan (**18**)   PD 123, 319 (**19**)

CP 96345 (**20**)   SR 38968 (**21**)

ity relationships for further designed activity enhancement. Although numerous attempts have been made to map peptides topologically with their nonpeptide mimics (Pierson & Freer, 1992; Portoghese et al., 1988), these analogies rarely prove general when multiple analogues are considered. Recently, elegant molecular biological experiments with mutated cloned receptors have provided a rationale for this shortcoming (Fong et al., 1992; Gether et al., 1993). Mutations that affect the binding of the peptide mimetic do not affect the binding of the endogenous peptide ligand. The lesson is that although both ligands do bind to the same receptor, they do so in different ways, contacting disparate amino acid residues and, at best, occupying only partially overlapping space. Most of the success in screening has been in finding antagonists to G-protein coupled serpentine receptors (7-helix transmembrane spanning). This may be due in part to the inherently lipophilic environment of the receptor, and the relatively limited contact surface involved in this particular class of ligand-receptor interactions.

What lessons can be learned from these screening leads? Importantly, they have validated many of the critical concepts which underlie the rational design of peptidomimetics, in that they prove that compounds lacking amide bonds, as well as obvious pharmacophore similarity and flexibility, can be potent and selective ligands for peptide receptors. However, a very sobering note is that these relationships provide very limited information with regard to the development of generic solutions for rationally traversing the pathway from peptides to mimetics.

Combinatorial Screening

Ligand discovery involving large arrays (libraries) has become an increasingly popular strategy for lead discovery. Initially, most of the effort was focused on short (6–7 amino acid) peptide libraries (Fodor et al., 1991; Geysen & Mason, 1993; Houghten et al., 1991; Lam et al., 1993). The appeal of these methods is that they allow for the selection of a compound(s) that binds to a target with the highest affinity from a large combinatorial pool, which typically might contain 10^6 to 10^7 members.

In essence, this is a renaissance of the traditional screening of natural and synthetic compounds for potential therapeutic agents that has been a mainstay of the pharmaceutical industry. However, as discussed previously, peptides are less than ideal candidates for pharmaceutical development. Cognizant of this shortcoming, chemists are now developing libraries of cyclic peptides, novel biopolymers and nonpeptides. In this type of library, one additional problem that arises is the question of identifying leads.

Kerr et al. (1993) developed a strategy to encode each nonnatural component in their library pool with a unique peptide sequence that can be read by convential peptide sequencing. An encoded library containing 200 nonnatural decapeptoids was synthesized by the alternating parallel synthesis of a branched polymer containing both a binding ligand and a coding peptide. Recently, clever coding techniques that allow for the tracking of much larger libraries of nonpeptides have been developed. Nielsen et al. (1993) and Needels et al. (1993) have described an oligonucleotide coding system that allows for the amplification of the coding sequence by PCR. Ohlmeyer et al. (1993) recently reported a technique for the synthesis of encoded libraries in which the tag can be read by electron capture capillary gas chromatography. An intriguing caveat to the usage of libraries was pointed out in a recent paper by Chen et al. (1993). Libraries containing two million hexapeptides and two million cyclic hexapeptides failed to yield ligands for the SH3 domain of PI3 kinase. However, the use of a biased library that incorporated a known polyproline motif was quite successful.

The combinatorial synthesis of nonpeptide "druglike" chemical structures offers an exciting opportunity to speed up the drug discovery process. A significant advantage of this type of library is that the compounds screened already possess the desirable traits of conformational rigidity and biostability. Bunin & Ellman (1992) have laid the groundwork by demonstrating a solid phase method for a general and efficient synthesis of 1,4-benzodiazepines. Similarly, Dewitt et al. (1993) reported the "diversomer" approach. They were able to synthesize 40 discreet hydantoins and 40 benzodiazepines by a solid phase approach. It is anticipated that significant activity in this type of library construction and screening will be seen in the near future. In particular, efficient syntheses of novel classes of compounds would be highly desirable as large numbers of hydantoins and benzodiazepines already reside in the collections of major pharmaceutical companies.

Design of Nonpeptide Mimetics

There has been an increasing effort to rationally design and synthesize highly active analogues of biologically significant peptides and proteins. It is anticipated that such drugs might possess greater selectivity, and fewer side effects, than their present day counterparts (Fauchère, 1986) The complex problems associated with the rational design of mimetics are being addressed increasingly due to advances in molecular biology, spectroscopy, and computational chemistry. The determination of the receptor bound conformation of a peptide or protein ligand is invaluable for the rational design of mimetics. However, with few exceptions this information is not readily available (de Vos et al., 1992; Milner-White et al., 1988).

Conformational constraints constitute one of the most promising avenues for a solution to this problem, particularly if the constraint is such that only one conformation of the ligand is significantly populated. Rigid analogues pay a lower entropy cost upon binding to their receptor and, therefore, should bind more avidly, assuming appositive placement of pharmacophoric residues (Miklavc et al., 1987). Proteolytic enzymes generally prefer conformationally adaptable substrates; therefore, constrained analogues are generally endowed with increased proteolytic stability. Additionally, selectivity can be enhanced by precluding the formation of conformers that produce undesired bioactivity (Veber et al., 1979).

The Secondary Structure Approach

One approach to the design of peptidomimetics has been guided by the simple elegance which nature has utilized in the molecular architecture of proteinaceous species (Kaiser & Kezdy, 1984). Three basic building blocks (α- helices, β-sheets and reverse turns) are utilized for the construction of all proteins. The design and synthesis of peptidomimetic prosthetic units to replace these three architectural motifs is affording an opportunity to dissect and investigate complex structure-function relationships in proteins through the use of small synthetic, conformationally restricted components. This is a critical step toward the rational design of low molecular weight nonpeptide pharmaceutical agents, devoid of the shortcomings of conventional peptides. A recent symposium-in-print is devoted to this topic (Kahn, 1993).

Reverse Turn (β-Turns and γ-Turns)

The surface localization of turns in proteins, and the predominance of residues containing potentially critical pharmacophoric information, has led to the hypothesis that turns play critical roles in a myriad of recognition events (Rose et al., 1985). Reverse turns are classified into γ-turns, consisting of three residues (sometimes referred to as a C7 conformation), and the more common β-turns (C10 conformation), formed by a tetrapeptide. Unlike the α-helix and the β-sheet, β-turns are highly irregular secondary structures and exhibit widely variable backbone conformations as shown through numerous statistical investigations of well-resolved X-ray crystal structures (Lewis et al., 1973; Perczel et al., 1993; Venkatachalam, 1968; Wilmot & Thornton, 1988, 1990). A β-turn is defined as a tetrapeptide sequence in which the α-carbon distance between first and fourth residue is less than 7 Å, and is contained within a nonhelical region (Lewis et al., 1973). Beta turns can be classified further according to the backbone dihedral angles of the second and third residues (Wilmot & Thornton, 1988).

However, a classification based on the peptide backbone conformation may be irrelevant, in that overwhelming evidence indicates that the appositive placement of the amino acid side chains is the most critical role of the β-turn in molecular recognition. An alternative β-turn classification has been proposed which utilizes a single torsional parameter β. Beta defines the spatial relationship between the peptide bonds as they enter and exit the reverse turn and the relative orientations of the intervening side chains (Ball et al 1990, 1993). An excellent review by Ball and Alewood (1990) summarized the progress to that point in reverse turn mimetics.

The endogenous tripeptide Pro-Leu-Gly-NH$_2$ (**22**) has a modulatory effect on the CNS dopamine receptor and is believed to adopt a Type II β-turn conformation based upon NMR (Higashijima et al., 1978) and X-ray (Reed & Johnson, 1973) investigations. The biologically active conformation was probed by introducing various constraints (**23–29**) (Genin et al., 1993a,b; Sreenivasan et al., 1993; Subashinghe et al., 1993). The initial analogue (**23**) (Yu et al., 1988), which incorporated a γ-lactam at the Leu-Gly position, showed a promising 10,000-fold enhancement of the binding of the dopamine agonist ADTN compared with the linear tripeptide. Surprisingly, variation in the size and chirality of the lactam ring (**23–29**) did not affect potency. The 5,5 and 6,5-bicyclic thiazolidine lactam systems (**30, 31** and **32, 33**, respectively), originally designed to mimic a Type II' β-turn (Nagai & Sato, 1985), showed only modest (2 to 5-fold) increase in activity compared with the linear PLG. Compound **31** did not show any activity. Conformational analysis of **30** and **32** using a random conforma-

22

23 X = CH$_2$CH$_2$ * = R
24 X = CH$_2$ * = R
25 X = CH$_2$CO   * = R
26 X = CH$_2$O   * = R
27 X = (CH$_2$)$_3$ * = R
28 X = (CH$_2$)$_4$ * = R
29 X = (CH$_2$)$_4$ * = S

30 X = CH$_2$ * = R, R, S
31 X = CH$_2$ * = R, S, R
32 X = CH$_2$CH$_2$ * = R, R, S
33 X = CH$_2$CH$_2$ * = S, S, R

tional search program (Ferguson & Raber, 1989) with MM2 force field (Allinger, 1977) indicated that both **30** and **32** possess nearly ideal ϕ_2 and ψ_2 torsional angles (ideal angles are 120° and 80°, respectively). The (R)-4.4 and 5.4-spirolactam systems were also used to induce a β-turn in the PLG sequence (**34** and **35**) (Genin et al., 1993a). The structures were analyzed by X-ray crystallography and found to form a Type II β-turn (Genin et al., 1993b). Previous studies by Ward et al. (1990) with a series of neurokinin antagonists had shown (from energy calculations, and supported by NMR spectroscopic analysis in DMSO) that the extended conformation is favored for the (R)-4.4-spiro lactam system. However, the (S)-4.4-spirolactam demonstrated a Type II' β-turn conformation in DMSO, as judged by NMR spectroscopy.

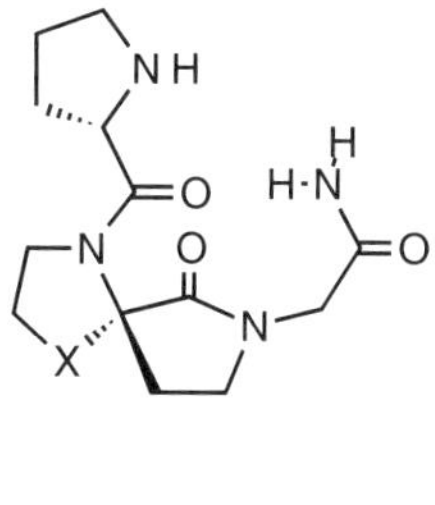

34 X=CH$_2$
35 X=CH$_2$CH$_2$

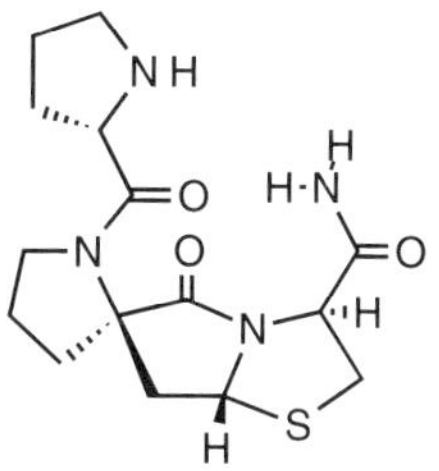

36

Further constrained analogues in this series were made by combining the bicyclothiazolidine lactam and spirolactam to form a spirobicyclic system (**36**) (Genin & Johnson, 1992). Molecular modeling and NMR studies (temperature coefficient of amide hydrogen and NOE) in CDCl$_3$ suggested the formation of a Type II β-turn through restriction to almost ideal values of three out of the four torsional angles (ϕ_2, ψ_2 and ϕ_3; root mean squared fit to an ideal Type II β-turn backbone atoms is 0.16 Å). However, in DMSO the mimetic exhibited a high temperature coefficient for the amide hydrogen, indicating that the β-turn is not stable, presumably due to flexibility at ψ_3. The modulatory activity of this analogue was tested and found to enhance the binding of ADTN to the dopamine receptor by only 40% at 1 μM concentration compared to 26% at 1 μM concentration for the linear PLG. The authors claim that these results clearly demonstrate that the bioactive conformation of PLG is a Type II β-turn; however, the failure to improve on the initially less constrained compound **23** by introducing additional constraints lead us to believe that although the β-turn conformation may be important, the addition of buttressing groups beyond the peptide backbone may be interfering with binding.

Baca et al. (1993) made use of Nagai's Type II' thiazolidine lactam β-turn mimetic (Nagai & Sato, 1985) to replace the type I' β-turn between Gly-16 and Gly-17 in HIV-1 protease. The β-turn mimetic-containing enzyme dimerized similarly to the native enzyme. It was fully active, possessed the same substrate specificity as the native enzyme, and showed enhanced thermal stability. Nicolaou et al. (1990) utilized 3-deoxy-β-D-glucose (**37**) as a scaffold to synthesize the first nonpeptidic analogue of the receptor recognizing β-turn (Phe7, Trp8, Lys9, Thr10) of somatostatin (**38**). This mimetic binds to the pituitary gland somatostatin receptor with an IC$_{50}$ value of 1.3 μM. In-

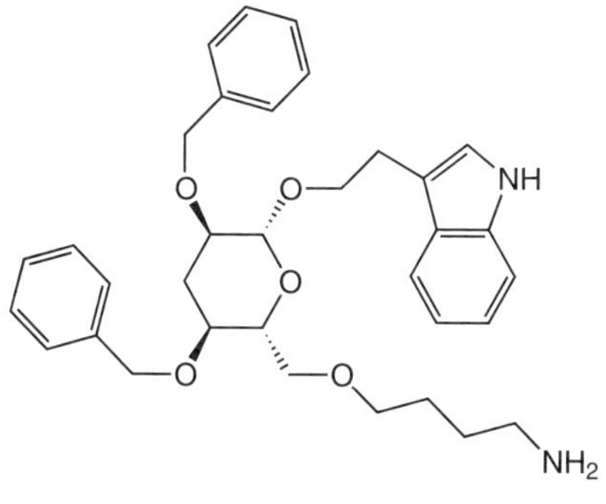

37

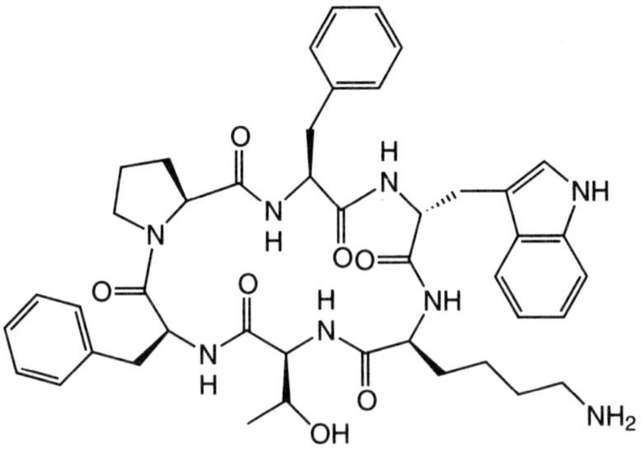

38

terestingly, in a functional assay this mimetic displayed agonist activity at 3 μM concentrations, a feature that is less frequently observed with peptidomimetics.

An interesting recent application involved the design of a benzodiazepine peptidomimetic (**39**) to inhibit the enzyme Ras farnesyl transferase. The benzodiazepine system is intended to mimic the proposed β-turn in the CAAX sequence of the enzyme substrate (James et al., 1993); it has an $IC_{50} < 1$ nM. Interestingly, compound **40**, a structurally related lipophilic, conformationally restricted mimic of the tripeptide (Z)Phe-His-Leu, is a potent inhibitor of angiotensin converting enzyme (Flynn et al, 1987). In this instance, by analogy to other metalloprotease substrate interactions (i.e., Borkakoti et al., 1994), the enzyme-bound substrate presumably adopts an extended structure.

γ-Turns

γ-turns are defined by a hydrogen bond between the carbonyl oxygen of amino acid residue i and the amide hydrogen of residue $i + 2$ (**41**). Two types of γ-turns exist: inverse γ-turns and classic γ-turns; their backbone structures are mirror images. Although rare, a recent analysis of 54 proteins with high resolution X-ray crystal structures has shown the existence of approximately ten classic γ-turns, with all but one producing chain reversal at the ends of β-hairpins. Inverse γ-turns are about six times more common than classic γ-turns in proteins, but produce chain kinks, and are usually not involved in chain reversal (one out of 61 inverse γ-turns) (Milner-White et al., 1988). However, γ-turns are well established as a relatively common feature in cyclic

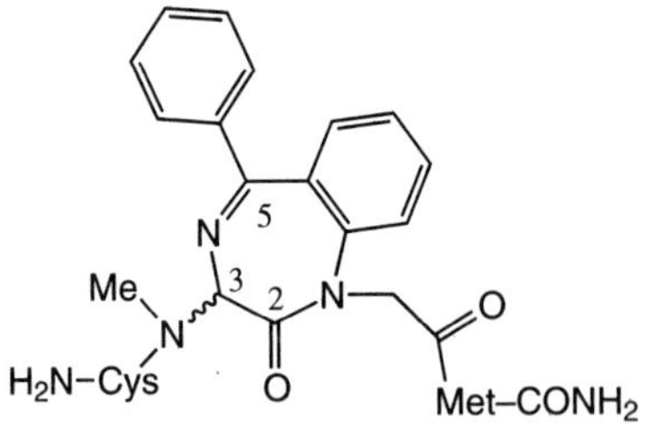

39 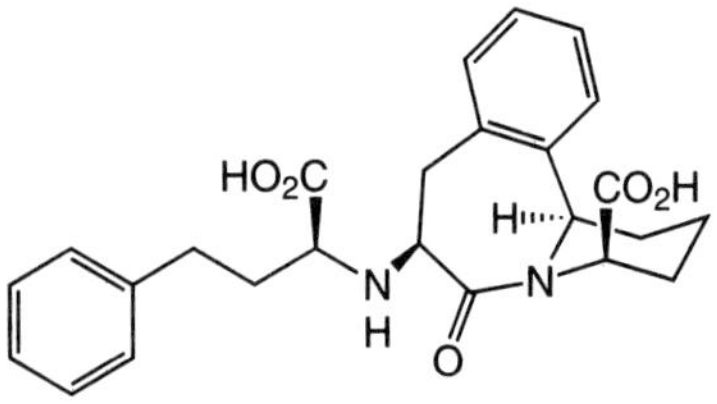   **40**

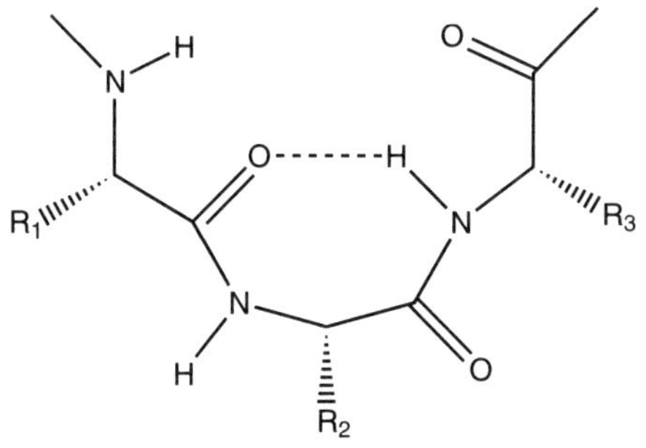

41

peptides that exhibit biological activity (see, e.g. Bienstock et al., 1993; Bogusky et al., 1993; Di Blasio et al., 1993; Pavone et al., 1992; Stroup et al., 1992). A *trans*-olefin C7 mimetic (**42**) was used to restrain a γ-turn conformation at the Asp residue in the RGD sequence (Callahan et al., 1992), which is known to inhibit the binding of fibrinogen to its receptor (gp-IIb/IIIa), thus resulting in inhibition of platelet aggregation and thrombus formation. NMR, X-ray crystallographic, and molecular modeling studies (Kopple et al., 1992) of the cyclic peptide Ac-c[Cys-NMeArg-Gly-Asp-Pen]-NH_2 had shown an extended Gly in the RGD region and a γ-turn at the aspartic acid residue. Two mimetic-containing RGD analogues (**42** and **43**, R = CH_3 or Ph) were synthesized and found to be potent platelet aggregation inhibitors *in vitro* and retained nanomolar affinity for the isolated human gp-IIb/IIIa receptor; this was comparable to the cyclic peptide and more than two orders of magnitude better than the corresponding linear peptide. A retro amide γ-turn mimetic (**44**) was also incorporated into the RGD peptide sequence to enforce the γ-turn conformation at the Asp residue. However, the retro amide mimetic **45** exhibited a significant reduction in binding to the human gp-IIIb/IIIa receptor, and *in vitro* platelet antiaggregatory activity (Callahan et al.,

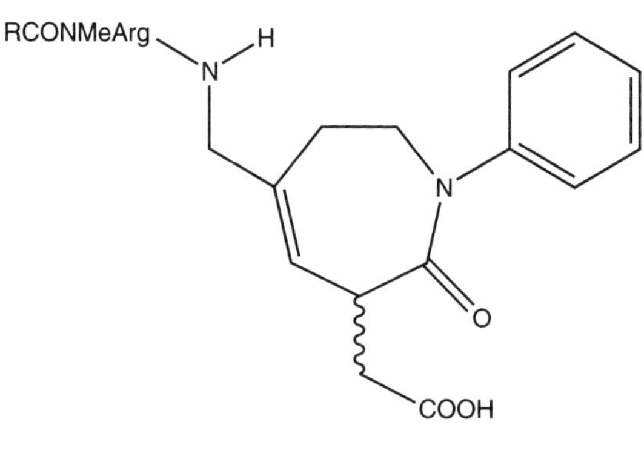

42 R = CH_3

43 R = C_6H_5

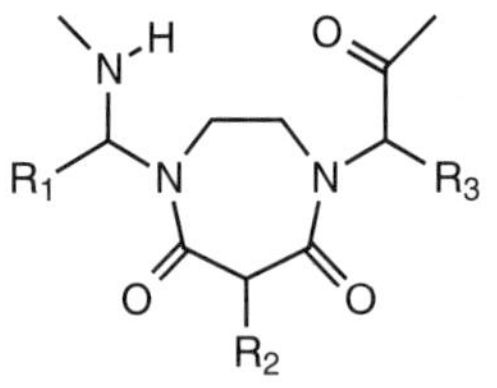

44

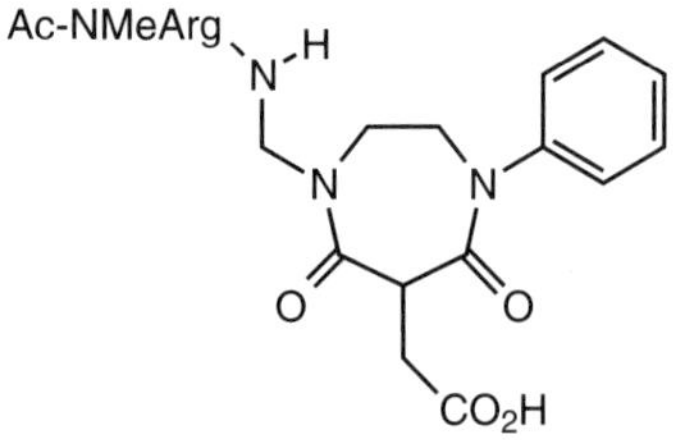

45

1993). A possible explanation for the loss of activity may be that the additional polar carbonyl group interferes with the binding of the mimetic.

However, the bioactive conformation of RGD is far from clear. The snake venom proteins echistatin and kistrin (Adler et al., 1991; Chen et al., 1991) contain the RGD sequence in a highly flexible loop. A β-turn has also been proposed as the bioactive conformation (Reed et al., 1988) and a series of bioactive peptidomimetics (e.g., **46**) have been designed based on this hypothesis (Barker et al., 1992; Hirschmann et al., 1992; McDowell & Gadek, 1992).

It may be that the binding of RGD analogues by gp-IIb/IIIa is primarily governed by the distance between the guanidinium and carboxylate moieties and is quite, although not completely, promiscuous in regard to the scaffolding involved.

A *trans* olefin mimic was also utilized (Tourwe et al., 1992) to study the active conformation of enkephalin (H-Tyr-Gly-Gly-Phe-Leu-OH). A γ-turn was introduced to replace either the Gly-Gly-Phe or Gly-Phe-Leu or Tyr-Gly-Gly sequences. The bindings and functional activities of the enkephalin analogues were tested at both the δ and μ opioid receptors. Only the analogue incorporating the Tyr-Gly-Gly mimetic demonstrated any activity, but this was only 1% of the affinity of leucine enkephalin for the δ-receptor. However, in a functional assay at the mouse vas deferens δ receptor, this analogue was found to be virtually inactive. The data suggests that the biologically active conformations of Leu-enkephalin at either the δ or the μ opioid receptors is not consistent with γ-turn formation.

A 6-membered lactam ring (**47**) (R_1 = H, R_3 = CH_2Ph) was incorporated into the nonapeptide bradykinin (Arg-Pro-Pro-Gly-Phe-Ser-Pro-Phe-Arg), in which a γ-turn formed between residues 6–8 has been hypothesized to be a bioactive conformation (Sato et al., 1992). One of the diastereomers was found to show very moderate (mi-

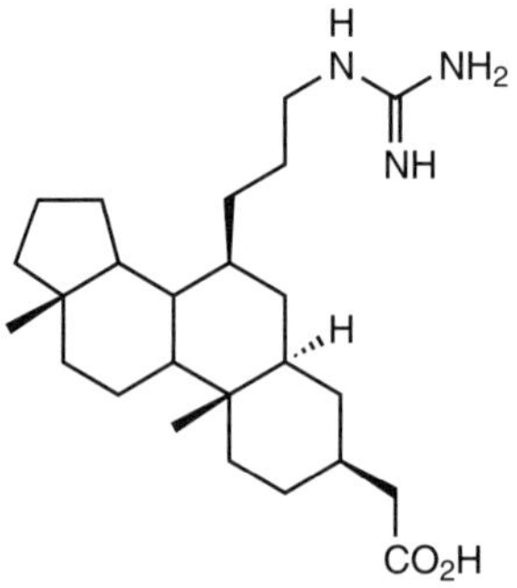

46

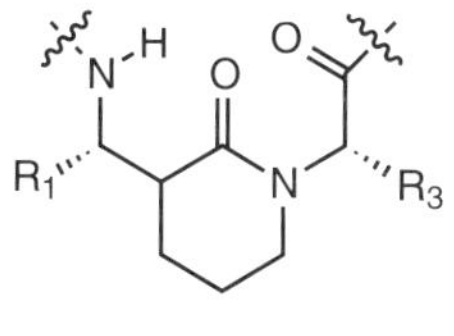

47

cromolar) affinity for the bradykinin receptor despite lacking the proper side chains of Ser-6 and Pro-7. The result tends to support the presence of a reverse turn (either a γ-turn or, more probably, β-turn) at this site.

α-Helix

Helices are the most common secondary structural element found in globular proteins, accounting for just over one-third of all residues (Barlow & Thornton, 1988). The key feature of an α-helix is the iterative pattern of backbone hydrogen bonding between the amide hydrogens and the carbonyl oxygens located four residues apart. The α-helix has been recognized as a hallmark of protein architecture since its conception by Pauling et al. (1951). It is postulated to play a key role in protein folding as the hydrogen bonds can be localized to intrasegment partners to form autonomous folding units in proteins (Kim & Baldwin, 1984; Presta & Rose, 1988; Shoemaker et al., 1987). Extensive effort has been directed toward an understanding of helix formation, its stability and amino acid propensities (Blaber et al., 1993; Chakrabartty et al., 1991,1993; Hermans et al., 1992; Komeiji et al., 1993; Padmanabhan et al., 1990; Padmanabhan & Baldwin, 1991). Yet as noted by Kemp and Curran (1988a, b), little attention has been given to the design of helical templates. This is due in large part to the inherently more difficult task of mimicking the approximately 12 amino acids (i.e., three turns of an α-helix) required to form a stabilized isolated helical peptide. The formation of an alpha helix involves two steps, initiation and propagation. To date, most of the effort in the design and synthesis of α-helix mimetics has centered around N-terminal initiation motifs.

Arrhenius and Satterthwait used a hydrazone-ethylene bridge to replace the 5 → 1 backbone hydrogen bond in one turn of an α-helix to afford the cyclic peptide shown in Figure 12-2 (Arrhenius et al., 1987; Arrhenius & Satterthwait, 1989). Conforma-

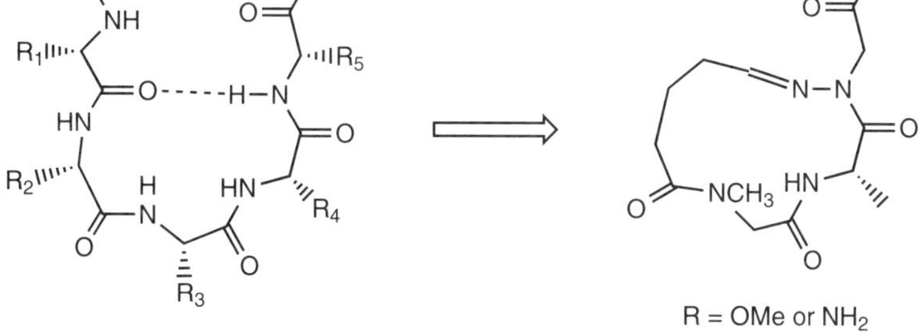

Figure 12-2. Use of a hydrazone-ethylene bridge to form a cyclic peptide from the 5 → 1 backbone H bond in an α-helix.

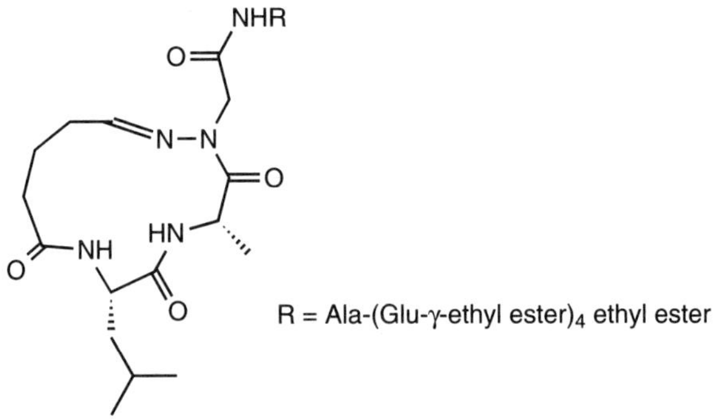

48

R = Ala-(Glu-γ-ethyl ester)$_4$ ethyl ester

tional analysis of both the methyl ester and amide were performed in CDCl$_3$ and DMSO-d_6 by NMR spectroscopy. Based upon 1D NOE measurements, both compounds seem to prefer the *cis-N*-methyl peptide bond conformation in DMSO and an equilibrium mixture in CDCl$_3$. Based on this analysis, the formation of an α-helix-inducing conformation in these macrocyclic compounds is inconclusive. A pentapeptide (Ala-(Glu-γ-ethylester)$_4$-ethyl ester) was subsequently added to the carboxy terminus of the macrocyclic template **48**. The conformations of the pentapeptide with and without the macrocyclic template were monitored by NMR utilizing ^3J coupling constants, and sequential NOE and H/D exchange rates of the amide protons in deuterated trifluoroethanol. The observed 3.8 Hz ^3J coupling constant (Pardi et al., 1984) observed for the alanine residue, together with the smaller exchange rates, tends to indicate the formation of an α-helical conformation.

The proper alignment of two or more hydrogen bond acceptors is crucial for the design of a successful helix nucleation template. By restraining two proline rings with a thiamethylene bridge, thereby forming a tricyclic template (**49**), Kemp and Curran (1988) intended to orient the three amide carbonyls at the proper pitch and spacing for a right-handed α-helix. Observation of an NOE from the acetyl methyl to the following proline αH in CDCl$_3$ (Kemp & Curran, 1988) leads one to believe that the carbonyl oxygen of the acetyl group at the N-terminus is improperly positioned (i.e., oriented *cis*), presumably to avoid the dipole-dipole repulsion of the aligned carbonyls. However, NMR investigation of polyalanine ($n = 1–6$) conjugated to this template in CDCl$_3$ indicated the existence of a conformational equilibrium between an intramole-

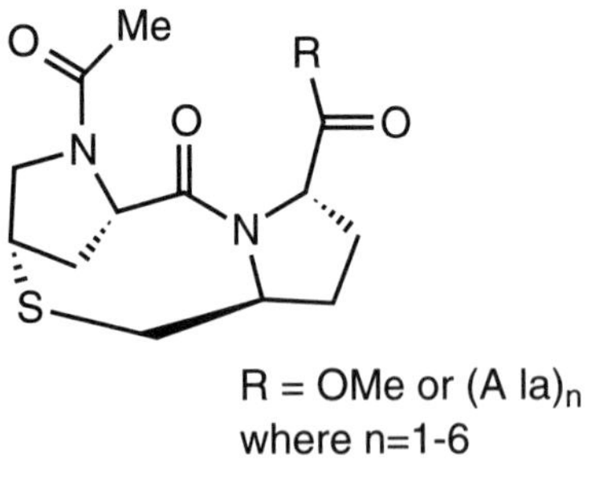

R = OMe or (Ala)$_n$ where n=1-6

49 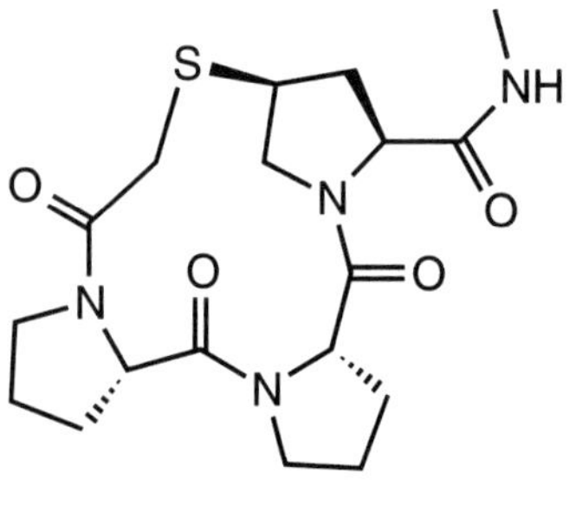   **50**

51 **52**

cularly hydrogen bonded structure with a *trans*-N-terminal acetamide bond, and a non-hydrogen bonded structure derived from the *cis* conformer (Kemp et al., 1991a). The *trans* to *cis* ratio, which correlates with the observed helix to random coil ratio, is found to be length and solvent dependent for the polyalanine oligomers (Kemp et al., 1991b).

In a subsequent design, the number of hydrogen bond donor sites was increased to four by using a cyclic triproline helix template (**50**) (Kemp & Rothman, 1992). An X-ray and NMR study determined that the template exists largely in a nonhelical conformation. This finding was in agreement with molecular mechanics calculations, which suggested that a helical conformation is approximately 4 kcal/mol less stable. Attachment of a polyalanine ($n = 1-3$) sequence to the template afforded a peptide that adopted largely a 3_{10} helical conformation.

A template which affords rigid alignment of three carbonyl oxygens was devised by Müller et al (1993). They utilized the cage compound (**51**) which is readily accessible via a series of electrocyclic addition reactions. A nonapeptide (mixture of Ala and Aib) coupled to this cage compound showed significantly increased α-helicity compared with the linear nonapeptide in 1:1 water-TFE solution as judged by CD spectra. On the other hand, the same nonapeptide coupled to the enantiomeric template (**52**), where the three carbonyl groups are aligned as in a left-handed helix, exhibited decreased helicity under the same conditions. Estimated α-helicities based on the CD elipticities at 222 nm are 40%, 70%, and 25% for the *N*-Boc protected linear, right-handed conjugate (**51**) and left-handed conjugate (**52**), respectively (Müller et al., 1993).

β-Sheet

Despite a wide array of potential applications for β-sheet mimetics, for example, enzyme inhibitors (Borkakoti et al., 1994), antigen presentation (Bjorkman et al., 1987), disruption of dimerization (Schramm, 1991, 1993), and second messenger signaling (Waksman et al., 1992), limited effort in these areas has been reported to date. The difficulty in forming a well defined β-sheet in unaggregated form has hindered our understanding of β-sheet properties. In proteins, β-sheets form as the result of an extensive hydrogen-bond network and side chain - side chain interactions. Providing hydrogen bond donors, acceptors, and side chain functional groups in the proper arrangement is a significant synthetic challenge (Kemp et al., 1991c).

Kemp and coworkers described the first β-sheet mimetic which utilizes a diacylaminoepindolidione template. This template was linked to a dipeptide Pro-D-Ala, which is presumed to adopt a β-turn conformation. Subsequent coupling of a urea, which perforce inverts the directionality of the peptide chain, permits the formation of an antipar-

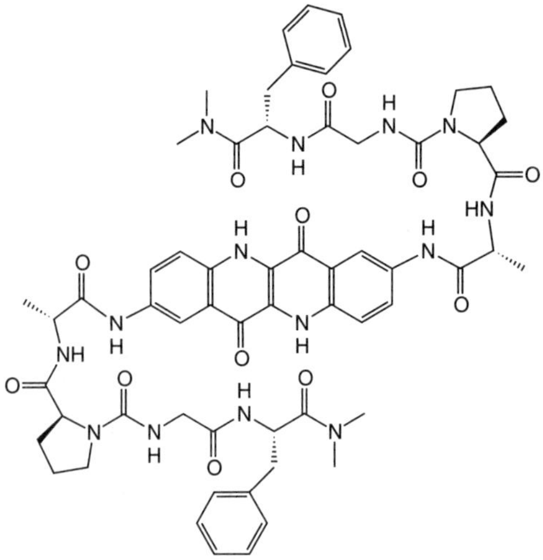

53

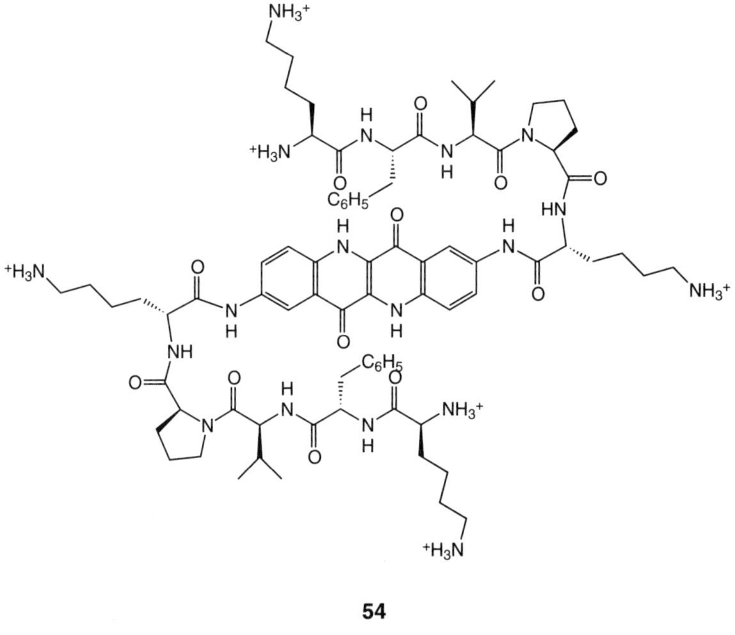

54

allel sheet structure (**53**) (Kemp & Bowen, 1988a, b, 1990; Kemp et al., 1991c). Removal of the urea allows for the formation of a parallel β-sheet (**54**) (Kemp et al., 1991c). The existence of a β-sheet conformation was confirmed by the observation of NOEs between the alpha hydrogens of glycine and H-1 of the epindolidione, and between the hydrogens of the *N*-methyl groups and H-10 of the epindolidione in DMSO. Additional

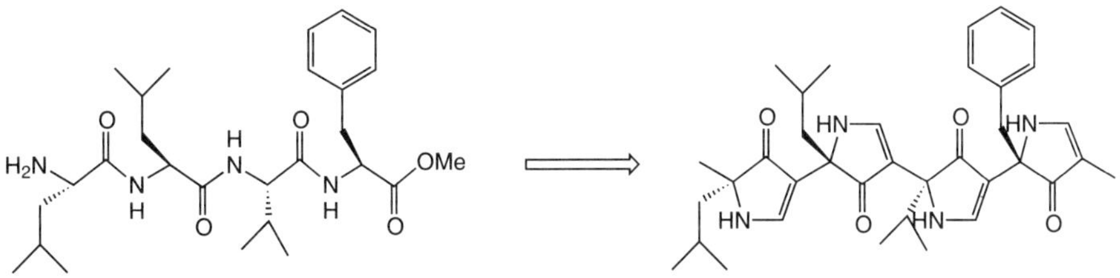

Figure 12-3. Use of 3.5-linked pyrrolin-4-ones to form a β-sheet.

supporting evidence was provided by the temperature dependence of amide proton chemical shifts, and geminal coupling constants. The β-turn forming sequence Pro-D-Ala was replaced by Sar-Gly to examine the effect of the β-turn on the formation of the antiparallel β-sheet. The CD spectrum for this substance, in solvents ranging in polarity from THF to DMSO, exhibited no band from 300 to 500 nm, indicating no significant interaction between the two amino acid asymmetric centers and the epindolidione, consistent with the absence of significant secondary structure (Kemp & Bowen, 1990). Although the epinodolidiones may provide valuable information on the nucleation and stability of parallel and antiparallel β-sheets, a significant shortcoming is the difficulty of incorporating side chain functionality into this template, which would be required for many biological applications. The use of 3,5-linked pyrrolin-4-ones (Smith et al., 1992, 1993b) overcomes this problem at the expense of a displaced NH group (Fig. 12-3). Initial modeling indicated that a β-strand would be the favored conformation. The X-ray crystal structure of this compound confirmed that it exists in a β-strand conformation and that the side chain orientations closely mimic a natural β-strand in angiotensin. It was also determined from examination of the unit cell that the mimetic dimer adopts an antiparallel β-pleated sheet. Aspartic proteinase inhibitor analogues (for renin and HIV-1) were constructed using the pyrrolinone template based upon previously reported inhibitors. The limited examples disclosed displayed encouraging binding affinities, and selectivities, as well as improved transport properties (Smith et al., 1994).

In an attempt to lock the peptide backbone into a β-sheet conformation, Martin et al. (1992) have used a trisubstituted cyclopropane in a renin inhibitor. Vinylogous polypeptides have also been used to constrain the peptide backbone conformation (Hagihara et al., 1992). They were found to adopt either antiparallel or parallel β-sheets as crystals depending on the side chain substitutents. Elongated vinylogous polypeptides seem to adopt a helical conformation. Flexibility and lack of hydrogen bond donor-acceptor capacity may hinder their generic application for enzyme inhibition, although cyclotheonamide B, a naturally occurring vinylogous polypeptide, is a relatively potent inhibitor of thrombin (Fusetani et al., 1990).

Using a dibenzofuran as a β-turn template (**55**) Kelly and coworkers were able to nucleate and stabilize a β-sheet conformation (Diaz et al., 1993; Tsang et al., 1994). Similarly, a tricyclic xanthene template was used to mimic the β-loop structure of the snake toxin flavoridin (**56**). This analogue displayed approximately 50-fold higher affinity for the fibrinogen receptor gp-IIb/IIIa than the corresponding linear peptide (Müller et al., 1993). It is interesting to note that in both cases, in addition to orienting the proper hydrogen bonding networks, hydrophobic clustering at the turn template seems to stabilize and nucleate the β-sheet (Tsang et al., 1994).

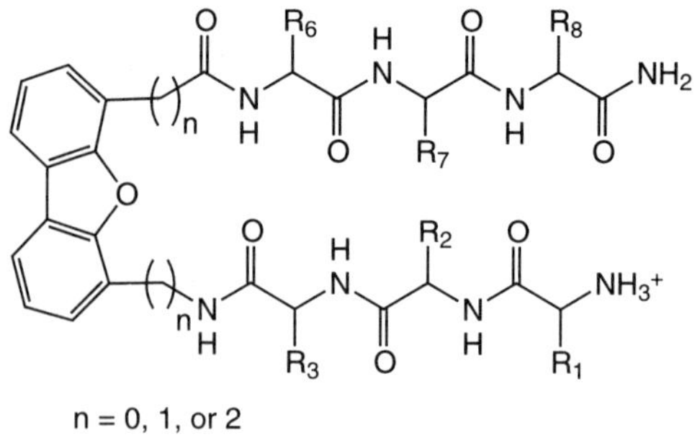

55

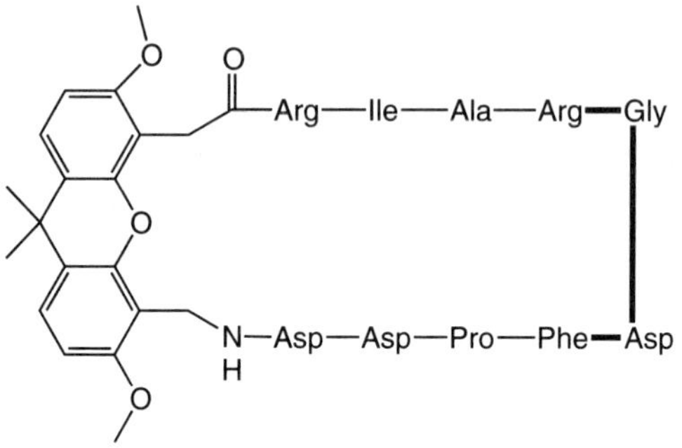

n = 0, 1, or 2

56

Structure 56: dibenzofuran-linked peptide with Arg—Ile—Ala—Arg—Gly and N(H)—Asp—Asp—Pro—Phe—Asp

Recent Advances from Our Laboratories
(Reverse Turn Mimetics)

In the design of reverse turn mimetic systems, there are a number of concerns and criteria that need to be addressed. β-Turns comprise a rather diverse group of structures. β-Turns are classified according to the ϕ and ψ angles of the $i + 1$ and $i + 2$ residues. In addition to a number of turn types (I, I', II, II', III, III', IV, V, Va, VIa, VIb, VII and VIII), the $C_\alpha i$ to $C_\alpha i+3$ distance varies from 4–7 Å (Wilmot & Thornton, 1988). From this cursory discussion, it should be readily apparent that no one structure can accurately mimic this diversity of turns. The interaction of the amino acid side chains with their complementary receptor groups is the critical determinant of biological specificity. A successful peptidomimetic must appositely position the appropriate functional groups on a relatively rigid framework. Therefore, an idealized mimetic design should incorporate the ability to accurately display critical pharmacophoric information in the same manner in which it is presented in native reverse turns. This is a far from trivial synthetic problem, in that it requires the stereo and enantiocontrolled introduction of a minimum of four noncontiguous asymmetric centers. Furthermore, the nonpeptidic character of these molecules promises to best overcome the inherent problems of peptides. Synthetic expediency is a major concern that should not be taken lightly, particularly at an early stage when the delineation of structure-activity relationships is critical, and requires the synthesis and evaluation of a series of related structures. Although we had achieved some success with our earlier generations of peptidomimetics, it became obvious to us that in order to harness the power of peptidomimetics for structure-function studies, they had to be readily accessible. The major breakthrough in peptide synthesis was the development of a solid phase format by Merrifield for which he was awarded the Nobel Prize in 1985 (Merrifield, 1985). It was obvious that we needed to capture the modular component nature and automated facility of solid phase peptide synthesis in our approach to peptidomimetics. This is additionally allowing us to develop libraries of conformationally constrained peptidomimetics. With this goal, it became relatively facile to propose an appropriate retrosynthetic strategy (Fig. 12-4).

The synthesis of the reverse turn mimetic can be performed in solution; however, it is designed to be and is fully compatible with solid phase synthesis protocols. In essence, it involves the coupling of the first modular component (**A**) to the amino terminus of a growing peptide chain (**B**). Coupling of the second component (**C**), removal of the protecting group P' and subsequent coupling of the third modular component (**D**) provides the nascent β-turn (**E**). The critical step in this sequence involves the use of an azetidinone as an activated ester to effect the macrocyclization reaction (Wasserman, 1987). Upon nucleophilic opening of the azetidinone by nucleophile X, a new amino terminus is generated for continuation of the synthesis. An important feature of this scheme is the ability to alter the X-group linker, both with regard to length and degree of rigidity or flexibility. The requisite stereogenic centers are readily derived, principally from the "chiral pool." The synthesis allows for the introduction of natural or unnatural amino acid side chain functionality in either L or D configuration. Additionally, deletion of the second modular component (**C**) provides access to γ-turn mimetics (Sato et al., 1992) (Fig. 12-5).

We have used our third generation mimetic system to explore structure-function relationships among molecules of the immunoglobulin gene superfamily. Immunoglobulins are constructed from a series of antiparallel β-pleated sheets connected by loops

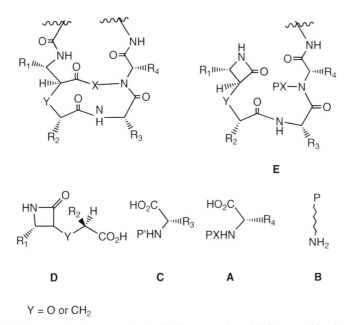

$Y = O$ or CH_2

Figure 12-4. Retrosynthetic strategy for the facile preparation of β-form peptidomimetics.

Figure 12-5. Retrosynthetic strategy for the elaboration of γ-turn mimetics.

(Amzel et al., 1979; Kabat, 1978). The specificity of these molecules is determined by the sequence and size of the canonical hypervariable complementarity determining regions (CDRs) (Chothia et al., 1989; Martin et al., 1989)

The monoclonal antibody 87.92.6 (mAb 87.92.6) is an antiidiotype antibody that binds to the cellular receptor of the type 3 reovirus. Sequence analysis revealed an intriguing homology between the two proteinaceous ligands which bind the receptor (Fig. 12-6) (Saragovi et al., 1991). In particular, a region within the CDR-2 of the light chain of mAb 87.92.6 and the hemagglutinin of the type 3 reovirus exhibited strong primary sequence homology. The V_L CDR-2 canonically exists in a reverse turn conformation (Bruck et al., 1986; Chotnia et al., 1987). Based on this analysis, we designed and synthesized a reverse turn mimetic (**57**), which incorporated the sequence Y, S, G, S, S. (Saragovi et al., 1991) Importantly, it displayed similar binding properties to the cellular reovirus receptor and to mAb 9BG5, and had the same inhibitory effect on cell proliferation as did the native antibody 87.92.6.

We have also designed a mimetic of the CDR-2 like region of human CD4. CD4 is a 55 kD glycoprotein, primarily found on the cell surface of the helper class of T cells. It binds the human immunodeficiency virus glycoprotein (HIV gp120) with high affinity ($K_d \approx$ 1-4 nM), and is an important route of cellular entry for the virus. Extensive mutagenesis (Ashkenazi et al., 1990; Landau et al., 1988) and peptide mapping experiments (Jameson et al., 1990) have shown that the region of amino acids 40–55 within the CDR-2-like domain of CD4 is critical for gp120 binding. X-ray crystallographic analysis showed that residues Gln^{40} through Phe^{43} reside on a highly surface exposed β-turn connecting the C' and C" β-strands. A first generation mimetic of this region (**58**) was designed and synthesized. NMR and molecular modeling analysis confirmed that the ten-membered ring system (**58**) closely mimicked the conformation of this loop. Importantly, this small molecule mimic abrogates the binding of HIV-1 (IIIB)

	Reo 323		I	V	S	Y	S	G	S	G	L	N	332
			O	O		♦	♦	♦	♦		♦	O	
mAb 87.92.6		46	L	L	I	Y	S	G	S	T	L	Q	55

♦ identical
O conservative substitution

Figure 12-6. Homology between monoclonal antibody 87.92.6 and hemagglutinin of type 3 retrovirus, both of which bind to the cellular reapter of the type 3 retrovirus.

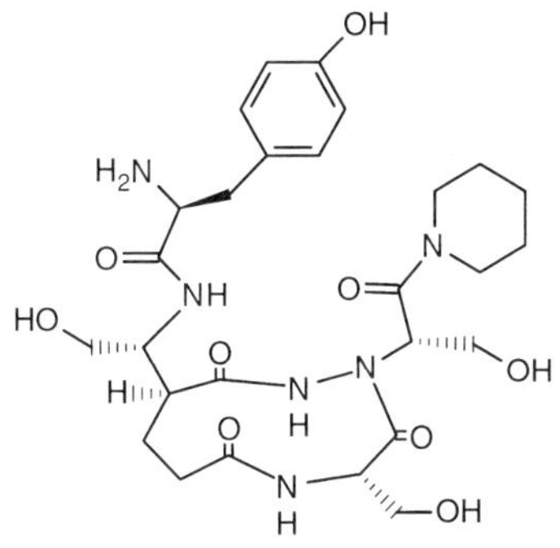

57

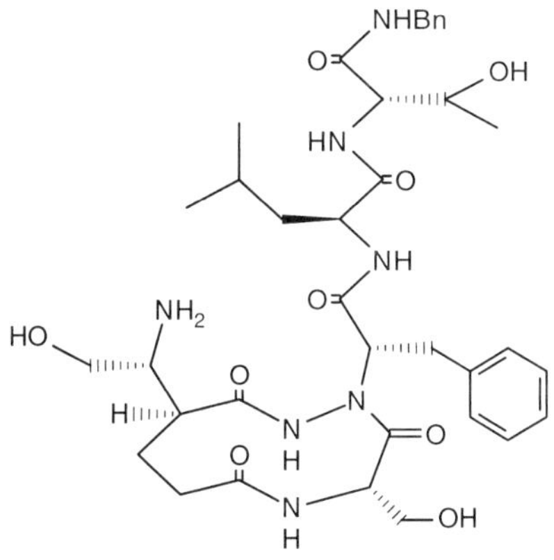

58

gp120 to CD4$^+$ cells at low micromolar levels and reduces syncytium formation 50% at 250 μg/ml (Chen et al., 1992).

Recently, we have been addressing a problem described previously concerning the relationship between morphine and enkephalin. The inherent mobility of the enkephalin framework, its rapid degradation *in vivo* (Roques et al., 1976) and the existence of multiple receptor subtypes (Mansour et al., 1988; Rapaka et al., 1986) have hampered the assessment of its bioactive conformations. Conformationally constrained peptides or peptidomimetics (Belanger et al., 1982; Hansen & Morgan, 1990; Schiller, 1990; Su et al., 1993) should facilitate this task. Several turn conformations have been proposed based upon computational models (Chew et al., 1991; Hassan & Goodman, 1986; Pettit et al., 1991; Smith et al., 1991), X-ray crystallography (Aubry et al., 1989; Griffin & Smith, 1988), and spectroscopic studies (Hruby et al., 1988; Mosberg et al., 1990; Picone et al., 1990).

In 1976, Bradbury et al. proposed a β-bend model stabilized by an intramolecular hydrogen bond between the NH of Phe^4 and the C=O of Tyr^1, which produces a spatial disposition between the Phe^4 aromatic ring and the tyramine segment of Tyr^1, analogous to that existing between the corresponding moieties in the potent morphine analogue PEO. Further support for the relevance of this conformation was provided by energy calculations on the potent [DAla^2, Met^5] enkephalin analogue, the lowest energy conformer (Balodis et al., 1978; Humblet & DeCoen, 1977; Manavalan & Momany, 1980) of which contained a folded structure with a turn centered on residues 2 and 3. However, evidence contrary to the biological signficance of a 4 → 1 β-turn has been presented by Freidinger (1981), and additionally by Schiller and Dimaio (1983) in the analysis of the conformations of 13- and 14-membered rigid cyclic analogues.

To further examine this hypothetical bioactive conformation, we have synthesized a family of 4 → 1 β-turn mimetics (**59–62**). The lowest energy conformer of the ten-membered ring system (**59**) is an excellent mimic of an idealized Type I′ β-turn (6 atom rms deviation 0.22 Å) and displays excellent overlap with the critical Phe^4 aromatic ring and tyramine moieties of PET (Fig. 12-7), yet it is essentially devoid of biological activity. Only the 14-membered ring analogue (**61**), which has a rather expanded loop structure, demonstrates any binding activity at the μ receptor, although even this was minimal. The results of this investigation can be interpreted as casting significant doubt on the

[Structures of compounds 59, 60, 61, 62 with X groups as shown]

59 X = $\begin{matrix} \text{N} \\ \text{H} \end{matrix}$ –CH(CH$_3$)–

60 X = N-H with isobutyl group

61 X = N-H with allyl group

62 X = NH

biological relevance of a 4 → 1 β-turn conformation for enkephalin. A similar series of experiments involving conformationally constrained analogues of 5 → 2 enkephalin β-turn mimetics is underway. It is hoped that this type of systematic approach to the synthesis of constrained reverse turn analogues, in conjunction with multiple peptide synthetic strategies, will clarify the situation as to the biological significance of these and other proposed receptor bound reverse turn conformations.

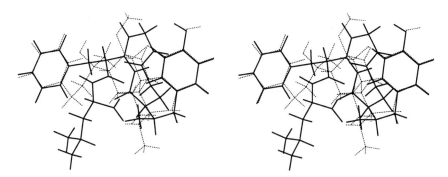

Figure 12-7. Overlay of 4 → 1 β-turn mimetic **58** with the critical Phe[4] aromatic ring and tyramine moieties of PET.

Summary

The advent of molecular biology, in particular cDNA cloning and monoclonal antibodies, has provided enormous opportunities for structural as well as functional analysis of a wide array of proteinaceous species. The critical roles that proteins play at all levels of biological regulation have opened virtually limitless potential for therapeutic intervention with recombinant proteins. However, with some notable exceptions (EPO, tPA, etc.), the therapeutic applications of proteinaceous species have been severely restricted. Proteins are subject to poor bioavailability, rapid proteolytic degradation and clearance, and are potentially antigenic. One approach to overcome these liabilities is to develop small molecule mimics.

Recently, there has been an increasing effort to rationally design and synthesize biologically active nonpeptide analogues of peptides and proteins. It is anticipated that these drugs of tomorrow will possess greater selectivity, and hence fewer side effects, than their present day counterparts. A number of approaches to this task have been outlined in this chapter. Perhaps the most fascinating feature to emerge in the field of peptidomimetics is the enormous structural diversity and creativity that is being utilized to mimic the ingenious simplicity of nature.

13

Use of Enzymes in Organic Synthesis

Zhen Yang and Alan J. Russell

The use of enzymes as biocatalysts in the transformation of organic compounds has a long history, dating back more than one hundred years to Pasteur's obervation that glycerol was produced by yeast during sugar fermentation. The manufacture of fine chemicals using isolated enzymes is now an established process. Enzyme-catalyzed organic reactions have now been extended from the synthesis of chiral synthons and low molecular weight substances, such as sugars and peptides, to more complex molecules such as oligosaccharides, polypeptides, nucleotides, and their conjugates. Enzymes are also used widely in the food industry (for example in the production of high fructose corn syrup), the pharmaceutical industry (for example in the modification of porcine insulin), and the detergent industry (for protein and fat degradation).

Enzymes are classified into six groups, according to the recommendations of the International Union of Biochemistry (Table 13-1). To date more than 2,000 enzymes have been identified, of which about 15% are commercially available. The catalytic versatility of hydrolases, combined with commercial availability and relatively low cost, make this class the most widely used. Some enzymes such as oxidoreductases need cofactors for their biocatalytic functions, and hence cofactor recycling should be considered when employing these enzymes in synthetic strategies (see later discussion).

Advantages and Disadvantages of Biocatalysts

As catalysts, enzymes are attracting significant attention from organic chemists because they provide a number of advantages over more conventional chemical catalysts, including effectiveness and efficiency, high specificity and selectivity, and the mild reaction conditions under which they operate.

Enzymes can accelerate a chemical reaction by a factor of 10^8 or more, which is often beyond the capabilities of most conventional chemical catalysts. Additionally, molar fractions of 10^{-3} to 10^{-4}% of an enzyme are sufficient to promote reaction rates, whereas chemical catalysts are often utilized at concentrations of 0.1 to 1% on a molar basis (Faber, 1992).

Enzymes often operate optimally under mild conditions: neutral pH, room temperature, and normal atmospheric pressure. Thus, undesired side reactions such as decomposition, isomerization, racemization and rearrangement are minimized in biocatalytic synthetic strategies. In addition, mild reaction conditions enable the use of relatively simple equipment in which the cost of generating high temperatures and pres-

Table 13-1 Classification of Enzymes

Class	Name	Functions	Total Number Identified[a]	Number Commercially Available[b]
1	Oxidoreductase	Oxidation-reduction: CH—OH↔C=O, CH—CH↔C=C, C—H↔C—OH	650	90
2	Transferase	Transfer of groups from one molecule to another: aldehyde, ketone, acyl, sugar, phosphoryl or methyl, from one molecule to another.	720	90
3	Hydrolases	Hydrolysis or formation of esters, amides, epoxides, nitriles, anhydrides, glycosides, etc.	636	125
4	Lyases	Addition or eleminination of small molecules (HX, X ≠ OH) on C=C, C=N, C=O bonds.	255	35
5	Isomerases	Isomerization such as racemization and epimerization.	120	6
6	Ligases	Connection of two molecules to form C—O, C—N, C—C, C—S bonds, with concomitant triphosphate cleavage.	80	6

[a,b]Data obtained from Davies et al. (1989).

sures, as often required by conventional chemical catalysts, is obviated. Furthermore, while organic chemists have viewed enzymes with some trepidation, the use of biocatalysts in nonaqueous media requires almost no special handling.

Enzymes display three major types of selectivities: *chemoselectivity, regioselectivity*, and *enantioselectivity*. Enzymes are chemoselective in that they recognize specific functional groups on a molecule. For example, an alcohol dehydrogenase can oxidize an alcohol to an aldehyde but cannot catalyze the hydrolysis of an ester group. Enzymes can distinguish between multiple copies of the same functional groups; hence they are regioselective. For instance, all positions of steroids and terpenes can be selectively hydroxylated by different enzymes. Enzymes are chiral catalysts, and they can exhibit enantioselectivity by reacting with one enantiomer of a racemic substrate pair, that is, they are able to effect kinetic resolution of racemates. Enzymes also distinguish enantiotopic functional groups in prochiral systems, producing optically active synthons, as discussed below.

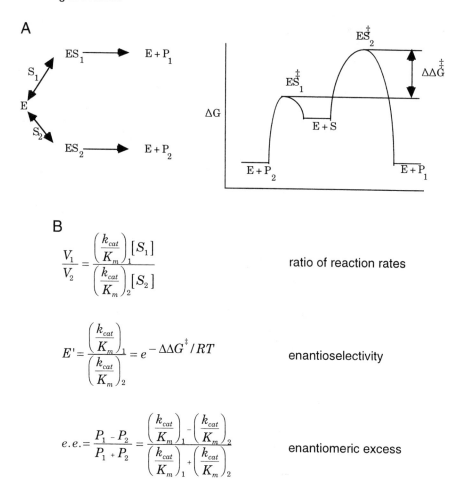

Figure 13-1. Kinetic analysis of an enantioselective reaction.

Enzymes also have other desirable properties such as biodegradability and broad applicability. Unlike conventional chemical catalysts, such as heavy metals, enzymes are natural catalysts and are completely biodegradable. The environmental acceptability of enzymes is another attractive feature of biocatalytic syntheses. In addition, enzymes can catalyze a broad variety of chemical reactions, as described later.

While the use of enzymes in organic synthesis can be of significant utility, there are also some potential drawbacks for biocatalytic processes, including cost and dependence on a narrow range of reaction conditions. As noted, the use of enzymes may be costly, especially when purified enzymes are employed or cofactors are required. It is important to stress, however, that enzyme costs can be as low as 1 to 2 cents per pound of low value product such as the cost of glucose isomerase in high fructose corn syrup production. Enzymes are sensitive to a variety of controllable variables and may be inactivated by high temperature or pressure, or extremes of pH. Enzymes are also subject to substrate or product inhibition, as indeed are many traditional chemical catalysts. One further limitation is that enzymes usually display optimal catalytic activity in water, while for organic synthesis organic solvents are the media of choice.

The development of biocatalysis as a viable alternative for conventional catalysis is dependent on successfully overcoming these perceived shortcomings. For example, microbiology and recombinant DNA technology have made the low-cost production of enzymes possible. Enzyme engineering is capable of altering intrinsic enzyme properties, and has even been used to design improved detergent enzymes, which now have an annual market of about $300 million. Enzymes can promote high activity in organic media, and offer many unusual properties that cannot be demonstrated in aqueous solution. Whole cell systems can be used instead of isolated enzymes to accomplish biotransformations; a comparison of these two methods has been published (Faber, 1992). Thus, enzymes can provide an attractive alternative to conventional chemical catalysts for organic synthesis.

Synthetically Important Enzyme-Based Reactions

Hydrolysis Reactions

Most enzymes that catalyze hydrolyses do not need cofactors, and a large number of "hydrolases" are commercially available. Hydrolases catalyze the hydrolysis of a wide range of substrates, such as carboxylic acid esters, amides, phosphate esters, epoxides and nitriles. Enzyme-catalyzed hydrolyses have been investigated intensively and used widely. In particular, due to their enantioselectivity, hydrolases have been used as a means of kinetic resolution of enantiomers.

A racemic mixture can be enzymatically resolved into two enantiomers. The resolution is kinetically controlled, which can be best understood by considering the steady-state kinetics of an enantioselective transformation where S_1 and S_2 compete for the same active site of the enzyme (Fig. 13-1). The enantioselectivity of the reaction is determined by the ratio of the specificity constants (k_{cat}/K_m) of the two competing reactions, and is related to the difference in free energy between the transition states of both the enzyme-substrate complexes ($\Delta\Delta G^{\ddagger}$). In an enzyme-catalyzed enantioselective transformation of a racemic mixture, the optical purity of the product is normally expressed as its enantiomeric excess (ee), and the relationship between ee, $\Delta\Delta G^{\ddagger}$ and the ratio of k_{cat}/K_m are given in Table 13-2. Examples of enantiomer resolutions are provided below.

Examples of enzymes that catalyze ester hydrolyses include esterases (such as pig and horse liver esterases), proteases (such as chymotrypsin, subtilisin, trypsin, pepsin

Table 13-2 Relationship Between Enantiomeric Excess (ee), $\Delta\Delta G^{\ddagger}$ and Specificity Constants for Enzyme-Catalyzed Racemic Resolutions.

ee (%)	$(k_{cat}/K_m)_1/(k_{cat}/K_m)_2$	$\Delta\Delta G^{\ddagger}$ (kcal/mol)
10	1.22	0.118
50	3	0.651
90	19	1.74
95	39	2.17
99	199	3.14
99.9	1999	4.50

Reprinted from Wong et al., 1991b, with permission.

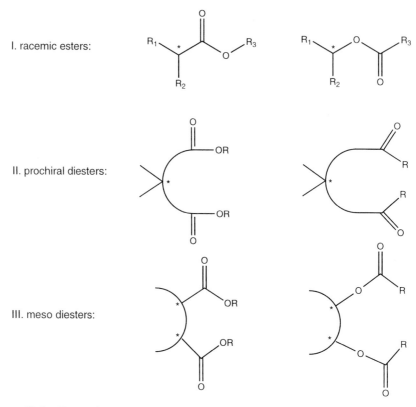

Figure 13-2. Types of esters that act as substrates for hydrolases. Reprinted from Faber, 1992, with permission.

and papain), and lipases (from various sources). Enzyme-catalyzed hydrolysis of a carboxylic acid ester is often employed to produce optically active materials in pharmaceutical chemistry (for a review see Margolin, 1993). Typical substrates involved in such biotransformations are shown in Figure 13-2. The chiral center of each substrate may, of course, be located in either the acid or alcohol portion of the molecule. Generally speaking, the Type I esters described in Figure 13-2 will meet several criteria (Faber, 1992). These include location of the chiral center as close as possible to the site of the reaction (i.e., to the carbonyl group of the ester). The substituents, R_1 and R_2, should differ in size and polarity to aid in chiral recognition by the enzyme and polar groups, such as -COOH, -CONH$_2$, or -NH$_2$, should be avoided or, if required, protected with a less polar group because esterases and lipases do not accept highly polar hydrophilic substrates. The R_3 group should be as short as possible, and the methine hydrogen atom is essential since α,α,α-trisubstituted carboxylates and esters of tertiary alcohols are usually not accepted by esterases and proteases.

Enzyme-catalyzed kinetic resolution of racemic esters is a valuable technique for obtaining optically active materials. When acting on a racemic pair of esters, proteases usually hydrolyze the enantiomer which most resembles the configuration of an S-amino acid (Bender & Killheffer, 1973). For example, N-acetyl-R,S-phenylalanine

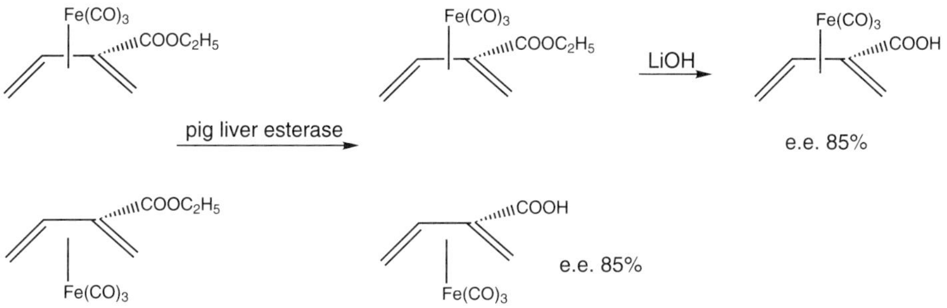

Figure 13-3. Enzymatic resolution of 2-ethoxycarbonylbuta-1,3-dienetricarbonyliron.

methyl ester has been resolved by treatment with α-chymotrypsin in aqueous methanol at 25°C for 2 hr; only the L-ester was hydrolyzed. Interestingly, the chirality does not necessarily need to be present on a tetrahedral carbon atom. It has been demonstrated that the helical chirality of a racemic organometallic ester was also recognized selectively by pig liver esterase (Fig. 13-3) (Alcock et al., 1988).

Hydrolase stereospecificity can also be utilized for the separation of two optically active acids or alcohols (Davies et al., 1989). If the chiral center of the ester is located in the acid portion of the substrate, hydrolysis of the racemic ester can afford an optically active ester and acid. Thus both the R and the S acids may be obtained if the unreacted optically active ester is further subjected to chemical hydrolysis. If the chiral center of the ester is present in the alcohol moiety, cleavage of the ester may result in an optically active alcohol; similarly, the unreacted optically active ester can be hydrolyzed chemically to produce the other optically active alcohol.

In addition, prochiral, or meso diesters can be hydrolyzed enzymatically to produce optically active monoesters (Fig. 13-2). The prochiral selectivity of an enzyme is affected by the structure of the α-substituent. Indeed, the preference of pig liver esterase varies from pro-S to pro-R upon the increase in the chain length of the substituent (Fig. 13-4) (Bjorkling et al., 1985). A similar phenomenon is observed in the desymmetrization of cyclic meso-1,2-dicarboxylates by the same enzyme (Sabbioni & Jones, 1987). The enzyme selectivity varies depending on the ring size of the substrate: pro-S ester is selected for the smaller ring while the R-counterpart reacts only when the ring is larger. Whether a pro-S or pro-R ester is selected is also dependent upon the specific enzyme used (Fig. 13-5) (Laumen & Schneider, 1984).

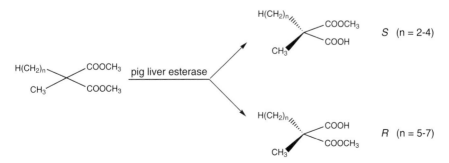

Figure 13-4. Enantioselective hydrolysis of a prochiral diester.

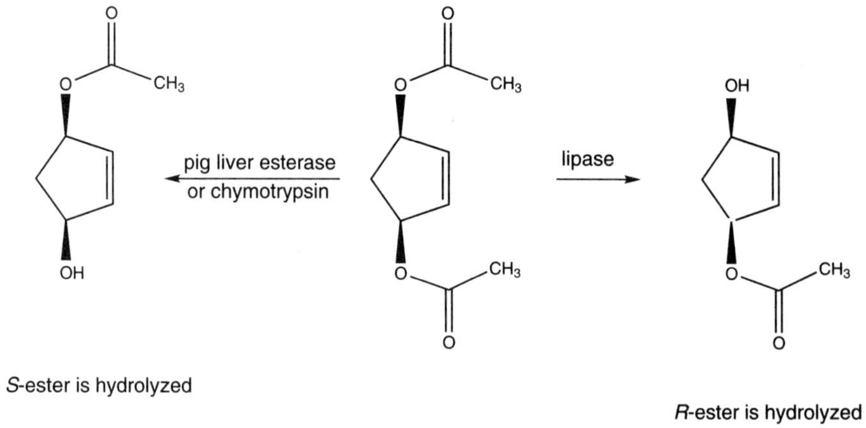

Figure 13-5. Enzymatic hydrolysis of prochiral cis-1,4-diacetyl-2-cyclopentenediol.

Enzymes that catalyze the cleavage or formation of amide linkages (Fig. 13-6) are referred to as amidases. This group includes carboxypeptidases, N-acylases, acid proteases and acyl transferases. These enzymes can be used in the resolution of amino acids (Jones & Beck, 1976), synthesis of S-amino acids (Wandrey, 1986), and the removal of the N-terminal protecting group from a dipeptide (Waldmann, 1988).

Amide bonds can also be formed in hydrolase-catalyzed reactions. A particularly important example of this is peptide synthesis. For example, Kullmann (1982) has successfully synthesized dynorphin with eight separate amino acid residues (Fig. 13-7). The enzyme-catalyzed process is preferable to chemical methods because the chemical preparation of such bonds is often accompanied by undesirable side reactions.

Redox Reactions

Enzymes that catalyze redox reactions are separately classified as dehydrogenases, oxygenases, or oxidases. Dehydrogenases are used in the reduction of carbonyl groups of aldehydes or ketones and of C=C bonds. They are particularly important because of their ability to convert a prochiral substrate to a chiral product (Fig. 13-8). Oxygenases catalyze the direct incorporation of molecular oxygen into an organic molecule, leading to the hydroxylation or epoxidation of C—H or C=C bonds. Oxidases, which are responsible for the transfer of electrons, are rarely employed in the biotransformation of non-natural organic compounds.

Almost all redox enzymes require cofactors, such as nicotinamide adenine dinucleotide (NAD(H)), (NADP(H)), flavin mononucleotide (FMN), flavin adenine dinucleotide (FAD), and pyrroloquinoline quinone (PQQ), which donate or accept the chemical equivalents for reduction or oxidation. Recycling of cofactors must be considered when using

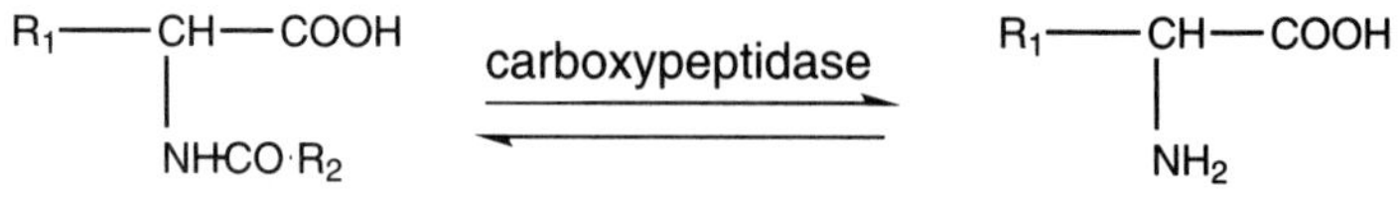

Figure 13-6. Cleavage and formation of the amide bond catalyzed by carboxypeptidase.

Use of Enzymes in Organic Synthesis / 427

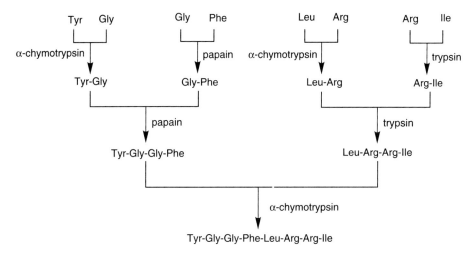

Figure 13-7. Enzymatic synthesis of dynorphin (1–8).

these enzymes in industrial applications. The efficiency of cofactor regeneration can be expressed in terms of total turnover number, which represents the total number of moles of product formed per mole of cofactor during the course of a complete reaction. Turnover numbers in the range of 10^3 to 10^4 are sufficient on a laboratory scale, while for commercial applications turnover numbers greater than 10^5 are necessary.

Since the addition of cofactors stoichiometrically is prohibitively expensive, whole microbial cells, which already contain cofactors, can be used instead of isolated enzymes to catalyze redox reactions. When a cell free enzyme is used, the cofactor can be regenerated by four methods: chemical, electrochemical, photochemical, and enzymatic. The last method is a particularly efficient one, and can itself be subdivided into two categories: coupled-substrate and coupled-enzyme (see Fig. 13-9).

Horse liver alcohol dehydrogenase, with its broad substrate specificity and narrow stereospecificity, is the most widely used enzyme in the reduction of aldehydes and ketones. Baker's yeast and whole microbial cells have also been used, the latter being advantageous because they contain the necessary cofactors and metabolic pathways for

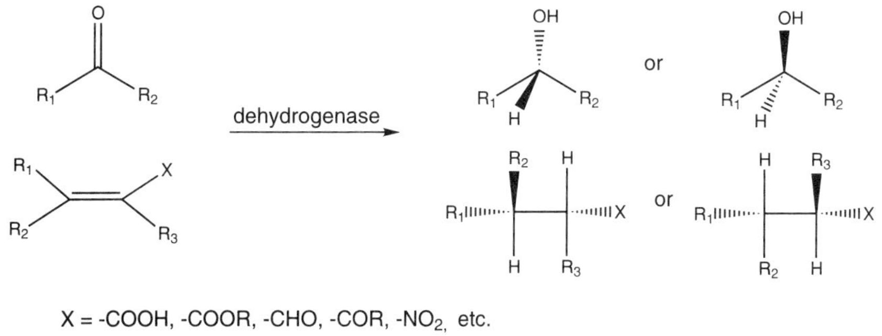

Figure 13-8. Reduction reactions catalyzed by dehydrogenases. Reprinted from Faber, 1992, with permission.

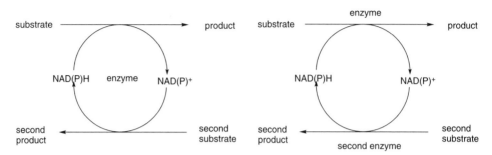

Figure 13-9. Enzymatic recycling of cofactor by coupled-substrate and coupled-enzyme method.

cofactor regeneration. A wide range of ketones can be reduced stereospecifically with dehydrogenases to give chiral secondary alcohols. For most of the commercially available dehydrogenases, the stereochemical course of the reaction may be related to the shape of the substrate, as introduced originally by Prelog (1964) (Fig. 13-10).

The use of enzymes to reduce ketones is well illustrated by the conversion of the α-keto acid (Et-CO-COOH) into (S)-2-hydroxybutanoic acid (99% yield, >99% ee) using lactate dehydrogenase (Kim & Whitesides, 1988). Kinetic resolution of mono-, bi-, and polycyclic ketones can also be achieved with horse liver alcohol dehydrogenase (see, e.g., Nakazaki et al., 1981). Horse liver alcohol dehydrogenase can also selectively reduce organometallic meso dialdehydes (Yamazaki & Hosono, 1988).

Specific reduction of C¨C double bonds is a challenging problem using conventional chemical methods, but can be achieved with high specificity, typically involving *trans*-addition of hydrogen across the C¨C bond, with an enzyme such as enoate reductase. Practically, such transformations are performed with whole microbial cells (Simon et al., 1985), mainly due to problems associated with cofactor recycling and the sensitivity of isolated enoate reductase to traces of oxygen. For effective reduction, the C¨C bond must be activated by an electron-withdrawing substituent, X (Fig. 13-8) (Fuganti & Grasselli, 1989). Therefore, α,f-unsaturated carboxylic acids and esters, aldehydes and ketones, and nitro compounds are all good substrates for the enzymatic reactions. When activated and conjugated 1,3-dienes are used as substrates, only the α,β-bond is reduced, while cumulated 1,2-dienes mainly yield the corresponding 2-alkenes.

In comparison to ketone reductions, enzyme-catalyzed alcohol oxidations are rarely used in organic synthesis because of the unfavorable reaction thermodynamics. The

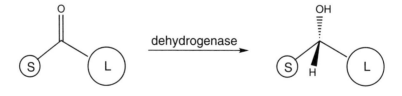

S = small group   L = large group

Figure 13-10. Asymmetric reduction of ketones.

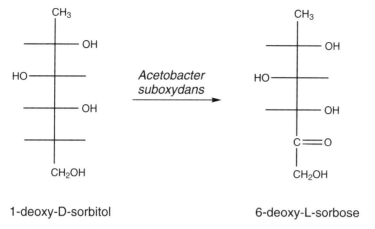

Figure 13-11. Regioselective oxidation of polyols.

relatively high pH necessary for enzyme activation can also be unsuitable for some substrates and cofactors. However, biocatalytic oxidation may be of practical interest in the regioselective oxidation of polyols (Fig. 13-11) (Kaufmann & Reichstein, 1967), the resolution of alcohols (Fig. 13-12) (Shimizu et al., 1987), and the desymmetrization of prochiral or meso-diols (Fig. 13-13) (Jones & Lok, 1979). Aldehydes and carboxylic acids can also be oxidized by enzymes such as alcohol dehydrogenases and aldehyde dehydrogenases.

Hydroxylation of saturated or unsaturated carbons is a useful type of biotransformation, given the difficulty of such reactions in traditional organic synthesis. Most biocatalytic hydroxylations involve whole microorganisms because the enzymes involved cannot be purified readily. An example of hydroxylation at saturated carbon is the preparation of stereoselectively hydroxylated steroids and terpenes. Indeed, all posi-

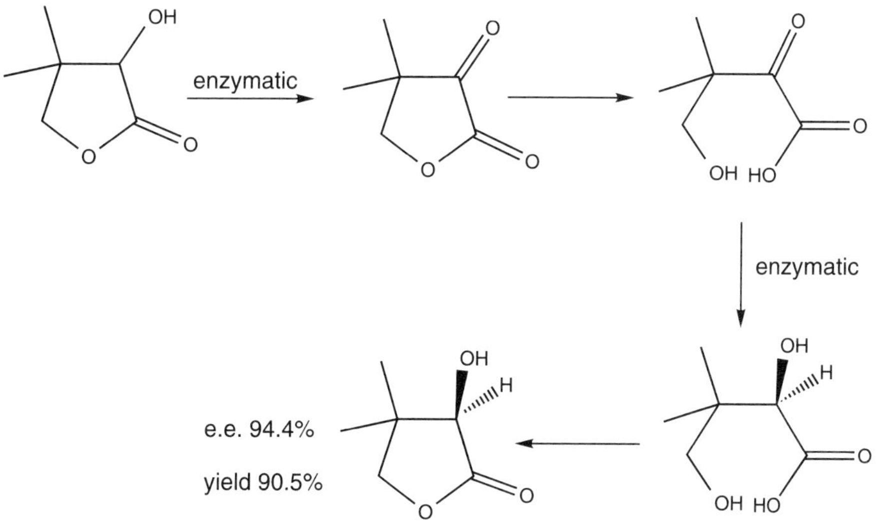

Figure 13-12. Enzymatic conversion of racemic pantoyl lactone to R-(-)pantoyl lactone.

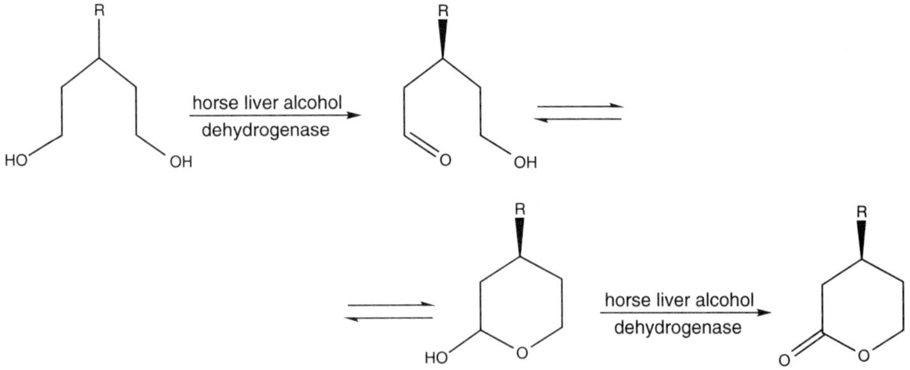

Figure 13-13. Asymmetrization of prochiral diols.

tions in the steroid nucleus, including the angular methyl groups C-18 and C-19, have been hydroxylated selectively by different microorganisms. For instance, hydroxylation of progesterone at the 11α-position can be achieved in a yield of 85 to 95% using *Rhizopus arrhizus* (Peterson et al., 1952), making 11α-hydroxyprogesterone available for therapy at a reasonable cost. Hydroxylations of terpenes and terpenoids are also of interest to organic chemists in view of their importance as fragrance or flavor agents. One example of this class of reactions is shown in Figure 13-14 (Hollinshead et al., 1983). Aromatic substrates can also be hydroxylated at benzylic positions. For example, epipodophyllotoxin, the glycoside of which is the antitumor agent etoposide, can be synthesized from the readily available natural deoxypodophyllotoxin in quantitative yield (Fig. 13-15) (Kondo et al., 1989).

Regiospecific hydroxylation of aromatic compounds is difficult utilizing conventional organic chemistry, but can be catalyzed by monooxygenases such as polyphenol oxidase, laccase and peroxidase. Polyphenol oxidase is able to catalyze the specific hydroxylation of a monophenol to its *o*-diphenol. The *o*-diphenol will be oxidized further by the enzyme to *o*-quinone, which is unstable in aqueous solution and polymerizes rapidly, thus limiting the enzyme's applicability in water-based reactions. Enzymatic generation of *o*-quinones can be useful in organic synthesis, as was demonstrated by Pandey et al. (1989) who prepared a series of coumestans in yields exceeding 90% using *o*-quinone obtained by polyphenol oxidase-catalyzed oxidation of catechol in the presence of 4-hydroxycoumarins (Fig. 13-16).

The Baeyer-Villiger reaction is a classical and useful chemical method for the production of esters or lactones from ketones (Baeyer & Villiger, 1899). This process can

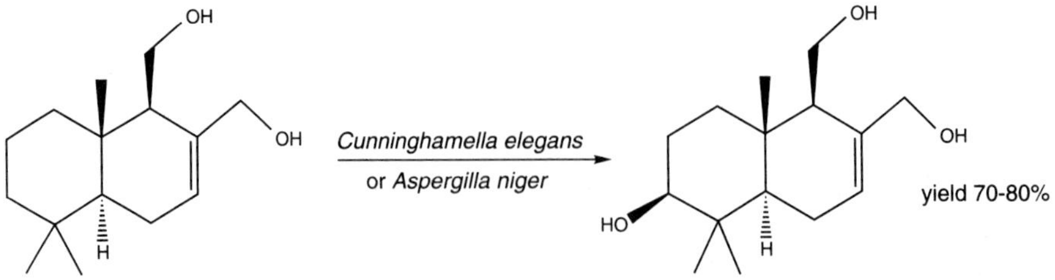

Figure 13-14. Microbial hydroxylation of terpene derivatives.

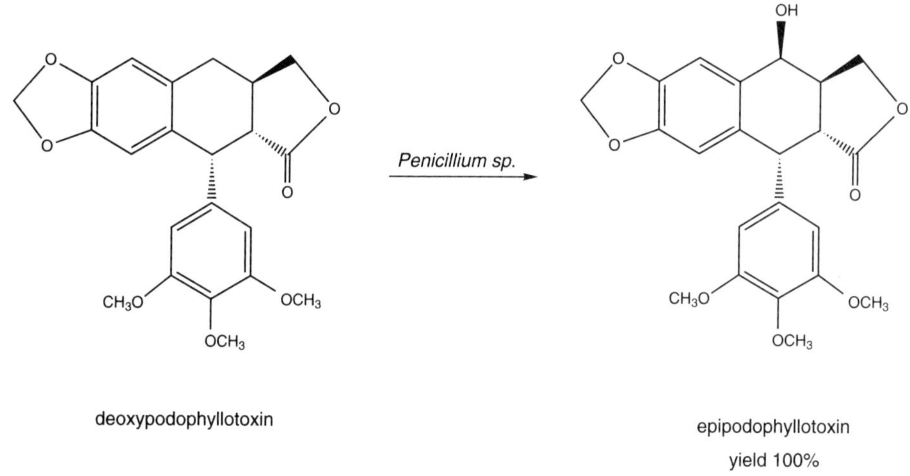

Figure 13-15. Benzylic hydroxylation, illustrated for the conversion of deoxypodophyllotoxin to epipodophyllotoxin.

be catalyzed using enzymatic methods. It is believed that the mechanism and regiochemistry of oxygen-insertion are similar for both conventionally and biologically catalyzed reactions (Fig. 13-17). The regiochemistry can be predicted by assuming that the carbon atom which is most capable of supporting a positive charge will migrate preferentially (Lee & Uff, 1967). A typical example was given by the formation of testolactone from progesterone via testosterone acetate, testosterone and androstenedione, by sequential enzymatic transformations including Baeyer-Villiger oxidations (Fig. 13-18) (Prairie & Talalay, 1963; Rahim & Sih, 1966).

There are some other oxidation reactions that can be performed biologically, including epoxidations, oxidations at heteroatoms such as sulfur and nitrogen, formation of peroxides, and dihydroxylation of aromatic compounds. Interested readers can refer to more complete texts on the use of enzymes in organic synthesis (Davies et al., 1989; Faber, 1992; Holland, 1992; Roberts, 1990).

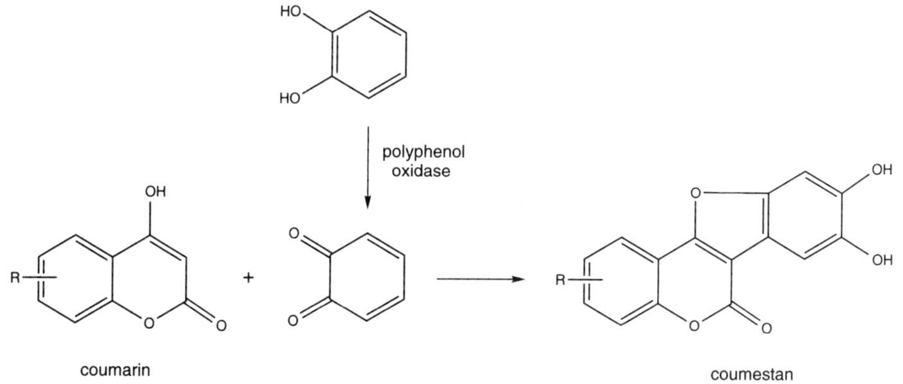

Figure 13-16. Polyphenol oxidase catalyzed synthesis of coumestan derivatives and their structural analogues.

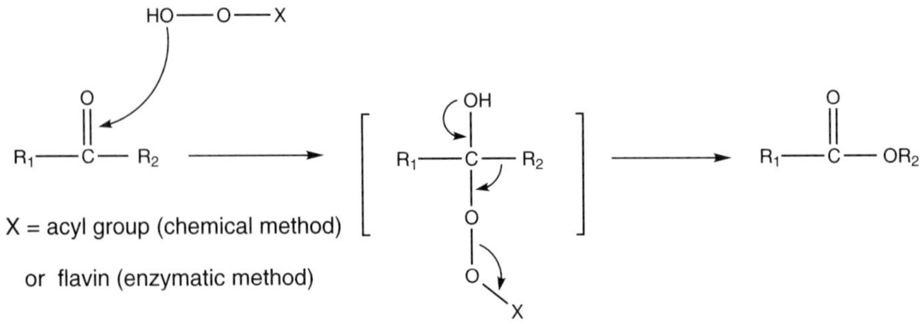

Figure 13-17. Mechanism of the Baeyer-Villiger reaction.

Carbohydrate Synthesis

Carbohydrates are one of the most important classes of compounds in nature; therefore, synthesis and modification of carbohydrates is a major target for biocatalysis in synthetic chemistry. While traditional chemical methods require selective protection and deprotection, enzymatic methods enable the regio and stereospecific synthesis of a target structure under mild conditions with a minimum of reaction and purification steps (Chen et al., 1992; Ichikawa et al., 1991; Toone et al., 1989; Whitesides & Wong, 1985; Wong et al., 1991a, 1992).

Asymmetric aldol condensation is one of the most effective methods for C—C bond formation in synthetic organic chemistry. Enzymatic aldol reactions catalyzed by aldolases provide a convenient means for effecting this reaction. More than 20 aldolases have been discovered (Wong et al., 1992), all of which share three common features (Wong, 1992). The first is that aldolases catalyze the condensation between a donor such as dihydroxyacetone phosphate (DHAP) and an acceptor (e.g., an aldehyde), lead-

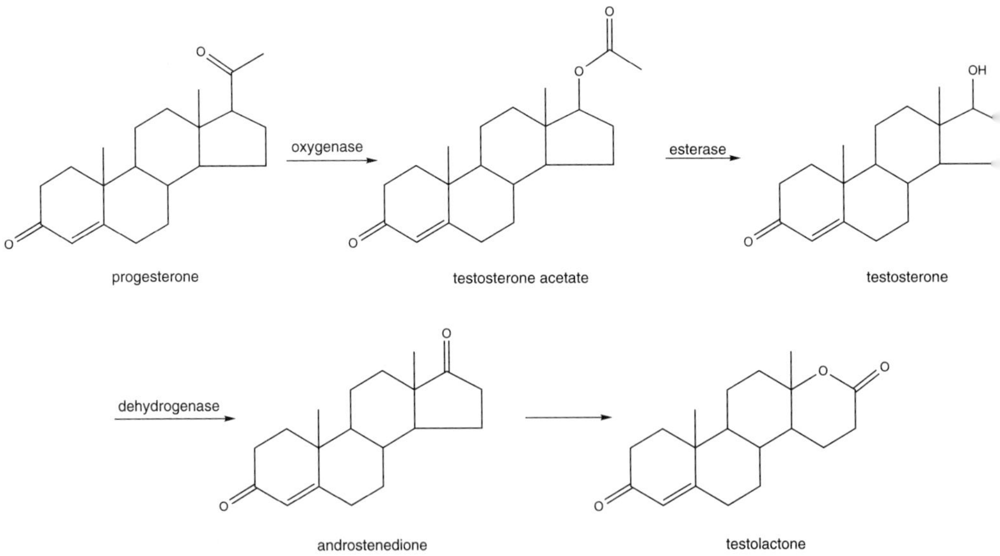

Figure 13-18. Enzymatic formation of testolactone from progesterone by sequential Baeyer-Villiger oxidations.

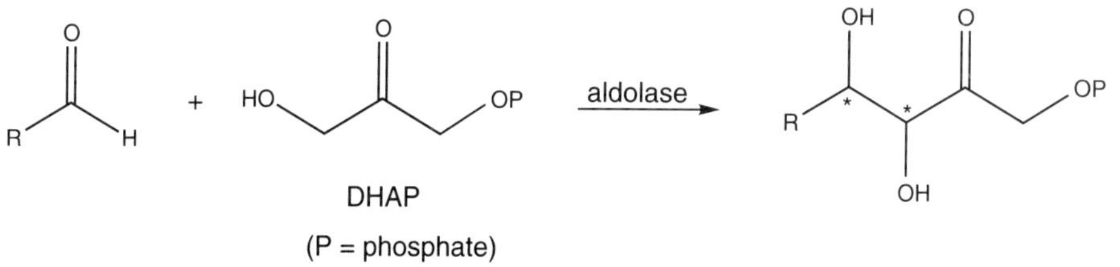

Use of Enzymes in Organic Synthesis / 433

Figure 13-19. Asymmetric aldol condensation catalyzed by aldolases.

ing to the formation of two new chiral centers and elongation of the aldehyde by a three-carbon unit (Fig. 13-19). Second, aldolases are highly specific for the donor substrate but exhibit flexibility for the acceptor component. Finally, the stereoselectivity of the reaction is controlled by the enzyme rather than by the substrate. The versatility of aldolases, and their utility in synthetic organic chemistry has been described in a number of review articles (Wong, 1992; Wong et al., 1992).

Aza sugars are nitrogen-containing unnatural carbohydrates which are useful for inhibition of glycoprocessing (Wong et al., 1992). Synthesis of aza sugars is a particularly important application that utilizes aldolases. The aldol condensation between DHAP and an azidoaldehyde is catalyzed by an aldolase; the azide group of the product is then reduced to an amine in the presence of a palladium catalyst. Intramolecular attack of the formed amine on the ketone then results in imine formation; the imine is further reduced by catalytic hydrogenation to give the aza sugar. Based on this general strategy, five aza-sugars have been prepared and are illustrated in Figure 13-20.

Several enzymes, including glycosidases, glycosyl transferases, transglycosidases,

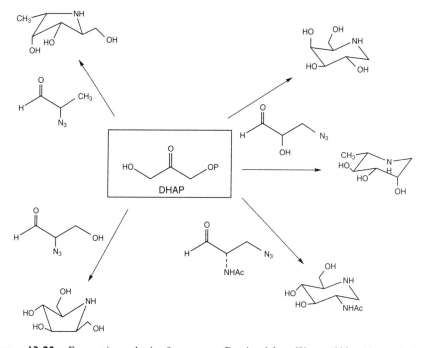

Figure 13-20. Enzymatic synthesis of aza sugars. Reprinted from Wong, 1992, with permission.

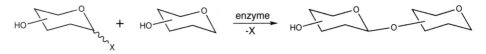

Figure 13-21. Enzymatic synthesis of oligosaccharides.

and phosphorylases, are involved in the synthesis of glycosides (Fig. 13-21) (Wong, 1992). An efficient multienzyme system has been developed successfully to catalyze the synthesis of a sialyl trisaccharide in a one step reaction using three nonactivated monosaccharides as starting materials (Ichikawa et al., 1991). It has also been demonstrated that the tetrasaccharide sialyl Lewis X (Le^x) can be synthesized using three different enzymes (Ichikawa et al., 1992). Thio and aza sugars can also be incorporated into oligosaccharides biocatalytically in a glycosyl transferase catalyzed reaction (Ichikawa et al., 1992).

Other Biotransformations

Enzymes catalyze a wide variety of other reactions which are also useful in organic synthesis, such as acyloin reactions, Michael-type addition reactions, halogenation and dehalogenation reactions, addition and elimination reactions, and O- and N-dealkylation. Space limitations prevent a complete listing of all classes of reactions, but readers can be referred to a standard compilation of enzyme reactions (Dixon & Webb, 1979).

Nonaqueous Enzymology in Organic Synthesis

Use of Enzymes in Organic Solvents

The significant potential of biocatalysis in organic synthesis has been widely recognized. However, most enzymes evolved under aqueous conditions; therefore, the use of enzymes in organic synthesis has been restricted because water is a poor solvent for the majority of organic reactants. Fortunately, extensive research conducted over the last decade has demonstrated that enzymes can also function effectively in nonaqueous media (for reviews see Brink et al., 1988; Deetz & Rozzell, 1988; Dordick, 1989; Halling, 1987; Khmelnitsky et al., 1988; Klibanov, 1989). Compared to aqueous conditions, biotransformations in organic media provide a number of features that are of particular value in organic synthesis. These include the ability to solubilize hydrophobic reactants, which enables increased substrate concentrations relative to water-based systems. The equilibria of reactions involving hydrolysis shift in the direction of synthesis. Enzymes in organic solvents can also catalyze novel reactions that are not feasible in water, for example, transesterifications. In addition, there is a reduction in water-dependent side reactions for nonaqueous systems, substrate specificity can be controlled by "solvent engineering," and enzymes exhibit enhanced thermostability in anhydrous systems. Finally, there is a reduced risk of microbial contamination of biocatalytic reactors in nonaqueous environments.

The various approaches developed for using enzymes in the presence of organic solvents are depicted in Figure 13-22. The following discussion focuses on the low-water solvent system. Water is absolutely necessary for enzyme catalysis in organic media. While enzymes are essentially inactive in completely "dry" systems, they can promote catalytic activity in the presence of sufficient water. As the degree of enzyme hydration increases the rate of enzymatic reactions accelerates, up to a point. Zaks and Klibanov (1988b) have demonstrated that water that is added to the solvent partitions between the solvent and enzyme. To a first approximation, the resulting enzyme activity depends only on the amount of water associated with the enzyme, and not on the water content in the entire system. As long as this "essential" or "bound" water is present, the enzyme retains activity in organic solvents. Although water is required for enzyme catalysis in organic solvents, the amount needed is dependent on the enzyme. Lipases appear to require very few molecules of water in order to exhibit activity (Valivety et al., 1992). Molecules of either chymotrypsin or subtilisin (proteases) are active when they are associated with less than 50 molecules of bound water (Zaks & Klibanov, 1988a). Alcohol dehydrogenase, alcohol oxidase, and polyphenol oxidase, in comparison, are active in organic solvents when several hundreds of water molecules are bound per enzyme molecule, that is, when there is enough water for a monolayer to form (Zaks & Klibanov, 1988b).

Presumably, enzyme catalysis occurs in the aqueous layer around the enzyme molecules and it is reasonable to speculate that hydration of the charged groups and some of the polar groups of the protein molecules is a prerequisite for enzyme catalysis. It

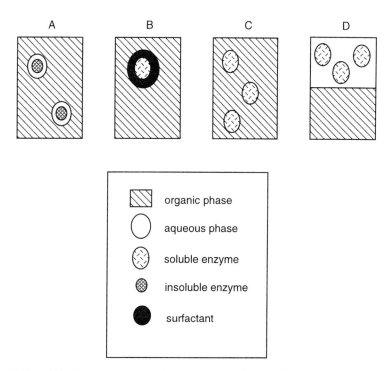

Figure 13-22. (A) The use of enzymes in a low-water solvent system. (B) A microemulsion. (C) A monophasic cosolvent system and (D) a biphasic organic-aqueous system. For (A) the enzyme is lyophilized, or immobilized, and suspended directly in an organic solvent; for (B) the enzyme is entrapped in the core of a spherical reversed micelle; for (C) the enzyme is dissolved in a mixture of water and water-miscible organic solvents.

has been suggested that in the dry condition, the charged and polar groups of the enzyme molecule interact with each other and produce an inactive "locked" conformation. The addition of water would make the enzyme molecule more flexible while maintaining its catalytically active conformation by assisting noncovalent interactions, that is, hydrogen bonding, electrostatic interactions, van der Waals and hydrophobic forces. This hypothesis is supported by data generated using polyphenol oxidase when water was replaced by various "hydrogen-bond formers" such as glycerol, ethylene glycol, and formamide; indeed, enzyme activity was enhanced (Zaks & Klibanov, 1988b). Addition of water may also increase enzyme activity by enhancing the polarity and flexibility of the active site of the enzyme. The addition of too much water, however, could give rise to the formation and growth of a water cluster within the active site, which could eventually offset the increased enzyme activity (Affleck et al., 1992).

The nature of the organic solvent is another fundamental factor that affects the behavior of enzymes suspended in organic media. It may not only affect the enzyme activity directly or indirectly, but also change the catalyst's substrate specificity (Wescott & Klibanov, 1993; Zaks & Klibanov, 1984, 1986), regioselectivity (Rubio et al., 1991), enantioselectivity (Fitzpatrick & Klibanov, 1991; Sakurai et al., 1988), and prochiral selectivity (Terradas et al., 1993). Briefly, the effect of solvents on enzyme activity can be summarized in the following ways. First, there is an influence of solvent on the water associated with the enzyme. Polar solvents are capable of "stripping" off the essential water from the enzyme surface, resulting in an insufficiently hydrated molecule and thereby leading to a decrease in enzyme activity, relative to activity in a nonpolar solvent. There is also an influence of solvent on the substrates and products. Solvents can affect the enzyme activity by direct reaction with the substrate or the product, or by indirectly altering the concentration of substrate or product in the aqueous layer around the enzyme. Concentration effects, which are the result of the changes of partitioning between aqueous and organic phases (Yang et al., 1992, Yang & Robb, 1994), are particularly important for those enzymes prone to substrate or product inhibition. Finally, there can be direct interactions between the enzyme and organic solvent. Solvent molecules dissolved in the microaqueous phase might inhibit or inactivate the protein. Alternatively, a direct contact between the organic-aqueous phase interface and the biocatalyst may lead to inactivation (Halling, 1987).

The search for a quantitative correlation between a solvent and the activity of an enzyme has been the subject of many papers, and a few groups have been bold enough to publish equations relating activity or specificity to solvent properties. Since hydrophobic solvents, compared to the hydrophilic ones, have a lower capacity for distorting the essential aqueous layer around the enzyme, the enzyme activity promoted in the former should be higher than in the latter. Therefore, the most popular guideline for solvent selection has been the "log P rule". Log P, which reflects the hydrophobicity of the solvent, is defined as the logarithm of the partition coefficient of a given solvent in an octanol-water mixture. Generally, biocatalytic activity is low in polar solvents having a log $P < 2$, moderate in solvents with a log P between 2 and 4, and high in apolar solvents having a log $P > 4$ (Laane et al., 1987). Many enzymes have been found to follow this rule, but occasional discrepancies have also been reported (Guinn et al., 1991; Yang & Robb, 1991; Zaks & Klibanov, 1988a). For example, the optimal solvent for polyphenol oxidase varies depending on the substrate used (Yang et al., 1992). Other criteria have also been proposed for prediction of activities in organic solvents, such as Hansen parameters (Schneider, 1991), denaturation capacity (Khmelnitsky et al., 1991) and partition coefficients of substrate and product (Yang et al., 1992; Yang & Robb, 1994). Ryu and Dordick (1992) have recently demonstrated that the effect of solvent on enzyme activity may result from three factors, including stabilization of the substrate ground state in the solvent, and penetration of the solvent into the active site of the enzyme, leading to a decrease in the local polarity and strengthening of the hydrogen bonding between the substrate and the enzyme. There may also be indirect effects on the structure of the active site of the enzyme as a result of interaction with a nonactive site region of the protein.

One of the advantages of nonaqueous biocatalysis is the ease of recovery and reuse of enzymes. Enzymes are insoluble in organic solvents, with the exception of dimethyl sulfoxide; hence, immobilization is usually unnecessary. If desired, however, adsorption of the enzyme onto an inert support such as celite and porous glass beads works well. Indeed, enzyme powders could be subject to rate-limiting internal diffusion; thus, for fast reactions, immobilization may be important. The selection and importance of support materials have thus received considerable attention. The effect of support on enzyme activity in organic solvents has been summarized (Adlercreutz, 1992), as described later.

As regards the effect of the support on the partitioning of substrate and product between the active site of the enzyme and the bulk organic phase, it has been found that a support, depending on its hydrophobicity, may have a significant impact on the local concentration of the substrate or product in the region of the enzyme active site. For example, Fukui and colleagues (Omata et al., 1979) have studied steroid conversion using cells immobilized in polymer gels, and discovered that the reaction rate increased with an increase in the hydrophobicity of the gel. This was interpreted as resulting from an enrichment of the hydrophobic substrate in the hydrophobic gel.

There is also an effect of the support on the water bound to the enzyme. When chymotrypsin was adsorbed on a series of supports, the enzyme activity in diisopropyl ether was dependent on the capacity of the support to adsorb water, in competition with the enzyme and the solvent (Reslow et al., 1988). Enzyme adsorbed on a hydrophobic support, such as Celite, promotes higher activity, simply because the enzyme can compete successfully for the water. In comparison, supports with a lower hydrophobicity adsorb significant amounts of water, thus resulting in insufficient hydration of the enzyme.

A support may have an effect on the kinetics of enzyme-catalyzed reactions as well. Even when the effect of support on water partitioning is minimized by controlling the amount of water associated with the enzyme, the activity of α-chymotrypsin and horse liver alcohol dehydrogenase still vary with different supports used (Adlercreutz, 1991). Hence, the choice of support directly affects enzyme kinetics for those enzyme-substrate pairs. A support can also affect the relative rates of different reactions catalyzed by the same enzyme. By using the polyamide support Accurel PA6 and low water activity, with α-chymotrypsin as the catalyst, substrate hydrolysis was suppressed while alcoholysis was maintained (Adlercreutz, 1991).

Adsorption of enzymes on an inert support may result in a decrease of specific activity, especially when a small amount of enzyme is immobilized on a support with a large surface area (Wehtje et al., 1993). However, this can be minimized by further addition of some adducts, such as proteins or polyethylene glycol, to the support.

Examples of Novel Properties Offered by Nonaqueous Enzymology

Structural analysis has revealed that enzyme powders suspended in organic solvents possess a more rigid conformation than in aqueous solution (Burke et al., 1989; Guinn et al., 1991). It has been hypothesized that this rigidity is responsible for some of the interesting properties that the enzyme acquires in nonaqueous surroundings.

Enzymes to be used in organic solvents are normally lyophilized or precipitated from aqueous solutions of the pH optimal for enzyme activity, as demonstrated by, for example, Zaks and Klibanov (1985, 1988a). This suggests that an enzyme can "remember" the pH of the last aqueous solution that it has been exposed to. The most plausible explanation is that the ionogenic groups of the enzyme acquire the ionization state corresponding to the pH of the aqueous solution, and this state is then retained during lyophilization and dispersal in the organic solvent. However, this is not always true. For example, Guinn et al. (1991) found that among the three pH values selected (2.0, 7.5, 11.0), alcohol dehydrogenase in water was most active at pH 7.5 while the enzyme in nearly anhydrous heptane showed highest activity when it was lyophilized from a solution having pH 2.0. Interestingly, the enzyme was virtually inactive in water at this pH. In a recent study conducted by Yang et al. (1994), it was observed that the apparent pH dependence of subtilisin was dependent upon the solvent used and the amount of water bound to the enzyme. The effect appeared to be related to the change in the polarity of the enzyme active site as a function of solvent and water content.

Structural rigidity, altered polarity, and altered structure of an enzyme in an organic solvent may lead directly to a change in substrate specificity. For instance, dry lipase, in contrast to its hydrated counterpart, is completely inactive when tertiary alcohols are used as substrates in transesterification reactions. Clearly, a possible explanation is that lipase lacks the conformational mobility needed to accommodate large substrate molecules in its active site when in the dry state (Zaks & Klibanov, 1984). Apart from mobility, solvent-induced changes in substrate specificity can result from changes in the ability of an enzyme to utilize the free energy of substrate binding to facilitate a reaction. For chymotrypsin, subtilisin, and carboxyl esterase, there is a complete reversal of substrate specificity upon replacement of water with octane as the reaction medium (Zaks & Klibanov, 1986). A conceivable explanation is that the main driving forces for enzyme-substrate binding in the case of many hydrolases are hydrophobic interactions that exist only in water. In aqueous solution, the polar groups in the active site of the enzyme and the substrate molecules form hydrogen bonds with water. Hydrophilic substrates form stronger bonds with water; therefore, more energy is required for the desolvation which leads to substrate-enzyme complex formation, and this leads to reduced reaction rates. In a nonhydrogen bonding solvent like octane, the substrate preference of the enzyme should be entirely distinct. When the solvent is changed, so will the partitioning behavior of substrate and product; thus, the specificity of an enzyme can be tailored by "solvent engineering." Wescott and Klibanov (1993) have recently proposed a thermodynamic model that attempts to predict substrate specificity in organic solvents.

Many enzymes have been found to be more stable in organic media. Dry lipase can not only withstand heating at 100°C for many hours, but also promotes higher activity at that temperature than at 20°C (Zaks & Klibanov, 1984). The thermostability decreases, however, when the water content of the solvent increases. Undoubtedly, the enhanced thermostability of the enzyme in anhydrous solvents results from its high conformational rigidity under dry conditions. Another benefit of using nonaqueous solvents is that water is required in a number of covalent processes involved in irreversible inactivation, such as deamination, peptide hydrolysis, and cysteine decomposition. All these irreversible processes are eliminated in zero-water media (Volkin et al., 1991).

Applications of Nonaqueous Enzymology

The realization that enzymes can function in nonaqueous media has greatly broadened their applicability in organic synthesis. To date, as a result of their catalytic versatility and commercial availability, most enzymes that are used for preparative conversions in organic solvents are hydrolases. However, other classes of enzymes, in particular oxidases, are also being exploited and should be commercialized in the next 5 to 10 years. Some of the more notable examples are discussed briefly below.

Enzymatic resolutions of racemic mixtures such as alcohols, amines, carboxylic acids and their esters can also be achieved in organic solvents with the aid of hydrolytic enzymes such as lipases and proteases. In nonaqueous systems, these enzymes catalyze esterification and transesterification (alcoholysis) rather than hydrolysis. A detailed discussion can be found in a number of reviews (for example, Klibanov, 1990).

Lipases have been used in a variety of reactions for the production of triglycerides, including esterifications, interesterifications, and transesterifications. An example is the retailoring of inexpensive oils (for instance, palm oil) through transesterification to produce cocoa butter equivalents (Fig. 13-23) (Harwood, 1989). It is worth noting that lipases are particularly interesting for use in organic media because they are active and stable in nonaqueous environments; they also have the remarkable ability to accommodate substrates of varying sizes and stereochemical complexities. Moreover, lipase-catalyzed esterification reactions in organic solvents are often more enantioselective than the corresponding hydrolytic reactions in water (Chen & Sih, 1989).

In water-restricted environments, the equilibrium for peptide bond hydrolysis is shifted toward synthesis. This has been demonstrated, for example, by the enzymatic synthesis of the dipeptide L-aspartyl-L-phenylalanine methyl ester (aspartame), which is widely used as a low-calorie sweetener. The aspartame precursor was synthesized by the thermolysin-catalyzed coupling of an N-protected aspartic acid to L-phenylalanine methyl ester in ethyl acetate. Many attractive features of biocatalysis (for instance, substrate specificity, enantioselectivity, and shifting of thermodynamic equilibrium using nonaqueous solvents) are exemplified in the synthesis of aspartame (Fig. 13-24) (Oyama & Kihara, 1984).

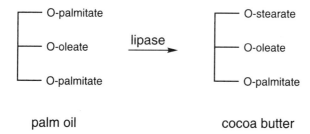

Figure 13-23. Production of cocoa butter equivalents from palm oil.

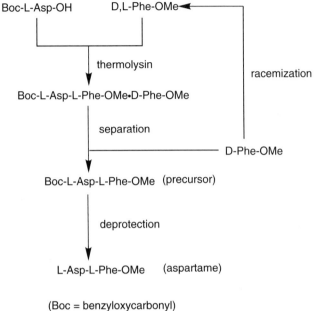

Figure 13-24. Enzymatic synthesis of aspartame.

Interestingly, it has been shown that a peptide can even be produced by chymotrypsin-catalyzed condensation of Boc-Ala-Phe-OMe and Leu-NH$_2$ with a salt hydrate (Na$_2$CO$_3$·10 H$_2$O) as a water donor in hexane, although all the chemicals involved were not dissolved in the solvent used (Kuhl et al., 1990). Presumably, the salt hydrate provides a thin layer of water surrounding the enzyme. Substrate will partition into the aqueous phase, and associate with the enzyme molecules dissolved in the microaqueous layer. The extremely low solubility of the product in hexane leads to its precipitation, which helps to shift the reaction equilibrium inside the aqueous layer toward condensation.

One of the most important applications using hydrolases in organic media is in the synthesis of polymers such as polyesters and polyacrylates (Dordick, 1992), because this is impossible in water using biocatalysts. For example, an optically active polyester (M_r 7900, ee >95%) can be produced from racemic monomers (Wallace & Morrow, 1989). The high selectivity of enzymes and high productivity of chemical catalysts can be combined for chemo-enzymatic synthesis of polymers. In this situation, a multistep process is involved, with an enzyme catalyzing a selective step, followed by nonselective chemical step(s) (see, e.g., Fig. 13-25) (Ghogare & Kumar, 1990).

Because of its high activity and stability in organic media, horseradish peroxidase is useful in a variety of applications. For instance, peroxidase can be used to catalyze the polymerization of phenols (Dordick et al., 1987), or to incorporate phenols into lignins, which has the potential to convert the modified lignins to higher value phenolic resins (Popp et al., 1991). In the former case, the molecular weight of biocatalytically synthesized polyphenols can be controlled by varying the water content in the reaction medium, and the use of organic solvents eliminates the drawback of the

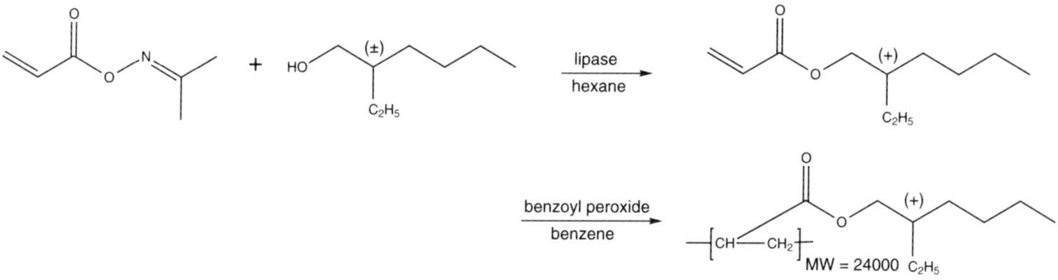

Figure 13-25. Chemo-enzymatic synthesis of poly(2-ethyl-1-hexanol acrylate).

low solubility of phenolic dimers and trimers in water, enabling phenolic resins to be produced without involving the use of the highly toxic formaldehyde.

The use of polyphenol oxidase in organic synthesis is limited by the low stabilities of both the enzyme and product in aqueous solution. However, both these drawbacks can be overcome by introducing the enzyme into an organic medium. It has been found that when the reaction is performed in chloroform, the o-quinone product is so stable that it can be converted back to the o-diphenol quantitatively with a chemical reductant such as ascorbic acid. This is illustrated well by the conversion of a tyrosine derivative to a derivative of L-dopa, a pharmaceutical agent used in the treatment of Parkinson's disease (Fig. 13-26) (Kazandjian & Klibanov, 1985).

The stability of o-quinone in chloroform has also been utilized in a two-stage synthesis of a sulfone product (Yang & Robb, 1991). The o-quinone was synthesized enzymatically by oxidation of 4-methylcatechol with polyphenol oxidase adsorbed on Celite in water-saturated chloroform; after separation from the enzyme, the product solution was added to a methanolic solution of sodium p-toluenesulfinate (Fig. 13-27). This also demonstrates the advantage of using enzymes in organic solvents when the coupling reagent necessary to form an adduct is inhibitory to the enzyme. The reaction process can be separated into two steps, hence avoiding the possibility of contact between the enzyme and the inhibitory reagent.

Alcohol dehydrogenase is not only catalytically active in organic media, but also maintains its stereoselectivity. A variety of optically active alcohols and ketones (ee

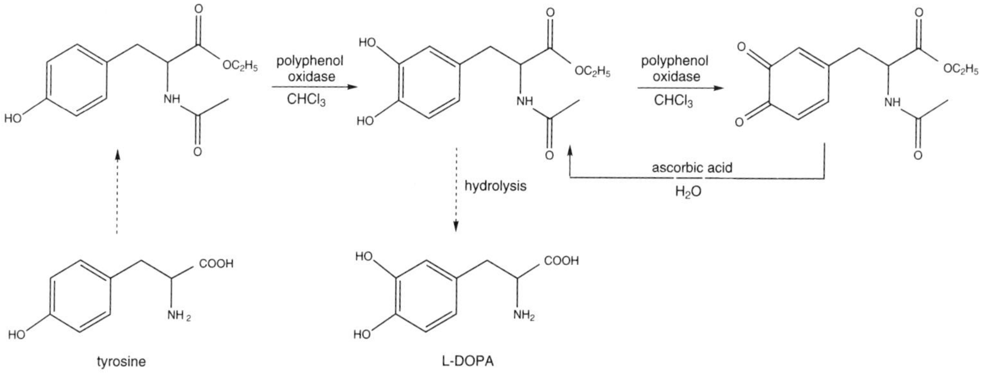

Figure 13-26. Enzymatic synthesis of L-dopa derivative in chloroform.

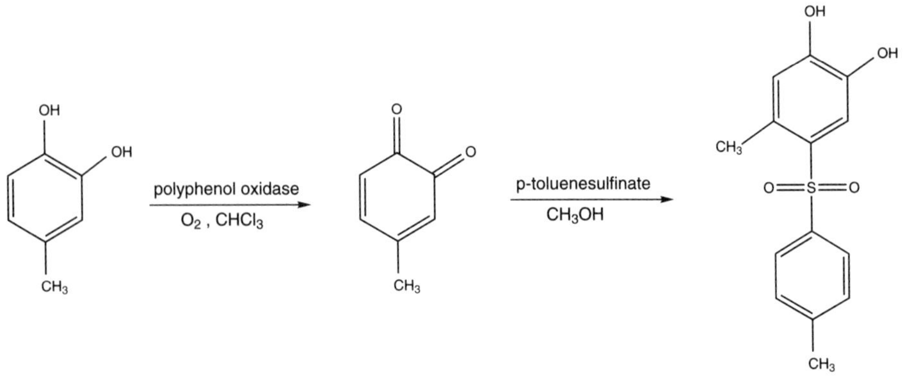

Figure 13-27. Synthesis of a sulfone product with polyphenol oxidase.

95 to 100%) have been prepared on a 1–10 mmole scale using horse liver alcohol dehydrogenase in organic solvents, and cofactor turnover numbers of 10^5–>10^6 have been obtained (Grunwald et al., 1986). Yang and Russell (1993) have successfully demonstrated the production of unsaturated aldehydes from unsaturated alcohols, which has significance in commercial polymer and fine chemical synthesis. Catalyzed by baker's yeast alcohol dehydrogenase (YADH), 3-methyl-2-buten-1-ol has been converted to its corresponding aldehyde in heptane. Since both enzyme and cofactor are insoluble in most organic solvents, they can be brought together by coprecipitation prior to placement in the organic medium; the recycling of the cofactor can be achieved by addition of a second substrate, acetone (Fig. 13-28). Interestingly, a higher turnover number of the cofactor was observed, accompanied by an increased enzyme activity and stability, when the enzyme and the cofactor were lyophilized in the presence of lactate dehydrogenase, which can bind the cofactor NAD but does not accept the substrate.

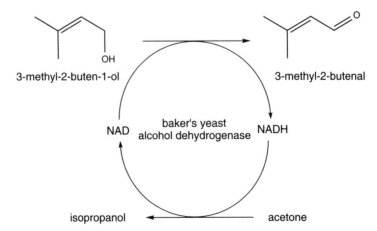

Figure 13-28. YADH-catalyzed reaction and cofactor regeneration in organic solvents.

Modification of Enzymes for Use in Organic Solvents

To overcome the drawbacks of heterogeneous systems resulting from the suspension of insoluble enzymes in organic solvents, enzymes have been solubilized in organic solvents by chemical modification (Inada et al., 1986). The concept involves covalent attachment of an amphiphilic polymer, usually polyethylene glycol (PEG), to the surface of proteins. This can be accomplished by linkage of the ϵ-amino groups of lysine residues to derivatives of PEG; some examples are provided in Figure 13-29.

PEG-modified enzymes are soluble up to concentrations of a few mg/ml in organic solvents such as benzene and chloroform. The solubility of modified catalase in benzene, for example, can be as high as 2 mg/ml depending on the degree of enzyme modification (Takahashi et al., 1984b). The specific activities of PEG enzymes are often comparable to, or even higher than, those in aqueous solutions. For instance, the maximum specific activity of PEG-modified catalase in benzene is 1.6 times higher than the activity of unmodified catalase in aqueous solution, and approximately five times greater than that of the modified enzyme in aqueous solution (Takahashi et al., 1984b). PEG-modified enzymes also exhibit high stability compared to the unmodified enzymes. The half-life of PEG-lipase in benzene is ten times greater than the native enzyme in water (Baillargeon & Sonnet, 1988), and the modified lipase retains about 50% of its activity in benzene after 3 to 5 months (Ajima et al. 1985). The optimal temperature for PEG-lipase dissolved in the reactants for ester exchange reaction (70°C)

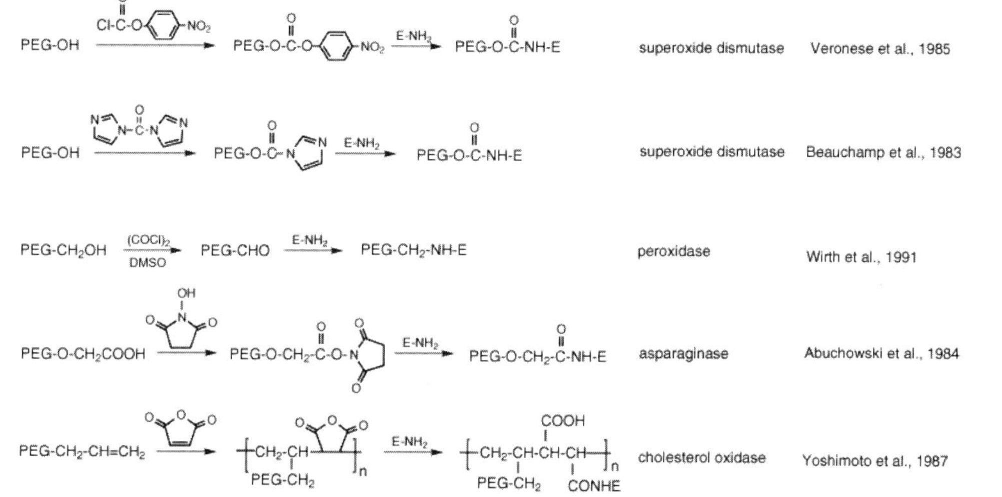

Figure 13-29. Modification of enzymes with polyethylene glycol.

is also higher than the temperature optimum for the hydrolysis of esters by the unmodified enzyme in an aqueous emulsified system (45°C) (Takahashi et al., 1985). Interestingly, a shift in optimal pH was observed when horseradish peroxidase was modified with PEG (pH 3.5 to 4.2) compared to the optimal pH for the unmodified enzyme (pH 5.2) (Souppe & Urrutigoity, 1989; Urrutigoity & Souppe, 1989; Wirth et al., 1991).

PEG-modified enzymes can be recovered simply from organic solvents by precipitation upon the addition of petroleum ether or hexane (Yoshimoto et al., 1984). Another easy and attractive way for the enzyme to be recovered is to link the PEG-enzyme conjugate to magnetite (Fe_3O_4) (Inada et al., 1988). Lipase, when modified in this way, can be recovered magnetically without loss of activity.

Use of Enzymes in Supercritical Fluids

A recent development in nonaqueous enzymology is the use of enzymes in a supercritical fluid. A supercritical fluid is a material above its critical temperature and pressure. Such a system is particularly attractive for enzymatic and chemical processing because it not only possesses the general advantages offered by the conventional solvent system but also exhibits some unique properties (Randolph et al., 1991). The viscosity, diffusivity and density of a solvent at supercritical conditions are intermediate between those of liquids and gases. Therefore, compared to enzymic reactions in conventional organic solvents, which can be subject to internal and external diffusion, the high diffusivity, low viscosity, and low surface tension inherent to supercritical fluids can facilitate increased reaction rates of diffusionally limited reactions due to the enhancement of mass transfer rates. The physical properties of a supercritical fluid can be tailored predictably over a wide range by modest variations in pressure or temperature. This "tunability" of solvent properties enables rational probing of enzyme activity, specificity, and stability in solvents. Due to this tunability, the solubility of a material can be controlled easily by adjusting the operation pressure, making product crystallization and solvent reuse and recycling possible. Since the critical temperatures of many fluids are below 100°C (Table 13-3), mild temperatures, which are necessary for thermally sensitive enzymes, can be used to generate the supercritical environment.

Since the first apparent demonstration of the activity of alkaline phosphatase in supercritical carbon dioxide (Randolph et al., 1985), a number of enzymes including polyphenol oxidase (Hammond et al., 1985), cholesterol oxidase (Randolph et al., 1988), lipase (Chi et al., 1988), subtilisin (Pasta et al., 1989), and *Aspergillus* protease (Kamat et al., 1993a) have been investigated in supercritical fluids such as carbon dioxide, ethane, ethylene, propane, fluoroform, and even sulfur hexafluoride (for reviews see Aaltonen & Rantakylä, 1991; Nakamura, 1990; Randolph et al., 1991). The pioneering work in this research area has demonstrated that enzymes not only retain their catalytic activity during exposure to supercritical fluids, but also exhibit high stability. For example, complete conversion was obtained within 30 minutes for the subtilisin-catalyzed transesterification between *N*-acetyl-L-phenylalanine chloroethyl ester and ethanol in supercritical CO_2 at 45°C and 150 bar (Pasta et al., 1989). A number of enzymes, such as lipase (Kamat et al., 1992) and subtilisin (Pasta et al., 1989), show higher activity in supercritical fluids than in conventional organic solvents. As far as stability is concerned, Dumont et al. (1992) have found that lipase retained 96% of its activity after five days of exposure to CO_2 at 15 MPa and 323 K. Therefore, the use

Table 13-3 Physical Properties of Several Supercritical Fluids[a]

Fluid	T_c (°C)	P_c (atm)	d_c (g/cc)
Carbon dioxide	31.0	72.8	0.468
Ethylene	9.2	49.7	0.218
Fluoroform	25.9	47.7	0.526
Ethane	32.28	48.16	0.21
Sulfur hexafluoride	45.6	37.1	0.734

[a]Data obtained from *CRC Handbook of Chemistry and Physics*, D. R. Lide, Ed., CRC Press, Boca Raton, 1990.

of supercritical fluids in biocatalysis provides an attractive means of improving the activity and utility of enzymes in anhydrous environments.

The use of enzymes in supercritical fluids is also attractive in biotransformations in that it allows the tuning of the enzyme activity and specificity by alterations in pressure. As mentioned, a change in pressure results in an alteration of the fluid physical properties. For instance, reducing pressure in supercritical fluoroform from 4000 to 850 psi resulted in a 10-fold increase in lipase activity, concomitant with a decrease of the dielectric constant of the solvent from 8 to 1 (Kamat et al., 1993b). Also, an increase of pressure can lead to an enhancement of the enantioselectivity of both subtilisin and *Aspergillus* protease (Kamat et al., 1993a). Correlations between the activity and enantioselectivity of an enzyme and the solvent dielectric can be explained in the same way as for conventional solvent systems (Affleck et al., 1992; Sakurai et al., 1988). In addition to changing the dielectric constant of the solvent used, it has been proposed that changes in pressure also affects enzyme activity by altering both the concentrations of reactants and products, and the rate of mass transfer. For example, the rate of a lipase-catalyzed interesterification increased as pressure increased from 83 to 111 bar, due to the enhanced substrate solubilities at higher pressure (Miller et al., 1990).

It is worth noting that as in conventional organic solvents, water is also important for biocatalysis in supercritical fluids. The initial rates of hydrolysis and interesterification catalyzed by lipase increased with the increase of water content in the reaction system (Chi et al., 1988), and the loss of enzyme activity due to the low water content can be regained by addition of more water, as demonstrated by Randolph et al. (1988) for cholesterol oxidase. In addition, changing the water content in the system can shift the equilibrium of lipase-catalyzed interesterification (Chi et al., 1988).

Summary

This chapter has discussed the use of enzymes in a myriad of synthetic processes; hopefully this will serve to convince readers that proteins can be used to catalyze many commercially relevant reactions. As the use of enzymes in "extreme" environments continues to expand, we believe that many chemical syntheses, which are currently prohibited by either cost or technology, will become viable.

14

Engineered Proteins in Materials Research

David A. Tirrell, Jane G. Tirrell,
Thomas L. Mason, and Maurille J. Fournier

Peptides and proteins have attracted scientific and technological interest largely because of their intriguing properties as catalysts, receptors, signalling molecules, and therapeutic agents. In attempts to understand and exploit these properties, protein engineering has been used primarily to obtain precious proteins in increased quantities, or to explore systematic alterations in protein sequence through site-directed mutagenesis. Design of protein structures *de novo* ("from scratch") has attracted less attention, and has been directed in the main toward studies of protein folding (Kamtekar et al., 1993). Such studies represent a key element in the current vigorous investigation of the connections between amino acid sequence and the three-dimensional structures of isolated protein chains in aqueous solution. This chapter describes protein engineering of quite another sort, in which the *proteinacious* nature of the product is less important than its *macromolecular* character.

Engineered Proteins

The chemical processes used to prepare the macromolecular materials (polymers) of commerce produce complex mixtures of products, in which chain length, sequence, and stereochemistry are characterized by broad statistical distributions. As an example, Figure 14-1 shows a matrix-assisted laser desorption mass spectrum of poly(methyl methacrylate) (PMMA), an important polymer sold under tradenames such as Lucite or Plexiglas. The sample represented in Figure 14-1 is far better defined in terms of molecular weight distribution than is the typical commercial PMMA, yet it is clear from the spectrum that no single molecular species constitutes more than a small percentage of the chain population. Similar complexity characterizes the distributions of *monomer sequences* in commercial copolymers, and the *stereochemistry* of vinyl polymers is generally discussed only in statistical terms. This complexity of structure has had profound consequences for polymer science and technology, because it means that this is a science and technology of *mixtures*. This is not necessarily an impediment to the development of useful properties, and conventional polymers are enormously important as practical, commercial materials. At the same time, because mixtures behave

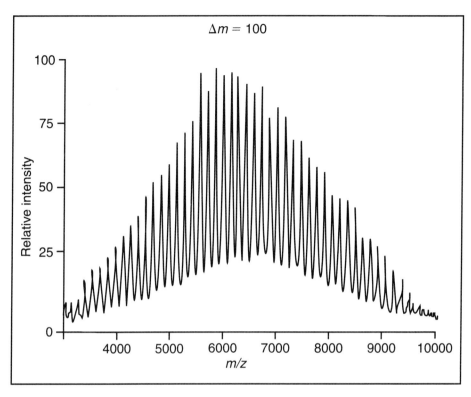

Figure 14-1. MALDI mass spectrum of poly(methyl methacrylate). Reprinted from Bahr et al., 1992, with permission.

differently from pure substances, one wonders what kinds of new materials properties might be achievable through better control of macromolecular architecture.

Protein engineering is now the most powerful means of controlling the architecture of macromolecular substances, and offers an important alternative to conventional polymerization processes as a route to new classes of polymeric materials. The exploitation of protein engineering to create new materials is outlined in schematic form in Figure 14-2. The process begins with design of a target structure, or more typically, a family of related target structures. The targets here are artificial proteins, amino acid copolymers designed to exhibit interesting *materials* properties, for example, interesting properties in the solid state, at interfaces, or in liquid crystal phases. The design of such proteins must take into account both "biological" and "materials" concepts in order to succeed. Once designed, the primary sequence of the target artificial protein is encoded into a complementary DNA, which is then made by solid phase synthesis and enzymatic ligation of oligonucleotide fragments. Conversion of the artificial gene to the artifical protein is then accomplished via the same kinds of genetic manipulations and biochemical engineering methods used to produce other classes of heterologous proteins in microbial hosts (Sambrook et al., 1989).

In our own laboratories, we have used this approach to address four issues in materials design and synthesis: (1) design and preparation of macromolecular crystals of

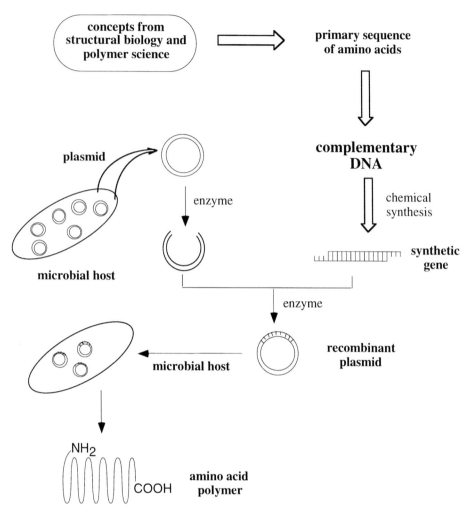

Figure 14-2. Outline of gene construction and protein synthesis.

controlled thickness and surface chemistry, (2) incorporation of nonnatural amino acids, (3) synthesis of rodlike helical macromolecules of uniform chain length, and (4) development of hybrid proteins comprising natural and artificial domains. We describe in this chapter the progress that has been made toward these objectives, and we hope thereby to illustrate some of the exciting opportunities that we perceive at this new interface of materials science and the biological sciences.

Crystal Design: The Solid-State Protein Folding Problem

A logical starting point for the design of macromolecular crystals is provided by the sequence-dependent secondary structures (helices, β-strands, and reverse turns) found in fibrous and globular proteins. For example, a repetitive polypeptide comprising β-strands followed by reverse turns would be expected to assemble into a lamellar aggregate like that shown schematically in Figure 14-3, in which lamellar thickness is controlled by the length of the strands, and surface chemistry is determined by functional groups contributed by amino acid residues confined to the turns. Such lamellar crystals are well known in polymer materials science (Keller, 1957), but are usually only metastable structures trapped by the kinetics of crystal growth. In the design shown in Figure 14-3, the lamellar structure should not only be stable, but should be predictable on the basis of the secondary structural elements used for its construction.

In order to test this approach to crystal design, we have prepared and expressed a family of artificial genes encoding polypeptides built from repeating units represented as sequence **1** (Deguchi et al., 1994; Krejchi et al., 1994, 1996).

$$\{-(AlaGly)_n ZGly-\}_x$$

1 $n = 3$ to 6

Z = Ala, Asn, Asp, Glu, Leu, Met, Phe, Ser, Tyr, Val, and ProGlu

In the crystal structures of synthetic aliphatic polyamides (nylons), kinetic factors limit the length of the crystal "stem" to that defined by six to eight lateral hydrogen bonds (Atkins et al., 1992; Dreyfuss & Keller, 1970a, b); the choice of $n = 3$ to 6 in **1** reflects that consideration. It has also been observed that the egg-stalk protein of *Chrysopa flava* (Geddes et al., 1968) folds at intervals of eight amino acid residues, giving rise to a lamellar thickness of about 3 nm. Bulky and polar amino acids in the Z position were chosen because such amino acids should be excluded from the interior and should therefore decorate the surfaces of the lamellar crystal. Glutamic acid in particular was chosen because it is the poorest β-sheet former of the twenty natural amino acids according to the Chou-Fasman predictions of protein structure and conformation (Chou & Fasman, 1974a, b).

Synthesis

The following paragraphs provide a brief description of the methods used in our laboratory to prepare polymers of repeating units such as **1** (Fig. 14-2).

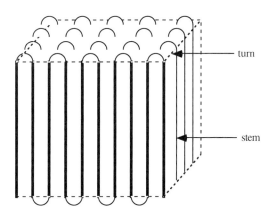

Figure 14-3. Schematic drawing of a chain-folded lamellar crystal. Reprinted from Parkhe et al., 1993, with permission.

```
                              Stop Gly Ala Gly Ala Gly Ala Gly Ala Gly Pro Glu Gly Ala Gly Ala Gly Ala Gly Pro Glu Gly Ala
AATTCG TAA GGT GCC GGC GCT GGT GCT GGG GCC GGT CCG GAA GGT GCA GGC GCT GGC GCG GGC GCG GGC CCG GAA GGT GCC G
       GC ATT CCA CGG CCG CGA CCA CGA CCC GGC CCA GGC CTT CCA CGT CCG CGA CCG CGC CCG CGC CCG GGC CTT CCA CGG CCTAG
  EcoRI        BanI                                                                                    ApaI       BanI    BamHI
```

2

Oligonucleotides encoding one or two repeats of the target repeating unit sequence are prepared via solid phase organic synthesis (McBride & Caruthers, 1983). After purification, the oligonucleotides are phosphorylated at the 5'-termini, annealed, and then ligated into appropriate restriction sites in a bacterial cloning vector. Oligonucleotides are typically designed according to the following rationale. First, as shown in fragment **2** (which encodes two copies of repeating unit **1** with $n = 4$ and $Z = $ ProGlu), a stop codon immediately following the 5' restriction site ensures disruption of the β-galactosidase α fragment encoded in common cloning vectors, and thus allows facile color screening for plasmids carrying the insert. Second, the oligonucleotides are designed so that the sequence encoding the repeating unit is flanked by two restriction sites which are used to isolate the fragments of interest after cloning and amplification (Cappello et al., 1990). Finally, the choice of codons for each amino acid in repeating unit **1** is determined by (1) the pattern of codon use in *E. coli* (Aota et al., 1988), (2) avoidance of strict sequence periodicity within the oligonucleotide, (3) the need to eliminate all *Ban*I sites except for those flanking the coding sequence, and (4) inclusion of an *Apa*I site to be used for screening transformants containing the insert.

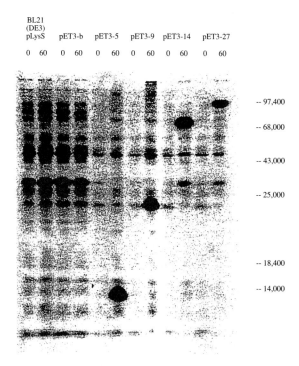

Figure 14-4. *In vivo* labeling of proteins containing (AlaGly)$_3$ProGluGly repeats in 12% polyacrylamide gels. pET3-5, pET3-9, pET3-14 and pET3-27 contain 5, 9, 14, and 27 repeats of the (AlaGly)$_3$ProGluGly sequence, respectively. Time points in minutes are relative to IPTG addition. Reprinted from McGrath et al., 1992, with permission.

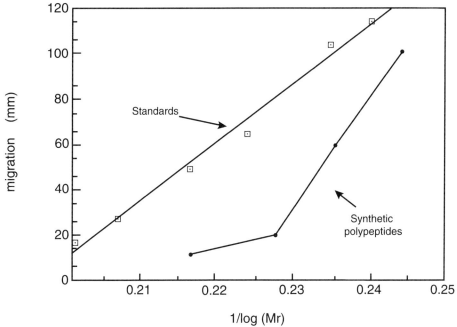

Figure 14-5. Plot of electrophoretic mobility versus reciprocal of log MW for a series of polypeptides containing 5, 9, 14, and 27 repeats of the (AlaGly)$_3$ProGluGly sequence. Standards: α-lactalbumin, β-lactoglobulin, carbonic anhydrase, egg albumin, bovine serum albumin, phosphorylase B (rabbit muscle). Reprinted from McGrath et al., 1992, with permission.

The presence of the insert in transformed bacterial cells can be confirmed by restriction analysis (for sequence **2**, for example, with the restriction enzymes *Apa*I and *Ban*I), and standard DNA sequencing methods are used to verify the integrity of the insert. The target DNA fragments are liberated by restriction digestion, purified, and self-ligated to produce a population of multimers. Inversion of repeats of the monomer sequences is suppressed by the fact that nonpalindromic ends are generated in *Ban*I digestion of the plasmid. Multimers of appropriate size are inserted (usually after an intermediate cloning step, which allows selection of coding sequences of preferred lengths) into an expression vector, which is chosen to direct bacterial synthesis of the target protein. At this point, the coding sequence may include N- and C-terminal extensions derived from the cloning and expression vectors; these are generally removed later by cleavage with CNBr or with appropriate enzymes.

Target protein synthesis can be demonstrated in a variety of ways. Because artificial proteins such as **1** bind conventional protein stains weakly, radiolabeling protocols or Western blots are most convenient. In a typical labeling experiment, transformed cells are grown in minimal medium supplemented with ^{3}H-glycine and a mixture of unlabeled amino acids lacking glycine. Following induction of protein synthesis, new protein bands corresponding to the target can be detected on autoradiograms of gels (Fig. 14-4). Protein products of sequence **1** may be present either in the soluble fraction of the cell lysate or in inclusion bodies, depending on charge density and the nature of residue(s) Z.

The apparent molecular weights of these artificial proteins as reported by gel electrophoresis are much higher than the expected molecular weights (Fig. 14-5). Highly acidic polypeptides such as **1** (Z = Glu or ProGlu) probably bind SDS weakly and may, therefore, adopt unusual micellar structures under the conditions of electrophoretic separation. The anomalously low average residue mass of these alanylglycine-rich polymers surely contributes further to this effect.

In our work with proteins such as **1**, neither the artificial genes nor the protein products have proven prohibitively unstable. Plasmids from transformed cells in some cases have been recovered and subjected to electrophoresis; even after 35 generations, no length polymorphism was observed. The protein products are sufficiently stable that they continue to accumulate in transformed cells for 2 to 3 hours (and perhaps longer) following induction (for an exception, see later). Typical yields from batch fermentation procedures have been in the range of 100 mg of protein per liter, and recent experiments using fed-batch fermentation methods have afforded yields of nearly 0.5 g of purified protein per liter of culture.

Molecular Structure

The primary structures of polymers of repeating units such as **1** are generally as expected. In one interesting exception, matrix assisted laser desorption (MALDI) mass spectrometry of the protein comprising 14 repeats of the sequence -(AlaGly)$_4$ProGluGly- (**3**), showed evidence of degradation of the product protein (Beavis et al., 1992) (Fig. 14-6). The spectrum of the isolated protein contained a series of signals that could best be explained by successive additions of amino acid residues, in the N- to the C-terminal direction, consistent with the anticipated amino acid sequence. MALDI mass spectrometry also gives an accurate measure of the mass of the polymer, and shows that mass determinations by SDS polyacrylamide gel electrophoresis are grossly in error.

Solid-State Structure

Proteins built up from repeating units of sequence **1** (Z = ProGlu) do not readily form β-sheets (McGrath et al., 1992). Under most of the crystallization conditions we have investigated, these polymers form amorphous solids, as indicated by wide-angle X-ray powder patterns that consist only of diffuse halos. Moreover, the amide I and amide II bands in the infrared spectrum are observed at 1653 and 1540 cm^{-1}, respectively, and not at the anticipated frequencies (1630 and 1525 cm^{-1}) characteristic of β-structures (Moore & Krimm, 1976). We have considered several explanations including (1) the odd number of amino acids in the repeating unit sequence precludes formation of hydrogen bonds over the full extent of the sheet (McGrath et al., 1992), (2) the geometry of the turn (presumably comprising the ProGlu dyad) is inconsistent with the parallel (or nearly parallel) trajectories required of the flanking strands, and (3) the steric bulk of the ProGlu dyad frustrates packing of sheets at a separation distance consistent with the small size of the Ala and Gly residues comprising the sheets. Whether or

Figure 14-6. Matrix-assisted laser desorption mass spectrometric analysis of {(AlaGly)$_4$ProGluGly}$_{14}$. (A) Spectrum of the target protein. RNAse A is included as an internal standard. (B) Spectrum expanded in the region of low molecular weight contaminants. (C) Mass spectrum of the protein sample after low molecular weight substances were removed by dialysis. Reprinted from Beavis et al., 1992, with permission.

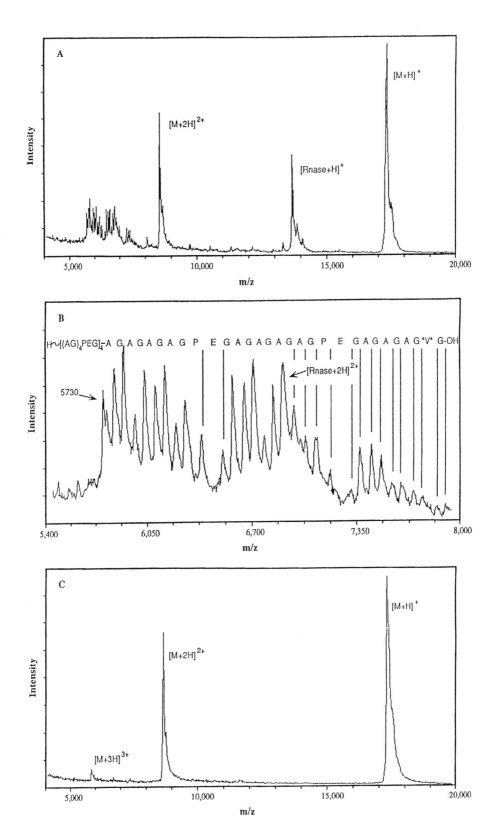

not any of these explanations is correct, it appears that the regularly folded conformation is destabilized to such an extent that the glassy state is preferred at room temperature or, alternatively, that crystallization is prohibitively slow.

In order to address these concerns, we explored a second generation of designed structures in which Pro was deleted from the repeating unit. We first considered polymers of repeating unit sequence -(AlaGly)$_n$GluGly- (**4**, $n = 3, 4, 5, 6$), and found that such polymers form β-sheet structures readily under conditions that failed to produce crystalline forms of the "ProGlu" polymer (Krejchi et al., 1994).

The infrared spectrum of {(AlaGly)$_3$GluGly}$_{36}$ (Fig. 14-7) shows amide I, II, and III vibrational modes at 1623, 1521, and 1229 cm^{-1}, respectively, all of which are characteristic of the β-sheet structure (Moore & Krimm, 1976), and a weak amide I vibration at 1698 cm^{-1} indicating the regularly alternating chain direction that defines antiparallel β-sheets (Miyazawa & Blout, 1961). Additional vibrational modes are thought to arise from reverse (β- or γ-) turn structures (Krimm & Bandekar, 1986).

Raman spectroscopy of {(AlaGly)$_3$GluGly}$_{36}$ further supports the proposed structure, showing the characteristic amide I band at 1664 cm^{-1} and splitting of the amide III band into two components at 1260 and 1228 cm^{-1} (Frushour & Koenig, 1975;

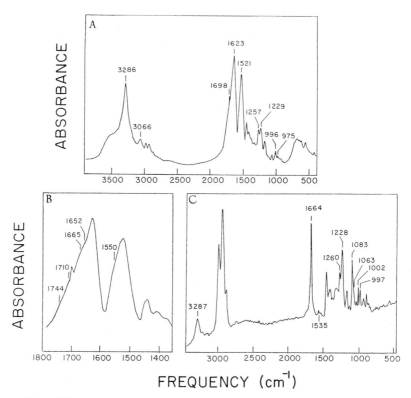

Figure 14-7. Vibrational spectra of {(AlaGly)$_3$GluGly}$_{36}$. (A) Fourier transform infrared spectrum of a KBr pellet of {(AlaGly)$_3$GluGly}$_{36}$ powder. (B) Expansion of the amide I region of the spectrum shown in A. (C) Raman spectrum of {(AlaGly)$_3$GluGly}$_{36}$. Reprinted from Krejchi et al., 1994, with permission.

Table 14-1 Chemical Shifts and Assignments for Selected Peaks Observed in CP/MAS ^{13}C NMR Spectra of {(AlaGly)₃GluGly}₃₆ and Poly(L-alanylglycine) (β-form)

	Chemical Shift (ppm)	
{(AlaGly)₃GluGly}₃₆	Poly(L-alanylglycine)	Assignment
49.9	48.5	Ala Cα
20.7	20.0	Ala Cβ
171.4	171.8	Ala C=O
43.6	43.3	Gly Cα
171.4	168.4	Gly C=O

Moore & Krimm, 1976). Weaker signals in the 1300 to 1330 cm^{-1} region have been attributed to turns (Krimm & Bandekar, 1986).

Cross-polarization magic angle spinning nuclear magnetic resonance (CP/MAS NMR) spectra provide additional confirmation that {(AlaGly)₃GluGly}₃₆ forms antiparallel β-sheets. Table 14-1 shows a comparison between the chemical shifts obtained for poly(L-alanylglycine) (Saito et al., 1984) and for {(AlaGly)₃GluGly}₃₆. The assignments are consistent with an antiparallel β-sheet structure, though there is evidence for other conformational states as well. For example, a shoulder at 16.8 ppm, assigned to the β-carbon of alanine, arises either from a fraction of the silk I structure (Ishida et al., 1990), or from amino acid residues in turn sequences (as proposed above to explain the infrared and Raman spectra).

Strong evidence for a chain-folded lamellar structure as the basic crystalline unit is provided by X-ray diffraction results obtained on the series of polymers, -{(AlaGly)ₙGluGly}ₓ- (**4**) (Krejchi et al., 1997). The dimensions of the corresponding orthorhombic unit cells are listed in Table 14-2, and small angle X-ray scattering gives a long-period spacing of 3.6 nm (for the polymer with $n = 3$), consistent with the anticipated lamellar thickness.

The unit cell parameters are consistent with those published for silks and for synthetic polypeptides known to adopt β-sheet architectures (Brown & Trotter, 1956; Warwicker, 1960). The value of the unit cell parameter a, 0.948 nm, was assigned on the basis of the second diffraction order spacing at 0.474 nm, which is characteristic of the

Table 14-2 Unit Cell Dimensions for Polymers of -{(AlaGly)ₙGluGly}ₓ-

Sample	Unit Cell Dimensions (nm)		
	a	b	c
{(AlaGly)₃GluGly}₃₆	0.948	1.060	0.695
{(AlaGly)₄GluGly}₂₈	0.948	1.028	0.695
{(AlaGly)₅GluGly}₂₀	0.957	0.970	0.695
{(AlaGly)₆GluGly}₁₄	0.964	0.962	0.695

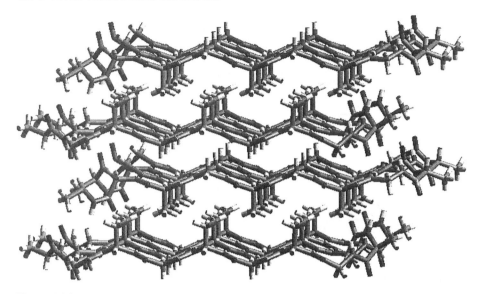

Figure 14-8. Computer-generated representation of the solid-state structure of {(AlaGly)$_3$GluGly}$_{36}$.

interchain distance in hydrogen-bonded, antiparallel β-sheets (Dreyfuss & Keller, 1970a, b; Hirichsen, 1973; Keller, 1959). Thus a is associated with the hydrogen bond direction. The value 1.060 nm observed for b represents twice the average periodicity of intersheet stacking. The intersheet spacing in β-sheet polypeptides is dependent on amino acid composition (Lucas et al., 1960), since the amino acid side chains protrude from the surfaces of the sheets and must be accommodated in the intersheet volume. The intersheet spacing for polyglycine has been reported as 0.34 nm (Nemethy & Printz, 1972), while that of *Nephila senegalensis* fibroin, which contains a high percentage of amino acid residues with bulky side chains, is 0.79 nm (Warwicker, 1960). The values reported for poly(L-alanylglycine), 0.44 nm (Fraser et al., 1965), and the β-form of poly(L-alanine), 0.54 nm (Brown & Trotter, 1956), are consistent with the results obtained for the series of polymers **4**.

Figure 14-8 shows a computer generated representation of the unit cell of polymer **4** ($n = 3$) in which the β-sheets are stacked along the b axis, the c axis extends horizontally in the plane of the image, and the hydrogen bond direction, a, is perpendicular to the plane of the page. The polymer chains in the crystals must fold back at the lamellar surfaces and reenter the crystalline lamellae, as indicated by the crystal dimensions observed, since the lamellar thickness is always shorter than the molecular length of the chains.

Since {(AlaGly)$_3$GluGly}$_{36}$ exhibits a well-defined β-sheet conformation and a uniform lamellar structure, it serves as a good model for studying the relationships between hydration and crystal structure. Infrared and Raman spectral studies of the {(AlaGly)$_3$GluGly}$_{36}$ polypeptide have shown that hydration is a stepwise process (Chen et al., 1995). At low water contents, the crystalline regions are not affected (Murthy et al., 1989; Tirrell et al., 1979; Vergalati et al., 1993). As the water content increases, the sheets become accessible to water and changes in intersheet spacing, and ultimately chain conformation, occur. Figure 14-9 shows the effects of humidity on the infrared spectrum of the carboxylate form of the sample. Increasingly marked effects on both the COO$^-$ and the amide I and II signals are observed as the water content increases. Figure 14-10 shows the intersheet distance as a function of humidity as determined by wide angle X-ray scattering, infrared spectra of the amide I and II regions at different relative humidi-

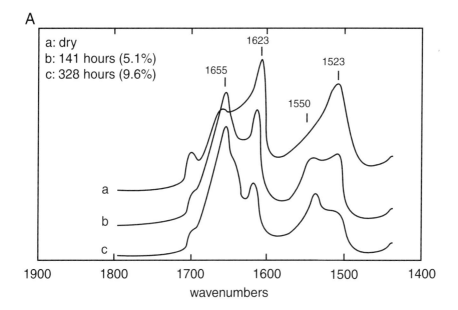

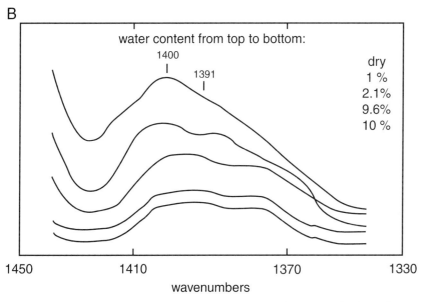

Figure 14-9. (A) Infrared spectra of the amide I and II region of {(AlaGly)$_3$GluGly}$_{36}$ in the carboxylate form as a function of hydration. (B) Effect of hydration on the symmetric COO$^-$ band in the infrared spectrum. Reprinted from Chen et al., 1995, with permission.

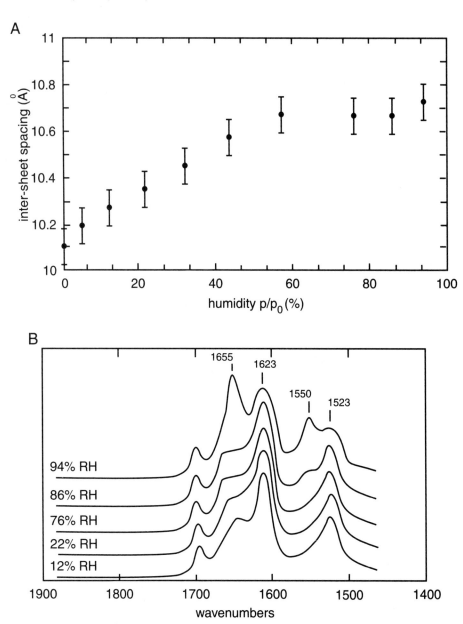

Figure 14-10. (A) Changes in intersheet spacing as a function of hydration. (B) Infrared amide I and II regions at different relative humidities. (C) Water content of hydrated samples in (A) and (B). Reprinted from Chen et al., 1995, with permission.

ties, and water contents of these same samples (from gravimetric measurements). The intersheet distance begins to change at low water contents, indicating hydration of the ionic groups at the lamellar surface; however, the infrared spectra reveal no changes in the chain conformation until the water content is quite high, that is, after the intersheet spacing has expanded, providing accessibility of the crystalline region to water.

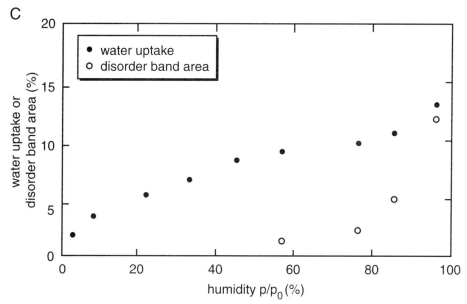

Figure 14-10. (Continued)

Artificial Proteins from Artificial Amino Acids

For some materials applications, it would be desirable to expand the set of monomer units available for construction of polymer chains beyond those normally used by organisms as substrates for protein synthesis. This section describes the incorporation of several non-natural amino acids—selenomethionine, *p*-fluorophenylalanine, 3-thienylalanine, 5′,5′,5′-trifluoroleucine, azetidine-2-carboxylic acid, and 3,4-dehydroproline—into genetically engineered artificial proteins expressed in bacterial hosts. Because of the relatively large sample size requirements of materials research, we have not yet exploited the intriguing approach of chemical misacylation of tRNAs and *in vitro* translation, a strategy developed and exploited successfully by others (Ellman et al., 1992; Hecht, 1992).

Selenomethionine

It has been known since the 1950s that selenomethionine (SeMet, **5**) can be substituted for methionine in the *in vivo* synthesis of bacterial proteins (Cowie et al., 1959; Cowie & Cohen, 1957; Hendrickson et al., 1990; Tuve & Williams, 1957). SeMet supports growth of *E. coli* methionine auxotrophs (Cowie & Cohen, 1957; Tuve & Williams, 1957) and has been shown to replace methionine virtually completely in thioredoxin of *E. coli* and phage T4 (Hendrickson et al., 1990). In addition, SeMet, with the bulky selenide group in the side chain, would be expected to be excluded from the interior of lamellar crystals, and thereby provide a population of reactive functional groups at the lamellar surface (Dougherty et al., 1993). With these ideas in mind, we undertook the synthesis of polymers of repeating unit sequence -(GlyAla)$_3$GlySeMet-.

The expression plasmid chosen for these experiments (pGEX-2T) (Smith & Johnson, 1988) encodes a 26 kD fragment of glutathione-S-transferase (GST) at the amino terminus of the target protein. The host methionine auxotroph was prepared from *E. coli* strain HB101 by mutagenesis with ethyl methanesulfonate and antibiotic selection in the absence of methionine (Miller, 1972). Auxotrophs were characterized by strict dependence on added methionine (or SeMet) for growth. As shown in Figure 14-11, substitution of SeMet for methionine allowed the mutant cells to grow, but with a two-fold increase in generation time.

To assess the level of SeMet substitution in the target protein, competition experiments were conducted using ^{35}S-methionine and unlabeled SeMet; labeled SeMet is not readily available. Cultures of transformed auxotrophs were grown to logarithmic phase in the presence of a small amount of methionine and with increasing amounts of SeMet.

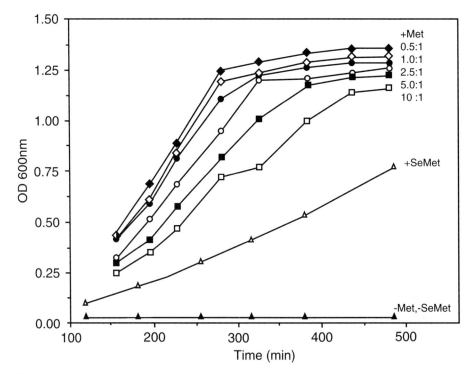

Figure 14-11. Growth kinetics of transformed *E. coli* cells grown in medium lacking methionine (−Met, −SeMet), with selenomethionine added without methionine (+SeMet), or methionine added with selenomethionine in different ratios (+Met). Reprinted from Dougherty et al., 1993, with permission.

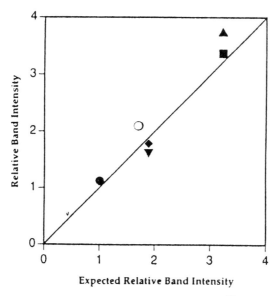

Figure 14-12. Effect of selenomethionine on the incorporation of ^{35}S-methionine into a polypeptide of repeating unit sequence -(GlyAla)$_3$GlySeMet-. Comparison of experimental and predicted incorporation values for the induced polypeptide at different ratios of selenomethionine to methionine. Experimental ratios were obtained by scanning autoradiographic signals. Each experimental value consists of intensity measurements taken from at least three separate time points and averaged to yield the experimental ratio. ●, 10:1/10X Met; ○, 2.5:1/5:1; ◆, 5:1/10X Met; ▼, 5:1/10:1; ■, 2:5:1/10:1; ▲, 2.5; 1/10X Met. 10X Met represents a control consisting of a 10-fold excess of unlabeled methionine. The theoretical ratios correspond to the intensities expected if the incorporation of selenomethionine is equivalent to that of methionine at each concentration tested. Reprinted from Dougherty et al., 1993, with permission.

^{35}S-Methionine was then added to the cultures, and protein synthesis was induced five minutes later by addition of isopropyl β-thiogalactopyranoside (IPTG). Samples were removed after varying periods of time, and analyzed by SDS polyacrylamide gel electrophoresis. Autoradiograms and stained gels were scanned densitometrically. Figure 14-12 illustrates the decrease in incorporation of radioactive methionine observed with increasing concentrations of SeMet in the growth medium; in fact, there is excellent agreement between the experimentally determined ratio of SeMet to methionine in the product, and the ratio of concentrations of the two amino acids in the medium. These data are consistent with complete or near complete substitution by SeMet.

p-Fluorophenylalanine

Polymers containing fluorinated amino acids are expected to exhibit many of the useful characteristics of conventional fluoropolymers, including low surface energy, low coefficient of friction, and excellent solvent resistance. In our first studies of genetically engineered fluoropolymers (Yoshikawa et al., 1994), *p*-fluorophenylalanine (pfF, **6**) was substituted for phenylalanine in a polymer comprising thirteen repeats of the octapeptide sequence -{(AlaGly)$_3$PheGly}-.

462 / Tirrell, Tirrell, Mason, and Fournier

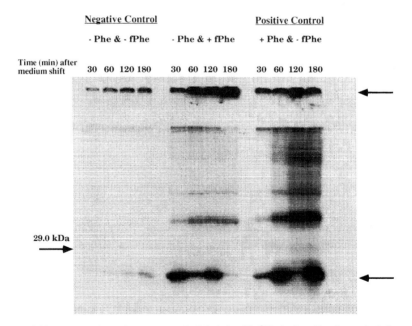

6

E. coli phenylalanine auxotrophs were generated by infecting a donor strain containing the *PheA*::Tn10 mutation with the transducing phage P1 and then transferring the transposon to the bacterial strain containing the expression vector pET3-b (Bochner et al., 1990; Miller, 1972; Rosenberg et al., 1987). This vector requires expression of bacteriophage T7 RNA polymerase for transcription of the target gene (Studier et al., 1990). In experiments directed toward incorporation of amino acid analogues, it might be anticipated that substitution of the analogue for the natural amino acid would result

In vivo Synthesis of -[(AlaGly)$_3$ fPheGly]$_{13}$-

Figure 14-13. Expression of target protein labeled with ^{3}H-glycine. For lanes 1–4 the culture medium lacks both phenylalanine and *p*-fluorophenylalanine; for lanes 5–8, the medium contains *p*-fluorophenylalanine but lacks phenylalanine; for lanes 9–12, the medium contains phenylalanine but lacks *p*-fluorophenylalanine. The arrows indicate the target protein; time points are relative to medium shift. Reprinted from Yoshikawa et al., 1994, with permission.

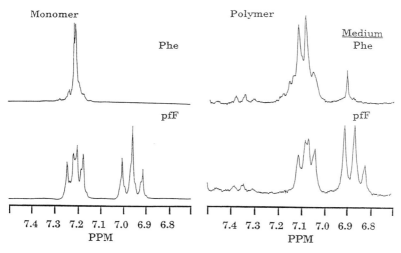

Figure 14-14. ^{1}H NMR spectra of -{(AlaGly)$_3$PheGly}- and -{(AlaGly)$_3$pfFGly}- in DCOOH. The aromatic regions of the spectra are compared with those for L-phenylalanine and p-fluoro-L-phenylalanine. The signal at 6.9 ppm for the protein containing phenylalanine is attributed to the solvent; this signal is obscured in the spectrum of protein containing p-fluorophenylalanine. Reprinted from Yoshikawa et al., 1994, with permission.

in an inactive polymerase. Therefore, we have adopted a protocol in which expression of the target protein begins with a ten minute induction in the presence of the twenty natural amino acids (but in the absence of pfF) followed by a shift to medium containing pfF after a pool of active polymerase has been established (Fig. 14-13).

Analysis of amino acid composition and integration of the relevant signals in the NMR spectra of target polymers made by such a protocol show a level of replacement of phenylalanine of 95 to 100% (Fig. 14-14, Table 14-3). The target proteins—{(AlaGly)$_3$PheGly} and {(AlaGly)$_3$pfFGly}—were also analyzed by Fourier transform

Table 14-3 Amino Acid Composition Analysis of Target Proteins Containing Phe and pfF

Amino acid	Mol% (theoretical)	Mol% (observed for proteins containing pfF)	Mol% (observed for proteins containing Phe)
Glycine	50.0	51.0	50.5
Alanine	37.7	37.0	36.4
Phenylalanine	12.3[a]	0.6	11.2
p-Fluorophenylalanine	12.3[b]	10.3	
Aspartic acid		1.1	1.0
Leucine			0.5
Valine			0.4

[a]Expected for proteins containing phenylalanine.
[b]Expected for proteins containing p-fluorophenylalanine.

7

infrared spectroscopy, which provides evidence for the antiparallel β-sheet architecture; the amide I bands are observed at 1625 cm^{-1} and the amide II bands at 1525 cm^{-1} (Krimm & Abe, 1972; Moore & Krimm, 1976). Evidence for β-sheet structure is also observed in wide angle X-ray scattering patterns of oriented crystal mats.

5',5',5'-Trifluoroleucine

5',5',5'-Trifluoroleucine (TfL, **7**) has been shown to be incorporated into proteins synthesized by *E. coli* leucine auxotrophs (Fenster & Anker, 1969), and incorporation of TfL into artificial proteins would be expected to allow placement of fluorinated residues at well-defined locations on the polymer chain.

The sequence of the protein in which we investigated the substitution of TfL for leucine was {(GlyAla)$_3$GlyLeu}$_{12}$GlyAla (Kothakota et al., unpublished). Gene synthesis, leucine auxotroph construction, and protein expression were carried out as described for expression of the protein containing pfF.

No target protein was produced by the host leucine auxotroph in the absence of leucine and TfL, while ^{1}H NMR spectra of proteins prepared in media supplemented with TfL indicated successful incorporation of the analogue. The level of substitution of TfL for leucine was estimated at a maximum of about 90% from NMR determination of the ratio of the methyl protons of TfL to those of alanine.

Of the stereoisomers of TfL (*2S,4S; 2S,4R; 2R,4R; 2R,4S*), only the *2S* isomers, or L-amino acids, are used in biological protein synthesis. In ^{19}F NMR spectra of monomeric TfL and {(GlyAla)$_3$GlyTfL}$_{12}$GlyAla, two sets of signals, attributable to the *2S,4S;2R,4R* and *2S,4R;2R,4S* isomers, respectively, were observed both for the monomer and for the polymer, indicating that both of the *2S* isomers are incorporated into protein to similar extents. The fact that *either* of the two diastereotopic methyl groups can be fluorinated without loss of translational activity, suggests that perhaps *both* can be fluorinated. Experiments with hexafluoroleucine are underway.

Infrared spectroscopy and X-ray diffraction studies indicate the presence of stacked antiparallel β-sheets in crystalline samples of both {(GlyAla)$_3$GlyLeu}$_{12}$GlyAla and {(GlyAla)$_3$GlyTfL}$_{12}$GlyAla. Vibrational modes similar to those observed for the periodic proteins discussed above (and characteristic of antiparallel β-sheets) are observed in the IR spectra.

The surface properties of thin films of {(GlyAla)$_3$GlyLeu}$_{12}$GlyAla and {(GlyAla)$_3$GlyTfL}$_{12}$GlyAla were assessed by measuring the advancing contact angles for water and hexadecane. The fluorinated form of the polymer exhibits decreased wettability by hexadecane, a sensitive indicator of surface fluorination.

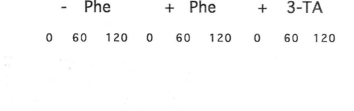

8

3-Thienylalanine

3-Thienylalanine (3-TA, **8**) was chosen for study because of its similarity to the 3-alkylthiophenes (Kothakota et al., 1995). Poly(3-alkylthiophene)s form excellent organic conductors, showing conductivities of about 2000 S cm^{-1} after doping (Roncali et al., 1988). The incorporation of 3-TA into engineered proteins may provide a means for electrodepositing such materials on electrodes or for fabrication of enzyme-based sensors or controlled delivery devices (Kothakota et al., 1995).

The target chosen for these experiments was a repeating polymer of sequence -{(GlyAla)$_3$GlyPhe}-$_{13}$, in which 3-TA was substituted for phenylalanine. Gel electrophoresis (Fig. 14-15) and amino acid composition analysis (Table 14-4) demonstrated efficient incorporation of 3-TA into recombinant proteins in place of phenylalanine.

Figure 14-15. Autoradiogram of ^{3}H-labeled proteins produced by transformed phenylalanine auxotrophs induced to synthesize protein in the presence of the 20 natural amino acids and shifted after ten minutes to media lacking Phe and 3-TA, containing Phe, or containing 3-thiemylalanine (3-TA). Reprinted from Kothakota et al., 1995, with permission.

Table 14-4 Amino Acid Compositions of Target Proteins Containing Phe or 3-Thienylalanine (3-TA)

Amino acid	Mol% (theoretical)	Mol% (observed for -{(GlyAla)$_3$GlyPhe}$_{-13}$	Mol% (observed for -{(GlyAla)$_3$Gly3-TA}$_{-13}$
Glycine	50.0	49.4	47.0
Alanine	37.7	35.6	34.1
Phenylalanine	12.3[a]	11.7	2.2
3-Thienylalanine	12.3[b]		10.8

[a]Expected for -{(GlyAla)$_3$GlyPhe}$_{-13}$.
[b]Expected for -{(GlyAla)$_3$Gly3-TA}$_{-13}$.

The relative amounts of phenylalanine and 3-TA in the product were also assessed quantitatively by ultraviolet spectroscopy (which takes advantage of the fact that the phenylalanine absorption maximum lies at 256 nm while that for 3-TA is at 233 nm), and by NMR spectroscopy. Figure 14-16 shows UV spectra for target proteins produced by cultures in which protein synthesis was induced in media supplemented either with phenylalanine or with 3-TA; the phenylalanine absorption maximum is absent from the spectrum of the target protein containing the electroactive analogue. Three sets of signals arising from the monosubstituted thiophene ring are observed in the NMR spectrum of -{(GlyAla)$_3$Gly3-TA}$_{-13}$, while two broad resonances attributed to the phenyl protons appear in the spectrum of -{(GlyAla)$_3$GlyPhe}$_{-13}$. Integration of the

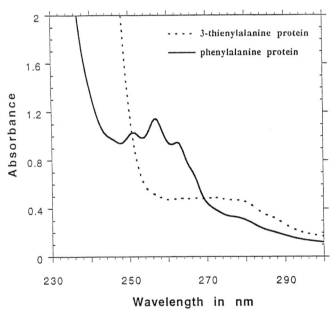

Figure 14-16. UV spectra of -{(GlyAla)$_3$GlyPhe}$_{-13}$ and -{(GlyAla)$_3$Gly3-TA}$_{-13}$. Reprinted from Kothakota et al., 1995, with permission.

spectrum of the 3-TA polymer indicates the extent of substitution of 3-TA for phenylalanine to be approximately 80%. Artificial proteins containing 3-TA have now been fabricated into composite films with poly(3-methylthiophene); studies of the electrochemical properties of such films are in progress.

Azetidine-2-carboxylic Acid

Earlier in this chapter, we showed that repeating polypeptides of the sequence -(AlaGly)$_n$ProGluGly- adopted disordered structures in the solid state. To explore the role of proline in promoting (or frustrating) the formation of β-sheets, azetidine-2-carboxylic acid (Aze, **9**) was substituted for proline in polypeptides of repeating sequence -(AlaGly)$_3$ProGluGly- (Deming et al., 1996). Previous studies have shown that the effects of incorporating azetidine-2-carboxylic acid into proteins can be striking (Mauger & Witkop, 1966); replacement of about 4% of the proline in collagen with Aze destabilizes the triple helix bundle (Lane et al., 1971). Computational studies have suggested that proteins containing Aze are more flexible than those containing proline, and more likely to adopt β-turn geometries (Zagari et al., 1990). Both characteristics should contribute to assembly of ordered β-sheets in periodic polypeptides.

Because Aze does not support bacterial growth (Peterson & Fowden, 1963), we grew transformed cells in medium supplemented with proline and then induced synthesis of the target protein in medium containing Aze. Autoradiograms of ^{3}H-labeled proteins produced in transformed proline auxotrophs show that more protein is produced in Aze-supplemented growth medium than in medium containing neither proline nor Aze. Direct determination of the relative amounts of proline and Aze in the product by amino acid analysis is precluded by the fact that Aze is degraded under the conditions used to hydrolyze and analyze peptides (Fowden, 1956). Nevertheless, the Aze content can be estimated reliably at about 25%, by determination of the extent of diminution of proline concentration (Table 14-5).

Also, ^{1}H NMR spectroscopy can be used to measure the level of Aze incorporation; the intensities of the proline resonances (3.62 ppm), compared to those of the alanine and glycine resonances (1.32 and 4.20 ppm, respectively) indicate that Aze is incorporated to a level of about 40%, somewhat higher than the estimate based on amino acid analysis.

The Pro and Aze forms of the polymers exhibit marked differences in pH-dependent solubility. Both polymers precipitate when the glutamate side chains are acid titrated, but partial replacement of proline with Aze results in a shift of the cloud point from pH 5.5 for the -(AlaGly)$_3$ProGluGly- polymers to pH 6.5 for those containing Aze.

Beta-sheet structure was indeed observed in -(AlaGly)$_3$ProGluGly- polymers containing Aze as shown by the typical amide vibrational modes in the infrared spectrum

9

Table 14-5 Amino Acid Compositions of -(AlaGly)₃ProGluGly-Repeating Proteins Made in Media Supplemented with Pro or Aze

Amino Acid	Mol% (theoretical)	Mol% (observed)[a]	Mol% (observed)[b]
Glycine	34.5	35.3	33.8
Alanine	29.0	27.9	28.1
Glutamic acid	9.5	10.6	10.9
Proline	10.0[c]	10.1	7.4

[a]Media supplemented with Pro.
[b]Media supplemented with Aze.
[c]Expected for proteins made in medium containing proline.

(Fig. 14-17). The Aze substituted polymer exhibited amide I and amide II absorptions at 1629 and 1522 cm^{-1}, respectively, as expected for proteins containing β-sheet structures (Moore & Krimm, 1976) while the corresponding vibrations observed for the proline form of the polymer (1654 and 1534 cm^{-1}) are not characteristic of β-sheets, as discussed previously (Bandekar & Krimm, 1986).

The changes observed in the structure of -(AlaGly)₃ProGluGly- upon low level substitution by Aze are remarkable, in view of the fact that only four to six residues, out of a total of 148 in the polymer chain, are modified. The factors giving rise to the conformational ordering of Aze-substituted polymers are currently under study.

3,4-Dehydroproline

3,4-Dehydroproline (Dhp, **10**) provides a means of introducing alkene groups into polypeptide chains at precisely determined locations. These groups are then available for subsequent site-specific chemical modifications, including oxidations, halogenations, and so forth.

Dhp is incorporated into protein much more efficiently than Aze (Mauger & Witkop, 1966; Rosenbloom & Prockop, 1970; Smith et al., 1962; Tristram & Thurston, 1966). In our experience, analysis of amino acid composition and ^{1}H NMR spectra gives levels of incorporation ranging from 90 to 100% (Deming et al., unpublished results).

Dhp in monomer form can be treated with H_2O_2 or with Br_2 in formic acid to produce 3,4-dihydroxyproline or 3,4-dibromoproline (Buku et al., 1980; Nakajima & Volcani, 1969), respectively, in virtually quantitative yields. Similar reactions can be carried out on Dhp polymers with essentially complete conversion of the Dhp residues, as indicated by the disappearance of the alkene proton resonances at 6.12 and 5.88 ppm in the ^{1}H NMR spectra.

Uniform Helical Rodlike Polymers

Poly(α,L-glutamic acid) (PLGA) and its derivatives, as produced by traditional synthetic methods, are characterized by the same kinds of broad molecular weight distributions that one finds in synthetic polymers generally (Block, 1983). In addition, PLGA

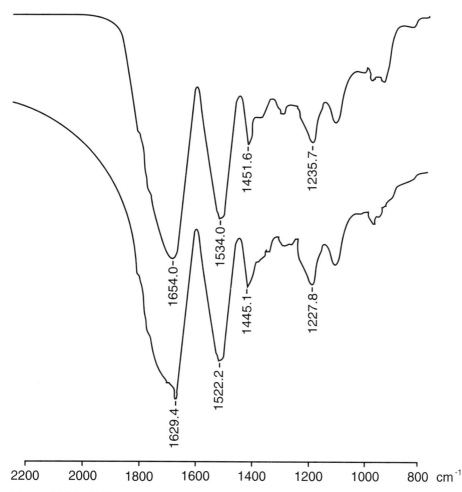

Figure 14-17. FTIR spectroscopy of unsubstituted -(AlaGly)₃ProGluGly- polymer (*top*) and -(AlaGly)₃ProGluGly- polymer containing Aze (*bottom*). Reprinted from Deming et al., 1996, with permission.

is usually synthesized in the ester form (e.g., as the benzyl ester, obtained by ring-opening polymerization of the corresponding N-carboxy-α-amino acid anhydride) and subsequent reactions that convert the ester to the parent acid can result in substantial racemization (Block, 1983). PLGA has been used in fundamental studies of polyelectrolytes and of the helix-coil transition (Poland & Scheraga, 1970), and poly(γ-benzyl α,L-glutamate) (PBLG), the most widely studied ester of PLGA, forms liquid crystalline solutions (Horton et al., 1990) and ordered monolayer films (McMaster et al., 1991). Because the heterogeneity of these materials complicates studies of their physical properties, production of monodisperse PLGA and various derivatives should provide significant new insights into the behavior of helical, rodlike macromolecules. Our own interest is directed in large part toward the development of novel smectic mesophases based on monodisperse PBLG.

We have used the biosynthetic strategy described in previous sections to prepare monodisperse derivatives of PLGA (Zhang et al., 1992). An oligonucleotide sequence encoding seventeen glutamic acid residues followed by a single aspartic acid unit, was synthesized using phosphoramidite chemistry (McBride & Caruthers, 1983). Codons for aspartic acid were included in the coding sequence to provide recognition and cleavage sites for the restriction endonuclease, *BbsI*, which was used to isolate the coding sequence after amplification. Although it is possible to devise coding sequences that use only glutamic acid codons, incorporation of flanking restriction sites into such a sequence would require that a *single* codon (either GAA or GAG) be used to construct the remainder of the oligonucleotide. We preferred to avoid such a design because of concern about genetic instability.

A population of multimers of the oligonucleotide sequence was produced in *E. coli* strain DH5αF', and a tetramer of this sequence was isolated and inserted into the expression vector, pGEX-3X, which yields the target protein as a fusion with a 26 kD fragment of glutathione-S-transferase at its amino terminus. The protein was purified by affinity chromatography on a glutathione-linked matrix, and subjected to CNBr cleavage to remove the GST fragment. Analysis by electrophoresis under nondenaturing conditions revealed a single tight protein band. Comparison of PLGA produced via the biosynthetic route, with commercially available samples of PLGA produced by conventional chemical synthesis (Zhang et al., 1992), is striking and serves as a powerful example of the precision of the biosynthetic strategy in the production of monodisperse polymeric materials. Studies of the benzylation of PLGA, and of the assembly behavior of the resulting poly(γ-benzyl α,L-glutamate)s, are underway.

Hybrid Artificial Proteins: Combining Materials Properties with Biological Function

Naturally occurring proteins often have two or more domains that are responsible for different activities such as recognition, binding, and catalysis (Baneyx et al., 1990; Hunger et al., 1990). Using the biosynthetic approach described in this chapter, it is possible to extend this idea to the construction of hybrid artificial proteins that combine the materials properties of synthetic polymers and the biological functions of natural proteins.

We have expressed hybrid proteins composed of alanylglycine-rich leader sequences ($\{(AlaGly)_3GluGly\}_{36}$ (Dong et al., 1994) or $\{(AlaGly)_3ProGluGly\}_{16}$) (Dong et al., 1996) attached at the N-terminus of the phosphotriesterase of *Pseudomonas diminuta*, which catalyzes efficient hydrolysis of organophosphorus compounds used as nerve agents and pesticides (Dumas & Raushel, 1989). The function of the artificial domains

Table 14-6 Catalytic Properties of Hybrid Protein and Phosphotriesterase

Parameter	Hybrid Protein	Phosphotriesterase
Specific activity (units/mg protein)	1976	7434
Michaelis constant (μM)	106	155
Catalytic rate constant (s^{-1})	2315	6540

^aOne unit of enzyme activity will hydrolyze 1 mmol of paraoxon per minute.

in these hybrids is to provide a simple and efficient means of immobilizing the phosphotriesterase on glass or on other basic surfaces. "Self-immobilizing" enzymes of this type might be expected to find application in detection of pesticides and in detoxification of nerve agents.

Polynucleotides encoding the artificial domains were constructed as described previously and ligated at the 5'-end of a plasmid-borne DNA sequence encoding the phosphotriesterase. The recombinant genes were expressed in *E. coli* and the hybrid proteins were purified by metal chelate affinity chromatography, taking advantage of an oligohistidine sequence encoded at the amino terminus.

Enzyme activity was detected in whole cell lysates and in soluble fractions subjected to metal chelate affinity chromatography, by monitoring the hydrolysis of the pesticide paraoxon to diethyl phosphate and the *p*-nitrophenolate anion, which absorbs strongly at 400 nm.

When cell lysates or purified proteins are loaded onto a DEAE-Sephadex anion-exchange column, the hybrid protein binds to the resin via the acidic, repetitive polypeptide domain. If paraoxon is then added to the column, the characteristic yellow color of the *p*-nitrophenolate anion is observed. Similar treatment of the unmodified phosphotriesterase reveals no activity on the column, as the enzyme has no strong affinity for the cationic resin. This experiment illustrates the functions of each of the two domains in the hybrid protein: binding activity arising from the artificial leader sequence, and catalytic activity from the phosphotriesterase domain.

The catalytic activity of the hybrid protein carrying the $\{(AlaGly)_3ProGluGly\}_{16}$ leader sequence is compared to that of unmodified phosphotriesterase in Table 14-6. While the specific activity and the catalytic rate constant of the hybrid protein are lower than those of the unmodified enzyme, the hybrid protein also has a lower Michaelis constant, indicating a higher affinity for substrate than that of the free phosphotriesterase. The pH profiles of catalytic activity of the hybrid protein and the unmodified phosphotriesterase are also very similar in the range of pH 6 to pH 11. Previous studies have shown that the bacterial phosphotriesterase has a pK_a of 6.1, indicating, together with chemical modification studies, that catalytic activity depends on the ionization of a histidine residue (Dumas & Raushel, 1990). We are currently exploring the immobilization of hybrid phosphotriesterases onto optical fibers for use as biosensors.

Summary

Protein engineering is opening important new opportunities in the design and synthesis of new kinds of polymeric materials and materials systems. By using a protein engineering approach, one can prepare new macromolecules with precise control of chain length, sequence and stereochemistry, and thereby control conformational properties and assembly behavior. Progress has been made in the engineering of macromolecular crystals (the "solid-state protein-folding problem"), and the prospects appear excellent for the design of new kinds of liquid crystal phases and surface arrays. The capacity of bacterial cells to accommodate nonnatural amino acids as substrates for protein biosynthesis extends the scope of the method, and has already led to artificial proteins with unusual surface properties and electrochemical behavior. Finally, the combination of artificial and natural protein domains provides a route to "hybrid proteins" in which both materials properties and biological function are subject to control.

References

Chapter 1

Anfinsen, C. B. (1973) *Science 181*, 223–230.
Dixon, M. & Webb, E. C. (1979) *Enzymes*, 3rd ed. Academic Press, New York.
Fasman, G. D. (1989) *Trends Biochem. Sci. 14*, 295–299.
Söll, D. (1988) *Nature 331*, 662–663.
Webb, E. C., preparer (1992), Enzyme Nomenclature 1992: Recommendations of the Nomenclature Committee of the IUBMB. Academic Press, New York.

Chapter 2

Akaji, K., Kuriyama, N., Kimura, T., Fujiwara, Y., & Kiso, Y. (1993) in *Peptides 1992* (Schneider, C. H. & Eberle, A. N., Eds.) pp 220–221, ESCOM Sci. Publ. B.V., Leiden.
Akiyama, M., Hasegawa, M., Takeuchi, H., & Shimizu, K. (1979) *Tetrahedron Lett. 28*, 2599–2602.
Al-Obeidi, F., Castrucci, A.-M. L., Hadley, M. E., & Hruby, V. J. (1989) *J. Med. Chem. 32*, 2555–2561.
Albericio, F. & Barany, G. (1984) *Int. J. Pept. Prot. Res. 23*, 342–349.
Albericio F. & Barany, G. (1985) *Int. J. Pept. Prot. Res. 26*, 92–97.
Albericio, F., Kneib-Cordonier, N., Biancanala, S., Gera, L., Masada, R. I. Hudson, D., & Barany, G. (1990) *J. Org. Chem. 55*, 3730–3743.
Anderson, G. W. (1970) in *Peptides: Chemistry and Biochemistry* (Weinstein, B., & Lande, S., Eds.) pp 255–266, Marcel Dekker, New York.
Anderson, G. W. & Callahan, F. M. (1958) *J. Am. Chem. Soc. 80*, 2902–2903.
Anderson, G. W., Callahan, F. M., & Zimmerman, J. E. (1967) *J. Am. Chem. Soc. 89*, 5012–5017.
Anderson, G. W. & McGregor, A. C. (1957) *J. Am. Chem. Soc. 79*, 6180–6183.
Anwer, M. K., Spatola, A. F., Bossinger, C. D., Flanigan, E., Liu, R. C., Olsen, D. B., & Stevenson, D. (1983) *J. Org. Chem. 48*, 3503–3507.
Arshady, R., Atherton, E., Gait, M. J., Lee, K., & Sheppard, R. C. (1979) *J. Chem. Soc., Chem. Commun.*, 423–425.
Atherton, E. & Sheppard, R. C. (1989) *Solid Phase Peptide Synthesis. A Practical Approach*, Oxford Univ. Press, Oxford.
Atherton, E., Glive, D. L. J., & Sheppard, R. C. (1975) *J. Am. Chem. Soc. 95*, 6584–6585.
Atherton, E., Caviezel, M., Over, H., & Sheppard, R. C. (1977) *J. Chem. Soc., Chem. Commun.*, 819–821.
Atherton, E., Brown, E., & Sheppard, R. C. (1981) *J. Chem. Soc., Chem. Commun.*, 1151–1152.
Bailey, J. L. (1950) *J. Chem. Soc.*, 3461–3466.
Barany, G., & Merrifield, R. B. (1979) in *Peptides* (Gross, E. & Meienhofer, J., Eds.) Vol. 2, pp 1–284, Academic Press, New York.
Barany, G., Kneib-Cordonier, N., & Mullen, D. G. (1987) *Int. J. Pept. Prot. Res. 30*, 705–739.
Barany G., Kneib-Cordonier, N., & Mullen, D. G., (1988) in *Encyclopedia of Polymer Science and Engineering*, Vol. 12, Second Ed., pp 811–858, Wiley, New York.

Barany, G. & Albericio, F. (1985) *J. Am. Chem. Soc. 107*, 4936–4942.
Barany, G. & Merrifield, R. B. (1977) *J. Am. Chem. Soc. 99*, 7363–7365.
Barany, G. & Merrifield, R. B. (1979) *Anal. Biochem., 95*, 160–166.
Barany, G. & Merrifield, R. B. (1980) *J. Am. Chem. Soc., 102*, 3084–3095.
Bayer, E., Dengler, M., & Hemmasi, B. (1985) *Int. J. Pept. Prot. Res. 25*, 178–186.
Beck-Sickinger, A. G., Dürr, H., & Jung, G. (1991) *Pept. Res. 4*, 88–94.
Berthot, J.-N., Loffet, A., Pinel, C., Reuther, F., & Sennyey, G. (1991) *Tetrahedron Lett., 32*, 1303–1306.
Birr, C. (1973) *Liebigs Ann. Chem.*, 1652–1662.
Birr, C. (1978) *Aspects of the Merrifield Peptide Synthesis*, Springer, Heidelberg.
Birr, C. (1990) in *Innovations and Perspectives in Solid Phase Synthesis* (Epton, R., Ed.) pp 155, SPCC Ltd, UK.
Blankemeyer-Menge, B., & Frank, R. (1988) *Tetrahedron Lett. 29*, 5871–5874.
Block, H., & Cox, M. F. (1963) in *Peptides, Proc. 5th European Symposium*, (Young, G. T., Ed.) pp 83–87, MacMillan, New York.
Bodanszky, M. (1956) *Acta Chim. Acad. Sci. Hung. 10*, 335–346.
Bodanszky, M. (1979) in *The Peptides: Analysis, Synthesis, Biology* (Gross, E. & Meienhoffer, J., Eds.) Vol. 1, pp 105–196, Academic Press, New York.
Bodanszky, M. (1993) *Principles of Peptide Synthesis*, Second Ed., Springer, New York.
Bodanszky, M., Klausner, Y.-S., & Ondetti, M. A. (1976) *Peptide Synthesis*, Second Ed., Wiley, New York.
Boisonnas, R.A. (1951) *Helv. Chim. Acta 34*, 874–879.
Bourdel, E., Califano, J. C., Llinares, M., Loffet, A., Rodriguez, M., & Martinez, J. (1993) *Poster Presented at the 13th American Peptide Symposium*, Edmonton, Canada.
Bouvier, M. & Taylor, J. W. (1992) *J. Med. Chem. 35*, 1145–1155.
Bray, A. M., Maeji, N. J., & Geysen, H. M. (1990) *Tetrahedron Lett. 31*, 5811–5814.
Bray, A. M., Maeji, N. J., Jhingran, A. G., & Valerio, R. M. (1991a) *Tetrahedron Lett. 32*, 6163–6166.
Bray, A. M., Maeji, N. J., Valerio, R. M., Campbell, R. A., & Geysen, H. M. (1991b) *J. Org. Chem. 56*, 6659–6666.
Brown, T. & Jones, J. H. (1981) *J. Chem. Soc. Chem. Commun.*, 648–649.
Brown, T., Jones, J. H., & Richards, J.D. (1982) *J. Chem. Soc., Perkins Trans. 1*, 1553–1561.
Butwell, F. G. W., Haws, E. J., & Epton, R. (1988) *Makromol. Chem. Macromol. Symp. 19*, 69–77.
Carpino, L. A., Chao, H. G., Beyermann, M., & Bienert, M. (1991) *J. Org. Chem. 56*, 2635–2642.
Carpino, L. A., Cohen, B. J., Stephens, Jr., K. E., Sadat-Aalaee, S. Y., Tien, J.-H., & Langridge, D. C. (1986) *J. Org. Chem. 51*, 3732–3734.
Carpino, L. A. & Han, G. Y. (1970) *J. Am. Chem. Soc. 92*, 5748–5749.
Carpino, L. A. & Han, G. Y. (1972) *J. Org. Chem. 37*, 3404–3409.
Carpino, L. A., Mansour, M. E., & Sadat-Alalaee, D. (1991) *J. Org. Chem. 56*, 2611–2614.
Castro, B. & Dormoy, J. R. (1972) *Tetrahedron Lett. 11*, 4747–4750.
Castro, B. & Dormoy, J. R. (1973a) *Tetrahedron Lett. 12*, 3243–3246.
Castro, B. & Dormoy, J. R. (1973b) *Bull. Soc. Chim. France*, 3359–3361.
Castro, B., Dormoy, J. R., Dourtoglou, B., Evin, G., Selve, C., & Ziegler, J. C. (1976) *Synthesis*, 751–752.
Castro, B., Dormoy, J. R., Evin, G., & Selve, C. (1975) *Tetrahedron Lett. 14*, 1219–1222.
Castro, B., Dormoy, J. R., Evin, G., & Selve, C. (1977) *J. Chem. Res.* 177–179.
Castro, B., Evin, G., Selve, C., & Seyer, R. (1977) *Synthesis*, 413–415.
Castro, B. & Selve, C. (1971) *Bull. Soc. Chim. France*, 2296–2298.
Chang, C.-D. (1980) *Int. J. Pept. Prot. Res. 15*, 485–494.
Chang, C.-D., Waki, M., Ahmad, M., Meienhofer, J., Lundell, E. O., & Haug, J. D. (1980) *Int. J. Pept. Prot. Res. 15*, 59–66.
Colombo, R. (1981) *J. Chem. Soc., Chem. Commun.*, 1012–1013.

Coste, J., Dufour, M.-H., Pantaloni, A., & Castro, B. (1991a) *Tetrahedron Lett. 31*, 669–672.
Coste, J., Frérot, E., Jouin, P., & Castro, B. (1991b) *Tetrahedron Lett. 32*, 1967–1970.
Coste, J., Le Nguyen, D., & Castro, B. (1990) *Tetrahedron Lett. 31*, 205–208.
Cox, M. T., Heaton, D. W., & Horbury, J. (1980b) *J. Chem. Soc., Chem. Commun.*, 799–800.
Cox, M. T., Gormley, J. J., Hayward, C. F., & Petter, N. N. (1980a) *J. Chem. Soc., Chem. Commun.*, 800–802.
Coy, D. H., Heinz-Erian, P., Jiang, N.-Y., Sasaki, Y., Taylor, J., Moreau, J.-P., Wolfrey, W. T., Gardner, J. D., & Jensen, R. T. (1988) *J. Biol. Chem. 263*, 5056–5060.
Curtius, T. (1902) *Chem. Ber. 35*, 3226–3228.
Dangles, O., Guibé, F., Balavoine, G., Lavielle, S., & Marquet, A. (1987) *J. Org. Chem. 52*, 4984–4993.
Daniels, S. B., Bernatowicz, M. S., Coull, J. M., & Köster, H. (1989) *Tetrahedron Lett. 30*, 4345–4348.
DeGrado, W. F. & Kaiser, E. T. (1980) *J. Org. Chem. 45*, 1295–1300.
DeGrado, W. F. & Kaiser, E. T. (1982) *J. Org. Chem. 47*, 3258–3261.
DeTar, D. F., Silverstein, R., & Rogers, Jr., F. F. (1966) *J. Am. Chem. Soc. 88*, 1024–1030.
Doulut, S., Rodriguez, M., Lugrin, D., Vecchini, F., Kitabgi, P., Aumelas, A., & Martinez, J. (1992) *Peptide Res. 5*, 30–38.
Dourtoglou, V., Gross, B., Lambropoulou, V., & Zioudrou, C. (1984) *Synthesis*, 572–574.
Dower, W. J., & Fodor, S. P. A. (1991) *Annu. Rep. Med. Chem. 26*, 271–280.
Dryland, A., & Sheppard, R. C. (1986) *J. Chem. Soc., Perkin Trans. 1*, 125–137.
Dutta, A. S., Furr, B. J. A., Giles, M. B., Valcaccia, B., & Walpole, A. L. (1978) *Biochem. Biophys. Res. Commun. 81*, 382–390.
Dutta, A. S., Gormley, J. J. Hayward, C. F., Morley, J. S., Shaw, J. S., Stacey, G. J., & Turnbull, M. T. (1977) *Life Sci. 21*, 559–562.
Erickson, B. W., & Merrifield, R. B. (1976) in *The Proteins*, Third Ed., Vol. 2, (Neurath, H. & Hill, R. L., Eds.) pp 255–577, Academic Press, New York.
Fehrentz, J.-A. & Castro, B. (1983) *Synthesis*, 676–678.
Felix, A. M., Wang, C. T., Heimer, E. P., & Fournier, A. (1988) *Int. J. Pept. Prot. Res. 31*, 231–238.
Fields, C. G., Lloyd, D. H., MacDonald, R. L., Ottenson, K. M., & Noble, R. L. (1991) *Pept. Res. 4*, 95–101.
Fields, G. B. & Noble, R. L. (1990) *Int. J. Pept. Prot. Res. 35*, 161–214.
Findeis, M. A. & Kaiser, E. T. (1989) *J. Org. Chem. 54*, 3478–3482.
Fisher, E. & Otto, E. (1903) *Chem. Ber. 36*, 2106–2117.
Fodor, S. P. A., Read, J. L., Pirrung, M. C., Stryer, L., Lu, A. T., & Solas, D. (1991) *Science 251*, 767–773.
Forest, M., & Fournier, A. (1990) *Int. J. Pept. Prot. Res. 35*, 89–94.
Four, P. & Guibé, F. (1982) *Tetrahedron Lett. 23*, 1825–1828.
Fournier, A., Danho, W., & Felix, A. M. (1989) *Int. J. Pept. Prot. Res. 33*, 133–139.
Fournier, A., Wang, C. T., & Felix, A. M. (1988) *Int. J. Pept. Prot. Res. 31*, 86–97.
Frérot, E., Coste, J., Pantaloni, A., Durfour, M.-H., & Jouin, P. (1991) *Tetrahedron Lett. 47*, 259–270.
Fujii, N., Hayashi, Y., Akaji, K., Shimamura, M., Yuguchi, S., Lazarus, L. H., & Yajima, H. (1987) *Chem. Pharm. Bull. 35*, 1266–1269.
Fuller, W. D., Cohen, M. P., Shabankareh, M., Blair, R. K., Goodman, M., & Naider, F. R. (1990) *J. Am. Chem. Soc. 112*, 7414–7416.
Fuller, W. D., Krotzer, N. J., Naider, F. R., Xue, C.-B., & Goodman, M. (1993a) in *Peptides 1992* (Schneider, C. H. & Eberle, A. N., Eds.) pp 229–230, ESCOM Sci. Publ. B.V., Leiden.
Fuller, W. D., Krotzer, N. J., Swain, P. A., Anderson, B. L., Comer, D., & Goodman, M. (1993b) in *Peptides 1992* (Schneider, C. H. & Eberle, A. N. Eds.) pp 231–232, ESCOM Sci. Publ. B.V., Leiden.

Furka, A., Sebestyen, F., Asgedom, M., & Dibo, G. (1988) *Abst. 14th Int. Congr. Biochem. 5*, 47.
Furka, A., Sebestyen, F., Asgedom, M., & Dibo, G. (1991) *Int. J. Pept. Prot. Res. 37*, 487–493.
Furr, B. J. A., Valcaccia, B. E., & Hutchinson, F. G. (1983) *Brit. J. Cancer 48*, 140–146.
Gacel, G., Zajac, J.-M., Delay-Goyet, P., Daugé, V., & Roques, B. P. (1988) *J. Med. Chem. 31*, 374–383.
Gante, J. (1989) *Synthesis*, 405–413.
Geiger, R. & König, W. (1981) in *The Peptides: Analysis, Synthesis, Biology* (Gross, E. & Meienhofer, J., Eds.) Vol. 3, pp 31–32, Academic Press, New York.
Geysen, H. M., Barteling, S., & Meloen, R., (1985) *Proc. Natl. Acad. Sci. U.S.A. 82*, 178–182.
Geysen, H. M., Meloen, R., & Barteling, S. (1984) *Proc. Natl. Acad. Sci. U.S.A. 81*, 2998–4002.
Giralt, E., Albericio, F., Andreu, D., Eritja, R., Martin, P., & Pedroso, E. (1981) *An. Quim. 77*, 120–126.
Giralt, E., Albericio, F., Pedroso, E., Granier, C., & Van Rietschoten, J. (1982) *Tetrahedron 38*, 1193–1208.
Gish, D. T., Katsoyannis, P. G., Hess, G. P., & Stedman, R. J. (1956) *J. Am. Chem. Soc. 75*, 5954–5955.
Gobbo, M., Biondi, L., Filira, F., & Rocchi, R. (1991) *Int. J. Pept. Prot. Res. 38*, 417–427.
Goodman, M., Spencer, J. R., Swain, P. A., Antonenko, V. V., Delaet, N. G., Naider, F. R., Xue, C.-B., & Fuller, W. D. (1993) in *Peptides 1992* (Schneider, C. H., & Eberle, A. N., Eds.) pp 29–31, ESCOM Sci. Publ. B.V., Leiden.
Gross, E. & Meienhofer, J., Eds. (1981) *The Peptides: Analysis, Synthesis, Biology* Vol. 3, Academic Press, New York,
Guibé, F., Dangles, O., & Balavoine, G. (1986a) *Tetrahedron Lett. 27*, 2365–2368.
Guibé, F., Dangles, O., Balavoine, G., & Loffet, A. (1986b) *Tetrahedron Lett. 30*, 2641–2644.
Guibé, F., & Saint M'Leux, Y. (1981) *Tetrahedron Lett. 22*, 3591–3594.
Guichard, G., Briand, J.-C., & Friede, M. (1993) *Peptide Res. 6*, 121–124.
Gutte, B. & Merrifield, R. B. (1971) *J. Biol. Chem. 246*, 1922–1941.
Hann, M. M., Sammes, P. G., Kennewell, P. D., & Taylor, J. B. (1982) *J. Chem. Soc., Perkin Trans. 1* 307–314.
Herranz, R., Syarez-Gea, M. L., Vinuesa, S., Garcia-Lopez, M. T., & Martinez, A. (1991) *Tetrahedron Lett. 32*, 7579–7582.
Hess, H. J., Moreland, W. T., & Laubach, G. D. (1963) *J. Am. Chem. Soc. 85*, 4040–4041.
Hey, H. & Arpe, H. J. (1973) *Angew. Chem. Int. Ed. Engl. 12*, 928–929.
Hill, P. S., Smith, D. D., Slaninova, J., & Hruby, V. J. (1990) *J. Am. Chem. Soc., 112*, 3110–3113.
Hirschmann, R., Strachan, R. G., Schwam, H., Schoenewaldt, E. F., Joshua, H., Barkemeyer, B., Veber, D. F., Paleveda, W. J., Jacob, T. A., Beesley, T. E., & Denkewalter, R. G. (1967) *J. Org. Chem. 32*, 3415–3425.
Hiskey, R. G. (1981) in *The Peptides*, Vol. 3, (Gross, E. & Meienhofer, J., Eds.) pp 137–167, Academic Press, New York.
Ho, P. T., Chang, D., Zhong, J., & Musso, G. F. (1993) *Peptide Res. 6*, 10–12.
Ho, T. L., Nestor, J. J., McCrae, G. I., & Vickery, B. H. (1984) *Int. J. Pept. Prot. Res. 24*, 79–83.
Hocart, S. J., Murphy, W. A., & Coy, D. H. (1990) *J. Med. Chem. 33*, 1954–1958.
Hofmann, K., Woolner, M. E., Sp̧hler, G., & Schwartz, E. T. (1958) *J. Am. Chem. Soc. 80*, 1486–1489.
Holladay, M. W. & Rich, D. H. (1983) *Tetrahedron Lett. 24*, 4401–4404.
Houghten, R. A., (1985) *Proc. Natl. Acad. Sci. U.S.A. 82*, 5131–5135.
Houghten, R. A. Pinilla, C., Bondelle, S. E., Appel, J. R., Dooley, C. T., & Cuervo, J. H., (1991) *Nature 354*, 84–86.
Hruby, V. J. (1996) in *The Practice of Medicinal Chemistry* (Wermuth, C. G., Ed.) Academic Press, London, pp 135–151.
Hruby, V. J., Al-Obeidi, F., Sanderson, D. G., & Smith, D. D. (1990) in *Innovations and Perspectives in Solid Phase Synthesis* (Epton, R., Ed.) pp 197–203, SPCC Ltd, Birmingham, UK.

Hruby, V. J., Wilke, S., & Al-Obeidi, F. (1997, in preparation)
Huse, W. D., Sastry, L., Iverson, S. A., Kang, A. S., Alting-Mees, M., Burton, D. R., Benkovic, S. J., & Lerner, R. A. (1989) *Science 246*, 1275–1281.
Itoh, M., Hagiwara, D., & Kamiya, T. (1974) *Tetrahedron Lett.*, 4393–4394.
Jones, D. A. (1977) *Tetrahedron Lett.*, 2853–2856.
Jones, J. H. (1991) *The Chemical Synthesis of Peptides*, Oxford Univ. Press, Oxford.
Jung, G., & Beck-Sickinger, A. G. (1992) *Angew. Chem. Int. Ed. Engl. 31*, 367–377.
Katakai, R. G. (1975) *J. Org. Chem. 40*, 2697–2702.
Kawasaki, A. K., Knapp, T. H., Walton, A., Wire, W. S., Hashimoto, S., Yamamura, H. I., Porreca, F., Burks, T. F., & Hruby, V. J. (1993) *J. Med. Chem. 36*, 750–757.
Kemp, D. S. (1979) in *The Peptides: Analysis, Synthesis, Biology* (Gross, E. & Meienhofer, J., Eds.) Vol. 1, pp 342–350, Academic Press, New York.
Kimura, T., Chino, N., Kumaguye, S., Kuroda, N., Emura, J., & Sakakibara, S. (1991) *Biochem. Soc. Trans. 18*, 1307–1309.
King, D. S., Fields, C. G., & Fields, G. B. (1990) *Int. J. Pept. Prot. Res. 36*, 255–266.
Kisfaludy, L., Ceprini, M. Q., Rakoczy, B., & Kovacks, J. (1967) in *Peptides, Proc. of 8th European Peptide Symposium* (Beyerman, H. C., van de Linde, A., & Maassen van den Brink, W., Eds.) pp 25–27, North Holland Publ., Amsterdam.
Kisfaludy, L., Roberts, J. E., Johnson, R. H., Mayer, G. L., & Kovacs, J. (1970) *J. Org. Chem. 35*, 3563–3565.
Kneib-Cordonier, N., Albericio, F., & Barany, G. (1990) *Int. J. Pept. Prot. Res. 35*, 527–538.
Knorr, R., Trzeciak, A., Bannwarth, W., & Gillessen, D. (1989) *Tetrahedron Lett. 30*, 1927–1930.
König, W. & Geiger, R. (1970) *Chem. Ber. 103*, 788–798.
Kricheldorf, H. R. (1977) *Makromol. Chem. 178*, 905–939.
Kricheldorf, H. R. (1987) in *α-Aminoacid-N-Carboxy-Anhydrides and Related Heterocycles Syntheses, Properties, Peptide Synthesis, Polymerization*, pp 22–23, Springer, Berlin.
Kunz, H. (1987) *Angew. Chem. Int. Ed. Engl. 26*, 294–308.
Kunz, H., & Dombon, B. (1988) *Angew. Chem. Int. Ed. Engl. 27*, 711–712.
Kuroda, H., Chen, Y.-N., Kimura, T., & Sakakibara, S. (1992) *Int. J. Pept. Prot. Res. 40*, 294–299.
Kuroda, H., Chen, Y.-N., Watanabe, T. X., Kimura, T., & Sakakibara, S. (1993) in *Peptides 1992* (Schneider, C.H. & Eberle, A. N., Eds.) pp 154–156, ESCOM Sci. Publ., Leiden.
Lam, K. S., Salmon, S. E., Hersh, E. M., Hruby, V. J., Al-Obeidi, F., Kazmierski, W. M., & Knapp, R. J. (1992) *Peptides Chem. Biol., Proc. 12th Amer. Pep. Symp.* (Smith, J. A. & Rivier, J. E., Eds.) pp 492–494, ESCOM Sci. Publ., Leiden.
Lam, K. S., Salmon, S. E., Hersh, E. M., Hruby, V. J., Kazmierski, W. M., & Knapp, R. J. (1991) *Nature 354*, 82–84.
Le Nguyen, D., Heitz, A., & Castro, B. (1985a) *J. Chem. Soc., Perkin Trans. 1*, 1915–1919.
Le Nguyen, D., Seyer, R., Heitz, A., & Castro, B. (1985b) *J. Chem. Soc., Perkin Trans. 1*, 1025–1031.
Lloyd-Williams, P., Jpu, G., Albericio, F., & Giralt, E. (1991) *Tetrahedron Lett. 32*, 4207–4210.
Loffet, A. & Zhang, H. X. (1992) in *Innovations and Perspectives in Solid Phase Synthesis* (Epton, R., Ed.) pp 72–82, SPCC Ltd, UK.
Loffet, A. & Zhang, H. X. (1993) *Int. J. Pept. Prot. Res. 42*, 346–351.
Loffet, A., Zhang, H. X., & Sennyey, G. (1993a) in *Peptides 1992* (Schneider, C. H., & Eberle, A. N., Eds.) pp 27–28, ESCOM Sci. Publ. B.V., Leiden.
Loffet, A., Zhang, H. X., Swain, P. A., Krotzer, N. J., Fuller, W. D., & Goodman, M. (1993b) in *Peptides 1992* (Schneider, C. H. & Eberle, A. N., Eds.) pp 33–36, ESCOM Sci. Publ. B.V., Leiden.
Lukas, T. J., Prystowsdy, M. B., & Erikson, B. W. (1981) *Proc. Natl. Acad. Sci. U.S.A. 78*, 2791–2795.
Marlowe, C. K. (1993) *Bioorg. Med. Chem. Lett. 3*, 437–440.
Martinez, J., Bali, J.-P., Rodriguez, M., Castro, B., Magous, R., Laur, J., & Lignon, M.-F. (1985) *J. Med. Chem. 28*, 1874–1879.

McKay, F. C. & Albertson, N. F. (1957) *J. Am. Chem. Soc. 79*, 4686–4690.
Meienhofer, J. (1979) in *The Peptides: Analysis, Synthesis, Biology* (Gross, E. & Meienhofer, J., Eds.) Vol. 1, pp 197–314, Academic Press, New York.
Meldal, M. (1992) *Tetrahedron Lett. 33*, 3077–3080.
Mergler, M., Nyfeler, R., Tanner, R., Gosteli, J., & Grogg, P. (1988a) *Tetrahedron Lett. 29*, 4005–4008.
Mergler, M., Tanner, R., Gosteli, J., & Grogg, P. (1988b) *Tetrahedron Lett. 29*, 4009–4012.
Merrifield, R. B. (1963) *J. Am. Chem. Soc. 85*, 2149–2153.
Merrifield, R. B. (1986) *Science 232*, 341–347.
Milton, R. C., Becker, E., Milton, S. C., Baxter, J. E., & Elsworth, J. F. (1987) *Int. J. Pept. Prot. Res. 30*, 431–432.
Mitchell, A. R., Kent, S. B. H., Engelhard, M., & Merrifield, R. B. (1978) *J. Org. Chem. 43*, 2845–2852.
Miyazawa, T., Otomatsu, T., Fukui, Y., Yamada, T., & Kuwata, S. (1992) *Int. J. Pept. Prot. Res. 39*, 308–314.
Munson, M. C., Lebl, M., Slaninova, J., & Barany, G. (1993) *Peptide Research 6*, 155–159.
Nishino, N., Mihara, H., Izumi, N., Fujimoto, T., Ando, S., & Ohba, M. (1993) *Tetrahedron Lett. 34*, 1295–1298.
Olah, G. A., Nohma, M., & Kerekes, I. (1973) *Synthesis*, 487–488.
Osapay, G. & Taylor, J. W. (1990) *J. Am. Chem. Soc. 112*, 6046–6051.
Osapay, G. & Taylor, J. W. (1992) *J. Am. Chem. Soc. 114*, 6966–6973.
Pavia, M. R., Sawyer, T. K., & Moos, W. H. (1993) *Bioorg. Med. Chem. Lett. 3*, 387–396.
Pietta, P. G., Cavallo, P. F., Takahashi, K., & Marshall G. R. (1974) *J. Org. Chem. 39*, 44–48
Pipkorn, R., Beyermann, M., & Henklein, P. (1993) in *Peptides 1992* (Schneider, C. H. & Eberle, A. N., Eds.) pp 222–223, ESCOM Sci. Publ. B.V., Leiden.
Poulos, C., Pasalimaniotou, P., Manolopoulou, A., & Tsegenidis, T. (1991) *Int. J. Pept. Prot. Res. 38*, 308–313.
Qian, J.-M., Coy, D. H., Jiang, N.-Y., Gardner, J. D., & Jensen, R. T. (1989) *J. Biol. Chem. 264*, 16667–16671.
Ragnarsson, U., Karlsson, S., & Lindeberg, G. (1970) *Acta Chem. Scand. 24*, 2821–2825.
Ramage, R. & Green, J. (1987) *Tetrahedron Lett. 28*, 2287–2290.
Ressler, C. (1956) *J. Am. Chem. Soc. 78*, 5956–5957.
Rich, D. H. & Gurwara, S. K. (1975) *J. Am. Chem. Soc. 97*, 1575–1579.
Rich, D. H. & Singh, J. (1979) in *The Peptides: Analysis, Synthesis, Biology* (Gross, E., & Meienhofer, J., Eds.) Vol. 1, pp 241–261, Academic Press, New York.
Rink, H. (1987) *Tetrahedron Lett. 28*, 3787–3790.
Rivaille, P., Gautron, J. P., Castro, B., & Milhaud, G. (1980) *Tetrahedron 36*, 4313–4419.
Rodriguez, M., Aumelas, A., & Martinez, J. (1990a) *Tetrahedron Lett. 31*, 5153–5156.
Rodriguez, M., Califano, J. C., Loffet, A., & Martinez. J. (1993) in *Peptides 1992* (Schneider, C. H. & Eberle, A. N., Eds.) pp 233–234, ESCOM Sci. Publ. B.V., Leiden.
Rodriguez, M., Heitz, A., & Martinez, J. (1990b) *Tetrahedron Lett. 31*, 7319–7321.
Rodriguez, M., Lignon, M.-F., Galas, M.-C., Fulcrand, P., Mendre, C., Aumelas, A., Laur, J., & Martinez, J. (1987) *J. Med. Chem. 30*, 1366–1373.
Sakakibara, S. & Shimonishi, Y. (1965) *Bull. Chem. Soc. Jpn. 38*, 1412–1413.
Schnabel, E. (1967) *Liebigs Ann. Chem. 702*, 188–196.
Scott, J. K. (1993) *TIBS 17*, 241–243.
Scott, J. K. & Smith, G. P. (1990) *Science 249*, 386–390.
Scott, R. P. W., Zolty, S., & Chan, K. K. (1972) *J. Chromatogr. Sci. 10*, 385–391.
Semenov, A. N. & Lomonosova, I. V. (1993) *Bioorg. Khim. 19*, 182–189.
Seyer, R., Aumelas, A., Caraty, A., Rivaille, P., & Castro, B. (1990) *Int. J. Pept. Prot. Res. 35*, 465–472.
Sheehan, J. C. & Hess, G. P. (1955) *J. Am. Chem. Soc. 77*, 831–835.

Sheehan, J. C. & Hlavka, J. J. (1956) *J. Org. Chem. 21*, 439–441.
Shih, H. (1993) *J. Org. Chem. 58*, 3003–3008.
Sieber, P. (1987) *Tetrahedron Lett. 28*, 2107–2110.
Sigler, G. F., Fuller, W. D., Chaturvedi, N. C., Goodman, M., & Verlander, M. (1983) *Biopolymers 22*, 2157–2162.
Slomczynaska, U., Zabrocki, J., Kaczmarek, K., Leplwy, M. T., Beusen, D. D., & Marshall, G. R. (1992) *Biopolymers 32*, 1461–1470.
Smith, G. P. (1985) *Science 228*, 1315–1317.
Sparrow, J. T. (1976) *J. Org. Chem. 41*, 1350–1353.
Spatola, A. F. (1983) in *Chemistry and Biochemistry of Amino Acids, Peptides and Proteins* (Weinstein, B., Ed.) Vol. 7, pp 267–357, Marcel Dekker, New York.
Spatola, A. F. (1993) *Methods Neurosci. 13*, 19–42.
Spencer, J. R., Antonenko, V. V., Delaet, N. G. J., & Goodman, M. (1992) *Int. J. Pept. Prot. Res. 40*, 282–293.
Stevens, C. M. & Watanabe, R. (1950) *J. Am. Chem. Soc. 72*, 725–727.
Stewart, J. M. & Young, J. D. (1984) *Solid Phase Peptide Synthesis*, Second Ed., Pierce Chemical Co., Rockford, IL.
Steward, F. H. C. (1965) *Aust. J. Chem. 18*, 887–901.
Swain, P. A., Anderson, B. L., Goodman, M., & Fuller, W. D. (1993) *Pept. Res. 6*, 147–154.
Tam, J. P., Heath, W. F., & Merrifield, R. B. (1983) *J. Am. Chem. Soc. 105*, 6442–6455.
Tam, J. P., Heath, W. F., & Merrifield, R. B. (1986) *J. Am. Chem. Soc. 108*, 5242–5251.
Tam, J. P., Tjoeng, F. S., & Merrifield, R.B. (1980) *J. Am. Chem. Soc. 102*, 6117–6127.
Tarbell, D. S., Yamamoto, Y. & Pop, B. M. (1972) *Proc. Natl. Acad. Sci. U.S.A. 69*, 730–732.
Taylor, J. W. & Osapay, G. (1990) *Acc. Chem. Res. 23*, 338–344.
Ten Kortenaar, P. B. W., Van Dijk, B. G., Peeters, J. J., Raaben, B. J., Adams, P. J., & Tesser, G. I. (1985) *Int. J. Pept. Prot. Res. 27*, 398–400.
Tessier, M., Albericio, F., Pedroso, E., Grandas, A., Eritja, R., Giralt, E., Granier, C., & Van Rietschoten, J. (1983) *Int. J. Pept. Prot. Res. 22*, 125–128.
Tourwe, D., DeCock, E., VanMarsenille, M., Vanderauwera, L., VanBinst, G., Viville, R., Degelaen, J., & Scarso, A. (1989) *Peptides 1988*, (Jung, G. & Bayer, E., Eds.) pp 562–564, W. de Gruyter, Berlin.
Tourwe, D., Meert, D., Couder, J., Ceusters, M., Elseviers, M., VanBinst, G., Toth, G., Burks, T. F., Schook, J. F., & Yamamura, H. (1990) *Peptides: Chemistry, Structure and Biology*, (Rivier, J. E. & Marshall, G. R., Eds.) pp 331–335, ESCOM Sci. Publ., Leiden.
Trost, B. M. (1980) *Acc. Chem. Res. 13*, 385–393.
Tsuji, J. (1980) in *Organic Synthesis with Palladium Compounds*, pp 125–132, Springer, Berlin.
Ulm, E. H., Hichens, M., Gomez, H. J., Till, A. E., Hand, E., Vassil, T. C., Biollaz, J., Brunner, H. R., & Schelling, J. L. (1982) *Brit. J. Clin. Pharmacol. 14*, 357–362.
Vaughan, Jr., J. R. (1951) *J. Am. Chem. Soc. 73*, 3547–3548.
von dem Bruch, K. & Kunz, H. (1990) *Angew. Chem. Int. Ed. Engl. 29*, 1457–1459.
Wang, S.-S. (1973) *J. Am. Chem. Soc. 95*, 1328–1333.
Wang, S.-S. & Kulesha, I. D. (1975) *J. Org. Chem. 40*, 1227–1234.
Wang, S.-S., Tam, J. P., Wang, B. S. H., & Merrifield R. B. (1981) *Int. J. Peptide Protein Res. 18*, 459–467.
Wang, S.-S., Yang, C. C., Kulesha, I. D., Sonenberg, M., & Merrifield, R. B. (1974) *Int. J. Peptide Protein Res. 6*, 103–109.
Wieland, T. & Bernhard, H. (1951) *Liebigs Ann. Chem. 572*, 190–194.
Wieland, T., Heinke, B., Vogeler, K., & Morimoto, H. (1962) *Liebigs Ann. Chem. 655*, 189–194.
Wünsch, E. & Drees, F. (1966) *Chem. Ber. 99*, 110–120.
Xaus, N., Albericio, F., Jorba, X., Clapés, P., Torres, J.-L., & Valencia, G. (1992) *Int. J. Pept. Prot. Res. 39*, 528–532.
Xue, C.-B. & Naider, F. (1993) *J. Org. Chem. 58*, 350–355.
Zhang, H. X., Guibé, F., & Balavoine, G. (1988) *Tetrahedron Lett. 29*, 623–626.

Chapter 3

Alpert, A. J. (1991) in *High-Performance Liquid Chromatography of Peptides and Proteins: Separation, Analysis, and Conformation* (Mant, C. T. & Hodges, R. S., Eds.) pp. 187–194, CRC Press, Boca Raton, FL.
Amati, B., Brooks, M. W., Levy, N., Littlewood, T. D., Evan, G. I., & Land, H. (1993) *Cell 72*, 233–245.
Anfinsen, C. B. (1973) *Science 181*, 223–230.
Baca, M. & Kent, S. B. H. (1992) *J. Am. Chem. Soc. 114*, 3992–3993.
Baca, M. & Kent, S. B. H. (1993) *Proc. Natl. Acad. Sci. U.S.A. 90*, 11638–11642.
Baca, M., Muir, T. W., Schnolzer, M., & Kent, S. B. H. (1995) *J. Am. Chem. Soc. 117*, 1881–1887.
Bayer, E. (1991) *Angew. Chem. 30*, 113–129.
Berman, A. L., Kolker, E., & Trifonov, E. N. (1994) *Proc. Natl. Acad. Sci. U.S.A. 91*, 4044–4047.
Branden, C. & Tooze, J. (1991) *Introduction to Protein Structure*, Garland Publishing, New York.
Brenner, M. (1967) in *Peptides. Proceedings of the Eigth European Peptide Symposium* (Beyerman, H. C., Ed.) pp. 1–7, North Holland, Amsterdam.
Canne, L. E., Bark, S. J., & Kent, S. B. H. (1996) *J. Am. Chem. Soc. 118*, 5891–5896.
Canne, L. E., Ferre-D'Amare, A. R., Burley, S. K., & Kent, S. B. H. (1995) *J. Am. Chem. Soc. 117*, 1881–1887.
Carreno, C., Carrascosa, A. L., Vinuela, E., Giralt, E., & Andreu, D. (1995) in *Peptides 1994* (Maia, H. L. S., Ed.) pp. 807–808, ESCOM Science Publishers B. V., The Netherlands.
Chait, B. T. & Kent, S. B. H (1992) *Science 256*, 1445–1448.
Chait, B. T., Wang, R., Beavis, R., & Kent, S. B. H. (1993) *Science 262*, 89–92.
Clark-Lewis, I., & Kent, S. B. H. (1989) in *The Use of HPLC in Receptor Biochemistry* (Kerlavage, A. R., Ed.), Vol. 14 *Receptor Biochemistry and Methodology* (Venter, J. C. & Harrison, L. C., Eds.) pp. 43–79, Alan R. Liss, New York.
Cohen, G. B., Ren, R., & Baltimore, D. (1995) *Cell 80*, 237–248.
Davies, D. R. (1990) *Annu. Rev. Biophys. Chem. 19*, 189–215.
Dawson, P. E., Churchill, M., Ghadiri, R., & Kent, S. B. H. (1997a) *J. Am. Chem. Soc. 119*, 4325–4329.
Dawson, P. E., Muir, T. W., Clark-Lewis, I., & Kent, S. B. H. (1994) *Science 266*, 776–779.
Dawson, P. E., Muir, T. M., Fitzgerald, M. C., & Kent, S. B. H. (1997b) *J. Am. Chem. Soc. 119*, 7917–7927.
Elder, J., Schnolzer, M., Hassellkus-Light, C. S., Henson, M., Lerner, D. A., Philips, T. R., Wagaman, P. C., & Kent, S. B. H. (1993) *Virology 67*, 1869–1876.
Englebretsen, D. R., Garnham, B. G., Bergman, D. A., & Alewood, P. F. (1995) *Tetrahedron Lett. 36*, 8871–8874.
Fenn, J. B. (1989) *Science 246*, 64–71.
Ferre D' Amare, A. R., Prendergast, G. C., Ziff, E. B., & Burley, S. K. (1993) *Nature 363*, 38–45.
Fisher, E. (1906) *Ber. Chem. Ges. 39*, 530–610.
Fitzgerald, M. C., Chernushevich, I., Standing, K. G., Kent, S. B. H., & Whitman, C. P. (1995) *J. Am. Chem. Soc. 117*, 11075–11080.
Fitzgerald, M. C., Chernushevich, I., Standing, K. G., Whitman, C. P., & Kent, S. B. H. (1996) *Proc. Natl. Acad. Sci. U.S.A. 93*, 6851–6856.
Fotouhi, N., Gallackalos, N. G., & Kemp, D. S. (1989) *J. Org. Chem. 54*, 2806–2817.
Gooding, K. M. & Freiser, H. H. (1991) in *High-Performance Ligiud Chromatography of Peptides and Proteins: Separation, Analysis, and Conformation* (Mant, C. T. & Hodges, R. S., Eds.) pp. 135–144, CRC Press, Boca Raton, FL.
Gutte, B. & Merrifield, R. B. (1969) *J. Am. Chem. Soc. 91*, 501–502.
Hackeng, T.M., Mounier, C. M., Bon, C., Dawson, P. E., Griffin, J. H., & Kent, S. B. H. (1997) *Proc. Natl. Acad. Sci. U.S.A. 94*, 7845–7850.

Hearn, M. T. W. (1991) in *High-Performance Liquid Chromatography of Peptides and Proteins: Separation, Analysis, and Conformation* (Mant, C. T. & Hodges, R. S., Eds.) pp. 105–122, CRC Press, Boca Raton, FL.
Hendrickson, W. A. & Wuthrich, K., Eds. (1996) *Macromolecular Structures 1996*, Current Biology, London.
Hirschmann, R. (1991) *Angew. Chem., Int. Ed. Engl. 30*, 1278–1301.
Hirschmann, R., Nutt, R. F., Veber, D. F., Vitali, R. A., Varga, S. L., Jacob, T. A., Holly, F. W., & Denkewalter, R. G. (1969) *J. Am. Chem. Soc. 91*, 507–508.
Hodges, R. S., & Merrifield, R. B. (1975) *Anal. Biochem. 45*, 241–272.
Hyland, L. J., Tomaszek, T. A., & Meek, T. D. (1991b) *Biochemistry 30*, 8454–8463.
Hyland, L. J., Tomaszdek, T. A., Roberts, G. D., Carr, S. A., Magaard, V. W., Bryan, H. L., Fakkhoury, S. A., Moore, M. L., Minnich, M. D., Culp, J. S., DesJarlais, R. L., & Meek, T. D. (1991a) *Biochemistry 30*, 8441–8453.
Inui, T., Bodi, J., Kubo, S., Hishio, H., Kimura, T., Kojima, S., Maruta, H., Muramatsu, T., & Sakakibara, S. (1996) *J. Peptide Sci. 2*, 28–39.
Jaskolski, M., Tomaselli, A. G., Sawyer, T. K., Staples, D. G. Heinrikson, R. L., Schneider, J., Kent, S. B. H., & Wlodawer, A. (1991) *Biochemistry 30*, 1600–1609.
Jones, J. (1991) *The Chemical Synthesis of Peptides*, Clarendon Press, Oxford.
Kent, S. B. H. (1988) *Annu. Rev. Biochem. 57*, 957–984.
Kent, S. B. H., Alewood, D., Alewood, P., Baca, M., Jones, A., & Schnolzer, M. (1992) in *Innovations and Perspectives in Solid Phase Synthesis: Peptides, Polypeptides, and Oligonucleotides*: (Proceedings of the Second International Symposium on Innovation and Perspectives in Solid Phase Synthesis, August, 1991) (Epton, R., Ed.) pp. 1–22, Intercept, Andover, England.
Keutmann, H. T., Sauer, M. M., Hendy, G. N., O'Riordan, J. L. H., & Potts, Jr., J. T. (1978) *Biochemistry 17*, 5723–5729.
Kiyama, S., Fujii, N., Yajima, H., Moriga, M., & Takagi, A. (1984) *Int. J. Pept. Protein Res. 23*, 174–186.
Lahoz, E. G., Xu, L., Schreiberagus, N., & Depinho, R. A. (1994) *Proc. Natl. Acad. Sci. U.S.A. 91*, 5503–5507.
Liu, C. F. & Tam, J. P. (1994a) *J. Am. Chem. Soc. 116*, 4149–4153.
Liu, C. F. & Tam, J. P. (1994b) *Proc. Natl. Acad. Sci. U.S.A. 91*, 6584–6588.
Lu, W., Qasim, M. A., & Kent, S. B. H. (1996) *J. Am. Chem. Soc. 118*, 8518–8523.
Mandolini, L. J. (1978) *J. Am. Chem. Soc. 100*, 550–554.
Mann, M., Meng, C. K., & Fenn, J. B. (1989) *Anal. Chem. 61*, 1702–1708.
Merrifield, R. B. (1963) *J. Am. Chem. Soc. 85*, 2149–2154.
Merrifield, R. B. (1986) *Science 232*, 341–347.
Miller, M., Baca, M., Rao, J. K. M., & Kent, S. B. H. (1996) *J. Molec. Structure*, in press.
Miller, M., Sathyanarayana, B. K., Toth, M. V., Marshalll, G. R., Clawson, L., Selk, L., Schneider, J., Kent, S. B. H., & Wlodawer, A. (1989) *Science 246*, 1149–1152.
Milton, R. C. D., Milton, S. C. F., & Adams, P. A. (1990) *J. Am. Chem. Soc. 112*, 6039–6046.
Milton, R. C. D., Milton, S. C. F., & Kent, S. B. H. (1992) *Science 256*, 1445–1448.
Milton, R. C. D., Milton, S. C. F., Schnolzer, M., & Kent, S. B. H. (1993) in *Techniques in Protein Chemistry IV*, (Angeletti, R. H., Ed.) pp. 257–267, Academic Press, New York.
Muir, T. M., Dawson, P. E., Fitzgerald, M. C., & Kent, S. B. H. (1996) *Chemistry & Biology 3*, 817–825.
Muir, T. W. (1995) *Structure 3*, 649–652.
Muir, T. W. & Kent, S. B. H. (1993) *Curr. Opin. Biotech. 4*, 420–427.
Muir, T. W., Williams, M., & Kent, S. B. H. (1995) *Anal. Biochem. 224*, 100–109.
Oroszlan, S. (1989) in *Viral Proteinases as Targets for Chemotherapy* (Drausslich, H., Oroszllan, S., & Wimmer, E., Eds.) pp. 87–100, Cold Spring Harbor Laboratory Press, Plainview, NY.
Pearl, L. H. & Taylor, W. R. (1987) *Nature 329*, 351–354.

Rajarathnam, K., Clark-Lewis, I., & Sykes, B. D. (1994) Biochemistry *33*, 6623–6630.
Roper, D. I., Subramanya, H. S., Shingler, V., & Wigley, D. B. (1994) *J. Mol. Biol. 243*, 799–801.
Rose, K. (1994) *J. Am. Chem. Soc. 116*, 30–34.
Schneider, J. & Kent, S. B. H. (1988) *Cell 54*, 363–368.
Schnolzer, M., Alewood, P., Jones, A., Alewood, D., & Kent, S. B. H. (1992a) *Int. J. Pept. Protein Res. 40*, 180–193.
Schnolzer, M., Jones, A., Alewood, P. F., & Kent, S. B. H. (1992b) *Anal. Biochem. 204*, 335–342.
Schnolzer, M., & Kent, S. B. H. (1992) *Science 256*, 221–225.
Schnolzer, M., Rackwitz, H. R., Laco, G. S., Elder, J., & Kent, S. B. H. (1996) *Virology, 224*, 268–275.
Scopes, R. K. (1993) in *Protein Purification: Principles and Practice*, pp. 187–237, Springer, New York.
Sieber, P., Kamber, B., Hartmann, A., Johl, A., Riniker, B., & Rittel, W. (1974) *Helv. Chem. Acta 57*, 2617–2621.
Smith, R., Brereton, I. M., Chai, R. Y., & Kent, S. B. H. (1996) *Nature Structural Biology 3*, 946–950.
Stewart, J. M. & Young, J. D. (1984) *Solid Phase Peptide Synthesis, Second Ed.*, Pierce Chemical Co., Rockford, IL.
Stivers, J. T., Abeygunawardana, C., Mildvan, A. S., Hajipour, G., Whitman, C. P., & Chen, L. H. (1996a) *Biochemistry 35*, 803–813.
Stivers, J. T., Abeygunawardana, C., Mildvan, A. S., Hajipour, G., & Whitman, C. P. (1996b) *Biochemistry 35*, 814–823.
Stivers, J. T., Abeygunawardana, C., Mildvan, A. S., Hajipour, G., Whitman, C. P., & Chen, L. H. (1996c) *Biochemistry 35*, 803–813.
Swain, A. L., Miller, M., Green, J., Rich, D. H., Schneider, J., Kent, S. B. H., & Wlodawer, A. (1990) *Proc. Natl. Acad. Sci. U.S.A. 87*, 8805–8809.
Tam, J. P., Heath, W. F., & Merrifield, R. B. (1983) *J. Am. Chem. Soc. 105*, 6442–6455.
Vogel, A. I. (1989) *Vogel's Textbook of Practical Organic Chemistry, Fifth Ed.*, (Furniss, B. S., Hannaford, A. J., Smith, P. W. G., & Tatchell, A. R., Eds.) p. 12, Wiley, New York.
Wieland, T., Bokelmann, E., Bauer, L., Lang, H. U., & Lau, H. (1953) *Liebigs Ann. Chem. 583*, 129–149.
Williams, M., Muir, T. W., Ginsberg, M., & Kent, S. B. H. (1994) *J. Am. Chem. Soc. 116*, 10797–10798.
Wlodawer, A. & Erickson, J. W. (1993) *Annu. Rev. Biochem. 62*, 543–585.
Wlodawer, A., Miller, M., Jaskolski, M., Sathyanarayana, B.K., Baldwin, E., Weber, I. T., Selk, L. M., Clawson, L., Schneider, J., & Kent, S. B. H. (1989) *Science 245*, 616–621.
Wu, X., Knudsen, B., Feller, S. M., Zheng, J., Sali, A., Cowburn, D., Hanafusa, H., & Kuriyan, J. (1995) *Structure 3*, 215–226.
Yajima, H. & Fujii, N. (1981) *J. Am. Chem. Soc. 103*, 5867–5871.
Yue, K. & Dill, K. A. (1995) *Proc. Natl. Acad. Sci. U.S.A. 92*, 146–150.

Chapter 4

Abe, Y., Shirane, K., Yokosawa, H., Matsushita, H., Mitta, M., Kato, I., & Ishii, S.-I. (1993) *J. Biol. Chem. 268*, 3525–3529.
Aebersold, R., Bures, E. J., Namchuk, M., Goghari, M. H., Shushan, B., & Covey, T. (1992) *Protein Sci. 1*, 494–503.
Aebersold, R. H., Leavitt, J., Saavedra, R. A., Hood, L. E., & Kent, S. B. H. (1987) *Proc. Natl. Acad. Sci. U.S.A. 84*, 6970–6974.
Aebersold, R., Pipes, G. D., Wettenhall, R. E. H., Nika, H., & Hood, L. E. (1990) *Anal. Biochem. 187*, 56–65.
Amons, R. (1987) *FEBS Lett. 212*, 68–72.
Andrews, P. C., & Dixon, J. E. (1987) *Anal. Biochem. 161*, 524–528.

Atherton, D., Fernandez, J., & Mische, S. M. (1993) *Anal. Biochem. 212*, 98–105.
Barber, M., Bordoli, R. S., Sedgwick, R. D., & Tyler, A. N. (1981) *Nature 1981*, 270–275.
Bauw, G., Damme, J. V., Puype, M., Vandekerckhove, J., Gesser, B., Ratz, G. P., Lauridsen, J. B., & Celis, J. E. (1989) *Proc. Natl. Acad. Sci. U.S.A.. 86*, 6005–6008.
Beavis, R. C. & Chait, B. T. (1989) *Rapid Commun. Mass Spec. 3*, 432–435.
Beavis, R. C., Chaudhary, T., & Chait, B. T. (1992) *Org. Mass Spec. 27*, 156–158.
Biemann, K. (1990) *Methods Enzymol. 193*, 455–479.
Biemann, K. & Scoble, H. A. (1987) *Science 237*, 992–998.
Billeci, T. M. & Stults, J. T. (1993) *Anal. Chem. 65*, 1709–1716.
Botelho, L. H., Ryan, D. E., Yuan, P.-M., Kutney, R., Shively, J. E., & Levin, W. (1982) *Biochemistry 21*, 1152–1155.
Brauer, A. W., Oman, C. L., & Margolies, M. N. (1984) *Anal. Biochem. 137*, 134–142.
Bruins, A. P., Covey, T. R., & Henion, J. (1987) *Anal. Chem. 59*, 2642–2646.
Burkhart, W. A., Moyer, M. B., & Bodnar, W. (1995) in *Techniques in Protein Chemistry VI*, (Crabb, J. W., Ed.) Academic Press, San Diego, in press.
Calaycay, J., Rusnak, M., & Shively, J.E. (1991) *Anal. Biochem. 192*, 23–31.
Carr, S. A., Barr, J. R., Roberts, G. D., Anumula, K. R., & Taylor, P. B. (1990) *Methods Enzymol. 193*, 501–518.
Chait, B. T., Wang, R., Beavis, R. C., & Kent, S. B. H. (1993) *Science 262*, 89–92.
Crimmins, D. L., McCourt, D. W., Thoma, R. S., Scott, M. G., Macke, K., & Schwartz, B. D. (1990) *Anal. Biochem. 187*, 27–38.
Davis, M. T. & Lee, T. D. (1992) *Protein Sci.1*, 935–944.
Davis, M. T., Stahl, D. C., Hefta, S. A., & Lee, T. D. (1995) *Anal. Chem. 67*, 4549–4556.
Deeg, M. A., Humphrey, D. R., Yang, S. H., Fergerson, T. R., Reinhold, V. N., & Rosenberry, T. L. (1992) *J. Biol. Chem. 267*, 18573–18580.
Degani, Y., Neumann, H., & Patchornik, A. (1970) *J. Am. Chem. Soc. 92*, 6969–6971.
Dell, A. (1990) *Methods Enzymol. 193*, 647–660.
Doering, T. L., Masterson, W. J., Hart, G. W., & Englund, P. T. (1990) *J. Biol. Chem. 265*, 611–614.
Edge, A. B., Faltynek, C. R., Hof, L., Reichert, Jr., L. E., & Weber, P. (1981) *Anal. Biochem. 118*, 131–137.
Edman, P. & Begg, G. (1967) *Eur. J. Biochem. 1*, 80–91.
Eng, J. K., McCormack, A. L., & Yates J. R., III (1990) *J. Am. Soc. Mass Spectrom. 5*, 976–989.
Enghild, J. J., Salvesen, G., Thogerson, I. B., Valnickova, Z., Pizzo, S. V., & Hefta, S. A. (1993) *J. Biol. Chem. 268*, 8711–8716.
Falick, A.M. & Maltby, D.A. (1989) *Anal. Biochem. 182*, 165–169.
Farnsworth, V. & Steinberg, K. (1993a) *Anal. Biochem. 215*, 190–199.
Farnsworth, V. & Steinberg, K. (1993b) *Anal. Biochem. 215*, 200–210.
Fernandez, J., DeMott, M., Atherton, D., & Mische, S. (1992) *Anal. Biochem. 201*, 255–264.
Gibson, B. W. & Cohen, P. (1990) *Methods Enzymol. 193*, 480–501.
Glass, J. D. & Pelzig, M. (1978) *Biochem. Biophys. Res. Commun. 81*, 527–531.
Goodlett, D. R., Armstrong, F. B., Creech, R. J., & van Breemen, R. B. (1990) *Anal. Biochem. 186*, 116–120.
Gross, E. & Witkop, B. (1962) *J. Biol. Chem. 237*, 1856–1860.
Heinrikson, R. L. (1970) *Biochem. Biophys. Res. Commun. 41*, 967–972.
Hellman, U., Wernstedt, C., Góñez, J., & Heldin, C.-H. (1995) *Anal. Biochem. 224*, 451–455.
Hendersen, L. E., Oroszlan, S., & Konigsberg, W. (1979) *Anal. Biochem. 93*, 153–157.
Henschen, A. (1993) *Protein Sci. 2 S1*, abstract 555M.
Henzel, W., J., Billeci, T. M., Stults, J. T., Wong, S. C., Grimley, C., & Watanabe, C. (1993) *Proc. Natl. Acad. Sci. U.S.A. 90*, 5011–5015.
Hewick, R. M., Hunkapiller, M. W., Hood, L. E., & Dryer, W. J. (1981) *J. Biol. Chem. 256*, 7990–7997.
Hildebrandt, E. & Fried, V. A. (1989) *Anal. Biochem. 177*, 407–412.

Hong, H.-Y., Choi, J.-K., & Yoo, G.-S. (1993) *Anal. Biochem. 214*, 96–99.
Hunt, D. F., Yates, III, J. R. I., Shabanowitz, J., Winston, S., & Hauer, C. R. (1986) *Proc. Natl. Acad. Sci. U.S.A.. 83*, 6233–6237.
Inglis, A. S. (1991) *Anal. Biochem. 195*, 183–196.
Jauregui-Adell, J. & Marti, J. (1975) *Anal. Biochem. 69*, 468–473.
Jue, R. A. & Hale, J. E. (1993) *Anal. Biochem. 210*, 39–44.
Karas, M. & Hillenkamp, F. (1988) *Anal. Chem. 60*, 2299–2301.
Karlsson, H., Hansson, G. C., & Carlstedt, I. (1989) *Anal. Biochem. 182*, 438–446.
Kawasaki, H. & Suzuki, K.(1990) *Anal. Biochem. 186*, 264–268.
Klapper, D. G., Wilde, C. E. I, & Capra, J. D. (1978) *Anal. Biochem. 85*, 126–131.
Kurth, J. & Stoffel, W. (1990) *Hoppe-Seyler's Biol. Chem. 371*, 675–685.
Laemmli, U. K. (1970) *Nature 227*, 680–685.
Laursen, R. A. (1971) *Eur. J. Biochem. 20*, 89–102.
Lee, T. D. & Shively, J. E. (1990) *Methods Enzymol. 193*, 361–374.
Lee, T. D. & Vemuri, S. (1990) *Biomed. Environ. Mass Spec.*, 638–645.
Lee, Y. C. & Scocca, J. R. (1972) *J. Biol. Chem. 247*, 5753–5758.
Liang, S.-P. & Laursen, R. A. (1990) *Anal. Biochem. 188*, 366–373.
Mahoney, W. C. & Hermodson, M. A. (1979) *Biochemistry 18*, 3810–3814.
Mann, M., Mortensen, P., Horjup, P., & Roepstorff, P. (1993) *Proc. Am. Soc. Mass Spec. 41*, 159a–159b.
Matsuo, H., Fujimoto, Y., & Tatsuno, T. (1966) *Biochem. Biophys. Res. Commun. 22*, 69–74.
Matsudaira, P. (1987) *J. Biol. Chem. 262*, 10035–10038.
McConville, M. J., Homan, S. W., Thomas-Oates, J. E., Dell, A., & Bacic, A. (1990) *J. Biol. Chem. 265*, 7385–7394.
McFadden, P. N., & Clarke, S. (1986) *J. Biol. Chem. 261*, 11503–11511.
Meyer, H. E., Hoffman-Posorske, E., & Heilmeyer, Jr., L. M. G. (1991) *Methods Enzymol. 201*, 169–185.
Meyer, H. E., Hoffman-Posorske, E., Korte, H., & Heilmeyer, Jr., L. M. G. (1986) *FEBS Lett. 204*, 61–66.
Miyatake, N., Kamo, M, Satake, K., Uchiyama, Y., & Tsugita, A. (1993) *Eur. J. Biochem. 212*, 785–789.
Murata, H., Takao, T., Anahara, S., & Shimonishi, Y. (1993) *Anal. Biochem. 210*, 206–208.
Muramoto, K., Nokihara, K., Ueda, A., & Kamiya, H. (1993) in *Methods in Protein Sequence Analysis* (Imahori, K. & Sakiyama, F., Eds.) pp. 29–36, Plenum Press, New York.
Neufeld, E., Goren, H. J., & Boland, D. (1989) *Anal. Biochem. 177*, 138–143.
Omenn, G. S., Fontana, A., & Anfinsen, C. B. (1970) *J. Biol. Chem. 245*, 1895–1902.
Pappin, D. J. C., Coull, J. M., & Köster, H. (1990) *Anal. Biochem. 187*, 10–19.
Pappin, D. J. C., Hjrup, P., & Bleasby, A. J. (1993) *Current Biol. 3*, 327–332.
Patterson, S. D., Hess, D., Yungwirth, T., & Aebersold, R. (1992) *Anal. Biochem. 202*, 193–203.
Patthy, L. & Smith, E. L. (1975) *J. Biol. Chem. 250*, 557–564.
Paxton, R. J., Mooser, G., Pande, H., Lee, T. D., & Shively, J. E. (1987) *Proc. Natl. Acad. Sci. U.S.A. 84*, 920–924.
Reim, D. & Speicher, D (1993) *Protein Sci. 2 S1*, abstract 176S.
Reim, D. F. & Speicher, D. W. (1993) *Anal. Biochem. 214*, 87–95.
Rice, R. H., Means, G. E., & Brown, D. (1977) *Biochim. Biophys. Acta 492*, 316–321.
Roepstorff, P. & Fohlman, J. (1984) *Biomed. Mass. Spec. 11*, 601.
Rosenfeld, J., Capdevielle, J., Guillemot, J. C., & Ferrara, P. (1992) *Anal. Biochem. 203*, 173–179.
Schechter, Y., Patchornik, A., & Burstein, Y. (1976) *Biochemistry 15*, 5071–5075.
Sheer, D. G., Yamane, D. K., Hawke, D. H., & Yuan, P.-M. (1990) *Peptide Res. 3*, 97–104.
Shively, J. E. (1986) in *Methods of Protein Microcharacterization*, (Shively, J. E., Ed.) pp 41–87, Humana Press, Clifton, NJ.
Simpson, R. J., Moritz, R. L., Begg, G. S., Rubira, M. R., & Nice, E. C. (1989a) *Anal. Biochem. 177*, 221–236.

Simpson, R. J., Ward, L. D., Reid, G. E., Batterham, M. P., & Moritz, R. L. (1989b) *J. Chromatogr. 476*, 345–361.
Slattery, T. K. & Harkins, R. N. (1993) in *Techniques in Protein Chemistry IV*, (Crabb, J. W., Ed.) pp 443–452, Academic Press, San Diego.
Smith, R. D., Olivares, J. A., Nguyen, N. T., & Udseth, H. R. (1988) *Anal. Chem. 59*, 436–441.
Stone, K. L., McNulty, D. E., LoPresti, M. L., Crawford, J. M., DeAngelis, R., & Williams, K. R. (1992) in *Techniques in Protein Chemistry III*, (Hogue-Angeletti, R. Ed.) pp 23–34, Academic Press, San Diego.
Strydom, D. J. (1988) *Anal. Biochem. 174*, 679–686.
Swiderek, K. M., Pearson, C. S., & Shively, J. E. (1993) in *Techniques in Protein Chemistry III*, (Hogue-Angeletti, R., Ed.) pp 127–134, Academic Press, San Diego.
Tarr, G. E., Beecher, J. F., Bell, M. & McKean, D. J. (1978) *Anal. Biochem. 84*, 622–627.
Tarr, G. E., Black, S. D., Fujitta, V. S., & Coon, M. J. (1983) *Proc. Natl. Acad. U.S.A. 80*, 6552–6556.
Titani, K., Sasagawa, T., Resing, K., & Walsh, K. A. (1982) *Anal. Biochem. 123*, 408–412.
Totty, N. F., Waterfield, M. D., & Hsuan, J. J. (1992) *Protein Sci. 1*, 1215–1224.
Tsugita, A., Kamo, M., Jone, C. S., & Shikama, N. (1989) *J. Biochem. 106*, 60–95.
Tsunasawa, S., & Hirano, H. (1993) in *Methods in Protein Sequence Analysis*, (Imahori, K., & Sakiyama, F., Eds.), pp 45–53, Plenum, New York.
Wang, Y., Fiol, C. J., DePaoli-Roach, A. A., Bell, A. W., Hermodson, M. A., & Roach, P. J. (1988) *Anal. Biochem. 174*, 537–547.
Welinder, K.G. (1988) *Anal. Biochem. 174*, 54–64.
Wessel, D. & Flugge, U. I. (1984) *Anal. Biochem. 138*, 141–143.
Whitehouse, C. M., Dreyer, R. N., Yamashita, M., & Fenn, J. B. (1985) *Anal. Chem. 57*, 675–679.
Wold, F. & Krishna, R. G. (1993) in *Methods in Protein Sequence Analysis*, (Imahori, K., & Sakiyama, F., Eds.), pp 167–171, Plenum, New York.
Xu, Q. Y. & Shively, J. E. (1988) *Anal. Biochem. 170*, 19–30.
Yamauchi, K., Sugimae, T., & Kinoshita, M. (1977) *Tetrahedron Lett.* 13, 1199–1202.

Chapter 5

Abad-Zapatero, C., Griffith, J. P., Sussman, J. L., & Rossmann, M. G. (1987) *J. Mol. Biol. 198*, 445–462.
Abrahams, J. P., Leslie, A. G. W., Lutter, R., & Walker, J. E. (1994) *Nature 370*, 621–628.
Anderson, A. G. & Hermans, J. (1988) *Proteins 3*, 262–265.
Anderson, B. F., Baker, H. M., Norris, G. E., Rumball, S. V., & Baker, E. N. (1990) *Nature 344*, 784–786.
Anderson, W. F., Ohlendorf, D. H., Takeda, Y., & Matthews, B. W. (1981) *Nature 290*, 754-758.
Arents, G., Burlingame, R., W., Wang, B.-C., Love, W., & Moudrianakis, E. (1991) *Proc. Natl. Acad. Sci. USA. 88*, 10148–10152.
Arents, G., & Moudrianakis, E. (1993) *Proc. Natl. Acad. Sci. U.S.A. 90*, 10489–10493.
Bahar, I., Atilgan, A. R., Jernigan, R. L., & Erman, B. (1997) *Proteins: Structure, Function, and Genetics 29*, 172–185.
Baker, T. S., Caspar, D. L. D., & Murakami, W. T. (1983) *Nature 303*, 446–448.
Baldwin, J. & Chothia, C. (1979) *J. Mol. Biol. 129*, 175–200.
Banner, D. M., Kokkinides, M., & Tsernoglou, D. (1987) *J. Mol. Biol. 196*, 657–672.
Barford, D., Hu, S.-H., & Johnson, L. N. (1991) *J. Mol. Biol. 218*, 233–260.
Behe, M. J., Lattman, E. E., & Rose, G. D. (1991) *Proc. Natl. Acad. Sci. U.S.A. 88*, 4195–4199.
Betts, L., Xiang, S., Short, S., Wolfenden, R., & Carter, C. W., Jr. (1994) *J. Mol. Biol. 235*, 635–656.
Biou, V., Yaremchuk, A., Tukalo, M., & Cusack, S. (1994) *Science 263*, 1404–1410.
Blivens, R. A. & Tulinsky, A. (1985) *J. Biol.Chem. 260*, 4264–4271.
Bolduc, J. M., Dyer, D. H., Scott, W. G., Singer, P., Sweet, R. M., Koshland, D. E., Jr., & Stoddard, B. L. (1995) *Science 268*, 1312–1318.

Bourne, Y., Redford, S. M., Steinman, H. M., Lepock, J. R., Tainer, J. A., & Getzoff, E. (1996) *Proc. Natl. Acad. Sci. U.S.A. 93*, 12774–12779.
Bowie, J. U., Luthy, R., & Eisenberg, D. (1991) *Science 253*, 164–170.
Brändén, C.-I. (1980) *Quart. Rev. Biophys. 13*, 317–388.
Brändén, C.-I. & Jones, T. A. (1990) *Nature 343*, 687–689.
Brändén, C.-I. & Tooze, J. (1991). *Introduction to Protein Structure*, Garland Publishing, Inc., New York.
Brennan, R. G. & Matthews, B. W. (1989a) *J. Biol. Chem. 264*, 1903–1906.
Brennan, R. G. & Matthews, B. W. (1989b) *Trends Biochem. Sci. 14*, 286–290.
Brunger, A. (1992) *Nature 355*, 472–474.
Brunger, A. & Rice, L. M. (1996) *Methods Enzymol.*, in press.
Brunger, A. T., Clore, G. M., Gronenborn, A. M., Saffrich, R., & Nilges, M. (1993) *Science 261*, 328–331.
Buehner, M., Ford, G. C., Moras, D., Olsen, K. W., & Rossmann, M. G. (1973) *Proc. Natl. Acad. Sci. U.S.A. 70*, 3052–3054.
Bullough, P. A., Hughson, F. M., Skehel, J. J., & Wiley, D. C. (1994) *Nature 371*, 37–43.
Burling, F. T., Weis, W. I., Flaherty, K. M., & Brunger, A. T. (1996) *Science 271*, 72–77.
Butler, P. J. G. & Klug, A. (1978) *Sci. Am. 239*, 62–69.
Campbell, A. P., & Sykes, B. D. (1993) *Annu. Rev. Biophys. Biomol. Structure 22*, 99–122.
Caplow, M., Ruhlen, R. L., & Shanks, J. (1994) *J. Cell Biol. 127*, 779–788.
Caplow, M. & Shanks, J. (1996) *Mol. Biol. Cell 7*, 663–675.
Carr, C. M. & Kim, P. S. (1993) *Cell 73*, 823–832.
Carson, M. & Bugg, C. E. (1986) *J. Mol. Graphics 4*, 121–125.
Carter, C. W., Jr. (1978) *Proc. Natl. Acad. Sci. U.S.A. 75*, 3649–3653.
Carter, C. W., Jr. (1979) *J. Biol. Chem. 254*, 12219–12223.
Carter, C. W., Jr. (1995) *Biochimie 77*, 92–98.
Carter, C. W., Jr. & Kraut, J. (1974) *Proc. Natl. Acad. Sci. U.S.A. 71*, 283–287.
Carter, C. W., Jr., Kraut, J., Freer, S. T., Alden, R. A., Sieker, L. C., Adman, E., & Jensen, L. H. (1972) *Proc. Natl. Acad. Sci. U.S.A. 69*, 3526–3529.
Carter, C. W., Jr., Kraut, J., Freer, S. T., Xuong, N.-H., Alden, R. A., & Bartsch, R. G. (1974) *J. Biol. Chem. 249*, 4212–4225.
Carter, C. W., Jr., Levinger, L. F., & Birinyi, F. (1980) *J. Biol. Chem. 255*, 748–754.
Carter, C. W., Jr. (1993) *Annu. Rev. Biochem. 62*, 715–748.
Carter, C. W., Jr., Doublié, S., & Coleman, D. E. (1994) *J. Mol. Biol. 238*, 346–365.
Carter, C. W., Jr., Freer, S. T., Xuong, N. H., Alden, R. A., & Kraut, J. (1971) *Cold Spring Harbor Symp. Quant. Biol. 36*, 381–385.
Caspar, D. L. D., & Klug, A. (1962) *Cold Spring Harbor Symp. Quant. Biol. 27*, 1–24.
Chan, A. W., Hutchinson, E. G., Harris, D., & Thornton, J. M. (1993) *Prot. Sci. 2*, 1574–1590.
Chan, H. S. & Dill, K. A. (1990) *Proc. Natl. Acad. Sci. U.S.A. 87*, 6388–6392.
Chothia, C. (1973) *J. Mol. Biol. 75*, 295–302.
Chothia, C. (1974) *Nature 248*, 338–339.
Chothia, C., Levitt, M., & Richardson, D. (1977) *Proc. Natl. Acad. Sci. U.S.A. 74*, 4130–4134.
Chothia, C., Wodak, S., & Janin, J. (1976) *Proc. Natl. Acad. Sci. U.S.A. 73*, 3793–3797.
Chou, K.-C. (1995) *Proteins: Structure, Function, and Genetics 21*, 319–344.
Chou, K.-C. & Zhang, C.-T. (1995) *Crit. Rev. Biochem. Mol. Biol. 30*, 275–349.
Church, G. M., Sussman, J. L., & Kim, S. H. (1977) *Proc. Natl. Acad. Sci. U.S.A. 74*, 1458–1462.
Clore, G. M. & Gronenborn, A., M. (1994) *Prot. Sci. 3*, 372–390.
Clore, G. M. & Gronenborn, A. M. (1991a) *J. Mol .Biol. 217*, 611–20.
Clore, G. M. & Gronenborn, A. M. (1991b) *Annu. Rev. Biophys. Biophys. Chem. 20*, 29–63.
Clore, G. M., Robien, M. A., & Gronenborn, A. M. (1993) *J. Mol. Biol. 231*, 82–102.
Colman, P. M., Varghese, J. N., & Laver, W. G. (1983) *Nature 303*, 41–44.
Crawford, J. L., Lipscomb, W. N., & Schellman, C. G. (1973) *Proc. Natl. Acad. Sci. U.S.A. 70*, 538–542.

Creighton, T. E. (1995) *Proteins: Structure and Molecular Principles*, Second Ed., W. H. Freeman, New York.
Crick, F. H. C. (1953) *Acta Crystallogr. 6*, 689–697.
Crick, F. H. C. & Watson, J. D. (1956) *Nature, 177*, 473–475.
Crippen, G. M. & Maiorov, V. N. (1995) *J. Mol. Biol. 252*, 144–151.
Cusack, S., Berthet-Colominas, C., Härtlein, M., Nassar, N., & Leberman, R. (1990) *Nature 347*, 249–255.
Daggett, V. & Levitt, M. (1993) *Annu. Rev. Biophys. Biomol. Structure 22*, 353–380.
Deisenhofer, J., Epp, O., Miki, K., Huber, R., & Michel, H. (1984) *J. Mol. Biol. 180*, 385–398.
Deisenhofer, J. & Michel, H. (1991) *Annu. Rev. Biophys. Biophys. Chem. 20*, 247–266.
Diamond, R. (1966) *Acta Crystallogr. 21*, 253–266.
Dill, K. & Chan, H. S. (1997) *Nature Structural Biol. 4*, 10–19.
Dill, K. A., Bromberg, S., Yue, K., Fiebig, K. M., Yee, D. P., Thomas, P. D., & Chan, H. S. (1995) *Prot. Sci. 4*, 561–602.
Dodson, E., Kleywegt, G. J. & Wilson, K. (1996) *Acta Crystallogr. D52*, 228–234.
Doublié, S. (1993) Ph.D. Dissertation, University of North Carolina at Chapel Hill.
Doublié, S., Gilmore, C. J., Bricogne, G., & Carter, C. W., Jr. (1995) *Structure 3*, 17–31.
Edmundson, A. B., Ely, K. R., Abola, E. E., Schiffer, M., & Panagiotopoulos, N. (1975) *Biochemistry 14*, 3953–3961.
Eisenhaber, F., Frömmel, C., & Argos, P. (1996) *Proteins: Structure, Function, and Genetics 25*, 169–179.
Egelman, E. H., Francis, N., & DeRosier, D. J. (1982) *Nature 298*, 131–135.
Erickkson, A. E., Baase, W. A., Zhang, X.-J., Heinz, D. W., Blaber, M., Baldwin, E. P., & Matthews, B. W. (1992) *Science 255*, 178–183.
Farber, G. K. & Petsko, G. A. (1990) *Trends Biochem. Sci. 15*, 228–234.
Filman, D. G., Syed, R., Chow, M., Macadam, A. J., Minor, P. D., & Hogle, J. M. (1989) *EMBO J. 8*, 156-174.
Finch, J. T., Brown, R. S., Rhodes, D., Richmond, T., Rushton, B., Levitt, M., & Klug, A. (1981) *J. Mol. Biol. 145*, 29–36.
Fitzgerald, P. M. D., McKeever, B. M., VanMiddlesworth, J. F., Springer, J. P., Heimbach, J. C., Leu, C. T., Herber, W. K., Dixon, R. A., & Darke, P. L. (1990) *J. Biol. Chem. 265*, 14209–14219.
Fitzgerald, P. M. D., & Springer, J. P. (1991) *Annu. Rev. Biophys. Biophys. Chem. 20*, 299–320.
Fitzgerald, P. M. D. (1991) In *Crystallographic Computing* 5 (Moras, D., Podjarny, A. D., & Thierry, J.-C., Eds.). International Union of Crystallography/Oxford University Press, Oxford, UK., pp. 333–347
Flöckner, H., Braxenthaler, M., Lackner, P., Jaritz, M., Ortner, M., & Sippl, M. J. (1995) *Proteins: Structure, Function, and Genetics 23*, 376–386.
Freer, S. T., Alden, R. A., Carter, C. W., Jr., & Kraut, J. (1975) *J. Biol. Chem. 250*, 46–54.
Fremont, D. H., Stura, E. A., Matsumura, M., Peterson, P. A., & Wilson, I. A. (1995) *Proc. Nat. Acad. Sci. U.S.A. 92*, 2479–2483.
Gaykema, W. P. J., Hol, W. G. J., Vereijken, J. M., Soeter, N. M., Bak, H. J., & Beintema, J. J. (1984) *Nature 309*, 23–29.
Gernert, K. M. (1994) Ph. D. Dissertation, Duke University.
Gernert, K. M., Surles, M. C., Labean, T. H., Richardson, J. S., & Richardson, D. C. (1995) *Prot. Sci. 4*, 2252–2260.
Gerstein, M., Anderson, B. F., Norris, G. E., Baker, E. N., Lesk, A. M., & Chothia, C. (1993) *J. Mol. Biol. 234*, 357–372.
Gerstein, M., Lesk, A., M., & Chothia, C. (1994) *Biochemistry 33*, 6739–6749.
Gething, M.-J. & Sambrook, J. (1992) *Nature 355*, 33–45.
Getzoff, E. D., Tainer, J. A., Weiner, P. K., Kollman, P. A., Richardson, J. S. & Richardson, D. C. (1983) *Nature 306*, 287–290.

Goldgur, Y., Mosyak, L., Reshetnikova, L., Ankilova, V., Lavrik, O., Svetlana, L., & Safro, M. (1997) *Structure 5*, 59–68.
Goldstein, R. A., Luthey-Schulten, Z. A., & Wolynes, P. G. (1992) *Proc. Natl. Acad. Sci. U.S.A. 89*, 9029–9033.
Hadju, J., Machin, P. A., Campbell, J. W., Greenhough, T. J., Clifton, I. J., Zurek, S., Gover, S., Johnson, L. N., & Elder, M. (1987) *Nature 329*, 178–181.
Hajdu, J., & Andersson, I. (1993) *Annu. Rev. of Biophys. Biomol. Structure 22*, 467–498.
Harbury, P. B., Kim, P. S., & Alber, T. (1994) *Nature 371*, 80–83.
Harbury, P. B., Tidor, B., & Kim, P. S. (1995) *Proc. Natl. Acad. Sci. U.S.A. 92*, 8408–8412.
Harbury, P. B., Zhang, T., Kim, P. S., & Alber, T. (1993) *Science 262*, 1401–1407.
Harrison, S. C. (1978) *Trends Biochem. Sci. 3*, 3–7
Harrison, S. C., & Aggarwal, A. K. (1990) *Annu. Rev. Biochem. 59*, 933–969.
Harrison, S. C., Olson, A. J., Schutt, C. E., Winkler, F. K., & Bricogne, T. (1978) *Nature 276*, 368-373.
Hartley, B. (1970) *Biochem. J. 119*, 805–822.
Hemmingsen, J. M., Gernert, K. M., Richardson, J. S., & Richardson, D. C. (1994) *Prot. Sci. 3*, 1927–37.
Henderson, R., Baldwin, J. M., Ceska, T. A., Zemlin, F., Beckmann, E., & Downing, K. H. (1990) *J. Mol. Biol. 213*, 899–929.
Hendrickson, W. A., & Ward, K. B. (1977) *J. Biol. Chem. 252*, 3012–3018.
Hermans, J. (1993) *Curr. Opin. Struct. Biol. 3*, 270–276.
Hogle, J. M., Chow, M., & Filman, D. J. (1987) *Sci. Am. 256*, 42–49.
Hol, W. J. G., van Duijnen, P. T., & Berensen, H. J. C. (1978) *Nature 273*, 443–446.
Holm, L., Ouzounis, C., Sander, C., Tuparev, G., & Vriend, G. (1992) *Prot. Sci. 1*, 1691–1698.
Holm, L. & Sander, C. (1991) *J. Mol. Biol. 218*, 183–194.
Holm, L. & Sander, C. (1992a) *J. Mol. Biol. 225*, 93–105.
Holm, L. & Sander, C. (1992b) *Proteins 14*, 213–223.
Holm, L. & Sander, C. (1993a) *J. Mol. Biol. 233*, 123–138.
Holm, L. & Sander, C. (1993b) *FEBS Lett. 315*, 301–306.
Holm, L. & Sander, C. (1997) *Structure 5*, 165–171.
Hutchinson, E. G. & Thornton, J. M. (1994) *Protein Science 3*, 2207–2216.
Hyde, C. C., Ahmed, S. A., Padlan, E. A., Miles, E. W., & Davies, D. R. (1988) *J. Biol. Chem. 263*, 17857–17871.
Jacobson, R. H., Zhang, X.-J., Dubose, R. F., & Matthews, B. W. (1994) *Nature 369*, 761–766.
James, M. N. G., Sielecki, A. R., Brayer, G. D., Delbaere, L. T. J., & Bauer, C. A. (1981) in *Structural Aspects of Recognition and Assembly in Biological Macromolecules* (Balaban, M., Ed.), Vol. 1, pp. 3–18, Balaban International Science Services, Rehovot, Israel.
Janin, J., Wodak, S., Levitt, M., & Maigret, B. (1978) *J. Mol. Biol. 125*, 357–386.
Janin, J., & Wodak, S. J. (1993) *Proteins 15*, 1–4.
Jones, A. T., & Liljas, L. (1984) *J. Mol. Biol. 177*, 73-92.
Jones, D., & Thornton, J. (1993) *J. Comput. Aided Mol. Des. 7*, 439–456.
Jones, D. T., Taylor, W. R., & Thornton, J. M. (1992) *Nature 358*, 86–89.
Jurnak, F., Yoder, M. D., Pickersgill, R., & Jenkins, J. (1994) *Current Opinion Struct. Biol. 4*, 802–806.
Kabsch, W. & Sander, C. (1983) *Biopolymers 22*, 2577–2637.
Kamtekar, S., Schiffer, J. M., Xiong, H., Babik, J. M., & Hecht, M. H. (1993) *Science 262*, 1680–1685.
Kantrowitz, E. R. & Lipscomb, W. N. (1988) *Science 241*, 669–674.
Karpen, M. E., de Haseth, P. L., & Neet, K. E. (1992) *Prot. Sci. 1*, 1333–1342.
Karplus, M. & Petsko, G. A. (1990) *Nature 347*, 631–639.
Kidera, A. & Go, N. (1992) *J. Mol. Biol. 225*, 457–475.
Kidera, A., Inaka, K., Matsushima, M., & Go, N. (1992) *J. Mol. Biol. 225*, 477–486.
Kim, S.-H. (1992) *Science 255*, 1217–1218.

Kim, Y., Geiger, J. H., Hahn, S., & Sigler, P. B. (1993) *Nature 365*, 512–527.
Kitaigordosky, A. I. (1973) *Molecular Crystals and Moleculers*, Academic Press, New York.
Klotz, I. M., Klippenstein, G. L., & Hendrickson, W. A. (1976) *Science 192*, 335–344.
Konnert, J. H. (1976) *Acta Crystallogr. 32*, 614–617.
Kraulis, P. J. (1991) *J. Appl. Crystallogr. 24*, 946–950.
Kubena, B. D., Luecke, H., Rosenberg, H., & Quiocho, F. A. (1986) *J. Biol. Chem. 261*, 7995–8002.
Kuhn, L. A., Siani, M. A., Pique, M., E., Fisher, C. L., Getzoff, E. D., & Tainer, J. A. (1992) *J. Mol. Biol. 228*, 13–22.
Kyte, J. (1995) *Structure in Protein Chemistry*. Garland Publishing, Inc, NY.
Lahm, A. & Suck, D. (1991) *J. Mol. Biol. 222*, 645–658.
Lamy, J., Bijholt, M. M. C., Sizaret, P. Y., Lamy, J., & van Bruggen, E. F. J. (1981) *Biochemistry 20*, 1849–1856.
Lamzin, V. & Wilson, K. (1993) *Acta Crystallogr. 49*, 129–147.
Lamzin, V. S., & Wilson, K. S. (1996) *Methods Enymol.*, in press.
Lane, A. N., & Lefèvre, J.-F. (1994) *Methods Enzymol. 239*, 596–619.
Lapthorne, A. J., Harris, D. C., Littlejohn, A., Lustbader, J. W., Canfield, R. E., Machin, K. J., Morgan, F. J., & Isaacs, N. W. (1994) *Nature 369*, 455–461.
Levitt, M. (1981) *Nature 294*, 379–38.
Levitt, M. & Chothia, C. (1976) *Nature 261*, 552–558.
Lewis, P., Momany, F., & Scheraga, H. (1973) *Biochim. Biophys. Acta 303*, 211–229.
Liddington, R., Derwenda, Z., Dodson, E., Hubbard, R., & Dodson, G. (1992) *J. Mol. Biol. 228*, 551–574.
Liddington, R. C., Yan, Y., Mouai, J., Sahli, R., Benjamin, T. L., & Harrison, S. C. (1991) *Nature 354*, 278–284.
Lim, W. A., Farruggio, D. C., & Sauer, R. T. (1992) *Biochemistry 31*, 4324–4333.
Lim, W. A. & Sauer, R. T. (1989) *Nature 339*, 31–36.
Lim, W. A. & Sauer, R. T. (1991) *J. Mol. Biol. 219*, 359–376.
Linderstrøm-Lang, K. U. (1952) *The Lane Medical Lectures*, Stanford University Press, Stanford, CA.
Lolis, E. & Petsko, G. A. (1990) *Annu. Rev. Biochem. 59*, 597–630.
London, R. E. (1989) *Methods Enzymol. 176*, 358–375.
Lovejoy, B., Le, T. C., Luthy, R., Cascio, D., ONeil, K. T., DeGrado, W. F., & Eisenberg, D. (1992) *Prot. Sci. 1*, 956–957.
Lumb, K. J., Carr, C. M., & Kim, P. S. (1994) *Biochemistry 33*, 7361–7367.
Lumb, K. J. & Kim, P. S. (1994) *J. Mol. Biol. 236*, 412–420.
Lumb, K. J. & Kim, P. S. (1995a) *Biochemistry 34*, 8642–8648.
Lumb, K. J. & Kim, P. S. (1995b) *Science 268*, 436–439.
Lymn, R. W. & Taylor, E. W. (1970) *Biochemistry 9*, 2975–2983.
Lymn, R. W. & Taylor, E. W. (1971) *Biochemistry 10*, 4617–4624.
Ma, P. C., Rould, M. A., Weintraub, H., & Pabo, C. O. (1994) *Cell 77,* 451-459.
Maiorov, V. N. & Crippen, G. M. (1994) *J. Mol. Biol. 235*, 625–634.
Maiorov, V. N., & Crippen, G. M. (1995) *Proteins: Structure, Function, and Genetics 22*, 273–283.
Matthews, B. W. (1993) *Annu. Rev. Biochem. 62*, 139–160.
Matthews, D. A., Alden, R. A., Birktoft, J. J., Freer, S. T., & Kraut, J. (1977) *J. Biol. Chem. 252*, 8875–8883.
Matthews, D. A., Appelt, K., & Oatley, S. J. (1989) *J. Mol. Biol. 205*, 449–454.
Mathews, F. S., Argos, P., & Levine, M. (1972) *Cold Spring Harbor Symp. Quant. Biol. 36*, 387–395.
McDonald, I. K., & Thornton, J. (1994) *J. Mol. Biol. 238*, 777–793.
McLachlan, A. D., Stewart, M., & Smillie, L.B. (1975) *J. Mol. Biol. 98*, 281–291.
McLachlan, A. D. & Stewart, M. (1975) *J. Mol. Biol. 98*, 293–304.

McLachlan, A. D. & Stewart, M. (1976) *J. Mol. Biol. 103*, 271–298.
Michie, A. D., Orengo, C. A., & Thorton, J. M. (1996) *J. Mol. Biol. 262*, 168–185.
Milburn, M. V., Tong, L., DeVos, A. M., Brunger, A., Yamaizumi, Z., Nishimura, S., & Kim, S.-H. (1990) *Science 247*, 939–945.
Morris, A. L., MacArthur, M. W., Hutchinson, E. G., & Thornton, J. (1992) *Proteins: Structure, Function, and Genetics 12*, 345–364.
Moss, D. S., Tickle, I. J., Theis, O., & Wostrack, A. (1996). In *Macromolecular Refinement* (Dodson, E. J., Moore, M., Ralph, A. & Bailey, S., Eds.), pp 105–113.
Namba, K., Pattanayek, R., & Stubbs, G. (1989) *J. Mol. Biol. 208*, 307–321.
Navaratnam, N., Battacharya, S., Sherman, N., Fujino, T., Carter, C. W., Jr., & Scott, J. (1997) *J. Mol. Biol.*, submitted.
Navaratnam, N., Shah, R., Patel, D., Fay, V., & Scott, J. (1993) *Proc. Natl. Acad. Sci. U.S.A. 90*, 222–226.
Neidhart, D. J., Kenyon, G. L., Gerlt, J. A., & Petsko, G. A. (1990) *Nature 347*, 692–694.
Neidhart, D. J., Howell, P. L., Petsko, G. A., Powers, V. M., Li, R., Kenyon, G. L., & Gerlt, J. A. (1991) *Biochemistry 30*, 9264–9273.
Newcomer, M. E., Lewis, B. A. & Quiocho, F. A. (1981) *J. Biol. Chem. 256*, 13218–13222.
Nicholls, A., Sharp, K. A., & Honig, B. (1991) *Proteins: Structure, Function, and Genetics 11*, 281–296.
Noble, M. E. M., Verlinde, C. L. M. J., Groendijk, H., Kalk, K. H., Wierenga, R. K., & Hol, W. G. J. (1991) *J. Med. Chem. 34*, 2709–2716.
O'Connell, (1997)
O'Shea, E. K., Klemm, J. D., Kim, P. S., & Alber, T. (1991) *Science 254*, 539–544.
O'Shea, E. K., Rutkowski, R., & Kim, P. S. (1989) *Science 243*, 538–542.
Oldfield, T. J. & Hubbard, R. E. (1994) *Proteins: Structure, Function, and Genetics 18*, 324–337.
Orengo, C. A., Flores, T. P., Taylor, W. R., & Thornton, J. M. (1993) *Prot. Eng. 6*, 485–500.
Orengo, C. A. & Thornton, J. M. (1993) *Structure 1*, 105–120.
Orlova, E. V., Serysheva, I. I., van Heel, M., Hamilton, S. L., & Chiu, W. (1996) *Nature Structural Biol. 3*, 547–551.
Otwinowski, Z., Schevitz, R. W., Zhang, R.-G., Lawson, C. L., Joachimiak, A. J., Marmorstein, R., Luisi, B. F., & Sigler, P. B. (1988) *Nature 335*, 321-326.
Ouzounis, C., Sander, C., Scharf, M., & Schneider, R. (1993) *J. Mol. Biol. 232*, 805–825.
Pai, E. F., Krengel, U., Petsko, G. A., Goody, R. S., Kabsch, W., & Wittinghofer, A. (1990) *EMBO J. 9*, 2351–2359.
Pakula, A. A. & Sauer, R. T. (1989) *Proteins 5*, 202–210.
Pauling, L., Corey, R. B., & Branson, H. R. (1951) *Proc. Natl. Acad. Sci. U.S.A. 37*, 205–211.
Peng, J. W., & Wagner, G. (1994) *Methods Enzymol. 239*, 563–596.
Perutz, M. (1992) *Protein Structure: New Approaches to Disease and Therapy*, W. H. Freeman, New York.
Perutz, M. F. (1970) *Nature 228*, 726–739.
Perutz, M. F. (1978) *Sci. Am. 239*, 92–125.
Perutz, M. F. (1979) *Annu. Rev. Biochem. 48*, 327–386.
Phillips, G. N., Lattman, E.E., Cummins, P., Lee, K.Y., & Cohen, C. (1979) *Nature 278*, 413–417.
Philips, S. E. V. (1980) *J. Mol. Biol. 142*, 531–554.
Phillips, S. E. V. (1991) *Curr. Opin. Struct. Biol. 1*, 89–98.
Presta, L. G. & Rose, G. D. (1988) *Science 240*, 1632–41.
Quiocho, F. A. & Vyas, N. K. (1984) *Nature 310*, 381–386.
Quiocho, F. A., Wilson, D. K., & Vyas, N. K. (1989) *Nature 340*, 404–407.
Radzicka, A., Acheson, S. A., & Wolfenden, R. (1992) *Bioorg. Chem. 20*, 382–386.
Radzicka, A. & Wolfenden, R. (1994) *Science 265*, 936–937.
Ramachandran, G. N., Ramakrishnan, C., & Sasisekharan, V. (1963) *J. Mol. Biol. 7*, 95–99.
Rayment, I. (1993) *Science 261*, 50–58.
Rayment, I. (1996) *Structure 4*, 501–504.

Rayment, I., Baker, T. S., Casper, D. L. D., & Murakami, W. T. (1982) *Nature 295*, 110–115.
Rayment, I. & Holden, H. M. (1994) *Trends Biochem. Sci. 19*, 129–134.
Rayment, I., Holden, H. M., Whittaker, M., Yohn, C. B., Lorenz, M., Holmes, K. C., & Milligan, R. A. (1993) *Science 261*, 58–65.
Read, R. J. (1997) *Methods Enzymol. 277*, in press.
Rice, P. A., Yang, S.-W., Mizoguchi, K., & Nash, H. A. (1996) *Cell 87*, 1295-1306.
Richards, F. M. (1974) *J. Mol. Biol. 82*, 1–14.
Richards, F. M. (1977) *Annu. Rev. Bioph. Bioeng. 6*, 151–176.
Richards, F. M. (1985) *Methods Enzymol. 115*, 440–464.
Richards, F. M. (1991) *Sci. Am. 264*, 54–7, 60–63.
Richards, F. M. & Lim, W. A. (1993) *Quart. Rev. Biophys. 26*, 423–498.
Richardson, J. S. & Richardson, D. C. (1988) *Science 240*, 1648–1652.
Richardson, D. C. & Richardson, J. S. (1992a) *Protein Science 1*, 3–9.
Richardson, J. (1977) *Nature 268*, 495–500.
Richardson, J. S. (1973) *Proc. Nat. Acad. Sci. U.S.A. 73*, 2619–1623.
Richardson, J. S. (1981) *Adv. Prot. Chem. 34*, 167–339.
Richardson, J. S. (1985) *Methods Enzymol. 115*, 359–380.
Richardson, J. S., Getzoff, E. D., & Richardson, D. C. (1978) *Proc. Natl. Acad. Sci. U.S.A. 75*, 2574–2578.
Richardson, J. S., & Richardson, D. C. (1987) in *Protein Engineering* (Oxender, D. L., & Fox, C. F., Eds.), pp. 149–163, Alan R. Liss, New York.
Richardson, J. S. & Richardson, D. C. (1990) in *Prediction of Protein Structure and the Principles of Protein Conformation* (Fasman, G., Ed.), pp. 1–97, Plenum, New York.
Richmond, T. J., Finch, J. T., Rushton, B., Rhodes, D., & Klug, A. (1984) *Nature 311*, 532–538.
Ries-Kautt, M. & Ducruix, A. (1997) *Methods Enzymol. 276*, 23–59.
Ringe, D. & Petsko, G. A. (1986) *Methods Enzymol. 131*, 389–433.
Roe, S. M. & Teeter, M. M. (1993) *J. Mol. Biol. 229*, 419–427.
Rooman, M. J. & Wodak, S. J. (1988) *Nature 335*, 45–49.
Rossmann, M. G. & Argos, P. (1976) *J. Mol. Biol. 105*, 75–95.
Rossmann, M. G. & Argos, P. (1977) *J. Mol. Biol. 109*, 99–129.
Rossmann, M. G., Moras, D., & Olsen, K. W. (1974) *Nature 250*, 194–199.
Rost, B. & Sander, C. (1993) *J. Mol. Biol. 232*, 584–99.
Sablin, E. P., Kull, F. J., Cooke, R., Vale, R. D., & Fletterick, R. J. (1996) *Nature 380*, 555–559.
Sander, C. & Schneider, R. (1991) *Proteins 9*, 56–68.
Sander, C. & Schneider, R. (1993) *Nucleic Acids Res. 21*, 3105–3109.
Sander, C., Vriend, G., Bazan, F., Horovitz, A., Nakamura, H., Ribas, L., Finkelstein, A. V., Lockhart, A., Merkl, R., & Perry, L. J. (1992) *Proteins 12*, 105–110.
Schlichting, I., Almo, S. C., Rapp, G., Wilson, K., Petratos, K., Lentfer, A., Wittinghofer, A., Kabsch, W., Pai, E. F., Petsko, G. A., & Goody, R. S. (1990) *Nature 345*, 309–315.
Scully, J. & Hermans, J. (1994) *J. Mol. Biol. 235*, 682–694.
Shaanan, B., Gronenborn, A. M., Cohen, G. H., Gilliland, G. L., Veerapandian, B., Davies, D. R., & Clore, G. M. (1992) *Science 257*, 961–64.
Sheldrick, G. & Schneider, T. R. (1996) *Methods Enzymol.*, in press
Sheriff, S., Hendrickson, W. A., & Smith, J. L. (1987) *J. Mol. Biol. 197*, 273–296.
Shortle, D. (1995) *Nature Structural Biol. 2*, 91–93.
Sibanda, B. L. & Thornton, J. M. (1985) *Nature 316*, 170–174.
Singh, R. K., Tropsha, A., & Vaisman, I. I. (1996) *J. Comp. Biol., 3*, 213–221.
Skolnick, J. (1997) *J. Mol. Biol.*, in press.
Smith, C. A. & Rayment, I. (1996) *Biophys. J. 70*, 1590–1602.
Smith, F. R. & Simmons, K. (1994) *Proteins 18*, 295–300.
Smith, F. R., Lattman, E. E., & Carter, C. W., Jr. (1991) *Proteins: Structure, Function, and Genetics 10*, 81–91.
Somers, W. S., & Phillips, S. E. V. (1992) *Nature 359*, 387-391.

Srinivasan, R. & Rose, G. D. (1994) *Proc. Natl. Acad. Sci. U.S.A. 91*, 11113–11117.
Srinivasan, R. & Rose, G. D. (1995) *Proteins: Structure, Function, and Genetics 22*, 81–99.
Starich, M. R., Sandman, K., Reeve, J. N., & Summers, M. F. (1997) *J. Mol. Biol. 255*, 187–203.
Stehle, T., & Harrison, S. C. (1996) *Structure 4*, 165-182.
Stevens, R. C., Gouaux, J. E., & Lipscomb, W. N. (1990) *Biochemistry 29*, 7691–7699.
Stewart, D. E., Sarkar, A., & Wampler, J. E. (1990) *J. Mol. Biol. 214*, 253–260.
Stewart, M. & McLachlan, A. D. (1975) *Nature 257*, 331–333.
Stouten, P. F., Sander, C., Ruigrok, R. W., & Cusack, S. (1992) *J. Mol. Biol. 226*, 1073–1084.
Subramanian, S., Tcheng, D. K. & Fenton, J. M. (1996) In *Proceedings of the Fourth International Conference on Intelligent Systems in Molecular Biology* (States, D., Ed.), pp 218–229. AAAI Press, San Francisco, CA.
Sun, P. D. & Davies, D. R. (1995) *Annu. Rev. Biophys. Biomol. Structure 24*, 269–291.
Tainer, J. A., Getzoff, E. D., Alexander, H., Houghten, R. A., Olson, A. J., Lerner, R. A., & Hendrickson, W. A. (1984) *Nature 312*, 127–134.
Tainer, J. A., Getzoff, E. D., Richardson, J. S., & Richardson, D. C. (1983) *Nature 306*, 284–289.
Takasugawa, F. & Kamitori, S. (1996) *J. Am. Chem. Soc. 118*, 8945–8946.
Tan, S. Hunziker, Y., Sargent, D. F., & Richmond, T. J. (1996) *Nature 381*, 127-131.
Tropsha, A., Singh, R. K., Vaisman, I. I., & Zheng, W. (1996) in *Pacific Symposium on Biocomputing '96* (Hunter, L., & Klein, T. E., Eds.), pp. 614–623, World Scientific, Singapore.
van Gunsteren, W. F., Luque, F. J., Timms, D., & Torda, A. E. (1994) *Annu. Rev. Biophys. Biomol. Structure 23*, 847–863.
Van Holde, K. E. & Miller, K. I. (1982) *Quart. Rev. Biophys. 15*, 1–129.
Waksman, G., Kominos, D., Robertson, S. C., Pant, N., Baltimore, D., Birge, R. B., Cowburn, D., Hanafusa, H., Mayer, B. J., Overduin, M., Resh, M. D., Rios, C. B., Silverman, L., & Kuriyan, J. (1992) *Nature 358*, 646–651.
Wagenknecht, T., Grassucci, R., Berkowitz, J., Wiederrecht, G. J., Xin, H.-B., & Fleischer, S. (1996) *Biophysical J. 70*, 1709–1715.
Wagner, G., Hyberts, S. G. & Havel, T. F. (1992) *Annu. Rev. Biophys. Biomol. Struct. 21*, 167–198.
Walker, J. E., Saraste, E. M., Runswick, M. J., & Gay, N. J. (1992) *EMBO J. 1*, 945–951.
Wang, L., O'Connell, T., Tropsha, A., & Hermans, J. (1996) *J. Mol. Biol. 262*, 283–293.
Warrant, R. W. & Kim, S. H. (1978) *Nature 271*, 130–135.
Watenpaugh, K. D., Sieker, L. C., Herriott, J. R., & Jensen, L., H. (1973) *Acta Crystallogr. B29*, 943–956.
Weber, P. C., Ohlendorf, D. H., Wendeloski, J. J., & Salemme, F. R. (1989) *Science 243*, 85–89.
Weber, P. C., Wendoloski, J. J., Pantoliano, M. W., & Salemme, F. R. (1992) *J. Am. Chem. Soc. 114*, 3197–3200.
Westhof, E., Altschuh, D., Moras, D., Bloomer, A. C., Mondragon, A., Klug, A., & Regenmortel, M. H. V. V. (1984) *Nature 311*, 123–126.
Wiley, D. C., Wilson, I. A., & Skehel, J. J. (1981) *Nature 289*, 374–378.
Williams, M. A., Goodfellow, J. M., & Thornton, J. M. (1994) *Prot. Sci. 3*, 1224–1235.
Wilmanns, M., Hyde, C. C., Davies, D. R., Kirschner, K., & Jansonius, J. N. (1991) *Biochemistry 30*, 9161–9169.
Wilson, D. K. & Quiocho, F. A. (1993) *Biochemistry 32*, 1689–1694.
Wilson, D. K. & Quiocho, F. A. (1994) *Nature Struct. Biol. 1*, 691–694.
Wilson, D. K. Rudolph, F. B., & Quiocho, F. A. (1991) *Science 252*, 1278–1284.
Wilson, I. A., Skehel, J. J., & Wiley, D. C. (1981) *Nature 289*, 366–373.
Wlodawer, A., Miller, M., Jaskolski, M., Sathyanarayana, B. K., Baldwin, E., Weber, I. T., Selk, L. M., Clawson, L., Schneider, J., & Kent, S. B. (1989) *Science 245*, 616–621.
Wolfenden, R. (1976) *Annu. Rev. Biophys. Bioeng. 5*, 271–306.

Wolfenden, R. (1983) *Science 222*, 1087–1093.
Wu, H., Lustbader, J. W., Liu, Y., Canfield, R. E., & Hendrickson, W. A. (1994) *Structure 2*, 545–558.
Wyman, J. & Gill, S. J. (1990). *Binding and Linkage*, University Science Books, Mill Valley, CA.
Xiang, S., Carter, C. W., Jr., Bricogne, G., & Gilmore, C. J. (1993) *Acta Crystallogr. D49*, 193–212.
Xiang, S., Short, S. A., Wolfenden, R., & Carter, C. W., Jr. (1995) *Biochemistry 34*, 4516–4523.
Xiang, S., Short, S. A., Wolfenden, R., & Carter, C. W., Jr. (1996a) *Biochemistry 35*, 1335–1341.
Xiang, S., Short, S. A., Wolfenden, R., & Carter, C. W., Jr. (1997) *Biochemistry 36*,
Yang, W., Hendrickson, W. A., Kalman, E. T., & Crouch, R. J. (1990) *J. Biol. Chem. 265*, 13553–13559.
Yee, D. P., Chan, H. S., Havel, T. F., & Dill, K. A. (1994) *J. Mol. Biol. 241*, 557–573.
Yoder, M. D., Keen, N. T., & Jurnak, F. (1993a) *Science 260*, 1503–1507.
Yoder, M. D., Lietzke, S. E., & Jurnak, F. (1993b) *Structure 1*, 241–251.
Yonath, A. (1992) *Annu. Rev. Biophys. Biomol. Structure 21*, 77–93.
Zhang, K. Y. J., Cowtan, K., & Main, P. (1997) *Methods Enzymol. 277*, in press.
Zhang, L. & Hermans, J. (1994) *J. Am. Chem. Soc. 116*, 11915–11921.
Zhang, L. & Hermans, J. (1996) *Proteins: Structure, Function, and Genetics 24*, 433–438.

Chapter 6

Abagyan, R. A. (1993) *FEBS Lett. 325*, 17–22.
Alber, T. (1989a) *Annu. Rev. Biochem. 58*, 765–798.
Alber, T. (1989b) in *Prediction of Protein Strucutre and the Principles of Protein Conformation* (Fasman, G. D., Ed.) pp. 161–192, Plenum, New York.
Alonso, D. O., Dill, K. A., & Stigter, D. (1991) *Biopolymers 31*, 1631–1649.
Anderson, D. E., Becktel, W. J., & Dahlquist, F. W. (1990) *Biochemistry 29*, 2403–2408.
Anfinsen, C. B. (1973) *Science 181*, 223–230.
Anfinsen, C. B., Haber, E., Sela, M., & White, F. H., Jr. (1961) *Proc. Natl. Acad. Sci. USA 47*, 1309–1314.
Anson, M. L. & Mirsky, A. E. (1934a) *J. Gen. Physiol. 17*, 393–398.
Anson, M. L., & Mirsky, A. E. (1934b) *J. Gen. Physiol. 17*, 399–408.
Aune, K. C., Salahuddin, A., Zarlengo, M. H., & Tanford, C. (1967) *J. Biol. Chem. 242*, 4486–4489.
Baker, D., Sohl, J. L., & Agard, D. A. (1992) *Nature 356*, 263–265.
Baker, D. & Agard, D.A. (1994) *Biochemistry 33*, 7505–7509.
Baker, E. N. & Hubbard, R. E. (1984) *Prog. Biophys. Mol. Biol. 44*, 97–179.
Baldwin, R. L. (1991) *Chemtracts: Biochem. Mol. Biol. 2*, 379–389.
Baldwin, R. L. (1993) *Current Opinion Struct. Biol. 3*, 84–91.
Barlow, D. J. & Thornton, J. M. (1983) *J. Mol. Biol. 168*, 867–885.
Barrick, D. & Baldwin, R. L. (1993) *Protein Sci. 2*, 869–876.
Baum, J., Dobson, C. M., Evans, P. A., & Hanley, C. (1989) *Biochemistry 28*, 7–13.
Beasty, A. M., Hurle, M. R., Manz, J. T., Stackhouse, T., Onuffer, J. J., & Matthews, C. R. (1986) *Biochemistry 25*, 2965–2974.
Becktel, W. J. & Schellman, J. A. (1987) *Biopolymers 26*, 1859–1877.
Bernstein, F. C., Koetzle, T. F., Williams, G. J. B., Meyer, E. F., Jr, Brice, M. D., Rodgers, J. R., Kennard, O., Shimanouchi, T. & Tasumi, M. (1977) *J. Mol. Biol. 112*, 535–542.
Blundell, T., Carney, D., Gardner, S., Hayes, F., Howlin, B., Hubbard, T., Overington, J., Singh, D. A., Sibanda, B. L., & Sutcliffe, M. (1988) *Eur. J. Biochem. 172*, 513–520.
Bowie, J. U., Reidhaar-Olson, J. F., Lim, W. A., & Sauer, R. T. (1990) *Science 247*, 1306–1310.
Brandts, J. F., Halvorson, H. R., & Brennan, M. (1975) *Biochemistry 14*, 4953–4963.

Briggs, M. S. & Roder, H. (1992) *Proc. Natl. Acad. Sci. USA 89*, 2017–2021.
Broadhurst, R. W., Dobson, C. M., Hore, P. J., Radford, S. E., & Rees, M. L. (1991) *Biochemistry 30*, 405–412.
Carrell, R. W., Evans, D. L., & Stein, P. E. (1991) *Nature 353*, 576–578.
Chaffotte, A., Guillou, Y., Delepierre, M., Hinz, H. J., & Goldberg, M. E. (1991) *Biochemistry 30*, 8067–8074.
Chaffotte, A. F., Guillou, Y., & Goldberg, M. E. (1992) *Biochemistry 31*, 9694–9702.
Chan, H. S. & Dill, K. A. (1991) *Annu. Rev. Biophys. Biophys. Chem. 20*, 447–490.
Chen, B. L., Baase, W. A., Nicholson, H., & Schellman, J. A. (1992) *Biochemistry 31*, 1464–1476.
Chen, B. L., Baase, W. A., & Schellman, J. A. (1989) *Biochemistry 28*, 691–699.
Chen, B. L. & Schellman, J. A. (1989) *Biochemistry 28*, 685–691.
Christensen, H. & Pain, R. H. (1991) *Eur. Biophys. J. 19*, 221–229.
Clore, G. M. & Gronenborn, A. M. (1991) *Annu. Rev. Biophys. Biophys. Chem. 20*, 29–63.
Cornette, J. L., Cease, K. B., Margalit, H., Spouge, J. L., Berzofsky, J. A., & Delisi, C. (1987) *J. Mol. Biol. 195*, 659–685.
Creighton, T. E. (1974a) *J. Mol. Biol. 87*, 579–602.
Creighton, T. E. (1974b) *J. Mol. Biol. 87*, 563–577.
Creighton, T. E. (1990) *Biochem. J. 270*, 1–16.
Creighton, T. E. (1992a) *Science 256*, 111–112.
Creighton, T. E., Ed. (1992b) *Protein Folding*, W.H. Freeman, New York.
Creighton, T. E. (1992c) *Proteins*, Second Ed., W. H. Freeman, New York.
Daggett, V. & Levitt, M. (1993) *Annu. Rev. Biophys. Biomol. Struct. 22*, 353–380.
Daopin, S., Sauer, U., Nicholson, H., & Matthews, B. W. (1991) *Biochemistry 30*, 7142–7153.
Darby, N. J., van Mierlo, C. P., Scott, G. H., Neuhaus, D., & Creighton, T. E. (1992) *J. Mol. Biol. 224*, 905–911.
Dill, K. A. (1990) *Biochemistry 29*, 7133–7155.
Dill, K. A. & Shortle, D. (1991) *Annu. Rev. Biochem. 60*, 795–825.
Dobson, C. M. (1992) *Curr. Opin. Struct. Biol. 2*, 6–12.
Dobson, C. M. & Evans, P. A. (1984) *Biochemistry 23*, 4267–4270.
Dobson, C. M., Evans, P. A., & Radford, S. E. (1994) *Trends Biochem. Sci. 19*, 31–37.
Dolgikh, D. A., Gilmanshin, R. I., Brazhnikov, E. V., Bychkova, V. E., Semisotnov, G. V., & Venyaminov SYu (1981) *FEBS Lett. 136*, 311–315.
Donehower, L. A., & Bradley, A. (1993) *Biochim. Biophys. Acta 1155*, 181–205.
Dyson, H. J., Rance, M., Houghten, R. A., Lerner, R. A., & Wright, P. E. (1988a) *J. Mol. Biol. 201*, 161–200.
Dyson, H. J., Rance, M., Houghten, R. A., Wright, P. E., & Lerner, R. A. (1988b) *J. Mol. Biol. 201*, 201–217.
Dyson, H. J. & Wright, P. E. (1991) *Annu. Rev. Biophys. Biophys. Chem. 20*, 519–538.
Ellis, R. E. (1994) *Curr. Opin. Struct. Biol. 4*, 117–122.
Englander, S. W. & Mayne, L. (1992) *Annu. Rev. Biophys. Biomol. Struct. 21*, 243–265.
Eriksson, A. E., Baase, W. A., & Matthews, B. W. (1993) *J. Mol. Biol. 229*, 747–769.
Eriksson, A. E., Baase, W. A., Zhang, X. J., Heinz, D. W., Blaber, M., Baldwin, E. P., & Matthews, B. W. (1992) *Science 255*, 178–183.
Evans, P. A., Topping, K. D., Woolfson, D. N., & Dobson, C. M. (1991) *Proteins: Struct. Funct. Genet. 9*, 248–266.
Fasman, G. D., Ed. (1989) *Prediction of Protein Structure and the Principles of Protein Conformation*, Plenum, New York.
Fauchere, J.-L. & Pliska, V. (1983) *Eur. J. Med. Chem.-Chim. Ther. 18*, 369–375.
Fersht, A. R. (1972) *J. Mol. Biol. 64*, 497–509.
Fersht, A. R. (1993) *FEBS Lett. 325*, 5–16.
Fersht, A. R., Shi, J. P., Knill-Jones, J., Lowe, D. M., Wilkinson, A. J., Blow, D. M., Brick, P., Carter, P., Waye, M. M., & Winter, G. (1985) *Nature 314*, 235–238.
Fersht, A. R., Matouschek, A., & Serrano, L. (1992) *J. Mol. Biol. 224*, 771–782.

Fink, A. L., Anderson, W. D., & Antonino, L. (1988) *FEBS Lett. 229*, 123–126.
Frank, H. S. & Evans, M. W. (1945) *J. Chem. Phys. 13*, 507–532.
Garnier, J. (1990) *Biochimie 72*, 513–524.
Georgopoulos, C. & Welch, W. J. (1993) *Annu. Rev. Cell Biol. 9*, 601–634.
Gething, M. J. & Sambrook, J. F. (1992) *Nature 355*, 33–45.
Ghelis, C. & Yon, J. (1982) *Protein Folding*, Academic Press, New York.
Gierasch, L. M. & King, J., Eds. (1990) *Protein Folding*, American Association for the Advancement of Science, Washington, D.C.
Godzik, A., Kolinski, A., & Skolnick, J. (1993) *J. Compt. Aided Mol. Des. 7*, 397–438.
Goldenberg, D. P. (1989) in *Protein Structure: A Practical Approach* (Creighton, T. E., Ed.) pp. 225–250, IRL Press, Oxford University Press, New York.
Goldenberg, D. P. & Creighton, T. E. (1985) *Biopolymers 24*, 167–182.
Green, S. M. & Shortle, D. (1993) *Biochemistry 32*, 10131–10139.
Greer, J. (1991) *Methods Enzymol. 202*, 239–252.
Griko, Y. V., Privalov, P. L., Sturtevant, J. M., & Venyaminov, S.Yu (1988a) *Proc. Natl. Acad. Sci. U.S.A. 85*, 3343–3347.
Griko, Y. V., Privalov, P. L., Venyaminov, S. Y., & Kutyshenko, V. P. (1988b) *J. Mol. Biol. 202*, 127–138.
Haass, C., Koo, E. H., Mellon, A., Hung, A. Y., & Selkoe, D. J. (1992a) *Nature 357*, 500–503.
Haass, C., Schlossmacher, M. G., Hung, A. Y., Vigopelfrey, C., Mellon, A., Ostaszewski, B. L., Lieberburg, I., Koo, E. H., Schenk, D., Teplow, D. B., & Selkoe, D. J. (1992b) *Nature 359*, 322–325.
Haber, H. & Anfinsen, C. B. (1962) *J. Biol. Chem. 237*, 1839–1844.
Harrison, S. C. & Durbin, R. (1985) *Proc. Natl. Acad. Sci. U.S.A. 82*, 4028–4030.
Hartl, F. U., Hlodan, R., & Langer, T. (1994) *Trends Biochem. Sci. 19*, 20–26.
Haynie, D. T. & Freire, E. (1993) *Proteins: Struct. Funct. Genet. 16*, 115–140.
Hendrick, J. P. & Hartl, F. U. (1993) *Annu. Rev. Biochem. 62*, 349–384.
Hirs, C. H. W., Moore, S., & Stein, W. H. (1960) *J. Biol. Chem. 235*, 633–647.
Hughson, F. M., Barrick, D., & Baldwin, R. L. (1991) *Biochemistry 30*, 4113–4118.
Hughson, F. M., Wright, P. E., & Baldwin, R. L. (1990) *Science 249*, 1544–1548.
Jaenicke, R., Ed. (1980) *Protein Folding*, Elsevier/North-Holland Biochemical, New York.
Jeffrey, D. G. & Saenger, W., Eds. (1991) *Hydrogen Bonding in Biological Structures*, Springer, New York.
Jeng, M. F., Englander, S. W., Elove, G. A., Wand, A. J., & Roder, H. (1990) *Biochemistry 29*, 10433–10437.
Jennings, P. A. & Wright, P. E. (1993) *Science 262*, 892–896.
Jennings, P. A., Finn, B. E., Jones, B. E., & Matthews, C. R. (1993) *Biochemistry 32*, 3783–3789.
Karplus, M. & Weaver, D. L. (1976) *Nature 260*, 404–406.
Kato, S., Shimamoto, N., & Utiyama, H. (1982) *Biochemistry 21*, 38–43.
Katta, V. & Chait, B. T. (1993) *J. Am. Chem. Soc. 115*, 6317–6321.
Kauzmann, W. (1954) in *The Mechanism of Enzyme Action* (McElroy, W. D., & Glass, B., Eds.) pp 70–120, John Hopkins Press, Baltimore, MD.
Kauzmann, W. (1959) *Adv. Protein Chem. 14*, 1–63.
Kellis, J. T., Jr., Nyberg, K., & Fersht, A. R. (1989) *Biochemistry 28*, 4914–4922.
Kim, P. S. & Baldwin, R. L. (1990) *Annu. Rev. Biochem. 59*, 631–660.
Kleywegt, G. J., Bergfors, T., Senn, H., Le Motte, P., Gsell, B., Shudo, K., & Jones, A. T. (1994) *Structure 2*, 1241–1258.
Kuwajima, K. (1977) *J. Mol. Biol. 114*, 241–258.
Kuwajima, K. (1989) *Proteins: Struct. Funct. Genet. 6*, 87–103.
Kuwajima, K., Nitta, K., Yoneyama, M., & Sugai, S. (1976) *J. Mol. Biol. 106*, 359–373.
Kuwajima, K., Yamaya, H., Miwa, S., Sugai, S., & Nagamura, T. (1987) *FEBS Lett. 221*, 115–118.
Labhardt, A.M. (1982) *J. Mol. Biol. 157*, 331–355.

Landry, S. J., Jordan, R., McMacken, R., & Gierasch, L. M. (1992) *Nature 355*, 455–457.
Landry, S. J., Zeilstra-Ryalls, J., Fayet, O., Georgopoulos, C., & Gierasch, L. M. (1993) *Nature 364*, 255–258.
Levinthal, C. (1969) in *Mossbauer Spectroscopy in Biological Systems* (Degennes, P., Ed.) pp 22–24, University of Illinois Press, Urbana, IL.
Lim, W. A. & Sauer, R. T. (1989) *Nature 339*, 31–36.
Liu, Z. P., Rizo, J., & Gierasch, L. M. (1994) *Biochemistry 33*, 134–142.
Logan, T. M., Olejniczak, E. T., Xu, R. X., & Fesik, S. W. (1993) *J. Biomol. NMR 3*, 225–231.
Marqusee, S., Robbins, V. H., & Baldwin, R. L. (1989) *Proc. Natl. Acad. Sci. U.S.A. 86*, 5286–5290.
Matouschek, A., Kellis, J. T., Jr., Serrano, L., & Fersht, A. R. (1989) *Nature 340*, 122–126.
Matsumura, M., Becktel, W. J., & Matthews, B. W. (1988) *Nature 334*, 406–410.
Matthews, B. W. (1993a) *Annu. Rev. Biochem. 62*, 139–160.
Matthews, C. R. (1993b) *Annu. Rev. Biochem. 62*, 653–683.
Matthews, C. R. & Hurle, M. R. (1987) *Bioessays 6*, 254–257.
McCaldon, P. & Argos, P. (1988) *Proteins: Struct. Funct. Genet. 4*, 99–122.
Michalovitz, D., Halevy, O., & Oren, M. (1990) *Cell 62*, 671–680.
Miller, S., Hanin, J., Lesk, A. M., & Chothia, C. (1987) *J. Mol. Biol. 196*, 641–656.
Milner, J. (1991) *Proc. R. Soc. Lond. B 245*, 139–145.
Milner, J. & Medcalf, E. A. (1991) *Cell 65*, 765–774.
Miranker, A., Radford, S. E., Karplus, M., & Dobson, C. M. (1991) *Nature 349*, 633–636.
Miranker, A., Robinson, C. V., Radford, S. E., Aplin, R. T., & Dobson, C. M. (1993) *Science 262*, 896–900.
Mottonen, J., Strand, A., Symersky, J., Sweet, R. M., Danley, D. E., Geoghegan, K. F., Gerard, R. D., & Goldsmith, E. J. (1992) *Nature 355*, 270–273.
Nall, B. T., Zuniga, E. H., White, T. B., Wood, L. C., & Ramdas, L. (1989) *Biochemistry 28*, 9834–9839.
Nelson, R. J., Ziegelhoffer, T., Nicolet, C., Wernerwashburne, M., & Craig, E. A. (1992) *Cell 71*, 97–105.
Neri, D., Billeter, M., Wider, G., & Wüthrich, K. (1992a) *Science 257*, 1559–1563.
Neri, D., Wider, G., & Wüthrich, K. (1992b) *Proc. Natl. Acad. Sci. U.S.A. 89*, 4397–4401.
Neupert, W., Hartl, F. U., Craig, E. A., & Pfanner, N. (1990) *Cell 63*, 447–450.
Northrop, J. H. (1932a) *J. Gen. Physiol. 16*, 323–337.
Northrop, J. H. (1932b) *J. Gen. Physiol. 16*, 339–348.
Ohgushi, M., & Wada, A. (1983) *FEBS Lett. 164*, 21–24.
Orengo, C. A., Flores, T. P., Taylor, W. R., & Thornton, J. M. (1993) *Protein Eng. 6*, 485–500.
Pace, C. N. (1990) *Trends Biochem. Sci. 15*, 14–17.
Pace, C. N. (1992) *J. Mol. Biol. 226*, 29–35.
Pace, C. N. (1993) *Chemtracts: Biochem. Mol. Biol. 4*, 102–105.
Pace, C. N. & Grimsley, G. R. (1988) *Biochemistry 27*, 3242–3246.
Pace, C. N., Shirley, B. A., & Thomson, J. A. (1989) in *Protein Structure: A Practical Approach* (Creighton, T. E., Ed.) pp 311–330, IRL Press, Oxford University Press, New York.
Page, M. I. & Jencks, W. P. (1971) *Proc. Natl. Acad. Sci. U.S.A. 68*, 1678–1683.
Pauling, L. & Corey, R. B. (1951) *Proc. Natl. Acad. Sci. U.S.A. 37*, 252–256.
Pauling, L., Corey, R. B., & Branson, H. R. (1951) *Proc. Natl. Acad. Sci. U.S.A. 37*, 205–211.
Pfeil, W. & Privalov, P. L. (1976a) *Biophys. Chem. 4*, 23–32.
Pfeil, W. & Privalov, P. L. (1976b) *Biophys. Chem. 4*, 33–40.
Pfeil, W. & Privalov, P. L. (1976c) *Biophys. Chem. 4*, 41–50.
Privalov, P. L. (1979) *Adv. Protein Chem. 33*, 167–241.
Privalov, P. L. (1992) in *Protein Folding* (Creighton, T. E., Ed.) pp 83–126, W.H. Freeman, New York.
Privalov, P. L. & Gill, S. J. (1988) *Adv. Protein Chem. 39*, 191–234.

Privalov, P. L. & Griko Yu. V. (1986) *J. Mol. Biol. 190*, 487–498.
Privalov, P. L. & Khechinashvili, N. N. (1974) *J. Mol. Biol. 86*, 665–684.
Privalov, P. L. & Potekhin, S. A. (1986) *Methods Enzymol. 131*, 4–51.
Provencher, S. W. & Gl:ockner, J. (1981) *Biochemistry 20*, 33–37.
Ptitsyn, O. B. (1987) *J. Protein Chem. 6*, 273–293.
Ptitsyn, O. B. (1992) in *Protein Folding* (Creighton, T. E., Ed.), W.H. Freeman, New York.
Richards, F. M. (1977) *Annu. Rev. Biophys. Bioeng. 6*, 151–176.
Richardson, J. S. & Richardson, D. C. (1989) in *Prediction of Protein Structure and the Principles of Protein Conformation* (Fasman, G. D., Ed.), Plenum, New York.
Richardson, J. S., Richardson, D. C., Tweedy, N. B., Gernert, K. M., Quinn, T. P., Hecht, M. H., Erickson, B. W., Yan, Y. B., McClain, R. D., Donlan, M. E., & Surles, M. C. (1992) *Biophys. J. 63*, 1186–1209.
Ring, C. S. & Cohen, F. E. (1993) *FASEB J. 7*, 783–790.
Rizo, J., Liu, Z. P. & Gierasch, L. M. (1994) *J. Biomol. NMR 4*, 741–760.
Robertson, A. D. & Baldwin, R. L. (1991) *Biochemistry 30*, 9797–9714.
Rose, G. D. & Wolfenden, R. (1993) *Annu. Rev. Biophy. Biomol. Struct. 22*, 381–415.
Sanger, F., Tompson, E. O. P., & Tuppy, H. (1952) *2nd Congr. Intern. Biochim. Chim. Biol. IV*, 26–38.
Sauer, R. T., Jordan, S. R., & Pabo, C. O. (1990) *Adv. Protein Chem. 40*, 1–61.
Schellman, J. A. (1987) *Annu. Rev. Biophys. Biophys. Chem. 16*, 115–137.
Schmid, F. X. (1992) in *Protein Folding* (Creighton, T. E., Ed.) pp 197–241, W.H. Freeman, New York.
Schmid, F. X. (1993) *Annu. Rev. Biophys. Biomol. Struct. 22*, 123–142.
Segawa, S. & Sugihara, M. (1984a) *Biopolymers 23*(11 Pt 2), 2473–2488.
Segawa, S. & Sugihara, M. (1984b) *Biopolymers 23*(11 Pt 2), 2489–2498.
Sela, M., White, F. H., & Anfinsen, C. B. (1957) *Science 125*, 691–692.
Semisotnov, G. V., Rodionova, N. A., Razgulyaev, O. I., Uversky, V. N., Gripas, A. F., & Gilmanshin, R. I. (1991) *Biopolymers 31*, 119–128.
Serrano, L., Matouschek, A., & Fersht, A. R. (1992) *J. Mol. Biol. 224*, 805–818.
Sharp, K. A., Nicholls, A., Friedman, R., & Honig, B. (1991) *Biochemistry 30*, 9686–9697.
Shin, H. C., Merutka, G., Waltho, J. P., Tennant, L. L., Dyson, H. J., & Wright, P. E. (1993) *Biochemistry 32*, 6356–6364.
Shirley, B. A., Stanssens, P., Hahn, U., & Pace, C. N. (1992) *Biochemistry 31*, 725–732.
Shortle, D. (1993) *Curr. Opin. Struct. Biol. 3*, 66–74.
Shortle, D., Stites, W. E., & Meeker, A. K. (1990) *Biochemistry 29*, 8033–8041.
Skolnick, J. & Kolinski, A. (1990) *Science 250*, 1121–1125.
Spolar, R. S., Livingstone, J. R., & Record, M. T., Jr. (1992) *Biochemistry 31*, 3947–3955.
Stahl, N. & Jencks, W. P. (1986) *J. Am. Chem. Soc. 108*, 4196–4205.
Stickle, D. F., Presta, L. G., Dill, K. A., & Rose, G. D. (1992) *J. Mol. Biol. 226*, 1143–1159.
Sturtevant, J. M. (1987) *Annu. Rev. Phys. Chem. 38*, 463–488.
Sun, Y., Spellmeyer, D., Perlman, D.A., & Kollman, P. (1992) *J. Am. Chem. Soc. 114*, 6798–6801.
Tanford, C. (1968) *Adv. Protein Chem. 23*, 121–282.
Tanford, C. (1970) *Adv. Protein Chem. 24*, 1–95.
Tanford, C. (1980) *The Hydrophobic Effect*, Wiley, New York.
Thomas, P. J., Ko, Y. H., & Pedersen, P. L. (1992) *FEBS Lett. 312*, 7–9.
Tsuji, T., Chrunyk, B. A., Chen, X. W., & Matthews, C. R. (1993) *Biochemistry 32*, 5566–5575.
Utiyama, H., & Baldwin, R. L. (1986) *Methods Enzymol. 131*, 51–70.
Varley, P., Gronenborn, A.M., Christensen, H., Wingfield, P.T., Pain, R.H., & Clore, G.M. (1993) *Science 260*, 1110–1113.
Weissman, J. S., & Kim, P. S. (1991) *Science 253*, 1386–1393.
Weissman, J. S., & Kim, P. S. (1992a) *Proc. Natl. Acad. Sci. U.S.A. 89*, 9900–9904.

Weissman, J. S., & Kim, P. S. (1992b) *Science 256*, 112–114.
Williams, R. P. (1989) *Eur. J. Biochem. 183*, 479–497.
Wu, H. (1929) *Am. J. Physiol. 90*, 562–563.
Wu, H. (1931) *Chinese J. Physiol. 5*, 321–344.
Wüthrich, K. (1994) *Curr. Opin. tStruct. Biol. 4*, 93–99.
Xi, Y. G., Ingrosso, L., Ladogana, A., Masullo, C., & Pocchiari, M. (1992) *Nature 356*, 598–601.
Yutani, K., Ogasahara, K., & Kuwajima, K. (1992) *J. Mol. Biol. 228*, 347–350.
Zhang, J., Liu, Z. P., Jones, T. A., Gierasch, L. M., & Sambrook, J. F. (1992) *Proteins: Struct. Funct. Genet. 13*, 87–99.

Chapter 7

Abbadi, A., Mcharfi, M., Aubry, A., Prenilat, S., Boussard, G., & Marroud, M. (1991) *J. Am. Chem. Soc. 113*, 2729–2735.
Adams, M. W. W. (1990) *FEMS Microbiobiol. Rev. 75*, 219–237.
Adman, E., Watenpaugh, K. D., & Jensen, L. H. (1975) *Proc. Natl. Acad. Sci. U.S.A. 72*, 4854–4858.
Archambault, J., Lacroute, F., Rluet, A., & Friesen, J. D. (1992) *Mol. Cell. Biol. 12*, 4142–4152.
Baleja, J. D., Marmorstein, R., Harrison, S. C., & Wagner, G. (1992) *Nature 356*, 450–453.
Barlow, P. N., Luisi, B., Milner, A., Elliott, M., & Everett, R. (1994) *J. Mol. Biol. 237*, 201–211.
Blake, P. R., Day, M. W., Hsu, B. T., Joshua-Tor, L., Park, J. B., Hare, D. R., Adams, M. W. W., Rees, D. C., & Summers, M. F. (1992a) *Protein Sci. 1*, 1508–1521.
Blake, P. R., Lee, B., Park, J.-B., Zhou, Z. H., Adams, M. W. W., & Summers, M. F. (1994) *New J. Chem. 18*, 387–395.
Blake, P. R., Lee, B., Summers, M. F., Adams, M. W. W., Park, J.-B., Zhou, H. Z., & Bax, A. (1992b) *J. Biomol. NMR 2*, 527–533.
Blake, P. R., Park, J. B., Adams, M. W. W., & Summers, M. F. (1992c) *J. Am. Chem. Soc. 114*, 4931–4933.
Blake, P. R., Park, J. B., Bryant, F. O., Aono, S., Magnuson, J. K., Eccleston, E., Howard, J. B., Summers, M. F., & Adams, M. W. W. (1991) *Biochemistry 30*, 10885–10895.
Blake, P. R., Park, J. B., Zhou, Z. H., Hare, D. R., Adams, M. W. W., & Summers, M. F. (1992d) *Protein Sci. 1*, 1508–1521.
Blake, P. R. & Summers, M. F. (1994) *Adv. Biophys.Chem.*, Vol. 4, (Bush, C. A., Ed.), pp. 1–30, JAI Press, Greenwich, CT.
Bobsein, B. R., & Myers, R. J. (1981) *J. Biol. Chem. 256*, 5313–5316.
Brevard, C. & Granger, P. (1981) *Handbook of High Resolution Multinuclear NMR*, Wiley, New York.
Clemens, K. R., Wolf, V., McBryant, S. J., Zhang, P., Liao, X., Wright, P. E., & Gottesfeld, J. M. (1993) *Science 260*, 530–533.
Day, M. W., Hsu, B. T., Joshua-Tor, L., Park, J. B., Zhou, Z. H., Adams, M. W. W., & Rees, D. C. (1992) *Protein Sci. 1*, 1494–1507.
Everett, R. D., Barlow, P., Milner, A., Luisi, B., Orr, A., Hope, G., & Lyon, D. (1993) *J. Mol. Biol. 234*, 1–10.
Fourmy, D., Dardel, F., & Blanquet, S. (1993a) *J. Mol. Biol. 231*, 1078–1089.
Fourmy, D., Meinnel, T., Mechulan, Y., & Blanquet, S. (1993b) *J. Mol. Biol. 231*, 1068–1077.
Freemont, P. S. (1993) *Ann. N.Y. Acad. Sci. 684*, 174–192.
Gardner, K. H., Pan, T., Narula, S., Rivera, E., & Coleman, J. E. (1991) *Biochemistry 30*, 11292–11302.
Green, L. M. & Berg, J. M. (1990) *Proc. Natl. Acad. Sci. U.S.A. 87*, 6403–6407.
Hard, T., Kellenbach, E., Boblens, R., Maler, B. A., Dahlman, K., Freedman, L. P., Carlstedt-Duke, J., Yamamoto, K. R., Gustafsson, J.-A., & Kaptein, R. (1990) *Science 249*, 157–160.
Helmann, J. D., Ballard, B. T., & Walsh, C. T. (1990) *Science 247*, 946–948.

Johnston, M. (1987) *Nature 328*, 353–355.
Karplus, M. (1959) *J. Chem. Phys. 30*, 11–15.
Klevit, R., Herriott, J. R., & Horvath, S. J. (1990) *Proteins: Struct., Funct., & Genet. 7*, 215–226.
Knegtel, R. M. A., Katahira, M., Schilthuis, J. G., Bonvin, A. M. J. J., Boelens, R., Eib, D., van der Saag, P. T., & Kaptein, R. (1993) *J. Biomol. NMR 3*, 1–17.
Kochoyan, M., Havel, T. F., Nguyen, D. T., Dahl, C. E., Keutmann, H. T., & Weiss, M. A. (1991a) *Biochemistry 30*, 3371–3386.
Kochoyan, M., Keutmann, H. T., & Weiss, M. A. (1991b) *Proc. Natl. Acad. Sci. U.S.A. 88*, 8455–8459.
Kraulis, P.J. (1991) *J. Appl. Crystallogr. 24*, 946–950.
Kraulis, P. J., Raine, A. R. C., Gadhavi, P. L., & Laue, E. D. (1992) *Nature 356*, 448–450.
Lee, M. S., Gippert, G. P., Soman, K. V., Case, D. A., & Wright, P. E. (1989) *Science 254*, 635–637.
Lee, M. S., Kliewer, S. A., Provencal, J., Wright, P. E., & Evans, R. M. (1993) *Science 260*, 1117–1120.
Luisi, B. F., Xu, W. X., Otwinowski, Z., Freedman, L. P., Yamamoto, K. R., & Sigler, P. B. (1991) *Nature 352*, 497–505.
Ma, J. & Ptashne, M. (1987) *Cell 48*, 847–853.
Marmorstein, R., Carey, M., Ptashne, M., & Harrison, S. C. (1992) *Nature 356*, 408–414.
Martin, D. I. K. & Orkin, S. H. (1990) *Genes Dev. 4*, 1886–1898.
Mayaux, J.-F. & Blanquet, S. (1981) *Biochemistry 20*, 4647–4654.
Miller, J., McLachlan, A. D., & Klug, A. (1985) *EMBO J. 4*, 1609–1615.
Miller, W. T., Hill, K. A., & Schimmel, P. (1991) *Biochemistry 30*, 6970–6976.
Morellet, N., Julian, N., de Rocquigny, H., Maigret, B., Darlix, J. L., & Roques, B. P. (1992) *EMBO J. 11*, 3059–3065.
Omichinski, J. G., Clore, G. M., Appella, E., Sakaguchi, K., & Gronenborn, A. M. (1990) *Biochemistry 29*, 9324–9334..
Omichinski, J. G., Clore, G. M., Robien, M., Sakaguchi, K., Appella, E., & Gronenborn, A. M. (1992) *Biochemistry 31*, 3907–3917.
Omichinski, J. G., Clore, G. M., Sakaguchi, K., Appella, E., & Gronenborn, A. M. (1991) *FEBS Lett. 292*, 25–30.
Omichinski, J. G., Clore, G. M., Schaad, O., Felsenfeld, G., Trainor, C., Appella, E., Stahl, S. J., & Gronenborn, A. M. (1993) *Science* 261, 438–446.
Pan, T. & Coleman, J. E. (1991) *Biochemistry 30*, 4212–4222.
Pappaport, J., Cho, K., Saltzman, A., Prenger, J., Golomb, M., & Weinmann, R. (1988) *Mol. Cell. Biol. 8*, 3136–3142.
Pavletich, N. P. & Pabo, C. O. (1991) *Science 252*, 809–817.
Posorske, L. H., Cohn, M., Yanagisawa, N., & Auld, D. S. (1979) *Biochim. Biophys. Acta 576*, 128–133.
Qian, X., Gozani, S. N., Yoon, H., Jeon, C., Agarwal, K., & Weiss, M. A. (1993) *Biochemistry 32*, 9944–9959.
Rajavashisth, T. B., Taylor, A. K., Andalibi, A., Svenson, K. L., Karen, L., & Lusis, A. J. (1989) *Science 245*, 640–643.
Rees, D. C., Lewis, M., & Lipscomb, W. N. (1983) *J. Mol. Biol. 168*, 367–787.
Reines, D. & Mote, Jr., J. (1993) *Proc. Natl. Acad. Sci. U.S.A., 90*, 1917–1921.
Richardson, J. S. (1981) *Advances in Protein Chemistry 34*, 167–339.
Santos, R. A., Gruff, E. S., Koch, S. A., & Harbison, G. S. (1991) *J. Am. Chem. Soc. 113*, 469–475.
Sato, S. M. & Sargent, T. D. (1991) *Development 112*, 747–753.
Schwabe, J. W. R., Neuhaus, D., & Rhodes, D. (1990) *Nature 348*, 458–461.
Siva-Raman, L., Reines, D., & Kane, C. M. (1990) *J. Biol. Chem. 265*, 14554–14560.
South, T. L., Blake, P. R., Sowder III, R. C., Arthur, L. O., Henderson, L. E., & Summers, M. F. (1990) *Biochemistry 29*, 7786–7789.

Stetter, K. O., Fiala, G., Huber, G., Huber, R., & Segerer, G. (1990) *FEMS Microbiobiol. Rev.* 75, 117–124.
Summers, M. F. (1988) *Coordination Chemistry Review* 86, 43–134.
Summers, M. F. (1991) *J. Cell. Biochem.* 45, 41–48.
Summers, M. F., Henderson, L. E., Chance, M. R., Bess, J. W., South, T. L., Blake, P. R., Sagi, I., Perez-Alvarado, G., Sowder III, R. C., Hare, D. R., & Arthur, L. O. (1992) *Prot. Sci.* 1, 563–574.
Summers, M. F., South, T. L., Kim, B., & Hare, D. R. (1990) *Biochemistry* 29, 329–340.
Watenpaugh, K. D., Sieker, L. C., Herriott, J. R., & Jensen, L. H. (1973) *Acta Crystallog. B29*, 943–956.
Webb, J. R. & McMaster, W. R. (1993) *J. Biol. Chem.* 268, 13994–14002.
Whyatt, D. J., deBoer, E., & Grosveld, F. (1993) *EMBO J.* 12, 4993–5005.
Wright, J. G., Tsang, H.-T., Penner-Han, J. E., & O'Halloran, T. V. (1990) *J. Am. Chem. Soc.* 112, 2434–2435.
Yang, H.-Y. & Evans, T. (1992) *Mol. Cell. Biol.* 12, 4562–4570.
Zerbe, O., Pountney, D. L., von Philipsborn, W., & Vasak, M. (1994) *J. Am. Chem. Soc.* 116, 377–378.

Chapter 8

Albery, W. J. & Knowles, J. R. (1976a) *Biochemistry* 15, 5627–5631.
Albery, W. J. & Knowles, J. R. (1976b) *Biochemistry* 15, 5631–5640.
Banner, D. W., Bloomer, A. C., Petsko, G. A., Phillips, D. C., Pogson, C. I., & Wilson, I. A. (1975) *Nature* 255, 609–614.
Bernasconi, C. F. (1992) *Adv. Phys. Org. Chem.* 27, 119–238.
Blacklow, S. C. & Knowles, J. R. (1990) *Biochemistry* 29, 4099–4108.
Breslow, R. (1991) *Acc. Chem. Res.* 24, 317–324.
Bruice, T. C. & Schmir, G. L. (1959) *J. Am. Chem. Soc.* 81, 4552–4556.
Cardinale, G. J. & Abeles, R. H. (1968) *Biochemistry* 7, 3970–3978.
Chiang, Y. & Kresge, A. J. (1991) *Science* 253, 395–400.
Chiang, Y., Kresge, A. J., Pruszynski, P., Schepp, N. P., & Wirz, J. (1990) *Angew. Chem., Int. Ed. Engl.* 29, 792–794.
Coulson, A. F. W., Knowles, J. R., Priddle, J. D., & Offord, R. E. (1970) *Nature* 227, 180–181.
Creighton, D. J. & Murthy, N. S. R. K. (1992) *Enzymes* 20, 323–421.
Davenport, R. C., Bash, P. A., Seaton, B. A., Karplus, M. A., Petsko, G. A., & Ringe, D. (1991) *Biochemistry*, 30, 5821–5826.
Fee, Judith A., Hegeman, G. D. & Kenyon, G. L. (1974) *Biochemistry* 13, 2533–2538.
Fisher, L. M., Albery, W. J., & Knowles, J. R. (1976) *Biochemistry* 15, 5621–5626.
Fletcher, S. J., Herlihy, J. M., Albery, W. J., & Knowles, J. R. (1976) *Biochemistry* 15, 5612–5617.
Gallo, K. A. & Knowles, J. R. (1993) *Biochemistry* 32, 3981–3990.
Gerlt, J. A. & Gassman, P. G. (1992) *J. Am. Chem. Soc.* 114, 5928–5934.
Gerlt, J. A. & Gassman, P. G. (1993a) *Biochemistry* 32, 11943–11952.
Gerlt, J. A. & Gassman, P. G. (1993b) *J. Am. Chem. Soc.* 115, 11552–11568.
Gerlt, J. A., Kenyon, G. L., Kozarich, J. W., Neidhart, D. J., Petsko, G. A., & Powers, V. M. (1992) *Curr. Opin. Struct. Biol.* 2, 736–742.
Gerlt, J. A., Kozarich, J. W., Kenyon, G. L., & Gassman, P. G. (1991) *J. Am. Chem. Soc.* 113, 9667–9669.
Guthrie, J. P. (1974) *J. Am. Chem. Soc.* 96, 3608–3615.
Guthrie, J. P. (1977) *J. Am. Chem. Soc.* 99, 3991–4001.
Hammond, G. S. (1955) *J. Am. Chem. Soc.* 77, 334–338.
Hartman, F. C. (1971) *Biochemistry* 10, 146–154.

Hartman, F. C. & Ratrie, H. (1977) *Biochem. Biophys. Res. Commun.* 77, 746–752.
Herlihy, J. M., Maister, S. G., Albery, W. J., & Knowles, J. R. (1976) *Biochemistry 15*, 5601–5606.
Hibbert, F. & Emsley, J. (1990) *Adv. Phys. Org. Chem. 26*, 255–379.
Jencks, W. P. (1981) *Chem. Soc. Rev. 10*, 345–375.
Kallarakal, A. T., Mitra, B., Kozarich, J. W., Gerlt, J. A., Clifton, J. R., Petsko, G. A., & Kenyon, G. L. (1995) *Biochemistry 34*, 2788–2797.
Kenyon, G. L. & Hegeman, G. D. (1979) *Adv. Enzymol. Relat. Areas Mol. Biol. 50*, 325–360.
Kluger, R. (1990) *Chem. Rev. 90*, 1151–1169.
Knowles, J. R. (1976) *Crit. Rev. Biochem. 4*, 165–173.
Komives, E. A., Chang, L. C., Lolis, E., Tilton, R. F., Petsko, G. A., & Knowles, J. R. (1991) *Biochemistry 30*, 3011–3019.
Landro, J. A., Gerlt, J. A., Kozarich, J. W., Koo, C. W., Shah, V. J., Kenyon, G. L., Neidhart, D. J., Fujita, S., & Petsko, G. A. (1994) *Biochemistry 33*, 14213–14220.
Landro, J. A., Kallarakal, A., Ransom, S. C., Gerlt, J. A., Kozarich, J. W., Neidhart, D. J., & Kenyon, G. L. (1991) *Biochemistry 30*, 9274–9281.
Lin, D. T., Powers, V. M., Reynolds, L. J., Whitman, C. P., Kozarich, J. W., & Kenyon, G. L. (1988) *J. Am. Chem. Soc. 110*, 323–324.
Lodi, P. J. & Knowles, J. R. (1991) *Biochemistry 30*, 6948–6956.
Maister, S. G., Pett, C. P., Albery, W. J., & Knowles, J. R. (1976) *Biochemistry 15*, 5607–5612.
Mitra, B., Kallarakal, A. T., Kozarich, J. W., Gerlt, J. A., Clifton, J. R., Petsko, G. A., & Kenyon, G. L. (1995) *Biochemistry 34*, 2777–2787.
Neidhart, D. J., Distefano, M. D., Tanizawa, K., Soda, K., Walsh, C. T., & Petsko, G. A. (1987) *J. Biol. Chem. 262*, 15323–15326.
Nickbarg, E. B., Davenport, R. C., Petsko, G. A., & Knowles, J. R. (1988) *Biochemistry 27*, 5948–5960.
Plaut, B. & Knowles, J. R. (1972) *Biochem. J. 129*, 311–320.
Powers, V. M., Koo, C. W., Kenyon, G. L., Gerlt, J. A., & Kozarich, J. W. (1991) *Biochemistry 30*, 9255–9263.
Rieder, S. V. & Rose, I. A. (1959) *J. Biol. Chem. 234*, 1007–1010.
Schmidt, D. J. & Westheimer, F. H. (1971) *Biochemistry 10*, 1249–1253.
Schwab, J. M. & Henderson, B. S. (1990) *Chem. Rev. 90*, 1203–1245.
Strauss, D., Raines, R., Kawashima, R., Knowles, J. R., & Gilbert, W. (1985) *Proc. Nat. Acad. Sci. U.S.A. 82*, 2272–2276.
Tanner, M. E., Gallo, K. A., & Knowles, J. R. (1993) *Biochemistry 32*, 3998–4006.
Thibblin, A. & Jencks, W. P. (1979) *J. Am. Chem. Soc. 101*, 4963–4973.
Waley, S. G., Miller, J. C., Rose, I. A., & O'Connell, E. L. (1970) *Nature 227*, 181.
Whitman, C. P., Hegeman, G. C., Cleland, W. W., & Kenyon, G. L. (1985) *Biochemistry 24*, 3936–3942.

Chapter 9

Atkins, W. M., & Sligar, S. G. (1989) *J. Am. Chem. Soc. 111*, 2715–2717.
Beck von Bodman, S., Schuler, M. A., Jollie, D. R., & Sligar, S. G. (1987) *Proc. Natl. Acad. Sci. U.S.A. 83*, 9443–9447.
Egeberg, K. D., Springer, B. A., Sligar, S. G., Carver, T. E., Rohlfs, R. J., & Olson, J. S. (1990) *J. Biol. Chem. 265*, 11788–11795.
Hemsley, A., Arnheim, N., Toney, M. D., Cortopassi, G., & Galas, D. J. (1989) *Nucleic Acids Res. 17*, 6545–6551.
Hernan, R. A., Hui, H. L., Andracki, M. E., Noble, R. W., Sligar, S. G., Walder, J. A., & Walder, R. Y. (1992) *Biochemistry 31*, 8619–8628.
Higuchi, R., Krummel, B., & Saiki, R. (1988) *Nucleic Acids Res. 16*, 7351–7367.

Ho, S. N., Hunt, H. D., Horton, R. M., Pullen, J. K., & Pease, L. R. (1989) *Gene 77*, 51–59.
Innis, M. A., Gelfand, D. H., Sninsky, J. J., & White, T. J. (1990) *PCR Protocols*, Academic Press, New York.
Kadowaki, H., Kadowaki, T., Wondisford, F. E., & Taylor, S. I. (1989) *Gene 76*, 161–166.
Kunkel, T. A., Roberts, J. D., & Zakour, R. A. (1987) *Methods Enzymol. 154*, 367–382.
Loida, P. J. & Sligar, S. G. (1993a) *Protein Eng. 2*, 207–212.
Loida, P. J. & Sligar, S. G. (1993b) *Biochemistry 32*, 11530–11538.
Olsen, D. B., Sayers, J. R., & Eckstein, F. (1993) *Methods Enzymol. 217*, 189–217.
Rodgers, K. K., & Sligar, S.G. (1991a) *J. Am. Chem. Soc. 113*, 9419–9421.
Rodgers, K. K., & Sligar, S.G. (1991b) *J. Mol. Biol. 221*, 1453–1460.
Rohlfs, R. J., Mathews, A. J., Carver, T. E., Olson, J. S., Springer, B. A., Egeberg, K. D. & Sligar, S.G. (1990) *J. Biol. Chem. 265*, 3168–3176.
Sambrook, J., Fritsch, E. F., & Maniatis, T. (1989) *Molecular Cloning: A Laboratory Manual*, Cold Spring Harbor Press, NY.
Sligar, S. G. & Murray, R. I. (1986) in *Cytochrome P-450: Structure, Mechanism, and Biochemistry* (Ortiz de Montellano, P. R., Ed.) pp 429–503, Plenum, New York.
Sligar, S. G., Filipovic, D., & Stayton, P. S. (1991) *Methods Enzymol. 206*, 31–49.
Sonveaux, E. (1986) *Bioorg. Chem. 14*, 274–325.
Zhao, L. J., Zhang, X. Q., & Padmanabhan (1993) *Methods Enzymol. 217*, 218–227.

Chapter 10

Addadi, L., Jaffe, E. K., & Knowles, J. R. (1983) *Biochemistry 22*, 4494–4501.
Alt, F. W., Blackwell, T. K., & Yancopoulos, G. D. (1987) *Science 238*, 1079–1087.
Andrews, P. R., Smith, G. D., & Young, I. G. (1973) *Biochemistry 12*, 3492–3498.
Arevalo, J. H., Stura, E. A., Taussig, M. J., & Wilson, I. A. (1993a) *J. Mol. Biol. 231*, 103–108.
Arevalo, J. H., Taussig, M. J., & Wilson, I. A. (1993b) *Nature 365*, 859–863.
Ashley, J. A., Lo, C.-H. L., McElhaney, G. P., Wirsching, P., & Janda, K. D. (1993) *J. Am. Chem. Soc. 115*, 2515–2516.
Baldwin, E. & Schultz, P. G. (1989) *Science 24*, 1104–1107.
Barbas, C. F., Rosenblum, J. S., & Lerner, R. A. (1993) *Proc. Natl. Acad. Sci. U.S.A.. 90*, 6385–6389.
Bartlett, P.A. & Marlow, C.K. (1983) *Biochemistry 22*, 4618–4624.
Bartlett, P. A., Nakagawa, Y., Johnson, C. R., Reich, S. H., & Luis, A. (1988) *J. Org. Chem. 53*, 3195–3210.
Benkovic, S. J., Adams, J. A., Borders, C. L., Janda, K. D., & Lerner, R. A. (1990) *Science 250*, 1135–1139.
Braisted, A. C. & Schultz, P. G. (1990) *J. Am. Chem. Soc. 112*, 7430–7431.
Braisted, A. C. & Schultz, P. G. (1994) *J. Am. Chem. Soc. 116*, 2211–2212.
Branden, C. & Tooze, J. (1991) *Introduction to Protein Structure*, Garland Publishing, New York.
Brunger, A. T., Leahy, D. J., Hynes, T. R., & Fox, R. O. (1991) *J. Mol. Biol. 221*, 239–256.
Bryan, P., Pantoliano, M. W., Quill, S. G., Hsiao, H.-Y., & Poulos, T. (1986) *Proc. Natl. Acad. Sci. U.S.A. 83*, 3743–3745.
Campbell, P. A., Tarasow, T. M., Massefski, W., Wright, P. E., & Hilvert, D. (1993) *Proc. Natl. Acad. Sci. U.S.A. 90*, 8663–8667.
Carter, P. & Wells, J. A. (1987) *Science 237*, 394–399.
Carter, P. & Wells, J. A. (1988) *Nature 332*, 564–568.
Charbonnier, J.-B., Carpenter, E., Gigant, B., Golinelli-Pimpaneau, B., Tawfik, D., Eshhar, Z., Green, B. S., & Knossow, M. (1995) *Proc. Natl. Acad. Sci. U.S.A. 92*, 11721–11725.
Chook, Y. M., Gray, J. V., Ke, H., & Lipscomb, W. N. (1994) *J. Mol. Biol. 240*, 476–500.
Chook, Y. M., Ke, H., & Lipscomb, W. N. (1993) *Proc. Natl. Acad. Sci. U.S.A. 90*, 8600–8603.
Clark, B. R., & Engvall, E. (1980) *Enzyme-Immunoassay*, CRC Press, Boca Raton, FL.

Cochran, A. G. & Schultz, P. G. (1990a) *Science 249*, 781–783.
Cochran, A. G. & Schultz, P. G. (1990b) *J. Am. Chem. Soc. 112*, 9414–9415.
Collet, T. A., Roben, P., O'Kennedy, R., Barbas, C. F. I., Burton, D. R., & Lerner, R. A. (1992) *Proc. Natl. Acad. Sci. U.S.A. 89*, 10026–10030.
Colman, P. M. (1988) *Adv. Immunol. 43*, 99–132.
Copley, S. D., & Knowles, J. R. (1985) *J. Am. Chem. Soc. 107*, 5306–5308.
Copley, S. D., & Knowles, J. R. (1987) *J. Am. Chem. Soc. 109*, 5008–5013.
Cravatt, B. F., Ashley, J. A., Janda, K. D., Boger, D. L., & Lerner, R. A. (1994) *J. Am. Chem. Soc. 116*, 6013–6014.
Davies, D. R., Padlan, E. A., & Sheriff, S. (1990) *Annu. Rev. Biochem. 59*, 439–473.
Erlanger, B. F. (1980) *Methods Enzymol. 70*, 85–104.
Ferrin, T.E., Huang, C.C., Jarvis, L.E., & Langridge, R. (1988) *J. Mol. Graphics 6*, 13–27.
Fersht, A. (1985) *Enzyme Structure and Mechanism*, W.H. Freeman, New York.
Gajewski, J. J., Jurayj, J., Kimbrough, D. R., Gande, M. E., Ganem, B., & Carpenter, B. K. (1987) *J. Am. Chem. Soc. 109*, 1170–1186.
Gallacher, G., Jackson, C. S., Searcey, M., Goel, R., Mellor, G. W., Smith, C., & Brocklehurst, K. (1993) *Eur. J. Biochem. 214*, 197–207.
Garrard, L. J. & Zhukovsky, E. A. (1992) *Current Opinion Biotech. 3*, 474–480.
Gibbs, R. A., Benkovic, P. A., Janda, K. D., Lerner, R. A., & Benkovic, S. J. (1992a) *J. Am. Chem. Soc. 114*, 3528–3534.
Gibbs, R. A., Taylor, S., & Benkovic, S. J. (1992b) *Science 258*, 803–805.
Goding, J. W. (1983) *Monoclonal Antibodies: Principles and Practice*, Academic Press, New York.
Golinelli-Pimpaneau, B., Gigant, B., Bizebard, T., Navaza, J., Saludjian, P., Zemel, R., Tawfik, D. S., Eshhar, Z., Green, B. S., & Knossow, M. (1994) *Structure 2*, 175–183.
Görisch, J. (1978) *Biochemistry 17*, 3700–3705.
Gouverneur, V. E., Houk, K. N., Pascual-Teresa, B., Beno, B., Janda, K. D., & Lerner, R. A. (1993) *Science 262*, 204–208.
Gray, J. V., Eren, D., & Knowles, J. R. (1990) *Biochemistry 29*, 8872–8878.
Guilford, W. J., Copley, S. D., & Knowles, J. R. (1987) *J. Am. Chem. Soc. 109*, 5013–5019.
Guo, J., Huang, W., & Scanlan, T. S. (1994) *J. Am. Chem. Soc. 116*, 6062–6069
Guo, J., Huang, W., Zhou, G. W., Fletterick, R. J., & Scanlan, T. S. (1995) *Proc. Natl. Acad. Sci. U.S.A. 92*, 1694–1698.
Harlow, E. & Lane, D. (1988) *Antibodies: A Laboratory Manual*, Cold Spring Harbor Laboratory, Cold Spring Harbor, N.Y.
Haynes, M. R., Stura, E. A., Hilvert, D., & Wilson, I. A. (1994) *Science 263*, 646–652.
Hilvert, D., Carpenter, S. H., Nared, K. D., & Auditor, M.-T. M. (1988) *Proc. Natl. Acad. Sci. U.S.A. 85*, 4953–4955.
Hilvert, D., Hill, K. W., Nared, K. D., & Auditor, M.-T. M. (1989) *J. Am. Chem. Soc. 111*, 9261–9262.
Hilvert, D. & Nared, K. D. (1988) *J. Am. Chem. Soc. 110*, 5593–5594.
Iverson, B. L., Iverson, S. A., Roberts, V. A., Getzhoff, E. D., Tainer, J. A., Benkovic, S. J., & Lerner, R. A. (1990) *Science 249*, 659–662.
Iverson, B. L. & Lerner, R. A. (1989) *Science 243*, 1184–1188.
Jackson, D. Y., Jacobs, J. W., Sugasawara, R., Reich, S. H., Bartlett, P. A., & Schultz, P. G. (1988) *J. Am. Chem. Soc. 110*, 4841–4842.
Jackson, D. Y., Liang, M. N., Bartlett, P. A., & Schultz, P. G. (1992) *Angew. Chem. Int. Ed. Engl. 31*, 182–183.
Jackson, D. Y. & Schultz, P. G. (1991) *J. Am. Chem. Soc. 113*, 2319–2321.
Janda, H.D., Lo, L.-C., Lo, C.-H. L., Sim, M.-M., Wang, R., Wong, C.-H., & Lerner, R. A. (1997) *Science 275*, 945–948.
Janda, K. D., Schloeder, D., Benkovic, S. J., & Lerner, R. A. (1988) *Science 241*, 1188–1191.
Jencks, W. P. (1969) *Catalysis in Chemistry and Enzymology*, McGraw-Hill, New York.

Jencks, W. P. (1975) *Adv. Enzymol 43*, 219–410.
Kabat, E. A. (1976) *Structural Concepts in Immunology and Immunochemistry.* Holt, Rinehart, and Winston, New York.
Kabat, E. A., Wu, T. T., Reid-Miller, M., Perry, H. M., & Gottesman, K. S. (1987) *Sequences of Proteins of Immunological Interest*, Fourth Ed., National Institutes of Health, Bethesda, MD.
Kast, P., Asif-Ullah, M., Jiang, N., & Hilvert, D. (1996a) *Proc. Natl. Acad. Sci. U.S.A. 93*, 5043–5048.
Kast, P., Hartgerink, J. D., Asif-Ullah, M., & Hilvert, D. (1996b) *J. Am. Chem. Soc. 118*, 3069–3070.
Kraulis, P. J. (1991) *J. Appl. Crystallogr. 24*, 946–950.
Krebs, J. F., Sinzdak, G., & Dyson, H. J. (1995) *Biochemistry 34*, 720–723.
Lane, J. W., Hong, X., & Schwabacher, A. W. (1993) *J. Am. Chem. Soc. 115*, 2078–2080.
Lee, A. Y., Karplus, A. P., Ganem, B., & Clardy, J. (1995a) *J. Am. Chem. Soc. 117*, 3627–3628.
Lee, A. Y., Stewart, J. D., Clardy, J., & Ganem, B. (1995b) *Chemistry & Biology 2*, 195–203.
Lerner, R. A., Benkovic, S. J., & Schultz, P. G. (1991) *Science 252*, 659–667.
Lesley, S. A., Patten, P. A., & Schultz, P. G. (1993) *Proc. Natl. Acad. Sci. U.S.A. 90*, 1160–1165.
Levitt, M. & Perutz, M. F. (1988) *J. Mol. Biol. 201*, 751–754.
Lewis, C., Krämer, T., Robinson, S., & Hilvert, D. (1991) *Science 253*, 1019–1022.
Lewis, C., Paneth, P., O'Leary, M. H., & Hilvert, D. (1993) *J. Am. Chem. Soc. 115*, 1410–1413.
MacBeath, G. & Hilvert, D. (1994) *J. Am. Chem. Soc. 116*, 6101–6106.
MacBeath, G. & Hilvert, D. (1996) *Chemistry & Biology 3*, 433–445.
Mariuzza, R. A. & Poljak, R. J. (1993) *Curr. Opin. Immunol. 5*, 50–55.
Napper, A. D., Benkovic, S. J., Tramontano, A., & Lerner, R. A. (1987) *Science 238*, 1041–1043.
Novotny, J., Bruccoleri, R. E., & Saul, F. A. (1989) *Biochemistry 28*, 4735–4749.
Padlan, E. A. (1990) *Proteins 7*, 112–124.
Pauling, L. (1948) *Am. Sci. 36*, 51–58.
Pawlak, J. L., Padykula, R. E., Kronis, J. D., Aleksejczyk, R. A., & Berchtold, G. A. (1989) *J. Am. Chem. Soc. 111*, 3374–3381.
Phillips, M. A., Fletterick, R., & Rutter, W. J. (1990) *J. Biol. Chem. 265*, 20692–20698.
Phillips, M. A., Kaplan, A. P., Rutter, W. J., & Bartlett, P. A. (1992) *Biochemistry 31*, 959–963.
Pollack, S. J., Nakayama, G. R., & Schultz, P. G. (1988) *Science 242*, 1038–1040.
Pollack, S. J. & Schultz, P. G. (1989) *J. Am. Chem. Soc. 111*, 1929–1931.
Radzicka, A. & Wolfenden, R. (1995) *Science 267*, 90–93.
Rajewsky, K., Förester, I., & Cumang, A. (1987) *Science 238*, 1088–1094.
Reymond, J.-L., Jahangiri, G. K., Stoudt, C., & Lerner, R. A. (1993) *J. Am. Chem. Soc. 115*, 3909–3917.
Reymond, J.-L., Janda, K. D., & Lerner, R. A. (1991) *Angew. Chem. Int. Ed. Engl. 30*, 1711–1713.
Reymond, J.-L., Janda, K. D., & Lerner, R. A. (1992) *J. Am. Chem. Soc. 114*, 2257–2258.
Rini, J. M., Schulze-Gahmen, U., & Wilson, I. A. (1992) *Science 255*, 959–965.
Roberts, V. A., Stewart, J., Benkovic, S. J., & Getzoff, E. D. (1994) *J. Mol. Biol. 235*, 1098–1116.
Sauer, J. & Sustman, R. (1980) *Angew. Chem. Int. Ed. Engl. 19*, 779–807.
Schultz, P. G. & Lerner, R. A. (1993) *Acc. Chem. Res. 26*, 391–395.
Schwabacher, A. W., Weinhouse, M. I., Auditor, M.-T. M., & Lerner, R. A. (1989) *J. Am. Chem. Soc. 111*, 2344–2346.
Severence, D. L. & Jorgensen, W. L. (1992) *J. Am. Chem. Soc. 114*, 10966–10968.
Sheriff, S., Hendrickson, W. A., & Smith, J. L. (1987) *J. Mol. Biol. 197*, 273–296.
Shin, J. A., & Hilvert, D. (1994) *Biomed. Chem. Lett. 4*, 2945–2948.
Shokat, K. M., Leumann, C. J., Sugasawara, R., & Schultz, P. G. (1988) *Angew. Chem. Int. Ed. Engl. 27*, 1172–1174.
Shokat, K. M., Leumann, C. J., Sugasawara, R., & Schultz, P. G. (1989) *Nature 338*, 269–271.

Smiley, J. A. & Benkovic, S. J. (1994) *Proc. Natl. Acad. Sci. U.S.A. 91*, 8319–8323.
Sogo, S. G., Widlanski, T. S., Hoare, J. H., Grimshaw, C. E., Berchtold, G. A., & Knowles, J. R. (1984) *J. Am. Chem. Soc. 106*, 2701–2703.
Stewart, J. D. & Benkovic, S. J. (1995) *Nature 375*, 388–391.
Stewart, J. D., Krebs, J. F., Siuzdak, G., Berdis, A. J., Smithrud, D. B., & Benkovic, S. J. (1994a) *Proc. Natl. Acad. Sci. U.S.A. 91*, 7404–7409.
Stewart, J. D., Roberts, V. A., Thomas, N. R., Getzoff, E. D., & Benkovic, S. J. (1994b) *Biochemistry 33*, 1994–2003.
Stewart, J.D., Liotta, L.J., & Benkovic, S.J. (1993) *Acc. Chem. Res. 26*, 396–404.
Suga, H., Ersoy, O., Williams, S. F., Tsumuraya, T., Margolies, M. N., Sinskey, A. J., & Masamune, S. (1994) *J. Am. Chem. Soc. 116*, 6025–6026.
Tang, Y., Hicks, J. B., & Hilvert, D. (1991) *Proc. Natl. Acad. Sci. U.S.A. 88*, 8784–8786.
Tarasow, T. M., Lewis, C., & Hilvert, D. (1994) *J. Am. Chem. Soc. 116*, 7959–7963.
Tawfik, D., Zemel, R., Arad-Yelliln, R., Green, B. S., & Eshhar, Z. (1990) *Biochemistry 29*, 9916–9921.
Tawfik, D. S., Chap, R., Green, B. S., Sela, M., & Eshhar, Z. (1995) *Proc. Natl. Acad. Sci. U.S.A. 92*, 2145–2149.
Tawfik, D. S., Green, B. S., Chap, R., Sela, M., & Eshhar, Z. (1993) *Proc. Natl. Acad. Sci. U.S.A. 90*, 373–377.
Thorn, S. N., Daniels, R. G., Auditor, M.-T. M., & Hilvert, D. (1995) *Nature 373*, 228–230.
Uno, T. & Schultz, P. G. (1992) *J. Am. Chem. Soc. 114*, 6573–6574.
Wagner, J., Lerner, R. A., & Barbas, C. F., III (1995) *Science 270*, 1797–1800.
Weiss, U. & Edwards, J. M. (1980) *The Biosynthesis of Aromatic Amino Compounds*, Wiley, New York.
Wells, J. A., Cunningham, B. C., Graycar, T. P., & Estell, D. A. (1986) *Phil. Trans. Royal Soc. Lond. (A) 317*, 415–423.
Wells, J. A. & Estell, D. A. (1988) *TIBS 13*, 291–297.
Westheimer, F. H. (1962) *Adv. Enzymol. 24*, 441.
Wilmore, B. H. & Iverson, B. L. (1994) *J. Am. Chem. Soc. 116*, 2181–2182.
Wilson, I. A. & Stanfield, R. L. (1993) *Curr. Opin. Struct. Biol. 3*, 113–118.
Wolfenden, R. (1976) *Annu. Rev. Biophys. Bioeng. 5*, 271–306.
Xue, Y., Lipscomb, W. N., Graf, R., Schnappauf, G., & Braus, G. (1994) *Proc. Natl. Acad. Sci. U.S.A. 91*, 10814–10818.
Zemel, R., Schindler, D. G., Tawfik, D. S., Eshhar, Z., & Green, B. S. (1994) *Mol. Immunol. 31*, 127–137.
Zhou, G. W., Guo, J., Huang, W., Fletterick, R. J., & Scanlan, T. S. (1994) *Science 265*, 1059–1064.

Chapter 11

Acher, R. (1993) *Regulatory Peptides 45*, 1–13.
Al-Obeidi, F., de L. Caastrucci, A. M., Hadley, M. E., & Hruby, V. J. (1989) *J. Med. Chem. 32*, 2555–2561.
Aubry, A. & Marraud, M. (1989) *Biopolymers 28*, 109–122.
Bai, J. P. F. & Amidon, G. L. (1992) *Pharm. Res. 9*, 969–978.
Bertenshaw, S. R., Rogers, R. S., Stern, M. K., & Norman, B. H. (1993) *J. Med. Chem. 36*, 173–174.
Cody, W. L., Doherty, A. M., He, J. X., DePue, P. L., Rapundalo, S. T., Hingorani, G. A., Major, T. C., Panek, R. L., Dudley, D. T., Haleen, S. J., LaDouceur, D., Hill, K. E., Flynn, M. A., & Reynolds, E. E. (1992) *J. Med. Chem. 35*, 3301–3303.
Coy, D. H., Horvath, M. K., Nekola, M. V., Coy, E. J., Erchegyi, J., & Schally, A. V. (1982) *Endocrinology 110*, 1445–1447.
DeWied, D. (1971) *Nature 232*, 58–60.
Doherty, A. M. (1992) *J. Med. Chem. 35*, 1439–1508.

Doherty, A. M., Cody, W. L., DePre, P. L., He, J. X., Waste, L. A., Leonard, D. M., Leitz, N. L., Dudley, D. T., Rapundalo, S. T., Hingorani, G. P., Haleen, S. J., LaDouceur, D. M., Hill, K. E., Flynn, M. A., & Reynolds, E. E. (1993) *J. Med. Chem.* 36, 2587–2589.

Dutta, A. S. & Furr, B. J. A. (1985) in *Annu. Reports Med. Chem.* 20 (Bailey, D. M., Ed.) pp 203–214, Academic Press, Orlando.

du Vigneaud, V., Lawler, H. C., & Popenon, E. A. (1953a) *J. Am. Chem. Soc.* 75, 4880–4881.

du Vigneaud, V., Ressler, C., Swan, J. M., Roberts, C. W., Katsoyannis, P. G., & Gordon, S. (1953b) *J. Am. Chem. Soc.* 75, 4879–4880.

Engel, M. H., Sawyer, T. K., Hadley, M. E., & Hruby, V. J. (1981) *Anal. Biochem.* 116, 303–311.

Erspamer, V. & Melchiorri, P. (1980) *Trends Pharmacol. Sci.* 1, 391.

Erspamer, V., Melchiorri, P., Falconieri-Erspamer, G., Negri, L., Corsi, R., Severins, C., Barra, D., Simmaco, M., & Kreil, G. (1989) *Proc. Natl. Acad. Sci. U.S.A.* 86, 5188–5192.

Farmer, P. S. (1980) in *Drug Design* (Ariens, E. J., Ed.) pp 119–141, Academic Press, New York.

Freidinger, R. M., Veber, D. F., Perlow, D. S., Brooks, J. R., & Saperstein, R. (1980) *Science* 210, 656–658.

Fujino, M., Kobayashi, S., Obayash M., Shinagawa, S., Fukuda, T., Kitada, C., Nakayama, R., Yamazaki, J., White, W. F., & Rippel, R. H. (1972) *Biochem. Biophys. Res. Commun.* 49, 863–869.

Fukuroda, T., Fujikawa, T., Ozaki, S. , Ishikawa, K., Yano, M., & Nishikibe, M. (1994) *Biochem. Biophys. Res. Commun.* 199, 1461–1465.

Granner, D. K. (1993) in *Harper's Biochemistry*, Twenty-third Ed. (Murray, R. K., Granner, D. K., Mayes, P. A., & Rodwell, V. W., Eds.) Appleton & Lange, Norwalk, CT.

Gilman, A. G. (1987) *Annu. Rev. Biochem.* 56, 615–649.

Grant, G. A. & Chiappinelli, V. A. (1985) *Biochemistry* 24, 1532–1537.

Gray, W. R., Luque, A., Olivera, M. B., Barrett, J., & Cruz, L. J. (1981) *J. Biol. Chem.* 256, 4734–4740.

Hadley, M. E., Ed. (1989) *The Melanotropic Peptides*, Vols. 1, 2, 3, CRC Press, Boca Raton, FL.

Hadley, M. E. (1992) *Endocrinology*, Third Ed., Prentice Hall, Englewood Cliffs, NJ.

Hadley, M. E., Sharma, S. D., Hruby, V. J., Levine, N., & Dorr, R. T. (1993) in *The Melanotropic Peptides: Annals of the New York Academy of Sciences* Vol. 680 (Vaudry, H. & Eberle, A. N., Eds.) pp 424–439, and references therein.

Haviv, F., Fitzpatrick, T. D., Swenson, R. E., Nichols, C. J., Mort, N. A., Bush, E. N., Diaz, G. J., Bammert, G.,; Nguyen, A., Rhutasel, N. S., Nellans, H. N., Hoffman, D. J., Johnson, E. S., & Greer, J. (1993) *J. Med. Chem.* 36, 363–369.

Hill, P. S., Smith, D. D., Slaninova, J., & Hruby, V. J. (1990) *J. Am. Chem. Soc. 112*, 3110–3113.

Hirshmann, R. (1991) *Angew. Chem. Int. Ed. Engl.* 30, 1278 -1301.

Hirshmann, R., Nicolaou, K. C., Pietranico, S., Leahy, E. M., Salvino, J., Arison, B., Cichy, M. A., Spoors, P. G., Shakespeare, W. C., Sprengeler, P. A., Hamley, P., Smith, III, A. B., Reisine, T., Raynor, K., Maechler, L., Donaldson, C., Vale, W., Freidinger, R. M., Cascieri, M. R., & Strader, C. D. (1993) *J. Am. Chem. Soc.* 115, 12550–12568.

Houben, H. & Denef, C. (1994) *Peptides* 15, 547–582.

Hruby, V. J. (1982) *Life Sci.* 31, 189–199.

Hruby, V. J. (1992) in *Progress Brain Res.*, Vol. 92 (Joosse, J., Buijs, R. M., & Tilders, F. J. H., Eds.) pp 215–224, Elsevier, New York.

Hruby, V. J., Al-Obeidi, F., & Kazmierski, W. (1990) *Biochem. J.* 268, 249–262.

Hruby, V. J. , Cody, W. L., Wilkes, B. C., & Hadley M. E. (1984) in *Peptides, 1984* (Ragnarsson, U., Almquist, & Wilkes, Eds.) pp 505–508, Stockholm.

Hruby, V. J., Sharma, S. D., Toth, K., Jaw, J. Y., Al-Obeidi, F., Sawyer, T. K., & Hadley, M. E. (1993) *Ann. N.Y. Acad. Sci. 680*, 51–63.

Hruby, V. J., & Smith, C. W. (1987) in *The Peptides* Vol. 8, pp 77–207, Academic Press, Orlando.

Humphrey, M. J., & Ringrose, P. S. (1986) *Drug Metab. Rev. 17*, 283–310.
Ihara, M., Noguchi, K., Saeki, T., Fukurode, T., Tsuchido, S., Kimura, S., Fukami, T., Ishikawa, K., Nishibe, M., & Yano, M. (1991) *Life Sci. 50*, 247–255.
Ishikawa, K., Fukami, T., Nagase, T., Fujita, K., Hayama, T., Niijama, K., Mase, T., Ihara, M., & Yano, M. (1992) *J. Med. Chem. 35*, 2139–2142.
Ishikawa, K., Fukami, T., Nagase, T., Mase, T., Hayama, T., Niijama, K., Fujita, K., Urakawa, Y., Kumagai, U., Fukuroda, T., Ihara, M., & Yano, M. (1993) in *Peptides 1992* (Schneider, C. H. & Eberle, A. M., Eds.) pp 685, ESCOM, Leiden.
Ishikawa, K., Ihara, M., Noguchi, K., Mase, T., Mino, N., Saeki, T., Fukuroda, T., Fukami, T., Ozaki, S., Nagase, T., Nishikibe, M., & Yano, M. (1994) *Proc. Natl. Acad. Sci. U.S.A. 91*, 4892–4896.
Ito, A. S., del Castrucci, A. M., Hruby, V. J., Hadley, M. E., Krajcarski, D. T., & Szabo, A. G. (1993) *Biochemistry 32*, 12264–12272.
Jard, S., Elands, J., Schmidt, A., & Barberis, C. (1988) in *Progress in Endocrinology* (Imura, H. & Shizume, K., Eds.) pp 1183–1188, Elsevier, Amsterdam.
Jones, J. H., Ed. (1993) *Amino Acids and Peptides*, Vol. 24 (and 1–23), A Specialist Periodical Report, Royal Society of Chemistry, Cambridge.
Jost, K., Lebl, M., & Brtník, F. (1987) *Handbook of Neurohypophyseal Hormone Analogs*, Vols. 1 & 2, CRC Press, Boca Raton, FL.
Kangawa, K. & Matsuo, H. (1984) *Biochem. Biophys. Res. Commun. 118*, 131–139.
Karle, I. L. (1978) *J. Am. Chem. Soc. 100*, 1286–1289.
Karten, M. J. & Rivier, J. E. (1986) *Endocr. Rev. 7*, 44–66.
Kikuchi, T., Kubo, K., Ohtaki, T., Suzuki, N., Asami, T., Shimamoto, N., Wakimasu, M., & Fujimo, M.(1993) *J. Med. Chem. 36*, 4087–4093.
Lande, S. & Lerner, A. B. (1971) *Biochem. Biophys. Acta 251*, 246–253.
Li, C. H. (1964) *Nature 201*, 924.
Li, C. H. (1981) in *Hormonal Proteins and Peptides*, Vol. 10, (Li, C. H., Ed.) pp 1–34, Academic Press, Orlando.
Manning, M., Chan, W. Y., & Sawyer, W. H. (1993) *Regulatory Peptides 45*, 279–283.
Manning, M., Przybylski, J. P., Olma, A., Klis, W. A., Kruszynski, M., Wo, N. C., Pelton, G. H., & Sawyer, W. H. (1987) *Nature 329*, 839–840.
Manning, M. & Sawyer, W. H. (1993) *J. Receptor Res. 13*, 195–214.
Marks, N., Stern, F., & Sawyer, W. H. (1974) *Biochem. Biophys. Res. Commun. 61*, 1458–1463.
Matsumura, Y., Hisaki, K., Takaoka, M., & Morimoto, S. (1990) *European J. Pharmacol. 185*, 103–106.
McDowell, R. S., Blackburn, B. K., Gadek, T. R., McGee, L. R., Rawson, T., Reynolds, M. E., Robarge, K. D., Somers, T. C., Thorsett, E. D., Tischler, M., Webb II, R. R., & Venuti, M. C. (1994) *J. Am. Chem. Soc. 116*, 5077–5083.
McMahon, E. G., Paslomo, M. A., Moore, W. M., McDonald, J. F., & Stern, M. K. (1991) *Proc. Natl. Acad. Sci. U.S.A. 88*, 703–707.
Montecucchi, P. C., & Henschen, A. (1981) *Int. J. Peptide Protein Res. 18*, 113–120.
Morikawa, N., Hagiwara, K., & Nakajima, T. (1992) *Biochem. Biophys. Res. Commun. 189*, 184–190.
Mullins, L. S., Wesseling, K., Kuo, J. M., Garrett, J. B., & Raushel, F. M. (1994) *J. Am. Chem. Soc. 116*, 5529–5533.
Mutt, V. (1993) in *Peptides 1992, Proceedings of the 22nd European Peptide Symposium* (Schneider, C. H. & Eberle, A. N., Eds.) pp 3–20, ESCOM, Leiden.
Nestor, Jr., J. J., & Vickery, B. H. (1988) in *Annu. Reports Med. Chem.* (Allen, R. C., Ed.) pp 211–220, Academic Press, San Diego.
Noda, M., Teranishi, Y., Takahashi, H., Toyosato, M., Notake, M., Nakanishi, S., & Numa, S. (1982) *Nature 297*, 431–434.
Okada, K., Miyazaki, Y., Takada, J., Matsuyama, K., Yamaki, T., & Yano, M. (1990) *Biochem. Biophys. Res. Commun. 171*, 1192–1198.

Olson, G. L., Bolin, D. R., Bonner, M. P., Bös, M., Cook, C. M., Fry, D. C., Graves, B. J., Hatada, M., Hill, D. E., Kahn, M., Madison, V. S., Rusiecki, V. K., Sarabu, R., Sepinwall, J., Vincent, G. P., & Voss, M. E. (1993) *J. Med. Chem. 36*, 3039–3049.

Pease, L. G., & Watson, C. (1978) *J. Am. Chem. Soc. 100*, 1279–1286.

Qian, X., Kövér, K. E., Shenderovich, M. D., Loui, B.-S., Misika, A., Zalewska, T., Horváth, R., Davis, P., Bilsky, E. J., Porreca, F., Yamamura, H. I, & Hruby, V. J. (1994) *J. Med. Chem. 37*, 1746–1757.

Rivier, J., Rivier, C., Koerber, S. C., Kornreich, W. D., deMiranda, A., Miller, C., Galyean, R., Porter, J., Yamamoto, G., Donaldson, C. J., & Vale, W. W. (1992) in *Peptides, Chemistry and Biology,: Proceedings of the 12th American Peptide Symposium* (Smith, J. A. & Rivier, J. E., Eds.) p 33, ESCOM, Leiden.

Rizo, J. & Gierasch, L. M. (1992) *Annu. Rev. Biochem. 61*, 387–418.

Rizo, J., Koerber, S. C., Bienstock, R. C., Rivier, J., Gierasch, L. M., & Hagler, A. J. (1992a) *J. Am. Chem. Soc. 114*, 2860–2871.

Rizo, J., Koerber, S. C., Bienstock, R. C., Rivier, J., Hagler, A. J., & Gierasch, L. M. (1992b) *J. Am. Chem. Soc. 114*, 2852–2859.

Rodbell, M. (1985) *Trends Biochem. Sci. 10*, 461–464.

Rosamond, J. D., Comstock, J. M., Thomas, N. J., Clark, A. M., Blosser, J. C., Simmons, R. D., Gawlak, D. L., Loss, M. E., Augello-Vaisey, S. J., Spatola, A. F., & Benovitz, D. E. (1988) in *Peptides, Chemistry and Biology* (Marshall, G. R., Ed.) pp 610–612, ESCOM, Leiden.

Sakurai, T., Yanagisawa, M., & Masaki, T. (1992) *Trends Pharmacol. Sci. 13*, 103–108.

Sawyer, T. K., Sanfilippo, P. J., Hruby, V. J., Engel, M. H., Heward, C. B., Burnett, J. B., & Nadley, M. R. (1980) *Proc. Natl. Acad. Sci. U.S.A. 79*, 5754–5758.

Sawyer, T., Hruby, V. J., Darman, P. S., & Hadley, M. E. (1982) *Proc. Natl. Acad. Sci. U.S.A.* 1751–1755.

Schally, A. V., Nair, R. M. G., & Redding, T. W. (1971) *J. Biol. Chem. 246*, 7230–7236.

Schwyzer, R. (1977) *Ann. N.Y. Acad. Sci. 297*, 3–26.

Schwyzer, R. (1986) *Biochemistry 25*, 6335–6342.

Schwyzer, R. (1992) *ChemTracts 3*, 348–379.

Smith, J. A., & Pease, L. G. (1980) in *CRC Critical Reviews in Biochemistry* Vol. 8 (Fasman, G. D., Ed.) pp 315–399, CRC Press, Boca Raton, FL.

Spatola, A. F. (1983) in *Chemistry and Biochemistry of Amino Acids, Peptides, and Proteins*, Vol. 7 (Weinstein, B., Ed.) pp 267–357, Marcel Dekker, New York.

Spatola, A. F. & Agarwal, N. S. (1980) *Biochem. Biophys. Res. Commun. 97*, 1571–1574.

Spatola, A. F., Agarwal, N. S., Bettag, A. L., Yankeelov, Jr., J. A., Bowers, C. Y., & Vale, W. W. (1980) *Biochem. Biophys. Res. Commun. 97*, 1014–1023.

Spinella, M., Malik, A. B., Everitt, J., & Anderson, T. T. (1991) *Proc. Natl. Acad. Sci. U.S.A. 88*, 7443–7446.

Struthers, R. S., Tanaka, G., Koerber, S. C., Solmajer, T., Baniak, E. L., Gierasch, L. M., Vale, W., Rivier, J., & Hagler, A. T. (1990) *Proteins 8*, 295–304.

Toniolo, C. (1989) *Biopolymers 28*, 247–257.

Toniolo, C. (1990) *Int. J. Peptide Protein Res. 35*, 287–300.

Urry, D., & Walter, R. (1971) *Proc. Natl. Acad. Sci. U.S.A. 68*, 956–958.

Vale, W., Grant, G., Rivier, J., Monahan, M., Amoss, M., Blackwell, R., Burgus, R., & Guillemin, R. (1972) *Science 176*, 933–934.

Veber, D. F. (1992) in *Peptides, Chemistry and Biology* (Smith, J. A., & Rivier, J.E., Eds.) pp 3–14, ESCOM, Leiden.

Walter, R., Glickson, J. D., Schwartz, I. L., Havran, R. T., Meienhofer, J., & Urry, D. W. (1972) *Proc. Natl. Acad. Sci. U.S.A. 69*, 1920–1924.

Walter, R. (1977) *Fed. Proc. Am. Soc. Exp. Biol. 36*, 1872–1878.

Walter, R., Smith, C. W., Mehta, P. K., Boonjarern, S., Arruda, J. A. L., & Kurtzman, N. A. (1977) *Disturbances in Body Fluid Osmolality*, pp 1–36, American Physiol. Society, Bethesda, MD.

Williams, P. D., Anderson, P. S., Ball, R. G., Bock, M. G., Carroll, L. A., Chiu, S. H. L., Cli-

neschmidt, B. V., Culberson, J. C., Erb, J. M., Evans, B. E., Fitzpatrick, S. L., Freidinger, R. M., Kaufman, M. J., Lundell, G. F., Murphy, J. S., Pawluczyk, J. M., Perlow, D. S., Pettibone, D. J., Pitzenberger, S. M., Thompson, K. L., & Veber, D. F. (1994) *J. Med. Chem. 37*, 565–571.
Yamada, T. & Owyang, C. (1991) in *Encyclopedia of Human Biology*, Vol. 5, pp 713–724, Academic Press, Orlando.
Yamashiro, D. H., & Li, C. H. (1984) in *The Peptides* (Udenfriend, S. & Meienhofer, J., Eds.) Vol. 6, Academic Press, New York.
Yanagisawa, M., Kuribara, H., Kimura, S., Tomobe, Y., Kobayasshi, M., Mitsui, Y., Yazaki, Y., Goto, K., & Masaki, T. (1988) *Nature 332*, 411–415.
Zasloff, M. (1987) *Proc. Natl. Acad. Sci. U.S.A. 84*, 5449–5453.
Zatuchni, G. I., Shelton, J. D., & Sciarra, J. J., Eds. (1981) *LH-RH Peptides as Female and Male Contraceptives*, Harper & Row, Philadelphia.

Chapter 12

Adler, M., Lazarus, R. A., Dennis, M. S., & Wagner, G. (1991) *Science 253*, 445–448.
Advenier, C., Emonds-Alt, X., Vilain, P., Goulaouic, P., Proietto, V., Van Broeck, D., Naline, E., Neliat, G., Le Fur, G., & Breliere, J. C. (1992) *Br. J. Pharm. 105*, 77P.
Allinger, N. L. (1977) *J. Am. Chem. Soc. 99*, 8127–8134.
Amzel, L. M.; Poljak, R. J. (1979) *Annu. Rev. Biochem. 48*, 961–997.
Arrhenius, T., Lerner, R. A., & Satterthwait, A. C. (1987) in *Protein Structure, Folding and Design* (Oxender, D., Ed.) pp 453–465, Alan R. Liss, New York.
Arrhenius, T., & Satterthwait, A. C. (1989) in *The 11th American Peptide Symposium* (Rivier, J. E. & Marshall, G. R., Eds.) pp 870–872, ESCOM, Leiden, La Jolla, CA.
Ashkenazi, A., Presta, L. G., Marsters, S. A., Camerato, T. R., Rosenthal, K. A., Fendly, B. M., & Capon, D. J. (1990) *Proc. Natl. Acad. Sci. U.S.A. 87*, 7150–7154.
Aubry, A., Birlirakis, N., Sakarellos-Daitsiotis, M., Sakarellos, C., & Marroud, M. (1989) *Biopolymers 28*, 27–40.
Baca, M., Alewood, P. F., & Kent, S. B. H. (1993) *Protein Sci. 2*, 1085–1091.
Balaram, P. & Sudha, T. S. (1983) *Int. J. Pept. Prot. Res. 21*, 381–388.
Ball, J. G., Andrews, P. R., Alewood, P. F., & Hughes, R. A. (1990) *FEBS Lett. 273*, 15–18.
Ball, J. B. & Alewood, P. F. (1990) *J. Mol. Recognition 3*, 55–64.
Ball, J. B., Hughes, R. A., Alewood, P. F., & Andrews, P. R. (1993) *Tetrahedron 49*, 2467–2478.
Balodis, Y. Y., Nikiforovich, G. V., Grinsteine, I. V., Vegner, R. E. & Chipens, G.I. (1978) *FEBS Lett. 86*, 239–242.
Barker, P. L., Bullen, S., Bunting, S., Burdick, D. J., Chan, K. S., Deisher, T., Eigenbrot, C., Gadek, T. R., Gantzos, R., Lipari, M. T., Muri, C. D., Napier, M. A., Pitti, R. M., Padua, A., Quan, C., Stanley, M., Struble, M., Tom, J. Y. K., & Burnier, J. P. (1992) *J. Med. Chem. 35*, 2040–2048.
Barlow, D. J. & Thornton, J. M. (1988) *J. Mol. Biol. 201*, 601–619.
Belanger, P. C., Dufresne, C., Scheigetz, J., Young, R. N., Springer, J. P., & Dmitrienko, G. I. (1982) *Can. J. Chem. 60*, 1019–1029.
Benedetti, E., Toniolo, C., Hardy, P., Barone, V., Bavoso, A., Di Blasio, B., Grimaldi, P., Lelj. F., Pavone, V., Pedone, C., Nonora, G. M., & Lingham, I., (1984) *J. Am. Chem. Soc. 106*, 8146–8152.
Bienstock, R. J., Rizo, J., Koerber, S. C., Rivier, J. E., Hagler, A. T., & Gierasch, L. M. (1993) *J. Med. Chem. 36*, 3265–3273.
Birnbaum, S. & Mosbach, K. (1992) *Curr. Opin. Biotech. 3*, 49–54.
Bjorkman, P. J., Saper, M. A., Samraoui, B., Bennett, W. S., Strominger, J. L., & Wiley, D. C. (1987) *Nature 329*, 512–518.
Blaber, M., Zhang, X.-J., Matthews, B. W. (1993) *Science, 260*, 1637–1640.

Bock, M. G., DiPardo, R. M., Veber, D. F., Chang, R. S. L., Lotti, V.J., Freedman, S. B., & Freidinger, R. M. (1993) Bioorg. Med. Chem. Lett. *3*, 871–874.
Bogusky, M. J., Brady, S. F., Sisko, J. T., Nutt, R. F., & Smith, G. M. (1993) *Int. J. Pept. Prot. Res. 42*, 194–203.
Borkakoti, N., Winkler, F. K., Williams, D. H., D'Arcy, A., Broadhurst, M. J., Brown, P. A., Johnson, W. H., & Murray, E. J. (1994) *Struct. Biol. 1*, 106–110.
Bovy, P. R., Trapani, A. J., McMahon, E. G., & Palomo, M. (1989) *J. Med. Chem. 32*, 520–522.
Bruck, C., Co, M. S., Slaoui, M., Gaulton, G. N., Smith T., Fields, B. N., Mullins, J. I., & Greene, M. I. (1986) *Proc. Natl. Acad. Sci. U.S.A. 83*, 6578–6582.
Bunin, B. A. & Ellman, J. A. (1992) *J. Am. Chem. Soc. 114*, 10997–10998.
Callahan, J. F., Bean, J. W., Burgess, J. L., Eggleston, D. S., Hwang, S. M., Kopple, K. D., Koster, P. F., Nichols, A., Peishoff, C. E., Samanen, J. M., Vasko, J. A., Wong, A., & Huffman, W. F. (1992) *J. Med. Chem. 35*, 3970–3972.
Callahan, J. F., Newlander, K. A., Burgess, J. L., Eggleston, D. S., Nichols, A., Wong, A., & Huffman, W. F. (1993) *Tetrahedron 49*, 3479–3488.
Chakrabartty, A., Kortemme, T., Padamanabhan, S., & Baldwin, R. L. (1993) *Biochemistry 32*, 5560–5565.
Chakrabartty, A., Schellman, J. A., & Baldwin, R. L. (1991) *Nature 351*, 586–588.
Chen, J. K., Lane, W. S., Brauer, A. W., Tanaka, A., & Schreiber, S. L. (1993) *J. Am. Chem. Soc. 155*, 12591–12592.
Chen, S., Chrusciel, R. A., Nakanishi, H., Raktabutr, A., Johnson, M. E., Sato, A., Weiner, D., Hoxie, J., Saragovi, H. U., Greene, M. I., & Kahn, M. (1992) *Proc. Natl. Acad. Sci. U.S.A. 89*, 5872–5876.
Chen, Y., Pitzenberger, S. M., Garsky, V. M., Lumma, P. K., Sanyal, G., & Baum, J. (1991) *Biochemistry 30*, 11625–11636.
Chew, C., Villar, H. O., & Loew, G. H. (1991) *Mol. Pharmacol. 39*, 502–510.
Cho, C. Y., Moran, E. J., Cherry, S. R., Stephans, J. C., Fodor, S. P. A., Adams, C. L., Sundaram, A., Jacobs, J. W., & Schultz, P. G. (1993) *Science 261*, 1303–1305
Chorev, M. & Goodman, M. (1993) *Acc. Chem. Res. 26*, 266–273.
Chothia, C. & Lesk, A. M. (1987) *J. Mol. Biol. 196*, 901–917.
Chothia, C., Lesk, A. M., Tramontano, A., Levitt, M., Smith-Gill, S. J., Air, G., Sheriff, S., Padlan, E. A., Davies, D., Tulip, W. R., Colman, P. M., Spinelli, S., Alzara, P. M., & Poljak, R. J. (1989) *Nature 342*, 877–889.
Collins, J. L., Dambek, P. J., Goldstein, S. W., & Faraci, W. S. (1992) *Bioorg. Med. Chem. Lett. 2*, 915–918.
Cox, M. T., Heaton, D. W., Horbury, J. (1980) *J. Chem. Soc. Chem. Commun.*, 799–802.
Creighton, T. E. (1984) in *Proteins*, p. 163, W. H. Freeman, New York.
Crisma, M., Valle, G., Bonora, G. M., De Menego, E., Toniolog, C., Lelj., F., Barone, V., & Fraternali, F. (1990) *Biopolymers 30*, 1–11.
Damon, D. B. & Hoover, D. J. (1990) *J. Am. Chem. Soc. 112*, 6439–6442.
Degrado, W. F. (1988) *Adv. Protein Chem. 39*, 51–124.
Dewitt, S. H., Kiely, J. S., Stankovic, C. J., Shroeder, M. C., Reynolds Cody, D. M., & Pavia, M. R. (1993) *Proc. Natl. Acad. Sci. U.S.A. 90*, 6909–6913.
Diaz, H., Tsang, K. Y., Choo, D., Espina, J. R., Kelly, J. W. (1993) *J. Am. Chem. Soc. 115*, 3790–3791.
Di Blasio, F., Lombardi, A., Nastri, F., Saviano, M., Pedone, C., Yamada, T., Nakao, M., Kuwata, S., & Pavone, V. (1992) *Biopolymers 32*, 1155–1161.
Di Blasio, B., Lombardi, A., D'Auria, G., Saviano, M., Isernia, C., Maglio, O., Paolillo, L., Pedone, C., & Pavone, V. (1993) *Biopolymers 33*, 621–631.
Doughty, M. B., Chu, S. S., Misse, G. A., & Tessel, R. (1992) *Bioorg. Med. Chem. Lett. 2*, 1497–1502.
El Masdouri, L., Aubry, A., Sakarellos, C., Gomex, E. J., Cung, M. T., & Marraud, M. (1988) *Int. J. Pept. Prot. Res. 31*, 420–428.

Evans, B. E., Leighton, J. L., Rittle, K. E., Gilbert, K. F., Lundell, G. F., Hobbs, D. W., DiPardo, R. M., Veber, D. F., Pettibone, D. J., Anderson, P. S., & Freidinger, R. M. (1992) *J. Med. Chem. 35*, 3919–3927.
Evans, B. E., Rittle, K. E., Chang, R. S. L., Lotti, V. J., Freedman, S. B., Freidinger, R. M. (1993) *Bioorg. Med. Chem. Lett. 3*, 867–870.
Fauchère, J. L. (1986) *Adv. Drug. Res. 15*, 29–69.
Fauchère, J. L. (1987) in *QSAR in Drug Design and Toxicology* (Hadzi, D. and Jerman-Blazic, B., Eds.) 22, Amsterdam, The Netherlands.
Felder, E., Allmendinger, T., Fritz, H., Hugerbuler, E., & Keller, M. (1992) in *Proceedings of the Twelfth American Peptide Symposium* (Smith, J. A. & Rivier, J. E., Eds.) pp 161–162.
Ferguson, D.M. & Raber, D.J. (1989) *J. Am. Chem. Soc. 111*, 4371–4378.
Filatova, M. P., Dri, N. A., Komarova, N. A., Orfkkchovich, O. M., Reiss, V. M., Liepinya, I. T., & Nikiforovich, G. V. (1986) *Bioorg. Khim. 12*, 59–70.
Flynn, G. A., Giroux, E. L., & Dage, R. C. (1987) *J. Am. Chem. Soc. 109*, 7914–7915.
Fodor, S. P., Read, J. L., Pirrung, M. C., Stryer, L., Lu, A. T., & Solas., D. (1991) *Science 251*, 767–773.
Fong, T. M., Huang, R.-R. C., & Strader, C. D. (1992) *J. Biol. Chem. 267*, 25664–25667.
Franciskovich, J., Houseman, K., Mueller, R., & Chmielewski, J. (1993) *Bioorg. Med. Chem. Letts. 3*, 765–768.
Freidinger, R. M. (1981) in *Peptides, Synthesis, Structure, Function* (Rich, D. H. & Gross, E., Eds.) pp 673–683, Pierce Chemical Co., Rockford, IL.
Fusetani, N., Matsunaga, S., Matsumoto, H., & Takebayashi, Y. (1990) *J. Am. Chem. Soc. 112*, 7053–7054.
Genin, M. J. & Johnson, R. L. (1992) *J. Am. Chem. Soc. 114*, 8778–8783.
Genin, M. J., Mishra, R. K., & Johnson, R. L. (1993a) *J. Med. Chem. 36*, 3481–3483.
Genin, M. J., Ojala, W. H., Gleason, W. B., & Johnson, R. L. (1993b) *J. Org. Chem. 58*, 2334–2337.
Gether, U., Johansen, T. E., Snider, R. M., Lowe, J. A., Nakanishi, S., Schwartz, T. W. (1993) *Nature, 632*, 345–348.
Geysen, H. M. & Mason, T. J. (1993) *Bioorg. Med. Chem. Lett. 3*, 397–404.
Griffin, J. F., & Smith, G. D. (1988) *Opioid Peptides: An Update* (Rapaka, R. S. & Dhawan, B. N., Eds.) p 41, NIDA Research Monograph 87, Washington, D.C.
Grimaldi, P., Lelj, F., Pavone, V., Pedone, C., Nonora, G. M., & Lingham, I., (1984) *J. Am. Chem. Soc. 106*, 8146–8152.
Hagihara, M., Anthony, N. J., Stout, T. J., Clardy, J., & Schreiber, S. L. (1992) *J. Am. Chem. Soc. 114*, 6568–6570.
Hale, J. J., Finke, P. E., & MacCoss, M. (1993) *Bioorg. Med. Chem. Lett. 3*, 319–322.
Hansen, P. E., Morgan, B. A. (1990) *The Peptides: Analysis, Synthesis, Biology 6* (Udenfried, S. & Meienhofer, J., Eds.) pp 269–321.
Hassan, M. & Goodman, M. (1986) *Biochemistry, 25*, 7596–7606.
Hermans, J., Anderson, A. G., & Yun, R. H. (1992) *Biochemistry 31*, 5646–5653.
Higashijima, T., Tasumi, M., Miyazawa, T., & Miyoshi, M. (1978) *Eur. J. Biochem. 89*, 543–546.
Hirschmann, R., Sprengeler, P. A., Kawasaki, T., Leahy, J. W., Shakespeare, W. C., & Smith, A. B., III (1992) *J. Am. Chem. Soc. 114*, 9699–9701.
Houghten, R. A., Pinilla, C., Blondelle, S. E., Appell, J. R., Dooley, C. T., Cuervo, J. H. (1991) *Nature 354*, 84–86.
Howard, J. C., Momany, F. A., Andreatta, R. H., & Scheraga, H. A. (1973) *Macromolecules 6*, 535–541.
Howbert, J. J., Lobb, K. L., Britton, T. C., Mason, N. R., & Bruns, R. F. (1993) *Bioorg. Med. Chem. Lett. 3*, 875–880.
Hruby, V. J. (1987) *Trends Pharmacol. Sci. 8*, 336–339.
Hruby, V. J., Al-Oeidi, F., & Kazmierski, W. (1990a) *Biochem. J. 268*, 249–262.
Hruby, V. J., Kao, L. F., Pettitt, B. M., & Karplus, M. (1988) *J. Am. Chem. Soc. 110*, 3351–3359.
Hruby, V. J., Knapp, F. S., Kazmierski, W., Lui, G. K., Yamaura, H. I. (1990b) in *Peptides:*

Chemistry, Structure and Biology, Proceedings of the 11th American Peptide Symposium (Rivier, J. E. & Marshall, G. R., Eds.) pp 53–55, ESCOM, Leiden.

Hughes, J., Smith, T. W., Kosterlitz, H. W., Fothergill, L. A., Morgan, B. A., & Morris, H. R. (1975) *Nature 258*, 577–578.

Humblet, C. & DeCoen, J. L. (1977) in *Peptides: Proceedings of the Fifth American Peptide Symposium* (Goodrum, M. & Meienhofer, J., Eds.) pp 88–91, Wiley, New York.

James, G. L., Goldstein, J. L., Brown, M. S., Rawson, T. E., Somers, T. C., McDowell, R. S., Crowley, C. W., Lucas, B. K., Levinson, A. D., & Marsters, Jr., J. C. (1993) *Science 260*, 1937–1942.

Jameson, B. A., Rao, P. E., Kong. L. I., Hahn, B. H., Shaw, G. M., Hood, L. E., & Kent, S. B. H. (1990) *Science 240*, 1335–1339.

Jennings-White, C., & Almquist, R. G. (1982) *Tetrahedron Lett. 23*, 2533–2534.

Jones, Jr., W. C., Nestor, Jr., J. J., & du Vigneaud, V. (1973) *J. Am. Chem. Soc. 95*, 5677–5680.

Kabat, E. A. (1978) *Adv. Protein Chem. 32*, 1–75.

Kahn, M. Ed. (1993) *Peptide Secondary Structure Mimetics*. Tetrahedron Symposia-in-Print Number 50, Oxford, England.

Kaiser, E. T. & Kezdy, F. J. (1984) *Science 223*, 249–255.

Karle, I. L. & Balaram, P. (1990) *Biochemistry 29*, 6747–6756.

Kawai, M., Fukuta, N., Ito, N., Kagami, T., Butsugan, Y., Maruyama, M., & Kudo, Y. (1990) *Int. J. Pept. Protein Res. 35*, 452–459, and citations therein.

Kemp, D. S., Allen, T. J., & Oslick, S. L. (1991a) in *Peptides Chemistry and Biology: Proceedings of The 12th American Peptide Symposium*, (Smith, J. A. & Rivier, J. E., Eds.) pp 352–355, ESCOM, Leiden.

Kemp, D. S., Blanchard, D. E., & Muendel, C. C. (1991c) in *Peptides: Chemistry and Biology, Proceedings of the 12th American Peptide Symposium* (Smith, J.A., & Rivier, J.E., Eds.) pp 319–322, ESCOM, Leiden.

Kemp, D. S. & Bowen, B. R. (1988a) *Tetrahedron Lett. 29*, 5077–5080.

Kemp, D. S., & Bowen, B. R. (1988b) *Tetrahedron Lett. 29*, 5081–5082.

Kemp, D. S., & Bowen, B. R. (1990) in *Protein Folding: Deciphering the Second Half of the Genetic Code*. (Gierasch, L.M. & King, J., Eds.) pp 293–303, AAAS, Washington, D.C.

Kemp, D. S. & Curran, T. P. (1988a) *Tetrahedron Lett. 29*, 4931–4934.

Kemp, D. S. & Curran, T. P. (1988b) *Tetrahedron Lett. 29*, 4935–4938.

Kemp, D. S., Curran, T. P., Boyd, J. G., & Allen, T. J. (1991b) *J. Org. Chem. 56*, 6672–6682.

Kemp, D. S. & Rothman, J. H., (1992) in *Peptides: Chemistry and Biology, Proceedings of The 12th American Peptide Symposium*, (Smith, J. A. & Rivier, J. E., Eds.) pp 350–351, ESCOM, Leiden.

Kerr, J. M., Hanville, S. C., & Zuckerman, R. N. (1993) *J. Am. Chem. Soc. 155*, 2529–2531.

Kim, P.S. & Baldwin, R. L. (1984) *Nature 307*, 329–334.

Kobayashi, S. V., Caldwell, C. G., Springer, M. S., & Hagmann, W. K. (1992) *J. Med. Chem. 35*, 252–258.

Komeiji, Y., Uebayashi, M., Someya, J.-I., & Yamaota, I. (1993) *Proteins 16*, 268–277.

Kopple, K. D., Baures, P. W., Bean, J. W., Dambrosio, C. A., Hughes, J. L., Peishoff, C. E., & Eggleston, D. S. (1992) *J. Am. Chem. Soc. 114*, 9615–9623.

Lam, K. S., Hruby, V. J., Lebl., M., Knapp, R. J., Kazmierski, W. M., Hersh, E. M., & Salmon, S. E. (1993) *Bioorg. Med. Chem. Lett. 3*, 419–424.

Landau, N. R., Warton, M., & Littman, D. R. (1988) *Nature 334*, 159–162.

Lanza, T. J., Durette, P. L., Rollins, T., Siciliano, S., & Cianciarulo, D. N. (1992) *J. Med. Chem 35*, 252–258.

Lewis, P. N., Momany, F. A., & Sheraga, H. A. (1973) *Biochim. Biophys. Acta. 303*, 211–229.

London, R. E., Stewart, J. M., & Cann, J. R. (1990) *Biochem. Pharmacol. 40*, 41–48.

Lowe, J. A., Drozda, S. E., Snider, R. M., Longo, K. P., & Rizzi, J. P. (1993) *Bioorg. Med. Chem. Lett. 3*, 921–924.

Manavalan, P. & Momany, F. A. (1980) *Biopolymers 19*, 1943–1973.
Mansour, A., Khachturian, H., Lewis, M. E., Akil, H., & Watson, S. J. (1988) *Trends Neurosci. 11*, 308–314.
Mantlo, N.B., Kim, D., Ondeyka, D., Chang, R. S. L., Kivlighn, S. D., Siegl, P. K. S., & Greenlee, W. J. (1994) *Bioorg. Med. Chem. Lett. 4*, 17–22.
Marshall, G. R., Clark. J. D., Dunbar J. B., Smith, G. D., Zabrocki, J., Redlinski, A. S., & Leplawy, M. R. (1988) *Int. J. Pept. Prot. Res. 32*, 544–555.
Marshall, G. R., Gorin, F. A., & Moore M. L. (1978) *Annu. Rev. Med. Chem. 13*, 227–238.
Martin, A. C. R., Cheetham, J. C., & Rees, A. R. (1989) *RNAS 86*, 9628–9272.
Martin, S. F., Austin, R. E., Oalmann, C. J., Baker, W. R., Condon, S. L., Delara, E., Rosenberg, S. H., Spina, K. P., Stein, H. H., Cohen, J., Kleinert, H. D. (1992) *J. Med. Chem. 35*, 1710–1721.
McDowell, R. S. & Gadek, T. R. (1992) *J. Am. Chem. Soc. 114*, 9245–9253.
Merrifield, R. B. (1985) *Angew. Chem. Int. Ed. Engl. 24*, 799–810.
Miklavc, A., Kocjan, D., Avbelj, F., & Hadzi, D. (1987) in *QSAR in Drug Design and Toxicology* (Hadzi, D. and Jerman-Blazic, B., Eds.) pp 185–190, Amsterdam.
Milner-White, E. J., Ross, B. M., Ismail, R., Beinadj-Mostefa, K., & Poet, R. (1988) *J. of Mol. Biol. 204*, 777–782.
Morley, J. (1980) *Annu. Rev. Pharmacol. Toxicol. 20*, 81–110.
Mosberg, H. I., Sobczyk-Kojiro, K., Subramanian, P., Crippen, G. M., Ramalingam, K., & Woodard, R. W. (1990) *J. Am. Chem. Soc. 112*, 822–829.
Müller, K., Obrecht, D., Knierzinger, A., Stankovic, C., Spiegler, C., Bannwarth, W., Trzeciak, A., Englert, G., Labhardt, A. M., & Schönholzer, P. (1993) in *Perspectives in Medicinal Chemistry* (Testa, B., Kyburz, E., Fuhrer, W., & Giger, R., Eds.) pp 513–531, Verlag Helvetica Chimica Acta, Basel.
Nagai, U. & Sato, K. (1985) *Tetrahedron Lett. 26*, 647–650.
Nagaraj, R. & Balaram, P. (1978) *FEBS. Lett. 96*, 273–276.
Needels, M. C., Jones, D. G., Tate, E. H., Heinkel, G. L., Kochersperger, L. M., Dowler, W. J., Barrett, R. W., & Gallop, M. A. (1993) *Proc. Natl. Acad. Sci. U.S.A. 90*, 10700–10704.
Nicolaou, K. C., Salvino, J. M., Raynor, K., Pietranico, S., Reisine, T., Freidinger, R. M., & Hirschmann, R. (1990) *Pept. Chem. Struct. Biol. Proc. Am. Pept. Symp. 11th*, 881.
Nielsen, J., Brenner, S., & Janda, K. D. (1993) *J. Am. Chem. Soc. 115*, 9812–9813.
Nowick, J. S., Powell, N. A., Martinez, E. J., Smith. E. M., & Noronha, G. (1992) *J. Org. Chem. 57*, 3763–3765.
Ohlmeyer, M. H. J., Swanson, R. N., Dillard, L. W., Reader, J. C., Asouline, G., Kobayashi, R., Wigler, M., & Still, C. L. (1993) *Proc. Natl. Acad. Sci. U.S.A. 90*, 10922–10926.
Otsubo, K., Morita, S., & Uchida, M. (1993) *Bioorg. Med. Chem. Letts. 3*, 1633–1636.
Padmanabhan, S. & Baldwin, R. L. (1991) *J. Mol. Biol. 219*, 135–137.
Padmanabhan, S., Marqusee, S., Ridgeway, T., Laue, T. M., & Baldwin, R. L. (1990) *Nature 344*, 268–270.
Pardi, A., Billeter, M., & Wuthrich, K. (1984) *J. Mol. Biol. 180*, 741–751.
Paterson, Y., Rumsey, S. M., Benedetti, E., Nemethy, G., & Scheraga, H. A. (1981) *J. Am. Chem. Soc. 103*, 2947–2955.
Pauling, L., Corey, R. B., & Branson, H. R. (1951) *Proc. Natl. Acad. Sci. U.S.A. 37*, 205–211.
Pavone, V., Lambardi, A., D'Auria, G., Saviono, M., Nastri, F., Paolillo, L., DiBlasio, B., Pedone, C. (1992) *Biopolymers 32*, 173–183.
Perczel A., McAllister, M. A., Csaszar, P., & Csizmadia, I. G. (1993) *J. Am. Chem. Soc. 114*, 4849–4858.
Pettitt, B. M., Matsunaga, T., Al-Obeidi, F., Gehrig, C., Hruby, V. J., & Karplus, M. (1991) *Biophys. J. 60*, 1540–1544.
Picone, D., D'Ursi, A., Motta, A., Tacredi, T., & Temussi, P. A. (1990) *Eur. J. Biochem. 192*, 433–439.

Pierson, M. E. & Freer, R. J. (1992) *Pept. Res. 5*, 102–105.
Portoghese, P. S., Sultana, M., Nagase, H., & Takemori, A. E. (1988) *J. Med. Chem., 31*, 281–282.
Presta, L. G. & Rose, G. D. (1988) *Science*, 240, 1632–1641.
Rapaka, R. S., Barnett, G., & Hawks, R. L., Eds. (1986) in *Opioid Peptides: Medicinal Chemistry*, NIDA Research Monograph 69, Rockville, MD.
Raymond, M. & Gros, P. (1989) *Proc. Natl. Acad. Sci. U.S.A. 86*, 6488–6492.
Reed, J., Hull, W. E., von der Lieth, C. W., Kubler, D., Suhai, S., & Kinzel., V. (1988) *Eur. J. Biochem. 178*, 141–154.
Reed, L. L. & Johnson, P. L. (1973) *J. Am. Chem. Soc. 95*, 7523–7524.
Rodriguez, M., Heitz, A., & Martinez, J. (1990) *Tetrahedron Lett. 31*, 7319–7322.
Roques, B. P., Garbary-Jaureguiberry, C., Oberlin, R., Anteunis, M., & Lala, A. K. (1976) *Nature 262*, 778–779.
Rose, G. D., Gierasch, L. M., & Smith, J. A. (1985) *Adv. Protein Chem. 37*, 1–109.
Salituro, G. M., Pettibone, D. J., Clineschmidt, B. V., Williamson, J. M., & Zink, D. L. (1993) *Bioorg. Med. Chem. Lett. 3*, 337–340.
Salzmann, T. N., Ratcliff, R. W., Christensen, B. G., & Boufard, F. A. (1980) *J. Am. Chem. Soc. 102*, 6161–6163.
Samanen, J., Cash, T., Narindray, D., Brandeis, E., Adams, Jr., W., Weideman, H., & Yellin, T. (1991) *J. Med. Chem. 34*, 3036–3043.
Saragovi, H. U., Fitzpatrick, D., Raktabutr, A., Nakanishi, H., Kahn, M., & Greene, M. I. (1991) *Science 253*, 792–795.
Sato, M., Lee, J. Y. H., Nakanishi, H., Johnson, M. E., Chrusciel, R. A., & Kahn, M. (1992) *Biochem. Biophys. Res. Commun. 187*, 999–1006.
Schiller, P. W. (1990) in *The Peptides: Analysis, Synthesis, Biology*, 6. (Udenfired, S. & Meienhofer, J., Eds.) pp 219–268, Academic Press, Orlando, FL.
Schiller, P. W., & Dimaio, J. (1983) in *Peptides: Structure and Function* (Hruby, V. J. & Rich, D. H., Eds.) pp 269–278, Pierce Chemical Co., Rockford, Illinois.
Schramm, H. J., Nakashima, H., Schramm, W., Wakayama, H., & Yamamoto, N. (1991) *Biochem. Biophys. Res. Commun. 179*, 847–851..
Schramm, H.J., Billich, A., Jaeger, E., Rucknagel, K.-P., Arnold, G., & Schramm, W. (1993) *Biochem. Biophys. Res. Commun. 194*, 595–600.
Shoemaker, K. R., Kim, P. S., York, E. J., Stewart, J. M., & Baldwin, R. L. (1987) *Nature 326*, 563–567.
Simon, R. J., Kania, R. S., Zuckerman, R. N., Huebner, V. D., Jewell, D. A., Banville, S., Ng, S., Wang, L., Rosenberg, S., Marlowe, C. K., Spellmeyer, D. C., Tan, R., Frankel, A. D., Santi, D. V., Cohen, F. E., & Bartlett, P. A. (1992) *Proc. Natl. Acad. Sci. U.S.A. 89*, 9376–9371.
Smith, III., A. B., Hirschmann, R., Pasternak, A., Akaishi, R., Guzman, M. C., Jones, D. R., Keenan, T. P., Sprengeler, P. A., Darke, P. L., Emini, E. A., Holloway, M. K., & Schleif, N. A. (1994) *J. Med. Chem. 37*, 215–218.
Smith, III, A. B., Holcomb, R. C., Guzman, M. C., Kennan, T. P., Sprengeler, P. A., & Hirschman, R. (1993b) *Tetrahedron Lett. 34*, 63–66.
Smith, III, A. B., Keenan, T. P., Holcomb, R. C., Sprengler, P. A., Guzman, M. C., Wood, J. L., Carroll, P. J., & Hirschmann, R. (1992) *J. Am. Chem. Soc. 114*, 10672–10674.
Smith, P. E., Dana, L. X., & Pettitt, B. M. (1991) *J. Am. Chem. Soc 113*, 67–73.
Smith, R. G., Cheng, K., Schoen, W. R., Pong, S. S., Hickey, G., Jacks, T., Butler, B., Chan, W. W., Chaung, L. P., Judith, F., Taylor, J., Wyvratt, M. J., & Fisher, M. H., (1993a) *Science 260*, 1640–1643.
Snider, R. M., Pereira, D. A., Longo, K. P., Davidson, R. E., Vinick, F. J., Laitinen, K., Genc-Sehitoglu, E., & Crawley, J. N. (1992) *Bioorg. Med. Chem. Letts. 2*, 1535–1540.
Spaltenstein, A., Carpino, P. A., Miyake, F., & Hopkins, P. B. (1987) *J. Org. Chem. 52*, 3759–3766.
Spatola, A. F. & Edwards, J. V. (1986) *Biopolymers 25*, S229–244.

Sreenivasan U., Mishra, R. K., & Johnson, R. L. (1993) *J. Med. Chem. 36*, 256–263.
Subasinghe, N. L., Bontems, R. J., McIntee, E., Mishra, R. K., Johnson, R. L. (1993) *J. Med. Chem. 36*, 2356–2361.
Still, W. C., Tempczyk, A., Hawley, R. C., & Hendrickson, T. (1990) *J. Am. Chem. Soc. 112*, 6127–6129.
Stroup, A. N., Rockwell, A. L., Gierasch, L. M. (1992) *Biopolymers 32*, 1713–1725.
Su., T., Nakanishi, H., Xue, L., Chen, B., Tuladhar, S., Johnson, M. E., & Kahn, M. (1993) *Bioorg. Med. Chem. Lett 3*, 835–840.
Swain C. J., Seward E. M., Sabin, V., Owen, S., Baker, R., Cascieri, M. A., Sadowski, S., Strader, C., Ball, R. G. (1993) *Bioorg. Med. Chem. Lett. 3*, 1703–1706.
TenBrink, R. E. (1987) *J. Org. Chem. 52*, 418–422.
Thaisrivongs, S., Tomaselli, A. G., Moon, J. B., Hui, J., McQuade, T. J., Turner, S. R., Strohbach, J. W., Howe, W. J., Tarpley, W. G., & Heinrikson, R. L. (1991) *J. Med. Chem. 34*, 2344–2356.
Toniolo, C. (1990) *Int. J. Pept. Prot. Res., 35*, 287–300.
Toniolo, C. & Benedetti, E. (1991) *Trends Biochem. Sci. 16*, 305–353.
Toniolo, C., Crisma, M., Formaggio, F., Valle, G., Cavicchioni, G., Precigoux, G., Aubry, A., Kamphuis, J. (1993) *Biopolymers 33*, 1061–1072.
Tourwe, D., Couder, J., Ceusters, M., Meert, D., Burds, T. F., Kramer, T. H., Davis, P., Knapp, R., Yamamura, H. I., Leysen, J. E., & Van Binst, G. (1992) *Int. J. Pept. Prot. Res. 39*, 131–136.
Tsang, K. Y., Diaz, H., Graciani, N., & Kelly, J. W. (1994) *J. Am. Chem. Soc 116*, 3988–4005.
Tung, R. D., Dhaon, M. K., Rich, D. H. (1986) *J. Org. Chem. 51*, 3350–3354.
Valentine, J. J., Nakanishi, S., Hageman, D. L., Snider, M. R., Spencer, R. W., & Vinick, F. J. (1992) *Bioorg. Med. Chem. Lett. 2*, 333–338.
Vara Prasad, J. V. N. & Rich, D. H. (1990) *Tetrahedron Lett. 31*, 1803–1806.
Veber, D. F., Holly, F. W., Paleveda, W. J., Nutt, R. F., Bergstrand, S. J., Torchiana, M., Glitzer, M. S., & Saperstein, R. (1979) *Nature 280*, 512–514.
Venkatachalam, C. M. (1968) *Biopolymers 6*, 1425–1436.
de Vos, A. M., Ultsch, M., & Kossiakoff, A. A. (1992) *Science 255*, 306–312.
Waksman, G. (1992) *Nature, 358*, 646–653.
Ward, W. H. J., Timms, D., & Fersht, A. R. (1990) *Trends Pharmacol. Sci. 11*, 280–284.
Wasserman, H. H. (1987) *Aldrichim. Acta 20*, 63–74.
Watling, K. J. (1992) *Trends Pharmacol. Sci.* 13, 266–269.
William, R. M., Lee, B. H., Miller, M. M., & Anderson, O. P. (1989) *J. Am. Chem. Soc. 111*, 1063–1083.
Wilmot, C. M. & Thornton, J. M. (1988) *J. Mol. Biol. 203*, 221–232.
Wilmot, C. M. & Thornton, J. M. (1990) *Protein Eng. 6*, 479–493.
Yamazaki, T., Ro. S., Goodman, M., Chung, N. N., & Schiller, P. W. (1993) *J. Med. Chem. 36*, 708–719.
Yous, S., Andrieux, J., Howell, H. E., Morgan, P. J., Renard, P., Pfeiffer, B., Lesieur, D., & Guardiola-Lemaitre, B. (1992) *J. Med. Chem. 35*, 1484–1486.
Yu, K. L., Rajakumar, G., Srivastava, L. K., Mishra, R. K., & Johnson, R. L. (1988) *J. Med. Chem. 31*, 430–436.

Chapter 13

Aaltonen, O. & Rantakylä, M. (1991) *CHEMTECH April*, 240–248.
Abuchowski, A., Kazo, G. M., Verhoest, C. R., Es, T. V., Kafkewitz, D., Nucci, M. L., Viau, A. T., & Davis, F. F. (1984) *Cancer Biochem. Biophys. 7*, 175–186.
Adlercreutz, P. (1992) in *Biocatalysis in Non-conventional Media* (Tramper, J., Vermüe, M. H., Beeftink, H. H., & von Stockar, U., Eds.) pp 55–61, Elsevier, Amsterdam.
Adlercreutz, P. (1991) *Eur. J. Biochem. 199*, 609–614.

Affleck, R., Xu, Z.-F., Suzawa, V., Focht, K., Clark, D. S., & Dordick, J. S. (1992) *Proc. Natl. Acad. Sci. U.S.A. 89*, 1100–1104.
Ajima, A., Yoshimoto, T., Takahashi, K., Tamaura, Y., Saito, Y., & Inada, Y. (1985) *Biotechnol. Lett. 7*, 303–306.
Alcock, N.W., Crout, D. H. G., Henderson, C. M., & Thomas, S. E. (1988) *J. Chem. Soc., Chem. Commun.*, 746–747.
Baeyer, A. & Villiger, V. (1899) *Chem. Ber. 32*, 3625.
Baillargeon, M. W. & Sonnet, P. E. (1988) *Ann. N.Y. Acad. Sci.* 244–249.
Beauchamp, C. O., Gonias, S. L., Menapace, D. P., & Pizzo, S.V. (1983) *Anal. Biochem. 131*, 25–33.
Bender, M.L. & Killheffer, J. V. (1973) *Crit. Rev. Biochem. 1*, 149.
Bjorkling, F., Boutelje, J., Gatenbeck, S., Hult, K., Norin, T., & Szmulik, P. (1985) *Tetrahedron 41*, 1347–1352.
Brink, L.E.S., Tramper, J., Luyben, K. Ch. A. M., & Riet, K.Van't. (1988) *Enzyme Microb. Technol. 10*, 736–743.
Burke, P.A., Smith, S.O., Bachovchin, W. W., & Klibanov, A. M. (1989) *J. Am. Chem. Soc. 111*, 8290–8291.
Chen, L., Dumas, D. P., & Wong, C.-H. (1992) *J. Am. Chem. Soc. 114*, 741–748.
Chen, C.-S., & Sih, C. J. (1989) *Angew. Chem. Int. Ed. Engl. 28*, 695–707.
Chi, Y. M., Nakamura, K., & Yano, T. (1988) *Agric. Biol. Chem. 52*, 1541–1550.
Davies, H. G., Green, R. H., Kelly, D. R., & Roberts, S. M. (1989) *Biotransformations in Preparative Organic Chemistry: The Use of Isolated Enzymes and Whole Cell Systems*, Academic Press, London.
Deetz, J. S. & Rozzell, J. D. (1988) *Trends Biotechnol. 6*, 15–19.
Dixon, M. & Webb, E. D. (1979) *Enzymes*, Academic Press, New York.
Dordick, J. S. (1989) *Enzyme Microb. Technol. 11*, 194–211.
Dordick, J. S. (1992) *Trends Biotechnol. 10*, 287–293.
Dordick, J. S., Marletta, M. A., & Klibanov, A. M. (1987) *Biotechnol. Bioeng. 30*, 31–36.
Dumont, T., Barth, D., Corbier, C., Branlant, G., & Perrut, M. (1992) *Biotechnol. Bioeng. 39*, 329–333.
Faber, K. (1992) *Biotransformations in Organic Chemistry*, Springer-Verlag, Berlin.
Ferjancic, A., Puigserver, A., & Gaertner, H. (1990) *Appl. Microbiol. Biotechnol. 32*, 651–657.
Fitzpatrick, P. & Klibanov, A. M. (1991) *J. Am. Chem. Soc. 113*, 3166–3171.
Fuganti, C. & Grasselli, P. (1989) in *Biocatalysis in Agricultural Biotechnology, ACS Symp. Ser. 389* (Whitaker, J.R., & Sonnet, P.E., Eds.) pp 359, Am. Chem. Soc., Washington.
Gaertner, H. & Puigserver, A. (1989) *Eur. J. Biochem. 181*, 207–213.
Ghogare, A. & Kumar, G. S. (1990) *J. Chem. Soc., Chem. Commun.* 134–135.
Grunwald, J., Witz, B., Scollar, M. P., & Klibanov, A. M. (1986) *J. Am. Chem. Soc. 108*, 6732–6734.
Guinn, R. M., Skerker, P. S., Kavanaugh, P., & Clark, D. S. (1991) *Biotechnol. Bioeng. 37*, 303–308.
Halling, P. J. (1987) *Biotechnol. Adv. 5*, 47–84.
Hammond, D. A., Karel, M., & Klibanov, A. M. (1985) *Appl. Biochem. Biotechnol. 11*, 393–400.
Harwood, J. (1989) *Trends Biotechnol. 14*, 125–126.
Holland, H. L. (1992) *Organic Synthesis with Oxidative Enzymes*, VCH, Weinheim, Germany.
Hollinshead, D. M., Howell, S. C., Ley, S. V., Mahon, M. & Ratcliffe, N. M. (1983) *J. Chem. Soc., Perkin Trans. 1*, 1579–1689.
Ichikawa, Y., Lin, Y.-C., Dumas, D. P., Shen, G.-J., Garcia-Junceda, E., Williams, M. A., Bayer, R., Ketcham, C., Walker, L. E., Paulson, J. C., & Wong, C.-H. (1992) *J. Am. Chem. Soc. 114*, 9283–9298.
Ichikawa, Y., Lin, Y.-C., Shen, G.-J., & Wong, C.-H. (1991) *J. Am. Chem. Soc. 113*, 6300–6302.
Inada, Y., Takahashi, K., Yoshimoto, T., Ajima, A., Matsushima, A., & Saito, Y. (1986) *Trends Biotechnol. 7*, 190–194.

Inada, Y., Takahashi, K., Yoshimoto, T., Kodera, Y., Matsushima, A., & Saito, Y. (1988) *Trends Biotechnol. 6*, 131–134.
Jones, J. B. & Beck, J. F. (1976) *Tech. Chem. 10*, 107–401.
Jones, J. B. & Lok, K. P. (1979) *Can. J. Chem. 57*, 1025–1032.
Kamat, S., Barrera, J., Beckman, E. J., & Russell, A. J. (1992) *Biotechnol. Bioeng. 40*, 158–166.
Kamat, S. V., Beckman, E. J., & Russell, A. J. (1993a) *J. Am. Chem. Soc. 115*, 8845–8846.
Kamat, S., Iwaskewycz, B., Beckman, E. J., & Russell, A. J. (1993b) *Proc. Natl. Acad. Sci. U.S.A. 90*, 2940–2944.
Kaufmann, H. & Reichstein, T. (1967) *Helv. Chim. Acta 50*, 2280–2287.
Kazandjian, R. Z. & Klibanov, A. M. (1985) *J. Am. Chem. Soc. 107*, 5448–5450.
Khmelnitsky, Y. L., Lavashov, A. V., Klyachko, N. L., & Martinek, K. (1988) *Enzyme Microb. Technol. 10*, 710–724.
Khmelnitsky, Y. L., Mozhaev, V. V., Belova, A. B., Sergeeva, M. V., & Martinek, K. (1991) *Eur. J. Biochem. 198*, 31–41.
Kim, M.-J. & Whitesides, G. M. (1988) *J. Am. Chem. Soc. 110*, 2959–2964.
Klibanov, A. M. (1989) *Trends Biochem. Sci. 14*, 141–144.
Klibanov, A. M. (1990) *Acc. Chem. Res. 23*, 114–120.
Kondo, K., Ogura, M., Midorikawa, Y., Kozawa, M., Tsujibo, H., Baba, K., & Inamori, Y. (1989) *Agr. Biol. Chem. 53*, 777–782.
Kuhl, P., Halling, P. J., & Jakubke, H.-D. (1990) *Tetrahedron Lett. 31*, 5213–5216.
Kullmann, W. (1982) *J. Org. Chem. 47*, 5300–5303.
Laane, C., Boeren, S., Vos, K., & Veeger, C. (1987) *Biotechnol. Bioeng. 30*, 81–87.
Laumen, K. & Schneider, M. (1984) *Tetrahedron Lett. 25*, 5875–5878.
Lee, J. B. & Uff, B. C. (1967) *Quart. Rev. 21*, 429–457.
Lide, D. R., Ed. (1990) *CRC Handbook of Chemistry and Physics*, CRC Press, Boca Raton, FL.
Margolin, A. L. (1993) *Enzyme Microb. Technol. 15*, 266–280.
Miller, D. A., Blanch, H. W., & Prausnitz, J. M. (1990) *Ann. N.Y. Acad. Sci. 613*, 534–537.
Nakamura, K. (1990) *Trends Biotechnol. 8*, 288–292.
Nakazaki, M., Chikamatsu, H., Naemura, K., Suzuki, T., Iwasaki, M., Sasaki, Y., & Fujii, T. (1981) *J. Org. Chem. 46*, 2726–2730.
Omata, T., Iida, T., Tanaka, A., & Gukui, S. (1979) *Eur. J. Appl. Microbiol. Biotechnol. 8*, 143–155.
Oyama, K. & Kihara, K. (1984) *CHEMTECH* Feb. 100–105.
Pandey, G., Muralikrishna, C., & Bhalerao, U. T. (1989) *Tetrahedron 45*, 6867–6874.
Pasta, P., Mazzula, G., Carrea, G., & Riva, S. (1989) *Biotechnol. Lett. 2*, 643–648.
Peterson, D. H., Murray, H. C., Eppstein, S. H., Reineke, L. M., Weintraub, A., Meister, P. D., & Leigh, H. M. (1952) *J. Am. Chem. Soc. 74*, 5933–5936.
Popp, J. L., Kirk, T. K., & Dordick, J. S. (1991) *Enzyme Microb. Technol. 13*, 964–968.
Prairie, R. & Talalay, P. (1963) *Biochemistry 2*, 203–208.
Prelog, V. (1964) *Pure Appl. Chem. 9*, 119–130.
Rahim, M. A. & Sih, C. J. (1966) *J. Biol. Chem. 241*, 3615–3623.
Randolph, T. W., Blanch, H. W., & Clark, D. S. (1991) in *Biocatalysts for Industry* (Dordick, J. S., Ed.) pp 219–237, Plenum Press, New York.
Randolph, T. W., Blanch, H. W., Prausnitz, J. M., & Wilke, C. R. (1985) *Biotechnol. Lett. 7*, 325–328.
Randolph, T. W., Blanch, H. W., & Prausnitz, J. M. (1988) *AICHE J. 34*, 1354–1360.
Reslow, M., Adlercreutz, P., & Mattiasson, B. (1988) *Eur. J. Biochem. 172*, 573–578.
Roberts, S. M. (1990) in *Microbial Enzymes and Biotechnology*, (Second Ed., Fogarty, W. M. & Kelly, C. T., Eds.) pp 395–424, Elsevier, London.
Rubio, E., Fernandez-Mayorales, A., & Klibanov, A. M. (1991) *J. Am. Chem. Soc. 113*, 695–696.
Ryu, K., & Dordick, J. S. (1992) *Biochemistry 31*, 2588–2598.
Sabbioni, G. & Jones, J. B. (1987) *J. Org. Chem. 52*, 4565–4570.

Sakurai, T., Margolin, A. L., Russell, A. J., & Klibanov, A. M. (1988) *J. Am. Chem. Soc. 110*, 7236–7237.
Schneider, L. V. (1991) *Biotechnol. Bioeng. 37*, 627–638.
Shimizu, S., Hattori, S., Hata, H., & Yamada, H. (1987) *Appl. Environ. Microbiol. 53*, 519–522.
Simon, H., Bader, J., Günther, H., Neumann, S., & Thanos, J. (1985) *Angew. Chem. Int. Ed. Engl. 24*, 539–553.
Souppe, J. & Urrutigoity, M. (1989) *New J. Chem. 13*, 503–506.
Takahashi, K., Ajima, A., Yoshimoto, T., & Inada, Y. (1984b) *Biochem. Biophys. Res. Commun. 125*, 761–766.
Takahashi, K., Kodera, Y., Yoshimoto, T., Ajima, A., Matsushima, A., & Inada, Y. (1985) *Biochem. Biophys. Res. Commun. 131*, 532–536.
Takahashi, K., Nishimura, H., Yoshimoto, T., & Inada, Y. (1984a) *Biochem. Biophys. Res. Commun.* 121, 261–265.
Terradas, F., Teston-Henry, M., Fitzpatrick, P. A., & Klibanov, A. M. (1993) *J. Am. Chem. Soc. 115*, 390–396.
Toone, E., Simon, E. S., Bednarski, M. D., & Whitesides, G. M. (1989) *Tetrahedron 45*, 5365–5422.
Urrutigoity, M., & Souppe, J. (1989) *Biocatalysis 2*, 145–149.
Valivety, R. H., Halling, P. J., & Macrae, A. R. (1992) *FEBS Lett. 301*, 258–260.
Veronese, F. M., Largajolli, R., Boccu, E., Benassi, C. A., & Schiavon, O. (1985) *Appl. Biochem. Biotechnol. 11*, 141–152.
Volkin, D. B., Staubli, A., Langer, R., & Klibanov, A. M. (1991) *Biotechnol. Bioeng. 37*, 843–853.
Waldmann, H. (1988) *Tetrahedron Lett.* 29, 1131–1134.
Wallace, J. S. & Morrow, C. J. (1989) *J. Polym. Sci. Part A: Polym. Chem. 27*, 2553–2567.
Wandrey, C. (1986) in *Enzymes as Catalysts in Organic Synthesis* (Schneider, M. P. & Reidel, D., Eds.) pp 263, Dordrecht, Holland.
Wehtje, E., Adlercreutz, P., & Mattiasson, B. (1993) *Biotechnol. Bioeng. 41*, 171–178.
Wescott, C. R. & Klibanov, A. M. (1993) *J. Am. Chem. Soc. 115*, 1629–1631.
Whitesides, G. M. & Wong, C.-H. (1985) *Angew. Chem. Int. Ed. Eng. 24*, 617–638.
Wirth, P., Souppe, J., Tritsch, D., & Biellmann, J.-F. (1991) *Bioorg. Chem. 19*, 133–142.
Wong, C.-H. (1992) *Trends Biotechnol. 10*, 337–341.
Wong, C.-H., Ichikawa, Y., Krach, T., Narvor, C. G., Dumas, D. P., & Look, G. C. (1991a) *J. Am. Chem. Soc. 113*, 8137–8145.
Wong, C.-H., Liu, K. K.-C., Kajimoto, T., Chen, L., Zhong, Z., Dumas, D. P., Liu, J. L.-C., Ichikawa, Y., & Shen, G.-J. (1992) *Pure Appl. Chem. 64*, 1197–1202.
Wong, C.-H., Shen, G.-J., Pederson, R. L., Wang, Y.-F., & Hennen, W. J. (1991b) *Methods Enzymol. 202*, 591–620.
Yamazaki, Y. & Hosono, K. (1988) *Tetrahedron Lett.* 29, 5769–5770.
Yang, F. X. & Russell, A. J. (1993) *Biotechnol. Prog. 9*, 234–241.
Yang, Z. & Robb, D. A. (1991) *Biochem. Soc. Trans. 20*, 13S.
Yang, Z. & Robb, D. A. (1994) *Biotechnol. Bioeng. 43*, 365–370..
Yang, Z., Robb, D. A., & Halling, P. J. (1992) in *Biocatalysis in Non-conventional Media* (Tramper, J., Vermüe, M. H., Beeftink, H. H., & von Stockar, U., Eds.) pp 585–592, Elsevier, Amsterdam.
Yang, Z., Zacherl, D., & Russell, A. J. (1994) *J. Am. Chem. Soc. 115*, 12251–12257.
Yoshimoto, T., Takahashi, K., Nishimura, H., Ajima, A., Tamaura, Y., & Inada, Y. (1984) *Biotechnol. Lett. 6*, 337–340.
Yoshimoto, T., Ritani, A., Ohwada, K., Takahashi, K., Kodera, Y., Matsushima, A., Saito, Y., & Inada, Y. (1987) *Biochem. Biophys. Res. Commun. 148*, 876–882.
Zaks, A. & Klibanov, A. M. (1984) *Science 224*, 1249–1251.
Zaks, A. & Klibanov, A. M. (1985) *Proc. Natl. Acad. Sci. U.S.A. 82*, 3192–3196.

Zaks, A. & Klibanov, A. M. (1986) *J. Am. Chem. Soc. 108*, 2767–2768.
Zaks, A. & Klibanov, A. M. (1988a) *J. Biol. Chem. 263*, 3194–3201.
Zaks, A. & Klibanov, A. M. (1988b) *J. Biol. Chem. 263*, 8017–8021.

Chapter 14

Aota, S., Gojobori, T., Ishibashi, F., Maruyama, T, & Ikemura, T. (1988) *Nucleic Acids Res. 16*, R315–R402.
Atkins, E. D. T., Hill, M., Hong, S. K., Keller, A., & Organ, S. (1992) *Macromolecules 25*, 917–924.
Bahr, U., Deppe, A., Karas, M., Hillenkamp, F., & Giessmann, U. (1992) *Anal. Chem. 64*, 2866–2869.
Bandekar, J. and Krimm, S. (1986) in *Advances in Protein Chemistry* (Anfinsen, C. B., Edsall, J. T., Richards, F. M., Eds.) Vol. 38, pp 183–364, Academic Press, Orlando, FL.
Baneyx, F., Schmidt, C., & Georgiou, G. (1990) *Enzyme Microb. Technol. 12*, 337–342.
Beavis, R. C., Chait, B. T., Creel, H. S., Fournier, M. J. Mason, T. L., & Tirrell, D.A. (1992) *J. Am. Chem. Soc. 114*, 7584–7585.
Block H. (1983) *Poly(γ-benzyl-L-glutamate) and Other Glutamic Acid Containing Polymers*, Gordon and Breach, New York.
Bochner, B. R., Huang, H., Schieven, G. L., & Ames, B. N. (1990) *J. Bacteriol. 143*, 926.
Brown, L. & Trotter, I. F. (1956) *Trans. Faraday Soc. 52*, 537–548.
Buku, A., Faulstich, H., Wieland, T., & Dabrowski, J. (1980) *Proc. Natl. Acad. Sci. U.S.A. 77*, 2370–2371.
Cappello, J., Crissman, J., Dorman, M., Mikolajczak, M., Textor, G., Marquet, M., & Ferrari, F. (1990) *Biotechnol. Prog. 6*, 198–202.
Chen, C. C., Krejchi, M. T., Tirrell, D. A., & Hsu, S. L. (1995) *Macromolecules 28*, 1464–1469.
Chou, P. Y. & Fasman, G. D. (1974a) *Biochemistry 13*, 211–222.
Chou, P. Y. & Fasman, G. D. (1974b) *Biochemistry 13*, 222–245.
Cowie, D. B. & Cohen, G. N. (1957) *Biochim. Biophys. Acta 26*, 252–261.
Cowie, D. B., Cohen, G. N., Bolton, E. T., & De Robichon-Szulmajster, H. (1959) *Biochim. Biophys. Acta 34*, 39–46.
Deguchi, Y., Fournier, M. J., Mason, T. L., & Tirrell, D. A. (1994) *J. Macromol. Sci., -Pure Appl. Chem. A31*, 1691–1700.
Deming, T. J., Fournier, M. J., Mason, T. L., & Tirrell, D. A. unpublished results.
Deming, T. J., Fournier, M. J., Mason, T. L., & Tirrell, D. A. (1996) *Macromolecules 29*, 1442–1444.
Dong, W., Fournier, M. J., Mason, T. L., & Tirrell, D. A. (1994) *Polymer Preprints 35*, 419.
Dong, W., Fournier, M. J., Mason, T. L., & Tirrell, D. A. (1996) *Polym. Mat. Sci. Eng. 74*, 71–72.
Dougherty, M. J., Kothakota, S., Mason, T. L., Tirrell, D. A., & Fournier, M. J. (1993) *Macromolecules 26*, 1779–1781.
Dreyfuss, P. & Keller, A. (1970a) *J. Macromol. Sci.-Phys. B4*, 811–835.
Dreyfuss, P. & Keller, A. (1970b) *J. Polym. Sci. Part B., Polym. Phys. 8*, 253–258.
Dumas, D. P. & Raushel, F. M. (1989) *J. Biol. Chem. 264*, 19659–19665.
Dumas, D. P. & Rauschel, F. M. (1990) *J. Biol. Chem. 265*, 21498–1503.
Ellman, J. A., Mendel, D., & Schultz, P. G. (1992) *Science 255*, 197–200.
Fenster, E. D. & Anker, H. S. (1969) *Biochemistry 8*, 269–274.
Fowden, L. (1956) *Biochem. J. 64*, 323–332.
Fraser, R. D. B. MacRae, T. P., Stewart, H. C., & Suzuki, E. (1965) *J. Mol. Biol. 11*, 706–712.
Frushour, B. G. & Koenig, J. L. (1975) *Biopolymers 14*, 2115–2135.
Geddes, A. J., Parker, K. D., Atkins, E. D. T., & Beighton, E. (1968) *J. Mol. Biol. 32*, 343–358.
Hecht, S. M. (1992) *Acc. Chem. Res. 25*, 545–552.

Hendrickson, W. A., Horton, J. R., & LeMaster, D. M. (1990) *EMBO J. 9*, 1665–1672.
Hirichsen, G. (1973) *Makromol. Chem. 166*, 291.
Horton, J. C., Donald, A. M., & Hill, A. (1990) *Nature 346*, 44–45.
Hunger, H. D., Flachmeier, C., Schmidt, C., Behrendt, G., & Coutelle, C. (1990) *Anal. Biochem. 187*, 89–93.
Ishida, M., Asakura, T., Yokoi, M., & Saito, H. (1990) *Macromolecules 23*, 88–94.
Kamtekar, S., Schiffer, J. M., Xiong, H., Babik, J. M., & Hecht, M. H. (1993) *Science 262*, 1680–1685.
Keller, A. (1957) *Philos. Mag. 2*, 1171–1175.
Keller A. (1959) *J. Polym. Sci. 36*, 361–382.
Kothakota, S., Mason, T. L., Tirrell, D. A., Atkins, E. D. T., & Fournier, M. J., unpublished results.
Kothakota, S., Mason, T. L., Tirrell, D. A., & Fournier, M. J. (1995) *J. Am. Chem. Soc. 117*, 536–537.
Krejchi, M. T., Atkins, E. D. T., Waddon, A. J., Fournier, M. J., Mason, T. L., & Tirrell D. A. (1994) *Science 265*, 1427–1432.
Krejchi, M. T., Cooper, S. J. Deguchi, Y., Atkins, E. D. T. Fournier, M. J., Mason, T. L., & Tirrell, D. A. (1997) *Macromolecules 30*, 5012–5024.
Krimm, S. & Abe, Y. (1972) *Proc. Natl. Acad. Sci. U.S.A. 69*, 2788–2792.
Krimm, S. & Bandekar, J. (1986) *Adv. Protein Chem. 38*, 181–364.
Lane, J. M., Dehm, P., & Prockop, D. J. (1971) *Biochim. Biophys. Acta 236*, 517–527.
Lucas, F., Shaw, T. B., & Smith, S. G. (1960) *J. Mol. Biol. 2*, 339–349.
Mauger, A. B. & Witkop, B. (1966) *Chem. Rev. 66*, 47–86.
McBride, L. J. & Caruthers, M. H. (1983) *Tetrahedron Lett. 24*, 245–248.
McGrath, K. P., Fournier, M. J., Mason, T. L., & Tirrell, D. A. (1992) *J. Am. Chem. Soc. 114*, 727–733.
McMaster, T. C., Carr, H. J., Miles, M. J., Cairns, P., & Morris, V. J. (1991) *Macromolecules 24*, 1428–1430.
Miller, J. H. (1972) *Experiments in Molecular Genetics*, Cold Spring Harbor Laboratory, NY.
Miyazawa, T. & Blout, E. R. (1961) *J. Am. Chem. Soc. 83*, 712–719.
Moore, W. H. & Krimm, S. (1976) *Biopolymers 15*, 2465–2483.
Murthy, N. S., Stamm, M., Sibilia, J. P., & Krimm, S. (1989) *Macromolecules 22*, 1261–1267.
Nakajima, T., & Volcani, B. E. (1969) *Science 164*, 1400–1401.
Nemethy, G. & Printz, M. P. (1972) *Macromolecules 5*, 755–761.
Parkhe, A. D., Fournier, M. J., Mason, T. L., & Tirrell, D. A. (1993) *Macromolecules 26*, 6691–6693.
Peterson, P. J. & Fowden, L. (1963) *Nature 200*, 148–151.
Poland, D. & Scheraga, H. A. (1970) *Theory of Helix-Coil Transitions in Biopolymers*, Academic Press, New York.
Roncali, J., Yassar, A., & Garnier, F. (1988) *J. Chem. Soc., Chem. Commun.*, 581–582.
Rosenberg, A. H., Lade, B. N., Chui, D., Lin, S., Dunn, J. J., & Studier, F. W. (1987) *Gene 56*, 125–135.
Rosenbloom, J. & Prockop, D. J. (1970) *J. Biol. Chem. 245*, 3361–3368.
Saito, H., Tabeta, R., Asakura, T., Iwanaga, Y. Shoji, A., Ozaki, T., & Ando, I. (1984) *Macromolecules 17*, 1405–1412.
Sambrook, J., Fritsch, E. F., & Maniatis, T. (1989) *Molecular Cloning: A Laboratory Manual*, Second Ed., Cold Spring Harbor Laboratory Press, Cold Spring Harbor, NY.
Smith, D. B. & Johnson, K. S. (1988) *Gene 67*, 31–40.
Smith, L. C., Ravel, J. M., Skinner, C. G., & Shive, W. (1962) *Arch. Biochem. Biophys. 99*, 60–64.
Studier, F. W., Rosenberg, A. H., Dunn, J. J., & Dubendorff, J. W. (1990) *Methods Enzymol. 185*, 60–89.
Tirrell, D., Grossman, S., & Vogl, O. (1979) *Makromol. Chem. 180*, 721–736.

Tristram, H. & Thurston, C. F. (1966) *Nature 212*, 74–75.
Tuve, T. & Williams, H. (1957) *J. Am. Chem. Soc. 79*, 5830–5831.
Vergalati, C., Imberty, A., & Perez S. (1993) *Macromolecules 26*, 4420–4425.
Warwicker, J. O. (1960) *J. Mol. Biol. 2*, 350–362.
Yoshikawa, E., Fournier, M. J., Mason, T. L., & Tirrell, D. A. (1994) *Macromolecules 27*, 5471–5475.
Zagari, A., Nemethy, G., & Scheraga, H. A. (1990) *Biopolymers 30*, 951–959.
Zhang, G., Fournier, M. J., Mason, T. L., & Tirrell, D. A. (1992) *Macromolecules 25*, 3601–3603

Index

$\Delta G\ddagger_{int}$, 304–6, 308
$\Delta G°$, 304–6

A state, 231
Accessible surface, 227
Acetylation, 142
Actin, 18, 164, 165
Active site structure, triose phosphate
  isomerase (TIM), 293
  mandelate racemase (MR), 297
Active sites, 210
Acyl transfer, 282, 355
Acyl-antibody intermediate, 357, 361
Affinity chromatography, 17
Alanine, 2, 4
Alanine dipeptide, 169
Aldolases, 432, 433
α-Alkyl amino acid, 396
N^a-Allyloxycarbonyl type protection, 43
Alzheimer's disease, 224
Amide-to-sulfur, 259
Amide band, 452, 454, 464
Amide I vibration, 454
Amine bond replacement, 61
Amino acid, 1, 2
  21st, 5
  analysis, 13
  aromatic, 4
  bulky, 4
  charged, 4
  glycosylated, 6
  non-polar, 4
  non-standard, 5
  non-natural, 448, 459, 472
  phosphorylated, 4, 6
  polar, 4
  sequence, 7, 13
  serine, 4
  small, 4
  standard, 2, 4–6
  sulfur-containing, 4
  threonine, 4

tyrosine, 4
uncharged, 4
α-Amino isobutyric acid (Aib), 396
α-Amino protecting group, 28
Amino terminus, 13
Amphipathic helix, 8
Analytical ultracentrifugation, 13
Angiotensin, 396, 400, 406, 413
8–Anilino-1–naphthalene sulfonate (ANS), 247
Anilinothiazolinone (ATZ), 114, 116, 121,
  122, 123
ANS binding, 247
Antigenic determinant, 158
Antigenicity, 158
Antiparallel β-sheet structure, 455, 456
APOBEC-1 cytidine deaminase, 196
Apomyoglobin (ApoMb), 234, 235, 247, 249,
  252
Apo-α-lactalbumin (LA), 241
Apoenzyme, 10
Apoprotein, 10
Arabinose-binding protein, 216
Arginine, 4
Ascorbic acid, 4
Asn-Gly bond, 130
Asp N, 127
Asp-Pro bond, 130
Asparagine, 4
Aspartic acid, 3, 4
Aspartate carbamoyltransferse, 202
Assay, enzyme, 16
Autocrine, 367, 371
Aza-sugar, 433, 434
Azetidine-2–carboxylic acid, 459, 467

B-factor, 157
Backbone, 7
Bacteriophage T4 lysozyme, 238
Bacterial ferredoxins, 195
Barnase, 88, 238, 247
Bence-Jones protein, 221
β-Bend, 417

Benzodiazepine, 400, 402, 406
Beta sheet, 8, 13, 15
  antiparallel, 9
  parallel, 9
Beta galactosidase, 154
Big endothelin, 387–89
Biosensor, 471
Blot digest, 104
Boc chemistry, 38, 39, 41, 46, 67
N^α-Boc protected amino acid, 41
Boc/Fmoc strategy for on-resin lactam formation, 56
Bombesin, 400
Bound solvent water molecule, 183
Bovine pancreatic ribonuclease, 226
  trypsin inhibitor, BPTI, 228–30
BQ-123, 391, 393
Bradykinin, 396, 408, 409
Bragg's law, 158
Braunitzer's reagent, 128
Briggs, G., 21
Broken symmetry, 213, 221
p-(Bromomethyl)mandelate, 300, 301, 303
Brookhaven protein databank, 160
β-Bulge, 175, 176

C-terminal protection, 40
C-terminal group, 150
C3G, 93
C5a, 400
Calorimetric enthalpy, 235
Capillary columns, 124, 133
Capillary electrophoresis, 133
Capillary LC, 139
Carbodiimide, 120
Carbohydrate, 10, 432
Carbanions, 281
Carbon acid, 306
Carbonic anhydrase, 247
Carbonic anhydrase B, 234
Carboxy terminus, 7, 11
Cassette mutagenesis, 313, 316, 318
Catalytic antibodies, 25, 340–42, 344, 346, 354, 355, 358, 365, 366
Catalysis of elementary steps, 292
CBz chemistry, 38, 39
CCK, 379, 396, 400
cCrk, 93
  N-terminal SH3 domain, 92

CD4, 416
CDR-2, 416
Circular dichroism (CD), 13, 236, 248, 252, 253
  far-UV, 246
  near-UV, 247
  stopped flow, 242, 248, 249
Cellular retinoic acid binding protein (CRABP), 233
Chaperonins, 164
Chemoselective reaction, 73, 74, 78, 79
Chemical synthesis, 72, 74, 75, 98
  human parathyroid hormone, 71
  peptides, 67, 72, 73
Chemical ligation, 73, 74, 75, 78, 82, 83, 85, 98, 99
Chemical misacylation of tRNAs, 459
Chorismate mutase, 346–48, 350–54, 365, 366
Cis peptide bond, 183
Cis-enediolate, 288
Claisen condensation, 284
Cleavage/deprotection, 45–47, 50, 56, 58, 60, 63
cMyc, 80
cMyc-Max, 80
CNBr cleavage, 105, 125, 128
Cobalt, 10
Codon, 6
  stop, 6
  termination, 6
  use, 450
Coenzyme, 24, 25
  cobalamine, 10
Cofactor, 10, 14, 24
  recycling, 420, 426
Cold denaturation, 235
Collagen, 4, 18
Collapsed form, 234
Collision induced dissociation (CID), 141
Combinatorial chemistry, 51, 52
Compact intermediate, 234
Complementarity determining region (CDR), 336, 338, 350, 354, 358, 361, 362
Combinatorial screening, 401, 402
Complementarity determining regions, CDR, 416
Complementary packing surface, 171
Concerted process, 279
Conformational stability, 234, 236
  constraint, 395, 402
  entropy, 241, 256

Continuous flow synthesis, 63
Continuous flow reactor, 115
Coomassie Blue, 102, 109, 124
Cooperativity, 199, 218
Correlation spectroscopy, COSY, 245
Coupling, 38, 40, 44, 55
   hindered peptides or amino acids, 40
   methods, 29, 31
   reaction, 40
   reagents, 36,55
Covalent structure, 68–70, 89, 99
CRABP, 237, 252–54
   acid-unfolded form, 253, 254
Crevice, 184
$\beta\alpha\beta$ Crossover connection, 176
Cross-validation, 155
Crossing angles, 175
Crystallographic, 157
Crystal design, 448, 449
Curvature, 175
α-Cyano-4–hydroxycinnamic acid, 138
Cyanogen bromide, 120
Cyclization, 55
Cyclization of peptides, 55
Cyclization on a solid support, 56
Cyclization reaction, 58
Cyclization reagents, 58
1,2–Cyclohexanedione, 128
Cys-X-X-Cys knuckle, 262
Cysteine, 4, 5, 130
Cysteine residue, 110
Cystic fibrosis, 224
Cystine, 4
Cystine knot superfamily, 197
Cytidine deaminase, 195
Cytochrome, 233
   acid-unfolded form, 254
Cytochrome C, 234
Cytoskeleton, 202

D-alanine, 175
2D gel, 101, 103, 111
Database, 106
   search, 145
Debye-Waller, 155
Dehydrogenase, 421, 426–29, 435, 437, 438, 441, 442
3,4–Dehydroproline, 459, 468
Denaturation, 15, 224–26

DEPTU, 117
N^{α} Deprotection/coupling procedure, 44
Deprotection methods, 30, 41–44, 49, 58, 62, 63
Design of protein structures *de novo*, 27, 28, 446
Differential scanning microcalorimetry, 227, 235
Differential binding, 292
Difficult coupling, 40
Diffusion-collision model, 231
Diffusion controlled association/dissociation process, 291
Dihydrofolate reductase, 247
Diisopropylethylamine, 115
Dimer, 9
Dipeptide, 1
Diphenylthiourea (DPTU), 117
Diphenylurea (DPU), 117
Disproportionation, 221
Disulfide bond, 196, 226, 228–30, 247
DITC, 120
Dithioerythritol (DTE), 135
Dithiothreitol (DTT), 121, 135
Diverse peptide library, 51
DMPTU, 117
Dimethylsulfoxide (DMSO), 111, 112, 125, 126
DNA polymerase, deep vent, 328
Domain, 9, 73, 80, 81, 83, 85, 89, 96, 99
Dopamine, 403, 405
Double-reciprocal plot, 23
Double jump, 243
DSC, 235
DTT-Ser, 121

E1cb mechanism, 284
E2 mechanism, 284
EDC, 120
Edman degradation, 13, 105
   solid phase, 120
Effective enthalpy, 235
Electrophoresis, gel, 11
   denaturing, 12
   native gel, 12
   two dimensional, 12
Electron paramagnetic resonance (EPR), 14
Electrospray ionization mass spectrometry, 67–69, 71, 72, 74, 81, 85, 96, 134, 138, 139, 142, 143, 244, 245, 249, 250, 257

Electrostatic interaction, 234, 239, 240, 256
β-Elimination reaction, 284
Enantioselective, 341, 348, 356
Endo Lys C, 127
Endo Glu C, 127
Endo Arg C, 127
Endoglycosidases, 144
Endocrine, 367, 371
Endothelin, 371, 387, 388, 389, 390, 391
Enediolate intermediate, 292
Energy metabolism, 179
Energy transduction, 215
Enkephalin, 396, 399, 408, 417, 418
Enolate anion, 281, 285, 286
Enolic intermediate, 280, 300, 306, 307, 308, 309
Enthalpy, 226, 227, 235, 240,
Entropy, 226, 227, 235, 240
Entropy trap, 342, 343, 346, 347
Enzyme, 2
Enzyme nomenclature, 20
Enzyme kinetics, 21
Enzyme immobilization, 26
ES complex, 22
Ester hydrolysis, 355, 362
Esterase, 423, 424, 425, 438
Evolution of enzyme function, 292
Evolution of catalytic efficiency, 310
Expression vector, 451, 462, 470

F1–ATPase, 154
Fab fragment, 191
FAB, 105
Ferritin, 18
Fibrous, 18
Fluorescein isothiocyanate (FITC), 122
p-Fluorophenylalanine, 459, 461
FIV protease, 87, 88
Fluorescence spectroscopy, 236
Fmoc chemistry, 38, 39, 41, 46
Fold, 172, 190
Folded state, 235, 240
Folding intermediates, 244
Fourier analysis, 167
Framework model, 231
Free energy, 226, 235
  diagram, 287, 288
  native state, 238
  transfer, 227
  unfolded state, 238

Gel filtration, 13, 17
Genetic code, 6, 14
Gene duplication, 203
Gibbs free energy, 234
Glass fiber, 107
Globular, 18
Glucagon, 400
Glutamic acid, 3, 4, 449, 470
Glutamine, 4
Glutathione peroxidase, 5
Glutamic acid, 449, 470
Glycopeptide, 43
Glycosylphosphatidylinositol (GPI), 100, 101
Glycosylation, 146
Glycosaminoglycan, 151
Glyceraldehyde 3–phosphate, 287
Glycine, 2, 4, 175
Greek key, 195
Ground state, 339, 340, 353, 365
Growth hormone releasing peptide, 400
Guanidine HCl (Gdm HCl), 111, 225, 236

Haemagglutinin, 197, 218
Hairpin, 173
Haldane, J. B. S., 21
Haldane relationship, 296
Hammond postulate, 279, 292, 311
Handshake configuration, 209
Hapten, 338, 340, 341, 343–46, 348, 350, 353–59, 361–66
Heat capacity change, 227
Heat capacity, 235
Helicity, 15
Helical symmetry, 202
3_{10} Helix, 174
4_{14} Helix, 174
α-Helix, 8, 13, 174, 227, 232, 233, 240, 247, 252–54, 403, 409, 410
β-Helix, 179, 191
Helix-breakers, 15
Helix-turn-helix, 213
Helix-loop-helix, 213
Hemoglobin, 10, 19, 201, 220
Hemocyanin, 206
Heme, 10
Hen lysozyme, 233, 241, 245, 247, 249, 251, 252, 254
Heterotetramer, 9
Heteronuclear single quantum coherence (HSQC) spectroscopy, 245

Heteronuclear single quantum coherence
  spectra (HSQC), 253, 254
Hexafluoroacetone, 111
High performance liquid chromatography
  (HPLC), 18, 68, 69, 71, 74, 85, 102,
  103, 107, 114, 115, 121, 124, 132
  preparative, 74, 81
  preparative reverse phase, 71
  reversed phase, 67–69, 81
Hinge-bending, 215, 216
Histidine, 4
Histones, 18
  core particle, 208
  octamer, 208
HIV-1 protease, 83–85, 87–89, 216, 405
Holoprotein, 10
Holoenzyme, 10
Homology modeling, 361, 362
Human parathyroid hormone, 71
Human chlorionic gonadotropin, 196
Hybridization, 318
Hybridoma, 341
Hybrid protein, 448, 470–72
Hydrophobic-interaction chromatography, 17
Hydroxyproline, 4
Hydroxylamine, 128
Hydrophobic effect, 227, 234, 239, 256
  collapse, 231, 257
  core, 246
  interaction, 227, 231, 235, 238, 239, 241,
    257
Hydrophobicity, 227, 238, 239
Hydrogen bonding, 171, 227, 228, 232, 234,
  238–40, 256, 257, 259, 260
Hydrodynamic radius, 237
Hydrogen/deuterium exchange pulse labeling,
  244, 246, 248, 249, 257
Hydrolase, 420, 423, 425, 426, 438–40
Hydrolysis, 280, 421, 423–26, 437, 439, 443,
  445
Hydroxylation, 426, 429–31
Hydrophilicity, 158
Hypervariable sequence, 192

Icosohedral symmetry, 202
IgG molecule, 191
IL8, 78, 87
Imidazolium, 4
Imidazolate anion, 295
Imino acid, 4

Immobilon PSQ, 124
Immobilon CD, 125
Immunization, 335, 345, 350, 364, 366
Immune system, 335, 338–40, 342, 348, 364,
  365
Immune response, 335, 338–40, 345, 354,
  365
Immobilization, 437
In vitro translation, 459
In-gel digest, 104, 124
Indole, 4
Influenza neuraminidase, 200
Information transduction, 215
Infrared, 456–58, 464, 467
Insulin, 7, 18, 371, 372
Interferon, 18
Intermediate state, 228, 230, 234
Interatomic vectors, 153
Interactive computer graphic tool, 160
Interleukin, 18
Interleukin-1β, 254
Intersheet spacing, 456, 458
Intrinsic kinetic barrier, 304
Intramolecular interface, 185
Ion pairs, 238
Ion-exchange chromatography, 17
Iron, 10
Iso-Asp bond, 132
Isoelectric focusing, 12
Isomerization, 161, 247
Isopropyl β-thiogalactopyranoside (IPTG),
  461

Jigsaw puzzle model, 231

Kinetics, 21
Kinetic intermediate, 234, 238, 246, 247, 249
  competence, 285
  resolution, 421, 423, 424
K_m, 22, 23
Knotted protein, 190
α-Lactalbumin (α-LA), 231, 234, 247, 249

Lambda repressor, 238
Lamellar aggregate, 448
  crystals, 449
  thickness, 449, 455, 456
Laue method, 158

LC/MS, 105
LC/MS/MS, 105
Left-handed α-helical conformation, 175
Left-handed crossover connection, 178
Leucine, 4
Levinthal paradox, 228
Library, 401, 402
Linderstrøm-Lang's hierarchy, 162, 169
Lineweaver-Burke plot, 23
Linked equilibria and transduction, 222
Lipase, 424, 435, 438, 439, 444, 445
Lipid, 1, 10, 11
Liquid phase peptide synthesis, 35, 41
Loop, 232
   closure, 216
Luteinizing hormone-releasing hormone (LH-RH), 370, 373, 377–383, 385
Lysine, 4
α-Lytic protease, 230

Macromolecular character, 446
   crystals, 447, 448, 472
   material, 446
Mad, 80
Madelate racemase (MR), 202, 286, 295, 297, 304, 307, 309
MAGE, 160
Magainins, 19
Marcus formalism, 304, 311
Mass spectrometry, 14
   MALDI-TOF, 94, 96, 105, 109, 134, 137, 143, 144
   MS/MS, 106, 107, 141, 147
   SIMS, 105, 134
Matrix, 136, 137
Max, 80–82
Melano-Tan I, 387
α-Melanophore stimulating hormone (MSH), 383–86
Melatonin, 400
Membrane-bound protein, 11, 191
Membrane protein, 199
Menten, M., 21
Merrifield, R. B., 7
Mesomolecular assemblies, 206
Metastable state, 230
Methionine, 4
N-Methylated amino acid, 395, 396
Micellar structure, 452

Michaelis, L., 21
Michaelis-Menten, 22
   constant, 22
   kinetics, 21
Microstate, 228
Mixed α/β proteins, 246
Mixed α-helix, 257
Mixil, 80
Model bias, 155
Molecular chaperone, 226, 232
Molecular dynamics, 158, 174
Molten globule, 230, 231, 233, 247, 249
   model, 231
   state, 233, 234, 241, 247
Monoclonal antibodies, 341
Multiple synthetic method, 51
Muscle thin filament, 165
Mutational analysis, 238
Mutagenic primers, 328
Mutation, 188
Myc, 82
Myc-Max, 81
Myoglobin, 10, 247
Myosin, 18

N-glycans, 147
N-terminus, 13
N-terminal sequence, 104
N-terminal group, 150
Native state, 232, 233, 237
   conformation, 9
Native chemical ligation, 76
Neprilysin, 387
Neuraminidase, 144, 146
Neurocrine, 367
Neurokinin, 400, 404
Neuropeptide Y, 400
Neurotensin, 400
NH-O hydrogen bond, 260
NH-S hydrogen bond, 258, 260, 262–64, 267, 269, 272, 273, 276–78
NH-S(Cys) hydrogen bond, 273
Nitrocellulose, 125
NMR spectroscopy, 226, 227, 233, 245, 249, 250, 252, 253, 258–60, 262, 267–74, 276–78
   magic angle spinning, 455
   ^{15}N NMR, 295

Non-native secondary structure, 245, 246, 254, 257
  α-helix, 245, 253
  β-sheet, 245
Non-aqueous, 421, 434, 437–39, 444
NTCB, 130
Nucleoprotein, 11
Nucleic acid, 1, 10
Nucleic acid-protein complexes, 11

Oligonucleotide, 315, 316
Oligosaccharide, 420, 433, 434
Oligomeric protein, 199
On-blot digest, 124
One bead method, 52
One peptide method, 52
One-base mechanism, 298
Opioid, 396, 399, 408
Organic solvent, 422, 434–41, 443–45
Orthogonal synthesis, 28, 43, 45, 46, 49, 50, 59, 60
4-Oxalocrotonate tautomerase (OT), 88, 96, 97, 98
Oxazolone, 123
Oxidoreductase, 420
Oxidase, 426, 430, 435, 436, 440, 441, 444, 445
Oxidation, 426, 428, 429, 431, 441
Oxytocin, 373–76, 400

p53, 224
P-loop, 210
Packing densities, 186
Parameters, 155
Paracrine, 367, 371
Parallel β-structures, 176
  β-sheet, 195
Partially folded state, 241, 256
PEG, 443, 444
  lipase, 443
Pepsin, 127
Peptide, 1, 72
  bond, 7, 161
  coupling reagents, 37
  library, 50, 51, 54
  primary structure, 8, 13, 164
  synthesis, 28, 39, 40, 42, 54, 60, 62, 63, 67, 68

Peptide methyl esters, 143
Peptidyl prolyl cis-trans isomerase (PPI), 226, 242, 243
Peptoid, 398
Peroxidse, 430, 440
pH memory, 438
  dependent conformational change, 218
  optimum, 23
Phage 434–repressor, 226
Phenylalanine, 4, 14
Phenylisothiocyanate (PITC), 112, 115, 117, 121, 122
Phenythiohydantoin (PTH), 105
Phenylalanyl-tRNA synthetase, 169
Phosphotyrosine, 4, 122
Phosphoserine, 4, 122
Phosphothreonine, 4, 122
Phosphorylation, 148
Phosphorane intermediate, 280
Phosphoramidon, 388, 389, 390
Picornaviruses, 203
α-Phenylglycidate, 299, 302, 303
o-Phthaldialdehyde (OPA), 119
pKa, of the α-proton of a carbon, 283
  matched, 307
pK_E, 306
Plasminogen activator inhibitor-1 (PAI-1), 230
Plasmid preparation, 316
PLP-independent racemases, 296
Polybrene, 109, 116, 120
Polypeptide, 2, 51, 65, 66, 69–74, 76, 78–81, 83, 89, 91–94, 96–99
  synthesis, 5
  subunit, 9
Polymer, 446, 449, 450, 452, 454–56, 459, 461, 463–65, 467, 468
Polymerase chain reaction (PCR), 15, 313, 321, 325
Polymorphism, 215
Poly(L-alanine), 456
  (L-glutamic acid), 468
Polyacrylamide, 102
Post-translational modification, 5, 6
Pre-molten globule, 233, 254
Predominantly β-sheet protein, 237, 247
Proline bond, 4, 119, 247, 467, 468
Prosthetic group, 25
Protecting group, 28, 30, 40–44, 49, 50, 56, 62

Protected amino acid, 38, 43, 44, 67
  peptide, 49, 66, 74
  peptide segments, 74
Protein, 1, 2
  denaturation, 225, 226
  design, 169
  disulfide isomerase (PDI), 226
  engineering, 446, 447, 472
  folding, 224, 225, 228, 231
  membrane-bound, 17
  motion database, 216
  regulatory, 18
  secondary structure, 8, 169
  signature analysis, 89, 96, 98, 99
  storage, 18
  structural, 18
  synthesis, 7
  transport, 19
Protease, 423, 424, 426, 444, 445
Proteolysis, 203
1,2 and 1,3–Proton migration reaction, 284
1,1–Proton transfer reaction, 284, 295
1,2–Proton transfer reaction, 287
Proton transfer reactions, to and from carbon, 279, 282
Phenylthiocarbamyl (PTC), 112, 117, 121
Phenylthiohydantoin (PTH), 114–17, 121, 123
PTH-dehydroalanine, 122
Puzzle pieces, 185
Polyvinylidene difluoride (PVDF), 107–12, 115, 121, 124, 129, 133
Polyvinylpyrrolidone (PVP)-40, 109, 124, 125
  PVP-360, 125
Pseudopeptide/peptidomimetic library, 51
Pseudopeptides: amide bond replacement, 60
Pyridoxal phosphate, 296
Pyroglutamic acid, 101
Pyroglutamyl proteins, 110

Quarternary structure, 9, 14
Quasi equivalence, 203
Quaternary structure, 199
  polymorphism, 220
Quenched-flow, 242, 245
  H/D exchange, 244

R and T structures, 219
  interfaces, 220

R group, 8
Racemization, 31, 33–35, 39–42, 51, 55, 62, 280
Racemases, 296
Ramachandran plot, 169, 175
Raman, 454–56
RASMOL, 160
Rd-knuckle, 259, 268, 269, 276
Recommended name, 21
Redox reaction, 426, 427
Reduced triton X-100, 128
Reduction, 426, 427, 428, 434
Refinement, 154
Renaturation, 225
Residue, 2
Resin handle, 56
Resin, 44, 65–67, 90, 91, 93
Restriction site, 450, 470
Reverse translation, 15
Reverse turn, 232, 403, 414–16, 418
Reversibility, 226
R_{free}, 155
RGD peptide, 407
Ribbon representation, 159
Ribonuclease A, 233, 241
  T1, 238
Ribosome, 6
Right-handed twist, 175
RNase, 226
Rod-like helical macromolecules, 448
Rotation symmetry, 199
Rotational allomerism, 210
S-alkylation, 110, 122
S-atrolactate, 297

Sarcosylsarcosine, 396
Scanning microcalorimetry, 236
Scrapie, 224
Screw symmetry, 202
Screening, 317, 400, 401
Scurvy, 4
SDS gel, 12, 101, 103, 104, 107, 108, 110, 112, 124
Selenocysteine, 5, 6
Selenomethionine, 459–61
Serine, 4
SH2 domain, 93
SH3 domain, 93, 95, 96
Shear, 216

β-Sheet, 227, 232, 240, 245, 247, 254, 257, 403, 411–13, 455, 456, 464, 467, 468
β-Strand, 413, 416
Side chain, 2, 9, 162
  orthogonal, 29
  protection, 29, 43, 49
Side chain-to-side chain cyclization, 37
  lactam formation, 38
Side reactions, 32
Signal transduction, 87, 93
Silaic acid, 144
Silent mutation, 316
Silver stain, 102
Site-directed mutagenesis, 25
Site-directed mutants, 294
Size exclusion, 17
Solid phase peptide synthesis, 27, 28, 31, 32, 39, 40, 41, 43, 44–46, 51, 52, 54, 55, 63, 65–68, 71–73, 90, 91, 405
  fragment condensation, 45
Solid phase peptide synthesis, automated, 38, 54
Solid and solution phase synthesis, 36–38, 54
Solid support, 46
Somatostatin, 370, 373, 405
Specific activity, 16
Specificity, 421, 423, 425, 427, 428, 436, 438, 439, 444
Stabilized trypsin, 128
Staphylococcal nuclease, 235, 238
Steady-state, 22
  equation, 22
  theory, 21
Stepwise process, 279
Stereochemical complementarity and packing, 184, 186
Steric overlap, 186
Stopped flow (SF)-fluorescence, 242, 249
Stopped-flow (SF), 242, 244
Structural complementarity, 213
  intermediates, 244
  refinement, 180
Subunit, 2, 9
  interfaces, 209
Succinic anhydride, 128
Sugar, O-linked, 10
  N-linked, 10
Sulfation, 148
Sulfo-Tyr, 122
Supercritical fluid, 444, 445

Super R state, 221
Swiss roll, 195
Synthesis of peptides, 27, 39, 46, 49, 51, 52
  any peptide, 30
  difficult peptides, 29
  polypeptide ligands, 28
  polypeptides and proteins, 27
  short peptides, 38
Systematic name, 20

T4 lysozyme, 234, 247
T4 phage lysozyme, 235
Taq DNA polymerase, 328
Teflon tape, 125
Termini, 7
Tertiary structure, 9, 10, 15, 190
Tetrahedral intermediates, 280
Thermodynamics of protein folding, 234
  analysis, 234
  stability, 234
Thermal denaturation, 236
Thermodynamic barrier, 304
Thermostability, 434, 438, 439
3–Thienylalanine, 459, 465
Three dimensional orthogonal protection, 29, 60
Thyrotropin releasing hormone (TRH), 370, 373
Time of flight (TOF), 105, 137
Tobacco mosaic virus, 202
Tomato bushy stunt virus, 203
Topological switch points, 210
Total chemical synthesis of proteins, 65, 69, 72–74, 78, 80–85, 87–89, 96
Total synthesis of human IL8 by native ligation, 76
Total correlation spectroscopy, TOCSY, 245
Toxins, 367
Transcription, 6
  factor, 80–82, 87, 258, 259, 266, 267
  regulator, 19, 263
Translation, 6
Transverse urea gradient gel electrophoresis, 237
Transition state, 238, 246, 247, 335, 339–44, 347, 348, 352, 353, 355, 360–62, 364–66
  analogue, 212, 221, 340–42, 346–50, 352, 355, 356, 358, 364–66
  intermediate, 279, 280, 282
  theory, 285

Transporting viral RNA, 191
TRH, 396
Tributylphosphine (TBP), 110
Trichloroacetic acid (TCA), 107, 112
Triethylamine (TEA), 107, 115
Trifluoroacetic acid (TFA), 107, 108, 110, 116
5',5',5'-Trifluoroleucine, 459, 464
Trifluoromethanesulfonic acid, 147
Trimer, 9
Trimethylamine, 115
Trimethylphosphate, 130
Triose phosphate isomerase (TIM), 285, 287, 304, 307, 309
  three dimensional structure, 293
Tripeptide, 2
Triton X-100, 102, 127
Tropocollagen, 4
Tropomyosin, 164
Tryptophan, 3, 4, 14, 129
Tryptophan synthase α subunit, 247
Trypsin, 127
Tubulin, 18
γ-Turn, 403, 406, 407, 408, 415
Turnover number, 23
Turns, 15, 172
β-Turn, 173, 397, 403–6, 408, 411, 413, 414, 417, 418
Two-state, 230, 234, 235, 242, 244, 256 model, 249
Two-base mechanism, 298
Type I' turn, 175
Tyrosine, 4, 14

Ubiquitin, 243
Unfolded state, 232, 233, 235, 241, 243, 244, 253, 257
Uniform binding, 292

Universal genetic code, 6
Unprotected synthetic peptide segments, 73
Unprotected peptide, 69, 74, 75, 78, 83
Urea, 126, 128, 225, 226, 236, 253

Valine, 4
van der Waals, 239, 240
  interaction, 234, 239, 256
  interaction potential, 186
van't Hoff plot, 235
Vasopressin, 373–76, 400
Velcro, 172
Venus fly trap, 216
4-Vinylpyridine, 126
Vitamin
  B_{12}, 10
  C, 4
  C deficiency, 4
V_{max}, 22, 23
Voronoi polyhedra, 187

Water, 174
Western blot, 12, 13

X-ray crystallography, 14, 153
X-ray scattering, 455, 457, 464

Y quaternary state, 221
Ypsilanti mutant hemoglobin, 220

Zinc finger, 10, 258–64, 268–72, 277
Zinc, 10
Zitex, 109, 125
Zwitterion, 2